Isotopic Analysis

Edited by
Frank Vanhaecke and
Patrick Degryse

Related Titles

Nölte, J.

Fehlerfrei durch die ICP Emissionsspektrometrie

2012

ISBN: 978-3-527-31897-1

Michener, R., Lajtha, K. (eds.)

Stable Isotopes in Ecology and Environmental Science

2007

ISBN: 978-1-4051-2680-9

Voges, R., Heys, J.R., Moenius, T.

Preparation of Compounds Labeled with Tritium and Carbon-14

2009

ISBN: 978-0-470-51607-2

Hill, S. J. (ed.)

Inductively Coupled Plasma Spectrometry and its Applications

2007

ISBN: 978-1-4051-3594-8

Meier-Augenstein, W.

Stable Isotope Forensics: An Introduction to the Forensic Application of Stable Isotope Analysis

2010

ISBN: 978-0-470-51705-5

Edited by Frank Vanhaecke and Patrick Degryse

Isotopic Analysis

Fundamentals and Applications Using ICP-MS

WILEY-VCH Verlag GmbH & Co. KGaA

The Editors

Prof. Dr. Frank Vanhaecke
Ghent University
Dept. of Analytical Chemistry
Krijgslaan 281 - S12
9000 Ghent
Belgium

Prof. Dr. Patrick Degryse
Katholieke Universiteit Leuven
Center for Archaeol. Sciences
Celestijnenlaan 200 E
3001 Leuven
Belgium

Library of Congress Card No.: applied for

British Library Cataloguing-in-Publication Data
A catalogue record for this book is available from the British Library.

Bibliographic information published by the Deutsche Nationalbibliothek
The Deutsche Nationalbibliothek lists this publication in the Deutsche Nationalbibliografie; detailed bibliographic data are available on the Internet at <http://dnb.d-nb.de>.

Print ISBN: 978-3-527-32896-3
ePDF ISBN: 978-3-527-65051-4
ePub ISBN: 978-3-527-65050-7
mobi ISBN: 978-3-527-65049-1
oBook ISBN: 978-3-527-65048-4

Cover Design Adam-Design, Weinheim
Typesetting MPS Limited, Chennai
Printing and Binding Markono Print Media Pte Ltd, Singapore

Contents

Isotopic Analysis: Fundamentals and Applications Using ICP-MS,
First Edition. Edited by Frank Vanhaecke and Patrick Degryse.

Preface

Although several instrumental techniques allow information on the isotopic composition of target elements to be obtained, mass spectrometry is without doubt the most versatile and most powerful.

For isotopic analysis of metals and metalloids, thermal ionization mass spectrometry (TIMS) has been the "gold standard" for a long time. Although also serving other purposes, such as accurate and precise determination of element concentrations in the context of the production of reference materials and characterization of materials from the nuclear industry, TIMS was predominantly deployed in the domain of geo- and cosmochemistry. The introduction of single-collector inductively coupled plasma mass spectrometry (ICP-MS) in 1983 and, especially, of multi-collector (MC) ICP-MS about a decade later, however, had a tremendous impact on the field of "isotopic analysis." The much higher ionization efficiency of the inductively coupled plasma (ICP) ion source, the enhanced sample throughput, and the flexibility in terms of sample introduction are the reasons most quoted to explain the current success of MC-ICP-MS. In effect, MC-ICP-MS is gradually replacing/has gradually replaced TIMS as the "gold standard" for isotopic analysis by now.

ICP-MS was not immediately introduced as a tool for isotopic analysis, but as a very powerful technique for (ultra-)trace element determination, combining the multi-element capabilities of inductively coupled plasma optical emission spectrometry (ICP-OES) with an even higher detection power than atomic absorption spectrometry (AAS). As a result, the user community of ICP-MS was different from that of TIMS, although also here the geochemical community has often been working at the forefront of development, especially in the domains of isotopic analysis and of direct bulk and spatially resolved analysis by using laser ablation (LA) as a means of sample introduction. However, by using ICP-MS instead of spectrometric techniques such as AAS or ICP-OES that have only marginal capabilities for isotopic analysis, also scientists from other fields learned to appreciate the additional capabilities offered by the isotope-specific information that a mass spectrometric technique provides. In fact, all ICP-MS users rely on isotopic information, for example, for identifying element patterns in the mass spectrum and for revealing and correcting for spectral interferences via mathematical equations. A new community of users discovered the merits of isotope dilution as

Isotopic Analysis: Fundamentals and Applications Using ICP-MS,
First Edition. Edited by Frank Vanhaecke and Patrick Degryse.

Published 2012 by WILEY-VCH Verlag GmbH & Co. KGaA

a calibration approach, not only as a means for obtaining the highest accuracy and precision, but also for compensating for analyte losses during sample pretreatment and element speciation procedures, and more "adventurous" uses of isotope dilution were explored. Given the diversity of the ICP-MS community, it came as no surprise that the technique was also increasingly used for tracer experiments with stable isotopes aiming at obtaining a more profound insight into various physical processes and (bio)chemical reactions in the context of both fundamental physicochemical studies, for example, concerning reaction rates, and applied research, for example, in environmental and biomedical studies. The user community also started to use ICP-MS for determining isotope ratios that show variation in Nature as a result of the presence of one (e.g., Sr) or more (e.g., Pb) radiogenic isotopes although, as a result of the rather modest precision of single-collector ICP-MS, only the less demanding cases could be tackled in this way.

The gap in isotope ratio precision existing between TIMS and single-collector ICP-MS was bridged by the introduction of MC-ICP-MS, in which the noisy character of the ICP ion source is counteracted by simultaneous monitoring of the ion beams involved, a practice that was already established earlier in TIMS. The user community of MC-ICP-MS nowadays seems to consist of two groups: researchers from the geo- and cosmochemical domain have added MC-ICP-MS to their instrumentation and/or replaced (part of their) TIMS instruments by MC-ICP-MS devices, while single-collector ICP-MS users looking for an improved isotope ratio precision are also keen MC-ICP-MS users.

The introduction of MC-ICP-MS has not only facilitated the work otherwise done using TIMS, it also allows the study of elements that were previously not or barely accessible using TIMS. The most prominent example is Hg, the volatility of which precludes the use of TIMS. Furthermore, the introduction of MC-ICP-MS has revolutionized the field as its use contributed strongly to the current awareness that the isotopic composition not only shows variations as a result of isotope fractionation for the lighter elements, but actually varies for all of them, even for U, the heaviest naturally occurring element.

As the community of MC-ICP-MS users is more diverse than that of TIMS was/is for the reasons outlined above, there seemed to be a demand for a book on isotopic analysis via single- and multi-collector ICP-MS with a wider scope, one that addresses not only applications from the domains of geo- and cosmochemistry (but certainly also without denying the pioneering role that researchers in this research field have played and are still playing), but also from other areas, such as archeometry, forensics, and biomedical studies. The Editors were invited to realize such a book, a task that could only be completed with contributions from prominent scientists from diverse fields. It was the intention to provide a book that is accessible to the interested newcomer to the field, while also supplying the more experienced user with some new insights. To achieve this goal, both the very basics and also advanced topics are addressed, as detailed below.

Chapter 1 especially aims at providing newcomers to the field with general information on the origin of natural variations in the isotopic composition of metals and metalloids.

Chapters 2 and 3 present an overview of single- and multi-collector ICP-MS instrumentation and their respective capabilities. Also appropriate ways to overcome spectral overlap are addressed. As the ICP is a robust ion source operated at atmospheric pressure, there are various means of sample introduction. Although pneumatic nebulization of sample solution is the standard approach, LA of solid material avoids the need for digestion in bulk analysis, and also allows spatially resolved information to be obtained. Recent advances in LA are discussed in Chapter 4, in which the fundamental technical challenges associated with the handling of transient signals are also considered.

Although MC-ICP-MS permits isotope ratios to be measured with very high precision (down to 0.002% RSD), it needs to be realized that the raw data show a substantially larger bias (1% per mass unit order of magnitude at mid-mass) with respect to the corresponding true value. This bias is caused by preferential transmission of the heavier isotope during the extraction process, and obviously this instrumental mass discrimination needs to be corrected for. The methods used for this purpose are discussed into detail in Chapter 5.

Chapters 6 and 7 focus on the "quality" of isotope ratio measurement data by concentrating on isotopic reference materials (and the current shortage thereof) and on the total uncertainty accompanying all measurement data.

Chapter 8 provides the reader with the basic principles of isotope dilution mass spectrometry used for elemental analysis and also discusses more advanced features of this calibration approach, such as its use in direct solid sample analysis and in elemental speciation work, wherein not the total amount but that of various chemical species of a target element need to be determined.

Chapter 9 presents an accessible overview of methods for geochronological dating and illustrates the capabilities and limitations of MC-ICP-MS in this context with real-life applications reported in the literature.

Chapters 10–12 put geo- and cosmochemical applications in the spotlight. Chapter 10 clarifies the origin of cosmochemical isotopic variations and explains how these variations can be exploited for advancing insight into the universe and our solar system with its planetary bodies. Chapters 11 and 12 bring the reader back to Earth. Chapter 11 addresses paleoproxies in more detail, but instead of providing an overview of all the contexts in which it is attempted to reconstruct paleoconditions (e.g., pH, temperature, redox potential, salinity) prevailing during a physicochemical process in the past by investigating an isotopic signature systematically affected by the above-mentioned conditions, this chapter focuses on paleoredox proxies only and discusses into detail all factors that jeopardize a correct interpretation in this context. Chapter 12 then provides a more general description of how isotopes can be used as tracers of elements across the geosphere–biosphere interface.

Chapters 13 and 14 illustrate how the isotopic composition of selected target elements may be interpreted as a fingerprint, providing information on the provenance of materials and objects of relevance in an archeometric or forensic context.

In Chapter 15, nuclear applications are described and it is clarified in which contexts ICP-MS is superior to radiometric techniques that have to rely on the actual decay of the radionuclides present in the sample for data collection.

Isotopic analysis using ICP-MS can also advance our insight into the human metabolism, as illustrated extensively in Chapter 16. Various concepts in tracer studies with stable isotopes are delineated and case studies are reviewed on an element-by-element basis. Also, emerging applications based on natural variations in the isotopic composition of mineral elements are considered.

Finally, in Chapter 17, the relatively recent use of MC-ICP-MS in elemental speciation work – realized by coupling a chromatographic separation technique to an MC-ICP-MS device as an isotope-specific detector – is discussed. Also, the consequences of the transient nature of the signals thus obtained on the data collection are considered.

Considerable work both from the Editors and from the other contributors went into the making of this volume. We hope that these efforts have resulted in a book that is considered "useful" by our colleagues working in or interested in working in the still rapidly evolving and fascinating field of isotopic analysis using single- and multi-collector ICP-MS.

Ghent and Leuven, March 2012

Frank Vanhaecke
Patrick Degryse

List of Contributors

David Amouroux
LCABIE – Laboratoire de Chimie
Analytique Bio-inorganique et
Environnement
IPREM, CNRS-UPPA-UMR-5254
Hélioparc, 2 Avenue du Président
Pierre Angot
64053 Pau
France

Rasmus Andreasen
Imperial College London
Department of Earth Science and
Engineering
South Kensington Campus
Prince Consort Road
London SW7 2AZ
UK

Sylvain Berail
LCABIE – Laboratoire de Chimie
Analytique Bio-inorganique et
Environnement
IPREM, CNRS-UPPA-UMR-5254
Hélioparc, 2 Avenue du Président
Pierre Angot
64053 Pau
France

Patrick Degryse
Katholieke Universiteit Leuven
Center for Archaeological Sciences
Celestijnenlaan 200E
3001 Leuven
Belgium

Olivier F.X. Donard
LCABIE – Laboratoire de Chimie
Analytique Bio-inorganique et
Environnement
IPREM, CNRS-UPPA-UMR-5254
Hélioparc, 2 Avenue du Président
Pierre Angot
64053 Pau
France

Charles Douthitt
Thermo Fisher Scientific
8848 S Raven Ridge
Safford
AZ 85546
USA

Marlina A. Elburg
Ghent University
Department of Geology and Soil
Science
Krijgslaan 281-S8
9000 Ghent
Belgium

Isotopic Analysis: Fundamentals and Applications Using ICP-MS,
First Edition. Edited by Frank Vanhaecke and Patrick Degryse.

Published 2012 by WILEY-VCH Verlag GmbH & Co. KGaA

Vladimir N. Epov
LCABIE – Laboratoire de Chimie
Analytique Bio-inorganique et
Environnement
IPREM, CNRS-UPPA-UMR-5254
Hélioparc, 2 Avenue du Président
Pierre Angot
64053 Pau
France

Klaus G. Heumann
Johannes Gutenberg Universität
Mainz
Institut für Anorganische Chemie
und Analytische Chemie
Duesbergweg 10-14
55128 Mainz
Germany

Takafumi Hirata
Kyoto University
Graduate School and Faculty of
Science
Kitashirakawa Oiwake-cho
Sakyo-ku
Kyoto 606-8502
Japan

Michael E. Ketterer
Northern Arizona University
Department of Chemistry and
Biochemistry
Box 5698
Flagstaff
AZ 86011-5698
USA

Kurt Kyser
Queen's University
Department of Geological Sciences
and Geological Engineering
99 University Avenue
Kingston
ON K7L 3N6
Canada

Juris Meija
National Research Council Canada
NRC, Institute for National
Measurement Standards
1200 Montreal Road
Ottawa
ON K1A 0R6
Canada

Thomas Meisel
Montanuniversität Leoben
Department für Allgemeine,
Analytische und Physikalische
Chemie
Lehrstuhl Allgemeine und
Analytische Chemie
Franz-Josef-Strasse 18
8700 Leoben
Austria

Zoltán Mester
National Research Council Canada
NRC, Institute for National
Measurement Standards
1200 Montreal Road
Ottawa
ON K1A 0R6
Canada

Christophe Pécheyran
LCABIE – Laboratoire de Chimie
Analytique Bio-inorganique et
Environnement
IPREM, CNRS-UPPA-UMR-5254
Hélioparc, 2 Avenue du Président
Pierre Angot
64053 Pau
France

Wolfgang Pritzkow
BAM Bundesanstalt für
Materialforschung und -prüfung
Unter den Eichen 87
12205 Berlin
Germany

Mark Rehkämper
Imperial College London
Department of Earth Science and Engineering
South Kensington Campus
Prince Consort Road
London SW7 2AZ
UK

Martín Resano
University of Zaragoza
Facultad de Ciencias
Departamento de Química Analítica
Pedro Cerbuna 12
50009 Zaragoza
Spain

Maria Schönbächler
The University of Manchester
School of Earth,
Atmospheric and Environmental Sciences
Manchester M13 9PL
UK

Johannes Schwieters
Thermo Fisher Scientific
Hanna-Kunath-Strasse 11
28199 Bremen
Germany

Ralph E. Sturgeon
National Research Council Canada
NRC, Institute for National Measurement Standards
1200 Montreal Road
Ottawa
ON K1A 0R6
Canada

Scott C. Szechenyi
Pacific Northwest National Laboratory
P.O. Box 999
MSIN: J4-70
Richland
WA 99352
USA

Frank Vanhaecke
Ghent University
Department of Analytical Chemistry
Krijgslaan 281-S12
9000 Ghent
Belgium

Jochen Vogl
BAM Bundesanstalt für Materialforschung und -prüfung
Unter den Eichen 87
12205 Berlin
Germany

Thomas Walczyk
National University of Singapore
Department of Chemistry
3 Science Drive/S17-07-20
117543 Singapore
Singapore

Laura E. Wasylenki
Indiana University
Department of Geological Sciences
1001 East Tenth Street
Bloomington
IN 47405-1405
USA

Michael Wieser
University of Calgary
Department of Physics and Astronomy
2500 University Drive NW
Calgary
AB T2N 1N4
Canada

Lu Yang
National Research Council Canada
NRC, Institute for National Measurement Standards
1200 Montreal Road
Ottawa
ON K1A 0R6
Canada

1 The Isotopic Composition of the Elements

Frank Vanhaecke and Kurt Kyser

1.1 Atomic Structure

Early in the twentieth century, Rutherford realized that Thomson's late nineteenth century plain cake model for the atom, describing the atom as consisting of electrons floating around in a positive sphere, had to be replaced by a "Saturnian" model, describing the atom as consisting of a small central nucleus surrounded by electrons, rotating on rings [1]. This view was supported by a study of the behavior of a beam of α particles (see below, particles resembling the nucleus of an He atom, thus consisting of two protons and two neutrons) directed on to a very thin Au metal foil, known as the Geiger–Marsden experiment [2]. Since only a minor fraction of the α-particles were recoiled or deflected, and for the majority the path was not affected, it had to be concluded that for the largest part, an atom consists of empty space. According to Bohr's later model [3], the atom contains a nucleus composed of positively charged protons and neutral neutrons, having approximately the same mass. This nucleus is a factor of $\sim 10^4$ smaller than the size of the atom (although the concept of size itself is not self-evident in this context) and holds practically all of its mass. As they both reside in the nucleus, protons and neutrons are also referred to by the common term nucleon. The negatively charged electrons are substantially lighter (almost 2000 times) and rotate around the central nucleus in different orbits (also termed shells), corresponding to different energy levels. Subsequently, insight into the atomic structure has evolved tremendously and a multitude of other particles have been discovered, but for many chemical considerations – including a discussion of isotopes – the Bohr model still largely suffices.

As all protons carry the positive unit charge (1.602×10^{-19} C), they mutually repel one another. This electrostatic repulsion is overcome by the so-called nuclear force [4]. This is a very strong force, but effective only within a very short range. In fact, the very short range over which this force is effective even causes the largest nuclei (e.g., those of U) to be unstable (see below). Further clarification of the nature of this nuclear force requires a more thorough discussion of the atomic

Isotopic Analysis: Fundamentals and Applications Using ICP-MS,
First Edition. Edited by Frank Vanhaecke and Patrick Degryse.

Published 2012 by WILEY-VCH Verlag GmbH & Co. KGaA

structure, including a discussion on quarks, but this is beyond the scope of this chapter. Electrostatic attraction between the positive nucleus and the orbiting negative electrons provides the centripetal force required to keep the electrons from drifting away from the nucleus.

1.2 Isotopes

The chemical behavior of an atom is governed by its valence electrons (electron cloud) and, therefore, atoms that differ from one another only in their number of neutrons in the nucleus display the same chemical behavior (although this statement will be refined later). Such atoms are called isotopes and are denoted by the same chemical symbol. The term isotopes refers to the fact that different nuclides occupy the same position in the periodic table of the elements and was introduced by Todd and Soddy in the early twentieth century [5].

To distinguish between the isotopes of an element, the mass number A – corresponding to the sum of the number of protons and the number of neutrons (the number of nucleons) in the nucleus – is noted as a superscript preceding the element symbol: ^{A}X. The atomic number Z, corresponding to the number of protons in the nucleus, may be added as a subscript preceding the element symbol – $^{A}_{Z}X$ – but is often omitted as this information is already inherent in the element symbol.

As a result of their difference in mass, isotopes of an element can be separated from one another using mass spectrometry (MS), provided that they are converted into ions. In fact, this is exactly how isotopes were discovered: Thomson separated the ion beams of two Ne isotopes using a magnetic field, while their detection was accomplished with a photographic plate [6]. With a similar setup, typically referred to as a mass spectrograph, Aston was subsequently able to demonstrate the existence of isotopes for a suite of elements [7].

Although several techniques provide a different response for the isotopes of an element, for example, infrared (IR) spectroscopy, nuclear magnetic resonance (NMR) spectroscopy and neutron activation analysis (NAA), MS is the technique of choice for the majority of isotope ratio applications. The isotopic composition of the light elements H, C, N, O, and S is typically studied via gas source MS, and for ^{14}C dating, accelerator mass spectrometry (AMS) is replacing radiometric techniques to an increasing extent. For isotopic analysis of metals and metalloids, thermal ionization mass spectrometry (TIMS) and inductively coupled plasma mass spectrometry (ICP-MS) are the methods of choice. This book is devoted to the use of (single-collector and multi-collector) ICP-MS in this context and its basic operating principles, capabilities, and limitations are discussed in Chapters 2 and 3.

The relative abundance of one nuclide of the element M (^{1}M) is calculated as the amount (number of atoms N or number of moles n) of nuclide ^{1}M divided by the total amount (number of atoms or number of moles) of the element M:

$$\theta(^1\mathrm{M}) = \frac{N(^1\mathrm{M})}{\sum_{i=1}^{i=m} N(^i\mathrm{M})} = \frac{n(^1\mathrm{M})}{\sum_{i=1}^{i=m} n(^i\mathrm{M})}$$

for an element with m isotopes.

1.3
Relation Between Atomic Structure and Natural Abundance of Elements and Isotopes

Except for the lightest atoms, the binding energy per nucleon is remarkably stable (varying only from 7.6 to 8.8 MeV) for the naturally occurring elements (Figure 1.1). On the basis of this curve, it is understood that fission of a heavy nucleus

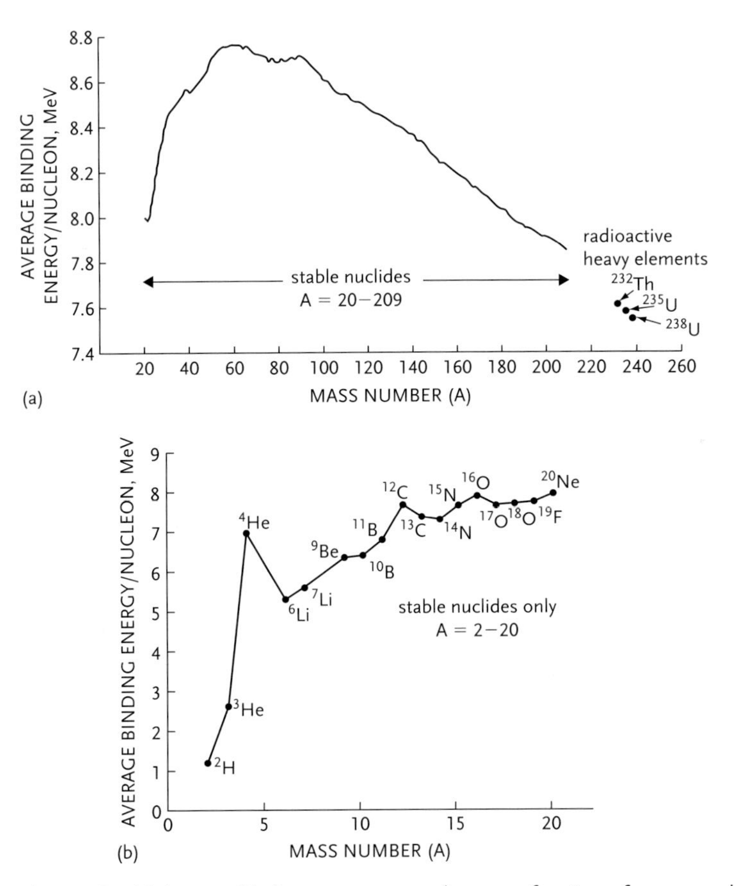

Figure 1.1 (a) Average binding energy per nucleon as a function of mass number for nuclides with a mass number from 20 to 238. (b) Average binding energy per nucleon as a function of mass number for nuclides with a mass number from 1 to 20. Reproduced with permission of John Wiley & Sons, Ltd., from [8].

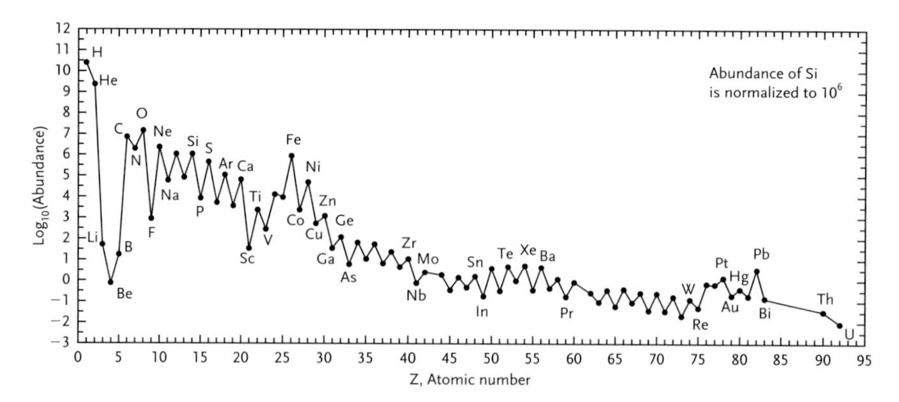

Figure 1.2 Natural relative abundance of the elements as a function of their atomic number. Reproduced with permission of John Wiley & Sons, Ltd., from [9]. The graph is based on the data published by Lodders [10].

into lighter nuclei (e.g., in a nuclear reactor) or fusion of two H atoms into He (the basis of solar energy) are exo-energetic because the process results in nuclei/ a nucleus characterized by a substantially higher binding energy per nucleon.

With the lighter atoms (see Figure 1.1b), nuclei with an even number of protons and an even number of protons show a higher binding energy per nucleon and thus higher stability (compare, e.g., the binding energies for ^{4}He and ^{3}He, ^{12}C and ^{13}C, and ^{16}O and ^{17}O). In addition, elements with an even number of protons, reflected by an even atomic number Z, are more abundant in Nature than those with an uneven number (Figure 1.2).

This variation in binding energy per nucleon also exerts a pronounced effect on the isotopic composition of the elements, especially for the light elements. "Even–even isotopes" for elements such as C and O (^{12}C and ^{16}O) are much more abundant than their counterparts with an uneven number of neutrons (^{13}C and ^{17}O). Despite the *overall* limited variation in binding energy per nucleon as a function of the mass number for the heavier elements, its variation among isotopes of an element may vary substantially, leading to a preferred occurrence of even–even isotopes, as illustrated by the corresponding relative isotopic abundances for elements such as Cd and Sn (Table 1.1). In both the lower (^{106}Cd through ^{110}Cd) and the higher (^{114}Cd through ^{116}Cd) mass ranges, only Cd isotopes with an even mass number occur. In addition, the natural relative isotopic abundances for ^{113}Cd and, to a lesser extent, ^{111}Cd are low in comparison with those of the neighboring Cd isotopes with an even mass number. Similarly, Sn, for which 7 out of its 10 isotopes are characterized by an even mass number, the isotopes with an odd mass number have a lower natural relative abundance than their neighbors.

Table 1.1 Isotopic composition of Cd and Sn according to Böhlke *et al.* [11].

Element	Atomic number *Z*	Isotopes and natural relative abundances (mol%)	
Cd	48	^{106}Cd	1.25
		^{108}Cd	0.89
		^{110}Cd	12.49
		^{111}Cd	12.80
		^{112}Cd	24.13
		^{113}Cd	12.22
		^{114}Cd	28.73
		^{116}Cd	7.49
Sn	50	^{112}Sn	0.97
		^{114}Sn	0.66
		^{115}Sn	0.34
		^{116}Sn	14.54
		^{117}Sn	7.68
		^{118}Sn	24.22
		^{119}Sn	8.59
		^{120}Sn	32.58
		^{122}Sn	4.63
		^{124}Sn	5.79

1.4 Natural Isotopic Composition of the Elements

As a first approximation, it can be stated that all elements have an isotopic composition that is invariant in Nature. This is the result of thorough mixing of most nuclides prior to the formation of the solar system some 4.6 billion years ago [12]. Addition of a stable isotopic tracer to a natural system induces a change in the isotopic composition of a target element. The use of isotope dilution for elemental assay (Chapter 8) and tracer experiments for monitoring a physical or a (bio)chemical process (Chapter 16) are based on measuring the induced changes in the isotopic composition of a target element thus obtained, as discussed in later chapters.

There are, however, various reasons why elements may show variations in their natural isotopic composition:

- **Radiogenic nuclides:** Some elements have one or several radiogenic nuclides, meaning that over time, such a nuclide is being produced as a result of the decay of a naturally occurring and long-lived radionuclide. The additional production of such a radiogenic nuclide has a pronounced effect on the temporal isotopic composition of the element with (the) radiogenic nuclide(s) and – as a result of the way in which isotopic abundances are defined – also affects the relative isotopic abundances of the other isotopes.

- **Extraterrestrial materials:** In some extraterrestrial material such as meteorites, elements may show isotopic compositions that are distinct from all terrestrial material investigated. This is related to decay of radionuclides that may already be extinct, due to half-lives which are very short compared with the age of the solar system of 4.6×10^9 years. Such variations are rare for terrestrial materials, in large part due to preferential sampling of the crust, whereas some extraterrestrial material, such as iron meteorites, resemble the Earth's core, in which parent to daughter element ratios may be much higher than in the crust.
- **Interaction between cosmic rays and terrestrial matter:** The Earth's atmosphere and, to a lesser extent, its surface are constantly bombarded with cosmic radiation which interacts with terrestrial material, resulting in isotopic variations in some elements. The best known example is the production of ^{14}C from ^{14}N by (n,p) reaction in the atmosphere, with the neutron involved created by cosmic ray-induced spallation. ^{14}C, a radionuclide with a half-life of 5730 years, is oxidized to CO_2 and enters the food chain via photosynthesis, thus affecting the isotopic composition of C in all living organisms.
- **Mass-dependent isotope fractionation:** The original theory that isotopes of an element are chemically identical has to be refined. As a result of their slight difference in mass, the isotopes of an element tend to participate in physical processes and (bio)chemical reactions with slightly different efficiencies. These differences in efficiency are related to a slight difference in equilibrium for each different isotopic molecule (thermodynamic effect) or in the rate with which the isotopes participate in a process or reaction (kinetic effect). This phenomenon is referred to as isotope fractionation and is well characterized for the lighter elements H, C, N, O, and S, the isotopic composition of which is typically studied via gas source isotope ratio MS. Especially the light elements that are redox sensitive show substantial variations in their isotopic composition, because different oxidation states correspond to substantially different bonding environments. In general, the extent of isotope fractionation is governed by the extent to which an element takes part in physical processes, such as diffusion, or chemical reactions wherein there is a change in bonding environment, and the relative difference in mass between the isotopes. Among the metals and metalloids, Li and B show significant natural isotopic variations as a result of isotope fractionation because of the large relative difference in mass of their isotopes. For the heavier elements – for which the isotopes show a much smaller relative mass difference – conventional wisdom held that there was minimal isotope fractionation, but the enhanced capabilities offered by state-of-art MS have demonstrated that all elements are prone to isotope fractionation, even an element as heavy as U [13].
- **Mass-independent fractionation:** Most cases of isotope fractionation are characterized by a linear relationship between the magnitude of the effect established and the difference in mass between the isotopes considered. For an increasing number of elements, however, an apparently aberrant behavior is established for some of their isotopes. This is currently a hot topic of research

and is attributed to a subtle difference in the interaction between the nucleus of those isotopes and the surrounding electron cloud, which affects the bonding environment for certain isotopes and results in mass-independent fractionation of these nuclides in chemical reactions. The interaction between the nucleus and the electron cloud in such instances is hypothesized to be influenced by variation in the volume of the nucleus or in its magnetic properties (see below). Mass-independent fractionation provides the elements with a distinct isotopic pattern that differs from that predicted by mass-dependent fractionation processes.

- **Anthropogenic effects:** Via a variety of processes, human-made changes in the isotopic composition of an element can be accomplished by enhancing the fractionation beyond those in normal reactions, or by producing specific isotopes. Production of enriched U for fueling nuclear reactors and production of enriched isotopic tracers for tracer experiments or for isotope dilution MS are examples of the effects of human intervention on isotopic compositions.

Various processes that result in isotopic variations will be discussed in more detail in the following sections. The way in which the quantification of such variations can be used in the context of various real-life applications will be demonstrated in later chapters.

1.4.1 Elements with Radiogenic Nuclides

1.4.1.1 Radioactive Decay

When considering all nuclides that occur in Nature, a distinction can be made between stable and radioactive nuclides (radionuclides). The nucleus of a radionuclide undergoes spontaneous radioactive decay, whereby it is converted into another nucleus.

Several types of radioactive decay can occur [4, 8, 14].

In the case of α-decay, an α-particle, containing two protons and two neutrons, is emitted from the nucleus, resulting in a reduction in its mass number by four units and in its atomic number by two mass units:

$$^{A}_{Z}X \rightarrow {}^{A-4}_{Z-2}Y + {}^{4}_{2}\alpha$$

α-Decay predominantly occurs for very heavy nuclides with an atomic mass number >200.

The term β-decay is used for those decay processes in which the mass number of the decaying nuclide remains the same, but the atomic number changes. This situation occurs in case of conversion of:

- a neutron into a proton, accompanied by emission of a β^- particle (i.e., an electron emitted by the nucleus, also called a negatron)
- a proton into a neutron, caused by the capture of an electron, usually coming from the K shell, or accompanied by emission of a β^+ particle (also called a positron).

Nuclides with a relative excess of neutrons are prone to β^--decay, nuclides with a relative excess of protons (or shortage of neutrons) to β^+-decay.

Whereas α-particles are characterized by specific energies, the energies of emitted β-particles show a continuous distribution. This puzzled scientists for a long time, but was understood when it was realized that β-decay is accompanied by the emission of a neutrino or anti-neutrino and the energy that is released is distributed over the β-particle and the (anti)neutrino. Hence the various forms of β-decay can be described as follows:

$$\begin{aligned} &\beta^- \text{ decay}: && {}^A_Z\text{X} \rightarrow {}_{Z+1}^{\ \ A}\text{Y} + {}_{-1}^{\ 0}\beta^- + \overline{\nu} \\ &\beta^+ \text{ decay}: && {}^A_Z\text{X} \rightarrow {}_{Z-1}^{\ \ A}\text{Y} + {}_{1}^{0}\beta^+ + \nu \\ &\text{electron capture}: && {}^A_Z\text{X} + {}_{-1}^{\ 0}\text{e}^- \rightarrow {}_{Z-1}^{\ \ A}\text{Y} + \nu \end{aligned}$$

In addition, when a nucleus is in an excited state, it can emit γ-radiation. In this case, the difference in energy between the higher and the lower energy states is emitted in the form of a γ-photon upon relaxation. Emission of this radiation does not affect either the atomic number or the mass number.

A nuclide can undergo spontaneous radioactive decay when the process is energetically favorable, which necessitates the sum of the masses of the resulting particles being smaller than the sum of the masses of the starting particles in all instances, except electron capture.

The radioactive decay process is normally a first-order reaction with a characteristic half-life $T_{1/2}$ – this is the time interval in which half of the nuclide population decays – and corresponding decay constant λ:

$$\lambda = \frac{\ln 2}{T_{\frac{1}{2}}}$$

As a result of this first-order behavior, radioactive decay is a process that can be described mathematically by an exponential decrease in the number of parent nuclides N as a function of time (Figure 1.3):

$$N_t = N_0 e^{-\lambda t}$$

where N_t is the number of parent nuclides at time t and N_0 is the original number of nuclides at time $= 0$.

Therefore, this phenomenon can be exploited for dating processes. Among the most widespread dating methods is radiocarbon dating, relying on the β^- decay of ^{14}C, which allows for age determination of the remains of living organisms (time passed since time of death), such as human remains or wood [15]. Because radiocarbon dating involves either radiometry or AMS, it will not be covered in this book. Other dating methods, such as U, Th–Pb, Pb–Pb, and Rb–Sr dating, however, can be carried out, relying on isotope ratio measurements via ICP-MS, and are discussed in detail in Chapter 9.

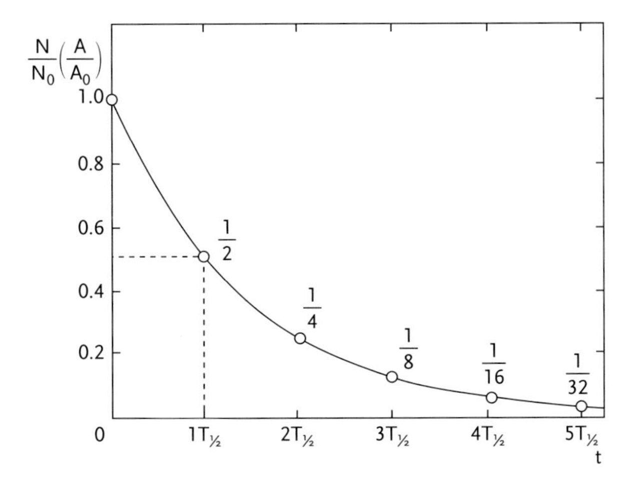

Figure 1.3 Decrease in the number of parent nuclides as a function of time *t* as a result of radioactive decay.
Reproduced with permission of Ellis Horwood from [14].

1.4.1.2 **Elements with Radiogenic Nuclides**

Some elements show natural variations in their isotopic composition because one or more of their isotopes are radiogenic [9, 16]. A radiogenic nuclide is continuously produced by decay of a naturally occurring radionuclide, thus leading to a steadily increasing relative abundance of this isotope, at least as long as the parent radionuclide and daughter radiogenic nuclide reside together. Because of the way in which relative isotopic abundances are calculated, the presence of one radiogenic nuclide affects the relative abundance of all isotopes of the daughter element.

The natural variation observed for elements with radiogenic isotopes is fairly pronounced and their isotopic analysis can serve various purposes. Once the half-life of the parent nuclide is known, then is its decay rate and thus the generation rate of the daughter (or progeny) nuclide and isotopic analysis can be used for dating (age determination) purposes, as described in Chapter 9. In addition to the geochronological application, isotopic analysis of these elements can also be used for provenance determination, that is, determination of the (geographic) origin of a material, object, or living species, as is discussed in detail in later chapters.

Examples of elements showing radiogenic isotope accessible via ICP-MS are provided below.

Strontium Strontium has four stable isotopes (^{84}Sr, ^{86}Sr, ^{87}Sr, and ^{88}Sr), one of which, ^{87}Sr, is radiogenic, as it is produced via the β^- decay of ^{87}Rb:

$$^{87}\mathrm{Rb} \rightarrow {}^{87}\mathrm{Sr} + \beta^- + \overline{\nu}$$

where β^- represents the particle emitted from the ^{87}Rb nucleus and $\overline{\nu}$ an antineutrino. The corresponding half-life is 48.8×10^9 years, which is more than 10 times the age of the solar system, such that only a limited fraction of ^{87}Rb has decayed so far.

Although the isotopic composition of Rb, which has two isotopes, ^{85}Rb and ^{87}Rb, varies slowly as a function of time, no fractionation of Rb isotopes is known to occur so that the isotopic composition of Rb currently is the same for all terrestrial material. In contrast, the isotopic composition of Sr does show natural variation as a result of the decay of ^{87}Rb and between rocks or minerals the isotopic composition of Sr will vary, depending on their Rb/Sr ratio and the time during which these elements have spent together. As the Rb/Sr elemental ratio and the Sr isotopic composition or, more specifically, the ratio of the radiogenic nuclide to a reference isotope, typically $^{87}Sr/^{86}Sr$, are measurable, this decay can be used for geochronological purposes (see Chapter 9). In addition, Sr isotopic analysis is also deployed in the context of the determination of the provenance of, among other things, agricultural products and human remains (in an archaeological or forensic context) and for providing insight into human and animal migration behavior [17] (see Chapters 13 and 14).

Lead Lead has four stable isotopes, three of which are radiogenic. The decay chains of ^{238}U ($T_{1/2} = 4.468 \times 10^9$ years), ^{235}U ($T_{1/2} = 0.407 \times 10^9$ years), and ^{232}Th ($T_{1/2} = 14.010 \times 10^9$ years) finally result in ^{206}Pb, ^{207}Pb, and ^{208}Pb as daughters, respectively (Figure 1.4), and only ^{204}Pb is not radiogenic.

For all three decay chains, the first step is much slower than the subsequent steps and, for many practical purposes, the entire chain can be described as a one-step process. As a result of these processes, the isotopic composition of Pb is governed by the U/Pb and Th/Pb elemental ratios and the times during which these elements have resided together. There is a substantial difference between the isotopic composition of crustal Pb and that in ores, because upon ore formation Pb was separated from U and Th and its isotopic composition was therefore "frozen" at that moment, while in the Earth's crust the decay of the parents continued to affect the isotopic composition of Pb. As a result, Pb isotopic analysis provides an excellent tool to distinguish between local (crustal) Pb and Pb pollution resulting from ore-derived Pb (e.g., used in anti-knock compounds previously added to petrol or used for other industrial purposes) [19–25]. Pb ores also show different isotopic signatures among them and therefore, Pb isotopic analysis is also widely deployed as a means of provenance determination of objects of art [17] or of tracing transport of dust [22, 23, 26].

For environmental samples, such as atmospheric aerosols, sediments, and snow, Pb from several sources may contribute to the total Pb concentration and therefore the isotopic signature will be a mixture of those of the various contributions. When there is mixing between Pb from two sources, the Pb isotope ratio results plotted on a three-isotope plot, that is, one isotope ratio of Pb, plotted as a function of another with a common denominator (Figure 1.5), will fall on a straight mixing line between the two end-member compositions, and the extent to which each of the two sources contributes to the sample's signature can be calculated [27]. If one Pb isotope ratio is plotted as a function of the Pb

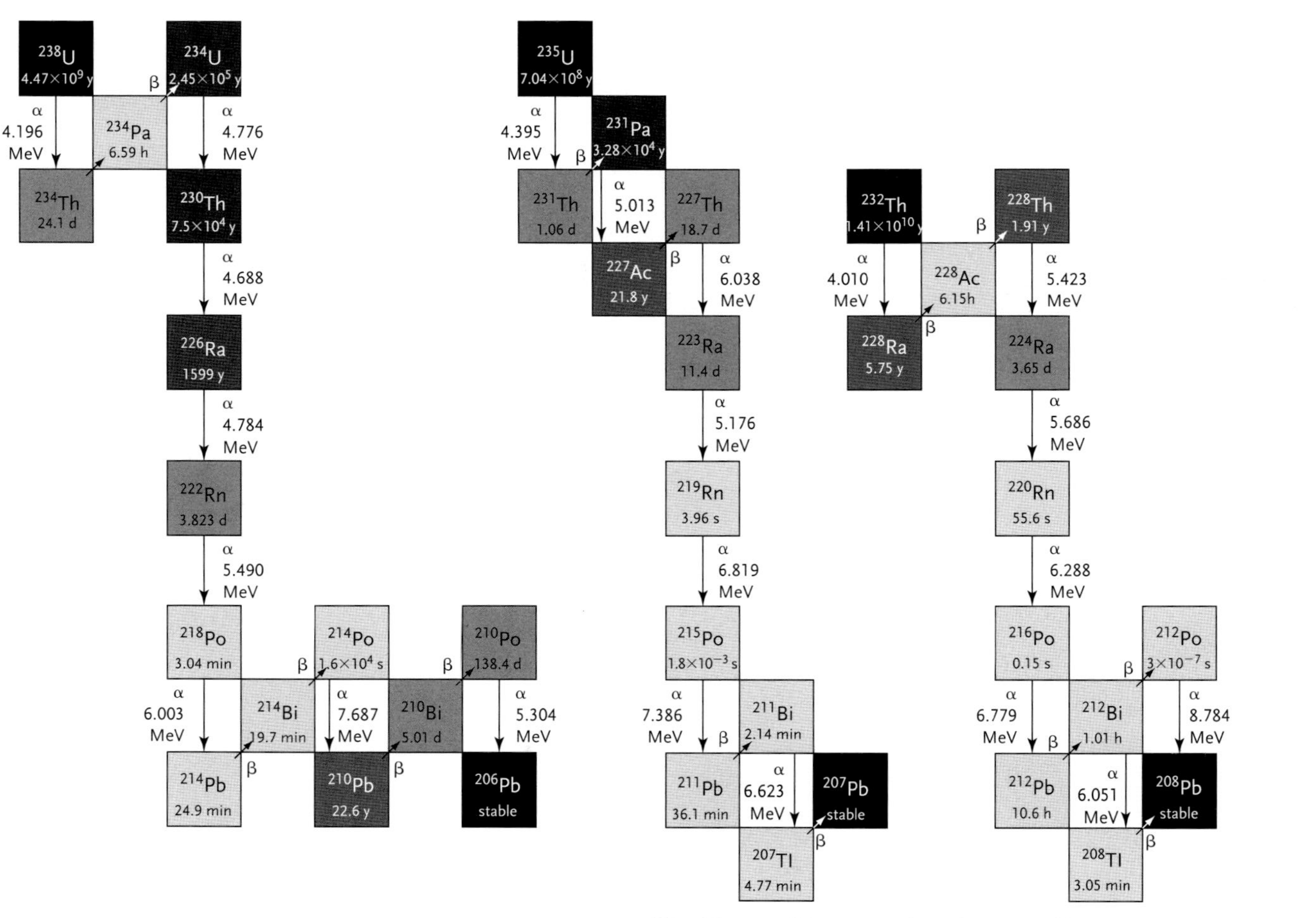

Figure 1.4 Decay of the naturally occurring and long-lived radionuclides ^{238}U, ^{235}U, and ^{232}Th, with ^{206}Pb, ^{207}Pb, and ^{208}Pb as stable end products, respectively. The gray scale of the nuclides in the decay chains gives an indication of the corresponding half-lives, with darker gray for longer values. Reproduced with permission of the Mineralogical Society of America – Geochemical Society from [18].

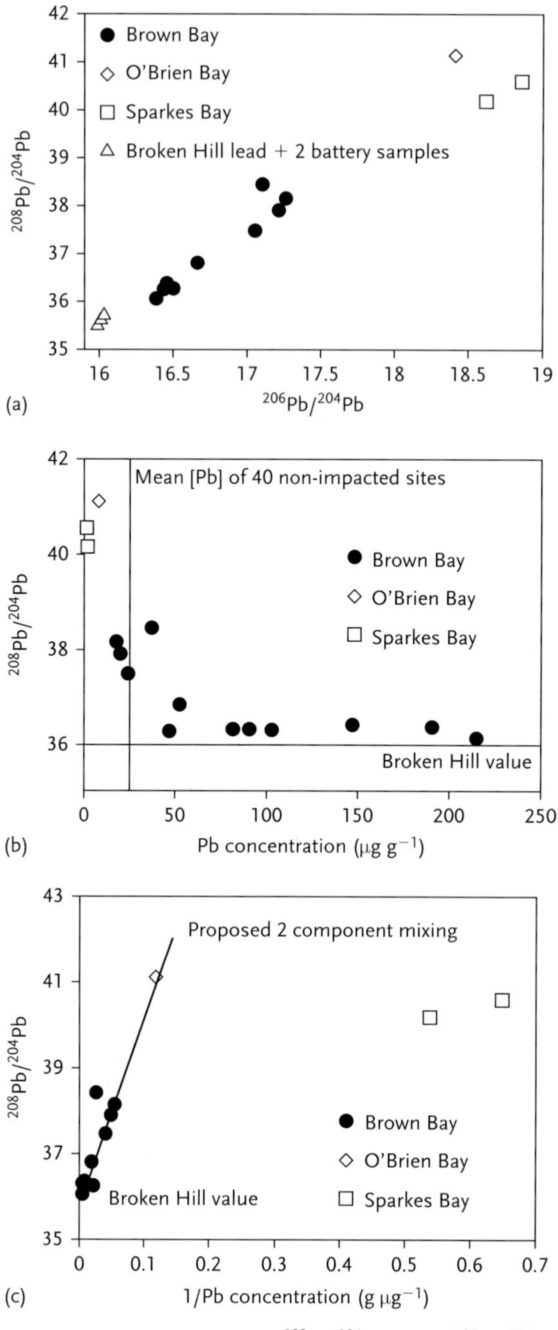

Figure 1.5 Three-isotope plot ($^{208}Pb/^{204}Pb$ versus $^{206}Pb/^{204}Pb$) for Antarctic sediments from the Pb-polluted Brown Bay (filled circles) and proxies for potential Pb sources – Broken Hill Pb ore and battery samples for anthropogenic Pb, on the one hand, and unpolluted sediments for local crustal Pb, on the other (O'Brien Bay and Sparkes Bay). (b) $^{208}Pb/^{204}Pb$ versus Pb concentration. (c) $^{208}Pb/^{204}Pb$ versus 1/Pb concentration.
Reproduced with permission of the Royal Society of Chemistry from [28].

concentration, the samples will plot on a hyperbolic line between the two end-members. This hyperbolic curve can be linearized for easier handling by plotting the isotope ratio as a function of the reciprocal Pb concentration (1/[Pb]).

If not two, but three end-members contribute to the final isotope ratios of the samples, the sample data will plot in the three-isotope plot in a triangular field, delimited by lines connecting the ratios of the end-members. As more end-members contribute to the final isotopic signature of a sample, it becomes increasingly more complex to reveal the extent to which the end-members contribute.

Other Parent–Daughter Pairs A summary of elements that contain one or more radiogenic nuclides and for which the isotopic composition is studied by (multi-collector) ICP-MS is provided in Table 1.2. Elements with radiogenic nuclides not included in Table 1.2 either are not amenable to ICP-MS analysis (e.g., Ar) or have such long half-lives that the variation in their isotopic compositions is too limited to be quantified using present-day ICP-MS instrumentation.

1.4.2 Effects Caused by Now Extinct Radionuclides

Throughout the history of our solar system, several radionuclides characterized by a half-life that is short with respect to the age of the solar system have become extinct. Variation in the isotopic composition of the element containing the corresponding daughter nuclide is sometimes only noticeable in extra-terrestrial material, such as meteorites. One of these now extinct radionuclides is ^{182}Hf that β^--decayed to ^{182}Ta with a half-life of ~9 million years. ^{182}Ta also undergoes β^--decay to ^{182}W with a half-life of only 144 days. As the character of Hf is a lithophile, preferring to reside in the Earth's crust, whereas W is a siderophile, preferring the Earth's core, Hf and W became separated from one another during segregation of the Earth's core. As a result, crustal material has a higher Hf/W elemental ratio than does core material. If differentiation occurred before ^{182}Hf became extinct, the decay of ^{182}Hf would have affected the isotopic composition of W in the crust more than if differentiation occurred after ^{182}Hf became extinct. The Earth's core is not accessible, but iron meteorites can be used as a proxy for planetary cores, including the Earth. Compared with crustal materials on Earth, iron meteorites indeed show lower relative abundances of ^{182}W. The ^{182}Hf–^{182}W chronometer can therefore be used to constrain the timing of planet differentiation [29].

Another example of an extinct isotope is ^{107}Pd that β^--decayed into ^{107}Ag with a half-life of 6.5 million years. The ^{107}Ag/^{109}Ag ratio is nearly constant at 1.08 for terrestrial materials, whereas iron meteorites have ratios as high as 9. No terrestrial materials have a Pd/Ag elemental ratio high enough to have resulted in such high ^{107}Ag/^{109}Ag ratios [30, 31].

1.4.3 Mass-Dependent Isotope Fractionation

As a result of their difference in mass, the isotopes of an element can participate in physical processes and/or chemical reactions with slight differences in efficiency,

Table 1.2 Elements with radiogenic nuclides that can be measured via (multi-collector) ICP-MS [9, 16].

Element containing radiogenic nuclide(s)	Isotopes (isotopic abundance as mole fraction [11]) with radiogenic nuclides indicated by the arrow	Parent radionuclide ($T_{1/2}$)	Radioactive decay
Sr	^{84}Sr (0.0055–0.0058)		
	^{86}Sr (0.0975–0.0999)		
	⇒ ^{87}Sr (0.0694–0.0714)	^{87}Rb (48.8 × 10^9 years)	$^{87}Rb \rightarrow {}^{87}Sr + \beta^- + \bar{\nu}$
	^{88}Sr (0.8229–0.8275)		
Nd	^{142}Nd (0.2680–0.2730)		
	⇒ ^{143}Nd (0.1212–0.1232)	^{147}Sm (1.06 × 10^{11} years)	$^{147}Sm \rightarrow {}^{143}Nd + \alpha$
	^{144}Nd (0.2379–0.2397)		
	^{145}Nd (0.0823–0.0835)		
	^{146}Nd (0.1706–0.1735)		
	^{148}Nd (0.0566–0.0678)		
	^{150}Nd (0.0553–0.0569)		
Hf	^{174}Hf (0.001619–0.001621)		
	⇒ ^{176}Hf (0.05206–0.05271)	^{176}Lu (3.57 × 10^{10} years)	$^{176}Lu \rightarrow {}^{176}Hf + \beta^- + \bar{\nu}^a$
	^{177}Hf (0.18593–0.18606)		
	^{178}Hf (0.27278–0.27297)		
	^{179}Hf (0.13619–0.13630)		
	^{180}Hf (0.35076–0.35100)		
Os	$^{184}Os^b$		
	^{186}Os		
	⇒ ^{187}Os	^{187}Re (4.161 × 10^{10} years)	$^{187}Re \rightarrow {}^{187}Os + \beta^- + \bar{\nu}$
	^{188}Os		
	^{189}Os		
	^{190}Os		
	^{192}Os		
Pb	^{204}Pb (0.0104–0.0165)		
	⇒ ^{206}Pb (0.2084–0.2748)	^{238}U (4.468 × 10^9 years)	See Figure 1.4
	⇒ ^{207}Pb (0.1762–0.2365)	^{235}U (0.407 × 10^9 years)	
	⇒ ^{208}Pb (0.5128–0.5621)	^{232}Th (14.010 × 10^9 years)	

^{a} There is also a smaller fraction (3%) of ^{176}Lu that decays to ^{176}Yb via electron capture.
^{b} No information provided in [11].

leading to mass-dependent isotope fractionation. In addition to mass-dependent fractionation, mass-independent isotope fractionation has also been observed for metalloids and metallic elements (see below), but their occurrence is much less common and their effect often less substantial. As a result, mass-independent fractionation effects were discovered later and are still less understood. When in a text or scientific paper there is no indication of whether the isotope fractionation is mass-dependent or mass-independent, it is implicitly assumed to be mass-dependent.

Because the changes in the isotopic composition due to isotope fractionation are small and a difference in isotope ratios relative to one another can be determined more easily than an absolute isotope ratio, an isotope ratio for a sample is usually expressed as the difference between the specific isotope ratio and that of a selected standard:

$$\delta = \frac{R_{\text{sample}} - R_{\text{standard}}}{R_{\text{standard}}} \times 1000(‰)$$

The difference is multiplied by 1000, and thus expressed in units of permil (‰), to obtain values that can be dealt with easily. Because increasingly small differences in isotope ratios can be measured, a multiplication factor of 10 000 is sometimes used (ε values):

$$\varepsilon = \frac{R_{\text{sample}} - R_{\text{standard}}}{R_{\text{standard}}} \times 10\,000$$

Mass-dependent isotope fractionation effects provide insight into the physical processes and chemical reactions during which they occurred and the prevailing conditions (such as pH, temperature, salinity, and oxidation potential). As a result, determination of isotope ratios affected by mass-dependent fractionation in samples that are chronological archives, such as speleothems, corals, or forams, can be used as paleoproxies for the conditions mentioned above, as will be discussed for paleoredox proxies in Chapter 11.

1.4.3.1 Isotope Fractionation in Physical Processes

A very relevant example of mass-dependent isotope fractionation involves the slow evaporation of water out of a glass. The water vapor produced is slightly isotopically lighter than the liquid water remaining in the glass because the phase with the stronger bonds, in this case liquid water, preferentially takes up the heavier isotope. This is because it requires less energy for a water molecule containing the lighter ^{16}O to be transferred from the liquid to the gas phase than it does for a water molecule containing ^{18}O. When the glass is half empty, the remaining water will be slightly enriched in ^{18}O compared with the water originally present in the full glass. The variation in the isotopic composition of O remaining in the glass can be described by the Rayleigh equation [32], which describes distillation or condensation under equilibrium conditions:

$$R_t = R_0 f_t^{(\alpha-1)}$$

where R_t is the $^{18}O/^{16}O$ ratio in the liquid phase at time $= t$, R_0 the same isotope ratio at time $= 0$, f_t the fraction of water remaining in the glass at time t and α the fractionation factor. In the case of evaporation of liquid water, the fractionation factor α is defined as

$$\alpha = \sqrt{\frac{(^{18}O/^{16}O)_{vapor}}{(^{18}O/^{16}O)_{liquid}}}$$

This fractionation factor varies as a function of temperature, approaching 1 at high temperatures for most processes and reactions.

Rayleigh's law can be applied in a physicochemical process in which the substrate is "consumed" and the product is removed. It is clear that the denser of the two phases considered will become enriched in the heavier of the two isotopes considered.

1.4.3.2 **Isotope Fractionation in Chemical Reactions**

For metals and metalloids, the isotope fractionation accompanying chemical transformations is of scientific interest. In the late 1940s, the pioneers Bigeleisen and Urey published papers describing the theoretical origin of isotope fractionation effects [33–36] and investigations into the mechanisms of isotope fractionation are still ongoing. Mass-dependent isotope fractionation in chemical reactions is a quantum mechanical phenomenon, into which basic or intuitive insights can be obtained from potential energy curves and vibrational energy levels for the molecules involved in the reaction.

Consider the simple example of the dissociation of a diatomic molecule into the corresponding atoms. The potential energy curve for a diatomic molecule can be approximated as a harmonic oscillator, for which the vibrational energy (E_{vib}) is quantized (Figure 1.6) [37, 38]. The molecule vibrates with a specific frequency, and the maximum displacement of the constituent atoms with respect to the

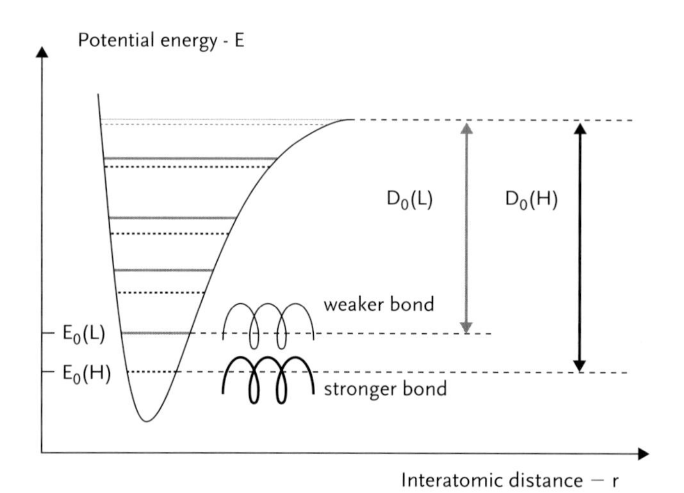

Figure 1.6 Potential energy as a function of interatomic distance in a diatomic molecule (Morse curve) showing vibrational energy levels for molecules containing the heavier (H) and lighter (L) of the two isotopes considered. The dissociation energies D_0 for these two types of molecules are also indicated.
Reproduced with permission of Springer from [37].

center can only adopt specific values. As a result, the diatomic molecule considered can only occupy discrete vibrational energy levels, described by

$$E_{\mathrm{vib}} = \left(n + \frac{1}{2}\right) h\nu$$

where n is the vibrational quantum number, h Planck's constant, and ν the vibrational frequency, which is determined by the force constant k (corresponding to the bond strength within the molecule) and the reduced mass of the diatomic molecule μ:

$$\nu = \frac{1}{2\pi}\sqrt{\frac{k}{\mu}}$$

The reduced mass is defined as

$$\frac{1}{\mu} = \frac{1}{m_1} + \frac{1}{m_2}$$

or

$$\mu = \frac{m_1 m_2}{m_1 + m_2}$$

where m_1 and m_2 are the masses of the constituent atoms.

From the above equations, it is clear that a diatomic molecule containing a heavier isotope will show a higher reduced mass than one with a lighter isotope. As a result, the latter will vibrate at a higher frequency. Vibrational energies for molecules containing the heavier and the lighter isotope, respectively, will be shifted with respect to one another.

When external energy such as heat or light is provided to the molecule, its vibrational energy can increase stepwise and ultimately the molecule can dissociate into the constituent atoms. The energy difference between the lowest vibrational energy level (with $n = 0$) and the level at which there is no more attraction between the atoms is termed the dissociation energy, which will differ depending on whether the light or the heavy isotope is present in the molecule. As a result, a molecule containing the light isotope will dissociate more readily than one containing the heavy isotope, resulting in mass-dependent fractionation whereby the molecule will be enriched in the heavier isotope and the separated atoms in the lighter isotope.

For more complex chemical reactions, consider that the reactants interact and form an activated complex, which can then be converted into the reaction products. The corresponding changes in potential energy are depicted for an exothermic reaction (assuming the contribution of the change in entropy ΔS to be negligible compared with that in enthalpy, such that $\Delta G = \Delta H - T\Delta S \approx \Delta H$) in Figure 1.7. If an element with more than one isotope is present, potential energy curves and vibrational energy levels for each of the stages involved, namely reagent, activated complex, and reaction product, will exist. A distinction can be

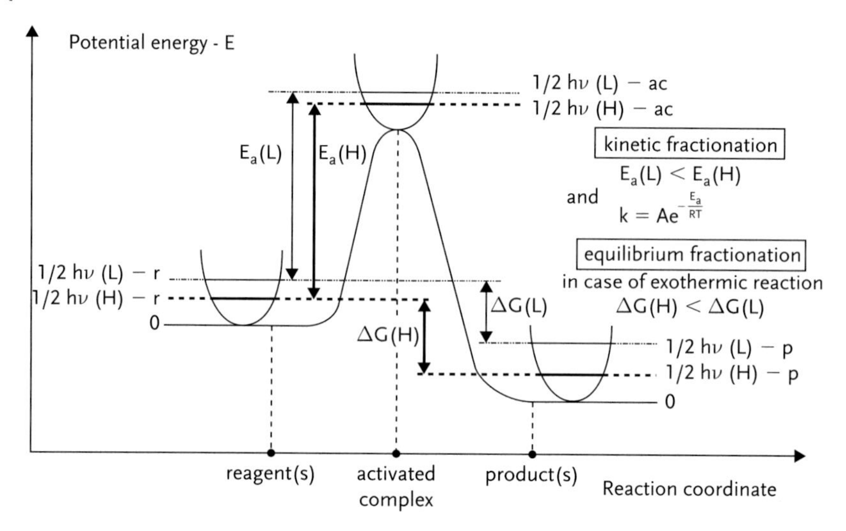

Figure 1.7 Variation of the potential energy as a function of the reaction coordinate. The potential energy curves for the reagent, activated complex, and reaction product containing the element considered are indicated, together with the vibrational energy levels, which vary according to the mass of the isotope present. Differences in ΔG govern thermodynamic isotope fractionation, whereas differences in E_a govern kinetic isotope effects. The latter occur when equilibrium cannot be reached.

made between two conditions: (i) one where chemical equilibrium is established and (ii) one where chemical equilibrium is not reached within the time frame considered.

If equilibrium can be established, the fractionation is of thermodynamic origin and its extent is governed by the difference in vibrational energy levels of both reagent and reaction product as a function of isotope mass. ΔG values differ depending on whether the light or the heavy isotope is present and this affects the equilibrium constant, as:

$$K = e^{-\frac{\Delta G}{RT}}$$

where R is the universal gas constant and T the prevailing temperature. The reagent or reaction product characterized by the strongest chemical bond will be enriched in the heavier isotope.

If within the time frame considered, however, chemical equilibrium is not attained, such as when the reaction is unidirectional (as in enzymatic reactions), when the reaction is proceeding at a relatively low temperature, or when the reaction products are removed, the fractionation is of kinetic origin and its extent is governed by the difference in vibrational energy levels of both reagent and activated complex. Activation energies (E_a) differ depending on whether the light or the heavy isotope is present and this affects the reaction rate k according to

$$k = Ae^{-\frac{E_a}{RT}}$$

where A is a constant, characteristic of the reaction considered.

The extent of isotope fractionation can be quantitatively expressed via the fractionation factor α:

$$\alpha = \frac{\left(\frac{N_2}{N_1}\right)_{\text{reaction product}}}{\left(\frac{N_2}{N_1}\right)_{\text{reagent}}} = \frac{\left(\frac{n_2}{n_1}\right)_{\text{reaction product}}}{\left(\frac{n_2}{n_1}\right)_{\text{reagent}}}$$

where N represents the number of nuclides and n the number of moles of the two isotopes considered.

If an element shows three isotopes or more, the mass dependence (usually – see below) of the isotope fractionation observed is expressed by the mass fractionation law:

$$\alpha_2 = \frac{\left(\frac{N_2}{N_1}\right)_{\text{reaction product}}}{\left(\frac{N_2}{N_1}\right)_{\text{reagent}}} \quad \text{and} \quad \alpha_3 = \frac{\left(\frac{N_3}{N_1}\right)_{\text{reaction product}}}{\left(\frac{N_3}{N_1}\right)_{\text{reagent}}}$$

with $\alpha_2 = \alpha_3^{\beta}$. It can be shown [39] that if the isotope fractionation is purely thermodynamically controlled, the mass fractionation factor β corresponds to

$$\beta = \frac{\frac{1}{m_1} - \frac{1}{m_2}}{\frac{1}{m_1} - \frac{1}{m_3}}$$

where m is the atomic mass.

If, on the other hand, the isotope fractionation is kinetically controlled, the mass fractionation factor β corresponds to

$$\beta = \frac{\ln(m_1) - \ln(m_2)}{\ln(m_1) - \ln(m_3)}$$

where m is either the atomic mass or the molecular mass of the compound containing the element under consideration for a transport process or the reduced mass for a process involving breaking bonds.

Maréchal *et al.* [40] defined a generalized law for the mass fractionation factor β:

$$\beta = \frac{m_1^n - m_2^n}{m_1^n - m_2^n}$$

in which $n = -1$ in the case of purely thermodynamic isotope fractionation, and approaches 0 for kinetically governed isotope fractionation.

Calculation of the mass fractionation factor β from experimental data provides insight into whether thermodynamic or equilibrium as opposed to kinetic effects are at the origin of the mass-dependent isotope fractionation, although often the fractionation is shown to be of "mixed" origin. For example, Wombacher *et al.*

studied the evaporation of molten Cd to determine the nature of the isotope fractionation thereby occurring [41]. The data plotted on a three-isotope plot with one isotope ratio of Cd plotted as a function of another one with a common denominator showed a linear relationship. The slope of the best-fitting straight line is the experimental value for β, which was compared with the predicted values based on the assumption of either purely thermodynamic or purely kinetic fractionation. The authors concluded that both kinetically and thermodynamically governed fractionation accompanied the process, resulting in a value of -0.35 for the exponent n.

The extent to which isotope fractionation is observed for a given element is determined by both the relative mass difference between its isotopes and the extent to which the element participates in physical processes and/or chemical reactions. For the light elements H, C, N, O, and S, variations in their isotopic composition caused by mass-dependent isotope fractionation have been extensively studied using gas source isotope ratio mass spectrometry. For most of the metallic and metalloid elements (except Li and B), the relative mass difference between the isotopes is more limited, such that the variation in isotopic composition thus created is considerably more limited. The high precision with which isotope ratios can be measured nowadays however, not only allows the small isotope fractionations to be revealed, but also quantified. Even for the heaviest naturally occurring element U, variation in its isotopic composition due to the occurrence of isotope fractionation has been demonstrated [13]. As a general rule, elements that can occur in the environment in several oxidation states tend to show more pronounced isotope fractionation.

The introduction of multi-collector ICP-MS has given rise to a breakthrough in this field and the corresponding studies are providing useful information in a variety of fields, as will be illustrated in later chapters. For a more detailed description of isotope fractionation in a variety of processes, the reader is referred to [42].

1.4.4 Mass-Independent Isotope Fractionation

All of the isotope fractionation effects discussed so far are "mass-dependent." This means that there is a linear relationship between the extent of fractionation observed and the mass difference between the isotopes considered. Some processes, however, give rise to an apparently aberrant behavior, whereby one or more isotopes display an additional effect on top of the well-understood mass-dependent fractionation. This apparently aberrant behavior is referred to as "mass-independent" or "anomalous" isotope fractionation. This effect is illustrated in Figure 1.8, where the extent of isotope fractionation experimentally observed for methylation of inorganic Sn using methylcobalamine is plotted as a function of the mass difference between the Sn isotope considered and the reference isotope ^{116}Sn. When this reaction proceeds in the dark, only mass-dependent fractionation is observed for all isotopes. However, under UV radiation, the two odd-numbered isotopes of Sn studied show a considerably more pronounced fractionation effect

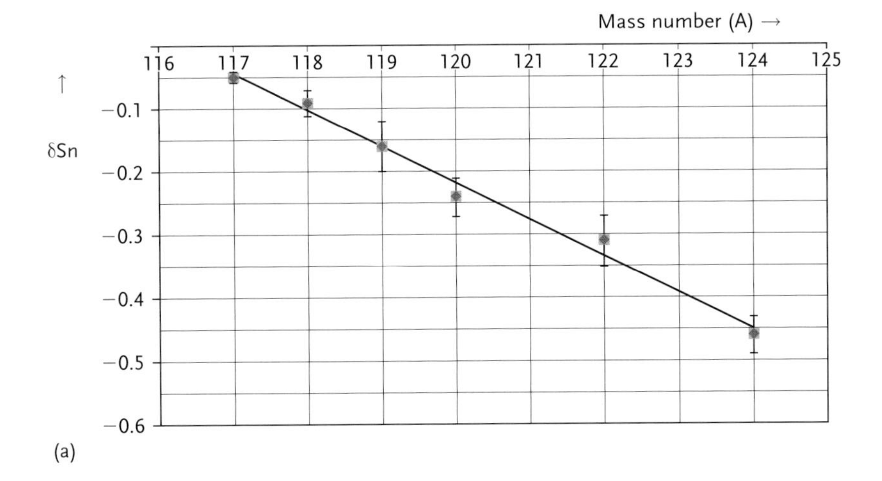

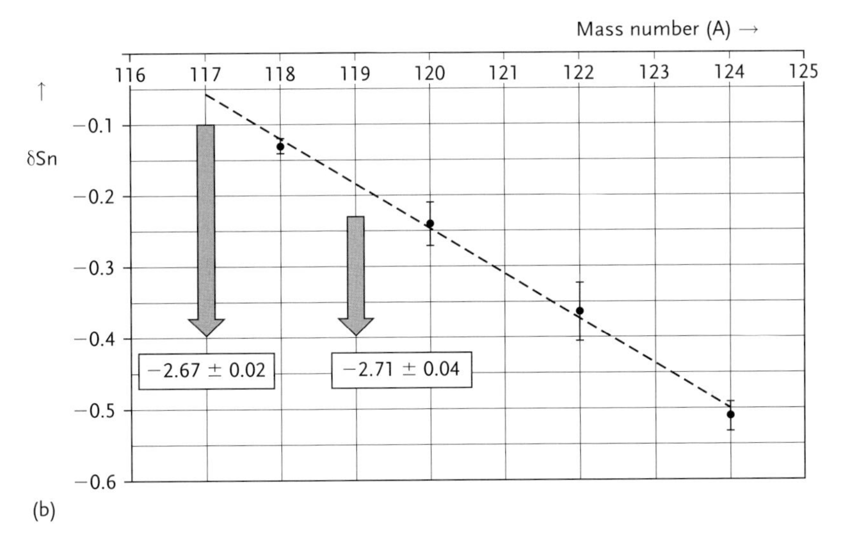

Figure 1.8 Isotope fractionation, expressed as $\delta_{Sn} = [(^{nnn}Sn/^{116}Sn)_{MeSn}/(^{nnn}Sn/^{116}Sn)_{inorganic\ Sn}] - 1 \times 1000‰$, for methylation of inorganic Sn using methylcobalamine as a function of the mass difference between the isotopes and the reference isotope ^{116}Sn. (a) Reaction in the dark, wherein only mass-dependent fractionation is observed. (b) Reaction under UV irradiation, which shows mass-independent fractionation for the odd-numbered isotopes ^{117}Sn and ^{119}Sn in addition to mass-dependent fractionation. Reproduced with permission of the Royal Society of Chemistry from [43].

and the corresponding results deviate from the linear correlation observed on the basis of the data for the other, even-numbered Sn isotopes. Below, such effects will be consistently termed mass-independent isotope fractionation.

The extent of *mass-independent* fractionation is indicated by the capital delta (Δ) value and is calculated by subtraction of the contribution from mass-dependent fractionation, as estimated from the behavior of the isotopes only displaying mass-dependent fractionation, from the experimentally observed fractionation [44]. For the example of Sn, the mass-independent contribution is calculated as follows [39, 45]:

$$\Delta^{117}\mathrm{Sn}/^{116}\mathrm{Sn} = \delta^{117}\mathrm{Sn}/^{116}\mathrm{Sn} - \left[\ln\left(\frac{m^{117}\mathrm{Sn}/m^{116}\mathrm{Sn}}{m^{124}\mathrm{Sn}/m^{116}\mathrm{Sn}}\right) \times \delta^{124}\mathrm{Sn}/^{116}\mathrm{Sn}\right]$$

$$\Delta^{118}\mathrm{Sn}/^{116}\mathrm{Sn} = \delta^{118}\mathrm{Sn}/^{116}\mathrm{Sn} - \left[\ln\left(\frac{m^{118}\mathrm{Sn}/m^{116}\mathrm{Sn}}{m^{124}\mathrm{Sn}/m^{116}\mathrm{Sn}}\right) \times \delta^{124}\mathrm{Sn}/^{116}\mathrm{Sn}\right]$$

$$\Delta^{119}\mathrm{Sn}/^{116}\mathrm{Sn} = \delta^{119}\mathrm{Sn}/^{116}\mathrm{Sn} - \left[\ln\left(\frac{m^{119}\mathrm{Sn}/m^{116}\mathrm{Sn}}{m^{124}\mathrm{Sn}/m^{116}\mathrm{Sn}}\right) \times \delta^{124}\mathrm{Sn}/^{116}\mathrm{Sn}\right]$$

$$\Delta^{120}\mathrm{Sn}/^{116}\mathrm{Sn} = \delta^{120}\mathrm{Sn}/^{116}\mathrm{Sn} - \left[\ln\left(\frac{m^{120}\mathrm{Sn}/m^{116}\mathrm{Sn}}{m^{124}\mathrm{Sn}/m^{116}\mathrm{Sn}}\right) \times \delta^{124}\mathrm{Sn}/^{116}\mathrm{Sn}\right]$$

$$\Delta^{122}\mathrm{Sn}/^{116}\mathrm{Sn} = \delta^{122}\mathrm{Sn}/^{116}\mathrm{Sn} - \left[\ln\left(\frac{m^{122}\mathrm{Sn}/m^{116}\mathrm{Sn}}{m^{124}\mathrm{Sn}/m^{116}\mathrm{Sn}}\right) \times \delta^{124}\mathrm{Sn}/^{116}\mathrm{Sn}\right]$$

Only for ^{117}Sn and ^{119}Sn and when the reaction proceeds under UV irradiation do the Δ values differ from zero.

Nuclear volume effects, sometimes referred to as nuclear field shift effects, are believed to be one cause of mass-independent isotope fractionation [46]. Nuclei of isotopes differ from one another only in their number of neutrons. Self-evidently, this provides the isotopes with a different mass, but this may also give rise to differences in the size and shape of the nuclei among the isotopes. The nuclei of nuclides with an odd number of neutrons are often smaller than they should be based on the mass difference relative to those of the neighboring nuclides with an even number of neutrons [47]. These differences in nuclear shape and size, and thus charge density, affect the interaction between the nucleus and the surrounding electron cloud. The resulting difference between the isotopes in terms of density and shape of the electron cloud results in slight differences in the efficiency with which they participate in chemical reactions [48].

The magnetic character of some nuclei also plays an important role in mass-independent fractionation effects. Nuclides characterized by an odd number of protons or odd number of neutrons are characterized by a non-zero nuclear spin. This is what makes these nuclides amenable to investigation by nuclear magnetic resonance (NMR) spectroscopy. A non-zero nuclear spin, however, also affects the interaction between the nucleus and the surrounding electron cloud via hyperfine nuclear spin–electron spin coupling, and thus also the behavior of these nuclides in chemical reactions [49, 50].

The methylation of inorganic Sn using methylcobalamine discussed above showed mass-independent fractionation related to radicals taking part in the reaction under UV irradiation [45, 51]. Conditions leading to a reduced occurrence of radicals, for example the presence of radical scavengers such as particles or OH^- ions, were found to decrease the extent of or even remove the mass-independent fractionation observed for ^{117}Sn and ^{119}Sn. Under conditions

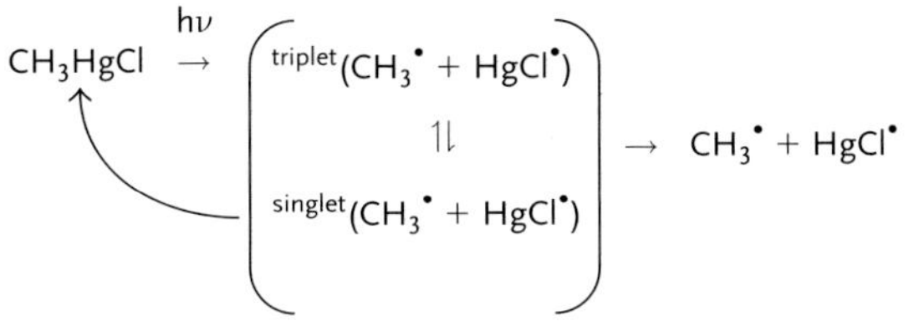

$$CH_3HgCl \xrightarrow{h\nu} \left(\begin{array}{c} {}^{triplet}(CH_3^{\bullet} + HgCl^{\bullet}) \\ \rightleftharpoons \\ {}^{singlet}(CH_3^{\bullet} + HgCl^{\bullet}) \end{array} \right) \rightarrow CH_3^{\bullet} + HgCl^{\bullet}$$

Figure 1.9 Radicalar reaction process for demethylation of $MeHg^{+}$ under UV radiation. Reproduced with permission of the Royal Society of Chemistry from [51].

favoring the existence of radicals, however, interaction between the nuclear spin of the odd-numbered Sn isotopes and their electron cloud is hypothesized to make these isotopes more amenable for the radicalar reactants, thus leading to an enrichment of the odd-numbered isotopes in methylated Sn.

Another reaction studied by Malinosvky *et al.* is the demethylation of methylmercury MeHgX under UV radiation, another process accompanied by mass-independent fractionation [51]. As shown in Figure 1.9, under the influence of UV radiation, the MeHgCl molecule is split and a short-lived so-called caged radical pair is formed. This caged radical pair is in triplet electron spin state – both electrons show the same electron spin. These radicals can either diffuse away from one another and then the Hg becomes demethylated, or a triplet–singlet electron spin conversion may occur, such that the MeHgCl can be formed again. Owing to hyperfine coupling between nuclear and electron spins, triplet–singlet conversion is believed to proceed faster for the odd-numbered Hg isotopes, so that these become enriched in the methylated form of Hg. Model calculations show that the experimental observations could not be explained solely by nuclear volume effects.

Although currently very much an "academic" topic, the study of mass-independent fractionation of metals and metalloids is also believed to have practical applications. Mass-independent fractionation provides the element with a very specific isotopic signature. For environmentally important elements, such as Hg and Sn, this can be exploited to reveal their sources and understand conversions, thereby enhancing our understanding of their biogeochemical cycles. Such fingerprints have already been demonstrated for Hg in real-life samples [52–54].

Epov *et al.* recently reviewed the literature on mass-independent fractionation for metals and metalloids [55], showing that in this emerging field most of the research so far has been devoted to Hg. However, in addition to Hg, mass-independent fractionation has been reported for Ba, Cd, Cr, Gd, Mo, Nd, Pb, Sm, Sn, Sr, Ru, Te, Ti, Zn, U, and Yb [48, 55].

1.4.5 Interaction of Cosmic Rays with Terrestrial Matter

The best known example of interactions between cosmic rays and terrestrial matter affecting the isotopic composition of an element is the production of ^{14}C from ^{14}N by (n,p) reactions in the atmosphere. The neutrons involved in the nuclear reaction are produced by spallation of nuclides in the atmosphere under the influence of cosmic radiation. ^{14}C, a radionuclide with a half-life of 5730 years, is oxidized to CO_2 and enters the food chain via photosynthesis, thus affecting the isotopic composition of C in all living organisms. As long as an organism is alive, the supply and the decay of ^{14}C are in dynamic equilibrium and the ^{14}C fraction of the carbon present in living organisms is on the order of 10^{-10}%. Once the organism dies, ^{14}C is no longer taken up and the ^{14}C present decays. Measurement of the remaining fraction of ^{14}C can provide information on the time of death for human or animal remains and materials from plants, such as wood. This forms the basis for radiocarbon or ^{14}C dating [15]. For many decades, ^{14}C dating has been performed using radiometric techniques, but AMS is now the method of choice because radiometric techniques can only use those ^{14}C atoms that decay during the duration of the measurement, whereas AMS can detect each of the ^{14}C atoms present [56–59].

Although the cosmic radiation is predominantly absorbed in the atmosphere, a small fraction reaches the surface of the Earth and both stable and radioactive cosmogenic nuclides can be produced. Owing to the very long half-lives of some of the radionuclides produced, such as ^{10}Be ($\sim 1.6 \times 10^6$ years) and ^{26}Al ($\sim 720 \times 10^3$ years), which are much longer than the half-life of ^{14}C, radiometric detection is not possible. Because of high sensitivity and minimal spectral interferences, AMS can be used to measure the corresponding isotope ratios, despite their extreme values (e.g., $<10^{-12}$). Based on knowledge of the production and decay rates of such cosmogenic radionuclides and determination of the corresponding isotope ratios, the duration of surface exposure to cosmic rays can be deduced.

Determination of the extreme isotope ratios encountered in this context is beyond the capabilities of ICP-MS techniques. However, multi-collector ICP-MS has also proven its utility in this context as it was used to obtain a more accurate value for the decay rate of ^{10}Be [60]. The radioactivity of a 100 mg l^{-1} solution with a $^{10}Be/^{9}Be$ ratio of ~ 1.4 (natural Be is monoisotopic) was measured using liquid scintillation counting, and the $^{10}Be/^{9}Be$ ratio was accurately determined via multi-collector ICP-MS. The combined results provided a half-life of $(1.388 \pm 0.018) \times 10^6$ years, permitting more accurate AMS-based data to be obtained.

1.4.6 Human-Made Variations

The best known example of human-made variations in the isotopic composition of a metal is that of U enrichment. Almost all natural uranium has (nearly) the same isotopic composition, with 99.27% ^{238}U, 0.72% ^{235}U, and 0.006% ^{234}U. The Oklo natural reactor [61] located in Gabon is the only known natural U deposit that shows a significantly different isotopic composition. Because of the earlier occurrence of a natural and self-sustaining fission process, the current relative abundance of ^{235}U has become significantly lower than 0.720% at Oklo. A fraction of the ^{235}U has been "consumed" during the fission process, as is the case in nuclear fission reactors used to produce energy. This natural fission process could take place because at the time it occurred (~2 billion years ago), the fraction of ^{235}U was high enough to sustain the fission reaction over thousands of years. Evidence of fission in U deposits with much higher grades than Oklo but having ages about 300–400 million years younger has been sought, but was never found.

For energy production in a light-water nuclear plant or for nuclear weapons, the fraction of ^{235}U in natural uranium is too low. Therefore, several approaches have been developed to increase the relative abundance of this isotope. For the production of the first atomic bomb at Oak Ridge, TN (USA), ^{235}U and ^{238}U were separated from one another by "preparative" mass spectrometry using large magnetic sectors (calutrons) after conversion to UCl_4 and electron ionization [62]. Isotope fractionation for ^{235}U enrichment is also possible via gaseous diffusion, gas centrifugation, or nozzle separation after conversion of U into UF_6 [63]. The level of enrichment required is dependent on the final use. Reactor-grade uranium intended for use as "fuel" in a nuclear fission reactor is often called low-grade uranium as an enrichment to 3–4% of ^{235}U is sufficient for this purpose. For the production of nuclear weapons, on the other hand, high-grade uranium with an isotopic abundance of $^{235}U \geq 90\%$ is used.

When enriched uranium is produced, depleted uranium (DU) with a ^{235}U isotopic abundance of $<0.720\%$ is also produced as a "waste product." DU is used in the manufacture of ammunition and projectiles that are capable of penetrating armored steel because of the high density of U metal [64]. Upon impact of such projectiles, the temperature rises due to friction and causes the uranium to catch fire and burn. Their use in recent wars in the Middle East and the former Yugoslavia has raised considerable concern because of the chemical toxicity of uranium [65]. DU is also used for the manufacture of counterweights located in the tail and the wings of airplanes with the purpose of increasing stability. The advantage of using DU in this context lies again in its very high density, enabling these counterweights to be compact, thus leaving more space for fuel. Isotopic analysis of U is carried out for a multitude of purposes. Applications based on U isotope ratio measurements are discussed in Chapter 15.

As a result of the industrial use of U, commercially acquired U-containing chemicals or U standard solutions intended for elemental assay purposes will not necessarily have the natural isotopic composition, but will most often be depleted in ^{235}U [66]. An unnatural isotopic composition may also be encountered for Li in chemicals, as it may be depleted in ^{6}Li, as a result of its use as ^{6}LiD as fusion fuel in thermonuclear weapons and in nuclear fusion [67].

Another example of an element for which human intervention is required to modify the isotopic composition, thus rendering the element more useful in specific applications, is boron. Natural B is composed of roughly 20% ^{10}B and 80% ^{11}B. As ^{10}B shows a thermal neutron cross-section (the probability that it will capture a thermal neutron) that is almost six orders of magnitude higher than that of ^{11}B, ^{10}B-enriched boron is used to control the chain reaction in nuclear fission reactors [68]. This enrichment requires human intervention. Upon neutron capture, ^{10}B undergoes (n,α) reaction, thus producing ^{7}Li. The production of these short-range α-particles also forms the basis for the use of ^{10}B in an experimental anti-cancer therapy, boron neutron capture therapy or BNCT [69]. In BNCT, the patient is first administered a ^{10}B-containing drug that selectively accumulates in the tumor tissue. Upon radiation with thermal neutrons, the nuclear reaction mentioned above occurs predominantly in tumor cells. The α-particles emitted typically travel a distance of only one cell, thus creating a far higher level of cell destruction for the neoplastic tissue than for the surrounding healthy tissue.

Finally, in the context of elemental assay via isotope dilution and tracer experiments using stable isotopes, an isotopically enriched spike or tracer is added to a sample wherein the element of interest typically shows the natural isotopic composition. These isotopically enriched spikes are another example of human-made variations.

References

1 Rutherford, E. (1911) The scattering of α and β particles by matter and the structure of the atom. *Philos. Mag.*, **21**, 669–688.

2 Geiger, H. and Marsden, E. (1909) On a diffuse reflection of the alpha-particles. *Proc. R. Soc. Lond. A*, **82**, 495–500.

3 Bohr, N. (1921) Atomic structure. *Nature*, **106**, 104–107.

4 Cunninghame, J.G. (1964) *Introduction to the Atomic Nucleus*, Elsevier, Amsterdam.

5 Budzikiewicz, H. and Grigsby, R.D. (2006) Mass spectrometry and isotopes: a century of research and discussion. *Mass. Spectrom. Rev.*, **25**, 146–157.

6 Thomson, J.J. (1913) Rays of positive electricity. *Proc. R. Soc. Lond. A*, **89**, 1–20.

7 Aston, F.W., The mass-spectra of chemical elements. *Philos. Mag.*, **39**, 611–625.

8 Ehmann, W.D. and Vance, D.E. (1991) *Radiochemistry and Nuclear Methods of Analysis*, John Wiley & Sons, Ltd., Chichester.

9 Faure, G. and Mensing, T.M. (2005) *Isotopes – Principles and Applications*, 3rd edn., John Wiley & Sons, Ltd., Chichester.

10 Lodders, K. (2003) Solar system abundances and condensation temperatures of the elements. *Astrophys. J.*, **591**, 1220–1247.

11 Böhlke, J.K., de Laeter, J.R., De Bièvre, P., Hidaka, H., Peiser, H.S., Rosman, K.J.R., and Taylor, P.D.P. (2005) Isotopic compositions of the elements, 2001. *J. Phys. Ref. Data.*, **34**, 57–67.

12 Faure, G. and Mensing, T.M. (2007) *Introduction to Planetary Science – the Geological Perspective*, Springer, Berlin.

13 Weyer, S., Anbar, A.D., Gerdes, A., Gordon, G.W., Algeo, T.J., and Boyle, E.A. (2008) Natural fractionation of ^{238}U/^{235}U. *Earth Planet. Sci. Lett.*, **72**, 345–359.

14 Navrátil, O., Hàla, J., Kopunec, R., Macásek, F., Mikulaj, V., and Lešetický, L. (1992) *Nuclear Chemistry*, Ellis Horwood, Chichester.

15 Taylor, R.E. (2000) Fifty years of radiocarbon dating. *Am. Sci.*, **88**, 60–67.

16 Dickin, A.P. (2005) *Radiogenic Isotope Geology*, 2nd edn., Cambridge University Press, Cambridge.

17 Balcaen, L., Moens, L., and Vanhaecke, F. (2010) Determination of isotope ratios of metals (and metalloids) by means of ICP-mass spectrometry for provenancing purposes – a review. *Spectrochim. Acta B*, **65**, 769–786.

18 Bourdon, B., Turner, S., Henderson, G.M., and Lundstrom, C.C. (2003) Introduction to U-series geochemistry. *Rev. Miner. Geochem.*, **52**, 1–19.

19 Chow, T.J. and Earl, J.L. (1970) Lead aerosols in the atmosphere: increasing concentrations. *Science*, **169**, 577–580.

20 Ault, W.U., Senechal, R.G., and Erlebach, W.E. (1970) Isotopic composition as a natural tracer of lead

in the environment. *Environ. Sci. Technol.*, **4**, 305–313.

21 Fachetti, S. (1988) Mass spectrometry applied to studies of lead in the atmosphere. *Mass Spectrom. Rev.*, **7**, 503–533.

22 Bollhofer, A. and Rosman, K.J.R. (2000) Isotopic source signatures for atmospheric lead: the southern hemisphere. *Geochim. Cosmochim. Acta*, **64**, 3251–3262.

23 Bollhofer, A. and Rosman, K.J.R. (2001) Isotopic source signatures for atmospheric lead: the northern hemisphere. *Geochim. Cosmochim. Acta*, **65**, 1727–1740.

24 Moor, H.C., Schaller, T., and Sturm, M. (1996) Recent changes in stable lead isotope ratios in sediments of Lake Zug, Switzerland. *Environ. Sci. Technol.*, **30**, 2928–2933.

25 Kersten, M., Garbe-Schönberg, D., Thomsen, S., Anagnostou, C., and Sioulas, A. (1997) Source apportionment of Pb pollution in the coastal waters of Elefsis Bay. *Greece Environ. Sci. Technol.*, **31**, 1295–1301.

26 Vallelonga, P., Gabrielli, P., Balliana, E., Wegner, A., Delmonte, B., Turetta, C., Burton, G., Vanhaecke, F., Rosman, K.J.R., Hong, S., Boutron C.F., Cescon, P., and Barbante, C. (2010) Lead isotopic compositions in the EPICA Dome C ice core and southern hemisphere potential source areas. *Quat. Sci. Rev.*, **29**, 247–255.

27 Döring, T., Schwikowski, M., and Gaggeler, H.W. (1997) The analysis of lead concentrations and isotope ratios in recent snow samples from high alpine sites with a double focusing ICP-MS. *Fresenius' J. Anal. Chem.*, **359**, 382–384.

28 Townsend, A.T. and Snape, I. (2002) The use of Pb isotope ratios determined by magnetic sector ICP-MS for tracing Pb pollution in marine sediments near Casey Station, East Antarctica. *J. Anal. At. Spectrom.*, **17**, 922–928.

29 Jacobsen, S.B. (2005) The Hf–W isotopic system and the origin of the Earth and the Moon. *Annu. Rev. Earth Planet. Sci.*, **33**, 531–570.

30 Kaiser, T. and Wasserburg, G.J. (1983) The isotopic composition and concentration of Ag in iron meteorites and the origin of exotic silver. *Geochim. Cosmochim. Acta*, **47**, 43–58.

31 Chen, J.H. and Wasserburg, G.J. (1990) The isotopic composition of Ag in meteorites and the presence of ^{107}Pd in protoplanets. *Geochim. Cosmochim. Acta*, **54**, 1729–1743.

32 Rayleigh, W.S. (1896) Theoretical considerations respecting the separation of gases by diffusion and similar processes. *Philos. Mag.*, **42**, 493–498.

33 Bigeleisen, J. and Mayer, M.G. (1947) Calculation of equilibrium constants for isotopic exchange reactions. *J. Chem. Phys.*, **15**, 261–267.

34 Urey, H.C. (1947) The thermodynamic properties of isotopic substances. *J. Chem. Soc.*, 562–581.

35 Bigeleisen, J. (1949) The relative reaction velocities of isotopic molecules. *J. Chem. Phys.*, **17**, 675–678.

36 Bigeleisen, J. (1965) Chemistry of isotopes. *Science*, **147**, 463–471.

37 Hoefs, J. (2004) *Stable Isotope Geochemistry*, 5th edn., Springer, Berlin.

38 Zeebe, R.E. and Wolf-Glabrow, D. (2001) *CO_2 in Seawater: Equilibrium, Kinetics, Isotopes*, Elsevier, Amsterdam.

39 Young, E.D., Galy, A., and Nagahara, H. (2002) Kinetic and equilibrium mass-dependent isotope fractionation laws in Nature and their geochemical and cosmochemical significance. *Geochim. Cosmochim. Acta*, **66**, 1095–1104.

40 Maréchal, C.N., Telouk, P., and Albarède, F. (1999) Precise isotopic analysis of copper and zinc isotopic compositions by plasma-source mass spectrometry. *Chem. Geol.*, **156**, 251–273.

41 Wombacher, F., Rehkämper, M., and Mezger, K. (2004) Determination of the mass-dependence of cadmium isotope fractionation during evaporation. *Geochim. Cosmochim. Acta*, **68**, 2349–2357.

42 Wolfsberg, M., Van Hook, W.A., Paneth, P., and Rebelo, L.P.N. (2010) *Isotope Effects in the Chemical, Geological and Biosciences*, Springer, Berin.

43 Jakubowski, N., Prohaska, T., Vanhaecke, F., Roos, P.H., and Lindemann, T. (2011) Inductively coupled plasma- and glow discharge plasma-sector field mass spectrometry – Part II: applications. *J. Anal. At. Spectrom.*, **26**, 727–757.

44 Blum, J.D. and Bergquist, B.A. (2007) Reporting of variations in the natural isotopic composition of mercury. *Anal. Bioanal. Chem.*, **388**, 333–359.

45 Malinovsky, D., Moens, L., and Vanhaecke, F. (2009) Isotopic fractionation of Sn during methylation and demethylation reactions in aqueous solutions. *Environ. Sci. Technol.*, **43**, 4399–4404.

46 Bigeleisen, J. (1996) Nuclear size and shape effects in chemical reactions. Isotope chemistry of the heavy elements. *J. Am. Chem. Soc.*, **118**, 3676–3680.

47 Aufmuth, P., Heilig, K., and Steudel, A. (1987) Changes in mean-square nuclear charge radii from optical isotope shifts. *At. Data Nucl. Data Tables*, **7**, 455–490.

48 Fujii, T., Moynier, F., and Albarède, F. (2009) The nuclear field shift effect in chemical exchange reactions. *Chem. Geol.*, **267**, 139–158.

49 Buchachenko, A.L. (1995) MIE versus CIE: comparative analysis of magnetic and classical isotope effects. *Chem. Rev.*, **95**, 2507–2528.

50 Buchachenko, A.L. (2001) Magnetic isotope effect: nuclear spin control of chemical reactions. *J. Phys. Chem. A*, **105**, 9995–10011.

51 Malinovsky, D., Latruwe, K., Moens, L., and Vanhaecke, F. (2010) Experimental study of mass-independence of Hg isotope fractionation during photodecomposition of dissolved methylmercury. *J. Anal. At. Spectrom.*, **25**, 950–956.

52 Laffont, L., Sonke, J.E., Maurice, L., Hintelmann, H., Pouilly, M., Bacarreza, Y.S., Perez, T., and Behra, P. (2009) Mercury isotopic compositions of fish and human hair in the Bolivian Amazon. *Environ. Sci. Technol.*, **43**, 8985–8990.

53 Gantner, N., Hintelmann, H., Zheng, W., and Muir, D.C. (2009) Variations in stable isotope fractionation of Hg in food webs of Arctic lakes. *Environ. Sci. Technol.*, **43**, 9148–9154.

54 Feng, X.B., Foucher, D., Hintelmann, H., Yan, Y.H., He, T.R., and Qiu, G.L. (2010) Tracing mercury contamination sources in sediments using mercury isotope compositions. *Environ. Sci. Technol.*, **44**, 3363–3368.

55 Epov, V.N., Malinovsky, D., Sonke, J.E., Vanhaecke, F., Begue, D., and Donard, O.F.X. (2011) Modern mass spectrometry for studying mass-independent fractionation of heavy stable isotopes in environmental and biological sciences. *J. Anal. At. Spectrom.*, **26**, 1142–1156.

56 Elmore, D. and Phillips, F.M. (1987) Acclerator mass spectrometry for measurement of long-lived radioisotopes. *Science*, **236**, 543–550.

57 Vogel, J.S., Turteltaub, K.W., Finkel, R., and Nelson, D.E. (1995) Accelerator mass spectrometry: isotope quantification at attomole sensitivity. *Anal. Chem.*, **67**, 353A–359A.

58 Filfield, L.K. (1999) Accelerator mass spectrometry and its applications. *Rep. Prog. Phys.*, **62**, 1223–1274.

59 Helborg, R. and Skog, G. (2008) Accelerator mass spectrometry. *Mass Spectrom. Rev.*, **27**, 398–427.

60 Chmeleff, J., von Blanckenburg, F., Kossert, K., and Jakob, D. (2010) Determination of the ^{10}Be half-life by multicollector ICP-MS and liquid scintillation counting. *Nucl. Instrum. Methods B*, **268**, 192–199.

61 West, R. (1976) Natural nuclear reactors – the Oklo phenomenon. *J. Chem. Educ.*, **53**, 336–340.

62 Yergey, A.L. and Yergey, A.K. (1997) A brief history of the calutron. *J. Am. Soc. Mass Spectrom.*, **8**, 943–953.

63 Settle, F.A. (2009) Uranium to electricity: the chemistry of the nuclear fuel cycle. *J. Chem. Educ.*, **86**, 316–323.

64 Bleise, A., Danesi, P.R., and Burkart, W. (2003) Properties, use and health effects of depleted uranium (DU): a general overview. *J. Environ. Radioact.*, **64**, 93–112.

65 Priest, N.D. (2001) Toxicity of depleted uranium. *Lancet*, **357**, 244–246.

66 Richter, S., Alonso, A., Wellum, R., and Taylor, P.D.P. (1999) The isotopic composition of commercially available uranium chemical reagents. *J. Anal. At. Spectrom.*, **14**, 889–891.

67 Qi, H.P., Coplen, T.B., Wang, Q.Z., and Wang, Y.H. (1997) Unnatural isotopic composition of lithium reagents. *Anal. Chem.*, **69**, 4076–4078.

68 Choppin, G.R. and Rydberg, J. (1980) *Nuclear Chemistry – Theory and Applications*, Pergamon Press, Oxford.

69 Barth, R.F., Coderre, J.A., Vicente, M.G.H., and Blue, T.E. (2005) Boron neutron capture therapy of cancer: current status and future prospects. *Clin. Cancer Res.*, **11**, 3987–4002.

2
Single-Collector Inductively Coupled Plasma Mass Spectrometry

Frank Vanhaecke

2.1
Mass Spectrometry

For each mass spectrometry (MS) technique, three essential parts of the instrumentation can be identified: (i) the ion source, in which the ions are produced, (ii) the mass spectrometer itself, in which the ions are separated from one another as a function of their mass-to-charge ratio, and (iii) the detection system that converts the ion beam into a measurable electrical signal (Figure 2.1). In inductively coupled plasma mass spectrometry (ICP-MS), three types of mass spectrometers are used in commercial instrumentation: (i) a quadrupole filter that at any given time transmits ions with a mass-to-charge ratio within a narrow window (approximately one atomic mass unit wide) only, (ii) a double-focusing sector field mass spectrometer that separates the ions from one another in space as a function of their mass-to-charge ratio, and (iii) a time-of-flight analyzer that separates the ions from one another in time as a function of their mass-to-charge ratio.

Self-evidently, collisions between the ions and gas molecules disturb the ion beam and have to be avoided to the largest possible extent. As a result, in all MS instrumentation, the mass spectrometer and detection system are brought under

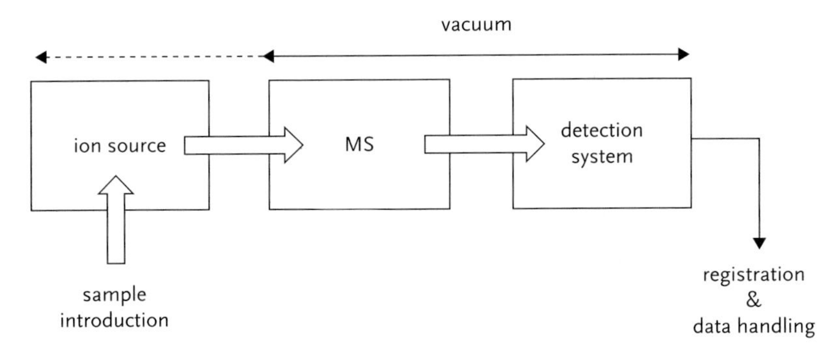

Figure 2.1 Schematic representation of the essential parts of mass spectrometric instrumentation.

Isotopic Analysis: Fundamentals and Applications Using ICP-MS, First Edition. Edited by Frank Vanhaecke and Patrick Degryse.

Published 2012 by WILEY-VCH Verlag GmbH & Co. KGaA

sufficiently high vacuum (sufficiently low pressure) by means of vacuum pumps. The necessity to use vacuum pumps adds to the cost of the instrumentation. In some MS techniques, also the ion source is operated under vacuum. This is the case, for example, in thermal ionization mass spectrometry (TIMS), where the sample (the isolated target element) is deposited on a metal filament, which is subsequently resistively heated in vacuum [1]. In ICP-MS, on the other hand, the plasma ion source is generated at atmospheric pressure. Finally, the MS instrument needs to be completed by a sample introduction system and a system for registration and handling of the analyte signals.

2.2
The Inductively Coupled Plasma Ion Source

The argon inductively coupled plasma (ICP) is a plasma ion source [2, 3]. A plasma is defined as a gas mixture at high temperature, containing ions and electrons in addition to neutral particles. The presence of these charged particles allows energy transfer into the plasma via induction.

An ICP is generated at the end of a quartz torch, consisting of three concentric tubes, the top end of which is surrounded by a water- or gas-cooled load coil. By sending a radiofrequency (either ~27.12, or ~40.68 MHz) current through this coil, a time-dependent magnetic field, varying with the same frequency, is generated. This varying magnetic field accelerates the electrons present in the plasma and forces them to move according to circular paths (Figure 2.2). Taking into account

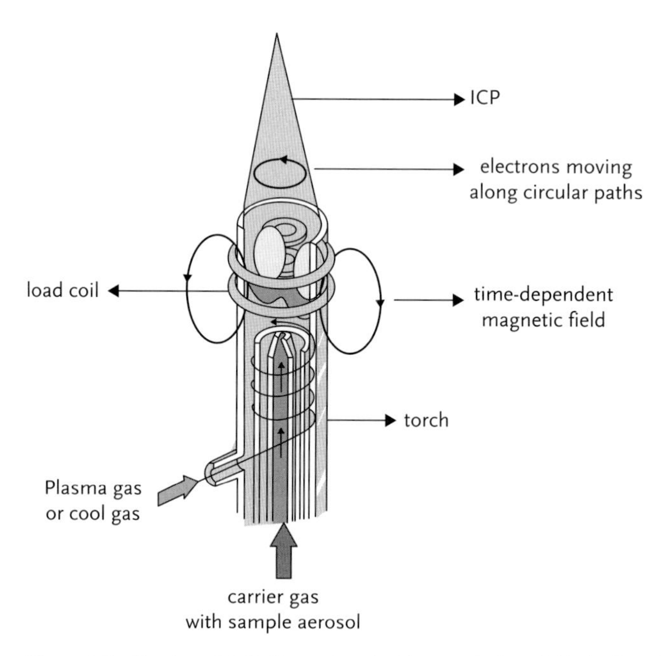

Figure 2.2 Torch and ICP. Reproduced with permission of Linde from [4].

the continuous downstream flow of Ar, the electrons are finally spiraling. As a result of the high kinetic energy that the electrons attain upon this acceleration, collisions with Ar atoms can result in (electron impact) ionization: $Ar + e^- \rightarrow Ar^+ + 2e^-$. As a result, the plasma is sustained as long as Ar is provided (typically at 15–20 l min^{-1}) and the r.f. current through the load coil is maintained. Originally, however, the Ar gas is neutral, such that "seed electrons" are required to start the process of plasma generation. Generation of seed electrons is accomplished by means of a high-voltage spark, created by the Tesla generator.

When sample aerosol (see below) is introduced into an ICP, the droplets are desolvated, molecules are broken down into the constituting atoms and these are ionized via electron impact and Penning ionization (ionization as a result of energy transfer from an excited Ar atom). Although an ICP is not in thermal equilibrium, its ionization efficiency can be decently estimated on the basis of the Saha equation [5]:

$$K_{ion} = \frac{n_i n_e}{n_a} = \left(\frac{2\pi m_e k T_{ion}}{h^2}\right)^{\frac{3}{2}} \frac{2Z_i}{Z_a} e^{-\frac{IE}{kT_{ion}}}$$

where K_{ion} is the ionization equilibrium constant, n_i, n_e, and n_a are the densities (number of particles per unit of volume) of ions, electrons, and atoms, respectively, m_e is the mass of the electron, k the Boltzmann constant (6.626×10^{-34} J s), T_{ion} the ionization temperature within the ICP, h Planck's constant (1.381×10^{-23} J K^{-1}), Z_i and Z_a the partition functions for the ionic and atomic state, respectively, and IE the ionization energy of the element considered.

The degree of ionization α can be deduced from the ionization constant:

$$\alpha = \frac{n_i}{n_a + n_i} = \frac{\frac{n_i n_e}{n_a}}{n_e + \frac{n_i n_e}{n_a}} = \frac{K_{ion}}{n_e + K_{ion}}$$

The results obtained upon carrying out this calculation, taking into account a typical ionization temperature of 7500 K and an electron density of 10^{15} cm^{-3}, demonstrate the high ionization efficiency of the ion source [6, 7]. These results are represented graphically in Figure 2.3 and illustrate that the ICP assures practically complete ionization ($\alpha \geq 0.9$) up to an ionization energy of 8 eV and even provides a reasonable ionization efficiency for metalloids ($\alpha \approx 0.3$–0.8) and nonmetals ($\alpha < 0.01$–0.3) [7]. This provides an important advantage over TIMS, where only elements characterized by an $IE < 7.5$ eV undergo sufficient conversion into M^+ ions [1].

The ICP torch consists of three concentric tubes through each of which Ar is flowing (Figure 2.2). Cool or plasma gas is flowing in between the outer tube and the middle tube and provides sufficient Ar to maintain the plasma, while providing a protective isolating sheathing between the quartz torch and the hot plasma, thus preventing the former from melting. Auxiliary gas is flowing in between the middle and the central tubes. In ICP-MS, the auxiliary gas plays only a modest role, although it can be used to optimize the position of the ICP flame with respect to the torch and the load coil. Finally, carrier, transport, or nebulizer gas is

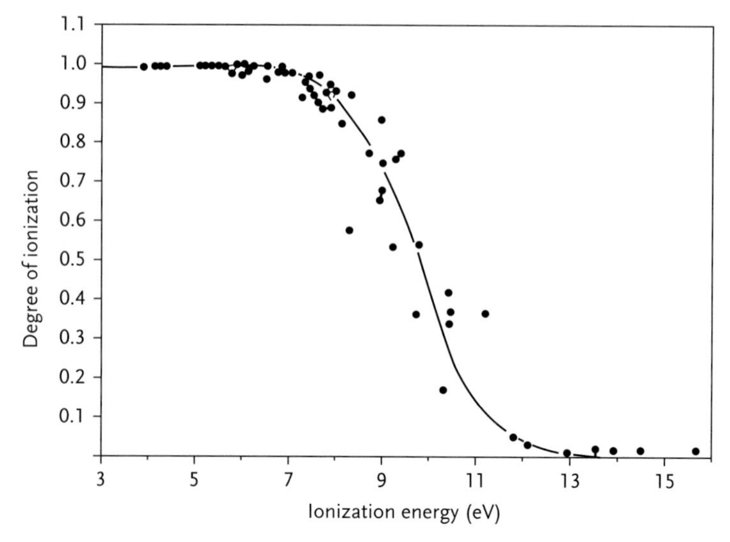

Figure 2.3 Degree of ionization in an ICP as a function of the element's ionization energy. Reproduced with permission of Wiley-VCH Verlag GmbH from [7].

flowing through the central tube and carries the sample aerosol (see below) into the ICP. This gas flow punctures the ICP, thus resulting in a toroidal plasma shape and giving rise to a dark channel in the center of the ICP, wherein the analyte ion formation takes place.

2.3
Basic Operating Principles of Mass Spectrometers

2.3.1
Mass Spectrometer Characteristics

Before discussing the basic operating principles of those mass spectrometer types used in ICP-MS instrumentation, some important characteristics need to be introduced: mass resolution, abundance sensitivity, mass range, and scanning speed [8].

2.3.1.1 Mass Resolution

The mass resolution of a mass spectrometer provides quantitative information on its capabilities to distinguish between ions showing a limited difference in mass only and thus to resolve neighboring spectral peaks. Two approaches for calculating mass resolution are used (Figure 2.4). Whereas one permits the user to determine the mass resolution on the basis of the experimentally observed width of a spectral peak at a given mass-to-charge ratio, the other one, the 10% valley definition, is suited to calculate the resolution required for resolving two

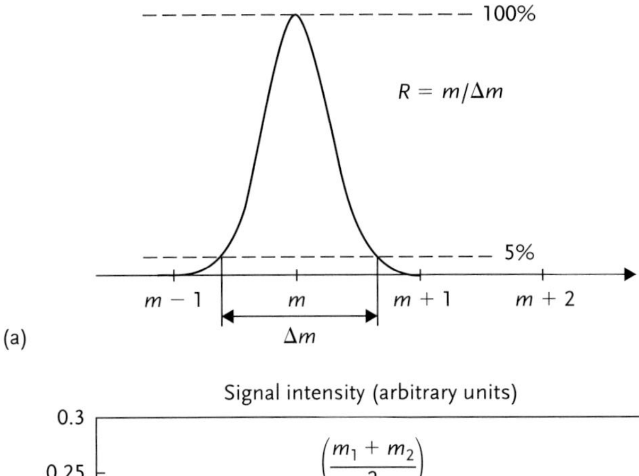

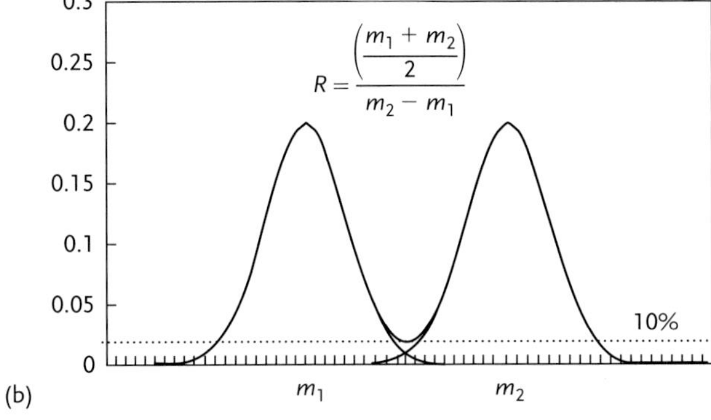

Figure 2.4 Mass resolution. (a) Calculation of the resolution offered by a mass spectrometer on the basis of the width of a spectral peak at 5% peak height (Δm) at mass m. (b) 10% valley definition: calculation of the resolution required to resolve the peaks of two ions showing a mass difference of $m_2 - m_1$.

neighboring spectral peaks on the basis of the exact masses of the corresponding ions. For the latter definition, two peaks of equal intensity are considered as resolved if the valley between the peak maxima does not exceed 10% of the peak heights. If there is a considerable difference in signal intensity between the two peaks considered, a higher mass resolution will be required to resolve the signal from the least abundant ion from the signal from the most abundant one to the same extent. It is important to stress that both approaches are equivalent as the 10% valley can be considered as the sum of two 5% contributions, originating from each of the spectral peaks.

2.3.1.2 **Abundance Sensitivity**

The abundance sensitivity also deals with the separation of peaks, but expresses the contribution of the tail of a neighboring peak to the signal intensity measured for the target element quantitatively. Abundance sensitivity is calculated as the contribution of the neighboring peak to the total intensity at the mass-to-charge ratio of interest divided by the intensity of the analyte signal (either signal height or signal integrated over a given mass interval is used). As spectral peaks are not necessarily symmetric, the effect of a lighter and a heavier neighboring ion on the analyte signal can be different. Narrow peaks and absence of "tails" provide a better abundance sensitivity. This aspect is especially important when dealing with "extreme" isotope ratios, characterized by a large difference between the relative isotopic abundances of the isotopes under consideration.

2.3.1.3 Mass Spectral Range

Although very important in organic MS, this aspect of a mass spectrometer is of lower relevance in ICP-MS, as the heaviest nuclide occurring in Nature is ^{238}U. As a result, a mass range of 0–250 u allows all signals of interest to be monitored.

2.3.1.4 Scanning Speed

The speed with which the mass spectrometer can scan (a part of) the mass spectrum or can switch from monitoring one nuclide to monitoring another is also important, especially in the context of isotopic analysis.

2.3.2 Quadrupole Filter

A quadrupole filter consists of four parallel cylindrical or hyperbolic rods, manufactured from or coated with conducting material. As the name suggests, it acts as a filter, transmitting ions with a mass-to-charge ratio within a narrow window (typically approximately one mass unit wide) only. These ions show a relatively stable trajectory through the quadrupole assembly (Figure 2.5), whereas those with a mass-to-charge ratio outside the window show unstable trajectories, resulting in their removal from the ion beam. The "location" of the mass window is determined by the (DC + AC) voltages applied to the rods, as is explained later.

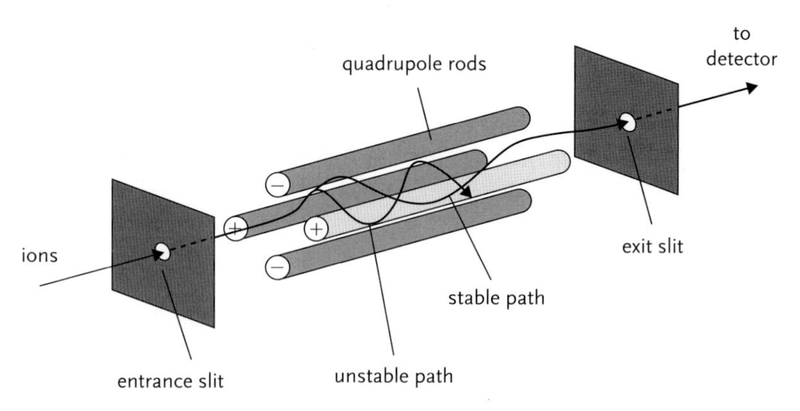

Figure 2.5 Operating principle of a quadrupole filter. Reproduced with permission of Dr. Paul Gates, School of Chemistry, University of Bristol, from [9].

The diametrically opposed rods of the quadrupole assembly are electrically connected, thus forming two electrode pairs (Figure 2.6).

To obtain an intuitive understanding of the operating principle of the quadrupole filter, the effect exerted on the ions has to be evaluated for each of these electrode pairs separately [10]. The voltage applied to the first pair of electrodes consists of a DC component, in this case +U, and an AC component, $V\sin\omega t$ (Figure 2.7). The positive DC component forces the ions to move towards the axis of the quadrupole assembly; the AC component will also focus the ions to the center during half a period, but will defocus them (attract them in the direction of the rods) during the other half period. Sufficiently heavy ions only feel the average potential, which is positive and thus focuses these ions on to the central axis. Lighter ions are removed from the beam due to the influence that the AC voltage exerts on their paths. This pair of electrodes thus acts as a high-mass filter. The voltage applied to the other pair of electrodes shows the same magnitude, but the opposite sign, such that the DC component in this case is $-U$, while the AC component shows a phase difference of 180°, $V\sin(\omega t + \pi)$ (Figure 2.7). As the average potential is now negative, this pair of electrodes acts as a low-mass filter. The heavier ions are defocused, and therefore removed, while only for the lighter ions does the focusing

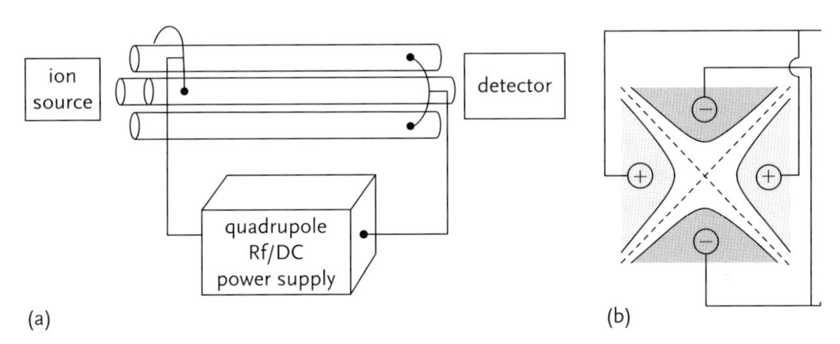

Figure 2.6 Schematic representation of a quadrupole filter (a) and the electrical connection of the diametrically opposed rods, leading to two electrode pairs (b).

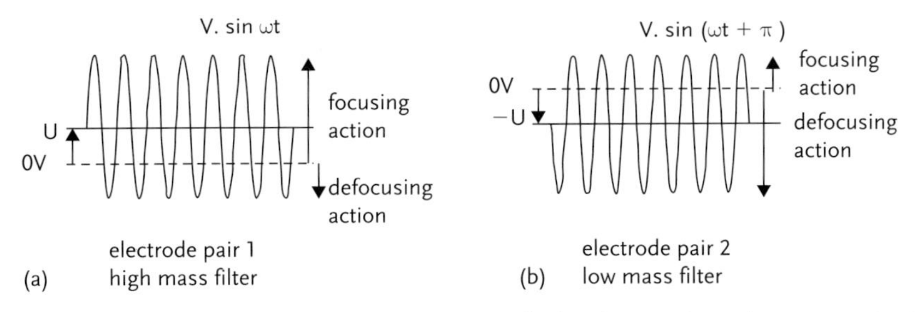

Figure 2.7 Voltages (DC and AC components) applied to the two electrode pairs constituting the quadrupole filter. (a) Electrode pair acting as high-mass filter; (b) electrode pair acting as low-mass filter.

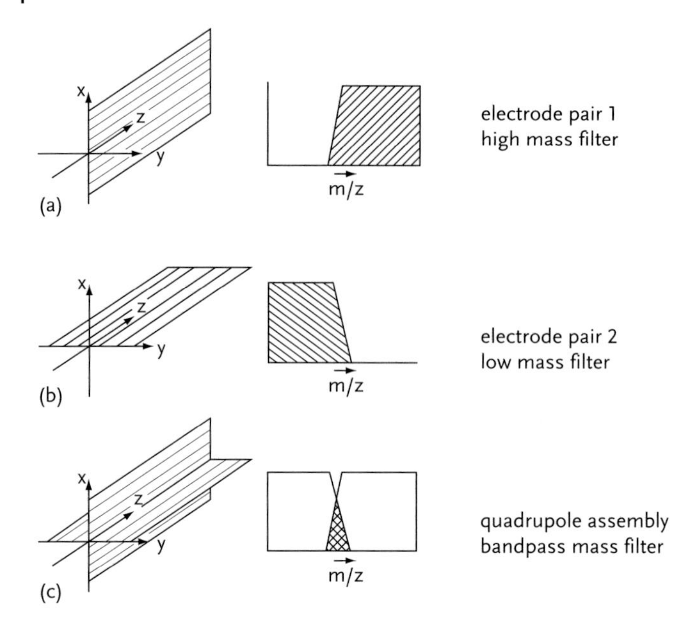

Figure 2.8 Combination of the high-mass filtering action of the first electrode pair with the low-mass filtering action of the second pair results in a narrow bandpass, typically ~1 mass unit wide. Reproduced with permission of the American Chemical Society from [10].

action of the AC component during half a period correct their path sufficiently. Self-evidently, the two electrode pairs function together and the combination of a high-mass and a low-mass filter results in a bandpass filter (Figure 2.8).

Continuous scanning of (a part of) the mass spectrum or "peak hopping" or "peak jumping" is accomplished by changing these voltages in such a way that the U/V ratio remains constant. The spectral peak width is constant over the mass range (mass resolution, on the other hand, varies as a function of the mass considered).

The most notable advantages of the quadrupole filter are its technical simplicity, the fact that it can be used at a somewhat higher pressure than a sector field mass spectrometer, and that there is a larger tolerance towards a spread in the kinetic energy of the incoming ions (no energy filter required). These aspects contribute to a lower purchase cost. Its major shortcoming is the limitation to unit mass resolution, at least in commercially available instrumentation [11, 12]. The quadrupole filter allows rapid scanning and is typically used when an exact determination of the ion masses or high mass resolution is not required.

2.3.3 Double-Focusing Sector Field Mass Spectrometer

In a magnetic sector, ions are separated from one another in space according to their mass-to-charge ratio [13–15]. When the ions coming from the source are

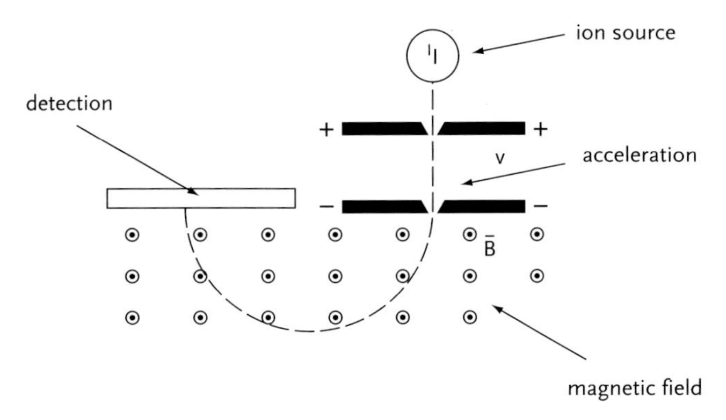

Figure 2.9 An ion introduced into a magnetic field (velocity $v \perp$ magnetic field B) is forced to move along a circular path, the radius of which is determined by its mass-to-charge ratio.

accelerated over a potential difference V and subsequently introduced into a magnetic field with the field lines perpendicular to the plane in which the ions move, they move according to a circular path (Figure 2.9).

The centripetal force required to move along a circular path is provided by the Lorentz force exerted by the magnetic field on the ion:

$$F = \frac{mv^2}{r} = qvB$$

where F is the force exerted on the ion, m the ion mass, v its velocity, r the radius of the circular path along which it moves, and q its charge (in ICP-MS, typically +1). As a result, the radius is described by

$$r = \frac{mv}{qB}$$

Relying on the fact that the kinetic energy that the ion obtains upon acceleration over a potential difference V is given by

$$E_{\text{kin}} = \frac{1}{2}mv^2 = qV$$

and thus

$$v = \sqrt{\frac{2qV}{m}}$$

the radius is also given by

$$r = \frac{\sqrt{2Vm}}{B\sqrt{q}}$$

In the oldest mass spectrometric, or more correctly mass spectrographic, experiments, an ion-sensitive emulsion was used to detect the ions at the exact location where they arrive. After development of the emulsion, a reproduction of the mass spectrum was obtained. The presence or absence of a specific element was then deduced from the presence or absence of the corresponding line on the film, while its concentration in the sample was deduced from the density of the line, as measured using a densitometer. Nowadays, electric detectors are used. When a single detector is used, the mass-to-charge ratio of the ion monitored can be selected by either adapting the magnetic field strength B (magnetic scanning or B-scanning) or the acceleration voltage V (electric scanning or E-scanning).

When the incoming ions show a spread in their kinetic energies and/or a higher mass resolution is required, the magnetic sector needs to be combined with an electrostatic sector. In TIMS instrumentation, a magnetic sector suffices as the sample pretreatment, typically consisting of isolation of the target element from the concomitant matrix, and the rather gentle ionization in vacuum lead to a limited risk of spectral overlap and a small energy spread. With an ICP as ion source, however, a double-focusing setup containing both a magnetic and an electrostatic sector is required. The basic operating principle of such an electrostatic sector is very straightforward (Figure 2.10).

As the ions move between a positively charged and a negatively charged bent plate, they are forced to move along a circular path. The centripetal force required to move along a circular path is provided by the force exerted by the electrical field on the ion:

$$F = \frac{mv^2}{r} = qE$$

where E is the strength of the electrical field. As a result, the radius of the circular path depends on the ion's kinetic energy:

$$r = \frac{mv^2}{qE} = \frac{2E_{\text{kin}}}{qE}$$

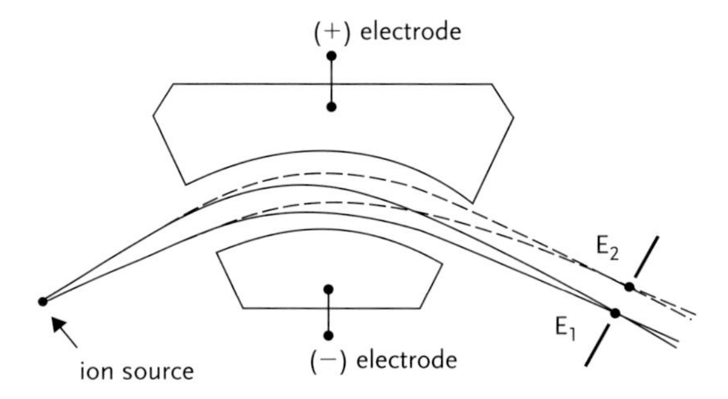

Figure 2.10 Use of electrostatic sector as energy filter. The radius of the circular path along which the ions are forced to move increases with higher ion kinetic energy ($E_2 > E_1$). Ions with an excessively high ($>E_2$) or an excessively low ($<E_1$) energy fall outside the slit width and will be removed from the ion beam.

If only singly charged ions M^+ are considered, this equation shows that they are dispersed as a function of their kinetic energy only. When a plate with a narrow slit is placed behind the electrostatic filter, only ions with an energy within an acceptable range are transmitted in the direction of the magnetic sector, whereas ions showing either too high or too low an energy are removed from the beam. The narrower the slit, the narrower is the energy spread, but the lower is the transmission efficiency. In order to avoid a huge loss in ion transmission efficiency, while maintaining the possibility of achieving high mass resolution, double-focusing setups have been developed.

In double-focusing setups, the dispersion of the ion beam as created by one of the sectors is exactly compensated for by the other sector. As a result, a wider slit width can be used between the two sectors, such that higher mass resolution can be attained without compromising ion transmission efficiency to an unacceptably large extent (Figure 2.11). The term "double focusing" refers in this context to the fact that both energy focusing and directional focusing are realized: ions of the same mass-to-charge ratio that show a difference in kinetic energy and/or direction are still focused by the double-focusing setup to one point. For realizing double focusing, both sectors and the way in which they are combined have to meet specific requirements [15]. Mattauch–Herzog, Nier–Johnson, and reverse Nier–Johnson geometries show these double-focusing properties.

With a Mattauch–Herzog geometry (Figure 2.12), the focal points of all ion beams are located in one focal plane. Therefore, it is the geometry of choice for

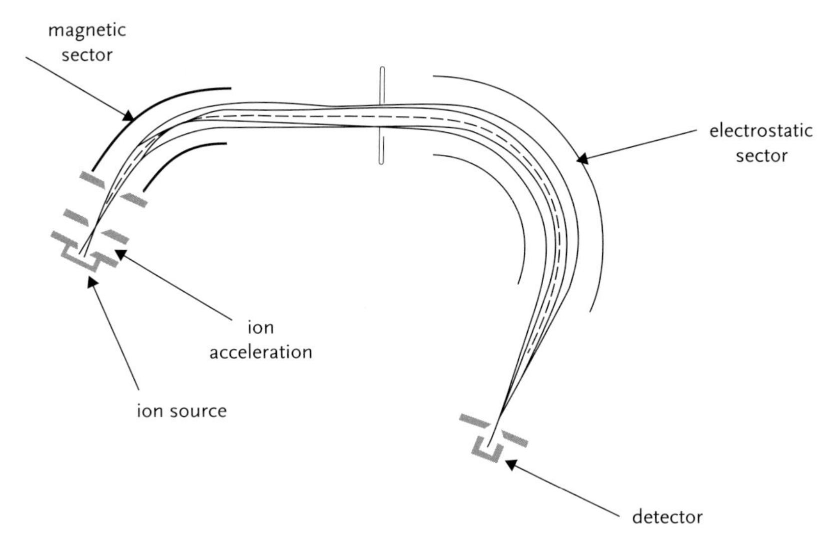

Figure 2.11 Double-focusing setup (reverse Nier–Johnson geometry; see below), illustrating how the dispersion of the first sector is compensated for by that of the second sector. Ions with the same mass-to-charge ratio, but a difference in kinetic energy and/or direction, are still focused to one point. Owing to the possibility of using a wider slit between the two sectors thus created, a lower loss in ion transmission efficiency is realized.

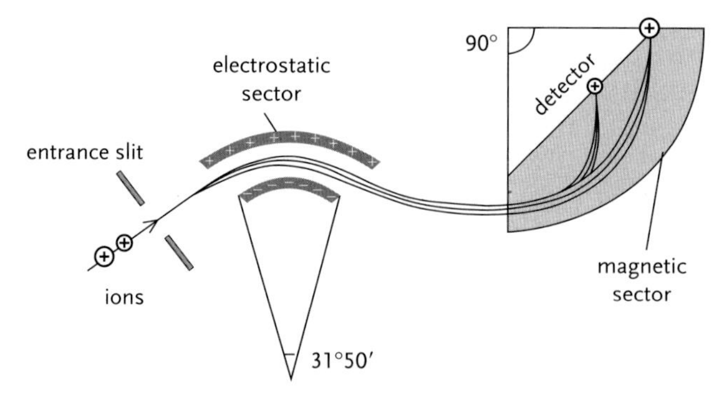

Figure 2.12 Mattauch–Herzog geometry providing directional and energy focusing for all ions in a focal plane. Reproduced with permission of FIZ Chemie Berlin from [16].

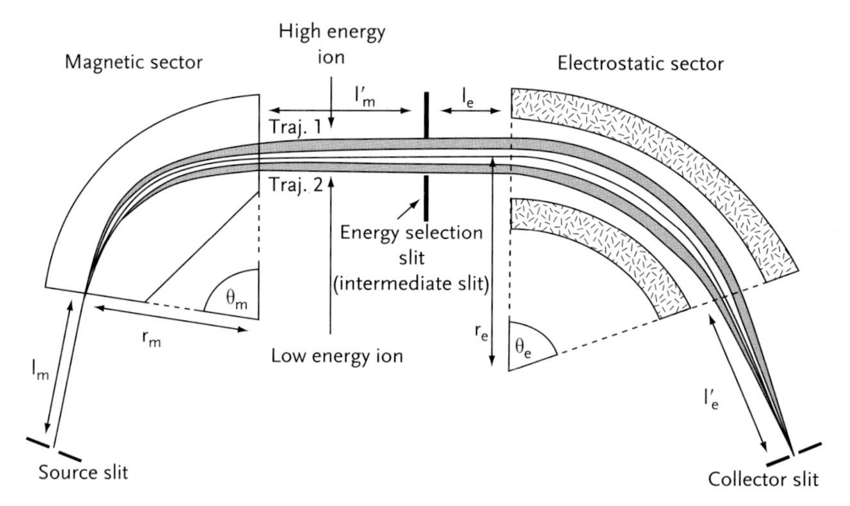

Figure 2.13 Nier–Johnson double-focusing geometry. Reproduced with permission of Wiley-VCH Verlag GmbH from [8].

simultaneous monitoring of the entire mass spectrum or a large part of it. In older days, this was realized by using an ion-sensitive emulsion; nowadays, it can also be accomplished by using a recently developed Faraday strip array detector [17, 18]. The first commercially available instrument with Mattauch–Herzog geometry, allowing simultaneous monitoring of the entire elemental spectrum (in 4800 sections), was commercially introduced at the 2010 Pittsburgh Conference & Exposition (Pittcon) [15].

The Nier–Johnson geometry (Figure 2.13) is used in multi-collector inductively coupled plasma mass spectrometry (MC-ICP-MS) instruments. Double focusing is not realized for all ions at the same time, but present-day MC-ICP-MS instruments show a mass range from m up to 1.3 m wherein the ion signals can be monitored under static conditions (no change of magnetic field strength or acceleration voltage).

Finally, the reverse Nier–Johnson geometry, in which the magnetic sector proceeds the electrostatic sector, is preferred for single-collector sector field ICP-MS instruments. The removal of most of the ions from the beam in the first sector ultimately leads to a lower background and a better abundance sensitivity, because at a lower beam density, fewer ions show an aberrant behavior as a result of collisions. For single-collector instruments, scanning or peak hopping is accomplished by changing the magnetic field strength B (magnetic scanning or B-scanning) or the acceleration voltage V (electric scanning or E-scanning). Because of magnet hysteresis, B-scanning is slower than E-scanning, but upon changing the acceleration voltage, the ion transmission efficiency is reduced. As a result, when ion signals spread over a large mass interval have to be monitored, a combination of changes in the magnetic field and subsequent E-scanning is typically used.

2.3.4
Time-of-Flight Analyzer

The basic operating principle of a time-of-flight (TOF) analyzer is quite straightforward [19, 20]. Consider a package of ions. These ions are accelerated over a potential difference V and subsequently introduced into a field-free flight tube (Figure 2.14). All singly charged ions obtain the same kinetic energy:

$$E_{\text{kin}} = \frac{1}{2}mv^2 = qV$$

As a consequence, the velocity with which an ion moves is determined by its mass:

$$v = \sqrt{2\frac{qV}{m}}$$

and hence so is the time t required to travel the distance to the detector, placed at the end of the flight tube with length L:

$$t = \frac{L}{v} = \frac{L\sqrt{m}}{\sqrt{2qV}}$$

Continuous registration of the incoming ions at the end of the flight tube thus provides a complete mass spectrum for the package of ions introduced.

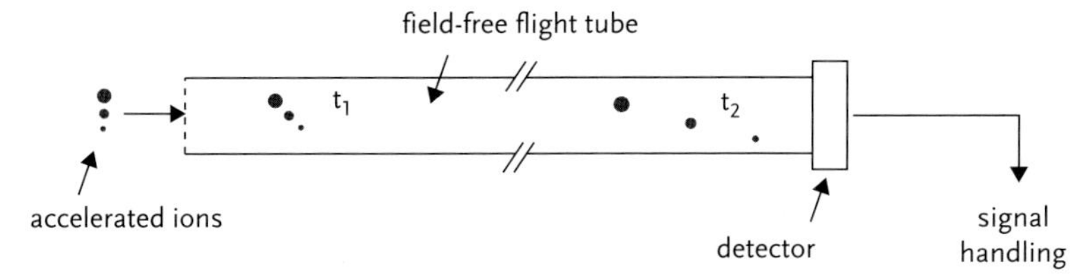

Figure 2.14 Schematic representation of the basic operating principle of a time-of-flight analyzer.

It is clear from the basic operating principle that ions cannot be introduced in a continuous way (otherwise a light ion, introduced later, could overtake a heavier ion, introduced earlier). For ion sources producing a continuous ion beam, such as the ICP, beam modulation is required. In the only ICP-TOF-MS instrument commercially available today, this is accomplished by orthogonal acceleration, the operating principle of which is illustrated in Figure 2.15. For this purpose, a repeller is used to which a voltage can be applied. If no voltage is applied to it, the ions cannot enter the TOF analyzer. If, on the other hand, a positive voltage is applied to it, the ions in front of the repeller are accelerated perpendicularly to the original beam and introduced into the TOF analyzer. Currently, up to 30 000 ion packages produced in this way can be analyzed per second, making the TOF analyzer particularly suited for monitoring transient signals of short duration.

Also for a TOF analyzer, the mass resolution is negatively influenced by a spread of the kinetic energy of the ions introduced. This can be remedied to some extent by means of a reflectron or ion mirror, consisting of a number of rings to which a decelerating voltage is applied (Figure 2.16). The ions are slowed, stopped, and accelerated in the opposite direction. As ions showing a higher kinetic energy penetrate deeper into the reflectron before they are stopped and accelerated in the other

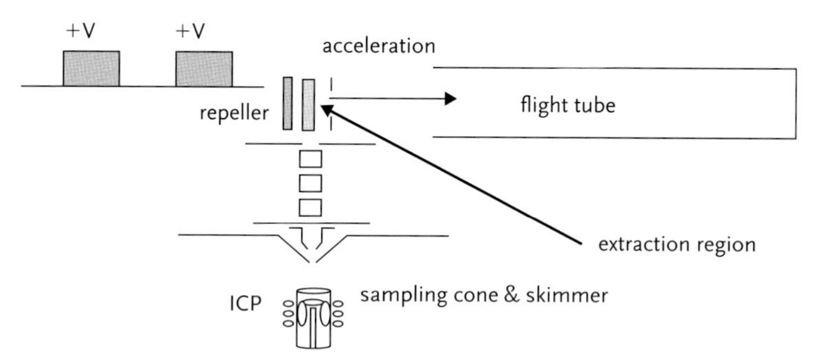

Figure 2.15 Orthogonal acceleration for beam modulation, as required when using a continuous ion source (in this case the ICP) with a TOF analyzer.

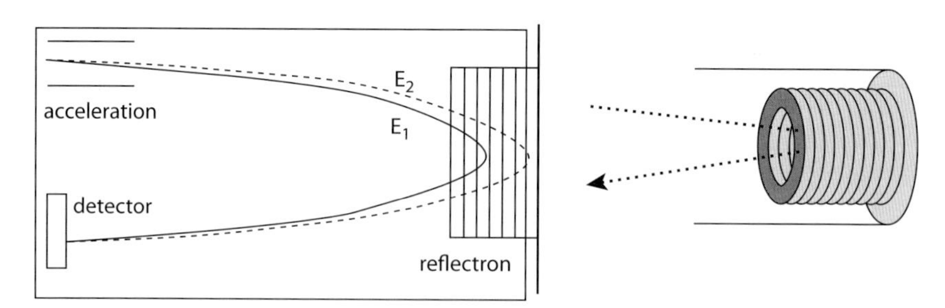

Figure 2.16 Use of reflectron to improve the mass resolution in a TOF analyzer. Two ions with the same mass-to-charge ratio, but a difference in kinetic energy ($E_2 > E_1$) travel a different distance and will reach the detector at the same time.

direction, they travel a somewhat longer distance than ions of the same mass-to-charge ratio, but with a somewhat lower kinetic energy, such that, finally, they reach the detector at the same time. Moreover, the ions are also forced to travel a considerably longer distance (tube length × 2) without having an effect on the footprint of the instrument. This increased pathlength is also beneficial for the mass resolution.

2.3.5 Comparison of Characteristics

Table 2.1 compares the characteristics of the three types of mass spectrometers used in commercially available ICP-MS instrumentation.

Table 2.1 Comparison of the mass resolution and scanning speed of the three types of mass spectrometers as used in ICP-MS instrumentation.

Type of mass spectrometer	Mass resolution	Scanning speed
Quadrupole filter	Unit mass resolution $R \approx 300$	$\sim$2500 u s^{-1} Full mass spectrum in 100 ms
Sector field	$R_{max} \approx 10\,000$	$\sim$1500 u s^{-1} Full mass spectrum in 150 ms
Time-of-flight	Unit mass resolution $R \approx 300$	$\sim$7 500 000 u s^{-1} Full mass spectrum in 0.033 ms

2.4 Quadrupole-Based ICP-MS

As in inductively coupled plasma optical emission (ICP-OES) spectra, in addition to atomic lines, intense ionic lines are also observed, the use of an ICP as an ion source for MS seemed logical, but overcoming the difference in pressure between the ICP (generated at atmospheric pressure) and the mass spectrometer (10^{-5}–10^{-9} mbar) proved difficult and had to be accomplished via the use of a two-cone interface. Despite the advantages that double-focusing sector field mass spectrometers (higher mass resolution) and TOF analyzers (high data acquisition speed) can offer, approximately 90% of the ICP-MS units used worldwide are equipped with a quadrupole filter for mass analysis.

Figure 2.17 provides a schematic overview of a quadrupole-based ICP-MS instrument. Typically, the sample solution is pumped to a nebulizer by means of a peristaltic pump. The nebulizer converts the sample solution into an aerosol. This primary aerosol is introduced into a spray chamber that filters out the droplets with diameter >10 μm. Although this process is highly inefficient – it reduces the analyte introduction efficiency by 1–2 orders of magnitude, depending on the actual type of nebulizer and spray chamber used – it is necessary to obtain a stable plasma

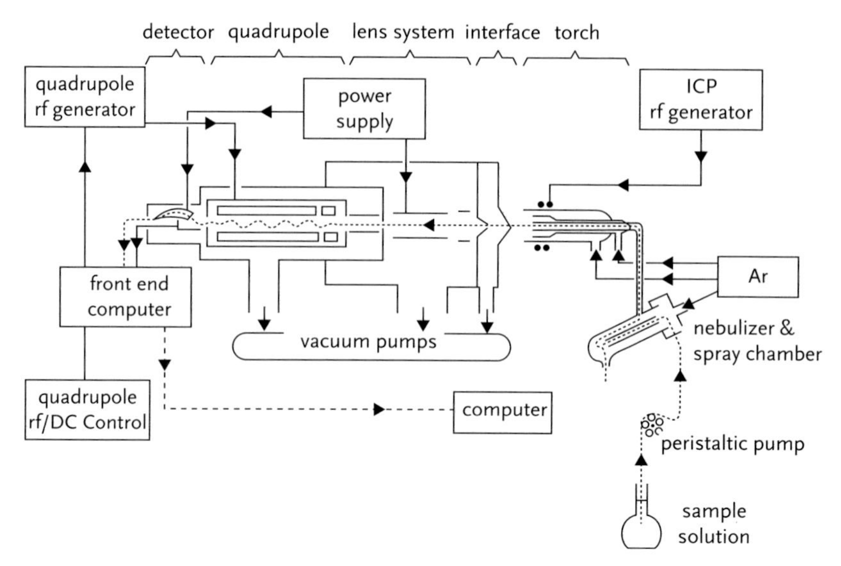

Figure 2.17 Schematic representation of a quadrupole-based ICP-MS setup.

and guarantee efficient desolvation, atomization, and ionization in the ICP. In the ICP, (i) the aerosol droplets are desolvated, resulting in small salt particles, (ii) individual molecules are released from these salt particles, (iii) the molecules thus formed are atomized, and (iv) these atoms are subsequently ionized. All of these processes occur within the 1–2 ms plasma residence time. The ions thus formed need to be introduced into the quadrupole filter for mass analysis, but the difference in pressure between the ICP (generated at atmospheric pressure) and the mass spectrometer (10^{-5}–10^{-9} mbar) needs to be overcome. The interface, which consists of two coaxially placed metal cones with a small central aperture (order of magnitude ~1 mm), plays a central role in this context [21]. The pressure in the expansion chamber between the two cones (the sampling cone and the skimmer) is of the order of 1 mbar. As a result of the difference in pressure between the ICP and the region between the two cones, part of the ICP is extracted into the expansion chamber and undergoes supersonic expansion. An important consequence of this expansion is that the entities present in the extracted plasma gas plume are moved away from one another, thereby strongly decreasing the possibility of interaction (e.g., ion–electron recombination). As a result, the plasma composition is "frozen" [22]. Most of the extracted plasma gas is removed by the vacuum system, but a central beam leaves the expansion chamber via the central aperture in the skimmer. At this point, the beam is still electrically neutral. Typically, the positive ions are then selectively extracted via a negatively charged extraction lens. The beam of positive ions is then transported to and introduced into the quadrupole filter. The ions transmitted by the quadrupole filter are detected by means of an electron multiplier, capable of detecting each transmitted ion individually. The operating principle and various aspects of this detector are discussed in section 2.7.2).

Self-evidently, the mass-to-charge ratio at which the signal intensity is monitored can be adapted by changing the voltages applied to the quadrupole rods. Although continuous scanning is possible, peak hopping or peak jumping approaches, realized by discontinuously changing the voltages applied, are often preferred, as in this way the measurement time is used more efficiently. For isotope ratio measurements, this is certainly the method of choice. Of course, both instrument operation and data handling are under computer control.

2.5 Sample Introduction Strategies in ICP-MS

To permit sample introduction into the ICP, the sample or a representative part of it has to be converted into a form that can be transported with a gas. Below, only those sample introduction strategies used most often in the context of isotopic analysis will be considered.

The standard sample introduction system consists of a pneumatic nebulizer (with which the aerosol is generated as a result of interaction of an accelerated gas flow on a liquid surface) mounted on a spray chamber [23–26]. Although various types of pneumatic nebulizer (Figure 2.18) exist, three types are most commonly

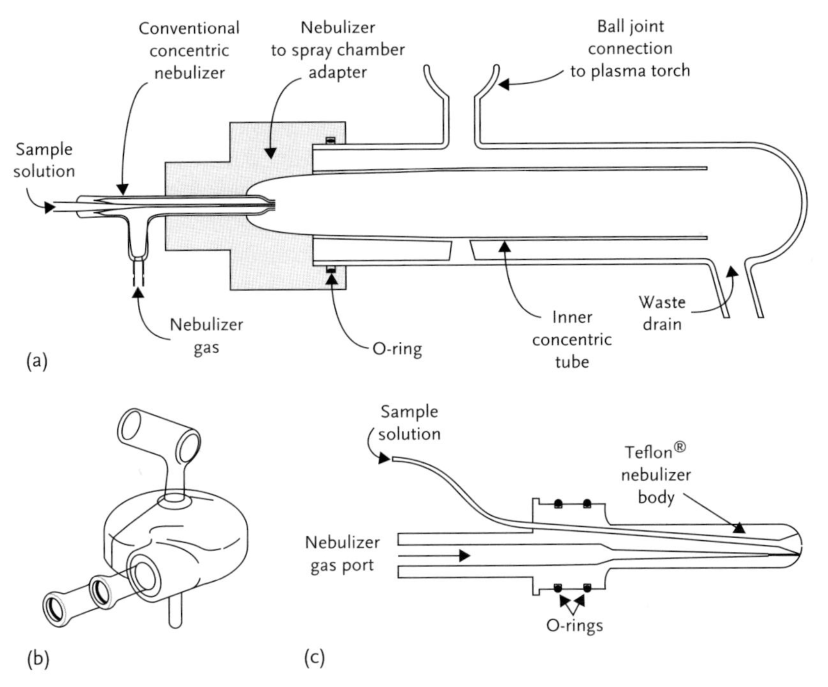

Figure 2.18 (a) Concentric nebulizer mounted on a Scott-type double-pass spray chamber. Reproduced with permission of Wiley-VCH Verlag GmbH from [25]. (b) Cyclonic spray chamber. Reproduced with permission of Advanstar Communications from [27]. (c) Burgener nebulizer. Reproduced with permission of Wiley-VCH Verlag GmbH from [25].

used: the traditional concentric nebulizer, typically operating at a sample uptake rate of ~1 ml min^{-1}, a microconcentric nebulizer, allowing stable sample introduction at considerably lower sample uptake rates (from <10 to several hundred μl min^{-1}), and the Burgener nebulizer (covering the entire range of sample uptake rates mentioned). Typically, sample solution is pumped to the nebulizer via a peristaltic pump, but when using a nebulizer showing spontaneous suction (the concentric nebulizers mentioned above), the peristaltic pump can be omitted [28]. Especially for isotope ratio measurements, this may be beneficial, as in this way the structured variation in the signal that can be observed when using a peristaltic pump can be avoided. Two types of spray chambers are typically used, the Scott-type double-pass and the cyclonic spray chamber (Figure 2.18). Both types aim at the removal of larger droplets by impaction and gravitational settling [26]. This is required to assure a stable plasma and guarantee efficient desolvation, atomization, and ionization in the ICP, but unfortunately leads to a 10–100-fold reduction in analyte introduction efficiency. The cyclonic spray chamber offers a somewhat higher analyte introduction efficiency, but a combination of both (also known as stable or tandem introduction system) provides the best signal stability [29, 30].

Spray chambers can be cooled via a water jacket or Peltier cooling to reduce the amount of solvent vapor introduced into the ICP [31, 32]. A further reduction in the amount of solvent introduced can be realized via a desolvation system. Traditionally, such a desolvation system consisted of a sequence of a heated and a cooled tube. In the heated tube, the solvent is vaporized, after which it condenses on the inner wall of the cooled tube and is thus removed. Nowadays, desolvation systems equipped with a membrane desolvator are often used [33, 34]. These basically consist of a tube manufactured from a semipermeable porous material, around which heated Ar gas is flowing in the opposite direction to the sample aerosol flow. The solvent is vaporized, and the gaseous solvent molecules leave the central tube via the pores and are carried off by the heated Ar flow. Desolvation of the sample aerosol can lead to an ~10-fold increase in signal intensity. For rather volatile analyte elements, (partial) analyte loss needs to be taken into account [35].

For some elements, sample introduction in the gaseous form presents an interesting alternative. This approach permits the separation of the element of interest from the concomitant matrix and provides (almost) quantitative analyte introduction. This approach has been used, for example, for Hg (reduction of Hg^{2+} to atomic Hg using $SnCl_2$) [36], Os (oxidation to OsO_4) [37–39], and hydride-forming elements, such as Se [40–42] (conversion into volatile hydrides using $NaBH_4$).

Solid samples can be analyzed directly with ICP-MS by using laser ablation (LA) as a means of sample introduction [43–45]. In this approach, a high-energy laser beam is focused onto the surface of the sample, placed in the ablation chamber. Upon impact of the laser pulse, a small amount of material is ablated and the dry aerosol thus formed is transported by means of a carrier gas (He or Ar) into the ICP (Figure 2.19).

The diameter of the laser beam can be typically varied between ~5 and ≥100 μm and, as the sample can be monitored via a video camera, the exact location to be examined can be selected. As a result, LA-ICP-MS is very well suited for the

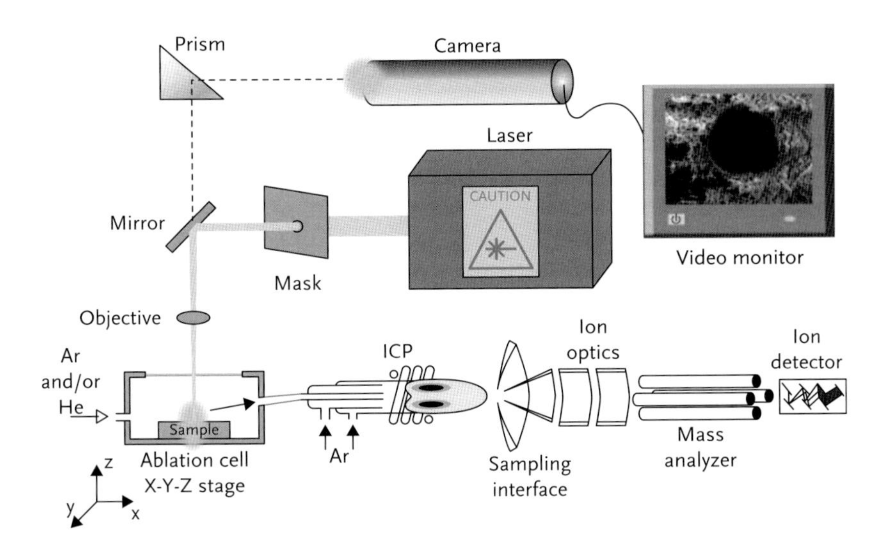

Figure 2.19 Schematic representation of the set-up for LA-ICP-MS. Reproduced with permission of Wiley-VCH Verlag GmbH from [46].

practically non-destructive elemental or isotopic analysis of valuable materials or artifacts. For trace element fingerprinting of gemstone-quality diamonds, for example, sample consumption could be limited to $\sim 2\,\mu g$ per sampling location [47], and when investigating objects of art [46], the damage inflicted (crater) on the sample is not visible to the naked eye when using a sufficiently small beam diameter (e.g., 15 μm). LA-ICP-MS is also being increasingly used for spatially resolved analysis, both lateral and in-depth [48]. With present-day LA systems, the ablation cell is mounted on a software-controlled translation stage that can be steered in all three spatial directions, also allowing two-dimensional rastering for studying and visualizing the spatial distribution of target elements in two dimensions [49].

Both solid-state (Nd:YAG-based) and gas (excimer-based) lasers are used. The fundamental wavelength of Nd:YAG lasers is 1064 nm (in the IR region), but by means of quadrupling or quintupling this fundamental frequency, UV wavelengths of 266 and 213 nm, respectively, are obtained [50]. Whereas the "standard wavelength" of Nd:YAG-based LA units used to be 266 nm, most systems sold nowadays provide laser radiation with a wavelength of 213 nm. Excimer (excited dimer)-based laser units provide even shorter wavelengths in the deep-UV region. ArF* excimer-based systems are the most widespread in this context and provide laser radiation with a wavelength of 193 nm [51]. Nowadays, laser radiation with a wavelength of 193 nm can also be produced by coupling an Nd:YAG based-laser to an optical parametric oscillator (OPO) system [52]. At this deep-UV wavelength, the coupling of the laser beam with UV-transparent materials, such as quartz, calcium fluoride, and carbonates, is more efficient and results in smaller particles that can be "digested" more efficiently by the ICP [53]. This also renders the ablation less matrix-dependent.

All lasers described so far work in the nanosecond regime with laser pulse widths varying between 3–5 ns (Nd:YAG-based lasers) and 10–15 ns (excimer lasers). With the more recently introduced femtosecond lasers, the energy is supplied to the solid sample within a much shorter time interval (<1 ps) [54–56]. This reduces the amount of heat generated and results in a cleaner, more explosion-like ablation [57], especially for metallic samples. In the latter case, avoiding heat transfer and thus melting also leads to an enhanced spatial resolution.

Also in the context of isotopic analysis, LA is often used as a means of sample introduction because of some of the advantages mentioned above, for example, quasi-nondestructive analysis in the case of valuable objects or spatially resolved analysis in the case of samples that can be considered as a chronological archive, such as speleothems [58]. Whereas the standard ablation chambers can house only relatively small samples, the development of larger and/or specific cells [59] and of alternative approaches, including the use of a non-contact cell [60] or in-air ablation [61], led to a continuous extension of the type of samples amenable to LA-ICP-MS analysis.

It should be taken into account, however, that the isotope ratio precision is typically worse than that with pneumatic nebulization as a means of sample introduction. Recently, Aramendía *et al.* provided a description of strategies to improve the isotope ratio precision attainable with the combination of LA and single-collector ICP-MS [62]. As the use of LA also affects the degree of mass discrimination, matrix-matched standards are required for adequate mass bias correction [63].

Recent progress in the use LA-MC-ICP-MS for isotopic analysis is discussed in Chapter 4.

Descriptions of other sample introduction systems, including ultrasonic nebulization (USN), direct injection nebulization (DIN), and electrothermal vaporization (ETV), can be found in the literature [64–66].

2.6 Spectral Interferences

As a result of the low ($\sim$unit) mass resolution characteristic of the quadrupole filter, the occurrence of spectral interferences is the most notorious disadvantage of ICP-MS. Spectral overlap occurs whenever two or more ions show the same *nominal* mass-to-charge ratio. Especially the occurrence of polyatomic ions, consisting of elements from the plasma gas, solvent, matrix, and/or entrained air, is at the origin of spectral interferences. For elements showing a second ionization energy lower than the first ionization energy of Ar, doubly charged ions can also be formed, further complicating the spectrum. Table 2.2 provides information on the types of interfering ions occurring in ICP-MS, although some overlap between the categories has to be taken into account. Self-evidently, for many *potentially*

Table 2.2 Classification of interfering ions in categories, with selected examples.

Type of interfering ion	Example/analyte nuclide affected
Isobaric nuclide	$^{40}Ar^+/^{40}Ca^+$ $^{58}Ni^+/^{58}Fe^+$ $^{87}Rb^+/^{87}Sr^+$ $^{204}Hg^+/^{204}Pb^+$
Ar-containing polyatomic ions	$^{40}ArC^+/^{52}Cr^+$ $^{40}ArO^+/^{56}Fe^+$ $^{40}Ar^{23}Na^+/^{63}Cu^+$ $^{40}Ar^{35}Cl^+/^{75}As^+$ $^{40}Ar_2^+/^{80}Se^+$
Oxide and hydroxide ions	$^{32}S^{16}O^+/^{48}Ti^+$ $^{35}Cl^{16}O^+/^{51}V^+$ $^{137}Ba^{16}O^+/^{153}Eu^+$ $^{136}Ba^{16}O^1H^+/^{153}Eu^+$
Doubly charged ions	$^{48}Ca^{2+}/^{24}Mg^+$ $^{206}Pb^{2+}/^{103}Rh^+$
Other types	$^{28}Si^{35}Cl^+/^{63}Cu^+$ $^{32}S^{16}O_2{}^+/^{64}Zn^+$ $^{32}S^{16}O_2{}^1H^+/^{65}Cu^+$ $^{23}Na^{23}Na^{16}O^+/^{62}Ni^+$

interfering ions, the concentrations of both the parent element and the element affected determine whether a meaningful interference does or does not occur.

Appropriate sample digestion, element–matrix separation techniques, aerosol desolvation [67, 68], and use of an alternative sample introduction system [66, 69, 70] are means to avoid (some cases of) spectral overlap. In present-day ICP-MS instrumentation, some more *general* strategies to tackle spectral overlap can also be deployed. In general, spectral interferences are an even greater nuisance in isotopic than in trace element analysis. A first reason for this lies in the fact that in isotopic analysis, at least two nuclides should display a signal free from spectral overlap. Moreover, whereas a limited extent of spectral overlap (e.g., 0.1–1%) may be negligible in trace element determination, this is definitely not the case in isotope ratio determination.

2.6.1 Cool Plasma Conditions

Cool plasma conditions are obtained at low r.f. power and high carrier gas flow rate. As these are conditions that favor the occurrence of secondary electrical discharges between the ICP and the sampling cone, some instruments require a grounded shield between the ICP torch and the load coil to decouple these components capacitively before adequate operation under cool plasma conditions is feasible [71]. Under cool plasma conditions, molecules such as NO and O_2 are no longer completely atomized. As a result of their respective ionization energies, electron transfer from these molecules to Ar-containing ions, giving rise to spectral overlap hampering the determination of various important elements (e.g., K, Ca, Cr, Fe), is a spontaneous process. As a result, the corresponding spectral overlap is avoided or at least very strongly reduced. This approach has been used in the isotopic analysis of various elements, including Ca [72], Cr [73], and Fe [74].

Drawbacks to the use of cool plasma conditions are a strongly reduced ionization efficiency for elements with a high ionization energy, a more pronounced occurrence of other types of polyatomic ions (oxide ions), and more pronounced matrix effects (matrix-induced signal suppression or enhancement). Plasma shielding for capacitive decoupling of the load coil and ICP has proven not only to be useful in the context of cool plasma conditions, but also to reduce the energy spread of the ions, thus enhancing their transmission efficiency [75].

2.6.2
Multipole Collision/Reaction Cell

A multipole collision/reaction cell consists of a multipole (consisting of $2n + 2$ parallel metallic rods) in an enclosed cell that can be pressurized with a gas [76–81]. It is located between the interface and the mass analyzer (Figure 2.20).

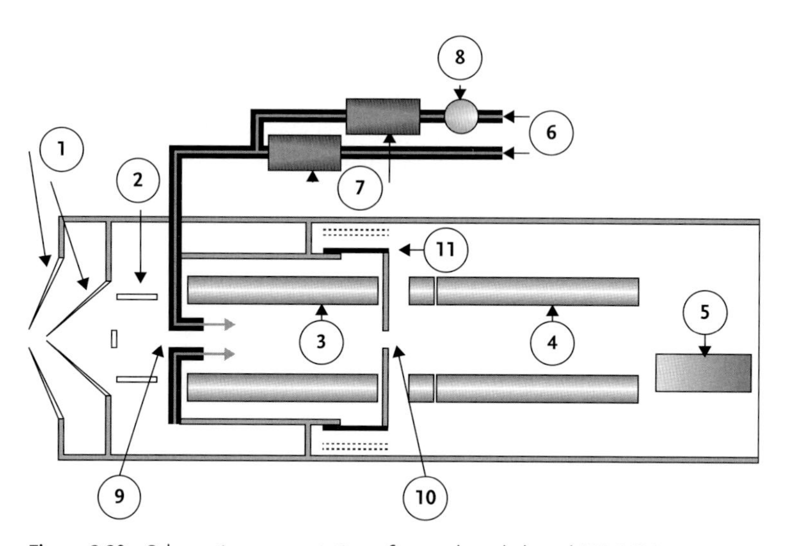

Figure 2.20 Schematic representation of a quadrupole-based ICP-MS instrument equipped with a collision/reaction cell: 1, sampling cone and skimmer; 2, extraction lens; 3, collision/reaction cell; 4, quadrupole filter; 5, detector; 6, collision or reaction gas; 7, mass flow controller; 8, getter (to remove traces of O_2); 9, cell entrance; 10, cell exit; 11, cell vent. Reproduced with permission of Elsevier from [82].

In commercially available single-collector ICP-MS instrumentation, either a quadrupole ($n = 1$), a hexapole ($n = 2$), or an octopole ($n = 3$) setup is used. An important differentiation has to be made, however, between a quadrupole assembly on the one hand and hexapole and octopole units on the other. Whereas the latter can only guide an ion from one point to another, a quadrupole assembly shows mass filtering capabilities [83].

2.6.2.1 **Overcoming Spectral Interference via Chemical Resolution**

When deploying chemical resolution to overcome spectral interferences, selective ion–molecule reactions are relied upon either to eliminate the contribution of an interfering ion at the nominal mass-to-charge ratio of the analyte nuclide or to convert the analyte ion into a molecular ion, which can be measured interference free at another mass-to-charge ratio.

One form of selective reaction between an unwanted ion and the reaction gas molecules is electron transfer. Spontaneous electron transfer from a reaction gas molecule to an ion occurs when the ionization energy of the former is lower than that of the neutral entity at the basis of the interfering ion (first ionization energy in the case of a singly charged ion, second in the case of a doubly charged ion). A few examples of this approach are presented below.

- Creating interference-free conditions at $m/z = 40$ for Ca monitoring [84]:

$$^{40}\mathrm{Ar}^{+} + \mathrm{NH_3} \rightarrow {}^{40}\mathrm{Ar} + \mathrm{NH_3^{+}}$$
$$^{40}\mathrm{Ca}^{+} + \mathrm{NH_3} \rightarrow \text{no reaction}$$

- Creating interference-free conditions at $m/z = 56$ for Fe monitoring [74]:

$$^{40}\mathrm{Ar}^{16}\mathrm{O}^{+} + \mathrm{NH_3} \rightarrow {}^{40}\mathrm{Ar}^{16}\mathrm{O} + \mathrm{NH_3^{+}}$$
$$^{56}\mathrm{Fe}^{+} + \mathrm{NH_3} \rightarrow \text{no reaction}$$

- Creating interference-free conditions at $m/z = 103$ for Rh monitoring [85]:

$$^{206}\mathrm{Pb}^{2+} + \mathrm{NH_3} \rightarrow {}^{206}\mathrm{Pb}^{+} + \mathrm{NH_3^{+}}$$
$$^{103}\mathrm{Rh}^{+} + \mathrm{NH_3} \rightarrow \text{no reaction}$$

"Real" chemical reactions, whereby atoms are exchanged, also occur. In this case, a reaction can only proceed when it is thermodynamically allowed, that is, when the reaction is accompanied by a decrease in Gibbs free energy, $\Delta G < 0$. Reactions with $\Delta G > 0$ can only proceed when energy is supplied and, in a collision/reaction cell, this is not the case. The change in Gibbs free energy accompanying a chemical reaction is determined by the changes in enthalpy (ΔH) and entropy (ΔS):

$$\Delta G = \Delta H - T\Delta S$$

The change in entropy ΔS accompanying an ion–molecule reaction in a collision/reaction cell is negligible, such that

$$\Delta G \approx \Delta H$$

As a result, an ion–molecule reaction *can only* proceed if it is exothermic ($\Delta H < 0$). Whether or not such a reaction *will also* proceed depends on the reaction rate. As a rule of thumb, it can be assumed that if a reaction *can* proceed (thermodynamics), it *will* actually proceed at a sufficient rate (kinetics). A few examples of this approach are presented below.

- Creating interference-free conditions at $m/z = 56$ for Fe monitoring [74]:

$$^{40}\mathrm{Ar}^{16}\mathrm{O}^+ + \mathrm{CO} \rightarrow {}^{40}\mathrm{Ar}^+ + \mathrm{CO_2}$$

$$^{56}\mathrm{Fe}^+ + \mathrm{CO} \rightarrow \text{no reaction}$$

- Creating interference-free conditions at $m/z = 106$ for Pd monitoring [86]:

$$^{90}\mathrm{Zr}^{16}\mathrm{O}^+ + \mathrm{O_2} \rightarrow {}^{90}\mathrm{Zr}^{16}\mathrm{O}_2^+ + \mathrm{O}$$

$$^{106}\mathrm{Pd}^+ + \mathrm{O_2} \rightarrow \text{no reaction}$$

In some cases, interference-free conditions are created by relying on selective reaction between the analyte ion and the reaction gas, allowing interference-free measurement of the analyte ion signal intensity at another mass-to-charge ratio. Examples of this approach are presented below.

- For S monitoring [87]:

$$^{32}\mathrm{S}^+ + \mathrm{O_2} \rightarrow {}^{32}\mathrm{SO}^+ + \mathrm{O}$$

$$^{16}\mathrm{O}_2^+ + \mathrm{O_2} \rightarrow \text{no reaction}$$

- For Sr monitoring [88]:

$$^{87}\mathrm{Sr}^+ + \mathrm{CH_3F} \rightarrow {}^{87}\mathrm{SrF}^+ + \mathrm{CH_3}$$

$$^{87}\mathrm{Rb}^+ + \mathrm{CH_3F} \rightarrow \text{no reaction}$$

An important facet of this approach, ignored so far, is the occurrence of side reactions, producing unwanted reaction products that may interfere with the determination of other target elements. In the case of a collision/reaction cell equipped with a quadrupole assembly, the formation of unwanted ions can be avoided by relying on its mass filtering capabilities. A mass window can be defined and all ions outside this window are removed, thus preventing them from taking part in further reactions [83]. As the mass window can be shifted in synchronicity with the scanning mass analyzer, the multi-element capabilities of ICP-MS can be preserved. As hexapole or octopole assemblies do not show mass filtering capabilities, the unwanted effects of newly created product ions are counteracted in another way, that is, by means of energy discrimination. As ions extracted from the ICP travel at higher speed than those formed in the reaction cell, a decelerating voltage can be relied upon to discriminate selectively against these newly formed ions. This approach is termed kinetic energy discrimination.

2.6.2.2 **Overcoming Spectral Interference via Collisional Deceleration and Kinetic Energy Discrimination**

This approach relies on the use of He as a non-reactive or inert collision gas in combination with energy discrimination. As the cross-section for collision with an He atom of a polyatomic ion is somewhat larger than that of an atomic ion, the former are slowed considerably more than the latter in the cell. As a result, a decelerating voltage may be applied to discriminate selectively against the polyatomic ions, such that only atomic ions are allowed into the mass analyzer.

2.6.3 High Mass Resolution with Sector Field ICP-MS

Approximately 10% of the ICP-MS instruments used worldwide are equipped with a double-focusing sector field mass spectrometer instead of the more traditional quadrupole filter. The use of this type of mass spectrometer requires the ions to be accelerated over a potential difference (e.g., 8000 V) before being introduced into the mass spectrometer [14, 15]. By reducing the width of an entrance and exit slit, the mass resolution can be increased (Figure 2.21). Whereas some instrument types allow continuous variation of the mass resolution, other offer a selection between a limited number of predefined resolution settings only.

At a sufficiently high mass resolution, ion signals that coincide at low mass resolution can be separated from one another, such that interference-free conditions are realized (Figure 2.22). With present-day instrumentation, a mass resolution of 10 000 (or slightly higher) can be realized, allowing most spectral interferences encountered to be resolved [89, 90].

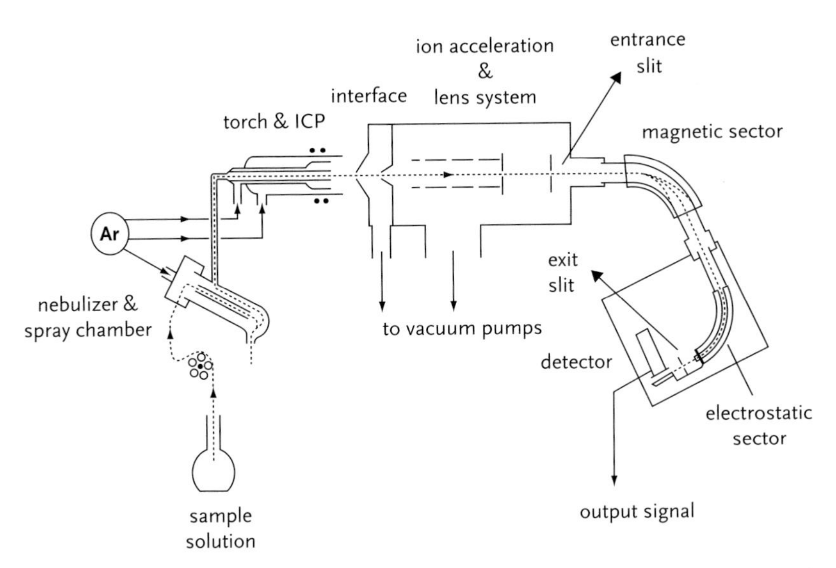

Figure 2.21 Schematic representation of sector field ICP-MS. The mass resolution can be increased by reducing the width of entrance and exit slits.

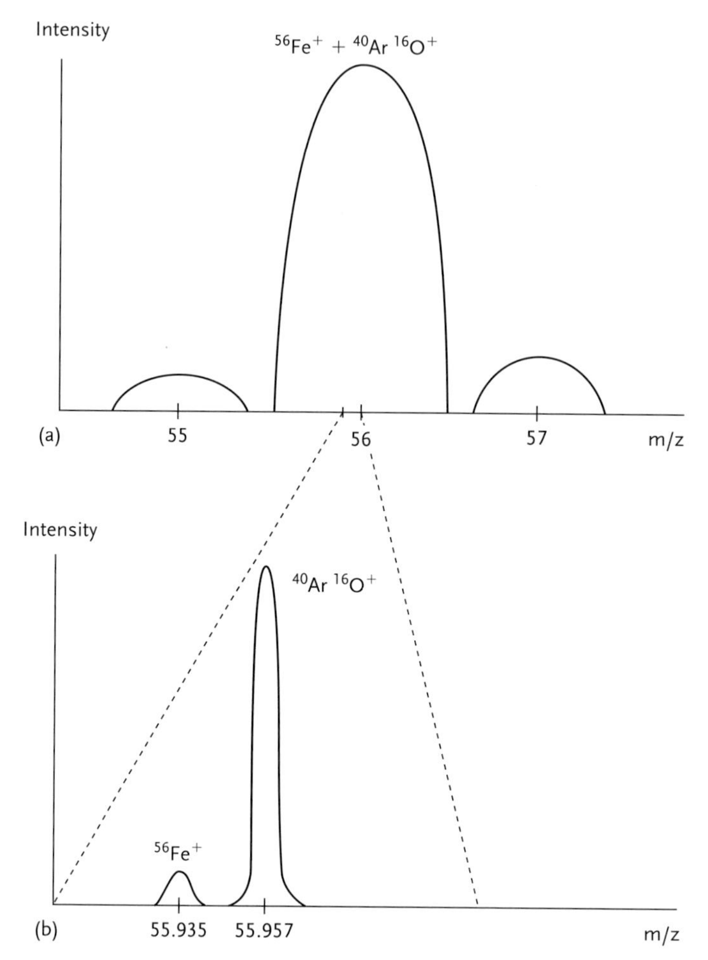

Figure 2.22 Schematic representation of the mass spectrum at $m/z = 56$, as obtained using a quadrupole-based ICP-MS instrument (a) and using a sector field ICP-MS instrument operated at higher mass resolution (b). Reproduced with permission of International Scientific Communications (UK) Ltd., from [91].

2.7
Measuring Isotope Ratios with Single-Collector ICP-MS

Owing to the rather noisy character of the ICP ion source and the sequential monitoring characteristic for ICP-MS equipped with a single detector only, single-collector ICP-MS does not provide sufficient precision for the most demanding isotope ratio applications. However, for applications involving measuring induced changes in the isotopic composition of a target element, as is the case in elemental assay via isotope dilution and in stable isotopic tracer experiments, the figures of merit can generally be considered as fit-for-purpose. In the context of studying natural variations, on the other hand, it can be said that single-collector ICP-MS can only be deployed successfully for studying those elements showing a relatively large variation in their isotopic composition, such as Li and B, the isotopic composition of which is affected due to isotope fractionation effects that are fairly pronounced as a result of the large relative differences in mass between their isotopes, and Sr and Pb, containing radiogenic isotopes. Moreover, even for the elements cited, single-collector ICP-MS is only suited for the less-demanding applications. For more challenging applications and for elements showing a smaller variation in their natural isotopic composition, isotopic analysis needs to be accomplished via multi-collector ICP-MS (see Chapter 3).

2.7.1 Isotope Ratio Precision

2.7.1.1 Poisson Counting Statistics

Before discussing the isotope ratio precision offered by various types of ICP-MS instrumentation and the factors affecting the precision, it needs to be pointed out that there is a "fundamental" limitation. When the variation in the arrival of ions at the detector can be considered as random, this fundamental limitation is governed by Poisson counting statistics [92]. According to Poisson counting statistics, the standard deviation on a signal corresponding to N counts (note that this is the *absolute* number of ions registered, not the count rate expressed in counts per second) is given by

$$s(N) = \sqrt{N}$$

Relying on the laws of error/uncertainty propagation, the standard deviation for an isotope ratio N_1/N_2 is then given by

$$s\left(\frac{N_1}{N_2}\right) = \frac{N_1}{N_2}\sqrt{\left[\frac{s(N_1)}{N_1}\right]^2 + \left[\frac{s(N_2)}{N_2}\right]^2}$$

This is the ultimate precision attainable and the isotope ratio precision observed in practice should be compared with that "predicted" by Poisson counting statistics to evaluate whether or not further progress can be made.

Gray *et al.* investigated this statement in practice and adapted their quadrupole-based ICP-MS instrument such that as many sources of noise as possible were eliminated, or their contribution minimized [93]. This included the use of a free expansion interface (without skimmer), a bonnet to shield the ICP from the surrounding atmosphere, and the introduction of the analyte (Xe) in gaseous form. Under these extraordinary conditions, they obtained a measurement precision equal to that calculated on the basis of Poisson counting statistics [down to 0.02% relative standard deviation (RSD)]. With a normal (two-cone) interface and sample introduction via pneumatic nebulization, they still obtained extraordinary values of around 0.05%, but at analyte concentrations that are unusually high (100 mg Ag l^{-1}), an isotope ratio close to 1 ($^{107}Ag/^{109}Ag$), and with the sheathing bonnet mentioned earlier.

As a result of the role of the number of counts registered for both isotopes in Poisson counting statistics, isotope ratio measurements are usually carried out at relatively high signal intensities (and thus at a sufficiently high target element concentration), while using sufficiently long measurement times. It has to be taken into account, however, that the higher the signal intensity, the larger is the influence of the detector dead time (see section 2.7.2.2). At higher signal intensities, the fraction of ions that arrive at the detector but do not contribute to the final signal intensity increases. As a result, the relative contribution of the correction for detector dead time to the total uncertainty will also be increased. Therefore, typically, a compromise signal intensity is aimed for. There is also no benefit in increasing the measurement time beyond reasonable values, as the experimentally obtained isotope ratio precision will typically no longer improve on extending total acquisition times beyond 30 min, due to issues with the long-term precision.

2.7.1.2 **Isotope Ratio Precision with Single-Collector ICP-MS**

When focusing attention on single-collector ICP-MS, the measurement precision, usually expressed as the RSD estimated on the basis of N subsequent measurements, typically significantly exceeds that predicted on the basis of Poisson counting statistics. This is a consequence of the combination of the ICP as a rather noisy ion source with sequential monitoring of the intensity of the ion beams of interest.

With a standard quadrupole-based ICP-MS instrument, the optimum isotope ratio precision attainable is $\geq 0.1\%$ RSD. This is the within-run or internal precision, expressed as the RSD (%) estimated on the basis of N (typically $N = 10$) replicate measurements. Some authors rather express the internal precision as the relative standard error, which corresponds to RSD $(\%)/\sqrt{N}$. As a result, care has to be taken when comparing isotope ratio precisions reported in the literature.

Values approaching 0.1% RSD internal precision can be obtained at sufficiently high signal intensities, while using a sufficiently long acquisition time and for an isotope ratio close to 1.

In theory, the variation in the signal intensities, in the form of both instability and drift, should not affect the isotope ratio measured and, thus, also the isotope ratio precision, when the intensities of the ion beams of interest are monitored simultaneously. This is accomplished in multi-collector ICP-MS, wherein an array of Faraday collectors (and/or ion-counting devices) is used for detection (see Chapter 3). In single-collector instrumentation, this can be approached by reducing (i) the number of spectral peaks monitored to those strictly required and (ii) the residence time per nuclide (and the number of acquisition points per spectral peak) [91–97]. As a result of the settling time of a quadrupole filter or a sector field mass spectrometer, a compromise setting needs to selected to combine rapid peak hopping or peak jumping with an acceptable duty cycle (a sufficiently high value for the ratio of the *actual* measurement time to the total time of the procedure). The effect of the residence time (also termed acquisition time or dwell time) per nuclide on the isotope ratio precision is shown in Figure 2.23.

Several approaches exist to improve the isotope ratio precision.

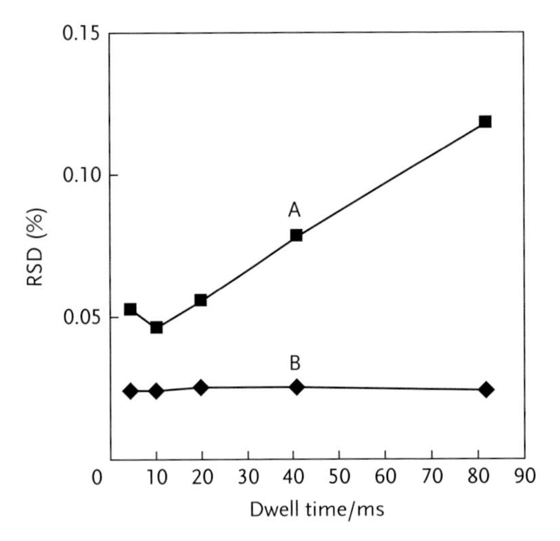

Figure 2.23 $^{107}Ag/^{109}Ag$ isotope ratio precision (RSD for 10 replicate measurements) as a function of the acquisition time per nuclide: (a) predicted on the basis of Poisson counting statistics; (b) observed experimentally. Reproduced with permission of the Royal Society of Chemistry from [28].

For isotope ratios that differ substantially from unity, for example, $^{204}Pb/^{208}Pb$, a longer residence time can be selected for the low-abundant nuclide (in this case ^{204}Pb). This results in an increase in the absolute number of counts registered for ^{204}Pb and therefore Poisson counting statistics demonstrate that a better isotope ratio precision is theoretically possible. This is counteracted, however, by the slower peak jumping or peak hopping that this results in, and therefore, again, compromise conditions need to be selected. Such compromise settings were determined via model calculations for Sr (monitoring of ^{86}Sr, ^{87}Sr, and ^{88}Sr) and Pb (monitoring of all Pb isotopes) isotopic analysis by Monna *et al.* [92]. The significant improvement in isotope ratio precision predicted by the model calculations were also reflected in the corresponding experiments, provided that conditions were chosen under which counting statistics form the most important contribution to the total variation.

Collisional damping in a collision/reaction cell provides a significant improvement in the isotope ratio precision [98, 99]. This effect is created by pressurizing the cell with a nonreactive collision gas, typically Ne. As a result, ions extracted from the ICP at slightly different moments in time are admixed in the cell, thereby damping the short-term variations in the ion beam to some extent. The effect of the use of Ne as a non-reactive collision gas on the isotope ratio precision observed in practice is illustrated in Figure 2.24.

By using collisional damping, the isotope ratio precision obtained can be improved to values of approximately 0.05% RSD (under optimum conditions). This gain in isotope precision, however, comes at the cost of a more pronounced instrumental mass discrimination [100–103], caused by preferential collisional losses of the lighter nuclide (Figure 2.25).

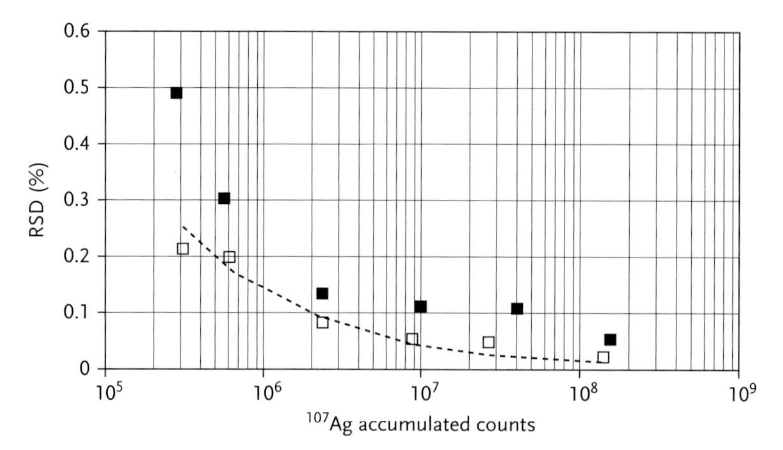

Figure 2.24 Internal isotope ratio precision (RSD) for $^{107}Ag/^{109}Ag$ as obtained using a quadrupole-based instrument equipped with a collision/reaction cell. Filled squares, vented cell; open squares, with Ne as an inert collision gas, introduced into the cell at a flow rate of 2 ml min^{-1}. The dotted line represents the precision as predicted by Poisson counting statistics. Reproduced with permission of the Royal Society of Chemistry from [99].

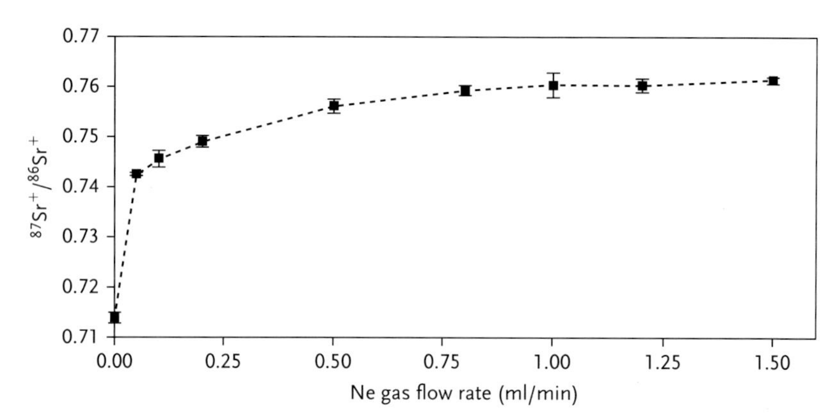

Figure 2.25 Raw $^{87}Sr/^{86}Sr$ isotope ratio result as obtained using a quadrupole-based ICP-MS instrument equipped with a collision/reaction cell as a function of the flow rate (~pressure) of Ne as an inert collision gas (corresponding true value: 0.710). Reproduced with permission of the Royal Society of Chemistry from [102].

With a sector field ICP-MS instrument operated at low mass resolution, a slightly better isotope ratio precision (better than 0.05% RSD) is obtained than with a quadrupole-based instrument [104, 105], at least when pure electrical scanning (E-scanning) is opted for (the magnetic field strength is not adapted during the measurement). This is a consequence of the flat-topped spectral peak shape observed at low mass resolution with sector field instrumentation, a result of the fact that, under these conditions, the beam diameter is smaller than the exit slit in front of the detector. Small differences in the location of the data acquisition point with respect to the spectral peak (mass calibration instability) will not result in a significant change in the signal intensity, thus providing more stable signals. On increasing the mass resolution, accomplished by narrowing both the entrance and the exit slits, the spectral peak rather becomes triangular (with a rounded peak), such that this advantage of sector field over quadrupole-based instrumentation no longer holds true. Also, the decrease in signal intensity accompanying an increase in mass resolution has a negative effect on the isotope ratio precision. Nevertheless, at medium mass resolution ($R = 4000$), the isotope ratio precision attainable is still similar to that for standard quadrupole-based ICP-MS [106]. At higher mass resolution, using a small acquisition window centered in the spectral peak only is reported to result in the best precision attainable.

When equipped with a TOF analyzer for mass analysis, the ions arriving at the detector all stem from the same package of ions extracted from the ICP and introduced into the TOF-analyzer at exactly the same time (or within a very short time interval), thus approaching the limit of simultaneous monitoring closer than is the case with quadrupole-based or sector field ICP-MS. As a result, an internal isotope ratio precision in the order of 0.05% RSD can be fairly easily accomplished [107–110]. Moreover, in contrast to the situation with quadrupole-based or sector field instrumentation, monitoring more ion signals does not have a negative effect on the isotope ratio precision obtained. With an ICP-TOF-MS instrument, the total measurement time does not have to be "divided" between all the ion signals of interest, whereas with quadrupole-based and sector field ICP-MS instruments, addition of another nuclide to the list of signals to be monitored while keeping the total measurement time constant results in a corresponding reduction in the total acquisition time for the other nuclides and in a larger time interval between two subsequent scans or sweeps.

Isotope ratio determination using an ICP-TOF-MS instrument can, however, be plagued by dependence of the isotope ratio result (at least when using the detector in analog mode) on the signal intensities [109, 111]. Rowland and Holcombe identified the attenuation of signals with an intensity below a threshold value (required for filtering out noise) as the origin of the problem, as this affects less intense signals to a more pronounced extent than more intense signals [111]. This problem can be mitigated by increasing the detector gain, although this shortens the detector lifetime, or by correcting for the differences in efficiency, or avoided by using the ion counting mode. For isotope ratio determination with ICP-TOF-MS, pulse counting detection is not recommended, however, because of the limited linear response of the detector (e.g., linear up to 20 000 cps only [111]), such that low analyte concentrations need to be deployed and the theoretically attainable isotope ratio precision (Poisson counting statistics) is relatively poor (Table 2.3).

Table 2.3 Isotope ratio precision attainable with commercially available ICP-MS instrumentation under optimum conditions.

Type of instrumentation	Isotope ratio precision (%) (RSD for $n = 10$)
Quadrupole-based ICP-MS	
Traditional	≥0.1
With inert collision gas in collision/reaction cell	~0.05
Sector field ICP-MS	
Low mass resolution	~0.05
Medium mass resolution[a]	≥0.1
Time-of-flight ICP-MS	~0.05

[a]One manufacturer of sector field ICP-MS instrumentation has recently introduced the use of a slit that allows combination of a higher mass resolution ($R \approx 2000$) with flat-topped peaks, thus maintaining the isotope ratio precision as typical for low mass resolution [15].

2.7.2 Detector Issues

2.7.2.1 Electron Multiplier Operating Principles

Both quadrupole-based and sector field ICP-MS instruments are equipped with an electron multiplier for ion detection [112, 113]. The operating principle of such a device is illustrated by means of the traditional continuous dynode electron multiplier (Figure 2.26). When an ion strikes the inner surface of this device, one or more electrons are set free. Owing to the potential difference between the front end of the detector (negatively charged at −2000 to −3000 V) and the back end (grounded), the electrons thus formed are accelerated towards the back end of the detector. This acceleration leads to multiple collisions with the inner surface. Whenever an electron collides with the inner surface, more electrons are set free. This multiplication effect leads to "an avalanche" of electrons, such that one ion arriving at the detector finally leads to 10^7–10^8 electrons and each ion is individually detected in pulse counting mode. Nowadays, electron multipliers with discrete dynodes are preferred (Figure 2.26). The operating principle of this type of electron multiplier is very similar, but the secondary electrons are accelerated from one dynode to the next as a result of a potential difference between each two successive dynodes.

2.7.2.2 Detector Dead Time

As the detection of an ion requires some time (due to the multiplication effect itself and the handling of the pulse by the associated electronics), an electron multiplier used in pulse counting mode is characterized by a dead time τ. During the time that a detector needs for handling one ion pulse, it cannot detect an ion arriving at the detector. Depending on the type of electron multiplier, τ is of the order of 5–100 ns. The higher the count rate, the larger is the fraction of ions that

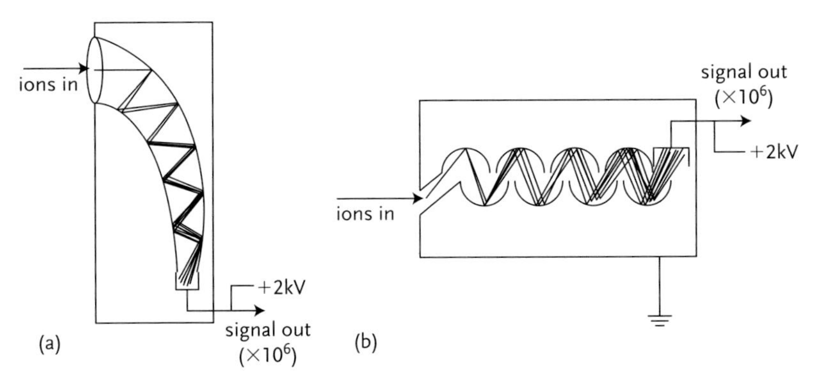

Figure 2.26 Basic operating principle of an electron multiplier. (a) Continuous dynode electron multiplier; (b) electron multiplier with discrete dynodes. Reproduced with permission of Dr. Paul Gates, School of Chemistry, University of Bristol, from [114].

is not detected. This is exactly why without adequate correction for detector dead time, an isotope ratio differing from 1 will show an additional bias (on top of that caused by instrumental mass discrimination; see section 2.7.3).

Correction for the detector dead time τ can be accomplished fairly easily and automatically via the instrument's software once the correct value is known. For isotopic analysis, one should never rely on a value predefined by the instrument manufacturer, but should determine the actual value experimentally. This is preferably done using the target element itself, as at least for some detector types, it has been demonstrated that the experimentally determined dead time may vary as a function of the element's mass number [115]. Dead time determination also has to be repeated regularly, as it changes upon aging of the detector. When during experimental dead time determination a very aberrant behavior is obtained (i.e., the results make no sense), this is often indicative of detector problems; the detector can, for example, be at the end of its lifetime.

The equation deployed for dead time correction depends on whether the multiplier behaves as non-paralyzable or as paralyzable. In the former case, the arrival of a second ion at the detector during the time that the detector needs for handling the first ion does not lead to an extension of the dead time. This behavior is often assumed, at least if the count rate is $\ll 1/\tau$.

For a non-paralyzable detector:

$$I_{\text{true}} = \frac{I_{\text{observed}}}{1 - (I_{\text{observed}}\tau)}$$

where I_{observed} and I_{true} (cps) are the observed and true count rates and τ (s) is the detector dead time.

There are several approaches for determining the dead time of the detector and its associated electronics. Most often, an approach relying on the fact that if the dead time is adequately corrected for, a selected isotope ratio (sufficiently different from unity) should show no systematic variation as a function of the element concentration is used. In the widely used Russ approach [116], a series of standard solutions of a given element, spanning a sufficiently wide concentration range, is measured. The value used by the instrument software for automatic dead time correction has to be set to zero prior to measuring the entire set of solutions. Subsequently, one selected isotope ratio for that element is calculated on the basis of signal intensities that are corrected for dead time losses assuming a number of detector dead time values in an appropriate range. The resulting isotope ratios are plotted as a function of the assumed dead time values (Figure 2.27). For each concentration level, the values thus obtained lie on a line and the lines for the various concentration levels (should) intersect in one point (x, y), The x-value provides the analyst with the detector dead time; the bias between the y-value and the corresponding "true" value, on the other hand, reflects the extent of mass discrimination.

Sag can interfere with the proper determination of the intersection point. The occurrence of sag or pulse pile-up [112, 117, 118] leads to a substantial loss of ions (greater than that caused by the detector dead time), thus increasing the slope of the corresponding line and shifting its intersection with the other lines to (apparently) higher dead time values. Sag is a result of the fact that, for avoiding "false" signals (e.g., resulting from ionization of residual gas molecules in the detector as a result of the electron multiplication effect), the magnitude of every individual ion pulse is compared with a threshold value by a "discriminator." At very high count rates, the gain of an electron multiplier is strongly reduced such that a substantial fraction of the incoming ions give rise to an output pulse that is actually below the threshold. As a result, much more pronounced signal losses

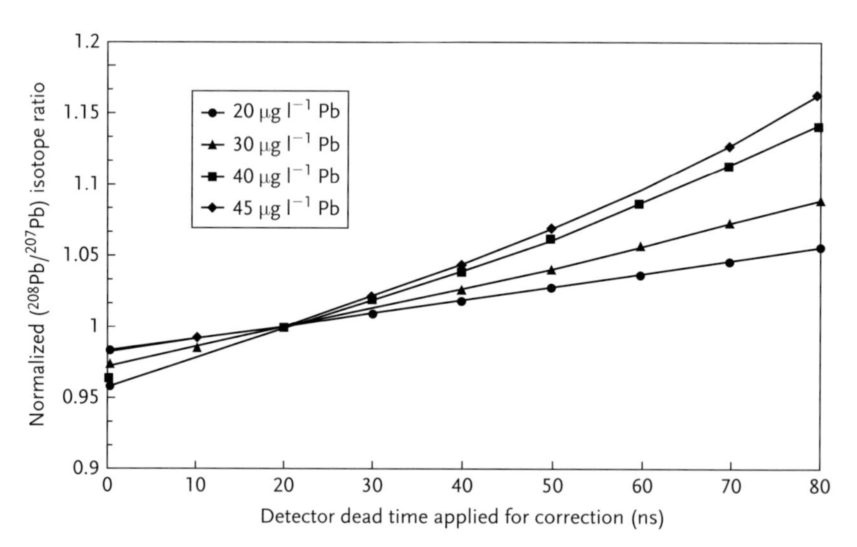

Figure 2.27 Experimental determination of dead time τ – lines corresponding to various analyte concentration levels (should) intersect in one point, the x-value of which corresponds to τ (in this specific case, 20 ns). Reproduced with permission of the Royal Society of Chemistry from [115].

than expected on the basis of the detector dead time occur and cannot be adequately corrected for. Self-evidently, sag should also be avoided in isotopic analysis.

The results from the measurements described in the previous paragraph can also be handled in another way [119], that is, by plotting the isotope ratio as a function of the element concentration for various assumed values of detector dead time (Figure 2.28a). The correct dead time should result in a line with a zero slope (horizontal line). A more practical way of establishing the correct dead time value is plotting the slopes of the lines thus obtained as a function of the assumed dead

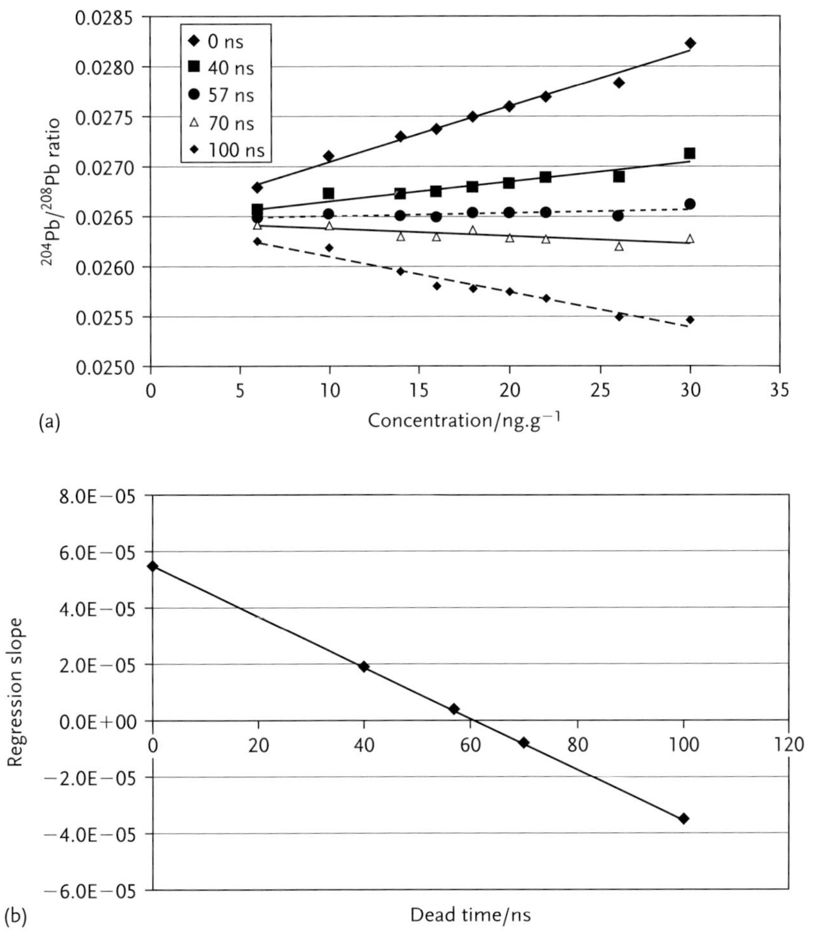

Figure 2.28 Experimental determination of detector dead time. (a) $^{204}Pb/^{208}Pb$ isotope ratio results as a function of Pb concentration, obtained assuming various values for detector dead time τ. That assumed value for τ leading to a horizontal line is the actual dead time. (b) Slopes of the straight lines depicted in (a) as a function of assumed value for τ. The intersection with the *x*-axis provides the actual dead time (in this specific case, 62 ns). Reproduced with permission of the Royal Society of Chemistry from [119].

time value (Figure 2.28b). The intersection of this line with the x-axis provides the analyst with the actual detector dead time. Other, but sometimes more complicated, methods for dead time determination have been described by, among others, Held and Taylor [120] and Appelblad and Baxter [121].

Finally, the detector dead time can also be electronically determined [122]. An interesting approach, potentially avoiding the need for dead time determination and reducing the related contribution to the total uncertainty, was described by Nygren *et al.* [123]. By changing the hardware configuration of the pulse detection system, they introduced an artificial and fixed dead time that is substantially longer than the actual dead time to mask temporal variations in the latter.

At higher count rates, the detector should be deployed in analog rather than in pulse counting mode. In analog mode, the individual ions arriving at the detector are no longer counted, but the electron multiplier is used as an analog amplifier instead. When deploying an electron multiplier in this mode, no correction for detector dead time is required. Most present-day detectors can be used in dual mode, whereby on the basis of the signal intensities encountered, an automatic choice between pulse counting and analog mode is made. Although this is very useful in the context of quantitative analysis (provided that both modes are adequately cross-calibrated), it has to be avoided in isotope ratio determination.

Sector field instruments can also be equipped with a triple-mode detector system. This means that in addition to an electron multiplier (mounted off-axis) that can be operated in dual mode, an additional Faraday cup is installed. When the signal intensity becomes so high that even when deployed in analog mode the electron multiplier can no longer handle the signal, the negative voltage at the front end of the detector is automatically switched off, such that the ion beam is no longer attracted to the multiplier, but enters the Faraday collector. This is a metal cup, grounded by a high resistance (typically $\sim 10^{11}\ \Omega$). When ions enter this detector, they are neutralized by electrons provided via the grounding. The potential difference thus created over the resistance is a measure of the ion signal intensity. This is the standard detector type in multi-collector ICP-MS and TIMS instrumentation, but is used in single-collector ICP-MS for the monitoring of very intense ion signals only. In the context of isotope ratio determination, this is of interest for measuring extreme isotope ratios, whereby the minor isotope can be measured with the electron multiplier in pulse counting mode and the major isotope using the Faraday collector. On the basis of its operating principle, it is clear that a Faraday collector does not suffer from dead time effects.

2.7.3 Instrumental Mass Discrimination

The beam leaving the expansion chamber via the skimmer aperture is electrically neutral and heavily dominated by neutral Ar atoms, forcing all entities to move with the same velocity [21, 22]. A negatively charged extraction lens selectively attracts the positive ions and repels the negatively charged entities (electrons and ions), which are removed by the vacuum system together with the neutral particles, which undergo neither attraction nor repulsion. The positive ions constituting the beam repel one another, which leads not only to defocusing, but also to a change in the beam composition [124]. As "lighter" and "heavier" ions move with the same velocity, their kinetic energy depends on their mass. As a result, the probability of removal from the beam as a result of mutual repulsion is far higher for a lighter than for a heavier ion. This phenomenon is referred to as the space-charge effect and it causes instrumental mass discrimination: heavier ions are transported more efficiently than are lighter ions [21, 22, 124]. One result of these space-charge effects is that ICP-MS is less sensitive for lighter than for heavier elements [125]. Another result is that these so-called space-charge effects cause a bias between a measured isotope ratio and the corresponding true value that needs to be corrected for [126]. Chapter 5 is entirely devoted to correction for instrumental mass discrimination and it makes clear that several approaches for correction exist. It needs to be stressed however that, as a result of the poorer isotope ratio precision, selection of a proper correction approach is less of an issue when working with single-collector instruments. Very often when working with single-collector instruments, these various approaches lead to final results that do not differ significantly from one another.

Also, a nozzle separation effect is expected to accompany the supersonic expansion in the extraction region [127]. This phenomenon is a result of collisions between the particles in the beam and the fact that only the central fraction of the beam passes the aperture of the skimmer and is characterized by a more pronounced loss of lighter nuclides than of heavier nuclides from the central ion beam. As a result, it works in the same direction as the space-charge effect and the contributions of both phenomena to the total instrumental mass discrimination cannot be distinguished from one another experimentally.

When measuring isotope ratios with a sector field ICP-MS instrument while using pure E-scanning (at a constant magnetic field strength), altering between the signals of interest is accomplished by changing the acceleration voltage. As changing the acceleration voltage has an effect on the ion transmission efficiency, this approach gives rise to an additional contribution to the mass discrimination [128].

Finally, it is important to note that mass fractionation such as occurs in TIMS and mass discrimination as occurs in ICP-MS are different phenomena (Figure 2.29).

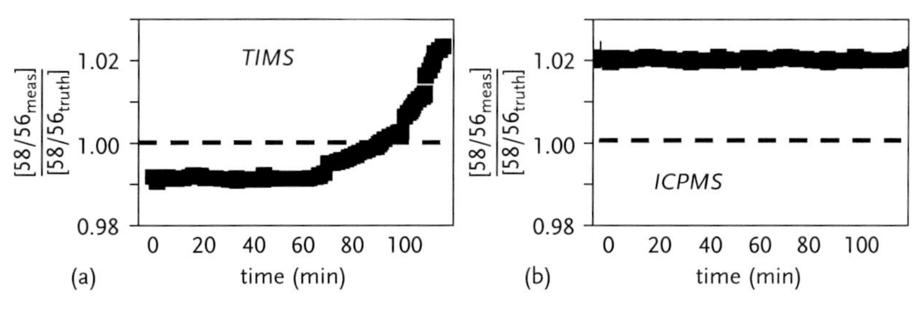

Figure 2.29 Comparison between mass fractionation as observed in TIMS (a) and mass discrimination as observed in ICP-MS (b). Modified and reproduced with permission of the Mineralogical Society of America from [129].

In the case of isotope ratio measurement via TIMS, a discrete amount of target element is deposited on a filament that is subsequently resistively heated in vacuum. The lighter isotope is ionized slightly more efficiently, such that the ion beam is initially slightly enriched in this isotope, while the target element remaining on the filament becomes slightly enriched in the heavier isotope. As a result of the latter enrichment, the isotopic composition of the ion beam changes progressively as a function of time. In ICP-MS, on the other hand, typically sample is introduced continuously (e.g., via nebulization) and the beam is always enriched in the heavier isotope. Although the extent of mass discrimination may show variations, there is no *systematic* trend as a function of measurement time. Instrumental mass discrimination in ICP-MS is typically of the order of 1% per mass unit, but is considerably larger for light elements (it can be as high as 10% per mass unit in the low-mass region).

References

1 Smith, D.H. (2000) Thermal ionization mass spectrometry, in *Inorganic Mass Spectrometry* (eds. C.M. Barshick, D.C. Duckworth, and D.H. Smith), Marcel Dekker, New York, pp. 1–30.

2 Turner, I.L. and Montaser, A. (1998) Plasma generation in ICP-MS, in *Inductively Coupled Plasma Mass Spectrometry* (ed. A. Montaser), Wiley-VCH Verlag GmbH, Weinheim, pp. 265–334.

3 Thomas, R. (2004) *Practical Guide to ICP-MS*, Marcel Dekker, New York.

4 The Linde Group, Inductively Coupled Plasma hiq.linde-gas.com/international/web/lg/spg/like35lgspg.nsf/docbyalias/anal_icp, last accessed on 05.02.2012.

5 Mermet, J.-M. (2007) Fundamental principlesof inductively coupled plasmas, in *Inductively Coupled Plasma Spectrometry and Its Applications*, 2nd edn. (ed. S.J. Hill), John Wiley & Sons, Ltd., Chichester, pp. 27–60.

6 Houk, R.S. (1986) Mass spectrometry of inductively coupled plasmas. *Anal. Chem.*, **58**, 97A–105A.

7 Taylor, H.E. and Garbarino, J.R. (1992) Analytical applications of inductively coupled plasma mass spectrometry, in *Inductively Coupled Plasmas in Analytical Atomic Spectrometry*, 2nd edn. (eds. A. Montaser and D.W. Golightly), Wiley-VCH Verlag GmbH, Weinheim.

8 Turner, P.J., Mills, D.J., Schröder, E., Lapitajs, G., Iacone, L.A., Haydar, D.A., and Montaser, A. (1998) Instrumentation for low- and high-resolution ICPMS, in *Inductively Coupled Plasma Mass Spectrometry* (ed. A. Montaser), Wiley-VCH Verlag GmbH, Weinheim, pp. 421–501.

9 University of Bristol (2011) Gas Chromatography Mass Spectrometry (GC/MS), www.bris.ac.uk/nerclsmsf/techniques/gcms.html (last accessed 21 December 2011).

10 Miller, P.E. and Denton, M.B. (1986) The quadrupole mass filter: basic operating concepts. *J. Chem. Educ.*, **63**, 617–622.

11 Ying, J.F. and Douglas, D.J. (1996) High resolution inductively coupled plasma mass spectra with a quadrupole mass filter. *Rapid Commun. Mass Spectrom.*, **10**, 649–652.

12 Chen, W., Collings, B.A., and Douglas, D.J. (2000) High-resolution mass spectrometry with a quadrupole operated in the fourth stability region. *Anal. Chem.*, **72**, 540–545.

13 Roboz, J. (1968) *Introduction to Mass Spectrometry – Instrumentation and Techniques*, John Wiley & Sons, Inc., New York.

14 Jakubowski, N., Moens, L., and Vanhaecke, F. (1998) Sector field mass spectrometers in ICP-MS. *Spectrochim. Acta B*, **53**, 1739–1763.

15 Jakubowski, N., Prohaska, T., Rottmann, L., and Vanhaecke, F. (2011) Inductively coupled plasma- and glow discharge plasma-sector field mass spectrometry. Part I: tutorial: fundamentals and instrumentation. *J. Anal. At. Spectrom.*, **26**, 693–726.

16 FIZ Chemie Berlin (2011) ChemgaPedia, www.chemgapedia.de (last accessed 21 December 2011).

17 Schilling, G.D., Andrade, F.J., Sperline, R.P., Denton, M.B., Barinaga, C.J., Koppenaal, D.W., and Hieftje, G.M. (2007) Continuous simultaneous detection in mass spectrometry. *Anal. Chem.*, **79**, 7662–7668.

18 Schilling, G.D., Ray, S.J., Rubinshtein, A.A., Felton, J.A., Sperline, R.P., Denton, M.B., Barinaga, C.J., Koppenaal, D.W., and Hieftje, G.M. (2009) Evaluation of a 512-channel Faraday-strip array detector coupled to an inductively coupled plasma Mattauch–Herzog mass spectrograph. *Anal. Chem.*, **81**, 5467–5473.

19 Mahoney, P.P., Ray, S.J., and Hieftje, G.M. (1997) Time-of-flight mass spectrometry for elemental analysis. *Appl. Spectrosc.*, **51**, A16–A28.

20 Goenaga-Infante, H. (2005) ICP-time-of-flight mass spectrometry, in *ICP Mass Spectrometry Handbook* (ed. S. Nelms), Blackwell, Oxford, pp. 69–84.

21 Niu, H. and Houk, R.S. (1996) Fundamental aspects of ion extraction in inductively coupled plasma mass spectrometry. *Spectrochim. Acta B*, **51**, 779–815.

22 Douglas, D.J. and French, J.B. (1988) Gas dynamics of the inductively coupled plasma mass spectrometry interface. *J. Anal. At. Spectrom.*, **3**, 743–747.

23 Browner, R.F. and Boorn, A.W. (1984) Sample introduction techniques for atomic spectroscopy. *Anal. Chem.*, **56**, 875A–888A.

24 Browner, R.F. and Boorn, A.W. (1984) Sample introduction – the Achilles heel of atomic spectroscopy. *Anal. Chem.*, **56**, 786A–798A.

25 Montaser, A., Minnich, M.G., McLean, J.A., Liu, H., Caruso, J.A., and McLeod, C.W. (1998) Sample introduction in ICPMS, in *Inductively Coupled Plasma Mass Spectrometry* (ed. A. Montaser), Wiley-VCH Verlag GmbH, Weinheim, pp. 83–264.

26 Todoli, J.L. and Vanhaecke, F. (2005) Liquid sample introduction and electrothermal vaporization for ICP-MS: fundamentals and applications, in *ICP Mass Spectrometry Handbook* (ed. S. Nelms), Blackwell, Oxford, pp. 182–227.

27 Thomas, R. (2007) A beginner's guide to ICP-MS. Part II: the sample introduction system. *Spectroscopy*, **16**, 56–60.

28 Begley, I.S. and Sharp, B.L. (1994) Occurrence and reduction of noise in inductively-coupled plasma mass spectrometry for enhanced precision in isotope ratio measurement. *J. Anal. At. Spectrom.*, **9**, 171–176.

29 Olofsson, R.S., Rodushkin, I., and Axelsson, M.D. (2000) Performance characteristics of a tandem spray chamber arrangement in double focusing sector field ICP-MS. *J. Anal. At. Spectrom.*, **15**, 727–729.

30 Krachler, M., Rausch, N., Feuerbacher, H., and Klemens, P. (2005) A new HF-resistant tandem spray chamber for improved determination of trace elements and Pb isotopes using inductively coupled plasma-mass spectrometry. *Spectrochim. Acta B*, **60**, 865–869.

31 Zhu, G.X. and Browner, R.F. (1988) Study of the influence of water-vapor loading and interface pressure in inductively coupled plasma mass spectrometry. *J. Anal. At. Spectrom.*, **3**, 781–789.

32 Pollmann, D., Pilger, C., Hergenroder, R., Leis, F., Tschöpel, P., and Broekaert, J.A.C. (1994) Noise power spectra of inductively-coupled plasma mass spectrometry using a cooled spray chamber. *Spectrochim. Acta B*, **49**, 683–690.

33 Botto, R.I. and Zhu, J.J. (1994) Use of an ultrasonic nebulizer with membrane desolvation for analysis of volatile solvents by inductively-coupled plasma

atomic emission spectrometry. *J. Anal. At. Spectrom.*, **9**, 905–912.

34 Tao, H. and Miyazaki, A. (1995) Decrease of solvent water loading in inductively-coupled plasma mass spectrometry by using a membrane separator. *J. Anal. At. Spectrom.*, **10**, 1–5.

35 Kahen, K., Jorabchi, K., and Montaser, A. (2006) Desolvation-induced non-linearity in the analysis of bromine using an ultrasonic nebulizer with membrane desolvation and inductively coupled plasma mass spectrometry. *J. Anal. At. Spectrom.*, **21**, 588–591.

36 Foucher, D. and Hintelmann, H. (2006) High-precision measurement of mercury isotope ratios in sediments using cold vapor generation multi-collector inductively coupled plasma mass spectrometry. *Anal. Bioanal. Chem.*, **384**, 1470–1478.

37 Sun, Y.L., Zhou, M.F., and Sun, M. (2001) Routine Os analysis by isotope dilution–inductively coupled plasma mass spectrometry: OsO_4 in water solution gives high sensitivity. *J. Anal. At. Spectrom.*, **16**, 345–349.

38 Schoenber, R., Nagler, T.F., and Kramers, J.D. (2000) Precise Os isotope ratio and Re–Os isotope dilution measurements down to the picogram level using multicollector inductively coupled plasma mass spectrometry. *Int. J. Mass Spectrom.*, **197**, 85–94.

39 Rodushkin, I., Bergman, T., Douglas, G., Engstrom, E., Sorlin, D., and Baxter, D.C. (2007) Authentication of Kalix (NE Sweden) vendace caviar using inductively coupled plasma-based analytical techniques: evaluation of different approaches. *Anal. Chim. Acta*, **583**, 310–318.

40 Janghorbani, M. and Ting, B.T.G. (1989) Comparison of pneumatic nebulization and hydride generation inductively coupled plasma mass spectrometry for isotopic analysis of selenium. *Anal. Chem.*, **61**, 701–708.

41 Rouxel, O., Ludden, J., Carignan, J., Marin, L., and Fouquet, Y. (2002) Natural variations of Se isotopic composition determined by hydride generation multiple collector inductively coupled plasma mass spectrometry. *Geochim. Cosmochim. Acta*, **66**, 3191–3199.

42 Layton-Matthews, D., Leybourne, M.I., Peter, J.M., and Scott, S.D. (2006) Determination of selenium isotopic ratios by continuous-hydride-generation dynamic-reaction-cell inductively coupled plasma mass spectrometry. *J. Anal. At. Spectrom.*, **21**, 41–49.

43 Günther, D., Jackson, S.E., and Longerich, H.P. (1999) Laser ablation and arc/spark solid sample introduction into inductively coupled plasma mass spectrometers. *Spectrochim. Acta B*, **54**, 381–409.

44 Vogt, C. and Latkoczy, C. (2005) Laser ablation ICP-MS, in *ICP Mass Spectrometry Handbook* (ed. S. Nelms), Blackwell, Oxford, pp. 228–258.

45 Longerich, H. (2008) Laser ablation–inductively coupled plasma mass spectrometry (LA-ICPMS), in *Laser Ablation ICP-MS in the Earth Sciences: Current Practice and Outstanding Issues* (ed. P. Sylvester), Short Course Series, vol. 40, Mineralogical Association of Canada, Quebec, pp. 11–18.

46 Resano, M., García-Ruiz, E., and Vanhaecke, F. (2010) Laser ablation-inductively coupled plasma mass spectrometry in archaeometric research. *Mass Spectrom. Rev.*, **29**, 55–78.

47 Resano, M., Vanhaecke, F., Hutsebaut, D., de Corte, K., and Moens, L. (2003) Possibilities of laser ablation-inductively coupled plasma mass spectrometry for diamond fingerprinting. *J. Anal. At. Spectrom.*, **18**, 1238–1242.

48 Balcaen, L., Lenaerts, J., Moens, L., and Vanhaecke, F. (2005) Application of laser ablation inductively coupled plasma (dynamic reaction cell) mass spectrometry for depth profiling analysis of high-tech industrial materials. *J. Anal. At. Spectrom.*, **20**, 417–423.

49 Becker, J.S., Zoriy, M., Wu, B., Matusch, A., and Becker, J.S. (2008) Imaging of essential and toxic elements in biological tissues by

LA-ICP-MS. *J. Anal. At. Spectrom.*, **9**, 1275–1280.

50 Jeffries, T.E., Jackson, S.E., and Longerich, H.P. (1998) Application of a frequency quintupled Nd:YAG source ($\lambda = 213$ nm) for laser ablation inductively coupled plasma mass spectrometric analysis of minerals. *J. Anal. At. Spectrom.*, **13**, 935–940.

51 Gunther, D., Frischknecht, R., Heinrich, C.A., and Kahlert, H.J. (1997) Capabilities of an argon fFluoride 193 nm excimer laser for laser ablation inductively coupled plasma mass spectrometry microanalysis of geological materials. *J. Anal. At. Spectrom.*, **12**, 939–944.

52 Horn, I., Günther, D., and Guillong, M. (2003) Evaluation and design of a solid-state 193 nm OPO–Nd:YAG laser ablation system. *Spectrochim. Acta B*, **58**, 1837–1846.

53 Guillong, M., Horn, I., and Günther, D. (2003) A comparison of 266 nm, 213 nm and 193 nm produced from a single solid state Nd:YAG laser for laser ablation ICP-MS. *J. Anal. At. Spectrom.*, **18**, 1224–1230.

54 Fernandez, B., Claverie, F., Pécheyran, C., and Donard, O.F.X. (2007) Direct analysis of solid samples by fs-LA-ICP-MS. *Trends Anal. Chem.*, **26**, 951–966.

55 Koch, J. and Günther, D. (2007) Femtosecond laser ablation inductively coupled plasma mass spectrometry: achievements and remaining problems. *Anal. Bioanal. Chem.*, **387**, 149–153.

56 Pisonero, J. and Günther, D. (2008) Femtosecond laser ablation inductively coupled plasma mass spectrometry: fundamentals and capabilities for depth profiling analysis. *Mass Spectrom. Rev.*, **27**, 609–623.

57 Hergenröder, R., Samek, O., and Hommes, V. (2006) Femtosecond laser ablation elemental mass spectrometry. *Mass Spectrom. Rev.*, **25**, 551–572.

58 Siklosy, Z., Demeny, A., Vennemann, T.W., Pilet, S., Kramers, J., Leel-Ossy, S., Bondar, M., Shen, C.C., and Hegner, E. (2009) Bronze Age volcanic event recorded in stalagmites by combined isotope and trace element studies. *Rapid Commun. Mass Spectrom.*, **23**, 801–808.

59 Feldmann, I., Koehler, C.U., Roos, P.H., and Jakubowski, N. (2006) Optimisation of a laser ablation cell for detection of hetero-elements in proteins blotted onto membranes by use of inductively coupled plasma mass spectrometry. *J. Anal. At. Spectrom.*, **21**, 1006–1015.

60 Asogan, D., Sharp, B.L., O'Connor, C.J.P., Green, D.A., and Hutchinson, R.W. (2009) An open, non-contact cell for laser ablation-inductively coupled plasma mass spectrometry. *J. Anal. At. Spectrom.*, **24**, 917–923.

61 Kovacs, R., Nishiguchi, K., Utani, K., and Günther, D. (2010) Development of direct atmospheric sampling for laser ablation-inductively coupled plasma mass spectrometry. *J. Anal. At. Spectrom.*, **25**, 142–147.

62 Aramendía, M., Resano, M., and Vanhaecke, F. (2010) Isotope ratio determination by laser ablation-single-collector-inductively coupled plasma-mass spectrometry. General capabilities and possibilities for improvement. *J. Anal. At. Spectrom.*, **25**, 390–404.

63 Resano, M., Marzo, P., Perez-Arantegui, J., Aramendía, M., Cloquet, C., and Vanhaecke, F. (2008) Laser ablation-inductively coupled plasma-dynamic reaction cell-mass spectrometry for the determination of lead isotope ratios in ancient glazed ceramics for discriminating purposes. *J. Anal. At. Spectrom.*, **23**, 1182–1191.

64 Poitrasson, F. and Dundas, S.H. (1999) Direct isotope ratio measurement of ultra-trace lead in waters by double focusing inductively coupled plasma mass spectrometry with an ultrasonic nebuliser and a desolvation unit. *J. Anal. At. Spectrom.*, **14**, 1573–1577.

65 McLean, J.A., Zhang, H., and Montaser, A. (1998) A direct injection high-efficiency nebulizer for inductively coupled plasma mass spectrometry. *Anal. Chem.*, **70**, 1012–1020.

66 Resano, M., Vanhaecke, F., and de Loos-Vollebregt, G. (2008) Electrothermal heating for sample introduction in atomic absorption, atomic emission and plasma mass spectrometry – a critical review with focus on solid sampling and slurry analysis. *J. Anal. At. Spectrom.*, **23**, 1450–1475.

67 Minnich, M.G. and Houk, R.S. (1998) Comparison of cryogenic and membrane desolvation for attenuation of oxide, hydride and hydroxide ions and ions containing chlorine in inductively coupled plasma mass spectrometry. *J. Anal. At. Spectrom.*, **13**, 167–174.

68 Kwok, K., Carr, J.E., Webster, G.K., and Carnahan, J.W., (2006) Determination of active pharmaceutical ingredients by heteroatom selective detection using inductively coupled plasma mass spectrometry with ultrasonic nebulization and membrane desolvation sample introduction. *Appl. Spectrosc.*, **60**, 80–85.

69 Klaue, B. and Blum, J.D. (1999) Trace analyses of arsenic in drinking water by inductively coupled plasma mass spectrometry: high resolution versus hydride generation. *Anal. Chem.*, **71**, 1408–1414.

70 Sturgeon, R.E. and Lam, J.W. (1999) The ETV as a thermochemical reactor for ICP-MS sample introduction. *J. Anal. At. Spectrom.*, **14**, 785–791.

71 Sakata, K. and Kawabata, K. (1994) Reduction of fundamental polyatomic ions in inductively-coupled plasma mass spectrometry. *Spectrochim. Acta B*, **49**, 1027–1038.

72 Fietzke, J., Eisenhauer, A., Gussone, N., Bock, B., Liebetrau, V., Nagler, T.F., Spero, H.J., Bijma, J., and Dullo, C. (2004) Direct measurement of $^{44}Ca/^{40}Ca$ ratios by MC-ICP-MS using the cool plasma technique. *Chem. Geol.*, **206**, 11–20.

73 Vanhaecke, F., Saverwyns, S., de Wannemacker, G., Moens, L., and Dams, R. (2000) Comparison of the application of higher mass resolution and cool plasma conditions to avoid spectral interferences in Cr(III)/Cr(VI) speciation by means of high-performance liquid chromatography – inductively coupled plasma mass spectrometry. *Anal. Chim. Acta*, **419**, 55–64.

74 Vanhaecke, F., Balcaen, L., de Wannemacker, G., and Moens, L. (2002) Capabilities of inductively coupled plasma mass spectrometry for the measurement of Fe isotope ratios. *J. Anal. At. Spectrom.*, **17**, 933–943.

75 Appelblad, P.K., Rodushkin, I., and Baxter, D.C. (2000) The use of Pt guard electrode in inductively coupled plasma sector field mass spectrometry: advantages and limitations. *J. Anal. At. Spectrom.*, **15**, 359–364.

76 Rowan, J.T. and Houk, R.S., (1989) Attenuation of polyatomic ion interferences in inductively coupled plasma mass spectrometry by gas-phase collisions. *Appl. Spectrosc.*, **43**, 976–980.

77 Feldmann, I., Jakubowski, N., and Stuewer, D. (1999) Application of a hexapole collision and reaction cell in ICP-MS. Part I: instrumental aspects and operational optimization. *Fresenius' J. Anal. Chem.*, **365**, 415–421.

78 Feldmann, I., Jakubowski, N., Thomas, C., and Stuewer, D. (1999) Application of a hexapole collision and reaction cell in ICP-MS. Part II: analytical figures of merit and first applications. *Fresenius' J. Anal. Chem.*, **365**, 422–428.

79 Tanner, S.D. and Baranov, V.I. (1999) Theory, design, and operation of a dynamic reaction cell for ICP-MS. *At. Spectrosc.*, **20**, 45–52.

80 Tanner, S.D., Baranov, V.I., and Bandura, D.R. (2002) Reaction cells and collision cells for ICP-MS: a tutorial review. *Spectrochim. Acta B*, **57**, 1361–1452.

81 Yamada, N., Takahashi, J., and Sakata, K. (2002) The effects of cell-gas impurities and kinetic energy discrimination in an octopole collision cell ICP-MS under non-thermalized conditions. *J. Anal. At. Spectrom.*, **17**, 213–1222.

82 Hattendorf, B. and Günther, D. (2003) Strategies for method development for an inductively coupled plasma mass spectrometer with bandpass reaction cell. Approaches with different reaction gases for the determination of selenium. *Spectrochim. Acta B*, **58**, 1–13.

83 Tanner, S.D. and Baranov, V.I. (1999) A dynamic reaction cell for inductively coupled plasma mass spectrometry (ICP-DRC-MS). II. Reduction of interferences produced within the cell. *J. Am. Soc. Mass Spectrom.*, **10**, 1083–1094.

84 Chen, K.L. and Jiang, S.J. (2002) Determination of calcium, iron and zinc in milk powder by reaction cell inductively coupled plasma mass spectrometry. *Anal. Chim. Acta*, **470**, 223–228.

85 Vanhaecke, F., Resano, M., Garcia-Ruiz, E., Balcaen, L., Koch, K.R., and McIntosh, K. (2004) Laser ablation-inductively coupled plasma-dynamic reaction cell-mass spectrometry (LA-ICP-DRC-MS) for the determination of Pt, Pd and Rh in Pb buttons obtained by fire assay of platiniferous ores. *J. Anal. At. Spectrom.*, **19**, 632–638.

86 Simpson, L.A., Thomsen, M., Alloway, B.J., and Parker, A. (2001) A dynamic reaction cell (DRC) solution to oxide-based interferences in inductively coupled plasma mass spectrometry (ICP-MS) analysis of the noble metals. *J. Anal. At. Spectrom.*, **16**, 1375–1380.

87 Bandura, D.R., Baranov, V.I., and Tanner, S.D. (2002) Detection of ultratrace phosphorus and sulfur by quadrupole ICP-MS with dynamic reaction cell. *Anal. Chem.*, **74**, 1497–1502.

88 Moens, L.J., Vanhaecke, F.F., Bandura, D.R., Baranov, V.I., and Tanner, S.D. (2001) Elimination of isobaric interferences in ICP-MS, using ion-molecule reaction chemistry: Rb/Sr age determination of magmatic rocks, a case study. *J. Anal. At. Spectrom.*, **16**, 991–994.

89 Vanhaecke, F. and Moens, L. (2004) Overcoming spectral overlap in isotopic analysis via single- and multi-collector ICP-mass spectrometry. *Anal. Bioanal. Chem.*, **378**, 232–240.

90 Jakubowski, N., Prohaska, T., Vanhaecke, F., Roos, P.H., and Lindemann, T. (2011) Inductively coupled plasma- and glow discharge plasma-sector field mass spectrometry. Part II: applications. *J. Anal. At. Spectrom.*, **26**, 727–757.

91 Lapitajs, G., Greb, U., Dunemann, L., Begerow, J., Moens, L., and Verrept, P. (1995) ICPMS in the determination of trace and ultratrace elements in the human body. *Int. Lab.*, (May), 21–27.

92 Monna, F., Loizeau, J.-L., Thomas, B. A., Guéguen, C., and Favarger, P.-Y. (1998) Pb and Sr isotope measurements by inductively coupled plasma mass spectrometer: efficient time management for precision improvement. *Spectrochim. Acta B*, **53**, 1317–1333.

93 Gray, A.L., Williams, J.G., Ince, A.T., and Liezers, M. (1994) Noise sources in inductively-coupled plasma mass spectrometry – an investigation of their importance to the precision of isotope ratio measurements. *J. Anal. At. Spectrom.*, **9**, 1179–1184.

94 Hinners, T.A., Heithmar, E.M., Spittler, T.M., and Henshaw, J.M. (1987) Inductively coupled plasma mass spectrometric determination of lead isotopes. *Anal. Chem.*, **59**, 2658–2662.

95 Koirtyohann, S.R. (1994) Precise determination of isotopic ratios for some biologically significant elements by inductively-coupled plasma mass spectroscopy. *Spectrochim. Acta B*, **49**, 1305–1311.

96 Furuta, N. (1991) Optimization of the mass scanning rate for the determination of lead isotope ratios using an inductively coupled plasma mass spectrometer. *J. Anal. At. Spectrom.*, **6**, 199–203.

97 Quétel, C.R., Thomas, B., Donard, O.F. X., and Grousset, F.E. (1997) Factorial optimization of data acquisition factors for lead isotope ratio determination by inductively coupled plasma mass spectrometry. *Spectrochim. Acta B*, 177–187.

98 Bandura, D.R. and Tanner, S.D. (1999) Effect of collisional damping in the dynamic reaction cell on the precision of isotope ratio measurements. *At. Spectrosc.*, **20**, 69–72.

99 Bandura, D.R., Baranov, V.I., and Tanner, S.D. (2000) Effect of collisional damping and reactions in a dynamic reaction cell on the precision of isotope ratio measurements. *J. Anal. At. Spectrom.*, **15**, 921–928.

100 Xie, Q.L. and Kerrich, R. (2002) Isotope ratio measurement by hexapole ICP-MS: mass bias effect, precision and accuracy. *J. Anal. At. Spectrom.*, **17**, 69–74.

101 Boulyga, S.F. and Becker, J.S. (2002) Comment on "Isotope ratio measurement by hexapole ICP-MS: mass bias effect, precision and accuracy" by Q. Xie and R. Kerrich, *J. Anal. At. Spectrom.*, 2002, **17**, 69. *J. Anal. At. Spectrom.*, **17**, 965–966.

102 Vanhaecke, F., Balcaen, L., Deconinck, I., de Schrijver, I., Almeida, C.M., and Moens, L. (2003) Mass discrimination in dynamic reaction cell (DRC)–ICP mass spectrometry. *J. Anal. At. Spectrom.*, **18**, 1060–1065.

103 Resano, M., Marzo, P., Perez-Arantegui, J., Aramendía, M., Cloquet, C., and Vanhaecke, F. (2008) Laser ablation-inductively coupled plasma-dynamic reaction cell-mass spectrometry for the determination of lead isotope ratios in ancient glazed ceramics for discriminating purposes. *J. Anal. At. Spectrom.*, **23**, 1182–1191.

104 Vanhaecke, F., Moens, L., Dams, R., and Taylor, P. (1996) Precise measurement of isotope ratios with a double-focusing magnetic sector ICP mass spectrometer. *Anal. Chem.*, **68**, 567–569.

105 Krachler, M., Le Roux, G., Kober, B., and Shotyk, W. (2004) Optimising accuracy and precision of lead isotope measurement (Pb-206, Pb-207, Pb-208) in acid digests of peat with ICP-SMS using individual mass discrimination correction. *J. Anal. At. Spectrom.*, **19**, 354–361.

106 Vanhaecke, F., Moens, L., Dams, R., Papadakis, I., and Taylor, P. (1997) Applicability of high resolution ICP-mass spectrometry for isotope ratio measurements. *Anal. Chem.*, **69**, 268–273.

107 Myers, D.P., Mahoney, P.P., Li, G., and Hieftje, G.M. (1995) Isotope ratios and abundance sensitivity obtained with an inductively-coupled plasma-time-of-flight mass spectrometer. *J. Am. Soc. Mass Spectrom.*, **6**, 920–927.

108 Vanhaecke, F., Moens, L., Dams, R., Allen, L., and Georgitis, S. (1999) Evaluation of the isotope ratio performance of an axial time-of-flight ICP-mass spectrometer. *Anal. Chem.*, **71**, 3297–3303.

109 Emteborg, H., Tian, X.D., Ostermann, M., Berglund, M., and Adams, F.C. (2000) Isotope ratio and isotope dilution measurements using axial inductively coupled plasma time of flight mass spectrometry. *J. Anal. At. Spectrom.*, **15**, 239–246.

110 Willie, S., Mester, Z., and Sturgeon, R.E. (2005) Isotope ratio precision with transient sample introduction using ICP orthogonal acceleration time-of-flight mass spectrometry. *J. Anal. At. Spectrom.*, **20**, 1358–1364.

111 Rowland, A. and Holcombe, J.A. (2009) Evaluation and correction of isotope ratio inaccuracy on inductively coupled plasma time-of-flight mass spectrometry. *Spectrochim. Acta B*, **64**, 35–41.

112 Kurz, E.A. (1979) Channel electron multipliers. *Am. Lab.*, **11**, 67–82.

113 Hunter, K. and Stresau, D. (2005) Ion detection, in *ICP Mass Spectrometry Handbook* (ed. S. Nelms), Blackwell Publishing, Oxford, pp. 117–146.

114 www.chm.bris.ac.uk/ms/images/emultiplier.gif (last accessed 21 December 2011).

115 Vanhaecke, F., de Wannemacker, G., Moens, L., Dams, R., Latkoczy, C., Prohaska, T., and Stingeder, G. (1998) Dependence of detector dead time on analyte mass number in inductively coupled plasma mass spectrometry. *J. Anal. At. Spectrom.*, **13**, 567–571.

116 Russ, G.P. III (1987) Isotope ratio measurements using ICP-MS, in *Applications of Inductively Coupled Plasma Mass Spectrometry* (eds. A.R.

Date and A.L. Gray), Chapman and Hall, London, pp. 90–114.

117 Dietz, L.A. (1965) Basic operating properties of electron multiplier ion detection and and pulse counting methods in mass spectrometry. *Rev. Sci. Instrum.*, **36**, 1763–1770.

118 Russ, G.P. and Bazan, J.M. (1987) Isotopic ratio measurements with an inductively coupled plasma source mass spectrometer. *Spectrochim. Acta B*, **42**, 49–62.

119 Nelms, S.M., Quetel, C.R., Prohaska, T., Vogl, J., and Taylor, P.D.P. (2001) Evaluation of detector dead time calculation models for ICP-MS. *J. Anal. At. Spectrom.*, **16**, 333–338.

120 Held, A. and Taylor, P.D.P. (1999) A calculation method based on isotope ratios for the determination of dead time and its uncertainty in ICP-MS and application of the method to investigating some features of a continuous dynode multiplier. *J. Anal. At. Spectrom.*, **14**, 1075–1079.

121 Appelblad, P.K. and Baxter, D.C. (2000) A model for calculating dead time and mass discrimination correction factors from inductively coupled plasma mass spectrometry calibration curves. *J. Anal. At. Spectrom.*, **15**, 557–560.

122 Rameback, H., Berglund, M., Vendelbo, D., Wellum, R., and Taylor, P.D.P. (2001) On the determination of the true dead-time of a pulse-counting system in isotope ratio mass spectrometry. *J. Anal. At. Spectrom.*, **16**, 1271–1274.

123 Nygren, U., Rameback, H., Berglund, M., and Baxter, D.C. (2006) The importance of a correct dead time setting in isotope ratio mass spectrometry: implementation of an electronically determined dead time to reduce measurement uncertainty. *Int. J. Mass Spectrom.*, **257**, 12–15.

124 Tanner, S.D. (1992) Space-charge in ICP-MS – calculation and implications. *Spectrochim. Acta B*, **47**, 809–823.

125 Ingle, C.P., Sharp, B.L., Horstwood, M.S.A., Parrish, R.R., and Lewis, D.J. (2003) Instrument response functions, mass bias and matrix effects in isotope ratio measurements and semi-quantitative analysis by single- and multi-collector ICP-MS. *J. Anal. At. Spectrom.*, **18**, 219–229.

126 Begley, I.S. and Sharp, B.L. (1997) Characterisation and correction of instrumental bias in inductively coupled plasma quadrupole mass spectrometry for accurate measurement of lead isotope ratios. *J. Anal. At. Spectrom.*, **12**, 395–402.

127 Heumann, K.G., Gallus, S.M., Rädlinger, G., and Vogl, J. (1998) Precision and accuracy in isotope ratio measurements by plasma source mass spectrometry. *J. Anal. At. Spectrom.*, **13**, 1001–1008.

128 Quetel, C.R., Prohaska, T., Hamester, M., Kerl, W., and Taylor, P.D.P. (2000) Examination of the performance exhibited by a single detector double focusing magnetic sector ICP-MS instrument for uranium isotope abundance ratio measurements over almost three orders of magnitude and down to pg g^{-1} concentration levels. *J. Anal. At. Spectrom.*, **15**, 353–358.

129 Albarède, F. and Beard, B. (2004) Analytical methods for non-traditional isotopes. *Rev. Miner. Geochem.*, **55**, 113–150.

3
Multi-Collector Inductively Coupled Plasma Mass Spectrometry

Michael Wieser, Johannes Schwieters, and Charles Douthitt

3.1
Introduction

The discovery of two isotopes of the element neon by J.J. Thomson in 1913 [1] heralded a significant research opportunity to study the origins and histories of atoms in the most revealing manner possible. Thomson arranged parallel electric and magnetic fields to separate ions of neon into two distinct parabolic traces on a photographic plate, each corresponding to a stable isotope of the gas. The intensity of the line on the plate corresponded to the abundance of the corresponding isotope in the gas sample. Following Thomson's lead, scientists including Aston, Dempster, and Bainbridge refined the original parabolic ray apparatus into instruments with increasing sensitivity and mass resolution, eventually used for measuring the isotopic compositions of several elements (including Li, C, O, Mg, Cl, K, Ca, Zn, Ge, and Te) and providing the atomic masses of the isotopes of these elements. In 1929, Aston demonstrated that lead was composed of four stable isotopes (^{204}Pb, ^{206}Pb, ^{207}Pb, and ^{208}Pb) and that relative isotope amount ratios (or, shorter, isotope ratios) could be used to calculate the age of the U-rich sample from which the lead was extracted [2]. In 1936, Dole identified a difference in the density of water formed using atmospheric oxygen compared with that of water from Lake Michigan [3]. He attributed this difference to an enrichment of ^{18}O in the oxygen in the water from Lake Michigan due to an equilibrium isotope exchange between the atmosphere and the lake water. Nier and Gulbransen undertook a systematic study of the ^{13}C/^{12}C isotope ratio in plant and mineral samples and found a variation as large as 5%, depending on the origin of the sample [4]. Meaningful variations in isotope ratios exist at the level of tens of parts per million. For example, in the case of carbon, the average ^{13}C/^{12}C isotope ratio ranges from 0.010 750 to 0.011 650, with biologically significant variations of $\pm$0.000 000 5. Insight into geological systems can also be obtained using isotope abundance variations. For example, Caro *et al.* [5] searched for evidence of anomalous amounts of ^{142}Nd in Archean minerals that were the result of alpha decay of the relatively short-lived (130 Myr) and thus now extinct ^{146}Sm. Caro and *et al.* measured relative enrichments of ^{142}Nd of 15 $\pm$ 4 ppm (2σ) in metamorphosed

Isotopic Analysis: Fundamentals and Applications Using ICP-MS,
First Edition. Edited by Frank Vanhaecke and Patrick Degryse.

Published 2012 by WILEY-VCH Verlag GmbH & Co. KGaA

sedimentary rocks from Isua, Greenland. This slight excess was evidence that the Earth's crust formed before the planet was 100 million years old and ^{146}Sm, which was still present then, was incorporated into the ancient minerals. Hence the goal of any instrument used to determine isotopic composition is to measure isotope ratios with the highest precision possible. The stability of modern instrumentation, and therefore the quality of the data produced, have led to important insights in many fields, including chemistry, physics, geology, biology, archeology, medicine, and environmental science, to name just a few.

3.2
Early Multi-Collector Mass Spectrometers

The isotopic composition of an element is typically determined by measuring isotope amount *ratios*, so that the absolute intensity of the ion current measured by the detector is not critical. With a single detector, the uncertainty of the isotope ratio is limited by the stability of the production of ions in the source and of the transmission of ions through the ion optical system. Of the analytical methods available, magnetic sector mass spectrometers offer an important advantage in their ability to separate ions spatially along a focal plane by the focusing action of the magnetic field. One can arrange several detectors to collect the ion currents of the isotopes of interest simultaneously. In this manner, any changes in the conditions in the ion source will affect all ions in a similar fashion and will therefore have minimal impact on the determination of the ratios of the measured ion currents. An additional feature of the magnetic sector mass spectrometer is that sufficient mass resolution can be combined with "flat-topped" peaks, such that the entire ion image is captured by the detector and small fluctuations in the position of the ion image in the focal plane do not result in variations in the measured ion current intensities. Since there are independent detectors for each isotope measured across the focal plane of the magnetic sector lens, the types of detectors can be tailored to the dynamic range required for accurate and precise isotope ratio measurements. For instance, in the measurement of thorium isotope abundances, the ratio of ^{230}Th to ^{232}Th can be very extreme and reach a value of just a few ppm. Here, the radiogenic ^{230}Th beam is measured by a high-sensitivity ion-counting detector, operated in parallel with an analog Faraday cup detector to measure the major ^{232}Th ion beam (see also Chapter 2 for a description of the basic operating principles of both detector types). This highlights the capability of large dynamic range and high sensitivity of sector field mass spectrometers.

The first commercially available multi-collector (MC) isotope ratio mass spectrometers offered by Finnigan MAT and VG Isotopes were fitted to thermal ionization sources. Thermal ionization of the purified target element from a metal filament, heated resistively in vacuum, is used for sensitive measurements of the alkali, alkaline earth and rare earth elements (REEs). Surface ionization techniques are typically element-specific and can produce relatively high ionization yields ($>10\%$) with low backgrounds. Early MC instruments were equipped with as many as nine individual Faraday cup detectors in fixed configurations for specific isotope systems. A system could be configured for the simultaneous measurement of ^{204}Pb, ^{206}Pb, ^{207}Pb, and ^{208}Pb, but could not be used to measure isotopes of strontium because the distance between adjacent lead detectors (one unit mass difference between ^{206}Pb and ^{207}Pb) is less than that between adjacent strontium masses (one unit mass difference between ^{86}Sr and ^{87}Sr).

3.3 Variable Multi-Collector Mass Spectrometers

A significant advance was the introduction of variable MC systems that allow the operator to position several detectors at different positions along the focal plane of the mass spectrometer (Figure 3.1). The locations of the detectors could be adjusted with micrometer precision to accommodate many isotope systems. The ability to position the detectors with fine precision is necessary for the measurement of isotope abundance ratios wherein interfering polyatomic (or

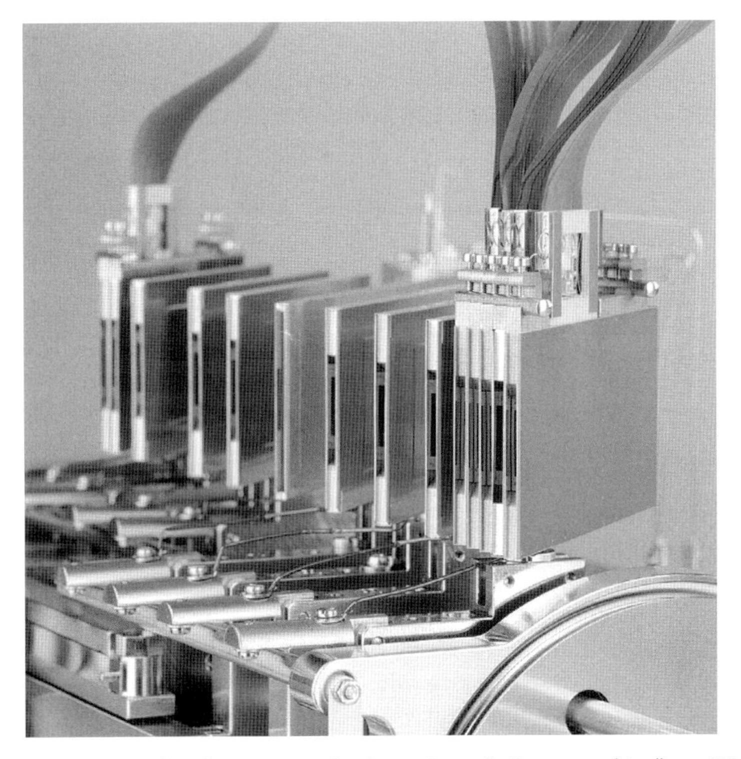

Figure 3.1 Multi-collector array in the Thermo Scientific Neptune multi-collector ICP-MS instrument. Miniaturized ion counters identical in size with Faraday detectors are mounted on the high-mass side (for U) and low-mass side (for Pb). Connections to the miniaturized ion counters are completely independent from the Faraday cup signal lines, protecting the integrity of the signals.

molecular) ions must be resolved from the target ions by high mass resolution. For example, in a standard-sized sector field mass analyzer, the spacing between interfering beams along the focal plane can be estimated to be typically on the order of 0.2 mm (or 200 μm) for polyatomic interferences in the majority of cases and the detector positions must be set to better than ± 20 μm to ensure good peak overlap of several target isotopes simultaneously. For some interferences requiring even higher mass resolution, the spacing may be much smaller, such that position control down to a few micrometer is required. Finally, there are interferences that cannot be resolved with a sector field instrument because the mass resolving power does not allow further physical separation of the ion beams.

Building on the high precision achieved with MC mass spectrometers for thermal ionization ion sources, Walder and Freedman [6] described an instrument that integrated an inductively coupled plasma (ICP) ion source to an MC platform. This instrument combined the versatility of an ICP source with the flexibility of a variable MC array. With this instrument, the isotope abundances of practically all elements across the periodic table were accessible for scientific research. While thermal ionization mass spectrometry (TIMS) is a powerful ionization technique for various elements, it has problems in ionizing elements with high first ionization potentials such as Hf. The power of the ICP as a more universal ion source overcomes this challenge and delivers high ionization efficiency for almost all elements across the periodic table. However, there are two significant design challenges to be overcome if an ICP ion source is used instead of a thermal ionization source. Because of the large spread in the kinetic energy of the ions created in the ICP source, an electrostatic sector lens had to be added to the magnetic sector in order to build a double-focusing geometry (see also Chapter 2). The geometry of the electrostatic sector is chosen such that the energy dispersion of the electrostatic sector and the energy dispersion of the magnetic sector match each other. The energy dispersion of the combination eliminates the energy dispersion to second order. Regardless of the larger energy spread of the ions generated in an ICP source, the double-focusing geometry preserves the resolving power of the mass analyzer in addition to the flatness of the peaks. The other design challenge to consider is that several stages of differential pumping are required to maintain an ultra-high vacuum in the mass spectrometer to ensure high abundance sensitivity since samples are introduced at atmospheric pressure into the plasma.

The first commercially available multi-collector inductively coupled plasma mass spectrometry (MC-ICP-MS) setup was the Plasma 54, introduced in 1992 by VG Elemental. This mass spectrometer incorporated the detector platform from the Sector 54 thermal ionization mass spectrometer and included an electrostatic analyzer before the entrance to the magnetic sector. This instrument featured seven Faraday cups and a Daly detector. The Daly detector [7] incorporates an Al knob maintained at +25 kV together with a scintillator screen and photomultiplier (Figure 3.2). Incoming ions are accelerated to the Al knob and large numbers of secondary electrons are produced as result of the impact of these ions on the aluminum surface. These electrons are then accelerated towards the scintillator,

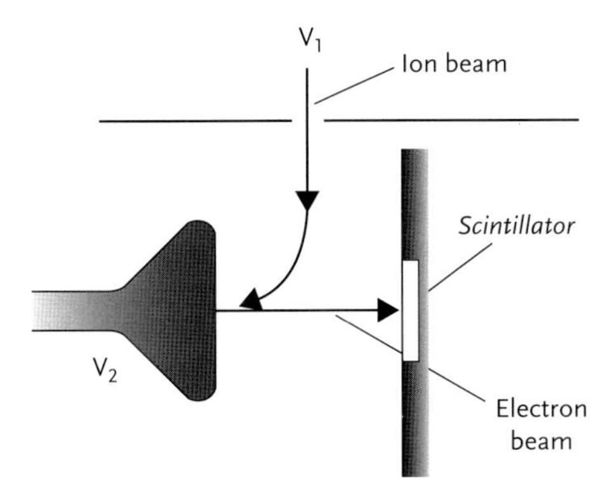

Figure 3.2 In the Daly detector, incoming ions are accelerated towards an Al anode maintained at high voltage (V_2). The secondary electrons produced by collision of the ions with the Daly knob are accelerated towards a scintillator. The subsequent light pulses are detected and amplified by a photomultiplier detector.
Modified from [7].

which is maintained at ground potential. The energies of the electrons are converted to flashes of light in the scintillator and the photons are detected by the photomultiplier. Early publications by users of this mass spectrometer generated considerable excitement in the isotope ratio community because of the high precision attainable in lead isotope abundance measurements, and also the ability to generate reliable data for elements with high ionization potentials that are difficult to measure by thermal ionization methods.

In 1996, VG Instruments introduced the Iso-PlasmaTrace multi-collector ICP–MS instrument. Nu Instruments introduced the Nu Plasma on to the market in 1997, which was followed by the large-dispersion Nu Plasma 1700 in 1999. Thermo Optek introduced the VG Axiom Plus MC-ICP-MS unit in 1998. This instrument was equipped with nine Faraday cup detectors and three electron multiplier detectors. Thermo Instruments introduced the Neptune in 2000. At present, there are three commercially available MC-ICP-MS instruments dedicated to isotope ratio analysis: the Neptune from ThermoFisher Scientific and the Nu Plasma II and Nu Plasma 1700 from Nu Instruments. Over the past two decades, the number of publications reporting on the use of MC-ICP-MS has grown significantly (Figure 3.3) [8].

3.4 Mass Resolution and Resolving Power

One of the challenges in measuring isotope ratios accurately is to achieve acceptable mass resolution to separate atomic ions from polyatomic ions, largely consisting of argide, hydride, and oxide ions. For example, in the measurement of Fe isotopes (^{54}Fe, ^{56}Fe, ^{57}Fe, and ^{58}Fe), common spectral interferences occur as a result of the presence of ArN^+, ArO^+, and $ArOH^+$ ions at masses 54, 56, 57,

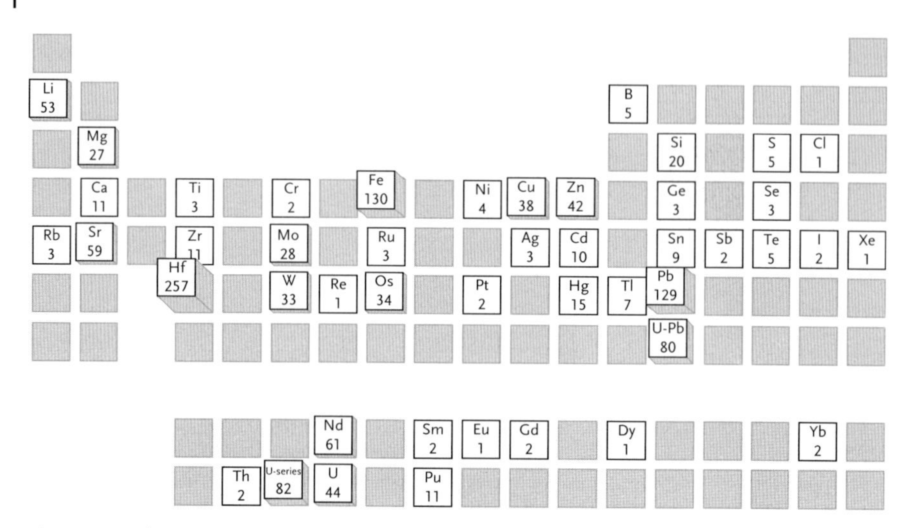

Figure 3.3 The number of papers on isotopic analysis based on the use of MC-ICP-MS instrumentation for each element, published between 1992 and publication of Douthitt's paper in 2008 [8].

and 58. Separating the ion beams of the target ions from those of the Ar-based ions with the same nominal mass-to-charge ratio is complicated by the large spread in the kinetic energy of the ions produced in the plasma, which is typically ~25 eV compared with ~2 eV for a thermal ionization source. So-called *double-focusing* ion optical designs couple electrostatic and magnetic sectors to achieve both energy filtering (velocity focusing) and mass separation of the ions (see also Chapter 2). Two designs that have been successfully applied to isotope ratio measurements are the Mattauch–Herzog and Nier–Johnson geometries. The Mattauch–Herzog mass spectrograph was first built in Vienna in 1935. Ions generated in the source are first accelerated into a 31.82° electrostatic analyzer, followed by a 90° magnetic sector. It is capable of double-focusing for all masses and has the unique ability to separate the ions along a flat focal plane, thus enabling simultaneous collection of multiple ion currents using a photographic plate. The Nier–Johnson geometry uses a 90° electrostatic sector immediately after the ion source for velocity focusing, then a 90° magnetic sector to separate the ions from one another according to their mass-to-charge ratio. Most commercially available MC-ICP-MS units employ Nier–Johnson geometry. The primary reason for the selection of this geometry for isotope ratio mass spectrometry is the need for mass dispersion along the focal plane. For achieving high mass resolution in single-collector sector field ICP-MS, the reverse Nier–Johnson geometry is preferable because of the higher abundance sensitivity. Because of the mass dispersion of the magnet, only the ions of choice enter the electrostatic analyzer, which further filters out stray ions by energy discrimination, leading to improved abundance sensitivity and reduced peak tailing in the mass spectrum.

The first generation of MC-ICP-MS instruments employed the relatively low mass resolution ion optics found in thermal ionization instruments. As users began to explore isotope abundance variations of Mg, Si, Ca, and Fe, it became necessary to improve the stability of the background and to suppress the influence of interferences. Ion–molecule interactions in the plasma source result in the production of polyatomic ion species that generate spectral interferences that cannot be resolved at low mass resolution. Also, trace amounts of elemental contaminants can interfere with minor isotopes of the element of interest. For resolving the overlap of isobaric nuclides, however, a very high mass resolution, far beyond the capabilities of present-day instrumentation, is required. Because of the small mass defect, hydrides also are difficult to resolve. In certain cases, doubly charged ions can also interfere. For example, in the case of calcium, Sr^{2+}, Ti^+, Ar^+, and hydrides of Ar and N interfere with almost every isotope of Ca (Table 3.1).

High mass resolution through spatial peak separation is the most robust way to eliminate spectral interferences. Peak flatness must be preserved to achieve high-precision ion current measurements. In practice, this is realized by adjusting the source slit width to generate an interference-free space between ion beams in the focal plane of the mass spectrometer. For high-precision isotope ratio measurements, it is advantageous to set the slits at the detector wider than the beam widths at the focal plane. In this manner, a wide plateau is generated as the relatively narrow image of the ion beam is swept across the wide detector opening. If the opening of the detector is wider than the separation between adjacent ion beams, the corresponding peaks will be unresolved. However, because of the mass defect of an isotope, polyatomic ions in the low and medium mass range always appear on the high mass side of elemental peaks. Therefore, one can position the detectors such that only the target ion currents are measured and polyatomic ions do not enter the detectors. This procedure does not just provide the widest peak plateau width, but also avoids detector slit assemblies that, in practice, would be a challenge to construct in a variable collector array. This approach (see Figure 3.4) is sometimes referred to as pseudo-high resolution [9].

Table 3.1 Elemental and polyatomic ions interfering with the measurement of calcium isotopes. Except for $^{40}Ar^+$, $^{88}Sr^{2+}$, and $^{46,48}Ti^+$, all interferences can be separated using higher mass resolution with a Thermo Scientific Neptune.

Isotope	Natural abundance (%)	Interferences	Resolution required
^{40}Ca	96.941	$^{40}Ar^+$	192 500
^{42}Ca	0.647	$^{40}Ar^1H_2^+$	2200
		$^{14}N_3^+$	830
^{43}Ca	0.135	$^{14}N_3{}^1H^+$	740
^{44}Ca	2.086	$^{12}C^{16}O_2^+$	1280
		$^{14}N_2{}^{16}O^+$	965
		$^{88}Sr^{2+}$	160 500
^{46}Ca	0.004	$^{46}Ti^+$	43 400
^{48}Ca	0.187	$^{48}Ti^+$	10 500

The practical mass resolution of this configuration is called the *resolving power* and calculated as shown in the following equation:

$$R_{\text{power}}(5\%, 95\%) = \frac{m}{m(5\%) - m(95\%)}$$

where $m(5\%)$ is the mass at which the signal intensity is 5% of the peak height, $m(95\%)$ is the mass at which the signal intensity is 95% of the peak height, and m is the mass of the peak (see Figure 3.4).

In contrast, in single-collector instruments, the electrostatic analyzer may be positioned after the magnetic sector. In this configuration, the energy filter will improve the abundance sensitivity of the instrument. This configuration cannot be used for the simultaneous detection of multiple isotopic ion currents since the magnetic field will select an ion beam of specific mass-to-charge ratio before energy filtering. The Thermo Scientific Neptune can achieve a resolving power of 10 000 with a transmission efficiency of $>5\%$. This is generally sufficient to overcome polyatomic interferences while preserving flat-topped peaks.

The large dispersion Nu Plasma 1700 from Nu Instruments shows full separation of the peaks while maintaining their flat-topped appearance at high mass resolution.

In another approach to reduce the energy spread of ions produced in the plasma source and remove polyatomic interferences, a hexapole collision cell (see also Chapter 2) is placed in front of the magnetic field. Argon, helium, or hydrogen gas is introduced into the cell, where ions from the plasma collide with the gas. Interaction between ions and the gas causes the ions to lose kinetic energy and polyatomic species formed in the plasma are decomposed as a result of ion–molecule collisions and/or, in the case of H_2, chemical reactions. The collision cell will attenuate most interferences by several orders of magnitude.

One way to reduce argon-based polyatomic interferences is to lower the power of the plasma, for example, to 300 W instead of 1200 W. A reduction in the plasma power (cool plasma conditions; see also Chapter 2) will lower the number of argon-containing ions generated in the ion source because the first ionization potential of Ar is relatively high compared with that of the other elements. The disadvantage is that a reduction in plasma power will make the ion source much more sensitive to matrix effects, that is, to changes in the composition of the aerosol, and the isotope ratios may become increasingly dependent on the concentration of the element in solution, even on the strength of the acid and the presence of other elements.

High-precision isotope ratio measurements require critical control of the background and interference suppression using high mass resolution is regarded as the most unambiguous way to eliminate spectral interferences.

3.5
Three-Isotope Plots for Measurement Validation

A particularly useful analytical tool is the "three-isotope plot," which allows one to confirm that measured isotope ratios are free from spectral interference from an

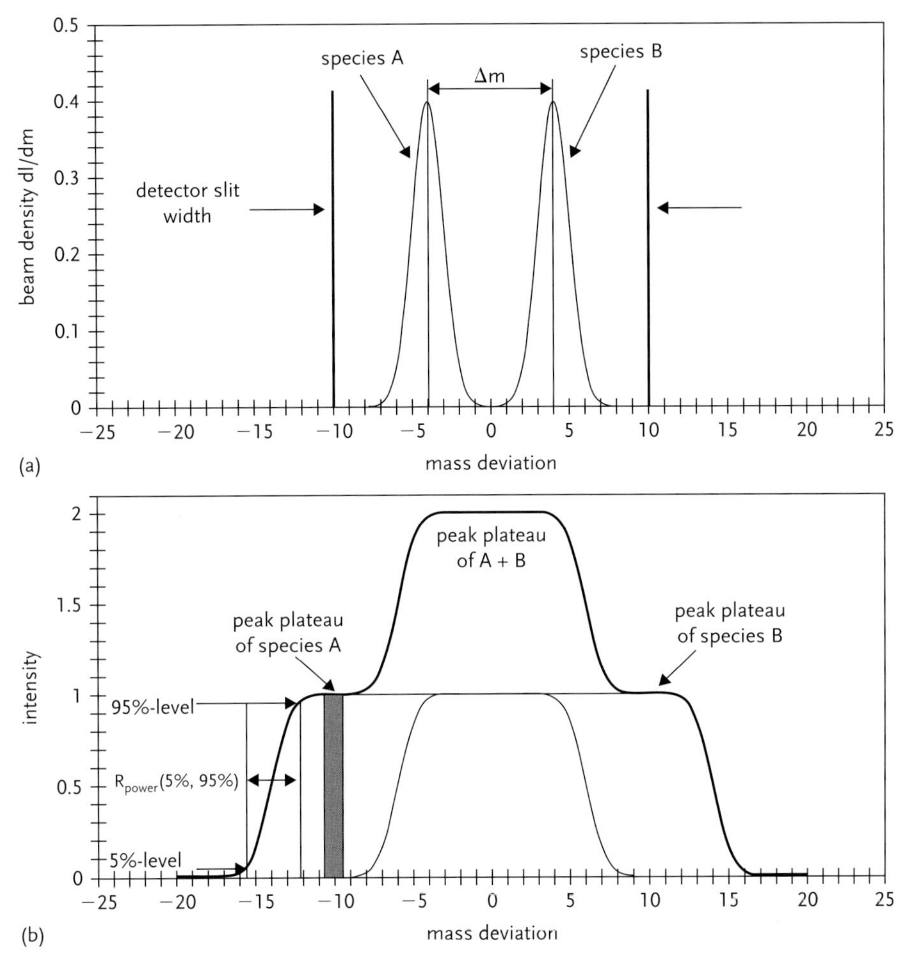

Figure 3.4 (a) Beam profiles for two ionic species (A and B) with a mass difference of Δm at the detector side of the mass spectrometer. (b) Recorded peak profile, when the adjacent ion beam profiles of ion species A and B with a mass difference of Δm are scanned across the low-resolution detector slit. The beams are separated by the use of a high-resolution entrance slit (narrow slit width). The detector slit, indicated by the two solid vertical lines in (a), is set at low mass resolution mode (wide slit width). First the lighter ion species A enters the detector slit and forms the first plateau section of the peak, which is indicated by the gray shadowed rectangle. At that time, the heavier ion species B is still clipped at the high-mass side of the detector slit. The second plateau section appears when both ion beams enter the detector slit. This is the situation as depicted in (a). Finally, the ion beam A is clipped on the low-mass side of the detector slit and only species B enters the cup and forms the third plateau section. The resolving power $R_{power}(5, 95\%)$ is given by the distance between the two vertical lines.
Reproduced from [10].

isobaric nuclide, a polyatomic ion, or doubly charged ion. For elements that have at least three stable isotopes, one can calculate the measured isotope ratios using a common reference isotope. If the measured isotope abundance data vary in nature because of mass-dependent isotope fractionation processes (see Chapter 1), then the values should plot along a straight line, the slope of which is consistent with the relative mass differences between the isotopes considered. However, if spectral interferences are impacting one of the measured isotopes, then the values will fall off the theoretical fractionation line (Figure 3.5). Finally, data that are affected by matrix effects caused by mismatched analytical conditions of the two samples will plot along a second line parallel to the theoretical mass-dependent curve. For example, in the case of Ca, one can plot $^{48}Ca/^{42}Ca$ isotope ratios versus $^{44}Ca/^{42}Ca$ data (Figure 3.6).

It is essential to demonstrate freedom from spectral interference from isobaric nuclides, doubly charged ions, and/or polyatomic ions when measuring systems that contain isotopes exhibiting limited isotopic abundance variations in Nature. One such example is magnesium, which has three stable isotopes, ^{24}Mg, ^{25}Mg, and ^{26}Mg, that have a large relative mass difference of 8% (^{24}Mg–^{26}Mg) and therefore one might expect reasonably large variations in natural isotopic composition. However, extensive variations have not been observed and previous investigations of Mg isotope abundance variations rather looked for anomalous amounts of ^{26}Mg that were the product of radiogenic decay [11, 12]. In this instance, the high stability of the mass bias in MC-ICP-MS allows the determination of magnesium isotope ratios with external precision of 0.01‰ once instrumental mass bias is corrected for. This

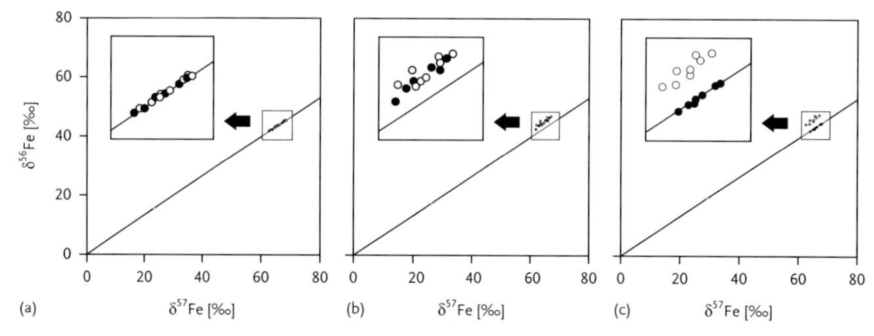

Figure 3.5 Use of three-isotope plots to check for spectral interferences in MC-ICP-MS. Each point represents the mean of an isotope ratio measurement of a standard (filled circles) or a sample (empty circle) of natural isotopic composition. Isotope ratios are plotted on the delta scale (δ) as relative deviations in parts per thousand from the known isotope ratio of an isotopic reference material of natural isotopic composition. The diagonal line represents the theoretical fractionation curve as defined by the isotopic masses and an exponential fractionation law. (a) Absence of isobaric interferences. Data points from standard and sample plot on the theoretical curve. (b) At least one isotopic signal in the mass spectrum of the standard and the sample is subject to spectral interference from an isobaric nuclide, polyatomic ion, or doubly charged ion. (c) Matrix differences between sample and standard result in an offset of the sample data points from the theoretical fractionation curve.
Reproduced from [13].

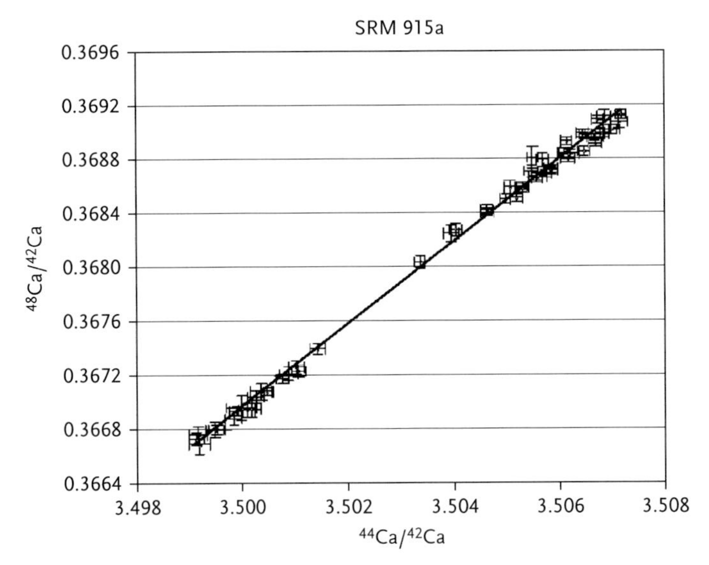

Figure 3.6 Three-isotope plot of $^{48}Ca/^{42}Ca$ versus $^{44}Ca/^{42}Ca$ isotope ratio measured for NIST SRM 915a $CaCO_3$ isotopic reference material. The data are in good agreement with mass fractionation and there is no evidence of spectral interferences from isobaric nuclides, polyatomic ions, and/or doubly charged ions affecting the results. The variability is the result of the drift in the extent of instrumental mass discrimination exhibited by the MC-ICP-MS instrument used. Reproduced from [14].

requires elimination of interfering ions by chemical separation prior to analysis and separation of polyatomic species formed in the plasma. In this case, the use of the three-isotope plot can demonstrate unambiguously that measured isotope ratios are only affected by mass-dependent processes. Chakrabarti and Jacobsen employed a GV Isoprobe-P multi-collector ICP-MS instrument to analyze a series of chondrite and differentiated meteorites, meteorites of Martian origin, and lunar and terrestrial basalts [15]. They observed that the $^{26}Mg/^{24}Mg$ isotope ratio of bulk silicate earth (BSE), carbonaceous and ordinary chondrites, Martian meteorites, pallasite olivines, and lunar materials were homogeneous to within 0.004%. These high-precision data enabled them to conclude that the magnesium stable isotopic composition of the inner solar system is constant to within a few parts per 10^5 (Figure 3.7) and can serve as reference points in understanding the partitioning of magnesium on the Earth in addition to the condensation and evaporation of this element in the early solar system.

3.6
Detector Technologies for Multi-Collection

Faraday cup detection is the most robust, linear, and accurate technology for the measurement of ion currents. Thin, deep buckets are precision-machined from

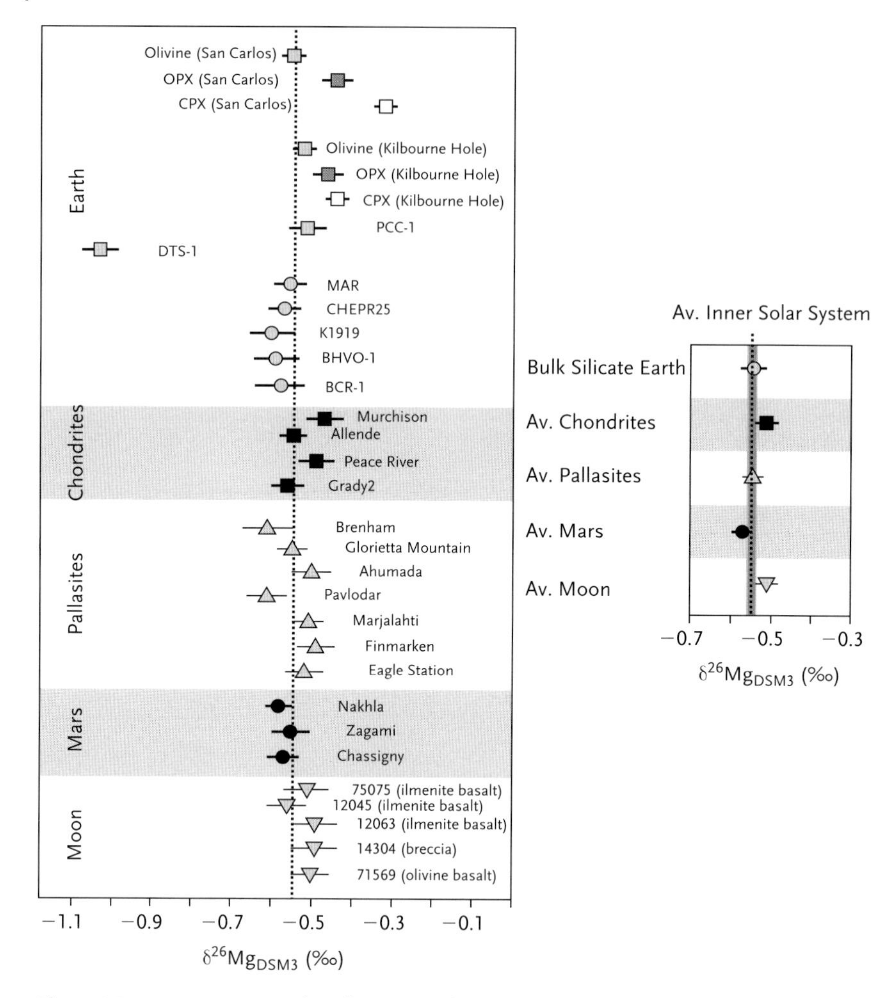

Figure 3.7 Mg isotope ratio data for terrestrial mantle-rocks and minerals, whole-rock chondrites, pallasite olivines, Martian meteorites, and lunar basalts and breccia. Error bars represent the external reproducibility in 2SE ($= 2\sigma/\sqrt{n}$, where n is the number of measurements per sample). Not considering the dunite DTS1 and the harzburgite PCC-1 (with 32% serpentine [15]), which are both likely affected by processes occurring after their emplacement within the Earth's crust, the average δ^{26}Mg of the BSE is -0.54 ± 0.04 (2SE), which overlaps with that of average chondrites (-0.52 ± 0.04), pallasites (-0.54 ± 0.04), Mars (-0.57 ± 0.02), and the Moon (-0.51 ± 0.03), showing that the stable Mg isotopic composition of the inner solar system (dotted lines, $\delta^{26}\text{Mg} = -0.54$) is homogeneous to within 0.05‰.
Reproduced from [15].

graphite sheets and fitted to the movable stages of the MC array. The buckets are light-tight and yield practically 100% collection of all incident ions. The large depth of the cup helps to ensure that any secondary electrons produced by energetic incident ions cannot escape the detector. Very often, the Faraday cups are made from graphite or graphite-coated metal. One reason for this is that carbon has a very low positive secondary ion yield under ion bombardment and therefore the negative secondary ions can be kept inside the Faraday cup using a secondary electron suppression aperture, which is usually a fundamental part of each Faraday cup. If there is significant production of positive ions sputtered by the incoming ion beam, these positive ions would be attracted by the negative secondary electron suppression aperture and extracted out of the Faraday bucket, resulting in an inaccurate current measurement of the incoming ion beam. The graphite cups are connected to current amplifiers with high-ohmic feedback resistors, typically on the order of 10^{11} Ω. High-ohmic feedback resistors have large temperature coefficients, usually in the range of ± 200 ppm $^{\circ}C^{-1}$. Therefore, the feedback resistors and current amplifiers are housed in an evacuated and temperature-stabilized (to better than $\pm 0.01^{\circ}C$) environment. The dominant source of noise is the random thermal motion of electrons inside the resistors, so-called Johnson noise. The magnitude of Johnson noise (ΔV) is given by the equation

$$\Delta V = \sqrt{\frac{4kRT}{t_m}}$$

where k is Boltzmann's constant, R is the value of the resistor, T is the temperature in units of Kelvin, and t_m is the integration time of the measurement.

The magnitude of the noise is inversely proportional to the integration time and quadrupling the integration time results in a decrease in the noise by a factor of two. In principle, the use of a 10 times higher value feedback resistor (10^{12} Ω) should achieve a 10-fold increase in the gain of the current amplifier and a $\sqrt{10}$ increase in the noise level. This is why there are attempts to increase the value of the feedback resistor further. At present, 10^{11} Ω resistors are typically used in most commercially available instruments. Some manufacturers go up to 10^{12} Ω amplifiers, which seek to improve the signal-to-noise ratio by a factor of $\sqrt{10}$. In practice, however, the signal-to-noise ratio increases by a factor of only two because the parallel current pathways in the amplifier, vacuum feedthroughs, and insulation of the Faraday cup are all of comparable resistance to the feedback resistor itself. Recently, even 10^{13} Ω amplifiers were explored and an improvement in signal-to-noise ratio by a factor of 2, compared with 10^{12} Ω, was observed [16]. Further increases in the value of the feedback resistor may not result in better signal-to-noise ratios. Also, increases in the value of the feedback resistor necessarily increase the time constant of the amplifier significantly.

Secondary electron multipliers are used to measure extremely low ion currents and extend the useful sensitivity by several orders of magnitude. Incoming ions strike a conversion dynode that ejects several electrons per incoming ion. These ions are accelerated towards a second conversion dynode where each electron again causes the ejection of several secondary electrons. Several cascading dynode stages result in gains of anywhere from 10^5- to 10^8-fold, depending on the voltage bias applied to the device (typically from 500 to 2000 V). In medium amplification or "analog" mode, the output signal is measured using conventional low-current amplifiers. An important disadvantage with this mode of operation is that the number of secondary electrons produced depends on the momentum of the incoming ion striking the first conversion dynode. For monoenergetic ions, the signal output is mass-dependent and there is a systematic error in the measured isotope ratio. Therefore, most secondary electron multipliers used for isotope ratio measurements are operated in "pulse-counting" mode, which offers higher amplifications in the range 10^6–10^8. The gain is large enough to detect and count these electron pulses by pulse-counting electronics. In this mode of operation, the measurement output (i.e., the number of pulses) is much less dependent on the momentum of the incoming ion and systematic biases are low enough that they can be neglected. The noise in ion-counting detectors arises from dark noise, dead time, linearity effects, and counting statistics (see also Chapter 2). The signal-to-noise ratio of these detectors is often much better than that of the Faraday cup designs. For high-precision isotope ratio measurements, the dynamic range of the ion-counting channels should be restricted to about 1–2 million counts per second in order to guarantee good linearity.

Secondary electron multipliers are bulky devices and cannot be placed adjacent to each other in an MC array to allow high-sensitivity ion current detection of neighboring masses, for example, ^{206}Pb, ^{207}Pb, and ^{208}Pb. In this case, miniaturized conversion dynodes (channeltrons) that are identical in size with the standard Faraday cups can be inserted directly in the collector array. In this way, one could design an MC system that incorporates a channeltron for the measurement of ^{235}U, a Faraday cup for the measurement of ^{208}Pb, and three additional channeltrons for ^{207}Pb, ^{206}Pb, and ^{204}Pb.

3.7 Conclusion

MC-ICP-MS shows several key features, including high sensitivity for all elements, high-precision isotope ratio measurements, and flexibility to analyze most isotopic systems (because of variable MC and detector technologies), and simultaneous detection has enabled many significant scientific advances to be achieved. Thus, many new and significant results have been generated in fields as diverse as cosmochemistry, geology, forensics, and human health. Simultaneous detection of the ion currents of the isotopes of interest also allows the analysis of transient signals generated by devices such as gas or liquid chromatographs (see also Chapter 17), laser ablation systems (see also Chapter 4), and on-line chemical reactors. Because all isotopic ion currents of interest are being measured at the same time, the absolute intensities of the ion beams do not need to remain constant over the course of the analysis (see Chapter 4 for information on the limitations in the monitoring of transient signals). The mass spectrometer "simply" functions as an on-line monitor of isotopic composition.

References

1 Thomson, J.J. (1913) Bakerian Lecture: rays of positive electricity. *Proc. R. Soc. Lond.*, **89**, 1–20.

2 Aston, F.W. (1929) The mass-spectrum of uranium lead and the atomic weight of protactinium. *Nature*, **123**, 313.

3 Dole, M. (1936) The relative atomic weight of oxygen in water and air II. A note on the relative atomic weight of oxygen in fresh water, salt water and atmospheric water vapor. *J. Chem. Phys.*, **4**, 778–780.

4 Nier, A.O. and Gulbransen, E.A. (1939) Variations in the relative abundance of the carbon isotopes. *J. Am. Chem. Soc.*, **61**, 697–698.

5 Caro, G., Bourdon, B., Birck, J.-L., and Moorbath, S. (2003) ^{146}Sm–^{142}Nd evidence from Isua metamorphosed sediments for early differentiation of the Earth's mantle. *Nature*, **423**, 428–432.

6 Walder, A.J. and Freedman, P.A. (1992) Isotopic ratio measurement using a double focusing sector mass analyzer with an inductively coupled plasma as an ion-source. *J. Anal. At. Spectrom.*, **3**, 571–575.

7 Daly, N.R. (1960) Scintillation type mass spectrometer ion detector. *Rev. Sci. Instrum.*, **31**, 264–267.

8 Douthitt, C.B. (2008) The evolution and applications of multicollector ICPMS (MC-ICPMS). *Anal. Bioanal. Chem.*, **290**, 437–440.

9 Vanhaecke, F. and Moens, L. (2004) Overcoming spectral overlap in isotopic analysis via single- and multi-collector ICP-mass spectrometry. *Anal. Bioanal. Chem.*, **378**, 232–240.

10 Weyer, S. and Schwieters, J.B. (2003) High precision Fe isotope measurements with high mass resolution MC-ICPMS. *Int. J. Mass Spectrom.*, **226**, 355–368.

11 Gray, C.M. and Compston, W. (1974) Excess Mg-26 in Allende meteorite. *Nature*, **251**, 495–497.

12 Lee, T. and Papanastassiou, D.A. (1976) Demonstration of Mg-26 excess in Allende and evidence for Al-26. *Geophys. Res. Lett.*, **3**, 109–112.

13 Walczyk, T. (2004) TIMS versus multicollector-ICP-MS: coexistence or struggle for survival?. *Anal. Bioanal. Chem.*, **378**, 229–231.

14 Wieser, M.E., Buhl, D., Boumann, C., and Schwieters, J. (2004) High precision calcium isotope ratio measurements using a magnetic collector inductively coupled plasma mass spectrometer. *J. Anal. At. Spectrom.*, **19**, 844–851.

15 Chakrabarti, R. and Jacobsen, S.B. (2010) The isotopic composition of magnesium in the inner Solar System. *Earth Planet. Sci. Lett.*, **293**, 349–358.

16 Tuttas, D., Schwieters, J.B., and Lloyd, N.S. (2010) Low noise Faraday cup measurements using multicollector mass spectrometers. *Geochim. Cosmochim. Acta*, **74**, A1061.

4
Advances in Laser Ablation–Multi-Collector Inductively Coupled Plasma Mass Spectrometry

Takafumi Hirata

4.1
Precision of Isotope Ratio Measurements

The combination of an inductively coupled plasma ion source and a magnetic sector-based mass spectrometer equipped with a multi-collector (MC) array [multi-collector inductively coupled plasma mass spectrometry (MC-ICP-MS)] offers precise and reliable isotope ratio data for many solid elements. In fact, MC-ICP-MS provides data, the trueness (accuracy) and precision of which is similar to, or, in some cases, even superior to, that achieved by thermal ionization mass spectrometry (TIMS), considered the "benchmark" technique for isotope ratio measurements of most solid elements [1]. The basic strength of ICP-MS lies in the ion source, which achieves extremely high ionization efficiency for almost all elements [2, 3]. Consequently, MC-ICP-MS is likely to become the method of choice for many geochemists, because it is a versatile, user-friendly, and efficient method for the isotopic analysis of trace elements [4–8]. The ICP ion source also accepts dry sample aerosols generated by laser ablation [9–16]. The combination of laser ablation (LA) with ICP-MS is now widely accepted as a sensitive analytical tool for the elemental and isotopic analysis of solid samples.

As a result of its continuous development, LA–ICP-MS provides ever more precise elemental and isotopic data. Enhancements in elemental sensitivity achieved for ICP-MS, together with a better understanding of the mechanism of the LA process, have led to successive improvements in the precision of isotope ratio measurements [17–19]. LA utilizing UV light – with a frequency-quadrupled (266 nm) or -quintupled (213 nm) Nd:YAG laser or an ArF* excimer laser (193 nm) – offers reduced elemental fractionation during ablation and better spatial resolution with a small ablation pit size, and is now the most widely used system for LA–ICP-MS [11, 18, 20–22]. Even with UV lasers, the precision of isotope ratio measurements achieved by LA–ICP-MS is commonly inferior to that obtained by conventional solution nebulization ICP-MS. MC-ICP-MS combined with conventional solution nebulization can achieve a precision of 0.005% or

Isotopic Analysis: Fundamentals and Applications Using ICP-MS, First Edition. Edited by Frank Vanhaecke and Patrick Degryse.

Published 2012 by WILEY-VCH Verlag GmbH & Co. KGaA

better for most elements [8, 23, 24]. MC-ICP-MS isotope ratio measurements using LA as a sample introduction technique achieve precisions ranging from 0.01 to 0.05%, suggesting that the precision with LA–MC-ICP-MS is slightly deteriorated in comparison with solution nebulization MC-ICP-MS [25]. This raises an interesting question: "Why are precisions achieved when using the laser ablation sample introduction technique poorer than those achieved when using solution introduction, even with a multi-collector system setup?" There are several possible explanations for the deterioration of analytical precision, the most obvious being (i) instability in the signal intensity profiles, (ii) isotope fractionation during the laser ablation event, and (iii) impact of the matrix effect on the mass bias factor (correction for instrumental mass discrimination). To obtain the most reliable and precise isotope ratio data, significant attention must be given to the contribution of these effects.

4.2 Stable Signal Intensity Profiles: Why So Important?

It has been widely recognized that a high and stable signal intensity is vital in achieving a better precision and realizing an improvement in the trueness of isotope ratio data. High signal intensity is very important because the contribution of the counting statistics $(1/\sqrt{N})$ can be minimized at high signal intensity (high N) [26]. However, even at high ion signal intensities, the precision of isotope ratio data can deteriorate when the signal intensity is unstable or spiky [27]. This is also true for solution nebulization. In the case of isotope ratio measurements using MC-ICP-MS, one may think that by ratioing the signals for two isotopes, the contribution of signal instability to the resulting isotope ratio data is removed. This might be true if the rate of the signal intensity changes was much slower than the detector response time. However, this is not the case for the changes in the signal intensity encountered when deploying LA for sample introduction. In fact, as mentioned above, the resulting precision of the isotope ratio measurements when using LA for sample introduction is typically poorer than that obtained with solution nebulization. Typical response times shown by conventional electron multipliers can be as short as 10 ns, whereas the response time of Faraday detectors, which are widely used for high-precision isotope ratio measurements, are typically in the region of 0.1 s [24, 28, 29]. If the signal intensity changes on a time scale shorter than the response time, the measured signal intensity cannot follow the changes in the actual signal intensity, and thus the measured intensities could deviate from the true intensities. Most of the commercially available MC instruments utilize Faraday collectors for ion detection [24]. The Faraday collector has remarkable advantages over other collector systems: (i) the measured and amplified ion current is finely proportional to the number of ions, (ii) there is no significant mass discrimination caused by differences in the chemical nature or mass of the entering ions [24], and (iii) the response or gain of the Faraday collector is virtually constant over long time periods of time. Therefore, the Faraday collector has been widely used for isotopic analysis, especially in an MC setup. The isotope ratio data obtained with Faraday collectors reported for many solid elements are considered "benchmark" data. A definitive disadvantage of the Faraday collector is the relatively long time constant (i.e., slow response) associated with the signal amplifier system [28]. Since the ion current of the isotopes monitored is typically in the region of 10^{-10} A (equivalent to about 10^6 ions per second), the ion current must be amplified, such that the voltage measured (Ohm's law: $V = RI$) is suitable for further processing. Figure 4.1a illustrates the principles of a simplified direct current amplifier. The positive ions arriving at the collector are neutralized by electrons arriving from ground through a high ohmic resistor, and the potential drop across this resistor is measured. This system, however, is not widely used as an amplification method for the Faraday collector because the entry of positively

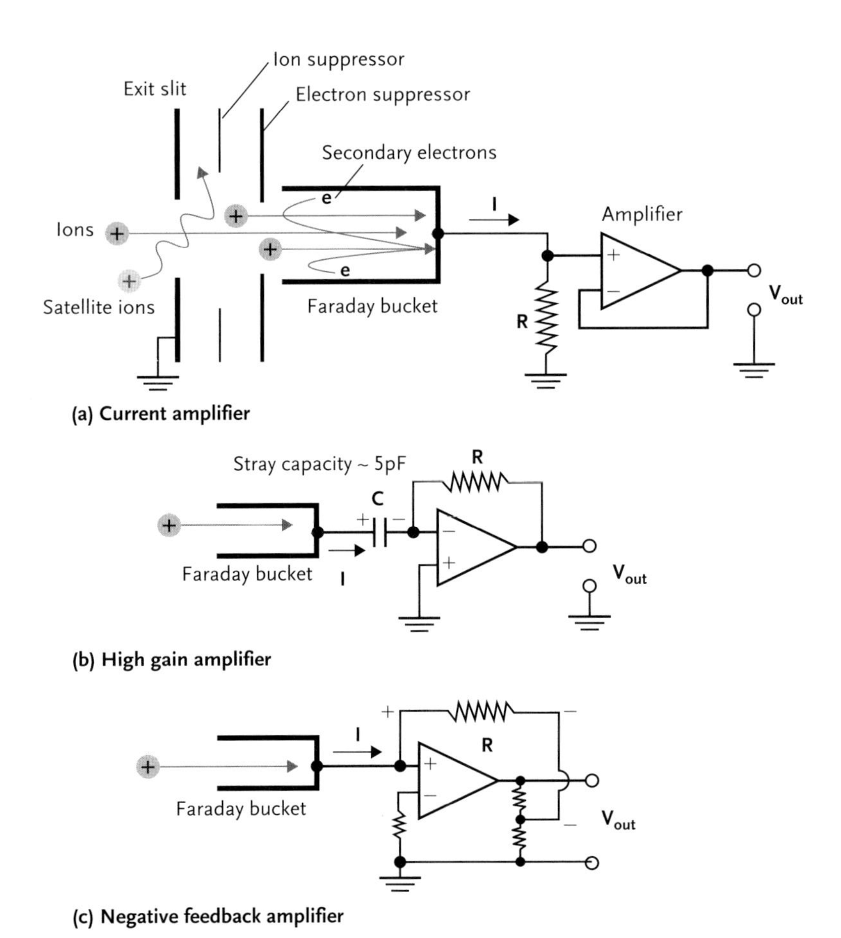

Figure 4.1 Schematic diagram of the amplifier circuit for the Faraday ion detector: (a) direct current amplifier; (b) high gain amplifier; (c) amplifier with negative feedback circuit. With the negative feedback circuit, a faster response can be achieved.

charged ions causes a potential increase in the Faraday bucket. This potential increase results in changes in collector efficiency with respect to the analytes. To overcome this, a high-gain amplifier shown in Figure 4.1b is preferred, because with this setup, the potential of the Faraday bucket is maintained at almost zero by the action of the amplifier. The time constant can be slightly improved by the high-gain amplifier system, but the time constants achieved by this amplifier system are still long compared with electron multipliers or the time range of the signal intensity changes. For precise isotope ratio measurements, relatively rapid changes in the ion current must be followed by these amplifier systems. The relationship between ion current and output voltage at the input of the electrometer can be defined by Kirchhoff's law as

$$RC\frac{\mathrm{d}V}{\mathrm{d}t} + V = RI \tag{4.1}$$

where C is the stray capacity associated with the Faraday collector, high ohmic resistor and feed lines between the detectors and amplifiers. If the ion current is suddenly changed from I_1 to I_2, the time profile of the output voltage (V) can be calculated by

$$V = RI_2 + R(I_1 - I_2)\mathrm{e}^{-\frac{t}{RC}} \tag{4.2}$$

This suggests that the measured voltages show an exponential change. $t = RC$ is known as the time constant of the circuit. In order to minimize the time constant, a reduction in the stray capacity or a high ohmic resistor (or both) is desirable. The input capacity can be significantly reduced by employing feedback amplification (Figure 4.1c). In the case of the negative feedback amplifier system, the polarity of the V_{out} is reversed to V_{in} through the pre-amplifier with a gain of G ($G = V_{\mathrm{out}}/V_{\mathrm{in}}$), and this results in negative feedback to the input voltage as defined by

$$V_{\mathrm{in}} - RI + \frac{R_\mathrm{b}}{R_\mathrm{a} + R_\mathrm{b}} V_{\mathrm{out}} = 0 \tag{4.3}$$

By replacing V_{in} by V_{out}/G in Eq. (4.3), we obtain

$$RI = V_{\mathrm{out}}\left(\frac{1}{G} + \frac{R_\mathrm{b}}{R_\mathrm{a} + R_\mathrm{b}}\right) \tag{4.4}$$

Assuming that

$$\frac{1}{G} \ll \frac{R_\mathrm{b}}{R_\mathrm{a} + R_\mathrm{b}} \tag{4.5}$$

Eq. (4.4) can be modified to

$$RI = V_{\mathrm{out}}\frac{R_\mathrm{b}}{R_\mathrm{a} + R_\mathrm{b}} \tag{4.6}$$

When $R_\mathrm{a} = 0$,

$$V_{\mathrm{out}} = RI \tag{4.7}$$

This indicates that the output voltage (V_{out}) is equal to GV_{in} ($=RI$), and the resulting total gain of the amplifier is not high ($\sim 10^5$). Despite this low gain of the negative feedback amplifier ($=V_{out}/RI$), zero stability and current linearity are remarkably improved because the output signal is fairly independent of fluctuations in gain (G). Moreover, the time constant is reduced, because the apparent capacitance (C) is reduced to $C/(1+G)$. The time constant is as low as 0.1 s with a $R = 10^{11}\ \Omega$ resistor. However, it should be noted that the stray capacity can cause a lowering of the time constant when a very high resistance (e.g., $R = 10^{11}\ \Omega$) is employed. The stray capacity depends strongly upon various parameters, such as instrumental setup and the length and geometry of the wiring between the Faraday collectors and the amplifiers (i.e., distance or angle between crossing cables), such that the residual capacitance cannot be reduced much below 10 pF. For example, if one uses a compensation capacitor of 30 pF and if the stray capacitance is 5 pF, the actual compensation capacitance will become 35 pF, and this results in a slower response of the amplification system than expected. More importantly, the stray capacitance can vary among the individual Faraday amplifiers, because of different geometric settings and, therefore, the time constant may vary between the Faraday collectors. This suggests that, even with multi-collection, the measured isotope ratio may vary when the signal intensity changes significantly. This is well demonstrated by the changes in the measured isotope ratios with sudden changes in signal intensity. Figure 4.2 shows the signal intensity of ^{63}Cu, obtained for a Cu standard solution prepared by dissolving NIST SRM 976 natural copper reference material. To obtain different signal intensities for Cu, two Cu solutions (with concentrations of $\sim$100 and $\sim$500 ng g^{-1}) were alternately introduced into the ICP ion source through solution nebulization. Because of the presence of a "bubble" between the two Cu solutions, mixing or diffusion of Cu between the two solutions was avoided. This means that no isotope fractionation could take place through the transport process of the solutions. Signal intensities for ^{63}Cu and ^{65}Cu were simultaneously monitored by two independent Faraday collectors. The resulting signal intensity increased rapidly for ^{63}Cu (increase rate of $> 5\ V\ s^{-1}$) when the high-concentration solution was introduced, and decreased when the low-concentration solution was introduced (Figure 4.2a). The time profile of the measured $^{65}Cu/^{63}Cu$ ratio is shown in Figure 4.2b, indicating that the measured $^{65}Cu/^{63}Cu$ ratio decreased by $\sim 2‰$ when the signal intensity increased from 1.2 to 5 V. In contrast, the measured $^{65}Cu/^{63}Cu$ ratio increased by $\sim 2‰$ when the signal intensity became low. It could be suggested that the mass bias changed because the "air bubble" was introduced into the ICP between the two sample solutions. The introduction of the air bubble could cause different ionization conditions, such as temperature or electron density in the ICP ion source, thereby resulting in changes in the measured $^{65}Cu/^{63}Cu$ ratio. However, this is not the case because the direction of change in the measured $^{65}Cu/^{63}Cu$ was the opposite with an increase or decrease in the signal intensity of Cu. Moreover, possible isotope fractionation through the nebulization or transport stage of the solution mist would be negligible. The most plausible explanation for the changes in the measured $^{65}Cu/^{63}Cu$ ratio would be the difference in the response of the

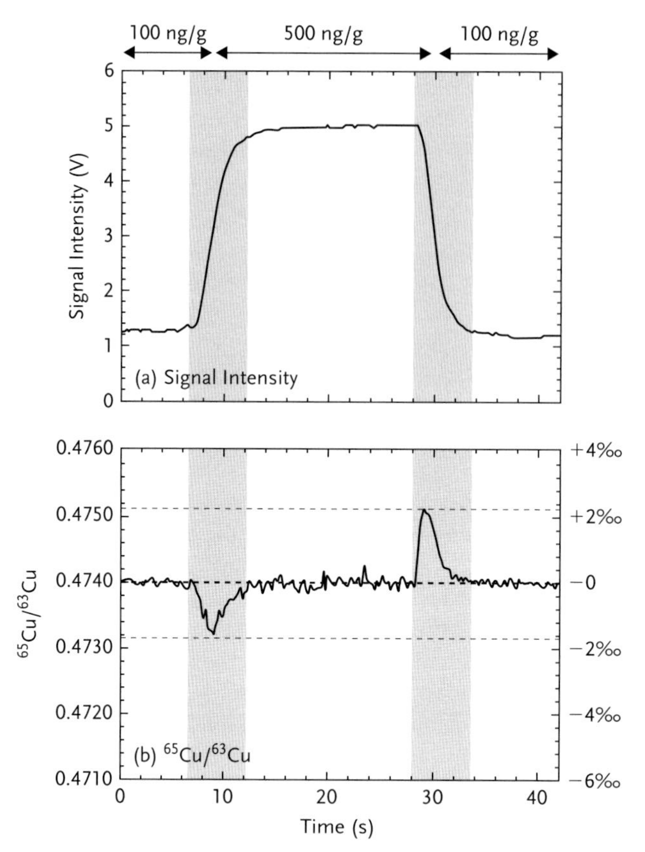

Figure 4.2 Time profile of signal intensity for ^{63}Cu (a) and the measured $^{65}Cu/^{63}Cu$ isotope ratio obtained upon introducing Cu standard solutions of different concentrations (100 and 500 ng g^{-1}) via conventional nebulization.

Faraday amplifiers. These data demonstrate clearly that sudden changes in signal intensity (~5 V s^{-1}) can cause fluctuations in the measured isotope ratios even in multi-collector systems. In the case of the isotopic analysis using ICP-MS, especially combined with LA as the sample introduction technique, signal intensities can change rapidly and, therefore, consideration must be given to erroneous measurements due to the slow response of Faraday amplifiers. To minimize erroneous measurements due to the contribution of the slow response of the Faraday amplifiers, stable or smoother signal intensity profiles are very important. In the case of isotope ratio measurements using TIMS, the slow response of the Faraday amplifiers may not be a major source of analytical error since the rate of change in signal intensities should be much slower than the time constant of the Faraday amplifiers. This is the main reason why the precision of isotope ratio measurements is higher (better) for TIMS than for MC-ICP-MS.

4.3
Signal Smoothing Device

Even with a multi-collector setup, stability of the signal intensities is still a key issue to obtain precise data and to improve the trueness of the results. In the case of LA, there are two major sources of signal instability: (i) the introduction of large-sized particles into the ICP and (ii) the introduction of a "pulsed" sample flow, induced by pulsed LA (when a low laser repetition rate is adopted). It is widely recognized that a spiky signal intensity profile can be due to the introduction of particles having a size larger than the "critical size," which is defined as the maximum size of the particles being completely atomized and ionized in the ICP ion source [30, 31]. Moreover, the introduction of large-sized particles into the ICP induces elemental and isotope fractionation during both the LA and the ionization process in the ICP. Therefore, it is desirable that large-sized particles are filtered out of the dry aerosol generated by LA [30–35]. However, since large-sized particles are the dominant contributor to the total mass of material ablated from the sample, their removal can cause a significant loss in the sample transport efficiency to the ICP ion source. To obtain a smooth signal intensity profile without significant loss of sample mass, the production of large-sized particles should be minimized through optimization of the LA process. The size distribution of the ablated particles can be modified towards smaller sizes by using a shorter wavelength [30, 36, 37] and/or a shorter pulse duration (laser emission duration) [38–41]. A recent development in the LA technique is the use of femtosecond (fs) lasers [38, 39, 41–43]. When using a fs-LA system, the laser-induced sample aerosol is characterized by smaller particle sizes, resulting in both reduced elemental fractionation and a higher transport efficiency of sample particles into the ICP ion source. The analytical capability can be further improved by employing a UV-fs-laser, operating at a wavelength of 260 or 196 nm [18, 38, 40, 43]. With LA using UV-fs-lasers, an even more stable and smoother signal intensity profile can be obtained from almost all materials, including metals and semiconductor samples. The difficulty in obtaining a stable signal intensity from metallic samples is discussed in the following section. Another serious cause of unstable signal intensity is the pulsed flow of sample aerosols induced by a pulsed laser operating at a low repetition rates <2 Hz (i.e., two laser shots per second). A lower repetition rate is advantageous because of the smaller aspect ratio (i.e., lower depth/diameter ratio) of the ablation pit that it typically produces. It is recognized that elemental fractionation becomes more serious as the aspect ratio increases (i.e., deeper crater depth). To reduce elemental fractionation during LA, elemental and isotopic analysis is sometimes carried out using a low laser repetition rate. Moreover, in the case of depth profiling analysis, a low laser repetition rate is preferred to improve the depth resolution. In these cases, the resulting signal intensity profile can become transient or oscillatory, reflecting the pulsation of the sample flow.

Figure 4.3a illustrates the signal intensity profile of ^{115}In, obtained upon laser ablation of NIST SRM 610 glass reference material with a laser repetition rate of 2 Hz. The resulting ^{115}In signal intensity is clearly oscillating at a $\pm 25\%$ level

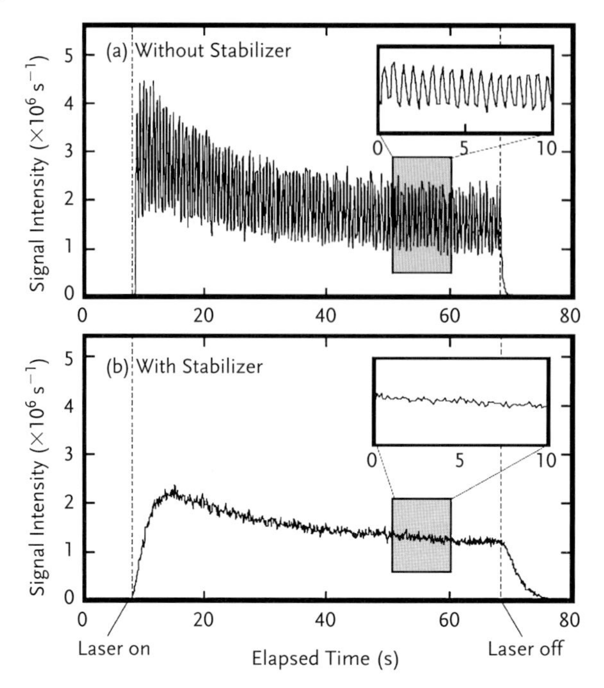

Figure 4.3 Signal intensity profile for ^{115}In (a) without and (b) with stabilizer.

from the average signal intensity. The enlarged scale of the resulting signal intensity profile reveals that the frequency of oscillation is 2 Hz, which is identical with the laser repetition rate employed in this measurement, and therefore the major source of oscillation. This oscillating signal intensity can be minimized effectively by inserting a signal stabilizer device between the sample cell and the ICP torch (Figure 4.4). With such a signal stabilizer device, the signal profile for ^{115}In can be modified dramatically toward a smoother profile (Figure 4.3b). The smoothing can be explained either by the mixing of the pulsed flow of sample aerosol or by removal of the large-sized particles through the stabilizer device. This suggests that the signal stabilizer device can also act as a particle filter, minimizing the introduction of large-sized particles into the ICP ion source [35]. In fact, the average signal intensity obtained with the signal stabilizer device was almost 15% lower than that obtained without the stabilizer. The lower signal intensity profile suggests that some fraction of the sample particles was removed by the stabilizer device. The larger volume of the stabilizer can provide a smoother signal intensity; however, the washout time becomes longer. The trade-off between signal smoothness and washout time is a matter of compromise, and should be carefully considered. The washout time typically exceeds 10 s when the stabilizer device is used. Moreover, the time required for the plateau intensity to establish is also increased when the stabilizer device is applied (Figure 4.3b). For elemental mapping studies (i.e., imaging mass spectrometry), the long washout time can

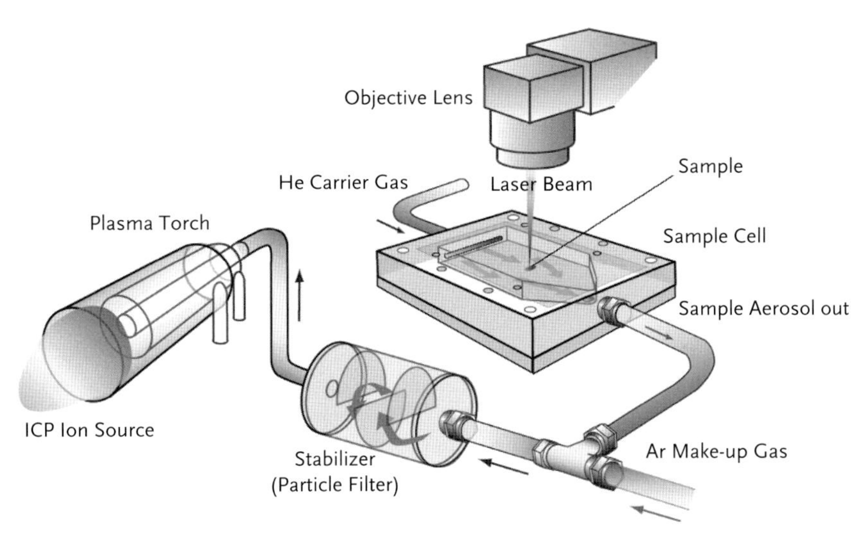

Figure 4.4 System setup for LA–ICP-MS equipped with a signal smoothing device. The internal volume of the stabilizer was 52 ml. With the stabilizer device, the signal intensity can be smoothed by both the buffering of the pulsed flow and the removal of the large-sized particles.

cause the lateral resolution to deteriorate. A shorter washout time allows faster laser scanning (or rastering), and therefore, a shorter analysis time [44, 45]. Despite the slow response achieved with the signal stabilizer device, its use is highly advisable for achieving higher analytical precision or better reliability of the elemental or isotope ratio measurements.

4.4 Multiple Ion Counting

MC arrays have been widely used in TIMS. The majority of the MC arrays consist of several Faraday collectors plus a single, fixed ion-counting device. The main advantage of multiple ion monitoring is that it obviates the need to cycle a number of narrow ion beams through a single collector, hence the collection efficiency of the ion beam is increased. Recently, continuous or discrete dynode electron multipliers have been developed that match the width of Faraday collectors and they therefore fit into the MC array, allowing isotope ratio measurements using either Faraday collectors, electron multipliers, or a combination of both. Ion-counting detectors can be utilized to achieve higher precision at lower analyte concentrations. Inter-collector bias and mass bias can be adequately corrected for based on the measured isotope ratio for an isotopic standard of known isotopic composition [46–49]. The major practical problem with the multiple ion-counting device is the large correction factor for the inter-collector biases, mainly due to large differences in the collector gain. The gain of a multiplier is seriously dependent upon both the high voltage used for amplification and the discrimination level used for to filter out "false" signals (see also Chapter 2). Moreover, the multiplier gain can change with time when the number of incoming ions hitting the first dynode is typically very high (i.e., high count rate signals). To avoid erroneous measurements due to changes of the multiplier gain, the corresponding correction factors must be calibrated regularly throughout the measurements. For achieving a smaller influence of the time-dependent changes of the multiplier gain or discrimination levels, very intense ion beams (i.e., high count rate signals) must be avoided. Despite the obvious success in obtaining a higher precision at lower target element concentrations, the typical precision of the isotope ratio measurements achieved by the multiple ion-counting technique is only about 0.05–0.2% [47], and therefore significantly poorer than that obtained by multi-collection using Faraday collectors. Nevertheless, many workers dealing with U–Pb chronology are increasingly interested in Pb isotopic analysis from a small area ($d < 20$ μm), hence they prefer to use the multiple ion-counting technique [with ICP-MS, TIMS, or secondary ion mass spectrometry (SIMS)]. For better precision and more reliable isotope ratio measurements, long-term stability in both the gain and noise level (i.e., discrimination level) is highly desirable.

4.5 Isotope Fractionation During Laser Ablation and Ionization

It is widely accepted that the precision or reproducibility of isotope ratio measurements can be improved dramatically by using an MC instrument. However, further correction or careful procedural protocols are required to also improve the trueness of the resulting isotopic data. There are two major considerations for better quality isotope ratio measurements using LA–ICP-MS analysis: the isotope fractionation during the LA and/or ionization process in the ICP, and the contribution of the matrix effect (non-spectral interference) to the mass bias effect. The latter is discussed in Section 4.6. First, the level of isotope fractionation during LA or ionization processes is discussed in this section.

With a better understanding of the mechanism of the LA processes and with the higher elemental sensitivity of the LA–ICP-MS technique, the precision of isotope ratio measurements has been improved successfully [19, 50, 51]. Laser ablation utilizing UV-light – with a frequency-quadrupled (266 nm) or -quintupled (213 nm) Nd:YAG laser or an ArF^* Excimer laser (193 nm) – can provide precise and reproducible signal intensity data from a small sampling area and is now the most widely used system for LA–ICP-MS [18, 20]. This is well demonstrated in zircon U–Pb chronology [15, 21, 47, 52–56]. The resulting precision of the ^{238}U–^{206}Pb age determination is now very similar to that achieved by the ion microprobe technique, which has been considered to provide "benchmark" age data [57]. However, even with UV lasers, serious elemental fractionation occurs during LA, mainly due to differences in elemental volatility [21, 52, 53, 56] or time-dependent changes in the particle size distribution [30, 31, 34, 36]. The reduction of elemental fractionation throughout the analysis is still a key issue in improving the data quality. It is widely recognized that the particle size distribution is a critical parameter, controlling both the analytical sensitivity (ionization efficiency) and the level of elemental fractionation in the ICP ion source. The size distribution of the sample aerosol is seriously dependent upon various conditions, including laser irradiance [40, 58, 59], cell geometry [39, 60], type of carrier gas [44, 61–63], and laser pulse length [42, 59, 60]. With the conventional nanosecond LA systems (e.g., Nd:YAG or ArF^* excimer lasers), high-speed camera imaging experiments revealed that the size of laser-generated particles from a glass sample was smaller than 1 μm, whereas the size of the particles released from an Si wafer or metallic sample was larger than the critical size ($\gg$1 μm). The difference in size distribution of laser-induced sample particles among sample materials can be attributed to differences in thermal conductivity ($1\,W\,m^{-1}\,K^{-1}$ for glass and $120\,W\,m^{-1}\,K^{-1}$ for Si) or to differences in the rate of thermal diffusion [64]. The rate of thermal diffusion of a metallic sample (or Si wafer) is in the order of 1 ps ($1\,ps = 10^{-12}\,s$) (Bogaerts, personal communication), which is much faster than that of nonmetallic samples [65–67]. In most cases, the typical pulse length of laser emission is 5–20 ns, much longer than the time scale for the cooling of metallic samples through thermal diffusion. Because of the large loss of laser energy due to faster thermal diffusion and the fact that the samples cannot be totally evaporated, a resulting molten phase is released from the ablation site by means of a laser-induced shockwave. The release of large-sized particles can be the main source of the signal spikes observed by ICP-MS monitoring [18]. Particle sizes larger than 150 nm for glass samples [68] and 400 nm for metals are not completely atomized and ionized in the ICP. Large-sized particles can survive their passage through the ICP (they are not totally evaporated) and solid material can therefore pass into the sampling orifice [30, 32]. This results in large elemental fractionation and production of signal spikes (Figure 4.5a).

This is particularly serious when metallic samples, such as Cu, Zn, Fe, or Si wafers, are analyzed via conventional nanosecond LA–ICP-MS [59, 61, 63]. For brass and glass materials, it is recognized that the total elemental composition of all transported particles represents the bulk composition of the solid sample, whereas individual particles may show different elemental compositions depending on their size [33, 34, 40]. Both the elemental fractionation and the spiky signal intensity profile can result in poor precision or reproducibility in isotope ratio measurements. One approach to reduce the size distribution of sample particles is to employ LA under low-pressure conditions. The effect of gas pressure in the sample cell on the data quality was well studied in laser-induced breakdown spectroscopy (LIBS), where enhancement of the signal intensity could be achieved by ablation under low-pressure conditions [69, 70]. The gain in signal intensity in LIBS is explained by better plasma expansion due to low local pressure or by a shift towards a smaller particle size distribution of the sample particles. Moreover, the level of sample redeposition after ablation can be reduced by low gas pressure conditions. The reduction in sample redeposition can be explained either by a smaller contribution of the down force, originating from the gas expansion

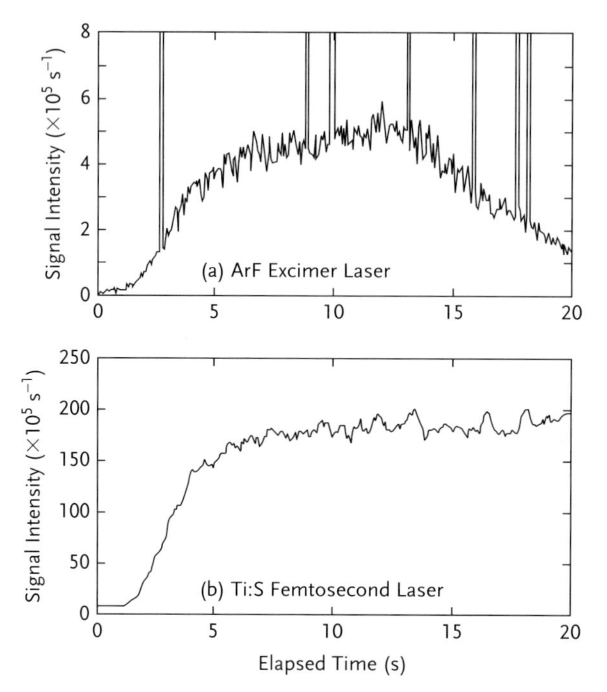

Figure 4.5 Signal intensity profile for ^{63}Cu obtained upon LA of metallic Cu (NIST SRM 976 reference material). With LA using an ArF* excimer laser, the signal intensity was very spiky and unstable (a). In contrast, with LA using the Ti:S fs-laser, no spiky signals were observed (b).

above the ablation pit, or by a relatively sharp release angle of the sample particles from the ablation pit. Both the smaller particle size distribution and the lower level of redeposition result in a higher transport efficiency and higher ionization efficiency in the ICP. In the case of ICP-MS, however, LA under low gas pressure cannot be applied because the ICP ion source is operated under atmospheric pressure. To overcome this, Fliegel and Günther [71] employed off-line LA under low and high gas pressure conditions (2–140 kPa). The LA was carried out under low (or high) gas pressure and the aerosol thus created was trapped in a closed cell; after the ablation, the cell was opened and the aerosol produced was flushed into the ICP. At lower gas pressure, a smaller particle size distribution was obtained and the resulting elemental fractionation effect was successfully minimized. However, as pointed out by the authors, with off-line LA, the transport efficiency of the sample particles became lower than that achieved by LA under atmospheric pressure due to the loss of sample particles within the ablation cell. Hirata reported on an on-line low gas pressure cell (LPC) device for elemental analysis using LA–ICP-MS [27]. Ambient gas in the sample cell was evacuated by a constant-flow diaphragm pump and the pressure in the sample cell was controlled by changing the flow rate of the He carrier gas. The sample aerosols produced were evacuated by the diaphragm pump, and the resulting sample particles were directly introduced into the ICP.

The degree of sample redeposition around the ablation pit could be reduced when the pressure of the ambient gas was lower than 50 kPa. The signal intensity profile for ^{63}Cu obtained upon LA of a metallic sample (NIST SRM 976 reference material) demonstrated that the spikes in the transient signal, which can become a large source of analytical error, were no longer observed. The resulting precision in the ^{65}Cu/^{63}Cu ratio measurements was 2–3% ($n = 10$, 2SD), half that obtained with LA under atmospheric pressure (6–10%). The newly developed LPC device provides easier optimization of the operating conditions, together with reduced sample redeposition and better stability of signal intensity, even for a metallic sample.

The advantage of the present technique is that on-line elemental and isotopic analyses can be made without any switching of the gas flow, obviating the risk of loss of ablated sample material in the cell. Despite the obvious success in obtaining both high transport efficiency and a smoothed signal intensity profile, a major source of analytical error originating from time-dependent changes in the measured isotope ratios still remains. Jackson and Günther [62] reported that the measured ^{65}Cu/^{63}Cu ratio increased systematically with prolonged ablation, and they demonstrated that the measured ^{65}Cu/^{63}Cu ratio showed large biases (1.5–4.8%) relative to the true ratio. To achieve reliable and reproducible isotope ratio data, time-dependent changes in the measured isotope ratios must be minimized. Several parameters, such as size distribution of the aerosols, aspect ratios, aerosol transport efficiencies, atomization/ionization efficiencies in the plasma and/or laser-induced isotope fractionation, must be considered to reduce time-dependent changes in the measured isotope ratios. A first cause of isotope fractionation during the LA event and subsequent handling of the laser-generated aerosol in the ICP is a result of the presence of larger sized particles. Also, redeposition of ablated sample material around the sample pit can generate isotope fractionation, as a result of preferential redeposition or ablation of specific elements or isotopes. To achieve both a small particle size distribution and a low level of sample redeposition, LA utilizing shorter pulse duration lasers (i.e., fs-lasers) is a very important trend. The efficiency of energy transfer from laser photons to electrons in solid samples can be dramatically improved by the high-energy irradiance achieved by fs-lasers. Even metallic samples, in which free electrons play an important role in diffusing the energy, can be effectively ablated without melting processes [39, 40]. The size distribution of the laser-induced sample particles can be modified dramatically towards smaller sizes. Less heating effects result in minimal elemental fractionation due to different elemental volatility and, more importantly, the production of small particles can reduce the size-related elemental fractionation, frequently revealed by changes in the U/Th ratio. Furthermore, the introduction of small particles into the ICP ion source results in stable signal intensities, which is very important for precise and reliable isotopic analysis. Figure 4.5b displays the signal intensity profile obtained upon LA of metallic Cu using an fs-laser. Both a high transmission efficiency (high sensitivity) and a smoothed signal intensity profile can be achieved when using fs-LA. The higher elemental sensitivity can be explained by the higher transmission efficiency of the sample particles, mainly due to their smaller size. With the stable signal intensity achieved with the fs-laser, the precision in the isotope ratio measurements can be improved dramatically. This can be well demonstrated by the long-term reproducibility of the $^{65}Cu/^{63}Cu$ isotope ratio measurements over a 6 month period. Figure 4.6 illustrates the measured $^{65}Cu/^{63}Cu$ ratios obtained upon LA of metallic Cu [JMC (Johnson Matthey Company) Cu and NIST SRM 976]. The resulting $\delta^{65}Cu/^{63}Cu_{NIST\ 976}$ value obtained for the JMC Cu was -0.217 ± 0.054‰ (Figure 4.6a). The $\delta^{65}Cu/^{63}Cu_{NIST\ 976}$ value obtained upon nebulization of a dissolved solution was -0.224 ± 0.051‰ (Figure 4.6b), which agreed well with the value obtained via LA within the analytical uncertainties [72]. More importantly, the resulting reproducibility of the $^{65}Cu/^{63}Cu$ ratio measurements achieved via fs-LA was very similar to that achieved by conventional solution nebulization. These data demonstrate clearly that stability of the signal intensity is a key parameter controlling the resulting precision and reliability of the results.

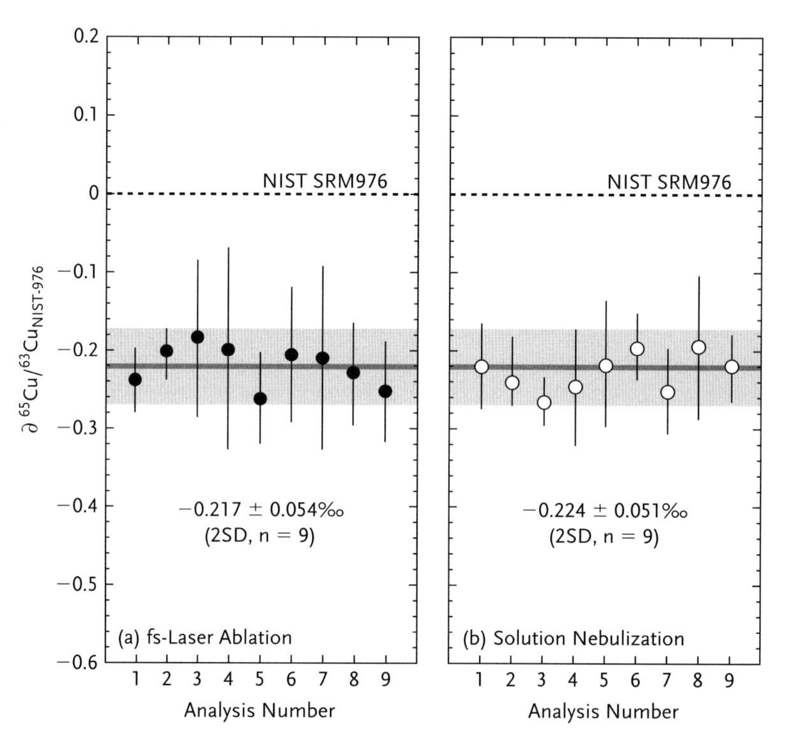

Figure 4.6 Long-term reproducibility of Cu isotope ratio measurements for pure copper metal (JMC Cu) obtained over a 6 month period. The measured $^{65}Cu/^{63}Cu$ ratios were normalized versus the Cu isotope ratio for NIST SRM 976 Cu reference material. Error bars represent analytical precision of individual solution and LA analyses at the 2σ level. Averaged $\delta^{65}Cu$ values obtained by solution nebulization and fs-LA were -0.224 ± 0.051 and -0.217 ± 0.054‰, respectively. This serves as a demonstration that there was no significant difference in the level of isotope fractionation between the two sample introduction techniques. Moreover, the resulting reproducibility in the $^{65}Cu/^{63}Cu$ ratio measurements achieved via fs-LA was very similar to that obtained via conventional solution nebulization.

4.6
Standardization of the Isotope Ratio Data

Copper occurs in Nature in various minerals with different chemical forms, including metal (native copper), sulfide (e.g., chalcopyrite and chalcocite), oxide (e.g., cuprite), and carbonate (e.g., malachite and azurite). The isotope ratio of Cu ($^{65}Cu/^{63}Cu$) varies significantly by as much as 12‰, reflecting the geochemical or physicochemical processes during the mineral's formation. Mason *et al.* [73] first reported the variation in the $^{65}Cu/^{63}Cu$ ratio at the sub-millimeter scale in a natural ore deposit. The resulting $^{65}Cu/^{63}Cu$ ratio was measured by means of TIMS for samples acquired via micro-drilling. With the micro-drilling technique, sampling of a specific area (<100 μm) can be realized. However, the resulting samples could contain several secondary minerals, and sampling of pure sulfide minerals is therefore difficult. Because such sea-floor sulfides are usually complexly inter-grown and very finely grained, sampling of specific small minerals can also be very difficult. To minimize the contribution of Cu from the secondary minerals, a new analytical technique that enables one to measure isotope ratios with better spatial resolution is highly desirable. For this purpose, the combination of LA and MC-ICP-MS is very powerful. With a combination of fs-LA and MC-ICP-MS, a precision of better than 0.1‰ can be achieved for most Cu-rich materials (metal, oxide, or sulfide). This is significantly lower than the variation in the $^{65}Cu/^{63}Cu$ ratio (caused by isotope fractionation) observed for natural Cu-rich minerals. Despite the high precision, it should be noted that correction for instrumental mass bias is still a key issue in obtaining reliable isotopic data due to matrix effects (non-spectral interferences). Figure 4.7a illustrates the comparison of the measured $^{65}Cu/^{63}Cu$ ratio obtained using either solution nebulization or fs-LA [72]. For the fs-LA technique, the

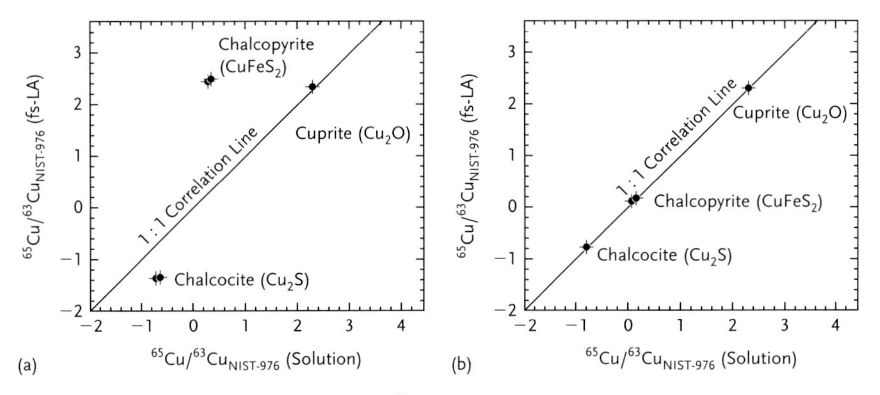

Figure 4.7 Comparison of the resulting $\delta^{65}Cu$ values obtained upon LA and nebulization of dissolved sample solutions. (a) All the $\delta^{65}Cu$ values were calculated based on the $^{65}Cu/^{63}Cu$ ratio for the NIST SRM 976 metallic Cu reference material; (b) all the $\delta^{65}Cu$ values were calculated based on the $^{65}Cu/^{63}Cu$ ratio for a matrix-matched standard. The gray line is the 1:1 correlation line.

measured $^{65}Cu/^{63}Cu$ ratio was normalized using the $^{65}Cu/^{63}Cu$ ratio obtained upon LA of metallic Cu (NIST SRM 976 reference material). For solution analysis, all solutions were prepared by dissolving the Cu-rich minerals using a mixture of acids, and subsequent dilution to an appropriate concentration. The $\delta^{65}Cu_{NIST\ 976}$ LA–ICP-MS values for cuprite (Cu_2O) agreed well with the data obtained by solution analysis. This suggests that the mass bias effect can be calibrated by monitoring the $^{65}Cu/^{63}Cu$ ratio for the metallic standard. In contrast, for the chalcocite (Cu_2S) or chalcopyrite ($CuFeS_2$) samples, the resulting $\delta^{65}Cu_{NIST\ 976}$ values do not fall on the 1:1 correlation line (gray line in Figure 4.7a). This indicates clearly that, in the case of chalcocite or chalcopyrite samples, the mass bias could not be adequately corrected for by relying on the metallic standard. Therefore, mass bias correction based on the $^{65}Cu/^{63}Cu$ ratio data for a matrix-matched standard is highly desired. The $\delta^{65}Cu_{NIST\ 976}$ values obtained upon calibration versus a matrix-matched standard were plotted against the $\delta^{65}Cu_{NIST\ 976}$ values obtained via solution analysis in Figure 4.7b. For all of the samples analyzed, the resulting $\delta^{65}Cu_{NIST\ 976}$ values determined by fs-LA using the matrix-matched standard fell close to the 1:1 correlation line (gray line in the Figure 4.7b), suggesting that the resulting $\delta^{65}Cu_{NIST\ 976}$ values obtained upon fs-LA agreed well with the data obtained by solution nebulization within the analytical uncertainties achieved in this study [72].

An ICP-MS instrument with an MC array system can easily improve the precision of the isotope ratio measurements. However, stability of the signal intensity profile, correction for isotope fractionation during LA and ionization, and correction of the mass bias effect using a matrix-matched calibration standard are also very important for improving the trueness of the isotopic data.

Acknowledgments

I am grateful to Dr. J. Andy Milton (National Oceanography Centre, University of Southampton, UK) for critical reading of the original manuscript and much technical advice. This work was partly supported by a Grant-in-Aid for Scientific Research from the Ministry of Education, Culture, Sports, Science, and Technology, Japan, and the Ministry of Agriculture, Forestry, and Fisheries of Japan.

References

1 Walczyk, T. (2004) TIMS versus multicollector-ICP-MS: coexistence or struggle for survival?. *Anal. Bioanal. Chem.*, **378**, 229–231.

2 Niu, H. and Houk, R.S. (1996) Fundamental aspects of ion extraction in inductively coupled plasma mass spectrometry. *Spectrochim. Acta B*, **51**, 779–815.

3 Becker, J.S. (2007) *Inorganic Mass Spectrometry : Principles and Applications*, Wiley-VCH Verlag GmbH, Weinheim.

4 Hirata, T. (1996) Lead isotopic analyses of NIST Standard Reference Materials using multiple collector inductively coupled plasma mass spectrometry coupled with a modified external correction

method for mass discrimination effect. *Analyst*, **121**, 1407–1411.

5 Belshaw, N.S., Zhu, X.K., Guo, Y., and O'Nions, R.K. (2000) High precision measurement of iron isotopes by plasma source mass spectrometry. *Int. J. Mass Spectrom.*, **197**, 191–195.

6 Beard, B.L., Johnson, C.M., Skulan, J.L., Nealson, K.H., Cox, L., and Sun, H. (2003) Application of Fe isotopes to tracing the geochemical and biological cycling of Fe. *Chem. Geol.*, **195**, 87–117.

7 Anbar, A.D. (2004) Iron stable isotopes: beyond biosignatures. *Earth Planet. Sci. Lett.*, **217**, 223–236.

8 Albarède, F., Telouk, P., Blichert-Toft, J., Boyet, M., Agranier, A., and Nelson, B. (2004) Precise and accurate isotopic measurements using multiple-collector ICPMS. *Geochim. Cosmochim. Acta*, **68**, 2725–2744.

9 Walder, J.A., Abell, I.D., and Platzner, I. (1993) Lead isotope ratio measurement of NIST 610 glass by laser ablation inductively coupled plasma mass spectrometry. *Specrochim. Acta B*, **48**, 397–402.

10 Thirlwall, M.F. and Walder, A.J. (1995) *In situ* hafnium isotope ratio analysis of zircon by inductively coupled plasma multiple collector mass spectrometry. *Chem. Geol.*, **122**, 241–247.

11 Longerich, H., Günther, D., and Jackson, S. (1996) Elemental fractionation in laser ablation inductively coupled plasma mass spectrometry. *Fresenius' J. Anal. Chem.*, **355**, 538–542.

12 Durrant, J.F. (1999) Laser ablation inductively coupled mass spectrometry: achievements, problems, prospects. *J. Anal. At. Spectrom.*, **12**, 1385–1403.

13 Woodhead, J.D. (2002) A simple method for obtaining highly accurate, Pb isotope data by MC-ICPMS. *J. Anal. At. Spectrom.*, **17**, 1381–1385.

14 Russo, R.E., Mao, X.L., Gonzalez, J., and Mao, S.S. (2002) Femtosecond laser ablation. *J. Anal. Atom. Spectrom.*, **17**, 1072–1075.

15 Tiepolo, M. (2003) *In situ* Pb geochronology of zircon with laser ablation-inductively coupled plasma-sector field mass spectrometry. *Chem. Geol.*, **199**, 159–177.

16 Hattendorf, B., Latkoczy, C., and Günther, D. (2003) Laser ablation–ICPMS. *Anal. Chem.*, **75**, 341A–347A.

17 Bogaerts, A., Chen, Z.Y., Gijbels, R., and Vertes, A. (2003) Laser ablation for analytical sampling: what can we learn from modeling?. *Spectrochim. Acta B*, **58**, 1867–1893.

18 Russo, R.E., Mao, X.L., Liu, C., and Gonzalez, J. (2004) Laser assisted plasma spectrochemistry: laser ablation. *J. Anal. At. Spectrom.*, **19**, 1084–1089.

19 Hirata, T. and Miyazaki, Z. (2007) High-speed camera imaging for laser ablation process: for further reliable elemental analysis using inductively coupled plasma-mass spectrometry. *Anal. Chem.*, **79**, 147–152.

20 Günther, D. and Hattendorf, B. (2005) Solid sample analysis using laser ablation inductively coupled plasma mass spectrometry. *Trends Anal. Chem.*, **24**, 255–265.

21 Hirata, T. and Nesbitt, R.W. (1995) U–Pb isotope geochronology of zircon: evaluation of the laser probe–inductively coupled plasma mass spectrometry technique. *Geochim. Cosmochim. Acta*, **59**, 2491–2500.

22 Jeffries, T., Pearce, N.G., Perkins, W., and Raith, A. (1996) Chemical fractionation during infrared and ultraviolet laser ablation inductively coupled plasma mass spectrometry: implications for mineral microanalysis. *Anal. Commun.*, **33**, 35–39.

23 Becker, J.S. (2005) Recent developments in isotope analysis by advanced mass spectrometric techniques. *J. Anal. At. Spectrom.*, **20**, 1173–1184.

24 Wieser, M.E. and Schwieters, J.B. (2005) The development of multiple collector mass spectrometry for isotope ratio measurements. *Int. J. Mass Spectrom.*, **242**, 97–115.

25 Hirata, T. (2002) *In-situ* precise isotopic analysis of tungsten using laser ablation multi-collector inductively coupled plasma mass spectrometry (LA-MC-ICP-MS) with time resolved data acquisition. *J. Anal. At. Spectrom.*, **17**, 204–210.

26 Frei, D. and Gerdes, A. (2009) Precise and accurate *in situ* U–Pb dating of zircon with high sample throughput by automated LA-SF-ICP-MS. *Chem. Geol.*, **261**, 261–270.

27 Hirata, T. (2007) Development of an on-line low gas pressure cell for laser ablation–ICP-mass spectrometry. *Anal. Sci.*, **23**, 1195–1201.

28 Beynon, J.H. and Brenton, A.G. (1982) *An Introduction to Mass Spectrometry*, University of Wales Press, Cardiff, p. 38.

29 Duckworth, H.E., Barber, R.C., and Venkatasubramanian, L.V.S. (1986) *Mass Spectroscopy*, 2nd edn., Cambridge University Press, Cambridge, p. 74.

30 Guillong, M. and Günther, D. (2002) Effect of particle size distribution on ICP-induced elemental fractionation in laser ablation inductively coupled plasma mass spectrometry. *J. Anal. At. Spectrom.*, **17**, 831–837.

31 Guillong, M., Kuhn, H., and Günther, D. (2003) Application of a particle separation device to reduce inductively coupled plasma-enhanced elemental fractionation in laser ablation–inductively coupled plasma-mass spectrometry. *Spectrochim. Acta B*, **58**, 211–220.

32 Bleiner, D., Gasser, P. (2004) Structural features of laser ablation particulate from Si target, as revealed by focused ion beam technology, Applied Physics Amaterials Science & Processing, **79**, 1019–1022.

33 Kuhn, H.-R. and Günther, D. (2003) Elemental fractionation studies in laser ablation inductively coupled plasma mass spectrometry on laser-induced brass aerosols. *Anal. Chem.*, **75**, 747–753.

34 Kuhn, H. and Günther, D. (2004) Laser ablation–ICP-MS: particle size dependent elemental composition studies on filter-collected and online measured aerosols from glass. *J. Anal. At. Spectrom.*, **19**, 1158–1164.

35 Tunheng, A. and Hirata, T. (2004) Development of signal smoothing device for precise elemental analysis using laser ablation–ICP-mass spectrometry. *J. Anal. At. Spectrom.*, **19**, 932–934.

36 Guillong, M., Horn, I., and Günther, D. (2003b) A comparison of 266 nm, 213 nm and 193 nm produced from a single solid state Nd:YAG laser for laser ablation ICP-MS. *J. Anal. At. Spectrom.*, **18**, 1224–1230.

37 Horn, I., Günther, D., Guillong, M. (2003) Evaluation and design of a solid-state 193 nm OPO-Nd : YAG laser ablation system, spectrochim. Acta B, **58**, 1837–1846.

38 Poitrasson, F., Mao, X., Mao, S.S., Freydier, R., and Russo, R.E. (2003) Comparison of ultraviolet femtosecond and nanosecond laser ablation inductively coupled plasma mass spectrometry analysis in glass, monazite, and zircon. *Anal. Chem.*, **75**, 6184–6190.

39 Koch, J., Bohlen, A., Hergenröder, R., and Niemax, K. (2004) Particle size distributions and compositions of aerosols produced by near-IR femto- and nanosecond laser ablation of brass. *J. Anal. At. Spectrom.*, **19**, 267–272.

40 Koch, J., Walle, M., Pisonero, J., and Günther, D. (2006) Performance characteristics of ultra-violet femtosecond laser ablation inductively coupled plasma mass spectrometry at ~265 and ~200 nm. *J. Anal. At. Spectrom.*, **21**, 932–940.

41 Hirata, T. and Kon, Y. (2008) Evaluation of the analytical capability of NIR femtosecond laser ablation–inductively coupled plasma mass spectrometry. *Anal. Sci.*, **24**, 345–353.

42 Horn, I. and von Blanckenburg, F. (2007) Investigation on elemental and isotopic fractionation during 196 nm femtosecond laser ablation multiple collector inductively coupled plasma mass spectrometry. *Spectrochim. Acta B*, **62**, 410–422.

43 Pisonero, J. and Günther, D. (2008) Femtosecond laser ablation inductively coupled plasma mass spectrometry: fundamentals and capabilities for depth profiling analysis. *Mass Spectrom. Rev.*, **27**, 609–623.

44 Iizuka, T. and Hirata, T. (2005) Improvements of precision and accuracy in *in situ* Hf isotope microanalysis of zircon using the laser ablation–MC-

ICPMS technique. *Chem. Geol.*, **220**, 121–137.

45 Autrique, D., Bogaerts, A., Lindner, H., Garcia, C.C., and Niemax, K. (2008) Design analysis of a laser ablation cell for inductively coupled plasma mass spectrometry by numerical simulation. *Spectrochim. Acta B*, **63**, 257–270.

46 Taylor, R.N., Warneke, T., Milton, J.A., Croudace, I.W., Warwick, P.E., and Nesbitt, R.W. (2003) Multiple ion counting determination of plutonium isotope ratios using multi-collector ICP-MS, *J. Anal. At. Spectrom.*, **18**, 480–484.

47 Simonetti, A., Heaman, L.M., Hartlaub, R.P., Creaser, R.A., MacHattie, T.G., and Böhm, C. (2005) U–Pb zircon dating by laser ablation-MC-ICP-MS using a new multiple ion counting Faraday collector array. *J. Anal. At. Spectrom.*, **20**, 677–686.

48 Snow, J.E. and Friedrich, J.M. (2005) Multiple in counting ICPMS double spike method for precise U isotopic analysis at ultra-trace levels. *Int. J. Mass Spectrom.*, **242**, 211–215.

49 Woodhead, J., Hergt, J., Meffre, S., Large, R.R., Danyushevsky, L., and Gilbert, S. (2009) *In situ* Pb-isotope analysis of pyrite by laser ablation (multi-collector and quadrupole) ICPMS. *Chem. Geol.*, **262**, 344–354.

50 Zeng, X., Mao, X.L., Greif, R., and Russo, R.E. (2004) Experimental investigation of ablation efficiency and plasma expansion during femtosecond and nanosecond laser ablation of silicon. *Appl. Phys. A*, **80**, 237–241.

51 Bogaerts, A. and Chen, Z.Y. (2004) Nanosecond laser ablation of Cu: modeling of the expansion in He background gas, and comparison with expansion in vacuum. *J. Anal. At. Spectrom.*, **19**, 1169–1176.

52 Horn, I., Rudnick, R.L., and McDonough, W.F. (2000) Precise elemental and isotope ratio determination by simultaneous solution nebulization and laser ablation-ICP-MS: application to U–Pb geochronology. *Chem. Geol.*, **164**, 281–301.

53 Jackson, S., Pearson, N., Griffin, W., and Belousova, E. (2004) The application of laser ablation–inductively coupled plasma-mass spectrometry to *in situ* U–Pb zircon geochronology. *Chem. Geol.*, **211**, 47–69.

54 Orihashi, Y., Nakai, S., and Hirata, T. (2008) U–Pb age determination for seven standard zircons using inductively coupled plasma-mass spectrometry coupled with frequency quintupled Nd-YAG ($\lambda = 213$ nm) laser ablation system: comparison with LA–ICP-MS zircon analyses with a NIST glass reference material. *Res. Geol.*, **58**, 101–123.

55 Cottle, J.M., Horstwood, M.S.A., and Parrish, R.R. (2009) A new approach to single shot laser ablation analysis and its application to *in situ* Pb/U geochronology. *J. Anal. At. Spectrom.*, **24**, 1355–1363.

56 Liu, H.C., Borisov, O.V., Mao, X.L., Shuttleworth, S., Russo, R.E. (2000) Pb/U fractionation during Nd : YAG 213 nm and 266 nm laser ablation sampling with inductively coupled plasma mass spectrometry. *Appl. spectrosc.*, **54**, 1435–1442

57 Ireland, T.R. and Williams, I. (2003) Considerations in zircon geochronology by SIMS. *Rev. Mineral. Geochem.*, **53**, 215–241.

58 Hergenröder, R., Samek, O., and Hommes, V. (2006) Femtosecond laser ablation elemental mass spectrometry. *Mass Spectrom. Rev.*, **25**, 551–572.

59 Koch, I., Feldmann, I., Jakubowski, N., and Niemax, K. (2002) Elemental composition of laser ablation aerosol particles deposited in the transport tube to an ICP. *Spectrochim. Acta B*, **57**, 975–985.

60 Vadillo, J.M. and Laserna, J.J. (2004) Laser-induced plasma spectrometry: truly a surface analytical tool. *Spectrochim. Acta B*, **59**, 147–161.

61 Eggins, S.M., Kinsley, L.P.J., and Shelley, J.M.G. (1998) Deposition and elemental fractionation processes during atmospheric pressure laser sampling for analysis by ICP-MS. *Appl. Surf. Sci.*, **127–129**, 278–286.

62 Jackson, S.E. and Günther, D. (2003) The nature and sources of laser-induced isotopic fractionation in laser ablation multicollector inductively coupled plasma mass spectrometry. *J. Anal. At. Spectrom.*, **18**, 205–212.

63 Iizuka, T. and Hirata, T. (2004) Simultaneous determinations of U–Pb age and REE abundances for zircons using ArF excimer laser ablation–ICPMS. *Geochem. J.*, **38**, 229–241.

64 Hirata, T., Hayano, Y., and Ohno, T. (2003) Improvements in precision of isotopic ratio measurements using laser ablation-multiple collector-ICP-mass spectrometry: reduction of changes in measured isotopic ratios. *J. Anal. At. Spectrom.*, **18**, 1283–1288.

65 Stuart, B.C., Feit, M.D., Rubenchik, A.M., Shore, B.W., and Perru, M.D. (1995) Laser-induced damage in dielectrics with nanosecond to subpicosecond pulses. *Phys. Rev. Lett.*, **74**, 2248–2251.

66 Chichkov, B.N., Momma, C., Nolte, S., von Alvensleben, F., and Tunnermann, A. (1996) Femtosecond, picosecond and nanosecond laser ablation of solids. *Appl. Phys. A*, **63**, 109–115.

67 Kozlov, B., Saint, A., and Skroce, A. (2003) Elemental fractionation in the formation of particulates, as observed by simultaneous isotopes measurement using laser ablation ICP-oa-TOFMS. *J. Anal. At. Spectrom.*, **18**, 1069–1075.

68 Kuhn, H., Guillong, M., and Günther, D. (2004) Size-related vaporisation and ionisation of laser-induced glass particles in the inductively coupled plasma. *Anal. Bioanal. Chem.*, **378**, 1069–1074.

69 Cristoforetti, G., Legnaioli, S., Palleschi, V., Salvetti, A., and Tognori, E. (2004) Influence of ambient gas pressure on laser-induced breakdown spectroscopy technique in the parallel double-pulse configuration. *Spectrochim. Acta B*, **59**, 1907–1917.

70 Iida, Y. (1990) Effects of atmosphere on laser vaporization and excitation processes of solid samples. *Spectrochim. Acta B*, **45**, 1353–1367.

71 Fliegel, D. and Günther, D. (2006) Low pressure laser ablation coupled to inductively coupled plasma mass spectrometry. *Spectrochim. Acta B*, **61**, 841–849.

72 Ikehata, K., Notsu, K., and Hirata, T. (2008) *In situ* determination of Cu isotope ratios in copper-rich materials by NIR femtosecond LA–MC-ICP-MS. *J. Anal. At. Spectrom.*, **23**, 1003–1008.

73 Mason, T.F.D., Weiss, D.J., Chapman, J.B., Wilkinson, J.J., Tessalina, S.G., Spiro, B. Horstwood, M.S.A., Spratt, J., and Coles, B.J. (2005) Zn and Cu isotopic variability in the Alexandrinka volcanic-hosted massive sulphide (VHMS) ore deposit, Urals, Russia. *Chem. Geol.*, **221**, 170–187.

5
Correction of Instrumental Mass Discrimination for Isotope Ratio Determination with Multi-Collector Inductively Coupled Plasma Mass Spectrometry

Juris Meija, Lu Yang, Zoltán Mester, and Ralph E. Sturgeon

5.1
Historical Introduction

From the advent of chemical science in the early nineteenth century, atomic weights of the elements were *the* royal measurands in chemistry. The global undertaking of the atomic weight measurements was, for nineteenth century chemistry, what the human genome project has been in the late twentieth century for biochemistry. High-precision determinations of the atomic weights of the elements were instrumental in many major chemistry debates of the time. For example, "Are the atomic weights of cobalt and of nickel the same?" [1], "Where does beryllium belong in the periodic table of the elements?" [2], and "Is it possible to express all atomic weights as multiples of hydrogen?" (Prout's hypothesis) [3, 4]. In the 1920s, the first understanding of the atomic weights of the elements, derived from high-precision atomic weight measurements, was realized. Aston received the 1922 Nobel Prize in Chemistry in part for the now-defunct "whole number rule," which viewed all isotope masses as integers in relation to each other (with the sole exception of hydrogen) [5]. Aston was the first to realize fully the value of isotopic composition measurements, which soon supplanted the classical methods of atomic weight determination [6]. Frederick Soddy, the "*inventor*" of the term *isotope*, and the man who stood next to Aston in Stockholm in 1922, was of the opinion that the determination of atomic weights "for the moment at least, appears as of little interest and significance as the determination of the average weight of a collection of bottles, some of them full and some of them more or less empty" [7]. For him, the future belonged to the measurement of isotope amount ratios (more commonly referred to as isotope ratios). In fact, one of the most controversial questions posed by humankind was soon answered with this tool when isotope measurements revealed the age of the Earth [8]. High-precision isotope amount ratio measurements continue to play a significant role in chemistry. For example, the recent high-profile determination of the Avogadro constant

Isotopic Analysis: Fundamentals and Applications Using ICP-MS,
First Edition. Edited by Frank Vanhaecke and Patrick Degryse.

Published 2012 by WILEY-VCH Verlag GmbH & Co. KGaA

from the 1 kg silicon sphere invokes measurement of the isotopic composition of silicon in that sphere [9]. Likewise, the discovery of a natural nuclear reactor in Gabon (see also Chapter 1) relied on high-precision measurements of the isotopic composition of uranium [10]. Although isotope amount ratio measurements can be performed using techniques other than mass spectrometry, such as gravimetry, nuclear magnetic resonance (NMR) spectroscopy or infrared (IR) spectroscopy, the use of these methods is rare. Since the mid-twentieth century, in fact, mass spectrometry has become the *de facto* method for isotope amount ratio measurements. Early measurements, however, were not always calibrated. In this regard, the isotope amount ratios measured were not always free from the biases incurred during the measurement process. As a tutorial example, consider the hypothetical measurement of the isotope amount ratios of hydrogen as the hydrogen gas escapes from a vessel through a pinhole: for an equimolar mixture of protium and deuterium, $N(^{2}H)/N(^{1}H) = 1$. In an oversimplified manner, the isotope amount ratios in the escaping gas, according to Graham's law of effusion, will not be $N(^{2}H)/N(^{1}H) = 1$, but $N(^{2}H)/N(^{1}H) \approx 1/\sqrt{2}$. Modern measurements involve more complex corrections, which form the basis for this chapter. Even with the advent of the twenty-first century, of the 62 elements having more than one stable isotope, there are no calibrated isotope amount ratio measurements for one-third of them [11]. Hence the need for calibration of isotope amount ratio measurements remains a state-of-the-art activity.

5.2 Mass Bias in MC-ICP-MS

The term *mass bias correction* is a popular metaphor for calibration of isotope amount ratios. Isotope amount ratio measurement results acquired through the use of inductively coupled plasma mass spectrometry (ICP-MS) can deviate from their true values by up to 25% (for lithium) [12]. The causes of this phenomenon within the ICP-MS instrument are not entirely understood yet, but likely arise during the supersonic expansion of ions passing through the sampler cone, and space-charge effects in the skimmer cone (i.e., based on mutual repulsion of positively charged ions, all moving with the same velocity) also play a significant role [13, 14]. Both processes favor the transmission of heavier isotopes and thus generate a bias in the measured isotope amount ratios. Along with the instrumental mass bias, matrix-dependent non-spectral discrimination can occur with changes in sample composition [15]. Fluctuations in mass bias through time also present a significant hindrance to the determination of accurate and precise isotope amount ratios, and must be dealt with. However, with the appropriate mathematical formalism, the temporal drift can be turned into an asset, as discussed in detail below. To achieve accurate isotope amount ratios, it is essential to correct for instrumental mass discrimination, which will further always be referred to as mass bias in this chapter. In fact, mass bias correction typically remains the single largest source of uncertainty in the measurement chain [15] (Figure 5.1).

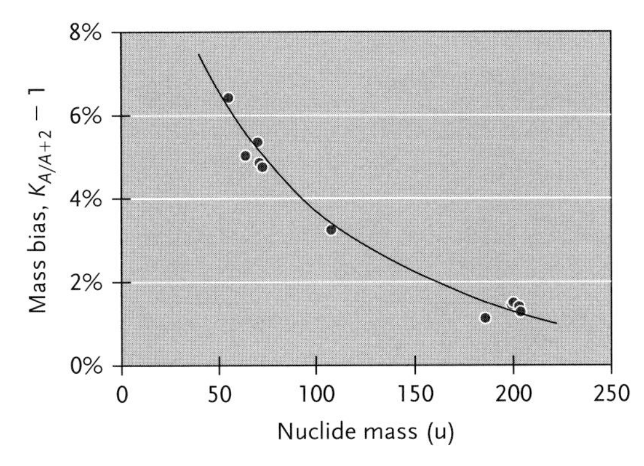

Figure 5.1 Typical bias in isotope amount ratio measurement results using a Neptune MC-ICP-MS instrument. Isotope ratios $N(^{A}E)/N(^{A+2}E)$ are considered here.

Thermal ionization mass spectrometry (TIMS) suffers from time-dependent mass bias, referred to as mass fractionation, as a result of the finite amount of sample on the source filament and the more efficient thermal ionization of the lighter isotope. Mass bias correction is more crucial with multi-collector (MC)-ICP-MS as the latter suffers significantly larger bias and, as noted earlier, it may not necessarily be constant over extended periods of time. Therefore, rigorous correction methods are required. Over the last few decades, several different mass bias correction methods have been successfully used for the determination of isotope amount ratios, as illustrated by Albarède *et al.* [16].

As a consequence of its high sample throughput, the use of MC-ICP-MS has resulted in a large number of publications. This deluge, however, has also created significant confusion and inconsistencies as to how isotope ratio data are evaluated [17]. Hence, the objective of this chapter is to assess the modern methodologies with a focus on mass bias correction methods. Some general rules for such measurements are suggested and discussed. In addition, accounting for all possible sources of uncertainty arising from the measurement process is of fundamental importance for producing an accurate value for the measurement results (see also Chapter 7). Calculation of combined uncertainty associated with the resulting ratio using a few commonly used mass bias correction models is therefore also outlined.

5.3 Systematics of Mass Bias Correction Models

Over the years, numerous methods have been proposed for calibrating isotope amount ratios [16–19]. At a glance, these methods can be formally classified using two key parameters that differ from method to method – whether or not data for the measurand and calibrant are acquired simultaneously, and whether or not the

Table 5.1 Approaches used to calibrate $N(^{87}Sr)/N(^{86}Sr)$ in mass spectrometry.

	Intra-elemental	Inter-elemental
Internal calibration	Sr–Sr internal correction	Zr–Sr, exponential model
	Double-spike ($^{88}Sr/^{84}Sr$)	Zr–Sr, regression model
External calibration	Gravimetric calibration	Zr/Sr bracketing

measurand and calibrant are isotope amount ratios of the same element. For the sake of clarity, we will resort to a case where the measurand is $N(^{87}Sr)/N(^{86}Sr)$. Although various calibration methods have evolved, often discriminated with the words *internal* or *external* being added to their name, these terms remain ambiguous. Here the terms *internal* and *external* are used synonymously with *simultaneous* and *sequential* analysis of the sample and calibrant and not to denote whether or not the measurand and calibrator are isotope ratios of the same element, such as $N(^{87}Sr)/N(^{88}Sr)$ and $N(^{86}Sr)/N(^{88}Sr)$, and $N(^{87}Sr)/N(^{88}Sr)$ and $N(^{91}Zr)/N(^{90}Zr)$, respectively. Four calibration strategies can be identified as frequently used (see Table 5.1).

5.3.1 External Gravimetric Calibration

The mass spectrometer is calibrated using synthetic mixtures of isotopes of high and known purity (^{87}Sr and ^{86}Sr). As the sample and calibrator runs are sequential, matrix matching and the stability of the instrument play significant roles in the veracity of the calibration [13, 20–22].

The ultimate question behind all calibration chains is how one can establish accurate isotopic composition of the enriched isotope materials with suitable precision. The measurements of the isotope amount ratio of the enriched materials, soon-to-be calibrators, initially cannot be calibrated. In a simplified example, the calibration can be done if the mass bias is identical when measuring both materials and their mixture [23]. Consider the silver system, $N(^{107}Ag)/N(^{109}Ag)$, and measurements of three isotope amount ratios: (i) of silver enriched in ^{107}Ag (r_A), (ii) of silver enriched in ^{109}Ag (r_B), and (iii) of the mixture of both of these materials (r_{mix}). If the chemical purity of both materials is known and the mass bias remains the same in all measurements, then the isotopic composition of both enriched materials can be obtained from the following three equations:

$$R_A = Kr_A,\ R_B = Kr_B,\ \text{and}\ R_{mix} = Kr_{mix} \tag{5.1}$$

Note that R_{mix} is a variable, $(n_{107,A} + n_{107,B})/(n_{109,A} + n_{109,B})$, that depends on K:

$$R_{mix} = \frac{x_{107,A}m_A/M_A + x_{107,B}m_B/M_B}{x_{109,A}m_A/M_A + x_{109,B}m_B/M_B} \tag{5.2}$$

where the relative isotopic abundance of ^{107}Ag in A is $x_{107,A} = Kr_A/(1 + Kr_A)$, that of ^{109}Ag is then $x_{109,A} = 1 - x_{107,A}$, and $M_A = m_{107}x_{107,A} + m_{109}x_{109,A}$, m being the sample mass and M the molar mass (and likewise for the other material). From these equations, the isotope amount ratio calibration factor is obtained:

$$K = (m_{109}/m_{107})\frac{m_A(r_{mix} - r_A) + m_B(r_{mix} - r_B)}{m_A r_B(r_A - r_{mix}) + m_B r_A(r_B - r_{mix})} \quad (5.3)$$

5.3.2 Internal Double-Spike Calibration

Calibration of the measured isotope amount ratio, $r_{87/86}$, is achieved by admixing known amounts of strontium that is enriched in ^{84}Sr and ^{88}Sr [24–27]. The isotopic composition of the "double-spike" strontium has to be known. Matrix matching does not have to be achieved *per se*. If the amount ratio of the non-radiogenic strontium isotopes, say $N(^{88}\text{Sr})/N(^{86}\text{Sr})$, is well known and does not vary significantly in Nature, it can be used to calibrate the measured ratio $r_{87/86}$ without any admixing of the double spike [28]. In both of its variations, this procedure involves the selection of an appropriate mass bias correction model, such as the exponential law (see below).

5.3.3 Internal Calibration (Inter-Element)

Arguably among the isotope ratio calibration strategies most frequently applied is the addition of a *foreign* element to the sample [29–31]. The calibration of the isotope amount ratio of the measurand is commonly achieved in three distinct ways. The exponential correction model is typically used with the assumption that both measurand and calibrant undergo the same mass discrimination. To avoid this assumption, the temporally correlated drift of both isotope amount ratio measurement results can be utilized. This is known as the regression model [32–35]. Alternatively, having possession of a sample of known isotope ratio of the measurand, this can be used to calibrate the isotope amount ratio in the foreign element [36]. This value then is in turn used to calibrate the isotope amount ratio of the measurand in the sample. Such an approach is similar to external gravimetric calibration with the exception that the foreign element serves as a proxy to eliminate the problems associated with sequential measurements.

5.3.4 External Bracketing Calibration (Inter-Element)

Albeit among the simplest mass bias correction models, this is the least favorable of all calibration strategies as it combines the shortcomings of all other techniques – the calibration is sequential and there is a need for a mass bias correction model and for matrix matching.

5.4
Logic of Conventional Correction Models

The need for a formal mass bias correction model arises when a specific isotope amount ratio is used to calibrate the amount ratio of *another* pair of isotopes, either of the same or of a different element, as for example when $N(^{88}Sr)/N(^{86}Sr)$ is used to calibrate $N(^{87}Sr)/N(^{86}Sr)$ or when $N(^{205}Tl)/N(^{203}Tl)$ is used to calibrate $N(^{202}Hg)/N(^{200}Hg)$. In cases where the measurand and calibrant are the same, there is no need to resort to mass bias correction models. In any situation, however, gravimetrically prepared synthetic mixtures of isotopes underpin the calibration.

As with TIMS, the core strength of MC-ICP-MS is its ability to register ion signals simultaneously. However, the traditional calibration approaches use *sequential* analysis of the measurand and the calibrant as a result of their temporally different introduction into the source. Hence a paradox of the choice: in contrast to sequential measurand–calibrator–measurand bracketing calibration, the calibrator can be admixed with the sample and both isotope amount ratios measured simultaneously. This, however, can only be done when the calibrator is a ratio of isotopes other than the measurand, which, in turn, requires a sound model for the mass bias transfer.

Mass bias in MC-ICP-MS varies with both time and nuclide mass. Since calibration is typically performed sequentially in a calibrator–measurand–calibrator bracketing mode, the effect of time has to be separated from that of mass in the mass bias correction model. Hence the logic of creating a mass bias discrimination model is to express the calibration factor, $K_{i/j}$, as a product of two functions: one that varies with time only and the other that varies with the nuclide masses only [16, 17, 19]. For example,

$$K_{i/j} = 1 + f(t)(m_i - m_j) \tag{5.4}$$

$$K_{i/j} = (m_i/m_j)^{f(t)} \tag{5.5}$$

or

$$\ln K_{i/j} = f(t)(m_i - m_j) \tag{5.6}$$

Clearly, there are many ways according to which one can express the calibration factor $K_{i/j}$ as functions of time and mass. Equation (5.4), for example, is commonly known as the linear correction law, Eq. (5.5) as Russell's law, and Eq. (5.6) as the exponential law. To add flexibility to the fitting function in the mass domain, it has been proposed that a discrimination exponent, n, be introduced into the mass bias correction factor [19]:

$$\ln K_{i/j} = f(t)(m_i^n - m_j^n) \tag{5.7}$$

When using any of these models, a calibrator with known isotope amount ratio is measured and from the difference between this known value and the corresponding measurement result, the value of $f(t)$ is obtained. This value then permits calculation of the isotope amount ratio correction factor ($K_{i/j}$) for the measurand.

The mass bias transfer with such models is attained using a ratio of these models for two sets of isotopes. Take, for example, Eq. (5.6):

$$\frac{\ln K_{i/j}}{\ln K_{k/l}} = \left[\frac{f_{i/j}(t)}{f_{k/l}(t)}\right]\left(\frac{m_i - m_j}{m_k - m_l}\right) \tag{5.8}$$

Since the logic of the model is such that the discrimination function is not nuclide mass dependent, rather it is time dependent only, the ratio $f_{i/j}(t)/f_{k/l}(t) = 1$. Hence

$$\frac{\ln K_{i/j}}{\ln K_{k/l}} = \frac{m_i - m_j}{m_k - m_l} \tag{5.9}$$

This can be rewritten as

$$K_{i/j}{}^{(m_k - m_l)} = K_{k/l}{}^{(m_i - m_j)} \tag{5.10}$$

The logic behind the last two expressions, especially the assumption that $f_{i/j}(t)/f_{k/l}(t) = 1$, is at the core of mass bias correction. Experimental tests of the latter, however, had to await the twenty-first century.

5.5 Pitfalls with Some Correction Models

5.5.1 Linear Law

The linear mass bias model, expressed as

$$R_{i/j} = r_{i/j}[1 + f(m_i - m_j)] \tag{5.11}$$

was first suggested by Dietz and others in 1962 for use with TIMS and soon became a popular tool in isotope amount ratio measurement science [37, 38]. This expression, albeit functional to a remarkably high degree of accuracy, is mathematically inconsistent for isotope ratios. In this vein, the widespread use of the linear correction model, in our opinion, illustrates the struggle in fundamental metrology of isotope amount ratio measurements. By definition,

$$R_{1/2} = R_{1/3}R_{3/2} \quad \text{and} \quad r_{1/2} = r_{1/3}r_{3/2} \tag{5.12}$$

Substituting this into the linear mass bias model, we obtain:

$$1 + f(m_1 - m_2) = [1 + f(m_1 - m_3)][1 + f(m_3 - m_2)] \tag{5.13}$$

which leads to

$$0 = f^2(m_1 - m_3)(m_3 - m_2) \tag{5.14}$$

This expression, however, can only be true if $f = 0$, which corresponds to a perfect (unbiased) measurement. The pitfall of the linear model illustrates that not all mathematical functions can be used to approximate the calibration factor K.

A coherent way of expressing a linear approximation of $K_{i/j}$ with respect to nuclide masses was put forward by Doherty *et al.* [39]:

$$R_{i/j} = r_{i/j} \frac{f + m_j}{f + m_i} \tag{5.15}$$

Note that this expression takes a form closely reminiscent of Padé approximants in mathematics. In reality, however, the poor performance of the linear law did not result from exposure of its underlying flaws; rather, it prompted the development of more complex, nonlinear, mass bias correction functions. In fact, the linear law was first *rationalized* as a first-order approximation of the power law [38]:

$$K_{i/j} = (1 + f)^{m_i - m_j} = 1 + f(m_i - m_j) + \cdots \tag{5.16}$$

5.5.2
Exponential Versus the Power Law

Consider the two most popular discrimination models: the exponential and the power law:

$$R_{i/j} = r_{i/j} e^{f_e(m_i - m_j)} \tag{5.17}$$

$$R_{i/j} = r_{i/j}(1 + f_p)^{m_i - m_j} \tag{5.18}$$

To date, these two laws are the most commonly used for high-precision isotopic analysis. However, these are not *two* different models, rather algebraic rearrangements of each other, much like expressions $y = e^{x \ln(a)}$ and $y = a^x$. Therefore, it is not at all surprising that publications aimed at statistical performance evaluation of these laws fail to report any difference between the two [18].

Russell's law [40], on the other hand, is *not* equivalent to the exponential mass bias correction model, although these two names are often used interchangeably. However, the difference in the correction factors for the two models is determined solely by the masses of all nuclides involved. Combining Eqs. (5.5) and (5.6) leads to the following:

$$\frac{\ln K^r_{i/j}}{\ln K^e_{i/j}} = \frac{(m_k - m_l)(\ln m_i - \ln m_j)}{(m_i - m_j)(\ln m_k - \ln m_l)} \tag{5.19}$$

where K^r and K^e are the correction factors according to the Russell and exponential laws, respectively.

5.6
Integrity of the Correction Models

5.6.1
Russell's Law

Non-conformity of the mass bias to Russell's law can be demonstrated in a simple experiment. From Russell's law [Eq. (5.5)], it follows that the slope of the linear regression between the logarithms of the measured isotope amount ratios is given by

$$b_{\mathrm{Hg/Tl}} = \frac{(f_{\mathrm{Hg}}/f_{\mathrm{Tl}})(m_{202} - m_{198})}{m_{205} - m_{203}} \tag{5.20}$$

From this, it is possible to obtain the ratio of both discrimination functions:

$$\frac{f_{\mathrm{Hg}}}{f_{\mathrm{Tl}}} = b_{\mathrm{Hg/Tl}} \frac{m_{205} - m_{203}}{m_{202} - m_{198}} \tag{5.21}$$

It is clear from Figure 5.2 that significant departures from the assumption $f_{\mathrm{Hg}} = f_{\mathrm{Tl}}$ are readily observed. Such conclusions have been drawn by many others in the past, but not always at a level of high statistical significance. In fact, such a conclusion is a logical extension of the observation that the magnitude of the mass bias increases as nuclide mass decreases (Figure 5.1).

It must be noted that the log-linear regression slope provides information about the value of $f_{\mathrm{Hg}}/f_{\mathrm{Tl}}$ in conjunction with the *assumed* discrimination law. Whether the discrimination law itself is appropriate needs more detailed consideration.

5.6.2 Discrimination Exponent

Inherent to all mass bias models stemming from Eq. (5.7) is the built-in variable (discrimination exponent) that distinguishes the various mass discrimination phenomena. Hence we have linear, exponential, equilibrium, power, and other discrimination models. This variable, in turn, is often used to identify the presence of a particular discrimination model [32, 41]. Consequently, considerable effort has been spent in extracting the numerical value of the mass bias discrimination

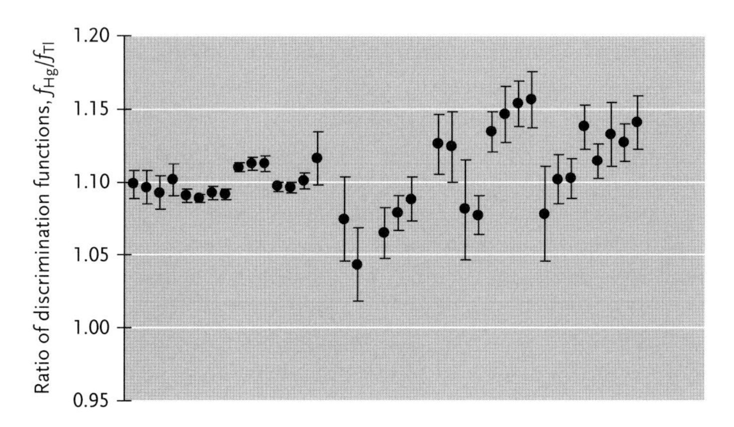

Figure 5.2 Departure from Russell's law of mass bias for isotope amount ratio measurement results using a Neptune MC-ICP-MS instrument. Isotope amount ratios $N(^{202}\mathrm{Hg})/N(^{198}\mathrm{Hg})$ and $N(^{205}\mathrm{Tl})/N(^{203}\mathrm{Tl})$ are considered here [Eq. (5.21)]. Agreement with Russell's law would result in $f_{\mathrm{Hg}}/f_{\mathrm{Tl}} = 1$. Uncertainties are quoted with the coverage factor $k = 1$.

exponent, n, from the experimentally determined uncorrected isotope amount ratios [42, 43]. Besides supposedly providing fundamental insights into the discrimination process, the value of n is deemed important as it affects the values of the mass bias-corrected isotope amount ratios.

To date, interrogation of the efficacy of mass bias correction models has largely resorted to attempts to determine the value of the discrimination exponent. In such experiments, the slope of the log-linear two-isotope ratio regression is used, which, in turn, leads to the discrimination exponent by solving the following expression [43–46]:

$$b = \frac{\ln K_{i/j}}{\ln K_{k/l}} = \left[\frac{f_{i/j}(t)}{f_{k/l}(t)}\right]\left(\frac{m_i^n - m_j^n}{m_k^n - m_l^n}\right) \tag{5.22}$$

Assuming $f_{i/j}(t) = f_{k/l}(t)$, one can solve for n. Iterative solutions are required to obtain n from b (and nuclide masses, m) since the above expression cannot be solved analytically. The fallacy of this approach is that one implicitly assumes that the ratio of discrimination functions is unity, that is, $f_{i/j} = f_{k/l}$ [17]. Without knowing this *a priori*, the discrimination exponent remains a dummy variable, the value of which cannot be independently evaluated by the currently proposed methods. To illustrate this, consider a system of $^{201}Hg/^{198}Hg$ and $^{202}Hg/^{200}Hg$. The experimentally obtained slope $b = 1.537(6)$ corresponds to $n = -3.2(4)$ if $f_1/f_2 = 1.00$ and $n = -0.5(4)$ if $f_1/f_2 = 1.02$. Therefore, even the slightest variations in the ratio of the discrimination functions greatly affect the perceived numerical value of n. Hence the "catch 22" situation: to estimate n one has to know f, but for that, one has to know n.

5.6.3
Discrimination Function

Consider a mass bias correction model, say $K_{i/j} = (m_i/m_j)^{f(t)}$ (Russell's law). According to the time–mass separation principle, $f(t)$ should be a function of time, not mass. Hence $f_{i/j} = f_{k/l}$. This statement can be evaluated experimentally. However, it is possible only when the aforementioned "catch 22" situation is circumvented because we have no knowledge regarding the value of the discrimination exponent. One way of doing so is to take advantage of the fact that for certain sets of isotopes, the value of the reduced mass function, $(m_i^n - m_j^n)/(m_k^n - m_l^n)$, does not depend on the value of n, or the dependence is extremely weak [17]. For mercury isotopes, for example, several sets of isotopes exist that fulfill this criterion (Figure 5.3). Among these is the set $^{198}Hg/^{201}Hg$ and $^{199}Hg/^{200}Hg$ for which $(m_{198}^n - m_{201}^n)/(m_{199}^n - m_{200}^n) = 3.004$ within a few parts per 10 000 ($-4 < n < +4$). For such a set of isotopes, therefore, it is possible to evaluate f with disregard to n. The results are shown in Figure 5.4, wherein significant departures from the assumption $f_{i/j} = f_{k/l}$ are apparent.

The above experimental evidence contradicts the mathematical framework used for the *general* mass bias correction laws. This, of course, does not mean that these

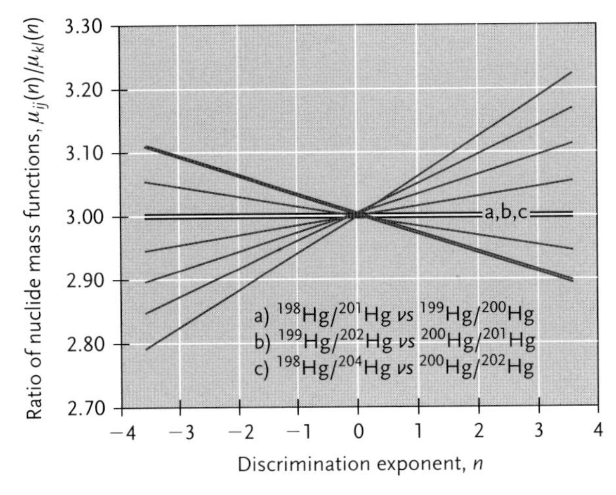

Figure 5.3 Ratio of nuclide mass functions ($\mu_{ij} = m_i^n - m_j^n$) as a function of the discrimination exponent (n) for selected isotopes of mercury. Adapted from [17].

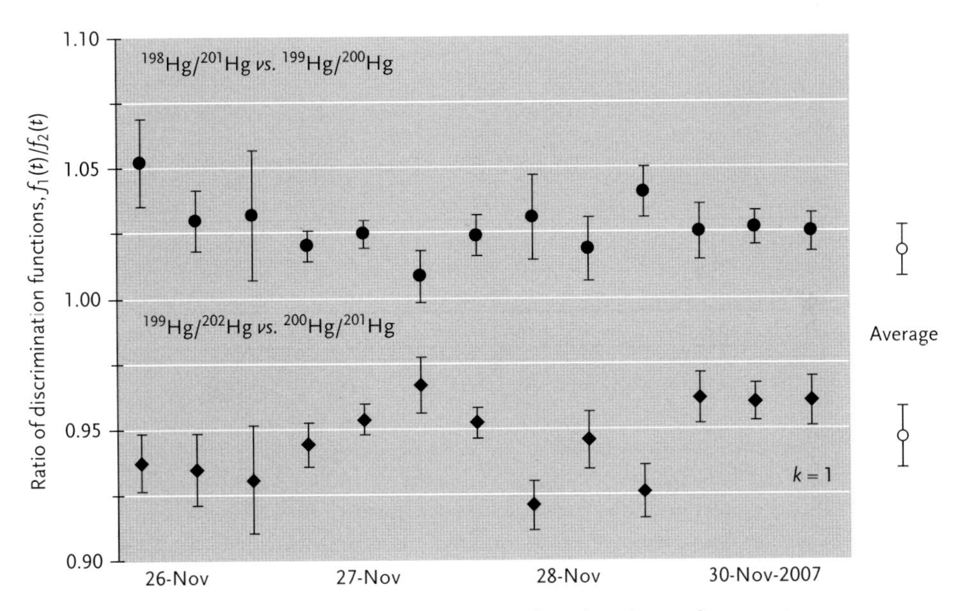

Figure 5.4 Ratio of mass bias discrimination functions for selected sets of mercury isotopes (Figure 5.3a,b). Adapted from [17].

correction laws are without use. Rather, it shows that the time–mass variable separation cannot be performed using the proposed equations. Considering the identified sources of bias, that is, the bias of the model, this must be acknowledged in the uncertainty of the reported results.

5.6.4
Second-Order Terms

A common explanation for (slight) deviations from the exponential correction law is that second-order terms have been neglected. Such reasoning is flawed. To demonstrate this, we trace below the reasoning that leads to the above conclusion.

The ion transmission efficiency for the nuclide of mass m_i is the ratio of the measured to true amount of that nuclide, $K_i = N_i^{\mathrm{meas}}/N_i^{\mathrm{true}}$. Hence the isotope ratio can be written as follows, based on the nuclide transmissions:

$$\frac{K_i}{K_j} = \frac{N_i^{\mathrm{meas}}/N_i^{\mathrm{true}}}{N_j^{\mathrm{meas}}/N_j^{\mathrm{true}}} = \frac{N_i^{\mathrm{meas}}/N_j^{\mathrm{meas}}}{N_i^{\mathrm{true}}/N_j^{\mathrm{true}}} \tag{5.23}$$

or

$$\frac{K_i}{K_j} = K_{i/j} = \frac{r_{i/j}}{R_{i/j}} \tag{5.24}$$

Although the true functional relation between K_i and K_j is unknown, it can be approximated using Taylor's theorem, which asserts that any sufficiently smooth function can locally be approximated by polynomials. This can be achieved by expanding $\ln K$ as a function of m^n, where n is an arbitrary discrimination exponent [19]:

$$\ln K_j \approx \ln K_j + \left[\frac{\mathrm{d} \ln K_j}{\mathrm{d}(m^n)}\right](m_i^n - m_j^n) + \left[\frac{\mathrm{d}^2 \ln K_j}{\mathrm{d}(m^n)^2}\right](m_i^n - m_j^n)^2 + \cdots \tag{5.25}$$

In the conventional form of the generalized power law, the first derivative of the transmission function (evaluated at m_j) is denoted as the discrimination function $f_{i/j}$ [19]. Hence trimming the Taylor expansion to the first degree gives the familiar generalized power law:

$$\ln K_{i/j} = \ln K_j + f_{i/j}(m_i^n - m_j^n) \tag{5.26}$$

If the discrimination function $f_{i/j}$ is mass independent, as postulated in the generalized power law, any higher order derivatives of $f_{i/j}$ with respect to m are zero. Therefore, deviations from the first-order Taylor expansion, that is, Eq. (5.26), indicate the mass dependence of the discrimination function.

5.7
The Regression Model

The ability to measure simultaneously the isotopes of two different elements with high precision has never been a particular challenge in mass spectrometry. The challenge has always been to ensure the accuracy of the mass bias transfer. Hence a fundamental limitation of the traditional mass bias correction models is their reliance on the equality between the discrimination functions of the calibrant and measurand and, to a much lesser extent, the value of the discrimination exponent. Both of these limitations can be obviated by taking advantage of the significant temporal drift of the mass bias when using MC-ICP-MS. Consider two isotopes of measurand A (^{i}A, ^{j}A) and two isotopes of the calibrant element B (^{k}B, ^{l}B):

$$R_{i/j} = K_{i/j} r_{i/j} \tag{5.27}$$

$$R_{k/l} = K_{k/l} r_{k/l} \tag{5.28}$$

Note that the above expressions are not merely definitions, as they also implicate the linearity in the isotope ratio response. In other words, the correction factors (K) do not depend on the magnitude of the isotope amount ratio (r). In the logarithmic scale, these two expressions become

$$\ln R_{i/j} = \ln K_{i/j} + \ln r_{i/j} \tag{5.29}$$

$$\ln R_{k/l} = \ln K_{k/l} + \ln r_{k/l} \tag{5.30}$$

These two expressions, when combined, become

$$\frac{\ln K_{i/j}}{\ln K_{k/l}} = \frac{\ln R_{i/j} - \ln r_{i/j}}{\ln R_{k/l} - \ln r_{k/l}} \tag{5.31}$$

From this, variable rearrangement leads to the following:

$$\ln r_{i/j} = \ln R_{i/j} - \left(\frac{\ln K_{i/j}}{\ln K_{k/l}}\right) \ln R_{k/l} + \left(\frac{\ln K_{i/j}}{\ln K_{k/l}}\right) \ln r_{k/l} \tag{5.32}$$

This expression forms a log-linear regression between the measured isotope amount ratios of the measurand and the calibrant. The intercept (a) and the slope (b) of this regression are

$$a - \ln R_{i/j} - \left(\frac{\ln K_{i/j}}{\ln K_{k/l}}\right) \ln R_{k/l} \tag{5.33}$$

$$b = \frac{\ln K_{i/j}}{\ln K_{k/l}} \tag{5.34}$$

These two equations can be combined:

$$a = \ln R_{i/j} - b \ln R_{k/l} \tag{5.35}$$

Since the values of the intercept (a) and slope (b) are known from best-fitting a straight line as provided by the least-squares method and the value of $R_{k/l}$ is known before the experiment, the isotope amount ratio of the measurand can be calculated [32]:

$$\ln R_{i/j} = a + b \ln R_{k/l} \tag{5.36}$$

which is equivalent to the following more elegant solution [17, 35]:

$$R_{i/j} = \mathrm{e}^{a} R_{k/l}^{b} \tag{5.37}$$

This is the model equation for the calibration of isotope amount ratios based on the log-linear temporal isotope amount ratio regression. Note that a and b are perfectly correlated ($\rho = +1$) if $R_{k/l} < 1$ ($\ln R_{k/l} < 0$) and perfectly anti-correlated ($\rho = -1$) if $R_{k/l} > 1$ ($\ln R_{k/l} > 0$). It is important to stress that this calibration method is fundamentally different from the conventional mass bias correction laws. Since the regression model does not invoke the principle of time–mass separation, it does not need either the discrimination exponent or the equality of the discrimination functions [17].

What is more important, this calibration model is *not* derived from either the exponential or Russell's laws as is commonly perceived (and originally presented) [15, 32]. Rather, it only requires the mass spectrometer response be linear [Eqs. (5.27) and (5.28)]. It is the *interpretation* of the slope and the intercept that can lead to the reliance on the exponential mass bias correction or even erroneous results. Consider, for example, the substitution of Eq. (5.34) in Eq. (5.32):

$$\ln r_{i/j} = \ln R_{i/j} - b \ln R_{k/l} + b \ln r_{k/l} \tag{5.38}$$

from which

$$R_{i/j} = r_{i/j} \left(R_{k/l} / r_{k/l} \right)^{b} \tag{5.39}$$

This expression has been used as the model equation for the regression model in the past. As pointed out by Baxter *et al.*[36], however, this is equivalent to assuming $f_{i/j} = f_{k/l}$. The pitfall of this approach is that the resultant value of the regression intercept is not used, rather its value is *assumed* from Eq. (5.35). To avoid the errors of Eq. (5.39), Baxter *et al.* proposed a revised regression model that combines two regressions: one for the reference material of the analyte and internal standard, and the other that combines the sample and the internal standard [36]. Reliance on the reference material for the measurand, however, is unnecessary, provided that the isotope amount ratio of the internal standard is known [Eq. (5.39)]. However, it is a viable option nevertheless.

Although this mass bias correction method is superior to all other currently known models, it demands large investments in measurement time and sample size. This cannot always be fulfilled in practice; hence there is a need for alternative calibration models, even at the cost of performance. Moreover, theoretical limitations of this calibration model are yet to be comprehended.

5.8 Calibration with Double Spikes

Isotope amount ratio calibration with double spikes involves measuring the relative amounts of at least four isotopes of an element (there are over 30 elements fulfilling the condition of availability of ≥ 4 isotopes). For two of these isotopes, the relative isotopic abundance is significantly enhanced by the addition of an enriched isotopic spike to the sample. Such an approach circumvents the bias incurred due to the mass bias calibration between different elements. The double-spike method was developed by Dietz *et al.* in 1962 for the determination of uranium isotope amount ratios [47]. Russell [24], Dodson [48], and Cumming [49] (1970–1973) later presented the mathematical framework and a more in-depth uncertainty propagation treatment for optimization of the amount of enriched isotope in the spike for both isotopes considered.

Like the traditional mass bias correction approaches, the double-spike method also relies on the *choice* of the mass bias model. The original formalism of the double spikes employed the linear mass bias law and, although double-spike calibration equations adapted for the exponential mass bias discrimination are available, linear models are still often used owing to their simplicity (see, for example, [50–52]). The caveat here is that erroneous results can be obtained when a linear correction is applied to data that do not follow such behavior. This is illustrated below.

Calibration using the double-spike method is an ingenious application of isotope dilution in mass spectrometry. Admixing the measurand and the spike (calibrant) results in a mixture, the isotopic composition of which is governed by the conservation of matter. Two sets of equations can be established: one that describes the conservation of matter during the mixing, and the other that expresses the relationship between the measured and true isotope amount ratios, that is, the discrimination law:

$$R_{i/k,\mathrm{m}} = qR_{i/k,\mathrm{s}} + (1-q)R_{i/k,\mathrm{a}} \quad \text{(mixing)} \tag{5.40}$$

$$R_{i/k} = r_{i/k} f(i/k) \quad \text{(discrimination)} \tag{5.41}$$

Note that the isotopic composition for the mixture (m) of measurand (analyte, a) and spike (s) can be expressed as a linear combination of the two. However, the mixing coefficient, q, is not the amount ratio of the analyte and spike (x); rather, these two variables are linked by the following expression [27]:

$$\frac{1-q}{q} = \left(\frac{1-x}{x}\right)\left(\frac{\Sigma R_{i/k,\mathrm{a}}}{\Sigma R_{i/k,\mathrm{s}}}\right) \tag{5.42}$$

Combination of the mixing and discrimination expressions leads to the following:

$$f_\mathrm{m} r_{i/k,\mathrm{m}} - qR_{i/k,\mathrm{s}} - (1-q)f_\mathrm{a} r_{i/k,\mathrm{a}} = 0 \tag{5.43}$$

If the discrimination law is known or assumed, then we have three unknowns: q and one variable for each of the two discrimination functions. Assume, for example, Russell's law:

$$\left(\frac{m_i}{m_k}\right)^{f_\mathrm{m}} r_{i/k,\mathrm{m}} - qR_{i/k,\mathrm{s}} - (1-q)\left(\frac{m_i}{m_k}\right)^{f_\mathrm{a}} r_{i/k,\mathrm{a}} = 0 \tag{5.44}$$

The set of three nonlinear equations has to be solved for f_a, f_m, and q. This can be done numerically (iteratively). A more common and significantly simplified approach for solving the above set of nonlinear equations is to use the first-order Taylor approximation of the nonlinear variable $(m_i/m_k)^{f_a}$, that is [27]:

$$\left(\frac{m_i}{m_k}\right)^{f_a} \approx 1 + f_a \ln\left(\frac{m_i}{m_k}\right) \tag{5.45}$$

With this, the above nonlinear equation becomes linear with respect to all three unknown variables:

$$\left[1 + f_m \ln\left(\frac{m_i}{m_k}\right)\right] r_{i/k,m} - qR_{i/k,s} - (1-q)\left[1 + f_a \ln\left(\frac{m_i}{m_k}\right)\right] r_{i/k,a} = 0 \tag{5.46}$$

or

$$q(r_{i/k,a} - R_{i/k,s}) + f_m \ln\left(\frac{m_i}{m_k}\right) r_{i/k,m} + (q-1) f_a \ln\left(\frac{m_i}{m_k}\right) r_{i/k,a} = r_{i/k,a} - r_{i/k,m} \tag{5.47}$$

This equation can be expressed using matrices, as F a = Y:

$$\left[(r_{i/k,a} - R_{i/k,s}); \ln\left(\frac{m_i}{m_k}\right) r_{i/k,m}; \ln\left(\frac{m_i}{m_k}\right) r_{i/k,a}\right] [q; f_m; (q-1) f_a]^T = [(r_{i/k,a} - r_{i/k,m})] \tag{5.48}$$

Solution of this equation, that is, the matrix $\boldsymbol{a} = [q;\ f_m;\ (q-1)f_a]$, is obtained using common matrix algebra:

$$a = F^{-1} Y \tag{5.49}$$

Although the mathematics of double-spike calibration is not new, it is not trivial. Owing to the complexity of nonlinear solutions, most opt for an approximate (linear) solution, which, in turn, may come at the expense of biased results.

In addition to the double-spike technique, an extension to triple spikes has been proposed by Galer [50]. However, as noted by Rudge *et al.* [27], no advantage has been found yet to justify the triple-spike calibration because the double-spike method always delivers smaller uncertainties than triple-spike calibration.

Practical disadvantages associated with use of the double-spike technique include the required availability of the high-purity enriched double spikes, the effort required to calibrate the isotopic composition of the spike, the need to avoid possible cross-contamination between the analysis of unspiked and spiked samples, the requirement for at least four interference-free isotopes of the analyte, and the need for two analyses of the sample – unspiked and spiked.

A major advantage of the double-spike technique is that the mass bias correction factor can be directly determined for each sample, thus eliminating the bias due to variations in the sample matrix. In this regard, double-spike calibration is an isotopic extension of the classical method of standard additions. This advantage has resulted in the rapid adoption of this technique for many elements other than lead, such as calcium, iron, chromium, germanium, selenium, cadmium, molybdenum, mercury, and uranium.

5.8.1 Caveat of the Model Choice

As mentioned earlier, double-spike correction relies on the choice of the mass bias correction model. This is arguably the theoretical weak spot of the method. Can this *choice*, however, lead to significant errors? It has been argued that such a question is moot as it has no impact on the corrected results given the small extent of fractionation observed in TIMS. MC-ICP-MS, however, experiences significantly larger mass bias; therefore, this question cannot be entirely dismissed. Consider this "textbook" analysis of analytical chemistry: the isotopic composition of natural lead is determined using the double-spike method based on the NIST equal-atom lead reference material SRM 982. To obtain the results, classical double-spike equations are used that rely on the linear discrimination law. Let us further have a mass spectrometer that follows Russell's law with bias of no more than 0.5% per atomic mass unit u. The sample and spike are both mixed in a equimolar ratio. It can be demonstrated that the results of such analysis will be significantly biased. Errors of up to 0.5% can be easily procured for the ratio of $N(^{206}\text{Pb})/N(^{208}\text{Pb})$, as illustrated in Table 5.2. For this very reason, SRM 982 has been deemed to be not a

Table 5.2 A *gedanken* experiment for double-spike calibration of natural lead using NIST equal-atom lead reference material SRM 982. (a) True isotope amount ratios in the sample ("natural" lead), calibrator (NIST SRM 982), and their equimolar mixture. (b) "Measured" isotope amount ratios subjected to Russell's law of less than 1% bias for the ratio $N(^{207}\text{Pb})/N(^{208}\text{Pb})$. (c) Estimated isotope amount ratios of natural lead using the traditional double-spike equations (assumed linear discrimination), showing a 0.5% bias in $N(^{207}\text{Pb})/N(^{208}\text{Pb})$.

A	Spike	1:1 mixture	Sample
(a) True isotope amount ratios, $N(^{A}\text{Pb})/N(^{208}\text{Pb})$			
204	0.0272	0.0281	0.0288
206	0.9998	0.6959	0.4621
207	0.4670	0.4446	0.4274
208	1.0000	1.0000	1.0000
(b) Measured isotope ratios, subjected to Russell's discrimination law			
		0.0292	0.0294
		0.7095	0.4671
		0.4489	0.4297
		1.0000	1.0000
(c) Lead isotope ratios in the sample calculated from the double-spike calibrations fit for linear discrimination law			
			0.0291
			0.4644
			0.4284
			1.0000

particularly suitable choice of double spike. This also illustrates the need for optimal double-spike design [27]. The purpose of this tutorial example, however, is to show how easy it is for the analytical results to be corrupted.

Note that even if we are to adopt a more fitting discrimination law for the double-spike calculus, we still have to accept the equality of the discrimination functions between the isotopes of the same element – an assumption that is now shown experimentally to be incorrect [17]. To date, this remains the breaking point in the chain of metrological traceability for double-spike calibration results.

5.9 Calibration with Internal Correction

5.9.1 Intra-Elemental Correction

Internal standardization refers to a variation of the double-spike method whereby the isotope amount ratio of an element is calibrated using a pair of other isotopes of the same element, the corresponding isotope ratio of which is believed to be invariant in Nature. In this method, both the sample and spiked sample are physically identical entities. Owing to its simplicity, this method has been widely applied to strontium and neodymium isotope analysis [53]. For example, the value of $N(^{88}\mathrm{Sr})/N(^{86}\mathrm{Sr})$ is assumed to be constant in Nature and is commonly used for the calibration of $N(^{87}\mathrm{Sr})/N(^{86}\mathrm{Sr})$. Fietzke and Eisenhauer, however, have reported $\delta(^{88}\mathrm{Sr}/^{86}\mathrm{Sr}) = +0.38(1)_{k=2}$ per mil for the IAPSO standard seawater relative to the NIST SRM 987 [54]. Likewise, Halicz *et al.* reported $\delta(^{88}\mathrm{Sr}/^{86}\mathrm{Sr}) = +0.28(14)_{k=2}$ per mil in marine sediments [55] and Yang *et al.* obtained $\delta(^{88}\mathrm{Sr}/^{86}\mathrm{Sr}) = +0.21(2)_{k=2}$ per mil in a fish liver tissue [56]. These observations suggest that the value of $N(^{88}\mathrm{Sr})/N(^{86}\mathrm{Sr})$ is not invariant in Nature and therefore the conventional isotope amount ratio value of $N(^{88}\mathrm{Sr})/N(^{86}\mathrm{Sr})$ should not be used in high-precision isotope amount ratio measurements.

Similarly, this raises questions regarding the determination of isotope amount ratios of neodymium or osmium – is it adequate to use the ratio for nonradiogenic isotopes for the mass bias correction of other isotope ratios? Are they indeed invariant in Nature? The 2009 IUPAC report on Atomic Weights addressed this question rather explicitly by abrogating the conventional notation of Standard Atomic Weights in favor of the interval of the Standard Atomic Weights [57]. Furthermore, concerning the mass bias correction, it is now established that mass bias correction factors are not identical for all isotopes of the same element. Hence the limitations of traditional mass bias correction methods must be accepted.

5.9.2 Inter-Elemental Correction

A major obstacle in implementing double-spike calibration is the difficulty in obtaining an appropriate isotopically enriched calibrant (in addition to the fact that many elements do not have the required four stable isotopes). The double spike can be substituted with another element, for which an isotope amount ratio is calibrated using traditional methods. This value can then be used to obtain mass bias corrected ratios for the measurand. As demonstrated by Weiss *et al.* [58], the NIST SRM 981 lead standard admixed with thallium was used to calibrate $N(^{205}\mathrm{Tl})/N(^{203}\mathrm{Tl})$, which was then successfully used for mass bias correction of Pb isotope amount ratios in samples. Similarly, strontium reference material (NIST SRM 987) was used to calibrate the isotope amount ratio of zirconium, which was then applied for mass bias correction of the isotope amount ratio of strontium in fish tissue samples [56].

The advantage of this approach for mass bias correction is that any temporal changes in the instrumental mass bias during the measurement process are accounted for. However, matrix-induced mass bias cannot be fully compensated with this type of mass bias correction and the corrected isotope amount ratios can show a dependence on the sample concentration [56]. Hence a near-perfect matrix separation, and close matching of analyte and calibrant concentrations are required to yield accurate isotope amount ratios.

Note that the calibrator element in this method serves as the mass bias correction proxy. Hence, for example, even though the value obtained for the Zr isotope ratio may itself be biased due to the limitations of the mass bias correction models employed (e.g., assumption $f_{\mathrm{Sr}} = f_{\mathrm{Zr}}$), this bias is largely negated in the second step of the calibration (Zr $\rightarrow$ Sr). Owing to error cancellation with this method, it is akin to the use of isotope dilution methods wherein reverse and direct isotope dilution protocols are performed in tandem. The error cancellation, however, occurs only when matrix separation and concentration matching are fully attained.

Related to the above-described inter-elemental correction method, the need for a proxy element can be obviated at the expense of resorting to *sequential* analysis of the sample and the calibrator, that is, with external gravimetric calibration. Although this technique is not favored by all, it has played a major role, for example, in the recent re-evaluation of the atomic weight of zinc [13].

5.10 Uncertainty Evaluation

An accurate uncertainty statement of the measurement result is arguably as important as the accuracy of the measurement result itself. However, since the process of isotope amount ratio measurement is far from trivial, so is also the evaluation of the uncertainty. Consider the following example as an illustration of current problems in conceptual understanding of the uncertainty evaluation of the measurement results.

Exponential mass bias correction of $N(^{202}\mathrm{Hg})/N(^{200}\mathrm{Hg})$ is performed using the NIST SRM 997 isotopic thallium standard with $N(^{205}\mathrm{Tl})/N(^{203}\mathrm{Tl}) = 2.38714$ $(102)_{k=2}$. The model equation of this measurement is

$$K_{202/200} \approx K_{205/203}^{(202-200)/(205-203)} \approx K_{205/203} \tag{5.50}$$

If we assume that the measurement is perfect, then the only variable contributing to the uncertainty of the measurement result is the uncertainty in the calibrator, that is, $R_{205/203}$, and the lowest relative uncertainty for $R_{202/200}$ (or $R_{200/198}$) therefore is 0.00102/2.38714 = 0.00043. This corresponds to an uncertainty of 0.00055 for $R_{202/200} = 1.29$ (0.00043×1.29) [35]. One can find reports, however, that claim uncertainties that are an order of magnitude lower than this. For example, Blum and Bergquist reported 0.00010 for $N(^{200}Hg)/N(^{198}Hg)$ [30], Xie *et al.* reported 0.00015 for $N(^{202}Hg)/N(^{200}Hg)$ [59], and Hintelmann and Lu reported 0.00020 for $N(^{202}Hg)/N(^{200}Hg)$ [60]. Such a practice of excluding the uncertainty of the calibrator is, for example, idiosyncratic to the geochemical community. It is important that all known uncertainty sources are included in the uncertainty statements. Comparable measurement results otherwise become a fleeting goal.

Given the function $y = f(x_1, \ldots, x_n)$, the combined uncertainty of a measurement result (y) can be evaluated from the standard uncertainties of all dependent variables $x_1, \ldots, x_n$. This is commonly done using the first-order Taylor expansion of the function $y = f(x_1, \ldots, x_n)$:

$$u_c(y) = JSJ^T \tag{5.51}$$

where J is the partial derivative vector (Jacobian) and S is the $n \times n$ covariance matrix of the dependent variables $x_1, \ldots, x_n$. The above expression is known as the law of uncertainty propagation [61]. If variables $x_1, \ldots, x_n$ are uncorrelated, the combined standard uncertainty of y can be approximated as

$$u_c(y) = (\partial y/\partial x_1)^2 u(x_1)^2 + \cdots + (\partial y/\partial x_n)^2 u(x_n)^2 \tag{5.52}$$

This generic expression, however, can lead to many errors since measurement results are, in fact, correlated in mass spectrometry [62]. As we shall see in the following section, ignoring isotope ratio correlations may lead to significant errors in the conclusions. Although the details of the uncertainty estimation are beyond the scope of this chapter, we present here an outline of calculation of combined standard uncertainty.

5.10.1 Uncertainty Modeling and the Double Spikes

Uncertainty evaluation is often perceived as a passive (*a posteriori*) part of the analysis process. The double-spike method of isotope ratio calibration, however, has demonstrated that uncertainty evaluation is an active research tool and the results obtained this way are significant in steering the entire field of mass bias correction.

It is known that the random error magnification in isotope dilution depends largely on the composition and the amount of the admixed isotopic spike (see also

Chapter 8). The same is true for the results from the double-spike method. Finding the optimal isotopic composition of the spike, however, has proved to be a challenge.

Both Galer [50] and Johnson and Beard [52] found that isotope amount ratio measurement uncertainty depends on which isotope is selected in the denominator. For example, it is argued that lower uncertainty is found in ^{206}Pb-denominator space than in ^{204}Pb-denominator space, and likewise that ^{57}Fe as the denominator isotope is superior to ^{54}Fe. Such a conclusion has prompted a preference for ^{206}Pb-enriched spikes over ^{204}Pb, among other things. If the above is true, then measuring $N(^{204}\text{Pb})/N(^{206}\text{Pb})$ must be superior to measuring $N(^{206}\text{Pb})/N(^{204}\text{Pb})$, clearly a misguided conclusion. This also implies that the isotope amount ratio covariance matrix is asymmetric–a conclusion that contradicts the associative property of real numbers under multiplication, that is, that $2 \times 3 \neq 3 \times 2$. It can be shown that this type of conclusion, that is, that the uncertainty of isotope amount ratios depends on the denominator isotope, is a result of ignoring the correlations between the isotopic signals and their ratios [27].

Virtually all published uncertainty evaluations of mass spectrometric methods assume independent ion currents or isotope amount ratios for all simultaneously measured isotopes [62, 63]. It is unclear why this paradigm has received such widespread acceptance. Isotopic signals in mass spectrometry are *always* correlated, provided that they are well above the background noise. The method of internal standardization is made possible precisely because mass spectrometric signals *are* correlated.

5.11 Conclusion

Metrology in mass spectrometry is a fairly new field and the principles presented in this chapter demonstrate that it plays an important role in assessing the quality of isotope amount ratio measurement results. Although all conventional mass bias correction procedures have their advantages and disadvantages, it is important to understand the fundamental assumptions associated with each model. Uncertainty evaluation of isotope amount ratio measurement results is undoubtedly in an *ad hoc* state-of-the-art. In other words, the often-claimed uncertainties of the isotope amount ratios in the order of parts per million are not full uncertainties as understood in ISO's *Guide to the Expression of Uncertainty in Measurement* (GUM) [64]. Terms and practices need to be revised, among other things. For example, uncertainty due to calibration must be included for meaningful comparison of results, and terms such as "internal" or "external" precision are urged for abrogation. As for the proper choice of mass bias correction, the major obstacle towards consistent and realistic isotope data lies with the fact that analysts are not fully acquainted with the nuances and assumptions of the mass bias correction models.

References

1 Richards, T.W. (1912) Atomic weights. *J. Am. Chem. Soc.*, **34**, 959–971.

2 Meija, J. (2010) Beryllium valence challenge. *Anal. Bioanal. Chem.*, **396**, 185–186.

3 Prout, W. (1815) On the relation between the specific gravities of bodies in their gaseous state and the weights of their atoms. *Ann. Philos.*, **6**, 321–330.

4 Budzikiewicz, H. and Grigsby, R.D. (2006) Mass spectrometry and isotopes: a century of research and discussion. *Mass Spectrom. Rev.*, **25**, 146–157.

5 Aston, F.W. (1966) Mass spectra and isotopes, in *Nobel Lectures, Chemistry*, Elsevier, Amsterdam, pp. 1922–1941.

6 Grayson, M.A. (ed.) (2002) *Measuring Mass: from Positive Rays to Proteins*, Chemical Heritage Foundation, Philadelphia, PA.

7 Soddy, F. (1932) *The Interpretation of the Atom*, John Murray, London.

8 Patterson, C., Tilton, G., and Inghram, M. (1955) Age of the Earth. *Science*, **21**, 69–75.

9 Flowers, J. (2004) The route to atomic and quantum standards. *Science*, **306**, 1324–1330.

10 de Laeter, J.R. and Hidaka, H. (2007) The role of mass spectrometry to study the Oklo–Bangombé natural reactors. *Mass Spectrom. Rev.*, **26**, 683–712.

11 Rosman, K.J.R. and Taylor, P.D.P. (1998) Isotopic compositions of the elements 1997. *Pure Appl. Chem.*, **70**, 217–235.

12 Millot, R., Guerrot, C., and Vigier, N. (2004) Accurate and high-precision measurement of lithium isotopes in two reference materials by MC-ICP-MS. *Geostand. Geoanal. Res.*, **28**, 153–159.

13 Ponzevera, E., Quetel, C.R., Berglund, M., Taylor, P.D.P., Evans, P., Loss, R.D., and Fortunato, G. (2006) Mass discrimination during MC-ICPMS isotopic ratio measurements: investigation by means of synthetic isotopic mixtures (IRMM-007 series) and application to the calibration of natural-like zinc materials (including IRMM-3702 and IRMM-651). *J. Am. Soc. Mass. Spectrom.*, **17**, 1413–1428.

14 Andren, H., Rodushkin, I., Stenberg, A., Malinovsky, D., and Baxter, D.C. (2004) Sources of mass bias and isotope ratio variation in multicollector ICP-MS: optimization of instrumental parameters based on experimental observations. *J. Anal. At. Spectrom.*, **19**, 1217–1224.

15 Yang, L. (2009) Accurate and precise determination of isotopic ratios by MC-ICP-MS: a review. *Mass Spectrom. Rev.*, **28**, 990–1011.

16 Albarède, F., Télouk, P., Blichert-Toft, J., Boyet, M., Agranier, A., and Nelson, B. (2004) Precise and accurate isotopic measurements using multiple-collector ICPMS. *Geochim. Cosmochim. Acta*, **68**, 2725–2744.

17 Meija, J., Yang, L., Sturgeon, R., and Mester, Z. (2009) Mass bias fractionation laws for multi-collector ICPMS: assumptions and their experimental verification. *Anal. Chem.*, **81**, 6774–6778.

18 Yang, L. and Sturgeon, R.E. (2003) Comparison of mass bias correction models for the examination of isotopic composition of mercury using sector field ICP-MS. *J. Anal. At. Spectrom.*, **18**, 1452–1457.

19 Kehm, K., Hauri, E.H., Alexander, C.M. O.D., and Carlson, R.W. (2003) High precision iron isotope measurements of meteoritic material by cold plasma ICP-MS. *Geochim. Cosmochim. Acta*, **67**, 2879–2891.

20 Pritzkow, W., Wunderli, S., Vogl, J., and Fortunato, G. (2007) The isotope abundances and the atomic weight of cadmium by a metrological approach. *Int. J. Mass Spectrom.*, **261**, 74–85.

21 Zhao, M., Zhou, T., Wang, J., Lu, H., and Xiang, F. (2005) Absolute measurements of neodymium isotopic abundances and atomic weight by MC-ICPMS. *Int. J. Mass Spectrom.*, **245**, 36–40.

22 Zhou, T., Zhao, M., Wang, J., and Lu, H. (2008) Absolute measurements and

certified reference material for iron isotopes using multiple-collector inductively coupled mass spectrometry. *Rapid Commun. Mass Spectrom.*, **22**, 717–720.

23 Qi, H.P., Berglund, M., Taylor, P.D.P., Hendrickx, F., Verbruggen, A., and de Bievre, P. (1998) Preparation and characterisation of synthetic mixtures of lithium isotopes. *Fresenius' J. Anal. Chem.*, **361**, 767–773.

24 Russell, R.D. (1971) The systematics of double spiking. *J. Geophys. Res.*, **76**, 4949–4953.

25 Woodhead, J.D. and Hergt, J.M. (1997) Application of the "double spike" technique to Pb-isotope geochronology. *Chem. Geol.*, **138**, 311–321.

26 Compston, W. and Obersby, V.M. (1969) Lead isotopic analysis using a double spike. *J. Geophys. Res.*, **74**, 4338–4348.

27 Rudge, J.F., Reynolds, B.C., and Bourdon, B. (2009) The double spike toolbox. *Chem. Geol.*, **265**, 420–431.

28 Boelrijk, N.A.I.M. (1968) A general formula for "double" isotope dilution analysis. *Chem. Geol.*, **3**, 323–325.

29 Longerich, H.P., Fryer, B.J., and Strong, D.F. (1987) Determination of lead isotope ratios by inductively coupled plasma-mass spectrometry (ICP-MS). *Spectrochim. Acta B*, **42**, 39–48.

30 Blum, J.D. and Bergquist, B.A. (2007) Reporting of variations in the natural isotopic composition of mercury. *Anal. Bioanal. Chem.*, **388**, 353–359.

31 Gallon, C., Aggarwal, J., and Flegal, A.R. (2008) Comparison of mass discrimination correction methods and sample introduction systems for the determination of lead isotopic composition using a multicollector-inductively coupled plasma mass spectrometer. *Anal. Chem.*, **80**, 8355–8363.

32 Maréchal, C.N., Télouk, P., and Albarède, F. (1999) Precise analysis of copper and zinc isotopic compositions by plasma-source mass spectrometry. *Chem. Geol.*, **156**, 251–273.

33 White, W.M., Albarède, F., and Télouk, P. (2000) High-precision analysis of Pb isotope ratios by multi-collector ICP-MS. *Chem. Geol.*, **167**, 257–270.

34 Woodhead, J. (2002) A simple method for obtaining highly accurate Pb isotope data by MC-ICP-MS. *J. Anal. At. Spectrom.*, **17**, 1381–1385.

35 Meija, J., Yang, L., Sturgeon, R.E., and Mester, Z. (2010) Certification of natural isotopic abundance inorganic mercury reference material NIMS-1 for absolute isotopic composition and atomic weight. *J. Anal. At. Spectrom.*, **25**, 384–389.

36 Baxter, D.C., Rodushkin, I., Engström, E., and Malinovsky, D. (2006) Revised exponential model for mass bias correction using an internal standard for isotope abundance ratio measurements by multi-collector inductively coupled plasma mass spectrometry. *J. Anal. At. Spectrom.*, **21**, 427–430.

37 Dietz, L.A., Pachucki, C.F., and Land, G.A. (1962) Internal standard technique for precise isotopic abundance measurements in thermal ionization mass spectrometry. *Anal. Chem.*, **34**, 709–710.

38 Dodson, M.H. (1963) A theoretical study of the use of internal standards for precise isotopic analysis by the surface ionization technique: Part I. General first-order algebraic solutions. *J. Sci. Instrum.*, **40**, 289–295.

39 Doherty, W., Gregoire, D.C., and Bertrand, N. (2008) Polynomial mass bias functions for the internal standardization of isotope ratio measurements by multi-collector inductively coupled plasma mass spectrometry. *Spectrochim. Acta B*, **63**, 407–414.

40 Russell, W.A., Papanastassiou, D.A., and Tombrello, T.A. (1978) Ca isotope fractionation on the Earth and other solar system materials. *Geochim. Cosmochim. Acta*, **42**, 1075–1090.

41 Vanhaecke, F., Balcaen, L., and Malinovsky, D. (2009) Use of single-collector and multi-collector ICP-mass spectrometry for isotopic analysis. *J. Anal. At. Spectrom.*, **24**, 863–886.

42 Young, E.D., Galy, A., and Nagahara, H. (2002) Kinetic and equilibrium mass-dependent isotope fractionation laws in Nature and their geochemical and cosmochemical significance. *Geochim. Cosmochim. Acta*, **66**, 1095–1104.

43 Wombacher, F., Rehkamper, M., and Mezger, K. (2004) Determination of the mass-dependence of cadmium isotope fractionation during evaporation. *Geochim. Cosmochim. Acta*, **68**, 2349–2357.

44 Hart, S.R. and Zindler, A. (1989) Isotope fractionation laws: a test using calcium. *Int. J. Mass Spectrom. Ion. Processes*, **89**, 287–301.

45 Wombacher, F. and Rehkamper, M. (2003) Investigation of the mass discrimination of multiple collector ICP-MS using neodymium isotopes and the generalized power law. *J. Anal. At. Spectrom.*, **18**, 1371–1375.

46 Malinovsky, D., Stenberg, A., Rodushkin, I., Andren, H., Ingri, J., Ohlander, B., and Baxter, D.C. (2003) Performance of high resolution MC-ICP-MS for Fe isotope ratio measurements in sedimentary geological materials. *J. Anal. At. Spectrom.*, **18**, 687–695.

47 Dietz, L.A., Pachucki, C.F., and Land, G.A. (1962) Internal standard technique for precise isotopic abundance measurements in thermal ionization mass spectrometry. *Anal. Chem.*, **34**, 709–710.

48 Dodson, M.H. (1970) Simplified equations for double-spiked isotopic analyses. *Geochim. Cosmochim. Acta*, **34**, 1241–1244.

49 Cumming, G.L. (1973) Propagation of experimental errors in lead isotope ratio measurements using the double spike method. *Chem. Geol.*, **11**, 157–165.

50 Galer, S.J.G. (1999) Optimal double and triple spiking for high precision lead isotopic measurement. *Chem. Geol.*, **157**, 255–274.

51 Hamelin, B., Manhes, G., Albarède, F., and Allègre, C.J. (1985) Precise lead isotope measurements by the double spike technique: a reconsideration. *Geochim. Cosmochim. Acta*, **49**, 173–182.

52 Johnson, C.M. and Beard, B.L. (1999) Correction of instrumentally produced mass fractionation during isotopic analysis of Fe by thermal ionization mass spectrometry. *Int. J. Mass Spectrom.*, **193**, 87–99.

53 Thirlwall, M.F. (1991) Long-term reproducibility of multicollector Sr and Nd isotope ratio analysis. *Chem. Geol.*, **94**, 85–104.

54 Fietzke, J. and Eisenhauer, A. (2006) Determination of temperature-dependent stable strontium isotope ($^{88}Sr/^{86}Sr$) fractionation via bracketing standard MC-ICP-MS. *Geochem. Geophys. Geosyst.*, **7**, 1–6.

55 Halicz, L., Segal, I., Fruchter, N., Stein, M., and Lazar, B. (2008) Strontium stable isotopes fractionate in the soil environments? *Earth Planet. Sci. Lett.*, **272**, 406–411.

56 Yang, L., Peter, C., Panne, U., and Sturgeon, R.E. (2008) Use of Zr for mass bias correction in strontium isotope ratio determinations using MC-ICP-MS. *J. Anal. At. Spectrom.*, **23**, 1269–1274.

57 Wieser, M.E. and Coplen, T.B. (2011) Atomic weights of the elements 2009 (IUPAC Technical Report). *Pure Appl. Chem.*, **83**, 359–396.

58 Weiss, D.J., Kover, B., Dolgopolova, A., Gallagher, K., Spiro, B., Roux, G., Mason, T.F.D., Kylander, M., and Coles, B.J. (2004) Accurate and precise Pb isotope ratio measurements in environmental samples by MC-ICP-MS. *Int. J. Mass Spectrom.*, **232**, 205–215.

59 Xie, Q., Lie, S., Evans, D., Dillon, P., and Hintelmann, H.J. (2005) High precision Hg isotope analysis of environmental samples using gold trap-MC-ICP-MS. *J. Anal. At. Spectrom.*, **20**, 515–522.

60 Hintelmann, H. and Lu, S.Y. (2003) High precision isotope ratio measurements of mercury isotopes in cinnabar ores using multi-collector inductively coupled plasma mass spectrometry. *Analyst*, **128**, 635–638.

61 Tellinghuisen, J. (2001) Statistical error propagation. *J. Phys. Chem. A*, **105**, 3917–3921.

62 Meija, J. and Mester, Z. (2007) Signal correlation in isotope ratio measurements with mass spectrometry: effects on uncertainty propagation. *Spectrochim. Acta B*, **62**, 1278–1284.

63 Meija, J. and Mester, Z. (2008) Uncertainty propagation of atomic weight measurement results. *Metrologia*, **45**, 53–62.

64 International Organisation for Standardisation (1995) *Guide to the Expression of Uncertainty in Measurement*, International Organisation for Standardisation, Geneva.

6
Reference Materials in Isotopic Analysis

Jochen Vogl and Wolfgang Pritzkow

6.1
Introduction

Isotopic variations are studied with the aim of providing answers to scientific questions. The relevant applications cover many fields, as the following examples demonstrate. Radioactive decay allows the dating of artifacts and minerals in archeometry (e.g., radiocarbon dating [1–3]) and geochronology (e.g., U–Pb dating [4–6]), based on the half-life of the parent nuclide. Studying isotopic variations caused by mass-dependent isotope fractionation accompanying processes such as evaporation of Cd contributes to a better understanding of the processes that occur during the impact of meteorites [7, 8]. Differences in the ground-state energy levels of $B(OH)_3$ and $B(OH)_4^-$ for the boron isotopes ^{10}B and ^{11}B combined with preferential incorporation of $B(OH)_4^-$ in rocks led to an isotope fractionation of boron in Nature that can be exploited as a paleoproxy [9, 10]. Human-made variations in the isotopic composition of an element by adding (an) enriched isotope(s) enable certain effects to be achieved, for example, an increase in the neutron shielding capacity of B by addition of ^{10}B, or processes to be studied, for example, the study of human and plant metabolism [11] (e.g., Mg absorption in humans [12, 13]), and in analytical chemistry quantification can be accomplished by applying isotope dilution mass spectrometry (IDMS) [14–16]. In recent years, the application of isotopic analysis to origin or provenance determination of food products, such as fruit juices, honey, milk, and wine [17–24], has strongly increased. Such applications are discussed in more detail in other chapters of this book, and this chapter rather deals with the interpretation of the numerical results obtained in this context.

In addition to the applications mentioned above, isotopic variations can in general be relied on to assess whether two samples are identical or of the same origin or not. However, it is only possible to determine indisputably that they are *not* identical, that is, when obtaining significantly different values for the isotope amount ratios – often referred to in short as isotope ratios – for the two samples. If both samples show the same isotope amount ratio(s), one can only assume that they are identical with a certain probability, because they could have been affected

Isotopic Analysis: Fundamentals and Applications Using ICP-MS,
First Edition. Edited by Frank Vanhaecke and Patrick Degryse.

Published 2012 by WILEY-VCH Verlag GmbH & Co. KGaA

by the same processes, leading to similar isotope amount ratios, or the measurement uncertainty can simply be too large to resolve the differences. Nowadays, isotopic analysis is increasingly used, for example, for checking the authenticity or for determining the origin of food, for provenance determination of archeological artifacts, and to establish a chain of evidence in forensic science.

In all these applications, isotope ratio data are produced, which are interpreted on an absolute or relative basis and which have an impact on our daily life, whether this is in science (e.g., age of an artifact), in society (e.g., provenance of food), or in public safety (e.g., neutron shielding in nuclear power plants). To ensure that these data are reliable and accurate, some specific requirements have to be fulfilled. The main requirement is that all these measurement results are *comparable*, which means that the corresponding results can be compared and differences between the measurement results can be used to draw further conclusions. This is only possible if the measurement results are traceable to the same reference [25]. This in turn can only be realized by applying isotopic reference materials (IRMs) for correction for bias and for validation of the analytical procedure. Whereas in earlier days only experts in mass spectrometry were able to deliver reproducible isotope ratio data, nowadays many laboratories, some of which may even have never been involved with mass spectrometry before, produce isotope ratio data using inductively coupled plasma mass spectrometry (ICP-MS). Especially for such users, IRMs are indispensable to permit proper method validation and reliable results. The rapid development and the broad availability of ICP-MS instrumentation have also led to an expansion of the research area and new elements are under investigation for their isotopic variations. In this context, all users require IRMs to correct for instrumental mass discrimination or at least to allow isotope ratio data to be related to a commonly accepted basis.

A broad overview of the application of ICP-MS in isotopic analysis was given in a tutorial review by Vanhaecke *et al.* [26].

6.2
Terminology

When talking about analytical issues, whether reference materials, measurement results, procedures, or something else, a common understanding of the relevant terms is required in order to prevent their misunderstanding or misuse. In analytical chemistry and also in isotopic analysis, often different terms are used to describe the same fact, property, or behavior. Frequently, even old and already rejected or replaced terms are still used. Several efforts have been made to harmonize terms in analytical chemistry. The most central publication in this context is the *International Vocabulary of Metrology – Basic and General Concepts and Associated Terms (VIM)* [25], prepared by a joint working group of several international commissions (BIPM – Bureau International des Poids et Mesures; ILAC – International Laboratory Accreditation Cooperation; ISO – International Organization for Standardization; IUPAC – International Union of Pure and

Applied Chemistry, and others). The VIM is mainly focused on basic expressions in metrology, the science of measurements. As the core subject of analytical chemistry and isotopic analysis is measurements, they belong to metrology and, consequently, the metrological terminology should be used.

Analytical terms can also be found in several ISO guides and in a very useful series of publications named *Glossary of Analytical Terms* (GAT) [27–29]. GAT is an ongoing project, run by the EURACHEM Education and Training Working Group, which translates analytical terms into more than 30 languages and publishes them together with their definitions. A list of defined terms and information on where they can be found was also given in Holcombe [30–32]. To avoid going beyond the scope of this book, only the most important terms are defined below.

Accuracy or measurement accuracy

Definition: "The closeness of agreement between a measured quantity value and a true quantity value of a measurand" [25].

Description: The term accuracy combines precision and trueness (i.e., the effects of random and systematic factors, respectively). "Suppose the results produced by the application of a method show zero or very low bias (i.e., are 'true'), their accuracy becomes equivalent to their precision" [27]. In general, accuracy is not a quantity and is not given a numerical value. Due to the worldwide concept of measurement uncertainty as expressed in ISO [33] and EURACHEM [34] documents, it is recommended to use the clearer and less ambiguous term "uncertainty" instead of "accuracy."

Bias or measurement bias

Definition: "Estimate of a systematic measurement error" [25].

Description: Bias is the total systematic error of a measurement result (in contrast to random error).

Conventional quantity value

Definition: "Quantity value attributed by agreement to a quantity for a given purpose" [25].

Description: Sometimes named "conventional true value," which is discouraged. As the "true value" is not accessible by definition, the best approach to it, the conventional quantity value, is used. The best available estimate of the value should be used. This strategy is often applied for reference material certification, when a reference quantity value is being established.

Error or measurement error

Definition: "Measured quantity value minus a reference quantity value" [25].

Description: Error is the sum of systematic and random error.

Precision or measurement precision

Definition: "The closeness of agreement between indications or measured quantity values obtained by replicate measurements on the same or similar objects under specified conditions" [25].

Description: Measurement precision is usually expressed numerically by measures of imprecision, such as standard deviation. The precision of a set of results, which have been performed under stated experimental conditions, can be quantified as standard deviation [27]. Precision is normally differentiated as repeatability and reproducibility.

Repeatability or measurement repeatability

Definition: "Measurement precision under a set of repeatability conditions of measurement" [25].

Description: Repeatability conditions are constant conditions and require the same measurement procedure, the same operator, the same measuring system, the same operating conditions at the same location, and replicate measurements on the same or similar samples over a short period of time. Often the term internal precision is used instead.

Reproducibility or measurement reproducibility

Definition: "Measurement precision under reproducibility conditions of measurement" [25].

Description: As soon as measurement conditions have been changed, it is a matter of reproducibility and no longer of repeatability. This is valid for the same individual sample measured under altered conditions (e.g., another day, another calibration, another operator), and also the measurement of different subsamples under the same or under varying conditions. Reproducibility may also be expressed in terms of dispersion characteristics of the result (e.g., standard deviation) and is often referred to as "external" standard deviation.

Measurement uncertainty

Definition: "Non-negative parameter characterizing the dispersion of the quantity values being attributed to a measurand, based on the information used" [25].

Description: "Uncertainty sets the limits within which a result is regarded as accurate, that is, precise and true" [27]. The combined measurement uncertainty is obtained using the individual (standard) measurement uncertainties associated with the input quantities. This means specifically that all calibrations or corrections have intrinsic uncertainty components, which have to be considered in the combined measurement uncertainty of a measurement result.

Quantity value

Definition: "Number and reference together expressing magnitude of a quantity" [25].

Description: It is the product of a number and the measurement unit, for example, 0.15 kg.

Reference quantity value

Definition: "Quantity value used as a basis for comparison with values of quantities of the same kind" [25].

Description: "A reference quantity value can be a true quantity value, in which case it is unknown, or a conventional quantity value, in which case it is known" [25].

A reference quantity value with associated measurement uncertainty is usually provided with reference to a material, for example, a certified reference material [25].

Trueness or measurement trueness

Definition: "The closeness of agreement between the average of an infinite number of replicate measured quantity values and a reference quantity value" [25].

Description: Trueness is not a quantity and cannot be expressed numerically [25]. It is inversely related to systematic measurement error or bias [25]. It is important to keep the difference from accuracy in mind. Accuracy is the closeness of agreement between a single test result and the conventional quantity value.

True value or true quantity value

Definition: "Quantity value consistent with the definition of a quantity" [25].

Description: The true value is a theoretical concept and would be obtained by a perfect measurement. In reality, it is indeterminate and not accessible, as all measurements are biased. The true value is unknowable.

Traceability or metrological traceability

Definition: "The property of a measurement result whereby the result can be related to reference through a documented unbroken chain of calibrations, each contributing to the measurement uncertainty" [25].

Description: For each analytical measurement, it should be possible to relate the result to an international standard or unit (e.g., international system of units – SI) through an unbroken chain of calibrations [27]. An international standard can also be realized by an internationally accepted reference material, as is the case for some δ-values (e.g., NIST SRM 951 for $\delta^{11}B$). Traceability allows comparability of measurement results.

Trackability

Definition: "The property of a measurement result whereby the result can be uniquely related to the sample" [27].

Description: Each step of an analytical procedure has to be documented, such that the result of a measurement can be linked unambiguously to the sample [27].

In the following, terms defining different types of reference materials are listed and described.

Reference material

Definition: "Material sufficiently homogeneous and stable with reference to specified properties, which has been established to be fit for its intended use in measurement or in examination of nominal properties" [25].

Description: Often reference materials (RMs) carry a reference quantity value, for example, mass fraction of Pb in soil, and are used as quality control material, as calibrator, or for validation purposes.

Certified reference material

Definition: "RM accompanied by documentation issued by an authoritative body and providing one or more specified property values with associated uncertainties and traceabilities, using valid procedures" [25].

Description: Certified reference materials (CRMs) may be used as calibrators or measurement trueness control material. When used as calibrator, a CRM permits traceable and thus comparable measurement results.

Isotopic reference material

Definition: RM certified for the isotopic composition of a specific element or compound contained in the isotopic reference material (IRM).

Description: The isotopic composition is characterized via isotope amount ratios or isotope amount fractions. These data have been obtained from measurements which have been corrected for all bias occurring in mass spectrometry. In terms of isotope ratio data, one can often read about "absolute" isotope ratio data or "absolute" measurements. Here it is meant that the measurements leading to isotope ratio data are a "best estimate for the true value." From a metrological point of view, all published isotope ratio data should be corrected for all bias and should be accompanied by an uncertainty budget. Isotope amount fractions and atomic weights should only be calculated from fully corrected isotope amount ratios. Anything else is observed data only. Thus, isotope amount fractions and molar masses are always "true" or "absolute" data [35].

Here a short remark on the designations "atomic weight" and "atomic mass" is necessary: They are atomic mass ratio numbers, since they are scaled according to the atomic mass of ^{12}C. They are relative atomic masses and therefore they are often given without the unit u. Ideally, these numbers should be used as molar mass of the specific element, $M(\mathrm{E})$, with the units g mol^{-1}. The conversion of relative atomic masses into molar masses is accomplished by using the Avogadro constant as given in Eq. (6.1). Details can be obtained elsewhere [36, 37].

$$M(\mathrm{E}) = \frac{1}{12} m(^{12}\mathrm{C}) N_{\mathrm{A}} A_{\mathrm{r}}(\mathrm{E}) \times 1000$$
$$\text{with } u = \frac{1}{12} m(^{12}\mathrm{C}) = 1.66053873(13) \times 10^{-27}\,\mathrm{kg} \qquad (6.1)$$

This conversion does not change the numerical value, but the uncertainty of the atomic mass increases by a factor of ~ 10. By no means can the atomic mass and the uncertainty from the literature be used and the unit of measure g mol^{-1} be appended.

Delta reference materials

Definition: RMs being certified for the deviation of a specific isotope amount ratio of a specific element, so called δ-value, from an internationally accepted reference defining the origin of the δ-scale [35].

Description: RMs certified only for isotope amount ratios relative to a specific standard, so-called δ-values, are also isotopic reference materials, but less specified. These materials will be characterized as delta reference materials (δ-RMs).

6.3
Determination of Isotope Amount Ratios

Each ion current, I_{obs}, measured with an ICP-MS instrument is biased as a result of mass discrimination, variations in detector gain and detector efficiency, and other effects. These other effects, such as background, detector dead time effects (counting devices), interferences, and matrix effects, will not be discussed here, because they vary strongly depending on the type of mass spectrometer. Discussions on these topics can be found in Chapters 2 and 3 and in the literature [38–42]. An observed isotope amount ratio, or in other words an ion current ratio, $R_{obs,i}$ [Eq. (6.2)], calculated for two isotopes a and b is therefore also biased. This means that every measured or observed isotope amount ratio is biased or, to put it bluntly, these observed isotope amount ratios are wrong, unless they have been corrected for all effects mentioned above.

$$R_{obs,i} = \frac{I_{obs,a}}{I_{obs,b}} \tag{6.2}$$

Assuming that no bias from interferences or drift effects occurs and that gain and efficiency factors of the detectors [e.g., Faraday cups in multi-collector (MC)-ICP-MS] have been determined and applied correctly, the ion current will be biased only by mass discrimination. To correct for this instrumental mass discrimination, a correction factor, the so-called *K*-factor, has to be determined, using Eq. (6.3):

$$R_{true,i} = K_i R_{obs,i} \tag{6.3}$$

where $R_{true,i}$ is the best estimate for the isotope amount ratio, which is commonly referred to as the "true" isotope amount ratio.

True values are unknowable by definition (see Section 6.2) and therefore require a replacement for practical work. The value which is closest to the "true" isotope amount ratio is the certified isotope amount ratio of an IRM, which in turn means that any correction for mass discrimination requires an IRM with known isotopic composition. With the ion current ratio and the certified isotope amount ratio of the IRM, $R_{cert,i}$, the *K*-factor can be calculated according to Eq. (6.4):

$$K_i = \frac{R_{cert,i}}{R_{obs,i}} \tag{6.4}$$

with $R_{true,i} = R_{cert,i}$, and can be used to correct the equivalent ion current ratio of a sample for mass discrimination, provided that the corresponding measurements are performed under identical conditions. For each isotope amount ratio, a separate K-factor has to be determined, which means that for an element with n isotopes, $n-1$ K-factors have to be determined. In the past, and still now, attempts have been made to describe mass discrimination with analytical terms. Different attempts or models, which are all based on Eq. (6.5):

$$R_{\text{true},i} = f(\varepsilon) R_{\text{obs},i} \rightarrow \varepsilon = f(R_{\text{true},i}, R_{\text{obs},i}, \Delta m) \quad (6.5)$$

can be found in the literature (see also Chapter 5) [43–47]. Several MC-ICP-MS studies claimed the exponential and power law models to be superior to the linear model in terms of accuracy [48, 49]. Other researchers found all models equally valid. Our own experience in terms of cadmium isotope amount ratio determinations revealed no differences between these models beyond their uncertainties [47]. An overview of the different findings is given in [50] and in Chapter 5.

In Eq. (6.5), the parameter ε is a correction factor per mass unit (Δm) and can be determined correctly only if an IRM is used for the correction. In many cases, it is assumed that a specific isotope amount ratio of an element is constant, whereas another isotope amount ratio of the same element is not due to natural radioactive decay, as is the case for strontium (^{87}Sr). By using an isotope amount ratio assumed to be constant (here ^{86}Sr/^{88}Sr), ε is calculated and is subsequently used to correct the isotope amount ratio containing the radiogenic nuclide (here ^{87}Sr/^{86}Sr). It is noted critically here that strictly, no constant isotope amount ratio exists. This can be demonstrated for a large number of elements, including strontium, owing to the very high precision of the present generation of mass spectrometers. In addition to this false assumption, it is common to assume the efficiency factor of detectors to be 1, which again is not correct. Different detectors have different efficiency factors, leading to an additional bias in isotope ratio determinations when using MC instruments. When determining ε for one specific isotope ratio with one specific pair of detectors and using it to correct another specific isotope ratio determined with another specific pair of detectors for mass discrimination, the difference in the detector efficiencies and gain will lead to a biased correction. Hence determined ε values carry uncertainties in the range of a few percent. This problem occurs, for example, when isotopic variations of Sr have to be determined. These problems do not occur when K-factors are used for the correction for mass discrimination, because the efficiency and gain factors are already included, provide that the same pair of detectors is used for sample and reference.

With the use of IRMs, these false assumptions can be avoided and traceable isotope amount ratios can be generated. At present, however, only for a limited number of elements is an IRM available, as can be seen from Tables 6.1 and 6.2. Natural-like IRMs (Table 6.1) can be used to correct for mass discrimination, whereas enriched IRMs (Table 6.2) are mainly used in tracer studies and isotope dilution experiments. The enriched IRMs offer sufficient quality, but the number of elements covered is too low. This has been discussed at several meetings with the summarized outcome that several enriched isotopes have been exhausted

Table 6.1 IRMs with natural-like isotopic composition; more details can be found on the websites of NIST [51], IRMM [52], ERM [53], and BAM [54].

Element	State	Composition	RM code[a]
Lithium	Solid	Li_2CO_3	IRMM-016a
Boron	Solid	H_3BO_3	IRMM-011, NIST SRM 951a
	Solution	Aqueous	IRMM-611
	Solution	Aqueous	ERM-AE101, -AE120, -AE121, -AE122
Magnesium	Solid	Mg chips	NIST SRM 980
	Solution	Nitric acid	IRMM-009, ERM-AE637
Silicon	Solid	Si single crystal	IRMM-017
		SiO_2	IRMM-018a
		Chips	NIST SRM 990[b]
Sulfur	Solution	Nitric acid	IRMM-643, -644, -645
Chlorine	Solid	NaCl	NIST SRM 975a
	Solution	Aqueous	ERM-AE641
Potassium	Solid	KCl	NIST SRM 985[b]
Chromium	Solid	$Cr(NO_3)_3 \cdot 9H_2O$	NIST SRM 979
	Solution	Hydrochloric acid	IRMM-012, -625
Iron	Solid	Fe cubes, wire	IRMM-014a, -014b
	Solution	Hydrochloric acid	IRMM-634
Nickel	Solid	Ni powder	NIST SRM 986
Copper	Solution	Nitric acid	ERM-AE633, -647
	Solid	Cu disk	NIST SRM 976[b]
Zinc	Solution	Nitric acid	IRMM-651, -3702
Gallium	Solid	Ga disk	NIST SRM 994
Bromine	Solid	NaBr	NIST SRM 977
Rubidium	Solid	RbCl	NIST SRM 984
	Solution	Nitric acid	IRMM-619
Strontium	Solid	$SrCO_3$	NIST SRM 987
Silver	Solid	$AgNO_3$	NIST SRM 978a
Cadmium	Solution	Nitric acid	BAM-I012
Rhenium	Solid	Re metal	NIST SRM 989[b]
Platinum	Solid	Pt wire	IRMM-010
Mercury	Solution	Hydrochloric acid	ERM-AE639
Thallium	Solution	Nitric acid	ERM-AE649
	Solid	Tl rod	NIST SRM 997
Lead	Solid	Pb wire	NIST SRM 981

[a]IRMM materials are available via www.irmm.jrc.be, NIST materials via www.nist.gov, ERM materials via www.erm-crm.org, and BAM materials via www.bam.de.
[b]Discontinued: this product is no longer produced.

and the supply of other is becoming critically low. This also demonstrates that the production of IRMs is too low and there is an undersupply of IRMs.

In the case of natural-like IRMs, however, the uncertainties of the certified isotope amount fractions are not sufficiently small for present isotope ratio determinations. The production and certification of IRMs is expensive and the

Table 6.2 Enriched IRMs; more details can be found on the websites of NIST [51], IRMM [52], and ERM [53].

Element	Enrichment	State	Composition	RM code[a]
Lithium	^{6}Li	Solid	Li_2CO_3	IRMM-15
		Solution	Hydrochloric acid	IRMM-615
Boron	^{10}B: 95%	Solid	H_3BO_3	NIST SRM 952
	^{10}B: 95%	Solution	Aqueous	IRMM-610
	^{10}B: 30%	Solution	Aqueous	ERM-AE102
	^{10}B: 50%	Solution	Aqueous	ERM-AE103
	^{10}B: 31.5%	Solution	Aqueous	ERM-AE104
Magnesium	^{26}Mg	Solution	Nitric acid	ERM-AE638
Sulfur	^{34}S	Solution	Nitric acid	IRMM-646
Chlorine	^{37}Cl	Solution	Aqueous	ERM-AE642
Calcium	^{41}Ca	Solution	Nitric acid	ERM-AE701
Chromium	^{50}Cr	Solution	Hydrochloric acid	IRMM-624
Iron	^{57}Fe	Solution	Hydrochloric acid	IRMM-620
Copper	^{65}Cu	Solution	Nitric acid	IRMM-632
Zinc	^{64}Zn	Solution	Nitric acid	IRMM-007-1 to -6, -652
	^{67}Zn	Solution	Nitric acid	IRMM-653
	^{68}Zn	Solution	Nitric acid	IRMM-654
Rubidium	^{87}Rb	Solution	Nitric acid	IRMM-618
Strontium	^{84}Sr	Solution	Nitric acid	IRMM-635
Cadmium	^{111}Cd	Solution	Nitric acid	IRMM-621, -622
Mercury	^{202}Hg	Solution	Hydrochloric acid	ERM-AE640
Lead	^{206}Pb	Solution	Nitric acid	NIST SRM 991[b]

[a]ERM materials are available via www.erm-crm.org, IRMM materials via www.irmm.jrc.be, and NIST materials via www.nist.gov.
[b]Material out of stock at the time this chapter was written.

producers bear the risk that the uncertainties obtained do not fall within the range intended (more details on the limitations of IRMs follow in Sections 6.4 and 6.5). For this reason, comparative measurements are a good alternative, as could be shown in the field of the so-called (traditional) stable isotopes (H, C, N, O, and S) for several years.

Here, the sample ion current ratio is simply related to the ion current ratio of a reference following Eq. (6.6) [21, 55]:

$$\delta(R) = \left[\left(\frac{R_{\text{sample}}}{R_{\text{reference}}}\right) - 1\right] \times 10^3(‰) \tag{6.6}$$

The difference in the ion current ratio compared with the ion current ratio of the reference is expressed as a so-called δ-value. This allows highly precise measurement results without requiring an IRM certified for its isotopic composition. The results are *relative* results only and the comparability of results is possible only when the same reference is used. Therefore, a system of international accepted references has been developed in the field of (traditional) stable isotope research, which is a historical expression for isotopic analysis of the elements H, C, N, O, and S. A very good and detailed overview on RMs for these stable isotopes was given by Gröning [56], which is still up-to-date and requires no amendment within this chapter. For all isotope measurements, IRMs are essential to obtain traceable and comparable isotope ratio data. When applying corrections using the *K*-factor, IRMs certified for their isotopic composition are necessary. For the determination of δ-values, RMs commonly accepted as the origin of the δ-scale are required. In any case, a second IRM of the same element is desirable, which is characterized by a different isotopic composition or δ-value. This second material can be used for method development, validation, and quality checks.

6.4 Isotopic Reference Materials

6.4.1 General

Metrologically "true" values do exist, but nobody will ever know them. What analysts and metrologists most commonly understand when talking about "true" values is "the best estimate for the true value." In terms of isotope ratio data, one can often read about "absolute" isotope ratio data or "absolute" measurements. Here it is meant that the measurements leading to isotope ratio data have been corrected for all bias occurring in mass spectrometry and therefore are a "best estimate for the true value." From a metrological point of view, all published isotope ratio data should have been corrected for all bias and should be accompanied by an uncertainty budget. Isotope amount fractions and molar masses should only be calculated from fully corrected isotope amount ratios. Anything else is observed data only. Thus, isotope amount fractions and molar masses are always "true" or "absolute" data. Whenever a differentiation between observed and corrected data is necessary or one of them should be emphasized, the authors prefer to use "observed" data and "true" data (see also Section 6.1).

6.4.2 Historical Development

The development of IRMs is strongly connected with the determination of atomic weights of the elements, because measurement techniques and procedures have primarily been developed for atomic weight determinations. A concise historical review of the determination of atomic weights was given by De Laeter *et al.* in the IUPAC technical report "Atomic weights of the elements: review 2000" [36]. This review also lists the latest IUPAC values for the isotopic composition of the elements. A later publication from 2005 presents the same data in a different form [57].

The first determination of the atomic weight of an element by mass spectrometric means was performed by Dempster in 1920 [58]. However, mass fractionation effects occurring in the mass spectrometers required further development. In the 1940s, the use of non-radioactive, stable isotopes increased, accelerated by the release of enriched isotopes by the US Atomic Energy Commission in 1946. In this phase of development, in 1950 Nier performed the first measurements of isotope amount fraction that were corrected for mass fractionation by means of synthetic mixtures of enriched isotopes [59]. This technology has been used since then for the determination of atomic weights and characterization of IRMs. Highly enriched and chemically purified isotopes were used to prepare gravimetrically synthetic isotope mixtures, for which theoretical isotope amount ratios can be calculated using the isotope amount fractions of the enriched isotopes. However, these isotope amount fractions are not known precisely and so they had to be determined by means of mass spectrometry. The observed isotope amount fractions of the enriched isotopes are affected by mass fractionation or discrimination and so are the theoretically calculated isotope amount ratios of the synthetic mixtures. This problem can only be solved by least-squares calculations, which yield the correction factor for the mass bias for the mass spectrometer applied. This correction is highly laborious and is therefore only used for isotope amount ratio determinations of the highest metrological quality, for example, aimed at refining the determination of the atomic weight of an element or assessing the isotopic composition of an IRM. Details on the application of synthetic isotope mixtures for the determination of molar masses and certification of IRMs can be found in the recent literature [47, 60–75]. Most of today's work is focused on the determination of the molar mass of isotopically enriched Si, which is being used in the international Avogadro Project, the intention of which is the redefinition of the SI unit of the mass – the kilogram – by the exact determination of fundamental constants: the Avogadro constant (N_A) and Planck's constant (h) [75–77]. These few references describe the worldwide work in this field within the last decade. Only part of this work was related to the production and certification of new IRMs.

In 1954, the National Bureau of Standards (NBS), now the National Institute of Standards and Technology (NIST), started a program to determine isotope amount fractions and to certify IRMs by mass spectrometric means [51]. In the 1960s, the Central Bureau for Nuclear Measurements (CBNM) in Geel, now the Institute for Reference Materials and Measurements (IRMM), also started a program to determine isotope amount fractions and to certify IRMs [52]. Up to 1993, much progress was made, resulting in a collection of IRMs covering 20 elements [78]. However, since 1993, the work on IRMs has declined strongly. In 2011, the status of IRMs is nearly the same as in 1993, and there are IRMs available for not more than 23 elements (Table 6.1). At the beginning of the new millennium, at the Bundesanstalt für Materialforschung und -prüfung (BAM) (the Federal Institute for Materials Research and Testing) in Germany, a small group started to work on IRMs [54]. Furthermore, many IRMs of IRMM and BAM are marketed as European Reference Materials (ERMs) [53]. With upcoming MC-ICP-MS, the worldwide need for IRMs increased dramatically. This has also been recognized by NIST, IUPAC, and BAM.

Radioactive IRMs are not considered separately. A list of the radioactive IRMs available is displayed in Table 6.3 to provide the reader with an overview. So far as IDMS applications or isotopic variations in Nature are concerned, they involve stable materials. IRMs for nuclear applications, especially those from IRMM, are of excellent quality and fulfill the needs of the users. Therefore, there is no strong need at present for further or better IRMs for use in this context.

Nearly all existing IRMs are searchable via two databases. The COMAR database is operated by BAM and lists several thousand CRMs including IRMs [79]. The GeoReM database lists a large number of materials including IRMs, but also materials that do not fulfill the requirements of an IRM (see the next section) [80]. The data attributed to the specific materials are based on a compilation of the existing literature, which is not complete in every case: BAM-I012 is a good example of this, because the basic reference [47] is not given in the database.

6.4.3 Requirements for Isotopic Reference Materials

The terms RM, CRM, and IRM have already been defined in Section 6.1. These internationally accepted definitions clearly make two important statements: the differentiation between RMs and CRMs and the definition of the basic requirements thereof. For an isotopic reference material, the requirements are as follows:

- An IRM must be sufficiently homogeneous.
- An IRM must be sufficiently stable for a long period of time.
- An IRM must be available to everyone for a long period of time.
- IRM values must be comprehensible and traceable.
- IRM values must be accompanied by an uncertainty statement.
- IRM values must be laid down in a certificate.

These are rather general requirements and need to be specified further (see below). Isotope amount ratios are used for different applications, such as the determination of elemental contents, determination of isotopic variations for the interpretation of specific incidents and processes in bio- or geochemistry, and characterization of materials. Interference corrections are also based on isotope amount ratios calculated from the IUPAC-tabulated isotope amount fractions [57].

In elemental analysis by ICP-MS, a difference in the isotopic composition between the sample and the calibration standard leads to a bias in the measurement result, that is, the elemental content determined. In serious routine analysis, the isotopic variation of certain elements therefore needs to be considered. For lead determination, often the ion currents of the isotopes ^{206}Pb, ^{207}Pb, and ^{208}Pb are summed and this sum is used for the calculations, because the sum of all isotope amount fractions is one and therefore is not susceptible to isotopic variations. However, two approximations are included: first, the ion current of ^{204}Pb is not used owing to its low isotope amount fractions and the interference from ^{204}Hg,

Table 6.3 Available radioactive IRMs; more details can be found on the websites of IRMM [52], NBL [81], and NIST [51].

Element	Isotope	Description	State	Composition	Code[a]
Iodine	^{129}I	High level	Solution	Alkaline	NIST SRM 3231
Lead	^{206}Pb	Radiogenic	Solid	Pb	NIST SRM 983[b]
	$^{206}Pb/^{208}Pb$	Equal atom	Solid	Pb	NIST SRM 982[b]
Thorium	^{230}Th		Solution	Nitric acid	IRMM-060, -061
	^{232}Th		Solution	Nitric acid	IRMM-035, -036
Uranium	^{234}U, ^{235}U, ^{236}U, ^{238}U		Solid	UF_6	IRMM-019 to -029 NBL 113-B
	^{234}U, ^{235}U, ^{236}U, ^{238}U		Solid	U_3O_8	NBL 125, 129-A, U002, U005-A, U010, U015, U020-A, U030-A, U100, U150, U200, U350, U500, U630, U750, U800, U850, U900, U930-D, U970
	^{233}U	Spike	Solution	Nitric acid	IRMM-040a, -051, -057, -058, -3660, -3660a, -3660b, NBL 111-A
	^{235}U	Spike	Solution	Nitric acid	IRMM-050, -054, NBL 135
	^{238}U	Spike	Solution	Nitric acid	IRMM-052, -053, -056
	^{233}U, ^{236}U	Double spike	Solution	Nitric acid	IRMM-3636, -3636a, -3636b
	^{233}U, ^{235}U, ^{238}U	Series	Solution	Nitric acid	IRMM-073, -074
	^{236}U, ^{238}U	Series	Solution	Nitric acid	IRMM-075
	^{234}U, ^{235}U, ^{236}U, ^{238}U		Solution	Nitric acid	IRMM-183 to -187, -3183 to -3187, EC NRM199, NBL 045
	^{233}U, ^{235}U, ^{236}U, ^{238}U	Series for γ-spectrometry	Solid	U_3O_8	IRMM-171, NBL 146, 969
Plutonium	^{239}Pu	Spike	Solution	Nitric acid	IRMM-081a, -086
			Solid	Pu	NBL 126-A
			Solid	PuO_2	NBL 122
	^{240}Pu	Spike	Solution	Nitric acid	IRMM-083
			–	–	NBL 136, 137, 138
	^{242}Pu	Spike	Solution	Nitric acid	IRMM-043, -044, -049c, -085, NBL 130
	^{244}Pu	Spike	Solution	Nitric acid	IRMM-084, ERM AE042
	^{239}Pu, ^{242}Pu	Series	Solid	Nitrate	IRMM-290, NBL 128
Mixed U	^{235}U, ^{239}Pu	Spike	Solid	Nitrate	IRMM-1027h, i, j, -1029a
and Pu	^{233}U, ^{242}Pu	Spike	Solution	Nitric acid	IRMM-046b, -090

[a] IRMM materials are available via www.irmm.jrc.be, ERM materials via www.erm-crm.org, and NBL materials via www.nbl.doe.gov.

[b] Radioactive due to ^{210}Pb present in the material.

and second, the molar mass is taken as invariant. In routine analysis, this approximation is allowed, because the uncertainties achievable are larger than a few percent. However, when performing elemental analysis at the highest level of accuracy and precision, such as an IDMS analysis for RM characterization, the isotopic variation has to be considered. This can be demonstrated by a theoretical IDMS experiment performed for the determination of the boron mass content in a sample. Three different cases are considered: (i) the sample shows an isotopic variation of $\delta^{11}B = 0‰$ versus NIST SRM 951 (boron IRM); (ii) the sample shows an isotopic variation of $\delta^{11}B = 30‰$; (iii) the sample shows an isotopic variation of $\delta^{11}B = +50‰$. In all cases, the spike has an invariant isotopic composition and the number of boron atoms in the sample is constant. The calculation of the boron mass content is carried out using the boron isotopic composition of $\delta^{11}B = 0‰$ for all cases, thus simulating the effect of ignoring isotopic variation. The boron mass content in case (a) is of course not biased versus the input data. Case (b) is biased by −0.8% and case (c) is biased by 1.2%. This error is metrologically not acceptable and isotopic variations have to be considered in IDMS analysis. Therefore, IRMs with "true" isotope amount fractions (and the corresponding molar mass) are required, offering expanded uncertainties in the range 0.1–0.01%, which is sufficient for metrological applications such as certification of a CRM by IDMS. Such IRMs are more than sufficient for routine elemental analysis, and also for interference corrections, which require expanded uncertainties between 1% and 0.1% only.

Even higher demands apply to measurements generating isotope amount ratios for different interpretations, for example, in geochemistry. By applying MC mass spectrometric instruments and highly sophisticated measurement procedures, including well-developed chemical separation procedures, reproducibilities of 0.001% can be achieved. This requires IRMs offering uncertainties of less than 0.001%. However, in many cases where interpretation of isotopic variations is important, no absolute data are required and relative or δ-values are sufficient.

6.5 Present Status, Related Problems, and Solutions

6.5.1 Present Status

Today, we face two opposite trends. Triggered by the rapid development of ICP-MS, the number of published isotope ratio data increased rapidly. This rapid development also caused an increasing number of wrong isotope ratio data to be published, such as the Sr isotope ratio data for ginseng by Choi *et al.* [82]. The extremely low $^{87}Sr/^{86}Sr$ data published by Choi *et al.* fall in a range that does not even exist today in terrestrial materials, as was demonstrated by Rosner [83]. The provision of suitable IRMs and their subsequent application can help to avoid such failings.

In contrast to the increasing need, the production and certification of IRMs stagnated, because most institutes focused on other research topics with higher visibility. Mostly, IRMs do not yield such visibility, because their production and certification at the required high level of metrological quality is tedious and uneconomical work, which is often scientifically not sufficiently acknowledged. Especially in the fields of geochemistry and food traceability, the research on isotopic variations is strongly increasing and is expanding to elements formerly not at the focus, such as copper, molybdenum, germanium, and selenium [84, 85]. Research on such elements requires RMs to permit reliable and comparable isotope ratio data to be obtained. These RMs should be provided by national metrology institutes such as NIST and BAM or international institutions such as IRMM, because they have the metrological background and the resources required to produce and certify such IRMs.

6.5.2
Related Problems

At present, the determination of isotope amount ratios is possible with reproducibilities of the order of 0.001% [47, 86], and future isotope research surely will go below this limit. This automatically means that the requirements for IRMs are of the same order or even higher. IRMs with uncertainties around 0.001% for the isotope amount ratio are currently not available. A representative selection of IRMs with uncertainties covering the existing range is presented in Table 6.4. The range

Table 6.4 Uncertainties of representative IRMs spanning the range of uncertainties achieved.

RM code	Element	Category	Year	State[a]	Relative uncertainty (%) ($k = 1$)		
					Isotope amount ratio[b]	Isotopic amount fraction[b]	Molar Mass
IRMM-010	Pt	Primary	1999	s	0.5	0.2	0.001
NIST SRM 980	Mg	Primary	1966	s	0.05[c]	0.02[c]	n.c.[d]
ERM-AE101	B	Secondary	2004	l	0.07	0.02	0.001
IRMM-011	B	Primary	1967	s	0.06	0.01	0.0009
NIST SRM 951a	B	Primary	1970	s	0.04[c]	0.008[c]	n.c.[d]
NIST SRM 981	Pb	Primary	1968	s	0.02[c]	0.008[c]	n.c.[d]
IRMM-651	Zn	Primary	2007	l	0.02	0.008	0.0002
BAM-I012	Cd	Primary	2006	l	0.007	0.005	0.00008
IRMM-643	S	Primary	2005	l	0.009	0.0005	0.00002

[a] s, solid material; l, liquid, aqueous or acidic solution.
[b] The isotope amount fraction and isotope amount ratio with the smallest uncertainty was selected.
[c] Only standard deviations are given in the certificate, no uncertainties.
[d] Not calculated in the certificate.

of the uncertainties or standard deviations for older materials reaches from around 0.5% to 0.007% for isotope amount ratios, stressing the lack of IRMs with uncertainties $\leq 0.001\%$.

At present, it is very difficult or even impossible to produce IRMs that meet these needs, because synthetic isotope mixtures have to be prepared under gravimetric control for the characterization of such IRMs. The weighing of solid materials in principle can be carried out with uncertainties of 0.0001%. However, it is the purity of the materials that is the bottleneck. This can be demonstrated for the example of Cd. Even after sublimation of the highly enriched isotopes under high vacuum, the remaining metal impurities of Cd are in the range between 1 and 10 mg kg^{-1} as shown by ICP-MS analysis [47]. The uncertainty achievable for the sum of impurities is therefore around 0.001%. The weighing uncertainties for preparing dilutions and mixtures are also around 0.001%, because evaporation of the solutions occurs to a limited extent during the weighing. This, of course, results in minimum uncertainties of several 0.001% for the isotope amount ratio determination of synthetic isotope mixtures. From our own work, we know that uncertainties of 0.01% are possible for single isotope amount ratios of an element. It is emphasized here that no pure reproducibilities, but real expanded uncertainties, are referred to, derived from a complete uncertainty budget with $k=2$ ($U=ku_c$). The uncertainties needed by the users and customers therefore currently cannot be achieved by applying this approach.

The next difficulty is the homogeneity of the IRMs. Most of the certified IRMs available are solid materials. In recent years, it was demonstrated that solutions prepared from different pieces of the same batch of pure metal show variations in their isotopic composition. These variations, falling within the relatively large uncertainties, can nowadays be reproducibly determined. Certified IRMs are produced from high-purity metals, where fractionation during winning or purification processes cannot be excluded. Therefore, IRMs must be prepared as solutions, which immediately resolves the homogeneity problem.

The homogeneity problem with solid materials can be shown nicely by the following two examples. The magnesium isotopic reference material NIST SRM 980 was produced in the mid-1960s and was a suitable material for Mg isotope ratio measurements for a long time [87]. An MC-ICP-MS study in 2003 indicated an isotopic heterogeneity of up to 4‰ for the $^{25}Mg/^{24}Mg$ ratio, which is far beyond the accuracy aimed at in current magnesium isotopic research [88]. Based on this and additional observations, there is a demand for new Mg materials with higher homogeneity and smaller uncertainties [89].

Similar observations could be made during a project on the atomic weight of Cd and the production of a Cd IRM in parallel. Fifteen single shots of the solid base material showed a maximum spread of ~1‰ for the ratio $^{114}Cd/^{110}Cd$. Therefore, no solid Cd IRM is provided. The Cd IRM (BAM-I012) is available only as a solution.

The other major problems are time and cost. A project for the certification of an IRM via synthetic isotope mixtures, of course including the molar mass determination, lasts a number of years and involves several scientists, who are, however, also able to work on other tasks in parallel. With MC-ICP-MS, however, the number of elements investigated for isotopic variations increases rapidly. The amount of IRMs required will soon cover nearly the entire periodic system. This need cannot be met by using synthetic mixtures. As a result, in recent years, often commercial solutions or solutions from uncharacterized solid materials have been installed by research groups as " isotopic reference materials," but these definitely do not fulfill the requirements for an IRM in terms of, for example, reliability, stability, sustainability, and accessibility. Often, personal interests play an important role, not always to the benefit of the community.

Therefore, scientists should contact the national metrology institutes such as NIST and BAM or international institutions such as IRMM when new materials are needed. The production of IRMs by these institutes ensures that the requirements for an IRM are fulfilled. Only with IRMs is the comparability of isotope ratio data enabled.

6.5.3
Solution

IRMs characterized by applying synthetic isotope mixtures currently cannot fulfill the present and future needs of the users. In many fields, however, where the interpretation of isotopic variations is of interest, δ-values are sufficient. Thus, the uncertainty of certified isotope amount ratios of the IRM is no longer limiting, because δ-RMs can be provided. The limits are thus set by homogeneity and accessibility. As described above, the problem with homogeneity can be avoided by providing solutions. The accessibility is guaranteed for these δ-RMs when they are provided by national institutes, such as NIST and BAM, or by international institutions, such as the IAEA (International Atomic Energy Agency), IRMM, or IUPAC. Even if a δ-RM runs short, as probably might happen for some materials within the next years, it should not be a problem to determine the δ-value for a new RM, provided that it is done in a timely manner and properly, as is guaranteed by a national metrology institute. These solutions will be suitable for defining the δ-scale of elements, as was done via SMOW (Standard Mean Ocean Water) for the δ-scale of hydrogen. Ideally, one δ-RM is provided for defining the origin of the δ-scale and another one for setting a second point with a well-defined distance to the origin. The δ-RM defining the origin of the δ-scale should be attributed a δ-value of zero, according to the wishes of the user. On the other hand, the base material for the δ-RM, preferably a pure element, should fulfill some requirements. The most important requirement is a high purity to avoid any impurities leading to interferences. With all requirements, it is obvious that not every material can be used for defining a δ-value of zero. Here the question arises of how the δ-scale should be defined. Assuming that the range of isotopic variations for an element is x‰, the origin of the δ-scale could be set at $x/2$‰. Unfortunately, among all the available materials suitable for producing a δ-RM, sometimes there is not a single one showing a δ-value of zero. However, it is not important for the determination of the δ-value in the sample and for the uncertainty whether the δ-RM carries a δ-value of zero or not. When the origin is set by a virtual δ-RM and the real δ-RM shows a specific δ-value versus the origin, the δ-value of the sample versus the origin can be calculated easily according to

$$\delta_{sa,0} = \delta_{sa,std} + \delta_{std,0} + (\delta_{sa,std}\delta_{std,0}) \times 10^{-3} \qquad (6.7)$$

where $\delta_{sa,0}$ and $\delta_{std,0}$ are the δ-values of the sample and the real δ-RM versus the origin, respectively, and $\delta_{sa,std}$ is the δ-value of the sample versus the real δ-RM. Although this calculation is very simple, the origin should be realized by a real δ-RM whenever possible.

RMs for defining the δ-scale are an essential transitional solution on the way to IRMs with certified isotope amount fractions. These IRMs certified for isotope amount fractions are essential for IDMS, for interference corrections, and for the correction for mass discrimination. Whenever possible, RMs defining the δ-scale should be additionally certified for their isotope amount fractions, even if the resulting uncertainties for the isotope amount fractions are too high for current research on isotopic variations. The aim is to obtain one δ-RM for every stable element, provided as a solution, fixing the origin of the δ-scale of the specific element and preferably certified for its isotope amount fractions, and a second δ-RM with a well-defined offset on the δ-scale. Research should continue to improve the capabilities for the determination of isotope amount fractions.

6.6 Conclusion and Outlook

The production of IRMs, which started in the 1960s, has stagnated during the last 15 years. On the other hand, the need for IRMs increased strongly with upcoming ICP-MS techniques, especially with MC instruments, in the beginning of the millennium. In parallel, the precision of the isotope amount ratio determinations reached a level which is beyond that of IRMs certified relying on synthetic isotope mixtures. Additionally, the number of elements for which IRMs are required has increased. This difference between the users' needs and the limitations of current IRMs can only be solved by providing IRMs defining a δ-scale for each element of interest. Such δ-RMs should be produced as solutions to avoid any complications with homogeneity issues. Whenever possible, isotope amount fractions should be provided additionally, even if the uncertainties are not sufficient for cutting-edge research on isotopic variations. Research should also focus on ways to provide isotope amount fractions with lower uncertainties.

For the coming years, around 27 elements have been selected by user communities as "of interest" in this context [35]. These elements are of interest to geo- and cosmochemists, and also to scientists working in the fields of human nutrition, human medicine, veterinary medicine, and food traceability. For 18 elements, such as Pb and U, IRMs or δ-RMs are available, provided as a solution or a solid material with sufficient homogeneity. Nevertheless, in some cases further materials are necessary, because the number of units is too small or the materials are too expensive to be widely accepted by users.

At BAM, IRMs and δ-RMs will be produced according to the requirements mentioned above. NIST has also started a program to produce new RMs for δ-values [90]. As discussed at the latest meeting of the IUPAC Commission on Isotopic Abundances and Atomic Weights (CIAAW) in July 2009 in Vienna, these activities will be supported by IUPAC.

References

1 Libby, W.F. (1946) Atmospheric helium three and radiocarbon from cosmic radiation. *Phys. Rev.*, **69** (11–12), 671–672.

2 Anderson, E.C., Libby, W.F., Weinhouse, S., Reid, A.F., Kirshenbaum, A.D., and Grosse, A.V. (1947) Natural radiocarbon from cosmic radiation. *Phys. Rev.*, **72** (10), 931–936.

3 Wohlfarth, B. (1996) The chronology of the last termination: a review of radio carbon-dated, high-resolution terrestrial stratigraphies. *Quat. Sci. Rev.*, **15** (4), 267–284.

4 Silver, L.T. and Deutsch, S. (1963) Uranium–lead isotopic variations in zircons – a case study. *J. Geol.*, **71** (6), 721–758.

5 Parrish, R.R. (1990) U–Pb dating of monazite and its application to geological problems. *Can. J. Earth Sci.*, **27** (11), 1431–1450.

6 Black, L.P., Kamo, S.L., Allen, C.M., Aleinikoff, J.N., Davis, D.W., Korsch, R.J., and Foudoulis, C. (2003) TEMORA 1: a new zircon standard for phanerozoic U–Pb geochronology. *Chem. Geol.*, **200** (1–2), 155–170.

7 Wombacher, F., Rehkamper, M., Mezger, K., and Munker, C. (2003) Stable isotope compositions of cadmium in geological materials and meteorites determined by multiple-collector ICPMS. *Geochim. Cosmochim. Acta*, **67** (23), 4639–4654.

8 Wombacher, F., Rehkamper, M., and Mezger, K. (2004) Determination of the mass-dependence of cadmium isotope fractionation during evaporation. *Geochim. Cosmochim. Acta*, **68** (10), 2349–2357.

9 Palmer, M.R., London, D., Morgan, G.B., and Babb, H.A. (1992) Experimental determination of fractionation of B-11/B-10 between tourmaline and aqueous vapor – a temperature-dependent and pressure-dependent isotopic system. *Chem. Geol.*, **101** (1–2), 123–129.

10 Palmer, M.R., Spivack, A.J., and Edmond, J.M. (1987) Temperature and pH controls over isotopic fractionation during adsorption of boron on marine clay. *Geochim. Cosmochim. Acta*, **51** (9), 2319–2323.

11 Stürup, S., Hansen, H., and Gammelgaard, B. (2008) Application of enriched stable isotopes as tracers in biological systems: a critical review. *Anal. Bioanal. Chem.*, **390** (2), 541–554.

12 Bohn, T., Walczyk, T., Davidsson, L., Pritzkow, W., Klingbeil, P., Vogl, J., and Hurrell, R.F. (2004) Comparison of urinary monitoring, faecal monitoring and erythrocyte analysis of stable isotope labels to determine magnesium absorption in human subjects. *Br. J. Nutr.*, **91** (1), 113–120.

13 Bohn, T., Davidsson, L., Walczyk, T., and Hurrell, R.F. (2004) Fractional magnesium absorption is significantly lower in human subjects from a meal served with an oxalate-rich vegetable, spinach, as compared with a meal served

with kale, a vegetable with a low oxalate content. *Br. J. Nutr.*, **91** (4), 601–606.

14 Heumann, K.G. (2004) Isotope-dilution ICP-MS for trace element determination and speciation: from a reference method to a routine method? *Anal. Bioanal. Chem.*, **378** (2), 318–329.

15 Vogl, J. (2007) Characterisation of reference materials by isotope dilution mass spectrometry. *J. Anal. At. Spectrom.*, **22** (5), 475–492.

16 Vogl, J. and Pritzkow, W. (2010) Isotope dilution mass spectrometry – a primary method of measurement and its role for RM certification. *MAPAN*, **25** (3), 135–164.

17 Kelly, S., Heaton, K., and Hoogewerff, J. (2005) Tracing the geographical origin of food: the application of multi-element and multi-isotope analysis. *Trends Food Sci. Technol.*, **16** (12), 555–567.

18 Padovan, G.J., De Jong, D., Rodrigues, L.P., and Marchini, J.S. (2003) Detection of adulteration of commercial honey samples by the C-13/C-12 isotopic ratio. *Food Chem.*, **82** (4), 633–636.

19 Calderone, G. and Guillou, C. (2008) Analysis of isotopic ratios for the detection of illegal watering of beverages. *Food Chem.*, **106** (4), 1399–1405.

20 Rossmann, A. (2001) Determination of stable isotope ratios in food analysis. *Food Rev. Int.*, **17** (3), 347–381.

21 Elflein, L. and Raezke, K.P. (2008) Improved detection of honey adulteration by measuring differences between C-13/C-12 stable carbon isotope ratios of protein and sugar compounds with a combination of elemental analyzer–isotope ratio mass spectrometry and liquid chromatography–isotope ratio mass spectrometry (delta C-13-EA/LC-IRMs). *Apidologie*, **39** (5), 574–587.

22 Almeida, C.M. and Vasconcelos, M. (2001) ICP-MS determination of strontium isotope ratio in wine in order to be used as a fingerprint of its regional origin. *J. Anal. At. Spectrom.*, **16** (6), 607–611.

23 Barbaste, M., Halicz, L., Galy, A., Medina, B., Emteborg, H., Adams, F.C., and Lobinski, R. (2001) Evaluation of the accuracy of the determination of lead isotope ratios in wine by ICP MS using quadrupole, multicollector magnetic sector and time-of-flight analyzers. *Talanta*, **54** (2), 307–317.

24 Knobbe, N., Vogl, J., Pritzkow, W., Panne, U., Fry, H., Lochotzke, H.M., and Preiss-Weigert, A. (2006) C and N stable isotope variation in urine and milk of cattle depending on the diet. *Anal. Bioanal. Chem.*, **386** (1), 104–108.

25 Bureau International des Poids et Mesures (BIPM) (2008) International Vocabulary of Metrology – Basic and General Concepts and Associated Terms (VIM), 3rd edn., JCGM 200:2008, http://www.bipm.org/en/publications/guides/vim.html (last accessed 23 December 2011); also published as ISO/IEC Guide 99–12:2007, International Organization for Standardization, Geneva.

26 Vanhaecke, F., Balcaen, L., and Malinovsky, D. (2009) Use of single-collector and multi-collector ICP-mass spectrometry for isotopic analysis. *J. Anal. At. Spectrom.*, **24** (7), 863–886.

27 Fleming, J., Neidhart, B., Tausch, C., and Wegscheider, W. (1996) Glossary of analytical terms (II to VI). *Accredit. Qual. Assur.*, **1**, 41–43; 87–88; 135; 190–191; 233–234; 277.

28 Fleming, J., Neidhart, B., Tausch, C., and Wegscheider, W. (1997) Glossary of analytical terms (VII to IX). *Accredit. Qual. Assur.*, **2**, 51–52; 160–161; 348–349.

29 Fleming, J., Neidhart, B., Tausch, C., and Wegscheider, W. (1998) Glossary of analytical terms (X). *Accredit. Qual. Assur.*, **3**, 171–173.

30 Holcombe, D. (1999) Alphabetical index of defined terms and where they can be found. Part I: A–F. *Accredit. Qual. Assur.*, **4**, 525–530.

31 Holcombe, D. (2000) Alphabetical index of defined terms and where they can be found. Part II: G–Q. *Accredit. Qual. Assur.*, **5**, 77–82.

32 Holcombe, D. (2000) Alphabetical index of defined terms and where they can be found. Part III: R–Z. *Accredit. Qual. Assur.*, **5**, 159–164.

33 ISO (1995) Guide to the Expression of Uncertainty in Measurement,

International Organization for Standardization, Geneva.

34 Ellison, S.L.R., Rosslein, M., and Williams, A. (eds.) (2000) Eurachem/CITAC Guide CG 4, Quantifying Uncertainty in Analytical Measurement, 2nd edn., http://www.measurementuncertainty.org/ (last accessed 23 December 2011).

35 Vogl, J. and Pritzkow, W. (2010) Isotope reference materials for present and future isotope research. *J. Anal. At. Spectrom.*, **25**, 923–932.

36 De Laeter, J.R., Bohlke, J.K., DeBièvre, P., Hidaka, H., Peiser, H.S., Rosman, K.J.R., and Taylor, P.D.P. (2003) Atomic weights of the elements: review 2000 (IUPAC technical report). *Pure Appl. Chem.*, **75** (6), 683–800.

37 Lide, D.R. (ed.) (2001) *Handbook of Chemistry and Physics*, 82nd edn., CRC Press, Boca Raton, FL

38 Vogl, J. (2005) Calibration strategies and quality assurance, in *ICP Mass Spectrometry*, 1st edn. (ed. S. Nelms), Blackwell, Oxford, pp. 147–181.

39 Montaser, A. (ed.) (1998) *Inductively Coupled Plasma Mass Spectrometry*, Wiley-VCH Verlag GmbH, New York.

40 Dulski, P. (1994) Interferences of oxide, hydroxide and chloride analyte species in the determination of rare-earth elements in geological samples by inductively-coupled plasma-mass spectrometry. *Fresenius' J. Anal. Chem.*, **350** (4–5), 194–203.

41 Mason, T.F.D., Weiss, D.J., Horstwood, M., Parrish, R.R., Russell, S.S., Mullane, E., and Coles, B.J. (2004) High-precision Cu and Zn isotope analysis by plasma source mass spectrometry – Part 1. Spectral interferences and their correction. *J. Anal. At. Spectrom.*, **19** (2), 209–217.

42 Nelms, S.M., Quetel, C.R., Prohaska, T., Vogl, J., and Taylor, P.D.P. (2001) Evaluation of detector dead time calculation models for ICP-MS. *J. Anal. At. Spectrom.*, **16** (4), 333–338.

43 Dodson, M.H. (1963) A theoretical study of use of internal standards for precise isotopic analysis by surface ionization technique. 1. General first-order algebraic solutions. *J. Sci. Instrum.*, **40** (6), 289–295.

44 Russell, W.A., Papanastassiou, D.A., and Tombrello, T.A. (1978) Ca isotope fractionation on Earth and other solar-system materials. *Geochim. Cosmochim. Acta*, **42** (8), 1075–1090.

45 Russ, G.P., and Bazan, J.M. (1987) Isotopic ratio measurements with an inductively coupled plasma source-mass spectrometer. *Spectrochim. Acta B*, **42** (1–2), 49–62.

46 Taylor, P.D.P., DeBièvre, P., Walder, A.J., and Entwistle, A. (1995) Validation of the analytical linearity and mass discrimination correction model exhibited by a multiple collector inductively-coupled plasma-mass spectrometer by means of a set of synthetic uranium isotope mixtures. *J. Anal. At. Spectrom.*, **10** (5), 395–398.

47 Pritzkow, W., Wunderli, S., Vogl, J., and Fortunato, G. (2007) The isotope abundances and the atomic weight of cadmium by a metrological approach. *Int. J. Mass Spectrom.*, **261** (1), 74–85.

48 Rehkämper, M. and Halliday, A.N. (1998) Accuracy and long-term reproducibility of lead isotopic measurements by multiple-collector inductively coupled plasma mass spectrometry using an external method for correction of mass discrimination. *Int. J. Mass Spectrom.*, **181**, 123–133.

49 Walder, A.J. (1997) Advanced isotope ratio mass spectrometry II: isotope ratio measurement by multiple collector inductively coupled plasma mass spectrometers, in *Modern Isotope Ratio Mass Spectrometry*, 1st edn. (ed. I.T. Platzner), John Wiley & Sons, Ltd., Chichester, pp. 11–82.

50 Ingle, C.P., Sharp, B.L., Horstwood, M.S.A., Parrish, R.R., and Lewis, D.J. (2003) Instrument response functions, mass bias and matrix effects in isotope ratio measurements and semi-quantitative analysis by single- and multi-collector ICP-MS. *J. Anal. At. Spectrom.*, **18** (3), 219–229.

51 NIST: National Institute for Standards and Technology (2011) Standard Reference Materials. http://www.nist.gov/srm/ (last accessed 08 February 2012).

52 IRMM: Institute for Reference Materials and Measurements, Joint Research Center (2011) Reference Materials. http://irmm.jrc.ec.europa.eu/reference_materials_catalogue/Pages/index.aspx (last accessed 08 February 2012).

53 ERM: European Reference Materials (2011) http://www.erm-crm.org/ (last accessed 08 February 2012).

54 BAM: Bundesanstalt für Materialforschung und -prüfung (Federal Institute for Materials Research and Testing) (2011) Reference Materials. http://www.bam.de/en/fachthemen/referenzmaterialien/index.htm (last accessed 08 February 2012).

55 Epstein, S. and Mayeda, T. (1953) Variation of O-18 content of waters from natural sources. *Geochim. Cosmochim. Acta*, **4** (5), 213–224.

56 Gröning, M. (2004) International stable isotope reference materials, in *Handbook of Stable Isotope Analytical Techniques*, 1st edn. (ed. P.A. de Groot), Elsevier, Amsterdam, vol. 1, pp. 874–906.

57 Bohlke, J.K., de Laeter, J.R., De Bievre, P., Hidaka, H., Peiser, H.S., Rosman, K.J.R., and Taylor, P.D.P. (2005) Isotopic compositions of the elements, 2001. *J. Phys. Chem. Ref. Data*, **34** (1), 57–67.

58 Dempster, A.J. (1920) Positive ray analysis of magnesium. *Science*, **52**, 559.

59 Nier, A.O. (1950) A redetermination of the relative abundances of the isotopes of carbon, nitrogen, oxygen, argon, and potassium. *Phys. Rev.*, **77** (6), 789–793.

60 Chang, T.L., Li, W.J., Qiao, G.S., Qian, Q.Y., and Chu, Z.Y. (1999) Absolute isotopic composition and atomic weight of germanium. *Int. J. Mass Spectrom.*, **189** (2–3), 205–211.

61 Chang, T.L., Li, W.J., Zhao, M.T., Wang, J., and Qian, Q.Y. (2001) Absolute isotopic composition and atomic weight of dysprosium. *Int. J. Mass Spectrom.*, **207** (1–2), 13–17.

62 Chang, T.L., Zhao, M.T., Li, W.J., Wang, J., and Qian, Q.Y. (2001) Absolute isotopic composition and atomic weight of zinc. *Int. J. Mass Spectrom.*, **208** (1–3), 113–118.

63 Ding, T., Valkiers, S., Kipphardt, H., De Bievre, P., Taylor, P.D.P., Gonfiantini, R., and Krouse, R. (2001) Calibrated sulfur isotope abundance ratios of three IAEA sulfur isotope reference materials and V-CDT with a reassessment of the atomic weight of sulfur. *Geochim. Cosmochim. Acta*, **65** (15), 2433–2437.

64 Briche, C., Held, A., Berglund, M., DeBièvre, P., and Taylor, P.D.P. (2002) Measurement of the isotopic composition and atomic weight of an isotopic reference material of platinum, IRMM-010. *Anal. Chim. Acta*, **460** (1), 41–47.

65 Chang, T.L., Zhao, M.T., Li, W.J., Wang, J., and Qian, Q.Y. (2002) Absolute isotopic composition and atomic weight of samariurn. *Int. J. Mass Spectrom.*, **218** (2), 167–172.

66 Tanimizu, M., Asada, Y., and Hirata, T. (2002) Absolute isotopic composition and atomic weight of commercial zinc using inductively coupled plasma mass spectrometry. *Anal. Chem.*, **74** (22), 5814–5819.

67 Russe, K., Valkiers, S., and Taylor, P.D.P. (2004) Synthetic isotope mixtures for the calibration of isotope amount ratio measurements of carbon. *Int. J. Mass Spectrom.*, **235** (3), 255–262.

68 Ding, T., Wan, D., Bai, R., Zhang, Z., Shen, Y., and Meng, R. (2005) Silicon isotope abundance ratios and atomic weights of NBS-28 and other reference materials. *Geochim. Cosmochim. Acta*, **69** (23), 5487–5494.

69 Valkiers, S., Russe, K., Taylor, P., Ding, T., and Inkret, M. (2005) Silicon isotope amount ratios and molar masses for two silicon isotope reference materials: IRMM-018a and NBS28. *Int. J. Mass Spectrom.*, **242** (2–3), 319–321.

70 Zhao, M.T., Zhou, T., Wang, J., Lu, H., Fang, X., Guo, C.H., Li, Q.L., and Li, C.F. (2005) Absolute isotopic composition and atomic weight of neodymium using thermal ionization mass spectrometry. *Rapid Commun. Mass Spectrom.*, **19** (19), 2743–2746.

71 Zhao, M.T., Zhou, T., Wang, J., Lu, H., and Xiang, F. (2005) Absolute measurements of neodymium isotopic

abundances and atomic weight by MC-ICPMS. *Int. J. Mass Spectrom.*, **245** (1–3), 36–40.

72 De Laeter, J.R. and Bukilic, N. (2006) The isotopic composition and atomic weight of ytterbium. *Int. J. Mass Spectrom.*, **252** (3), 222–227.

73 Ponzevera, E., Quétel, C.R., Berglund, M., Taylor, P.D.P., Evans, P., Loss, R.D., and Fortunato, G. (2006) Mass discrimination during MC-ICPMS isotopic ratio measurements: investigation by means of synthetic isotopic mixtures (IRMM-007 series) and application to the calibration of natural-like zinc materials (including IRMM-3702 and IRMM-651). *J. Am. Soc. Mass Spectrom.*, **17** (10), 1412–1427.

74 Valkiers, S., Varlam, M., Russe, K., Berglund, M., Taylor, P., Wang, J., Milton, M.J.T., and DeBièvre, P. (2007) Preparation of synthetic isotope mixtures for the calibration of carbon and oxygen isotope ratio measurements (in carbon dioxide) to the SI. *Int. J. Mass Spectrom.*, **264** (1), 10–21.

75 Mana, G., Rienitz, O., and Pramann, A. (2010) Measurement equations for the determination of the Si molar mass by isotope dilution mass spectrometry. *Metrologia*, **47** (4), 460–463.

76 Mana, G., and Rienitz, O. (2010) The calibration of Si isotope ratio measurements. *Int. J. Mass Spectrom.*, **291** (1–2), 55–60.

77 Rienitz, O., Pramann, A., and Schiel, D. (2010) Novel concept for the mass spectrometric determination of absolute isotopic abundances with improved measurement uncertainty: Part 1 – theoretical derivation and feasibility study. *Int. J. Mass Spectrom.*, **289** (1), 47–53.

78 DeBièvre, P., De Laeter, J.R., Peiser, H.S., and Reed, W.P. (1993) Reference materials by isotope ratio mass spectrometry. *Mass Spectrom. Rev.*, **12** (3), 143–172.

79 COMAR (2011) International Database for Certified Reference Materials, http://www.comar.bam.de/ (last accessed 09 February 2012).

80 Max Planck Institute (2011) GeoReM: Geological and Environmental Reference Materials, http://georem.mpch-mainz.gwdg.de (last accessed 09 February 2012).

81 NBL: New Brunswick Laboratories (2011) Certified Reference Materials, http://www.nbl.doe.gov (last accessed 23 December 2011).

82 Choi, S.-M., Lee, H.-S., Lee, G.-H., and Han, J.-K. (2008) Determination of the strontium isotope ratio by ICP-MS ginseng as a tracer of regional origin. *Food Chem.*, **108**, 1149–1154.

83 Rosner, M. (2010) Geochemical and instrumental fundamentals for accurate and precise strontium isotope data of food samples: comment on "Determination of the strontium isotope ratio by ICP-MS ginseng as a tracer of regional origin" (Choi *et al.*, 2008). *Food Chem.*, **121** (3), 918–921.

84 Wasylenki, L.E., Rolfe, B.A., Weeks, C. L., Spiro, T.G., and Anbar, A.D. (2008) Experimental investigation of the effects of temperature and ionic strength on Mo isotope fractionation during adsorption to manganese oxides. *Geochim. Cosmochim. Acta*, **72** (24), 5997–6005.

85 Moynier, F., Blichert-Toft, J., Telouk, P., Luck, J.M., and Albarede, F. (2007) Comparative stable isotope geochemistry of Ni, Cu, Zn, and Fe in chondrites and iron meteorites. *Geochim. Cosmochim. Acta*, **71** (17), 4365–4379.

86 Schmitt, A.D., Galer, S.J.G., and Abouchami, W. (2009) High-precision cadmium stable isotope measurements by double spike thermal ionisation mass spectrometry. *J. Anal. At. Spectrom.*, **24** (8), 1079–1088.

87 Catanzaro, E.J., Murphy, T.J., Garner, E.L., and Shields, W.R. (1966) Absolute isotopic abundance ratios and atomic weight of magnesium. *J. Res. Natl. Bur. Stand. A: Phys. Chem.*, **70A** (6), 453–458.

88 Galy, A., Yoffe, O., Janney, P.E., Williams, R.W., Cloquet, C., Alard, O., Halicz, L., Wadhwa, M., Hutcheon, I.D., Ramon, E., and Carignan, J. (2003) Magnesium isotope heterogeneity of the isotopic standard SRM980 and new reference materials for magnesium-isotope-ratio measurements. *J. Anal. At. Spectrom.*, **18** (11), 1352–1356.

89 Vogl, J., Pritzkow, W., and Klingbeil, P. (2004) The need for new SI-traceable magnesium isotopic reference materials. *Anal. Bioanal. Chem.*, **380** (7–8), 876–879.

90 Vocke, R. and Mann, J. (2010) Isotopic reference materials for the 21st century, presented at the FACSS Conference, 2008, Reno, NV, http://facss.org/contentmgr/showdetails.php/id/34733.

7
Quality Control in Isotope Ratio Applications

Thomas Meisel

7.1
Introduction

Although quality control has been accepted and implemented in many laboratories mainly dealing with routine commercial analysis, insufficient awareness in academia, in particular in research laboratories, still exists. Comparability of analytical results and traceability are issues that are also becoming prominent in the field of isotope ratio measurements of "non-traditional" elements with multi-collector (MC) inductively coupled plasma mass spectrometry (ICP-MS), especially as in many applications the limits of the technique are explored. Proficiency testing schemes (see below) and internationally organized comparative trials with expert laboratories reveal that even not all national metrological institutions (NMIs), such as the Bundesanstalt für Materialforschung und -prüfung (BAM) (the Federal Institute for Materials Research and Testing) in Germany and the National Institute of Standards and Technology (NIST) in the USA, and international organizations such as the International Atomic Energy Agency (IAEA) produce comparable results. Evidence is given by means of an example of a "traditional" isotope system, namely a uranium isotope ratio measurement carried out during a CCQM (Comité Consultatif pour la Quantité de Matière – Métrologie en Chimie; Consultative Committee for Amount of Substance – Metrology in Chemistry) pilot study, organized by the European Union's Institute for Reference Materials and Measurements IRMM [1] (Figure 7.1).

In the early days of mass spectrometry, research was focused on gas source isotope ratio mass spectrometry (IRMS) or thermal ionization mass spectrometry (TIMS), with as the main aim the determination of the isotopic composition and molar masses of the elements. Since the 1940s, isotope ratio measurements have also been used for the determination of isotope ratios involving a radiogenic nuclide and for quantitative element determination via isotope dilution (see also Chapter 8). The age of the solar system and the Earth were also of particular interest within isotope ratio science. With the establishment and improvement of TIMS instrumentation, the awareness of the importance of precision and accuracy and the need to be able to reproduce a result in another laboratory grew, while

Isotopic Analysis: Fundamentals and Applications Using ICP-MS,
First Edition. Edited by Frank Vanhaecke and Patrick Degryse.

Published 2012 by WILEY-VCH Verlag GmbH & Co. KGaA

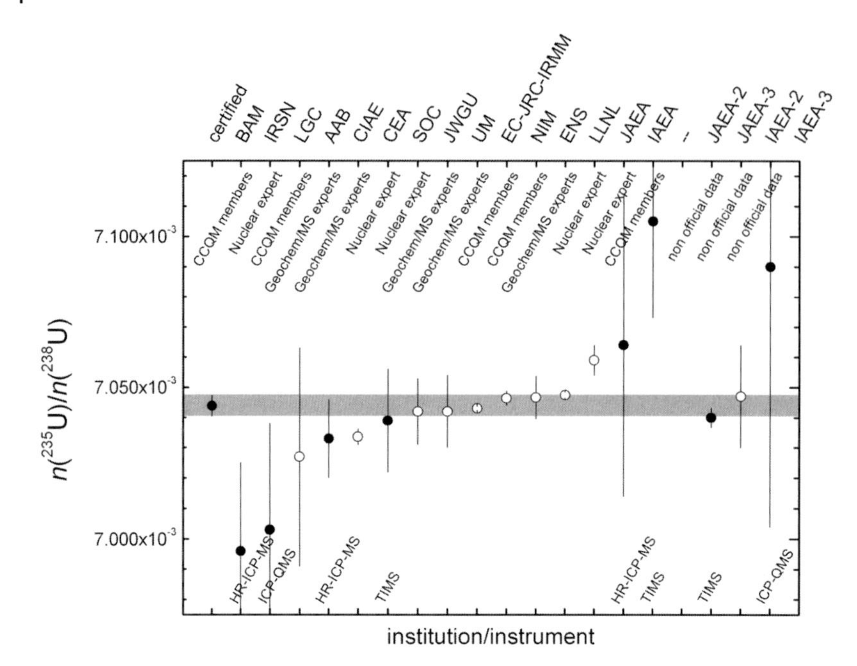

Figure 7.1 Results of $^{235}U/^{238}U$ isotope ratio measurements in a simulated biological/environmental material containing 5 mg kg^{-1} U, submitted by laboratories using different instruments and/or sample preparation techniques and with various levels of/areas of expertise [1]. Open symbols represent data obtained by MC-ICP-MS all other instruments are marked individually. The difference between the results obtained by expert laboratories is rather disillusioning. Laboratories were also asked to submit measurement uncertainties, which in some cases turned out to be underestimated and in others overestimated.

geologists/geochronologists realized the need to have an "error" on the age obtained for their models. As a result, different ways of assuring quality have been established, but independently. The methods for assuring quality of analytical data have typically been passed on within each laboratory.

While traditional radiogenic isotope systems, such as U–Pb and Rb–Sr, did not require very high levels of precision for samples that show a large spread of the parent/daughter ratio, the introduction of the Sm–Nd isotope system in the mid-1970s created a new challenge. As the isotopic variations were in the range of permil (‰) to percent (%) only, comparability became an issue. A reference material (the so-called La Jolla standard) was introduced by Wasserburg *et al.* [2] in 1981 and has been used ever since by most laboratories dealing with Nd isotope ratio measurements. This solution, so to speak, set a "standard" for making data comparable. However, the La Jolla standard was not used exclusively; other solutions, which were not traceable to the La Jolla standard, were also prepared as in-house or between-laboratory quality control materials. Comparability of data was not assured in any case, as laboratories introduced factors to correct for deviations between the Nd isotope ratios obtained with a particular TIMS instrument and the corresponding values published for the La Jolla solution. These factors were then used to correct individual sample results without understanding the reasons for the deviations observed and without taking the corresponding additional uncertainty component into account. It was clear from the beginnings of Nd isotope ratio measurements (for Sm–Nd geochronology) within the radiogenic isotope community that common solutions were necessary for quality control [1]. Although only data obtained for a homogeneous solution were compared and no data obtained for real (rock) samples were compared between laboratories, a first step to make data comparable was made.

With the advent of MC-ICP-MS, increasingly lower concentrations and higher precision levels were explored, leading to the discovery that not only light stable isotopes show mass-dependent and mass-independent isotope fractionation. As a result, new issues concerning quality control appeared. Owing to the minute, previously undiscovered and even disbelieved isotopic variations, it was not possible to express the "true" isotopic composition or isotopic abundances in absolute terms. As differences between MC-ICP-MS and TIMS results, between laboratories, and also between results obtained by different MC-ICP-MS instruments occurred, comparability had to be assured by other means. Isotope variations were only expressed relative to a reference value (similar to what was/is used for the O isotopes, i.e., using Standard Mean Ocean Water (SMOW) as a reference against which to report delta values). At the point where the limits of the methods are explored and the signal-to-background ratio becomes very low (the signal approaches the noise level), quality control becomes particularly important.

The importance of measurement uncertainty estimation as part of a quality control scheme needs to be recognized by all scientists working with isotope ratio measurements. Another important aspect is a common terminology. MC-ICP-MS turned out to be a very useful tool, both for geoscientists, who in most cases were previously TIMS users, and analytical chemists, who mostly were quadrupole-based ICP-MS or single-collector sector field ICP-MS users. Since both of these areas of science work on the same isotope systems, mainly that of non-traditional stable isotopes, exchange of information is possible only if both agree on the use of common terms and definitions. Here the use of VIM3 International Vocabulary of Metrology as the basis for a better communication is recommended and this document will be used in as practical a manner as possible throughout this chapter.

The definition of the measurand (the quantity that is being determined by measurement) is the heart of a result [3]. A correct measurand is needed to answer (fundamental) questions. However, this goal is dependent on the fitness-for-purpose of the currently available methods. Consider the following measurands relating to a 500 mg laboratory sample from one particular meteorite (the basic idea was taken from [3]):

1. the amount concentration of tungsten in a 100 mg test portion
2. the amount concentration of tungsten in the laboratory sample

3. the amount concentration of tungsten in the meteorite
4. the amount concentration of tungsten in a meteorite class (e.g., carbonaceous chondrites)
5. the isotope amount ratios (or, shorter, isotope ratios) of tungsten expressed as epsilon values (see also Chapter 1) in a 100 mg test portion
6. the isotope amount ratios of tungsten expressed as epsilon values in the meteorite
7. the solar abundance of tungsten
8. the "absolute" isotope amount ratios of tungsten in the meteorite
9. the "absolute" isotope amount ratios of tungsten of our solar system.

Nowadays we are able to obtain results for measurands 1–6, we might end up with a value for measurands 7–9, but a realistic measurement uncertainty will be far too large to identify any significant difference in the tungsten isotopic composition between parts of the solar system. However, values for measurands 1–8 or even 9 can be found in the literature, but the validity of the interpretations based on uncertain data needs to be questioned.

To make progress in science possible, it is important also to be able to rely on data produced by other laboratories and any particular analyst anywhere in the world should be able to reproduce reported data. Like trusting the weight balance at a farmers' market when buying fresh products, we should be able to trust all isotope ratio data obtained using MC-ICP-MS and published in a scientific journal. This goal can only be achieved by using *validated* analytical procedures that produce results that are *traceable* to a common reference that is fit for the purpose intended. This chapter will not provide recipes, neither for how to produce data that are fit-for-purpose, nor for how to establish a realistic uncertainty budget. The aim rather is to highlight areas where more awareness is needed in order to produce reliable and useful isotopic data.

7.2 Terminology and Definitions

All definitions in this section are taken directly from VIM3 [4] with some modifications, additions, and removal of notes that are not relevant for this chapter.

Measurand

Definition: Quantity intended to be measured.

Note 4: In chemistry, "analyte," or the name of a substance or compound, are terms sometimes used for "measurand." This usage is erroneous because these terms do not refer to quantities.

Description: Through this definition, measurand is mostly a concentration, but not an analyte or an element or a compound for which the term measurand is sometimes misused.

Measured quantity value

Definition: Quantity value representing a measurement result.

Note 1: For a measurement involving replicate indications, each indication can be used to provide a corresponding measured quantity value. This set of individual measured quantity values can be used to calculate a resulting measured quantity value, usually with a decreased associated measurement uncertainty.

Note 4: In the *Guide to the Expression of Uncertainty in Measurement* (GUM), published by the International Organization for Standardization ISO [5], the terms "result of measurement" and "estimate of the value of the measurand" or just "estimate of the measurand" are used for "measured quantity value" (Figure 7.2).

Measurement precision

Definition: The closeness of agreement between indications or measured quantity values obtained by replicate measurements on the same or similar objects under specified conditions.

Description: Measurement precision is usually expressed numerically by measures of imprecision, such as standard deviation, variance, or coefficient of variation, under the specified conditions of measurement. When a measurement precision is given, it is important to specify the conditions. These conditions can be, for example, repeatability condition of measurement, intermediate precision condition of measurement, or reproducibility condition of measurement (see ISO 5725-3:1994 and see below). The measurement precision is used to define measurement repeatability, intermediate measurement precision, and measurement reproducibility. In the VIM, it is mentioned that sometimes "measurement precision" is erroneously used to indicate measurement accuracy.

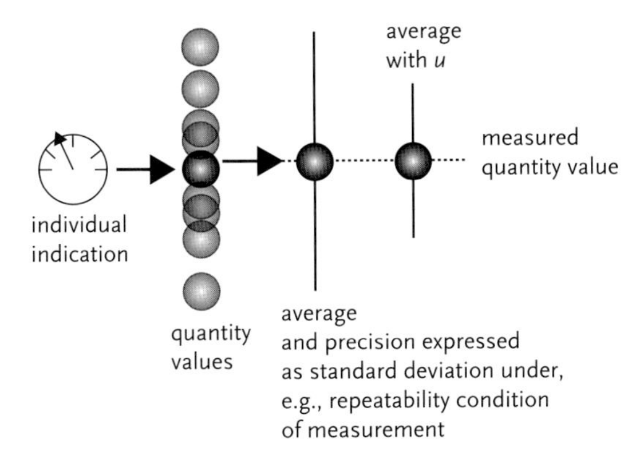

Figure 7.2 The measured quantity value is presented here as an average of quantity values which are based on replicate individual indications. Indications are quantity values provided by a measuring instrument or a measuring system.

Repeatability condition of measurement

Definition: Condition of measurement in a set of conditions that includes the same measurement procedure, same operators, same measuring system, same operating conditions, and same location, and replicate measurements on the same or similar objects over a short period of time.

Measurement repeatability

Definition: Measurement precision under a set of repeatability conditions of measurement (Figure 7.3).

Intermediate precision condition of measurement

Definition: Condition of measurement in a set of conditions that includes the same measurement procedure, same location, and replicate measurements on the same or similar objects over an extended period of time, but may include other conditions involving changes.

Description: The changes can include new calibrations, calibrators, operators and measuring systems. A specification of the conditions should contain the conditions changed and those unchanged, to the extent practical. In the isotope measurement community, the term "external reproducibility" (external precision) is sometimes used to designate this concept.

Intermediate measurement precision

Definition: Measurement precision under a set of intermediate precision conditions of measurement (Figure 7.3).

Reproducibility condition of measurement

Definition: Condition of measurement in a set of conditions that includes different locations, operators, measuring systems, and replicate measurements on the same or similar objects. The different measuring systems may use different measurement procedures.

Description: The associated precision can in most cases only be approximated and it will be the highest, thus the other extreme of precision under repeatability condition. It is important that in a publication, a statement concerning the conditions changed and unchanged, to the extent practical, is provided.

Measurement reproducibility

Definition: Measurement precision under reproducibility condition of measurement (Figure 7.3).

Metrological traceability

Definition: Property of a measurement result whereby the result can be related to a stated reference through a documented unbroken chain of calibrations, each contributing to the measurement uncertainty.

Note 1: For this definition, a "stated reference" can be a definition of a measurement unit through its practical realization, or a measurement procedure

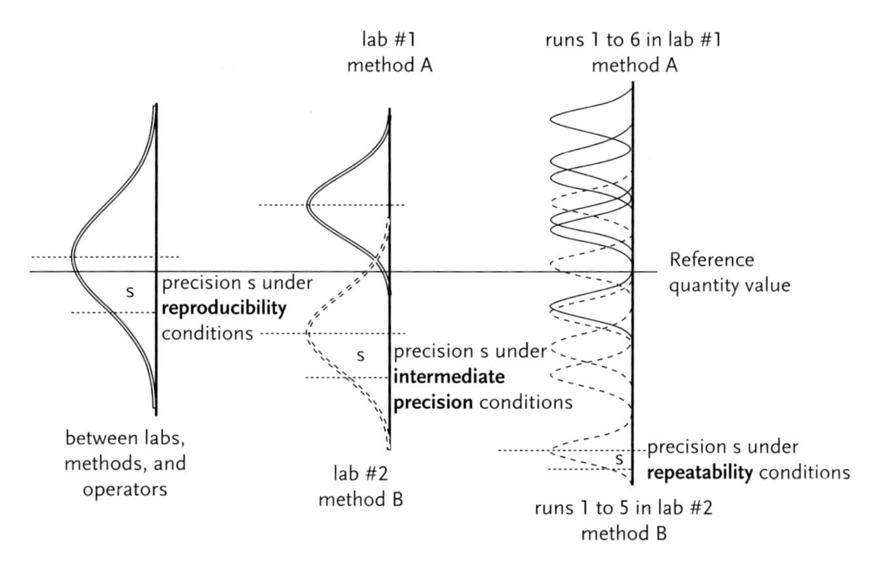

Figure 7.3 The three types of measurement conditions under which measurement precision can be determined are demonstrated via results obtained in two laboratories (1 and 2) and via two methods (A and B).

including the measurement unit for a non-ordinal quantity, or a measurement standard.

Note 2: Metrological traceability requires an established calibration hierarchy.

Metrological traceability to a measurement unit

Definition: Metrological traceability where the stated reference is the definition of a measurement unit through its practical realization'.

Note: The expression "traceability to the SI" means metrological traceability to a measurement unit of the International System of Units.

Metrological comparability of measurement results

Definition: Comparability of measurement results that are metrologically traceable to the same reference.

Description: Measurement results for the distances between the Earth and the Moon and between Paris and London are, for example, metrologically comparable when they are both metrologically traceable to the same measurement unit, for example, the meter.

Systematic measurement error

Definition: Component of measurement error that in replicate measurements remains constant or varies in a predictable manner. Synonyms are systematic measurement error, systematic error of measurement, and systematic error.

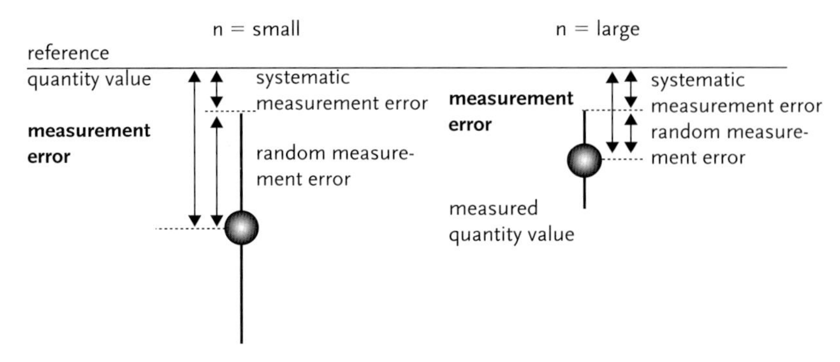

Figure 7.4 The systematic and the random errors add up to the measurement error. By increasing the number of measurements *n*, the random component can be reduced, but the systematic measurement error remains constant.

Note 1: The reference quantity value for a systematic measurement error is a true quantity value, or a measured quantity value of a measurement standard of negligible measurement uncertainty, or a conventional quantity value.

Note 2: Systematic measurement error, and its causes, can be known or unknown. A correction can be applied to compensate for a known systematic measurement error (Figure 7.4).

Measurement bias

Definition: Systematic measurement error or its estimate, with respect to a reference quantity value (also called bias).

Measurement uncertainty, uncertainty of measurement, uncertainty

Definition: Parameter characterizing the dispersion of the quantity values being attributed to a measurand, based on the information used.

Note 1: Measurement uncertainty includes components arising from systematic effects, such as components associated with corrections and the assigned quantity values of measurement standards, and also the definitional uncertainty. Sometimes known systematic effects are not corrected for but are instead treated as uncertainty components.

Note 2: The parameter may be, for example, a standard deviation called standard measurement uncertainty (or a specified multiple of it), or the half-width of an interval, having a stated coverage probability.

Note 3: Measurement uncertainty comprises, in general, many components. Some of these may be evaluated by Type A evaluation of measurement uncertainty from the statistical distribution of the quantity values from series of measurements and can be characterized by experimental standard deviations. The other components, which may be evaluated by Type B evaluation of measurement uncertainty, can also be characterized by standard deviations, evaluated from probability density functions based on experience or other information. The meaning of Type A and Type B will be clarified further in the text.

Detection limit, limit of detection (LOD)

Definition: Measured quantity value, obtained by a given measurement procedure, for which the probability of falsely claiming the absence of a component in a material is β, given a probability α of falsely claiming its presence.

Note 1: The International Union of Pure and Applied Chemistry (IUPAC) recommends default values for α and β equal to 0.05 (Figure 7.5).

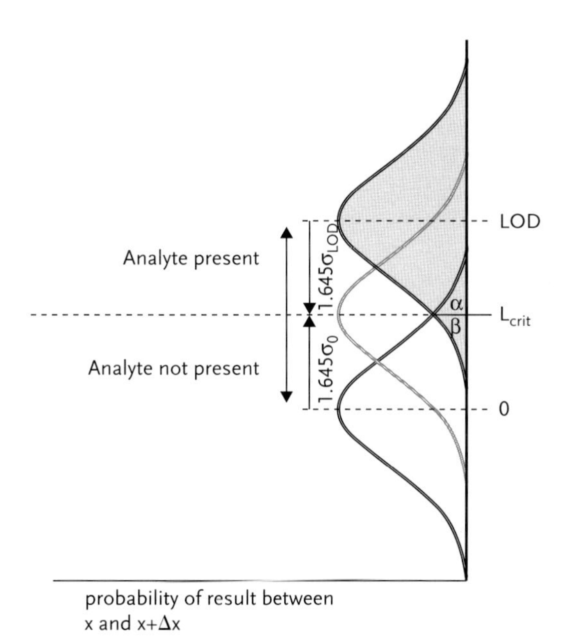

Figure 7.5 Limit of detection. For reasons of clarity, the distribution functions have been assumed to be the same. Following the definition of VIM3 and the IUPAC recommendation, which can also be found in a paper by Currie [6], the chance of positively detecting an analyte at the critical limit (L_{crit}) is only 50%. The distance between the mean of the blank solution and LOD is 2×1.645, which is 3.29 times the standard deviation of the blank solution. At this level, the risk of false non-detection of an analyte is in this case $\beta = 5\%$ and of false-positive detection $\alpha = 5\%$.

Reference quantity value, reference value

Definition: Quantity value, generally accepted as having a suitably small measurement uncertainty to be used as a basis for comparison with values of quantities of the same kind.

Note 1: A reference quantity value with associated measurement uncertainty is usually provided with reference to:

a. a material, for example, a certified reference material
b. a device, for example, a stabilized laser
c. a reference measurement procedure
d. a comparison of measurement standards.

7.3 Measurement Uncertainty

To make measurement results comparable and interpretable, each result needs to be associated with a statement of uncertainty of the measurement result. Although this is commonplace, ways of how to express uncertainties (erroneously called "errors") are diverse, unclear, and often without proper statements on how the uncertainty budget was derived (see also [1]). It should be noted that through the definitions it is clear that whereas measurement uncertainty defines a range in which the "true" value should lie with a given confidence, measurement error is a difference between two values.

The traditional way of expressing a measure of confidence in the measurement results is a statement of precision and accuracy, with precision (random component) being a kind of measure of how well data can be reproduced when repeated and with accuracy being some sort of confidence on how true the value (potentially affected by systematic error component or bias) might be. With this approach, the deviation from the true value is a combination of systematic and random errors. Here, these two kinds of errors are always considered to be distinguishable and are treated differently. There are no rules on how to combine these two types of error to obtain a total error on a measurement result. In some cases, no clear distinction between random and systematic error is made. Isotope geologists, for example, use the "external reproducibility," which is actually the intermediate measurement precision of a reference solution, as a measure of uncertainty. When the so-called "internal" precision, that is, the precision under repeatability condition, which is the standard error of the mean of block ratios (e.g., based on 10 means from 10 blocks, each comprising 10 isotope ratio measurements divided by the square root of 10 multiplied by 2), is larger than the intermediate measurement precision of the pure reference isotope solution, the "internal" precision of the measurement result of the isotope ratio determination is used as measure of uncertainty. In the case when the "internal" precision of the isotope amount ratio measurement is smaller than the intermediate precision of the pure reference solution, the intermediate precision is quoted as uncertainty of the measurement result. Although this may be practical, this approach does not take systematic errors into account. In addition to the problem of comparing standard errors of the mean with standard deviations, the intermediate measurement precision is only based on isotope ratio measurements of a matrix-free standard solution. Hence variance introduced through intrinsic inhomogeneity of most samples, sample preparation steps, matrix effects, and so on, are not included in the uncertainty budget.

The more useful and modern metrological approach to express the reliability of a measurement result is a statement of measurement uncertainty that is associated with every measurement result. It is not the goal of this method to approach the true value as close as possible. Rather, the objective is to assign an interval of reasonable values to the measurand. Instead of searching for systematic and random errors, components of measurement uncertainty are assigned and grouped into two categories, Type A and Type B, where Type A components

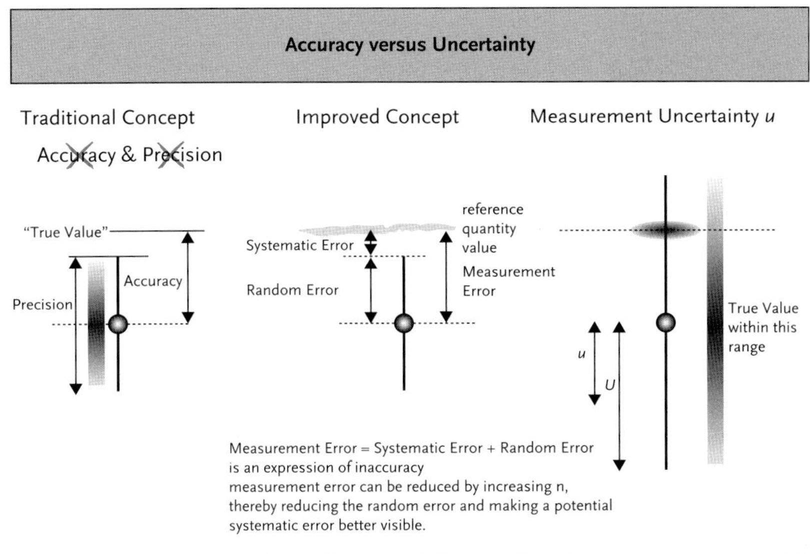

Figure 7.6 Comparison of the traditional and the metrological concepts. Whereas traditional concepts use differences, measurement uncertainty gives a range.

are evaluated by statistical means and Type B by other means (e.g., experience). The measurement uncertainty budget of a measurement result consists of a combination of variances caused by influence quantities. The combined standard uncertainty (u_c) for a typical isotope measurement can be described in a very simplified way, as shown in Eq. (7.1):

$$u_c{}^2 = u_{\text{sampling}}{}^2 + u_{\text{sample preparation}}{}^2 + u_{\text{isotope analysis}}{}^2 \qquad (7.1)$$

If the contributions of the single random effects are expressed as variances (the squared uncertainties), the most evident uncertainty components can be added linearly. From this equation, it is not evident which component of the uncertainty propagation has the largest influence on the combined standard uncertainty u_c. Only careful consideration of all possible influence quantities, taking into account correlations between them, can lead to a favorable measurement uncertainty that is fit-for-purpose. For the identification of possible influence quantities and their correlations, a cause–effect diagram (also termed fishbone diagram) is very useful. In a cause–effect diagram, all variables that have an influence on the measurement result are plotted as arrows, connected in such a way that their influence on a particular part of the measurement procedure is visualized. An example is given in Figure 7.7 for determination of a platinum group element (PGE) at low mass fractions ($<10\ \mu g\ kg^{-1}$) with ICP-isotope dilution mass spectrometry (IDMS).

The model [Eq. (7.2)] for the uncertainty calculations depicted in Figure 7.7 is based on the general isotope dilution equation of De Bièvre [8]:

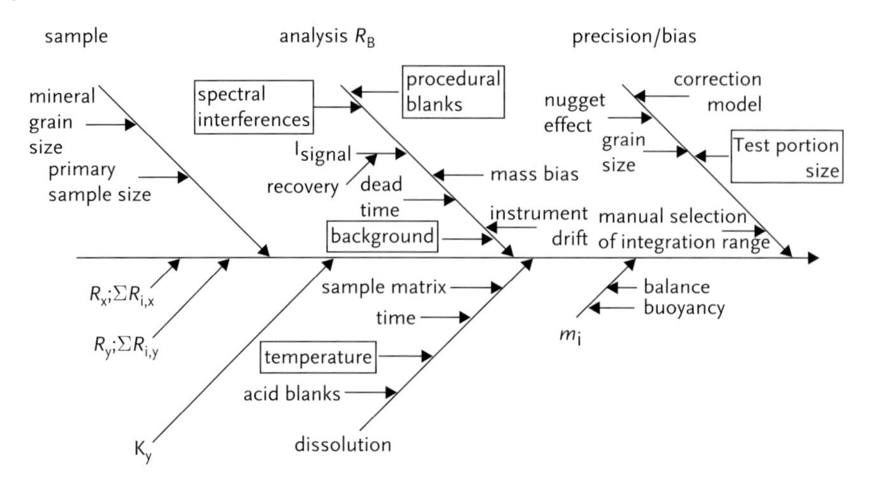

Figure 7.7 Cause–effect or fishbone diagram for the isotope dilution determination of the mass fraction of a platinum group element (Os) in geological reference materials [7]. The most important influence quantities are marked with a box. It takes into account natural variations in the $^{187}Os/^{188}Os$ ratio, which can contribute significantly to the total uncertainty in the Os concentration. R represents the isotope ratio (the artificially enriched isotope divided by the isotope used for normalization) for the blend (B), the spike (y) and the naturally occurring element (x). Masses of sample and spike are represented by m and are expressed in grams. $\Sigma R_{i,x}$ and $\Sigma R_{i,y}$ are the sum of all isotope ratios for the spike and the sample, respectively; k represents the amount contents in mol g^{-1}.

$$k_x = \frac{R_y - R_B}{R_B - R_x} \frac{\sum R_{ix}}{\sum R_{iy}} \frac{m_y}{m_x} k_y \tag{7.2}$$

As PGE-containing minerals are difficult to digest, the temperature at which the digestion step is carried through has a marked influence [9]. Blanks are important at a target element level lower than 1 µg kg^{-1}. Spectral interferences in ICP-MS cannot be neglected and in most natural matrices they can cause severe overestimations of concentrations. Hence this influence quantity has to be considered and its contribution to the budget estimated when mass-dependent and mass-independent isotope ratio fractionation need to be studied.

All other influence quantities cannot be neglected but are of lesser importance, but their magnitude of influence needs to be checked. Under normal conditions in an analytical laboratory, most influence quantities are under control, as is the case for, for example, mass determinations on balances or pipetting of solutions of known density for dilution purposes. In the following sections, the most significant influence quantities are discussed. These are typically important for most analytes and analytical procedures involving isotope ratio determinations with ICP-MS.

7.3.1 Influence Quantities

In a first step, influence quantities of isotope amount ratio measurements aiming at quantifying isotopic variations are discussed. In a second step, influence quantities also affecting isotope dilution measurements will be taken into account.

When considering the capabilities of modern analytical instruments, it is hard to imagine that through determination of isotope ratios in minute quantities of analyte at an incredibly high precision, it is possible to distill hypotheses on the first 50 Ma of our solar system [10–12]. As it turns out, some isotope amount ratio results that formed the basis for hypotheses, however, were not reproducible and therefore did not form a solid foundation for drawing conclusions [13]. Reasons for irreproducible isotope ratios are manyfold, but need to be understood. One way to identify sources of error is the study of all possible influence quantities individually through a bottom-up approach, as recommended by Eurachem (a network of organizations in Europe having the objective of establishing a system for the international traceability of chemical measurements and the promotion of good quality practices) [14]. This approach has the advantage of identifying the most important contributors to variance in the measurement result. Once identified, the influence of the source can be reduced or eliminated, leading to an improved combined standard uncertainty. Here, the most important influence quantities are described and their magnitude discussed.

7.3.1.1 Sampling

Sampling needs to be recognized as the first step in the measurement process and its contribution to the measurement uncertainty budget cannot be ignored [15]. Approaches on how to estimate realistic uncertainties arising from sampling can be found in the Eurachem/EUROLAB/CITAC/Nordtest/AMC guide [16].

7.3.1.2 Sample Preparation

Spectral interferences due to the occurrence of isobaric, polyatomic, and doubly charged ions, signal suppression, increased drift effects, and so on, are typically encountered when the target element or compound is not purified and not separated from its matrix. Several strategies for achieving proper separation have been developed. It must be emphasized that reliable isotope ratio data cannot be achieved with an MC-ICP-MS instrument without chemical analyte–matrix separation. Preferably, analyte isolation via a chromatographic process has to be accomplished with quantitative analyte recovery, as isotope fractionation on the chromatographic column has been reported.

7.3.1.3 Isotope Amount Ratio Determination

Single-Collector Versus Multi-Collector ICP-MS It was the combination of more than one detector with sector field ICP-MS instrumentation that allowed high-precision isotope amount ratio measurements reaching, in some cases, the previously unmatched precision of MC-TIMS instruments. As a result of the instability of the plasma-generated ion signal, single-collector quadrupole-based or single-collector sector field ICP-MS will never reach the precision that can be achieved with single-collector TIMS, but still can achieve precision levels that are fit-for-purpose for some isotope amount ratio determinations involving radiogenic isotopes. A successful application for single-collector instruments is, for example, the study of radiogenic lead isotopic variations as Pb isotope ratios ($^{207}Pb/^{206}Pb$, $^{208}Pb/^{206}Pb$) show natural variations exceeding the precision under reproducibility conditions of single-collector instruments. Another radiogenic isotope system that is accessible with single-collector instruments is the Re–Os isotope system, where the determination of the $^{187}Os/^{188}Os$ isotope amount ratio can be used for geochemical and oceanographic studies.

Single-collector instruments also prove very useful for mass content determinations via isotope dilution, as careful estimation of all quantities that influence the uncertainty budget demonstrates that the precision on the isotope amount ratio is typically not the dominant factor for high-precision measurements. Often, the accuracy of a mass content measurement will hardly improve through the use of MC-ICP-MS instruments as other influence quantities, such as uncontrolled spectral interferences and sample inhomogeneities, typically deserve more attention.

Mass Bias Correction and Drift Effects Instrumental mass discrimination and drift effects are two influence quantities having a major impact on the uncertainty budget if not properly taken into account. Instrumental mass bias affecting the isotope amount ratio result is considerably larger for ICP-MS than for TIMS. This in turn calls for careful validation of the mass bias correction by applying and comparing the effects of different correction procedures (see below).

Changes in the cone surface condition during a series of measurements can cause drift effects either through shift in the mass calibration or from changes in the signal intensities. A decrease in intensity will decrease the precision of the isotope amount ratio and possibly increase the uncertainty of the mass bias correction factor.

In any case, it is of paramount importance to separate the analyte from the matrix during sample preparation, as molecular and isobaric interferences can often be avoided in this way, mass bias becomes more stable and similar for samples and standards, and drift effects during ICP-MS analysis can be mitigated.

Internal Versus External Correction for Mass Bias If more than two isotopes of an element are available and one isotope ratio can be considered as constant, internal mass bias correction can be used. Let us consider an element with three isotopes, X, Y, and Z. If the X/Z ratio can be considered as constant in nature, the extent of instrumental mass discrimination can be determined by comparing the experimental value for X/Z with X_{ref}/Z_{ref} and the correction factor thus obtained can be used to correct the Y/Z isotope ratio for mass discrimination, based on a mass discrimination model. This approach is very common for the determination of $^{87}Sr/^{86}Sr$. Here, the $^{88}Sr/^{86}Sr$ ratio is assumed constant and monitored to correct subsequently the $^{87}Sr/^{86}Sr$ ratio (containing the radiogenic ^{87}Sr isotope) for instrumental bias. The assumption that $^{88}Sr/^{86}Sr$ is constant in nature has, however, been questioned recently [17–19]. Alternatively, if only two isotopes exist or if no constant ratio for normalization purposes is available, the so-called standard bracketing approach can be applied. Also in this case, a ratio for the standard used for mass bias correction needs to be known absolutely, but in most cases it is only a consensus or assumed value. In any case, an appropriate influence quantity contributing to the uncertainty budget needs to be considered.

7.3.1.4 **Data Presentation with Isotope Notation**

Even after mass bias and possible drift corrections have been carried through and their contributions to the uncertainty budget have been estimated, comparability is not assured. Minute variations in isotope ratios caused by fractionation can in most cases only be made visible and comparable when normalized to a ratio obtained for a common reference. Depending on the magnitude of deviation from a common reference isotope amount ratio, different notations are used.

The delta (δ) notation (see also Chapter 1) was introduced in the early 1950s when developing the ^{18}O thermometer. As the absolute isotopic abundance of ^{18}O in a sample could not be reliably determined, the $^{18}O/^{16}O$ ratio of the sample was compared with that of ocean water, measured in the same laboratory under the same measurement conditions, which was assumed to be constant. The measurement results were then not isotope amount ratios, but differences expressed as delta units (expressed in ‰). Through the introduction of MC-ICP-MS for isotope amount ratio measurements for non-traditional isotope systems, a similar situation as at the beginning of the ^{18}O thermometer occurred. For the development of repeatable analytical procedures, scientists used their in-house ICP-MS solutions as a normalization reference. Since several laboratories worked independently, different in-house standard solutions were used. Attempts were made to reach agreements, which were not always ruled by scientific arguments. For availability and homogeneity reasons, for example, it was proposed to use mean ocean molybdenum (MOMo) as common reference for the delta notation ($\delta^{98/95}Mo = [(^{98/95}Mo_{sample}/^{98/95}Mo_{MOMo}) - 1] \times 10^3$) for Mo isotope fractionation [20]. Ultimately, this idea did not succeed and rather a different commercially available Mo solution was used, which turned out to have an almost identical isotopic composition. Based on a thorough study of different reference solutions, a NIST solution is now proposed [21]. Even after a decade of Mo isotope studies, there has been no agreement on what to use to make data comparable. Also, no agreement has been reached between the groups working on Cd isotope fractionation.

The epsilon (ε) notation is also in common use and, for the example of Cd, ε is calculated as $\varepsilon^{114/110}Cd = [(^{114/110}Cd_{sample}/^{114/110}Cd_{reference}) - 1] \times 10^4$. The gamma (γ) notation, on the other hand, has been used for radiogenic isotope systems with much larger spreads in isotope ratios, as encountered in the context of the Re–Os parent–daughter pair. Here, the $^{187}Os/^{188}Os$ ratio (radiogenic isotope to a stable isotope) is of interest and the isotopic composition at a given time (t) is expressed as $\gamma Os_t = [(^{187/188}Os_{sample,t}/^{187/188}Os_{chondrite,t}) - 1] \times 10^2$ (%). Due to the radiogenic ingrowth of ^{187}Os through the decay of ^{187}Re, it is essential to quote the isotope amount ratio for a given time, which can be any time between 4.55 Ga ago, the time of the formation of our solar system, and today. Here, the hypothetical average isotopic evolution trajectory, typical for the composition of chondrites, a primitive and well-characterized class of meteorites, is used for reference purposes [22]. As for other isotope systems, there is no agreement about the reference in the case of the γ notation. In some cases, the chondrite average or the hypothetical Earth's primitive upper mantle evolution (which is slightly different to the chondrite evolution) is relied upon.

Although the use of notations such as δ, ε, γ, and so on is useful to make sometimes minute differences more visible, in addition to making data better comparable between laboratories, owing to the reduction of systematic errors, agreement between scientists or laboratories is often not achieved. This can cause confusion and misinterpretation of published data if little care is taken about the selection and description of the common reference used for calculating the isotope notation.

7.3.2
Example of Uncertainty Budget Estimation When Using Isotope Dilution

A simplified "bottom-up approach" for estimating the measurement uncertainty of the IDMS result for an individual PGE measurement result was presented by Meisel and Moser [7]. This approach takes into account the significant uncertainty contributions. Uncertainty components related to weighing, detector dead time, gravimetric dilution, and so on, need to be considered if a complete uncertainty budget is required, but they are negligible at trace element levels. A systematic approach that takes into account correlations between the uncertainty components was presented by Moser *et al.* [23].

Uncertainties were calculated by applying the spreadsheet method developed by Kragten [24], but without taking correlations into account. The following standard uncertainties were considered, following the suggestions and rules from the Eurachem/CITAC Guide (2000) [14].

- Isotope ratio measurement: the observed standard uncertainties estimated following Moser *et al.* [23].
- Spike calibration: 2% relative standard deviation (RSD) was assumed. This value is very conservative, but it is based on a combination of replicate spike calibrations and an additional component to allow for uncertainties in the standard solutions used for calibration. When using three Ru solutions from three different companies, three significantly different Ru concentrations were obtained for the spike, despite the "certified" and "traceable" PGE concentrations (nominal 10 μg ml^{-1}) that they were supposed to have.

- The standard deviation of the repeated total procedural blank measurements. As the value was close to the detection limit, 100% RSD was used as a worst-case estimate.
- The isotope ratio of the spike R_y (0.5% RSD) and of the sample R_x (0.1% RSD) used for calculations. Both make a very small contribution to the budget.
- Uncertainties in the sum of all isotope ratios $\Sigma R_{i,y}$ and $\Sigma R_{i,x}$ from the spike (0.5%) and the sample (0.17%), respectively.
- Uncertainties introduced through weighing were also considered, but their contributions are negligible as at least 100 mg of spike is added to the blends.
- A component added through sampling (heterogeneity) was ignored as only the uncertainty of an individual analysis was considered.
- Uncertainties introduced through variances amongst different reference material batches are not considered, as these are expected to be small or currently undetectable.

As expected, depending on the concentration level, the contribution of the blank increases with decreasing concentration.

7.3.3
Alternative Approach

Although the GUM/Eurachem bottom-up approach for measurement uncertainty estimation is very useful in identifying the most prominent influence quantities, which leads to an improvement of the combined measurement uncertainty, getting there is very tedious. Alternative approaches have been described which are based on a very pragmatic approach. To simplify the calculations, only two components arc considered: a random component that can be easily estimated through repeated measurements (including sampling, digestion, etc.) and a systematic effect. The systematic effect is estimated through a bias component. The bias can be derived from the comparison with a reference quantity value, for example, a certified value of a certified reference material. This approach is also known as the Nordtest calculation of measurement uncertainty [25]. The difficulty in this approach is the combination of these two components. A correction to the observed value can be made if considered significant and the uncertainty of such a correction needs to be included in the uncertainty budget. A bottom-up approach would be to treat the bias component as a variance and include it in the uncertainty budget. A discussion on the consequences of both approaches was given by O'Donnell and Hibbert [26].

7.3.4
How to Establish Metrological Traceability

Isotope ratio measurements are not only the foundation of new discoveries and models, but also form a basis for decisions in, for example, global trading, environmental monitoring, and food safety issues. In many cases, decisions have financial consequences and therefore need to be based on firm measurement results. In a global market, trade needs to rely on the measurement results obtained in other laboratories and this can only be assured through an unbroken chain of calibrations traced back to a common reference. In addition, data need not only to be comparable with other laboratories, but also to be comparable to data produced in the past. Hence data need to be comparable throughout space, time, and scientific disciplines.

Chemical amount of substance or concentration measurements are ideally traceable to the mole as a common reference (see the definition of metrological traceability to a measurement unit). Isotope ratios do not have a unit when the amount of the individual isotopes in the ratio is expressed in the same unit (e.g., mol mol^{-1}). In an ideal situation, isotope (amount) ratio measurements should be traceable to an isotopic reference material that has been calibrated against a synthetic isotope mixture. In this case, measurements can be made traceable to the SI unit. Since this is a rare case, most isotope amount ratio measurement are only traceable to the corresponding delta 0 reference material (see also Chapter 6). In the case of delta isotope measurements, the traceability chain ends at the delta unit which is 10^{-3} and multiples or fractions thereof. Therefore, the reference (the delta 0 reference solution) used for the delta calculations cannot be used for traceability purposes [27].

7.3.5 Method Validation

Validated methods allow useful measurement results to be produced, as needed for scientific purposes. Method validation is a planned set of experiments to determine the performance parameters of a method. Sufficient evidence needs to be collected to demonstrate that a method can be used routinely and is fit-for-purpose. Many textbooks in analytical chemistry deal with this topic. Here, only a few critical parameters and processes are discussed.

7.3.5.1 Limits of Detection, of Determination, and of Quantitation

The definition of the limit of detection (LOD) has been interpreted to be approximately three times the standard deviation derived from multiple measurements of a blank solution. This assumption is misleading, as it bears several problems. First, with the definition given, the LOD cannot be derived in the absence of known (or assumed) distributions of the concentration of a blank solution (or sample), and also of the solution containing the analyte. By multiplying the standard deviation of the blank solution by $k = 3$, it is assumed that the distribution of concentrations for the analyte-containing solution is identical with that of the blank. Second, all simplified calculations are based on a normal distribution, whereas a Poisson-like distribution of the concentrations of the blank (at level L_0) is more reasonably assumed. In a normal case, the standard deviation (σ) of at the blank level and at the limit of detection (L_0 and L_{LOD} respectively) are only estimated are only estimated (s_0 is an estimate for σ_0 based on a limited number of blank measurements) by a limited number of repeated measurements and, hence, the number of degrees of freedom ($\nu = n - 1$) needs to be taken into account. The calculation of the LOD for a normally distributed function can be approximated by LOD $\approx 2t_{95\%,\nu}\sigma_0$ (IUPAC and [6]), which for five replicates will give LOD $= 4.26\sigma_0$. However, in most cases, the distribution of σ_0 is only approximated by a limited number of replicates. In this case, the LOD needs to be increased owing to the uncertainty on the estimations of the distribution function. For example, if only five replicates are available, the LOD is $9.65s_0$, and thus more than three times higher than what is recommended in many textbooks. Hence, if the VIM3 LOD definition is used, the proper equations – taking into account the number of degrees of freedom (replicates – 1), α, and β (normally 0.05) – need to be selected and the influence of the distribution of the analyte concentration, matrix effects, interferences, and so on, also needs to be taken into account.

While the LOD is defined by VIM3, no definition can be found therein for any of the other limits, such as limit of quantification (LOQ), limit of quantitation, limit of determination, and so on, Several definitions for all of these limits can be found in the literature, hence there is plenty of space for new interpretations and definitions. Most definitions found in the literature are by no means comprehensive, as they do not give "recipes" on how actually to apply the concepts behind the definitions in real work. There seems to be some kind of consensus between all definitions for the LOQ, which is some kind of lower limit above which, roughly said, data can be reported with a certain confidence. Mermet [28] gave a good overview of existing definitions of LOQ, which are based on at least one of the following qualities: accuracy, trueness, precision, or uncertainty. Mermet grouped the existing LOQ definitions into categories. The first category contains definitions based on instrumental repeatability; here the assumption is made that the low background level signals are distributed normally. Second are LOQ definitions based on the RSD of the net signal (i.e., signal minus background) or the quantity level for which the RSD of repeated analysis reaches a level that is fit-for-purpose. Although a useful approach, it requires time-consuming repeated analysis with concentration levels approaching the range between the LOD and the expected LOQ. Another category of definitions is based on both accuracy and precision. Here, the incoherent use of accuracy and precision is not helpful for a better understanding. The fourth approach, based on the calibration graph, is not useful for isotope ratio measurements or isotope dilution. The method based on measurement uncertainty is metrologically the most correct approach. The estimation of measurement uncertainty for different concentration levels is a challenging task and is not easy to perform, in particular at very low analyte concentrations (see Figure 7.8).

The existing definitions are not applicable to isotope amount ratio measurements or isotope dilution measurements and give different results for the LOQ. These facts might hamper the actual use of the concept of LOQ and most scientists working in laboratories that are not required to implement quality systems rely on the limited information given by the LOD. Many literature examples demonstrate that the LOD in most cases is not properly defined and is considered to be a "magic" limit, above which numbers can be considered reliable.

No matter how these limits are defined, it is evident that the concentration range between LOD and LOQ provides measurement results with measurement

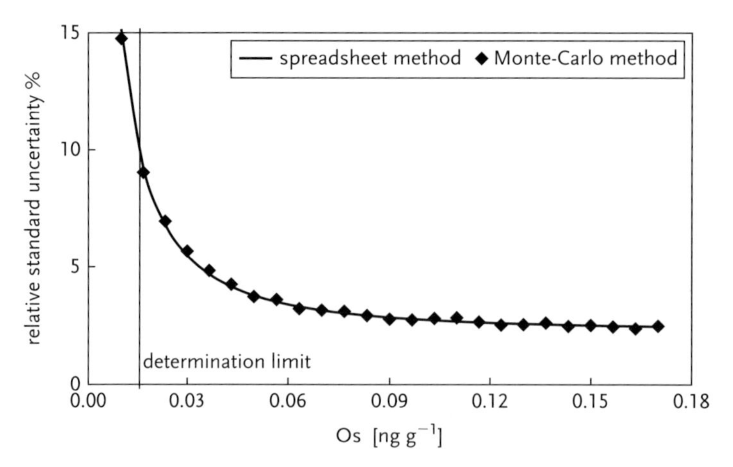

Figure 7.8 Uncertainty in Os concentration, determined via ICP-IDMS. The contribution of the spike calibration dominates the uncertainties at a higher concentration level. Blank contributions dominate the uncertainties at concentration levels approaching the determination limit. The standard deviations of the concentrations of replicate digestions were in most cases higher than the standard uncertainty estimated for the analysis of a single sample. It can therefore be assumed that in cases where the standard deviations of replicate measurements are very large, the nugget effect is dominant.

uncertainties well above those required for any useful application. Given the fact that LOD and LOQ are not well definable and not always applicable and that the LOD is of no use, the concept of LOD should be abandoned. The correct scientific approach, in particular in the range of the LOD and when minute isotope ratio differences are to be quantified, is going through a method validation process in the range of interest to demonstrate that the method applied produces measured quantity values that are fit for the use intended, or in other words that it is capable of verifying a hypothesis or answering the question posed.

7.3.5.2 **Inter-Laboratory Studies**

Even after a thorough method validation of the analytical procedure and establishment of a reasonable uncertainty budget, doubt about the confidence of the measurement results will remain. This is particularly the case when new isotope systems and fractionation processes are studied, as in these cases appropriate certified reference materials with measurands with uncertainties better than that suited for routine measurement are currently not available and will not become available in the near future. Inter-laboratory studies are a means to improve the performance of routine laboratories (proficiency tests), to test methods (any step from sampling to the isotope analysis), and to characterize test materials better (from in-house to certified reference materials). These studies can be organized at different levels of the method development steps, from the first observation of an isotope fractionation to routine performance of measurements by a group of scientists (the peers) and, later, by governmental organizations or international organizations (e.g., IRMM). In a first step, when isotope fractionation has been observed, standard solutions, for example ICP-MS solutions used for calibrations sitting around in the laboratory, can be sent out to colleagues to determine if the results are reproducible at all. In a further step, an inter-laboratory bias study can be carried out, wherein these solutions, or maybe even real samples, are used to determine method bias or laboratory bias. Further steps towards assigning consensus values of a test material (certification studies) are rarely performed in academia, as this is often regarded as less exciting and too time-consuming, and sometimes even not as something scientific. Nevertheless, there are some material certification studies that have been published, but this step is not sufficient to ensure traceability. Further steps need to be taken by international organizations or well-established commercial laboratories that are capable of following and implementing the recommendations and rules published by the ISO Committee on Reference Materials (REMCO) in ISO Guides 30–35.

The organization of and participation in inter-laboratory studies are essential for the advance of science and are therefore to be encouraged.

7.4 Conclusion

Instrument manufacturers are developing MC-ICP-MS instruments that allow better and better repeatability, thus permitting the detection of smaller and smaller isotope ratio variations. Scientists made/make use of these instrumental improvements and, through the use of better sample introduction systems and better control of and correction for instrumental mass bias and dead time (for ion-counting detectors), new useful discoveries of mass-dependent and mass-independent fractionation were/are made. For isotope variations to be detectable with MC-ICP-MS, other variances caused by sampling, sample inhomogeneities, sample preparation, and so on, need to be under control, as these variances can easily be larger than what state-of-art instruments are capable of detecting. Scientists should not be misled by data produced via high-precision instruments. A realistic measurement uncertainty budget and metrological traceability need to be established and a greater awareness of the importance of the LOQ needs to be created. Although less exiting than the discovery of new applications of isotope fractionation in low- and high-temperature geological processes, and also in biogeochemistry, forensic science, food safety, and environmental applications, the quality control instruments discussed need to be introduced and established for scientifically sound work.

References

1 Quétel, C.R., Lévêque, N., De Bolle, W., and Ponzevera, E. (2007) Final report on CCQM-P48: uranium isotope ratio measurements in simulated biological/environmental materials. *Metrologia*, **44**, (Tech. Suppl.), 08010.

2 Wasserburg, G.J., Jacobsen, S.B., DePaolo, D.J., McCulloch, M.T., and Wen, T. (1981) Precise determination of Sm/Nd ratios, Sm and Nd isotopic abundances in standard solutions. *Geochim. Cosmochim. Acta*, **45**, 2311–2323.
3 Hibbert, D.B. (2007) *Quality Assurance for the Analytical Chemistry Laboratory*, Oxford University Press, Oxford.
4 Bureau International des Poids et Mesures (BIPM) (2008) International Vocabulary of Metrology – Basic and General Concepts and Associated Terms (VIM), 3rd edn., JCGM 200:2008, http://www.bipm.org/en/publications/guides/vim.html (last accessed 23 December 2011); also published as ISO/IEC Guide 99–12:2007, International Organization for Standardization, Geneva.
5 ISO (1993) Guide to the Expression of Uncertainty in Measurement, International Organization for Standardization, Geneva.
6 Currie, L.A. (1999) Detection and quantification limits: origins and historical overview. *Anal. Chim. Acta*, **391**, 127–134.
7 Meisel, T. and Moser, J. (2004) PGE and Re concentrations in low abundance reference materials. *Geostand. Geoanal. Res.*, **28**, 233–250.
8 De Bièvre, P. (1990) Isotope dilution mass spectrometry: what can it contribute to accuracy in trace analysis? *Fresenius' J. Anal. Chem.*, **337**, 766–771.
9 Meisel, T., Reisberg, L., Moser, J., Carignan, J., Melcher, F., and Brügmann, G. (2003) Re–Os systematics of UB-N, a serpentinized peridotite reference material. *Chem. Geol.*, **201**, 161–179.
10 Schoenberg, R., Kamber, B.S., Collerson, K.D., and Eugster, O. (2002) New W-isotope evidence for rapid terrestrial accretion and very early core formation. *Geochim. Cosmochim. Acta*, **66**, 3151–3160.
11 Kleine, T., Münker, C., Mezger, K., and Palme, H. (2002) Rapid accretion and early core formation on asteroids and the terrestrial planets from Hf–W chronometry. *Nature*, **418**, 952–955.
12 Yin, Q.Z., Jacobsen, S.B., Yamashita, K., Blichert-Toft, J., Telouk, P., and Albarede, F. (2002) A short timescale for terrestrial planet formation from Hf–W chronometry of meteorites. *Nature*, **418**, 949–952.
13 Kleine, T., Touboul, M., Bourdon, B., Nimmo, F., Mezger, K., Palme, H., Jacobsen, S.B., Yin, Q.-Z., and Halliday, A.N. (2009) Hf–W chronology of the accretion and early evolution of asteroids and terrestrial planets. *Geochim. Cosmochim. Acta*, **73**, 5150–5188.
14 Ellison, S.L.R., Rosslein, M., and Williams, A. (eds.) (2000) Eurachem/CITAC Guide CG 4, Quantifying Uncertainty in Analytical Measurement, 2nd edn., http://www.measurementuncertainty.org (last accessed 23 December 2011).
15 Ramsey, M.H. (1998) Sampling as a source of measurement uncertainty: techniques for quantification and comparison with analytical sources. *J. Anal. At. Spectrom.*, **13**, 97–104.
16 Ramsey, M.H. and Ellison, S.L.R. (eds.) (2007) Eurachem/CITAC Guide: Measurement Uncertainty Arising from Sampling: a Guide to Methods and Approaches, produced jointly with EUROLAB, Nordtest and the RSC Analytical Methods Committee. Eurachem Secretariat, Prague.
17 Halicz, L., Segal, I., Fruchter, N., Stein, M., and Lazar, B. (2008) Strontium stable isotopes fractionate in the soil environments?. *Earth Planet. Sci. Lett.*, **272**, 406–411.
18 Yang, L., Peter, C., Panne, U., and Sturgeon, R.E. (2008) Use of Zr for mass bias correction in strontium isotope ratio determinations using MC-ICP-MS. *J. Anal. At. Spectrom.*, **23**, 1269–1274.
19 Fietzke, J. and Eisenhauer, A. (2006) Determination of temperature-dependent stable strontium isotope ($^{88}Sr/^{86}Sr$) fractionation via bracketing standard MC-ICP-MS. *Geochem. Geophys. Geosyst.*, **7**, Q08009.
20 Siebert, C., Nägler, T.F., von Blanckenburg, F., and Kramers, J.D.

(2003) Molybdenum isotope records as a potential new proxy for paleoceanography. *Earth Planet. Sci. Lett.*, **211**, 159–171.

21 Wen, H.J., Carignan, J., Cloquet, C., Zhu, X.K., and Zhang, Y.X. (2010) Isotopic delta values of molybdenum standard reference and prepared solutions measured by MC-ICP-MS: proposition for delta zero and secondary references. *J. Anal. At. Spectrom.*, **25**, 716–721.

22 Shirey, S.B. and Walker, R.J. (1998) The Re–Os isotope system in cosmochemistry and high-temperature geochemistry. *Annu. Rev. Earth Planet. Sci.*, **26**, 423–500.

23 Moser, J., Wegscheider, W., Meisel, T., and Fellner, N. (2003) An uncertainty budget for trace analysis by isotope dilution ICP-MS with proper consideration of correlation. *Anal. Bioanal. Chem.*, **377**, 97–110.

24 Kragten, J. (1994) Calculating standard deviations and confidence intervals with a universally applicable spreadsheet technique. *Analyst*, **119**, 2161–2165.

25 Magnusson, B., Naykki, T., Hovind, H., and Krysell, M. (2003) *Handbook for Calculation of Measurement Uncertainty*, 2nd edn., Nordtest, Espoo.

26 O'Donnell, G.E. and Hibbert, D.B. (2005) Treatment of bias in estimating measurement uncertainty. *Analyst*, **130**, 721–729.

27 Kipphardt, H. (2004) Traceability in isotopic measurements, in *Handbook of Stable Isotope Analytical Techniques*, Vol. 1 (ed. P.A. de Groot), Elsevier, Amsterdam, pp. 928–943.

28 Mermet, J.-M. (2008) Limit of quantitation in atomic spectrometry: an unambiguous concept?. *Spectrochim. Acta B*, **63**, 166–182.

8
Determination of Trace Elements and Elemental Species Using Isotope Dilution Inductively Coupled Plasma Mass Spectrometry

Klaus G. Heumann

8.1
Introduction

The determination of trace elements via isotope dilution mass spectrometry (IDMS) has been well known for a long time, but before 1990 it was more or less exclusively performed using thermal ionization mass spectrometry (TIMS) [1]. Most of this thermal ionization (TI)-IDMS work has been carried out in the context of nuclear technology, geochronology, and the certification of reference materials, because for these types of analysis the highest level of accuracy is required. IDMS is internationally accepted as a "primary method of measurement," providing high precision and accuracy and for which possible sources of error are well known and usually under control [2]. In the past, IDMS was also often called a "definitive method" owing to its highly accurate results, especially compared with more traditional calibration methods. However, as IDMS results are also associated with a corresponding uncertainty, it should no longer be qualified as a definitive method [3], even though it can provide the highest level of accuracy. After 1990, a continuously increasing number of papers reported on trace element determinations using inductively coupled plasma mass spectrometry (ICP-MS) in combination with isotope dilution. In addition, the first use of inductively coupled plasma isotope dilution mass spectrometry (ICP-IDMS) in the context of elemental speciation via high-performance liquid chromatography (HPLC)–ICP-MS was reported in 1994 [4, 5].

Even though the precision of isotope ratio measurements is usually better with TIMS than with ICP-MS and the mass bias is higher for ICP-MS, ICP-IDMS is increasingly replacing TI-IDMS for trace element determination, owing to the easier and much more time-efficient sample handling with ICP-MS. This difference is illustrated for iridium determination in photographic emulsions in Figure 8.1, which presents a comparison of the sample preparation steps required for ICP-IDMS and TI-IDMS [6].

Isotopic Analysis: Fundamentals and Applications Using ICP-MS,
First Edition. Edited by Frank Vanhaecke and Patrick Degryse.

Published 2012 by WILEY-VCH Verlag GmbH & Co. KGaA

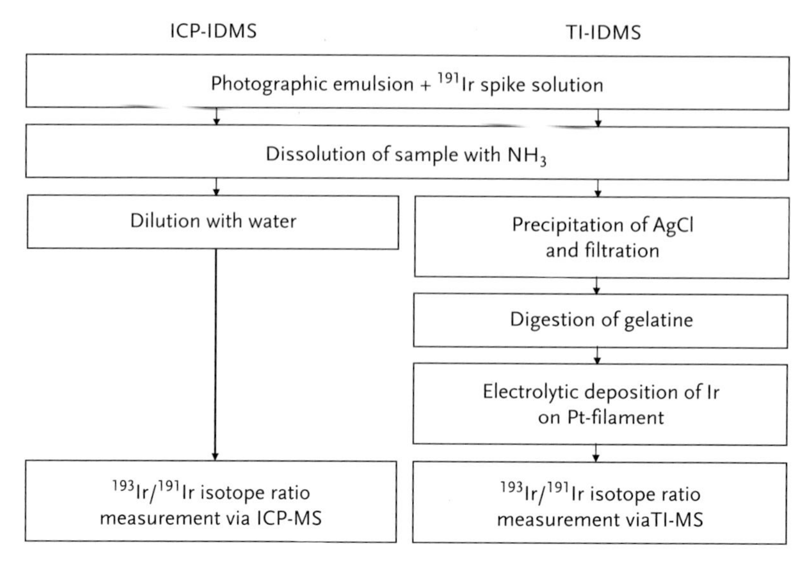

Figure 8.1 Comparison of sample pretreatment steps required for iridium determination in photographic emulsions via ICP–IDMS and TI-IDMS [6].

As a result of the time-efficient handling of samples by ICP-MS, isotope dilution techniques are now used within a broad range of applications. The unique advantages of ICP-IDMS in trace element determination compared with other analytical techniques also allow its use as a routine method in many fields of application, especially those in which high precision and accuracy are of crucial importance [7].

8.2 Fundamentals

8.2.1 Principles of Isotope Dilution Mass Spectrometry

The fundamental principles of IDMS have been extensively described in several textbooks [1, 8, 9], from which detailed information can be obtained. Here, those characteristics of ICP-IDMS that are specific in comparison with other analytical techniques are reviewed and attention is also paid to the fundamental topics to be considered carefully for successful analysis via ICP-IDMS.

In isotope dilution analysis, a sample with an unknown concentration of the target element, usually in its natural isotopic composition, is mixed and chemically equilibrated with a known amount of an isotope-enriched spike compound of the corresponding element. This approach self-evidently requires the element to be determined to have at least two isotopes measurable by ICP-MS. For poly-isotopic elements, such as lead with four stable isotopes, only two isotopes need to

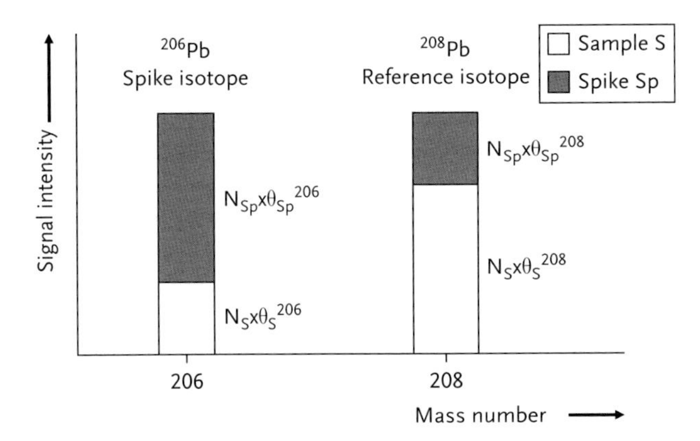

Figure 8.2 Schematic diagram representing isotope dilution analysis of lead (*N*, number of atoms; θ, relative isotopic abundance).

be selected for IDMS, one spike and one reference isotope. This is shown schematically in Figure 8.2, where ^{206}Pb is used as spike and ^{208}Pb as reference isotope. In the case of monoisotopic elements, isotope dilution can only be applied if a long-lived radioactive isotope is available for the spiking process. Radionuclides with a half-live of $>10^3$ years can usually be applied in trace element analysis without limitations. ^{129}I is an important example of a radionuclide (half-life 1.6×10^7 years) that was successfully used as a spike isotope [4].

One of the exceptional advantages of the isotope dilution technique compared with other methods is that a ratio (isotope ratio, R) is measured, whereas in all other analytical methods the *amount* of analyte needs to be measured, such that non-quantitative recovery during the sample pretreatment needs to be corrected for. For IDMS analysis, on the other hand, once the isotope dilution step has taken place, the recovery does not need to be known. This is because, for example, subsequent chemical separation procedures will usually not change the isotope ratio of the isotope-diluted sample to a significant extent. From this, it follows that sample loss does not affect the analytical result in IDMS analysis. An important precondition for successful element determination via isotope dilution is an equilibration of the sample and spike as early as possible in the analytical procedure. This prerequisite is best fulfilled in the case of solutions. Solid samples therefore need to be completely digested and thereby equilibrated with the spike, usually added in the form of an aqueous solution. If the analyte and the spike in the isotope-diluted solution are not equilibrated and exist in different chemical forms, sample preparation steps such as chromatographic separation will fractionate the analyte from the spike and the corresponding isotope ratio measurements will produce incorrect results.

Another advantage of IDMS is its independence of matrix effects, which present an important drawback for other calibration approaches. Even if matrix effects affect the signal intensity in ICP-MS, the reference and spike isotope are affected in the same way and to the same extent, so that the measured isotope ratio *R* is not affected. Also, instrument instability and signal drift, which are often a problem for external calibration methods, usually do not influence IDMS results because the reference and spike isotopes are subsequently measured within a very short time interval, in which drift can normally be neglected. Nevertheless, even such effects could be eliminated, if necessary, by using a multi-collector ICP-MS instead of a single-collector ICP-MS system. IDMS is a one-point internal calibration technique with a spike isotope as calibrant and therefore has tremendous advantages, especially in terms of accuracy, compared with other, more conventional, calibration methods. However, all these advantages cannot prevent ICP-IDMS from providing incorrect results if either one or both isotopes measured are subject to spectral interference. The occurrence of spectral interferences therefore needs to be carefully checked and excluded. If spectral interference hampers the measurement of the spike or reference isotope when using conventional quadrupole-based ICP-MS, the application of a quadrupole-based instrument equipped with a collision/reaction cell or of sector field ICP-MS can often solve the problem (see Chapter 2). Examples of the use of a quadrupole-based instrument with collision/reaction cell and of sector field ICP-MS in IDMS applications can be found in the literature, for example, [10, 11].

For the lead isotope ratio (R) measurement in the isotope-diluted sample, it follows from Figure 8.2 that

$$R\left(\frac{^{208}\mathrm{Pb}}{^{206}\mathrm{Pb}}\right) = \frac{N_\mathrm{S}\theta_\mathrm{S}{}^{208} + N_\mathrm{Sp}\theta_\mathrm{Sp}^{208}}{N_\mathrm{S}\theta_\mathrm{S}{}^{206} + N_\mathrm{Sp}\theta_\mathrm{Sp}^{206}} \tag{8.1}$$

where N_S and N_Sp are the number of analyte and spike atoms, respectively, and θ^{206} and θ^{208} are the relative isotopic abundance of the spike and reference isotope, respectively. It is important to point out that Eq. (8.1) can be easily transformed to be valid for any other element with, at least, two stable isotopes by simply replacing the index 208 by that corresponding to the heavier and the index 206 by that corresponding to the lighter isotope. If a radioactive spike isotope, not present in the sample, is used, Eq. (8.1) becomes even simpler because there is no natural contribution from the sample to the spike isotope. For example, for the mono-isotopic iodine (^{127}I), the first term of the numerator in Eq. (8.1) is omitted when using a radioactive ^{129}I spike. Based on Eq. (8.1), the concentration C_S of the element in the sample is given by

$$C_\mathrm{S}(\mu\mathrm{g\ g}^{-1}) = 1.66 \times 10^{-18}\left(\frac{M}{W_\mathrm{S}}\right) N_\mathrm{Sp}\frac{\theta_\mathrm{Sp}^{208} - R\theta_\mathrm{Sp}^{206}}{R\theta_\mathrm{S}{}^{206} - \theta_\mathrm{S}{}^{208}} \tag{8.2}$$

where M is the atomic or molecular weight of the analyte (g mol^{-1}) and W_S the sample weight (g). The factor 1.66×10^{-18} is required for conversion of the number of analyte atoms into the corresponding weight given in micrograms by using Avogadro's number ($N_\mathrm{A} = 6.022 \times 10^{23}$ atoms or molecules mol^{-1}).

Equations (8.1) and (8.2) are not corrected for instrumental mass bias (mass discrimination) and blank values. In the literature, one can therefore find much more complicated equations, which already include all necessary corrections for mass bias in the ICP-MS measurements and for the blank values [9, 12]. In these cases, it is often difficult to identify the exact meaning of all terms. It is therefore always important to trace back all complex equations to Eqs. (8.1) and (8.2), which are based on the fundamental principle of the isotope dilution technique, so that the different correction terms required can be identified more easily.

The isotopic composition of elements in samples is usually identical with the listed IUPAC information [13]. This is not the case when synthetic isotopic compositions appear, as is usually the case for nuclear samples. In some cases, the natural isotopic composition of an element shows variations that are so pronounced that the IUPAC values cannot be relied upon for accurate IDMS analysis. A prominent example is lead, for which the isotopic composition in natural samples shows a substantial range due to radiogenic contributions to ^{206}Pb, ^{207}Pb, and ^{208}Pb from the natural decay series of uranium and thorium (see Chapter 1). Natural variations in the isotopic composition should also be taken into consideration in ICP-IDMS analysis for the light elements lithium, boron, and sulfur, whereas for most of the other elements natural variations are relatively small [14] so that usually they can be neglected for calculation of an ICP-IDMS result. However, these variations should be considered in the uncertainty budget. Correction for instrumental mass bias (instrumental mass discrimination), however, is required. In principle, different correction approaches can be applied, for example, according to the linear law, the power law, or the exponential law [15], where the power law usually provides the best description of the mass discrimination process in ICP-MS (see also Chapter 5) [16]. For such a correction, the mass discrimination factor needs to be determined, ideally by measuring a corresponding certified isotopic standard of natural isotopic composition and comparing the experimental result with the certified value. However, in most cases, it is sufficient to measure a pure element standard of natural isotopic composition, assuming that the corresponding relative isotopic abundances are consistent with the IUPAC values. As an alternative, one can also directly use the uncorrected value for R and the experimentally determined uncorrected values for θ_{Sp} and θ_S. In both cases, identical and accurate results are obtained, as was demonstrated for the determination of trace amounts of copper [16]. In the latter procedure, the different uncorrected isotopic terms in Eq. (8.2) compensate for one another, such that an excellent approximation of the accurate result can be obtained without any mass bias correction.

A protocol for accurate determination of lead concentrations in water samples using ICP-IDMS was established by the National Institute of Standards and Technology (NIST), Gaithersburg, MD, USA [17]. A comparable scheme is presented in Figure 8.3, which summarizes schematically the different steps of isotope dilution analysis via ICP-MS. In the first line of this schematic diagram the initial solutions are listed, in the second the necessary isotope dilution processes, and in the third all isotope ratio measurements required. Reverse isotope dilution refers to exactly the same process, but in which the isotopic spike is considered as a sample and a standard solution of natural isotopic composition as a spike.

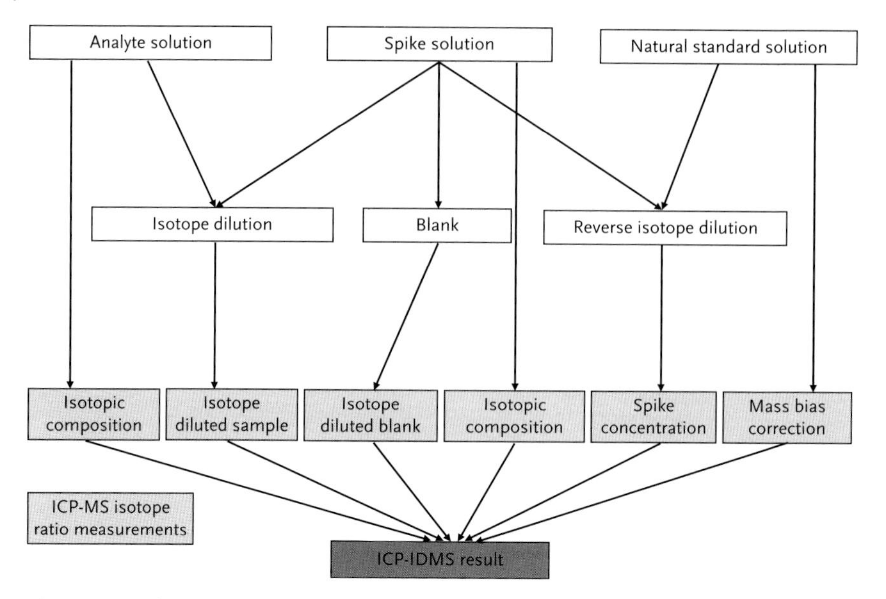

Figure 8.3 Schematic diagram showing the different analytical steps involved in trace element determination via ICP-IDMS.

8.2.2 Elements Accessible to ICP-IDMS Analysis

Figure 8.4 shows all elements in the periodic table which are accessible to ICP-IDMS analysis. All elements indicated in light gray are polyisotopic elements for which at least two stable isotopes exist. The elements indicated in dark gray are monoisotopic, but have one or more long-lived radioactive nuclide(s) (half-life $> 10^3$ years) which, in principle, can be applied as a spike, although their availability is sometimes a problem. The radioactive actinides thorium and uranium both have two long-lived natural isotopes, which usually allow unproblematic IDMS trace analysis. For the synthetic transuranium elements from neptunium to curium, corresponding radionuclides also exist, but in this case handling in special glove-boxes is usually recommended. All other elements, not highlighted (white) in Figure 8.4, usually cannot be determined via ICP-IDMS. This is mostly due to a lack of stable or long-lived radioactive nuclides (Na, P, Co, As, Au, etc.). Argon is used as plasma gas in ICP-MS and therefore cannot be determined. Nitrogen, oxygen, and the noble gases are theoretically determinable via ICP-IDMS, but contamination from the environment and from the corresponding impurities in the plasma gas usually prevents their determination in practice.

As one can see from Figure 8.4, most of the elements are determinable via ICP-IDMS. There are only two elements which are of high interest in the context of proteomics and environmental investigations which cannot be determined via ICP-IDMS: phosphorus and arsenic. Nevertheless, compared with TI–IDMS,

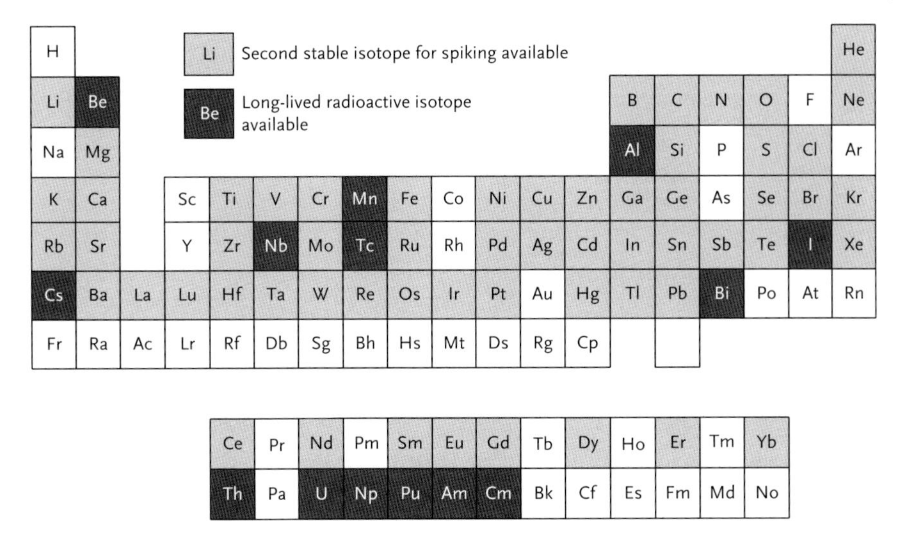

Figure 8.4 Elements accessible by ICP-IDMS.

more elements are covered by ICP-IDMS [1]. Elements such as silicon, platinum, and mercury are important examples which are now accessible via ICP-IDMS, but could not be determined using TI-IDMS.

8.2.3
Selection of Spike Isotope and Optimization of Its Amount

For most of the polyisotopic elements, isotopically enriched substances are commercially available. A calibrated pure aqueous spike solution is usually applied in the isotope dilution step, such that characterization of this solution needs to be carried out. After dissolution of the isotope-enriched substance, the concentration of such a stock solution can be determined most precisely and accurately via traditional (absolute) methods such as titration. This is possible as a result of the relatively high spike concentration and the normally negligible level of impurities. As an alternative to traditional methods, a reverse isotope dilution procedure can be applied (Figure 8.3). In this case, a certified standard solution of the corresponding element with natural isotopic composition is used for the isotope dilution step, but all other measurements and necessary corrections are similar to those explained in Section 8.2.1. In addition to its concentration, the isotopic composition of the spike solution also needs to be determined, which can be done easily by ICP-MS, at least on condition that the corresponding measurement is not hampered by spectral interferences. It is advisable that such a stock spike solution is stored at a relatively high concentration, because changes in concentration and isotopic composition with time are usually insignificant under these conditions. For the spiking process in trace analysis, a much lower concentration is required, which can be achieved easily by proper dilution shortly prior to the spiking step, taking into account optimization of the spike addition.

Meanwhile, metrological institutions, such as the Institute of Reference Materials and Measurements (IRMM) in Geel, Belgium, and the Bundesanstalt für Materialforschung und -prüfung (BAM) (the Federal Institute for Materials Research and Testing) in Berlin, Germany, distribute certified spike solutions [18, 19], for example, for boron, uranium, plutonium, copper, and zinc.

If an element has more than two stable isotopes, which is valid for more than 30 elements often measured via ICP-MS, one can select which isotope is to be used as spike and which as reference isotope. It is recommended that the following conditions are fulfilled:

1. Neither spike nor reference isotope should be subject to spectral overlap due to the occurrence of isobaric nuclides, doubly charged ions, or polyatomic ions.
2. Usually the isotope with the lowest natural isotopic abundance is selected as the spike isotope.
3. If the abundance of a natural isotope is extremely low (e.g., ^{204}Pb with a natural abundance of only $\sim 1.4\%$), the purchase cost for the corresponding enriched isotope is typically relatively high. In this case, selection of an isotope with a higher natural isotopic abundance is more cost-efficient. For lead determination, it is therefore much more convenient to select ^{206}Pb or ^{207}Pb instead of ^{204}Pb as spike isotope (Figure 8.2).

Most of the isotopically enriched substances of elements with natural isotopic abundances above 10%, cost $\leq$€2000 per 100 mg. For IDMS analysis, the amount of spike should be in the same range as the amount of the analyte, which means that usually not more than about 1 µg of a spike is consumed in one trace analysis; 1 µg of a spike will therefore cost $\leq$€0.02, making the spike costs negligible compared with the operating costs of ICP-MS. Only if milligram amounts of a spike or more are necessary does the spike contribute significantly to the total cost of an IDMS analysis. This is only acceptable in very special cases, where even small errors in the analytical result can become extremely expensive. In this context, the control of platinum and palladium concentrations at the mg g^{-1} level in automotive catalysts via ICP-IDMS was tested a couple of years ago because an accuracy better than 1% relative standard deviation (RSD) was desired, which could not be obtained via conventional analytical methods [7]. Due to spectral interferences hampering the accurate determination of palladium, the ICP-IDMS approach was not established in industry as a routine method, although excellent results were obtained for platinum. However, ICP-IDMS is usually not a method to be applied above µg g^{-1} levels, because many other methods exist which can provide similar precision and accuracy at higher concentrations.

An optimization of the ratio of the amount of spike to the amount of target element in the sample is recommended in IDMS analysis to minimize the error multiplication factor $f(R)$, which influences the precision of the result. From Eq. (8.3), it follows that the precision of an IDMS analysis is mainly dependent on the corresponding precision of the spike addition and the isotope ratio (R) measurement, assuming that the uncertainties on the relative isotopic abundances do not contribute significantly for either sample or spike. Under this assumption, the standard deviation s on N_S is described by the following approximation:

$$s^2(N_S) \sim s^2(N_{Sp}) + f^2(R)s^2(R) \tag{8.3}$$

Whereas the standard deviation on the amount of spike added $s(N_{Sp})$ is usually small, any statistical error on the isotope ratio (R) measurement is increased by the error multiplication factor $f(R)$, which is characterized mathematically by

$$f(R) = \frac{\left[\left(\frac{\theta^b}{\theta^a}\right)_S - \left(\frac{\theta^b}{\theta^a}\right)_{Sp}\right] R}{\left[R - \left(\frac{\theta^b}{\theta^a}\right)_S\right]\left[\left(\frac{\theta^b}{\theta^a}\right)_{Sp} - R\right]} \tag{8.4}$$

where $\theta^{a,b}$ are the isotopic abundances of the lighter and heavier isotopes, respectively. A detailed derivation of Eqs. (8.3) and (8.4) based on the law of propagation of errors can be found in the literature [1, 20]. By using Eqs. (8.1) and (8.4), one can calculate the error multiplication factor for different spike enrichments as a function of the atomic ratio N_S/N_{Sp}, as shown for iron in Figure 8.5 for various ^{57}Fe spike enrichments.

The curves in Figure 8.5 are relatively flat over a broad N_S/N_{Sp} range from about 10^{-1} to 10 for spike enrichments >90%, which means that under these conditions the precision of the IDMS result is nearly independent of the analyte to spike mixing ratio. It also demonstrates clearly that a highly enriched spike isotope (relative isotopic abundance approaching 100%) is usually not necessary for IDMS analysis. For other polyisotopic elements, the curves are similar to those in Figure 8.5, as can be seen from Figure 8.6 for boron, chlorine, and bromine for

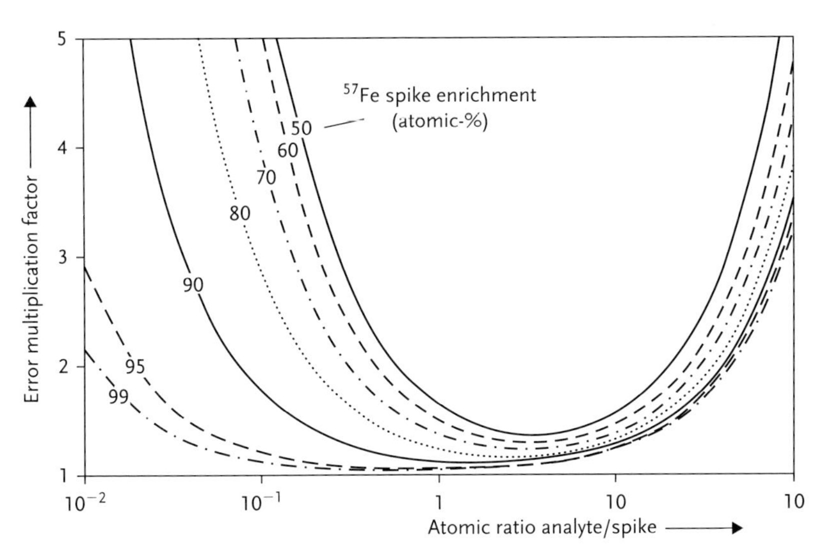

Figure 8.5 Error multiplication factors for IDMS determination of iron as a function of the analyte/spike atomic ratio for different ^{57}Fe spike enrichments.

commercially available isotopically enriched spikes with an isotopic enrichment of about 90% for the corresponding minor isotopes. Nevertheless, one should always calculate the corresponding optimum conditions for the actual IDMS analysis. The optimum conditions for precise IDMS analysis are reached at the minimum of the different curves, which correspond to an isotope ratio R_{opt} given by

$$R_{opt} = \sqrt{\left(\frac{\theta^b}{\theta^a}\right)_S \left(\frac{\theta^b}{\theta^a}\right)_{Sp}} \tag{8.5}$$

Only for spiking processes with a synthetic radionuclide, where the sample does not contribute to the intensity of this isotope in the isotope dilution process, do the curves for different spike enrichments show no minimum, as demonstrated in Figure 8.6 for iodine using an 86% enriched ^{129}I spike. Here, the curve of the error multiplication factor asymptotically reaches a constant minimum value at high analyte/spike ratios.

For optimization of the spike addition, it is necessary to know the approximate element concentration in the sample. As the precision of results is not significantly influenced over a wide range of analyte to spike mixing ratios, pre-analysis via a semiquantitative method is typically sufficient. If the expected amount of analyte is very low, it can also be advantageous to work with a distinct excess of spike to make chemical isolation of the isotope-diluted sample easier. This means that conditions to the left of the minimum in Figures 8.5 and 8.6 are used. Under extreme conditions of analyte to spike ratio, not only the enhanced error multiplication factor has to be taken into account, but also the deteriorating

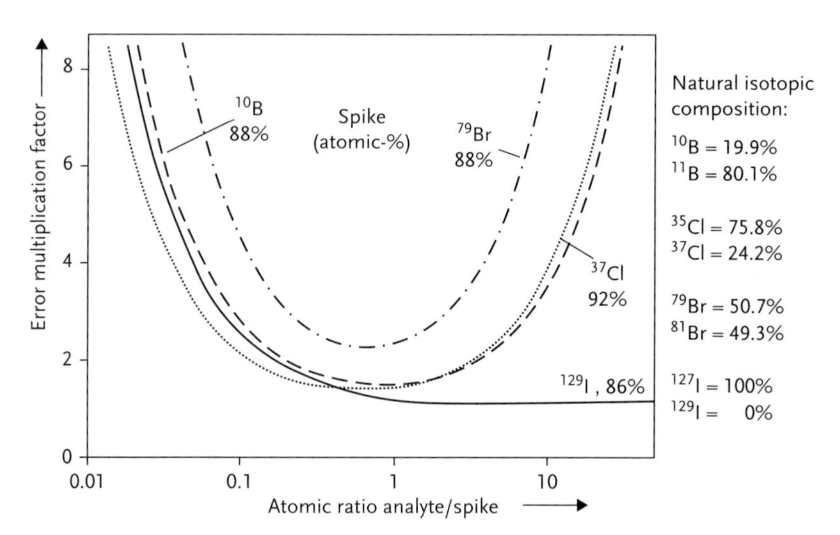

Figure 8.6 Error multiplication factors for IDMS determination of the polyisotopic elements boron, chlorine, and bromine and of the monoisotopic element iodine for specified spike isotope enrichments as a function of the analyte/spike atomic ratio.

precision of the isotope ratio (R) measurement, because the best precision in isotope ratio measurements is always obtained for ratios around 1:1.

8.2.4 Uncertainty Budget and Limit of Detection

Even though the isotope dilution technique is internationally accepted as a "primary method of measurement," the uncertainty of the results thus obtained should be evaluated via a corresponding uncertainty budget, a procedure more and more required for analytical data nowadays. The approaches described by Moser *et al.* [21] and Fortunato and Wunderli [22] for the determination of trace amounts of platinum group elements (PGEs) and rhenium in a certified geological reference material and for lead in wine via ICP-IDMS, respectively, are excellent models, demonstrating how to proceed for such an uncertainty budget (see also Chapter 7). An important item in the considerations is the fact that the estimation of uncertainty and the demonstration of traceability are interrelated. Traceability was proved via reliable uncertainty statements for the results and the uncertainty was evaluated based on the ISO/BIPM and Eurachem/Citac guidelines [23, 24].

Typical parameters which can influence the uncertainty of an IDMS analysis include heterogeneity of the sample, blank determination, spectral interferences, spectral background, mass bias, detector dead time, and spike concentration and its addition. Other parameters can also affect the uncertainty, but it is suitable from a practical point of view to reduce the number of uncertainty components to those which predominantly influence the uncertainty budget. In both examples mentioned, the measurement of the sample blank was one of the major contributions to the uncertainty budget, which agrees with common knowledge in IDMS [1, 11]. For the PGEs in the geological reference material, the heterogeneity also contributed strongly to the uncertainty budget. This agrees well with observations for other determinations of PGEs [25].

If the blank predominantly governs the limit of detection (LOD) in IDMS analysis, it follows that

$$\mathrm{LOD} = \mathrm{Bl} + 3s(\mathrm{Bl}) \tag{8.6}$$

where Bl is the blank value, expressed in concentration or amount of analyte, and s (Bl) is the standard deviation of the blank determination. Especially at low trace levels and for analytes with relatively high environmental abundances, the blank determination is the limiting factor in ICP-IDMS.

In any case, the isotope ratio R of the isotope-diluted sample needs to be significantly different from the corresponding isotope ratio of the spike R_{Sp}:

$$R \geq R_{\mathrm{Sp}} + 3s(R_{\mathrm{Sp}}) \tag{8.7}$$

where R_{Sp} and $s(R_{\mathrm{Sp}})$ are the spike isotope ratio and its standard deviation, respectively. Equation (8.7) is valid if the lighter isotope is the spike isotope and both R values are defined as the intensity ratio of the heavier to the lighter isotope [see Eq. (8.1)].

For elements with low natural abundance, it can sometimes be found that the isotope ratio measurement limits the LOD. But even for naturally low abundant elements such as uranium, plutonium, and the PGEs, the corresponding isotope intensities are often subject to interferences by small contributions from polyatomic ions [25, 26]. Such interferences simulate an additional blank value, which is still covered by the blank measurement, at least if the corresponding procedure with the blank solution gives rise to the same polyatomic ions as the real sample.

8.3
Selected Examples of Trace Element Determination via ICP-IDMS

8.3.1
Trends in ICP-IDMS Trace Analysis

In the past, ICP-IDMS for trace element determination was predominantly used in the context of the certification of standard reference materials, as a validation method for other analytical techniques, and as a routine method in nuclear technology and in geology, and it is still an important analytical tool in these areas. Today, for example, the production of SI-traceable reference values for certified reference materials largely relies on the use of ICP-IDMS [27, 28]. In 2006, Vogl reviewed the characterization of reference materials via the isotope dilution technique from the beginning of the 1990s onwards [29]. In geology, ICP-IDMS using either quadrupole-based or sector field ICP-MS instruments is often used today for accurate element determinations in rock samples and the successful use of ICP-IDMS for multi-element determinations of major, minor, and trace elements in silicate samples has been demonstrated [30]. ICP-IDMS is not only used for the determination of traces of uranium and plutonium in nuclear technology, but also increasingly in the context of environmental studies, for example, for rapid determination of traces of plutonium in environmental samples using ICP-IDMS combined with on-line column extraction [31].

The use of ICP-IDMS is especially advantageous if the accuracy of the results is crucial. Modern trends in ICP-IDMS analysis are therefore its use for the determination of elements with a “difficult behavior” during the sample handling, such as silicon [11], PGEs [21, 25, 32], and halogens [33], and also for the trace analysis of matrices that are difficult to digest, such as coal [34], and of organic fluids, such as petrol samples, which at higher concentrations are usually incompatible with an inductively plasma [35, 36]. In these cases, it is often difficult to produce accurate results via other analytical methods. In some of these trace analyses, direct laser ablation of isotope-diluted powder or of organic liquid samples was successfully applied.

However, ICP-IDMS also has high potential for routine analysis owing to its high accuracy and precision, matrix-independent and sensitive measurement, and its multi-element capability. The characterization of the spike solution is somewhat time consuming, but as more certified spike solutions become commercially available the isotope dilution method will become even more time effective, owing to its internal one-point calibration in contrast to other internal or external calibration procedures [7].

8.3.2 Direct Determination of Trace Elements in Solid Samples via Laser Ablation and Electrothermal Vaporization ICP-IDMS

Laser ablation–inductively coupled plasma mass spectrometry (LA–ICP-MS) has become increasingly important for the direct and fast multi-element trace analysis of compact and powdered solid samples over the last two decades. Moreover, LA–ICP-MS also allows spatially resolved analysis with good lateral resolution [37, 38]. However, calibration of LA–ICP-MS measurements for accurate determination of trace elements is still a problem, which is mainly due to fractionation processes and lack of certified standard reference materials with a matrix composition that is sufficiently similar to that of the various sample types that need to be analyzed [39]. A great variety of different calibration methods have therefore been developed, including matrix-matched and non-matrix-matched calibration techniques [39–42], but also techniques relying on the use of standard solutions, with corresponding aerosol introduction or direct ablation of the liquid, have been suggested [43, 44]. As a result of the different difficulties and limitations of these calibration methods, LA-ICP-IDMS is a promising alternative for calibration of measurements in powdered and liquid samples.

A first successful attempt to use the isotope dilution technique in LA–ICP-MS was carried out by Longerich and co-workers in 1999, with the aim of determining zirconium and hafnium in rock samples. A solid isotopic spike was homogeneously mixed with the sample through a combination of grinding and fusion [45]. A simpler sample pretreatment procedure without any fusion step was developed by Heumann and co-workers for a great variety of different powdered samples, and also for the determination of trace elements in petrol [25, 26, 46–48]. The sample preparation for powdered samples is represented in Figure 8.7. About

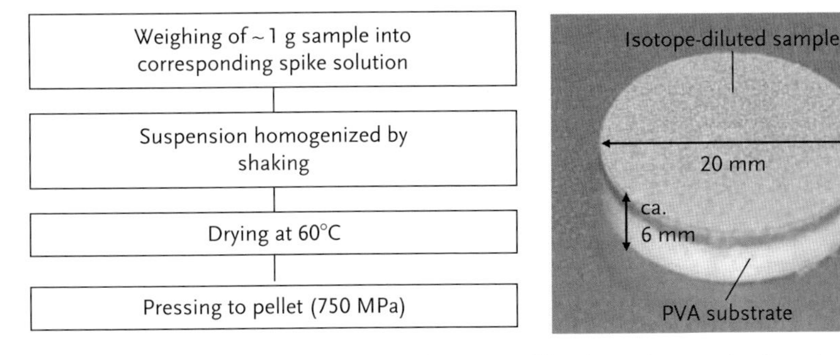

Figure 8.7 Sample preparation scheme for isotope-diluted powder samples for subsequent LA–ICP-IDMS analysis [46].

1 g of the sample is completely covered by the corresponding spike solution. For multi-element determination, a multi-spike solution prepared from single-spike stock solutions, in which the optimum spike additions were considered (see Section 8.2.3), is used. In cases where the sample is not moistened by the aqueous spike solution, a few milliliters of methanol are added. The suspension is then briefly shaken, dried, and the solid residue pressed with a laboratory press at about 750 MPa. If necessary, the mechanical stability of the pellets can be enhanced by using poly(vinyl alcohol) (PVA) powder, simultaneously fixed on the back of the sample (Figure 8.7).

The homogeneous adsorption of the spike on the powdered sample can be proved by measuring the corresponding isotope ratios at different locations within the pellet, where identical isotope ratios confirm this precondition. Also, the ablation process needs to be identical for the original chemical form of the element in the spike and in the sample, which means that equilibration must have occurred in the aerosol particles generated by the laser ablation process. At relatively high laser power densities of up to 4 GW cm^{-2} the aerosol particles are different in shape from the original ones showing that a melting process has taken place, which also confirms this precondition. Under these conditions, precisions of better than 10% RSD were obtained in multi-element LA–ICP-IDMS analysis and also the deviation between the experimental results and the corresponding certified values for reference materials was usually less than 10% [46]. Multi-element trace analysis via this LA–ICP-IDMS technique is even possible in matrices for which the capabilities of other analytical methods are often limited, and it can also be applied to volatile elements, otherwise suffering from partial losses during sample digestion. The simultaneous determination of the volatile elements chlorine, bromine, sulfur, and mercury in addition to seven other trace metals in coal, and also the accurate determination of 10 trace elements in alkaline fluoride samples are exceptional examples of this type of LA–ICP-IDMS analysis [34, 49]. When analyzing organic liquids, such as petroleum products, the sample pretreatment is extremely simple, because the spike only has to be mixed with the sample by shaking prior to the ablation of the isotope-diluted sample [47, 48]. However, in this case, the spike needs to be present in an organic solvent.

An alternative to the use of spike solutions for LA–ICP-IDMS analysis of powdered samples was developed by Donard and co-workers in 2008 by applying a matrix-matched multi-element solid spike for the determination of heavy metals in soil and sediment samples [50]. For production of such a powdered solid spike, the corresponding spike solutions, optimized for spike addition for the elements to be determined, were mixed with a soil reference material. After drying of the suspension thus obtained, this solid matrix-matched spike was mixed with the sample and homogenized in a mortar prior to pressing to pellets. Compared with the procedure in Figure 8.7, this approach has the advantage that the drying step, which takes some time, is eliminated and that the same spike can be used for all analyses of samples with similar matrix and element concentration. Hence this type of LA–ICP-IDMS will best fit the purpose of routine analyses when many samples of similar matrix and composition need to be analyzed. On the other hand, the procedure presented in Figure 8.7 is matrix independent and more flexible in terms of the number and concentration of elements to be determined.

Compact solid samples cannot be analyzed via LA–ICP-IDMS using the techniques mentioned above because a homogeneous mixture of the sample and spike – indispensable in isotope dilution analysis – cannot be realized for such a type of sample. To overcome this problem, an on-line solution-based calibration approach has been developed, which was called "on-line isotope dilution LA–ICP-MS," suggesting that a real IDMS analysis is carried out. However, this is by no means true. Here, the laser generated aerosol is continuously mixed with a nebulizer-generated aerosol of a spike solution in the ablation chamber [51, 52]. In addition to the potential occurrence of matrix and fractionation effects, this approach also requires the determination of a homogeneously distributed element in the sample and its subsequent use as an internal standard to correct for the different analyte introduction efficiencies of LA and solution nebulization. Real IDMS analyses of trace elements should only use the spike as an internal standard and no additional internal standard(s). In addition, one major advantage of LA–ICP-IDMS for the analysis of powdered samples is lost, namely the elimination of matrix and fractionation effects.

In principle, electrothermal vaporization (ETV) can also be used for the direct analysis of solid samples in combination with ICP-IDMS, even though ETV usually suffers more than laser ablation from several limitations, such as differences in vaporization temperature between the elements to be determined and reproducibility problems. Nevertheless, ETV–ICP-IDMS can be applied successfully under certain conditions, especially if a selective vaporization of the isotope-diluted element can be achieved. An interesting example is the determination of traces of boron in powdered biological samples by using $NH_4F \cdot HF$ (ammonium bifluoride) as a modifier. After the solid sample has first been subjected to a pyrolysis step at 550°C, a solution of the modifier and a ^{10}B-labeled spike solution are added in this approach. After drying, this mixture is heated to 1200°C and the $^{11}B/^{10}B$ isotope ratio of the isotope-diluted sample is measured [53]. Boron results obtained via this method agreed very well with the corresponding certified values for the reference materials analyzed and an LOD of 8 ng g^{-1} was obtained.

8.3.3 Representative Examples of Trace Element Determination via ICP-IDMS

8.3.3.1 Determination of Trace Amounts of Silicon in Biological Samples

The determination of trace amounts of silicon is presented as an example demonstrating an important advantage of IDMS: after the isotope dilution step, sample loss has no effect on the analytical result [11]. Figure 8.8 shows the sample preparation scheme for this analysis using a ^{30}Si-enriched spike solution. To profit from the above-mentioned advantage to the largest possible extent, the spike should be added as early as possible, followed by an equilibration between sample and spike. This was accomplished by adding the spike directly to the sample in a PFA vessel and, after addition of the appropriate concentrated acids and closing

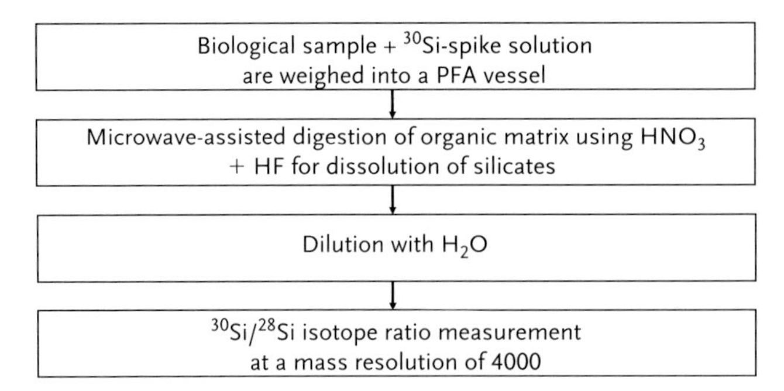

Figure 8.8 Sample preparation scheme for the determination of trace contents of silicon in biological materials via ICP-IDMS with a sector field ICP-MS instrument, operated at a mass resolution of 4000 [11].

the system to avoid loss of substance and also to prevent airborne contamination, subjecting the mixture to microwave-assisted acid digestion. Concentrated nitric acid under pressure in a closed vessel usually digests only the organic part of a biological sample. Therefore, if the biological sample also contains silicates, hydrofluoric acid also has to be added to dissolve this fraction.

The most critical step in most other analytical methods is opening the vessel after such a digestion procedure, because there is a high risk that volatile SiF_4 may be lost in an irreproducible way, so that the corresponding recoveries cannot be determined. On the other hand, loss of SiF_4 has no effect on the IDMS result as it is sufficient to isolate a (non-specified) amount of silicon for isotope ratio measurement using ICP-MS after simple dilution with highly pure water. The last analytical step is the $^{30}Si/^{28}Si$ isotope ratio measurement in the isotope-diluted sample, which needs to be carried out using sector field ICP-MS operated at a mass resolution of 4000 to avoid spectral interference.

Different biological samples were analyzed via the ICP-IDMS approach depicted in Figure 8.8 and the results obtained were compared with those of an inter-laboratory study in which 14 laboratories using various non-IDMS techniques participated. In this study, a silicon content of $5.3 \pm 1.0\ \mu g\ g^{-1}$ (mean $\pm$ standard deviation based on five independent analyses) was found for a homogeneous pork liver sample via ICP-IDMS, which agreed well with the result of the inter-laboratory study of $5.1 \pm 2.8\ \mu g\ g^{-1}$ (mean of the means of individual laboratories $\pm$ corresponding standard deviation). In contrast to this agreement, the results for a spinach powder were $333 \pm 11\ \mu g\ g^{-1}$ for ICP-IDMS and $176 \pm 189\ \mu g\ g^{-1}$ for the inter-laboratory study. The reason for this extreme difference between the ICP-IDMS result and the mean value, and also for the unacceptable standard deviation in the inter-laboratory study, is the presence of considerable amounts of silicate in spinach. Many laboratories had not adequately dealt with this silicate portion due to analytical difficulties with and errors in the sample preparation.

Similar problems are caused by losses of other analytes, which can be encountered when volatile compounds are produced. For example, losses of boron (as BF_3) can occur as a result of the use of hydrofluoric acid in the sample digestion, and in other circumstances, other volatile compounds, such as mercury, halogens, or OsO_4, can be (partly) lost. If the isotope dilution step and corresponding equilibration between analyte and spike take place in a closed system, subsequent loss of an unspecified amount of the isotope-diluted analyte, however, does not affect the corresponding isotope ratio and therefore also the accuracy of the analytical result.

8.3.3.2 Trace Element Analysis of Fossil Fuels

Combustion of fossil fuels generates emission of potentially toxic heavy metals and sulfur compounds, which is the primary source of anthropogenic sulfur in the atmosphere. In addition, elements such as nickel and vanadium can influence the activity of catalysts during oil cracking. These environmental and technical aspects are the underlying reason for the continuous reduction of the concentration of these elements in fossil fuels. As a result, improved analytical methods are also required to determine the corresponding element concentrations accurately, taking into account the continuously decreasing legal and technical limits. The use of ICP-IDMS methods in this field is therefore of increasing importance.

A more traditional technique was developed by Evans *et al.*, who determined the sulfur content in fossil fuels after microwave-assisted digestion of the sample, following equilibration with a ^{34}S-enriched spike in a closed vessel [35]. To avoid spectral overlap by the oxygen dimer ions (O_2^+), the $^{34}S/^{32}S$ isotope ratio of the isotope-diluted sample was measured at a mass resolution of 3000 using sector field ICP-MS. The results for certified reference materials showed excellent agreement with the corresponding certified values at the high $\mu g\ g^{-1}$ level. Yu *et al.* [54] also used a similar digestion method and a ^{34}S spike for the determination of sulfur in diesel fuel, but they avoided oxygen-based interferences by using ETV as a means of introducing the isotope-diluted sample into the ICP (dry plasma conditions), in combination with some nitrogen addition to the plasma gas (N_2 acts as additional oxygen scavenger), thus permitting the use of a quadrupole-based instrument for sulfur determination.

Digestion of fossil fuel samples is the most labor-intensive and time-consuming step in the analytical procedures described so far. Techniques with direct introduction of the isotope-diluted sample would therefore be much more attractive and suitable for use as routine methods. In this context, Heumann and co-workers developed two different methods which avoid the digestion step and rely on direct introduction of the isotope-diluted sample into the ICP. LA–ICP-IDMS was applied for the determination of sulfur and metals in petroleum products and crude oils [47, 48] and thermal vaporization (TV)–ICP-IDMS was used for the determination of trace amounts of sulfur in petroleum products [36]. To totally mix a petroleum or crude oil sample with the spike, the spike needs to be in an organic form. A ^{34}S-labeled dibenzothiophene spike in xylene and isotope-enriched organometallic compounds of the metals of interest in isobutyl methyl ketone were used for LA–ICP-IDMS purposes. For sulfur determination via TV–ICP-IDMS, the same ^{34}S spike was applied, but *n*-hexane was used as the solvent instead owing to the different requirements in terms of volatilization in both methods. Whereas for LA–ICP-IDMS a sector field mass spectrometer operated at medium mass resolution was necessary to avoid spectral interferences, oxygen-free injection of the isotope-diluted sample via thermal vaporization allowed the application of a quadrupole-based instrument in the case of TV–ICP-IDMS. In both cases, preparation of the isotope-diluted sample is very simple as it only requires mixing of an exactly known amount of the sample with a known amount of spike solution. The ^{34}S-labeled dibenzothiophene spike is currently not commercially available, but can be synthesized via a one-step synthesis from biphenyl and ^{34}S-enriched elemental sulfur [55].

For LA–ICP-IDMS, a volume of 10–20 μl of the isotope-diluted sample (there is no need to know the exact volume) was absorbed on a blank-free cellulose material, and subsequently covered with a thin Mylar foil to avoid evaporation of volatile compounds. After fixing this sample into a special sample holder in the ablation chamber, the laser ablation process was started and the transient ion signals of the spike and reference isotopes were measured [47, 48]. Figure 8.9 shows the transient ion signals of the spike and reference isotopes of sulfur and lead, respectively, obtained from isotope-diluted diesel fuel samples. As expected, the signal intensities vary strongly with time, but the corresponding isotope ratios remain constant, an essential precondition for applying the isotope dilution technique. The accuracy of the LA–ICP-IDMS results for sulfur and trace metals was demonstrated by very good agreement with certified values for reference materials and by comparison with ICP-IDMS analyses using microwave-assisted digestion, as described at the beginning of this section [35]. The LOD for sulfur was 0.04 μg g^{-1}, whereas for the metals it was in the range 0.02–0.2 μg g^{-1}, depending on the target element. Once the spike solutions are available, the time required for one LA–ICP-IDMS analysis is only ~10 min. The low LODs and the short analysis time also qualify this method for accurate routine trace element analysis of fossil fuel samples.

Thermal vaporization in combination with ICP-IDMS was also successfully applied to the determination of trace amounts of sulfur in petroleum products [36].

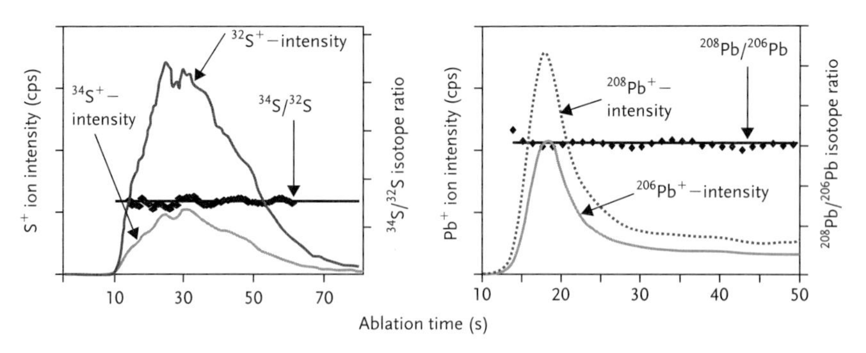

Figure 8.9 Signal intensities and corresponding sulfur and lead isotope ratios of an isotope-diluted diesel fuel sample as a function of time during LA–ICP-IDMS analysis. Reproduced with permission from [47,48].

Only a small volume of about 1 μl of the isotope-diluted sample was directly injected into a vaporizer unit, held at a constant temperature of 340°C, and the transient $^{34}S^+$ and $^{32}S^+$ signals were measured for the corresponding isotope ratio determinations. As a result of the oxygen-free injection using the vaporization unit (dry plasma), no spectral interference occurred, permitting the use of a quadrupole-based instrument. The very low amount of organic substance introduced into the ICP-MS system prevents the cones from clogging as a result of carbon deposition. The time between two injections could be less than 1 min. The accuracy of the results was confirmed at different concentration levels and for a variety of petroleum products by successful analysis of certified reference materials, and also by parallel analysis with ICP-IDMS after sample digestion. All these advantages, and also the low LOD of 0.04 μg g^{-1}, qualify TV–ICP-IDMS for the routine determination of trace amounts of sulfur, especially in modern "sulfur-free" petroleum products.

8.3.3.3 Trace Element Analysis via On-Line Photochemical Vapor Generation

Hydride generation (HG) using sodium tetrahydridoborate was often used in the past in atomic spectrometry as a selective sample introduction method for elements which form volatile compounds under these conditions. However, further improvements of this conventional HG method were not possible owing to many limitations [56]. The introduction of a new technique, ultraviolet photochemical vapor generation (UV-PVG), significantly extended the range of elements forming volatile elemental species. For example, iron, nickel, and several other transition and noble metals can be determined via this method [57, 58]. The main advantage of HG techniques is their much better sensitivity compared with sample introduction by conventional pneumatic solution nebulization. Sturgeon and co-workers therefore combined this advantage with the high precision and accuracy of isotope dilution by developing a UV-PVG–ICP-IDMS method, which was

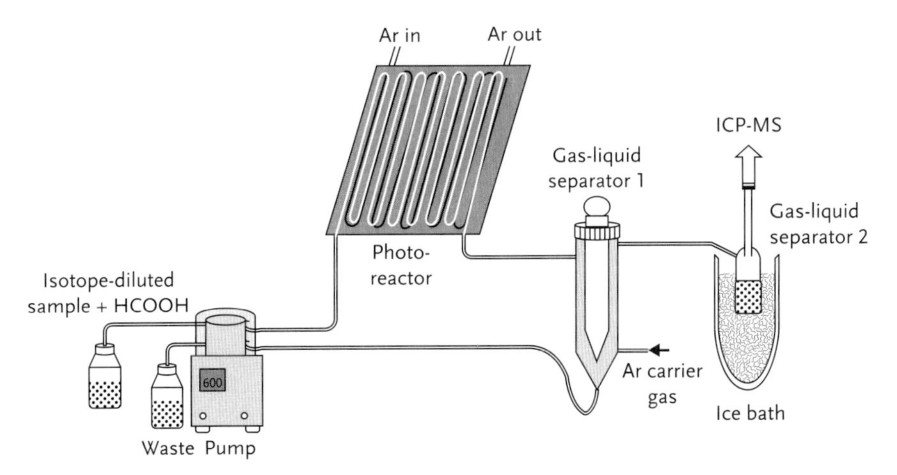

Figure 8.10 UV-PVG–ICP-IDMS system for determination of trace elements. Reproduced with permission from [59].

applied to the determination of trace amounts of selenium, iron, and nickel in biological tissues [59].

After microwave-assisted digestion of the sample with a mixture of nitric acid and hydrogen peroxide, during which equilibration with the isotopically enriched spikes added also took place, formic acid was used to convert the isotope-diluted elements into volatile compounds in a photo-reactor. In the case of iron and nickel, the corresponding volatile carbonyl compounds are most likely formed, whereas in the case of selenium, hydride compounds are obtained. A schematic representation of the UV-PVG–ICP-IDMS setup is shown in Figure 8.10. The good accuracy of results was demonstrated by the successful analysis of certified reference materials. Extremely low LODs of 0.18, 1.0, and 1.7 $pg\,g^{-1}$ were found for Ni, Fe, and Se, respectively, which are improvements of a factor of ~30 for the two metals and 150 for Se compared with conventional pneumatic solution nebulization.

8.3.3.4 **Determination of Trace Amounts of Platinum Group Elements**

An increasing number of PGE determinations via ICP-IDMS techniques have been reported during the last decade [21, 25, 60–63]. The reason for applying isotope dilution lies in the problem of other analytical methods accurately determining the usually very low concentrations at which PGEs occur in environmental and geological samples. This is demonstrated in Figure 8.11, which presents the results of palladium determinations in a homogenized road dust sample using different analytical methods in the context of an inter-laboratory study, carried out in 1999. Acceptable agreement between the results was only obtained by seven of

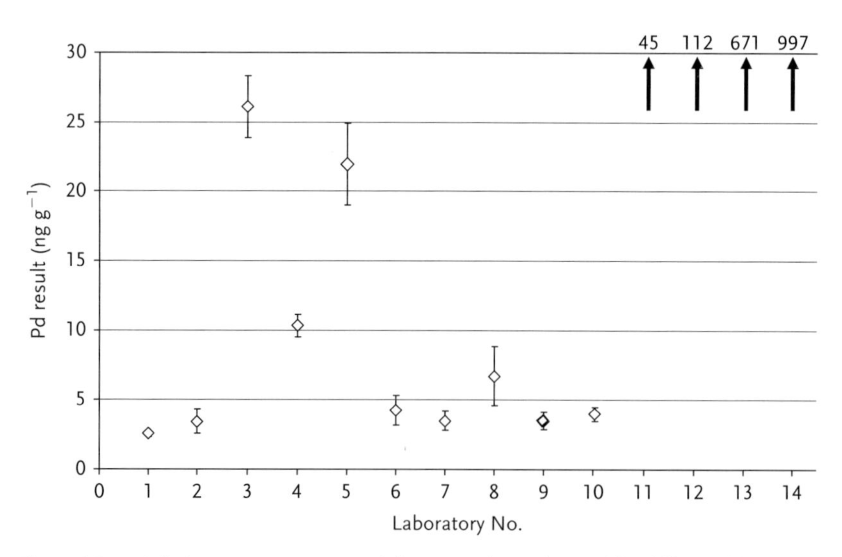

Figure 8.11 Palladium content in a road dust sample as obtained by different laboratories/analytical methods in the context of an inter-laboratory study (error bars represent standard deviations). Reproduced with permission from [60].

the 14 laboratories that reported Pd concentrations in the range 3–7 $ng\,g^{-1}$. Three laboratories reported values in the range 10–25 $ng\,g^{-1}$ and four laboratories reported much higher palladium contents of up to 997 $ng\,g^{-1}$.

In principle, all PGEs except rhodium can be determined via the isotope dilution technique because they have more than one stable isotope. However, it is difficult to determine osmium simultaneously with the other PGEs owing to the different properties of this element, especially as a result of the high volatility of OsO_4 usually formed during conventional digestion procedures. In addition, a careful separation of the PGEs from many other elements is required because otherwise several polyatomic ions that give rise to spectral overlap occur and thus hinder accurate PGE determination, even when using a sector field ICP-MS instrument operated at high mass resolution [60]. Mass resolutions in excess of 10 000, the current limit of commercially available sector field ICP-MS instruments, are often necessary, for example, to separate $^{66}Zn^{35}Cl^{+}$ ions from $^{101}Ru^{+}$ or $^{89}Y^{16}O^{+}$ ions from $^{105}Pd^{+}$. Most of the PGE determinations via ICP-IDMS were performed following microwave-assisted digestion of the sample, where the corresponding spike solutions were added, with an $HCl–HNO_3$ mixture and subsequent PGE separation from the other elements using anion- or cation-exchange chromatography. In some cases, osmium was determined via direct introduction of OsO_4, formed during digestion, into the ICP [61].

PGE traces were also determined via LA–ICP-IDMS, which has the important advantage that no complex and time-consuming digestion and separation are needed [25]. Owing to negligible oxide ion formation, which is otherwise an important source of PGE interferences, with laser ablation as a means of sample introduction, the isotope ratio measurements could be carried out using a sector field ICP-MS instrument operated at low mass resolution (300), thus providing maximum detection power. Sample preparation for laser ablation was carried out in the same way as shown in Figure 8.7. Pt, Pd, Ir, and Ru were determined in four reference materials of the Geological Survey of Canada. The LA–ICP-IDMS results are plotted against the corresponding certified or indicative values in Figure 8.12, demonstrating the excellent accuracy, and also the high sensitivity and linear dynamic range (from 10^{-1} to $>10^{3}$ $ng\,g^{-1}$) of this method.

On average, the uncertainties on the results obtained via LA–ICP-IDMS (in Figure 8.12 standard deviations are presented) are higher than those for the corresponding certified values. This can be explained by the very small amount of sample analyzed in one laser ablation run (about 3 mg), compared with that in the methods used for certification, where usually sample amounts of 0.2–2 g were taken. The analysis of small sample amounts is extremely sensitive to sample inhomogeneities, in the case of platinum, well known as nugget effects. On the other hand, laser ablation with its very small sample consumption is an excellent technique to reveal sample inhomogeneities, which is demonstrated in Figure 8.13 for the LA–ICP-IDMS determination of platinum in a road dust sample and in the geological reference material UMT-1 [25]. In both cases, ≥ 50 analytical runs with laser ablation were carried out (corresponding to the analysis of ≥ 50 sub-samples of the same isotope-diluted sample pellet) and the number of runs providing a platinum content within a given range is plotted in Figure 8.13.

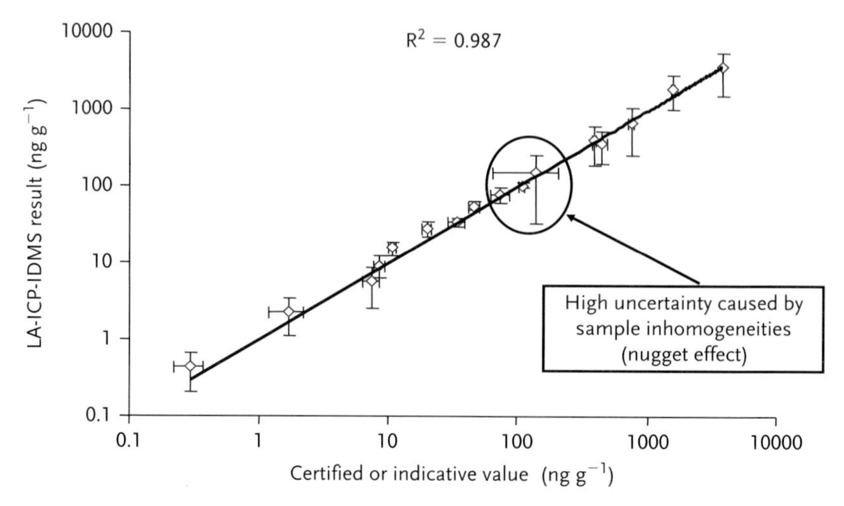

Figure 8.12 Results for Pt, Pd, Ir, and Ru as obtained via LA–ICP-IDMS for four reference materials of the Geological Survey of Canada compared with the corresponding certified or indicative values. Reproduced with permission from [25].

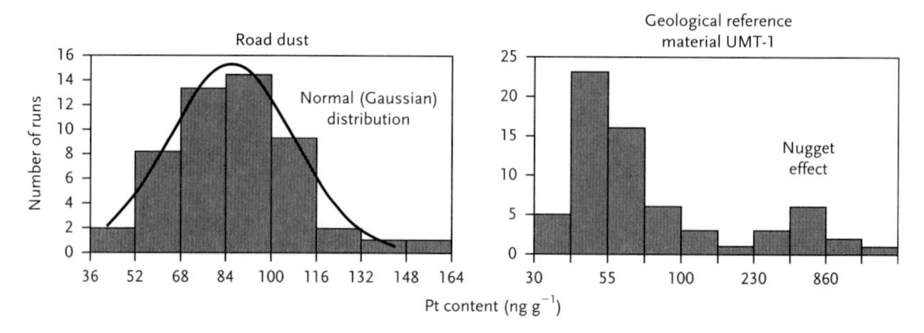

Figure 8.13 Distribution patterns of the platinum content in a road dust sample and the geological reference material UMT-1 determined via analysis of subsamples by LA–ICP-IDMS. Reproduced with permission from [25].

The platinum contents in the different sub-samples of road dust show a Gaussian distribution with its maximum almost at the mean value of the LA–ICP-IDMS analysis of 81.3 ± 3.3 ng g^{-1}, which means that platinum is homogeneously distributed. On the other hand, the LA–ICP-IDMS results for the UMT-1 material show two maxima, one at low and the other at a higher content, a situation produced by the presence of nuggets. From all sub-samples, a mean of 135 ± 71 ng g^{-1} (certified value 129 ± 5 ng g^{-1}) was calculated, a value which was only found directly in a minority of all the runs (Figure 8.13). Thus, the micro-analytical laser ablation technique demonstrates another capability of LA–ICP-IDMS, that is, to analyze distribution patterns quantitatively in powdered samples, which is of special importance for the characterization of certified reference materials.

8.3.3.5 Determination of Ultra-Trace Amounts of Transuranium Elements

As a consequence of weapons testing and nuclear accidents such as those in Chernobyl and Fukushima, contamination of the environment with transuranium radionuclides has become an important problem during the past few decades. Whereas the median level of the plutonium content in soils is only ~ 100 fg g^{-1} [64], in some regions of the world the plutonium content in upper soil layers and in living organisms exceeds the dangerous level of 1 pg g^{-1}. Accurate determination of such ultra-low transuranium contents is an important precondition for adequate risk assessment. ICP-IDMS is an ideal analytical technique for the determination of Pu at these low concentration levels owing to its high accuracy and sensitivity. Low LODs are the consequence of negligible abundances of these radioactive isotopes in the environment, such that they are usually only limited by potential spectral overlap. In addition, it is also important to determine not only the total content of transuranium elements, but also their isotopic composition, which allows the identification of the primary source.

Zheng and Yamada developed an ICP-IDMS technique for the determination of ^{241}Am ($T_{1/2} = 433$ years) in marine sediments because, as a result of its high particle affinity, this radioisotope is a useful tracer for understanding biogeochemical processes in the marine environment and for estimating mineral aerosol fluxes to the ocean [65]. ^{241}Am is a daughter product of ^{241}Pu, which is emitted into the atmosphere as a result of activities of the nuclear industry and weapons testing. A complex sample treatment procedure was applied, including a leaching step with HNO_3, followed by CaF_2 coprecipitation and chromatographic separation, to isolate americium from the sediment matrix and from possible substances which can interfere with the determination of ^{241}Am using ICP-MS (e.g., $^{209}Bi^{16}O_2^+$, $^{206}Pb^{35}Cl^+$). By applying a sector field ICP-MS instrument operated at low mass resolution (300) and ^{243}Am ($T_{1/2} = 7370$ years) as spike compound, an LOD of 0.32 fg g^{-1} sediment was achieved.

Ohtsuka *et al.* used a combination of on-line column separation and ICP-IDMS for the rapid determination of plutonium in environmental samples [66]. After leaching of plutonium from soil or sediment samples with 8 mol l^{-1} HNO_3, followed by chromatographic separation, the plutonium isotopes ^{239}Pu ($T_{1/2} = 2.4 \times 10^4$ years) and ^{240}Pu ($T_{1/2} = 6550$ years) were determined using a quadrupole-based ICP-MS instrument and a ^{242}Pu ($T_{1/2} = 3.8 \times 10^5$ years) spike, providing absolute LODs of 9.2 and 4.3 fg, respectively. The accuracy of the results was demonstrated by the good agreement between the sums of the ^{239}Pu and ^{240}Pu contents as obtained via ICP-IDMS and the corresponding certified values in certified reference materials. An interesting alternative to this method, combining plutonium leaching prior to its separation, is direct analysis of the isotope-diluted samples via LA–ICP-IDMS. Here, the environmental sample (soil or sediment) is mixed with a ^{244}Pu ($T_{1/2} = 8.3 \cdot 10^7$ years) spike according to the procedure described in Section 8.3.2 (Figure 8.7). When using sector field ICP-MS at a mass resolution of 4000 to avoid spectral interferences, especially at a mass-to-charge ratio of 239 ($^{207}Pb^{16}O_2^+$ and $^{238}UH^+$), LODs for ^{239}Pu and ^{240}Pu were found to be 300 fg g^{-1}, even for samples containing uranium at levels up to a few µg g^{-1} [26]. Also in this case, good accuracy was demonstrated by the successful analysis of certified reference materials. An important advantage of the use of LA–ICP-IDMS is the possibility of also revealing "hot spots" of plutonium, which often occur, as such analyses of soil samples from the Chernobyl region have shown.

8.3.4
ICP-IDMS in Elemental Speciation

8.3.4.1 Principles of ICP-IDMS in Elemental Speciation

Hyphenated techniques are the analytical methods most frequently used today for elemental speciation. For that purpose, a separation technique, such as high-performance liquid chromatography (HPLC), gas chromatography (GC), capillary electrophoresis (CE), or gel electrophoresis (GE), is coupled on-line with an elemental detection method, such as ICP-MS [67]. Hyphenated ICP-MS techniques can also be combined with the isotope dilution method for quantification of elemental species. Two different spiking modes are possible, the species-specific and species-unspecific mode. Rottmann and Heumann were the first to present a setup for HPLC–ICP-IDMS enabling the use of these two spiking modes [68] (Figure 8.14).

For species-specific ICP-IDMS analysis, the sample solution is first spiked with a solution containing the elemental species of interest, but in which the target element is isotopically enriched. After mixing the spike with the sample, usually by simple shaking, the isotope-diluted sample is analyzed for the corresponding elemental species. During mixing, it is important that no isotope exchange between different elemental species occurs, as this would lead to incorrect

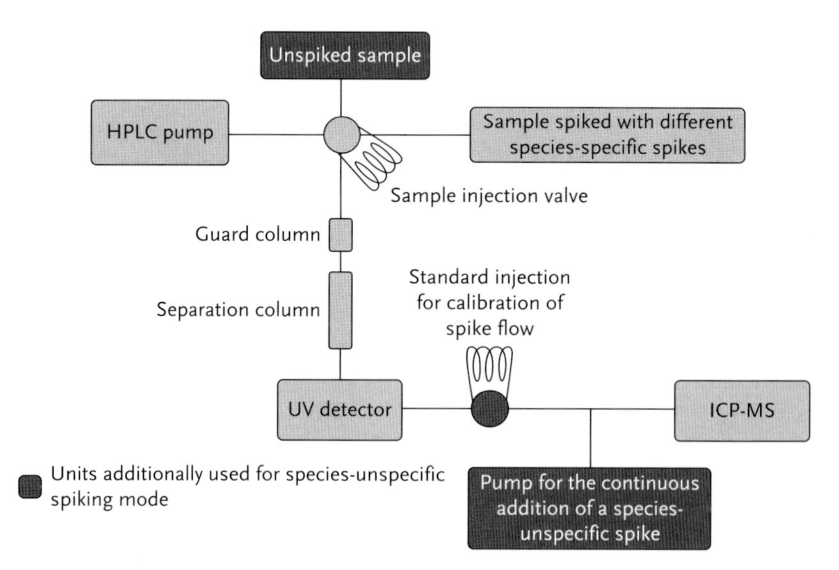

Figure 8.14 Elemental speciation via HPLC–ICP-IDMS using the species-specific or species-unspecific spiking mode. Reproduced with permission from [68].

results. This is especially a risk in the case of elemental species in which the target element is weakly bound, for example, in metal complexes or organic iodine compounds [7]. On the other hand, early mixing of the spike and the sample allows the analyst to rely on one of the most prominent advantages of the isotope dilution technique, namely that loss of analyte in the subsequent separation step has no effect on the final analytical result. After passing a flow-through UV detector, which is useful for the identification of organic elemental species, the separated species of interest is directly introduced into the ICP-MS instrument for measuring the transient signal intensities for the spike and reference isotopes. In species-specific mode, the isotope ratio of the corresponding isotope-diluted elemental species is constant over the entire peak. This is true if different elemental species do not overlap, a necessary precondition for ICP-MS detection, because this type of detector cannot differentiate between various species of the same element.

In species-unspecific spiking mode, the different elemental species are first separated from one another one and calibration by isotope dilution is performed by subsequent admixing a continuous flow of a spike solution of known concentration with the column effluent. Therefore, in the literature, this spiking mode is also often called "post-column" spiking. After mixing the separated elemental species with the spike, which can be in any chemical form of the element of interest, this mixture is introduced into the ICP-MS instrument and the signal intensities of the spike and reference isotopes are measured. A consequence of carrying out the spiking step after the separation is the necessity to separate the elemental species from one another without any analyte loss. On the other hand, this type of spiking has the advantage that the spike compound does not have to be identical with the elemental species to be determined and that, as a consequence, the same spike can be used for all separated elemental species. Indispensable for successful use of species-unspecific spiking is that no fractionation between different species occurs in the ICP introduction system and that the instrumental sensitivity does not depend on the chemical form. Whereas the latter precondition was found to be realized for all elemental species investigated due to the extremely high plasma temperature, the first precondition should always be proved. For most of the commonly used nebulizer systems, such as a cross-flow nebulizer, no difference in analyte introduction efficiency was found for different elemental species, but in some cases, for example, when using an ultrasonic nebulizer with membrane desolvator, significant differences in the analyte introduction efficiency were established for organic and inorganic elemental species of the same element [7]. In addition, the spike flow needs to be calibrated and this is carried out by reverse isotope dilution, using a known amount of a standard solution of natural isotopic composition, introduced via a loop located after the separation column (see Figure 8.14). Even when the critical parameters of the species-unspecific spiking mode can be controlled, the species-specific spiking mode has a higher level of accuracy assurance.

In contrast to the species-specific spiking mode, in the species-unspecific mode the isotope ratio of an elemental species peak varies with time because a constant spike flow is mixed with an elemental species of natural isotopic composition,

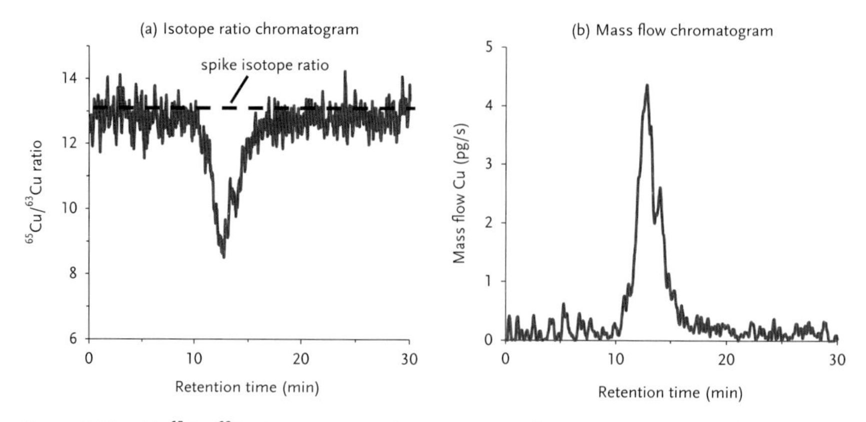

Figure 8.15 (a) $^{65}Cu/^{63}Cu$ isotope ratio chromatogram of a copper-containing elemental species by species-unspecific ICP-IDMS and conversion into (b) the corresponding copper mass flow chromatogram. Reproduced with permission from [67].

showing a time-dependent concentration profile. Figure 8.15a illustrates an isotope ratio chromatogram for the measurement of the $^{65}Cu/^{63}Cu$ isotope ratio in a peak of a copper-containing elemental species.

As can be seen in Figure 8.15a, one measures the isotope ratio of the spike solution ($^{65}Cu/^{63}Cu = 13$) as long as no copper-containing elemental species leaves the column. If a copper species appears, the isotope ratio shifts in the direction of the natural isotope ratio of copper ($^{65}Cu/^{63}Cu = 0.446$) to come back to the spike isotope ratio after the corresponding copper peak. Based on an analogous equation as 8.2 but correlated to pg amounts of the analyte instead of $\mu g\ g^{-1}$ concentrations (for every individual pair of data points), this isotope ratio chromatogram can be converted into the corresponding mass flow chromatogram (Figure 8.15b). This is most easily accomplished by using a computer program.

An interesting alternative to the more frequently applied "post-column" spiking mode was developed for HPLC–ICP-MS by Schaumlöffel *et al.*, based on "pre-column" isotope dilution [69]. In this case, the species-unspecific spike was directly added to the eluents used for separation, which is only possible, of course, if the spike compound does not react with the different analytes and if the column does not fractionate the spike from the analyte, which always needs to be carefully assessed. Successful application of this type of species-unspecific isotope dilution has been demonstrated for the quantification of sulfur-containing peptides using a ^{34}S-labeled sulfate spike.

8.3.4.2 Species-Specific ICP-IDMS

General Remarks For species-specific ICP-IDMS, the elemental species needs to be well defined in terms of composition and structure. This is feasible for most of the low molecular weight elemental species of environmental or biochemical importance, such as the inorganic compounds chromate and iodate, but also for organo-elemental species, such as monomethylmercury (MMM) ($MeHg^+$), trimethyllead (TML) (Me_3Pb^+), monobutyltin (MBT), dibutyltin (DBT), and tributyltin (TBT) [$SnBu_n^{(4-n)+}$, with $n = 1$, 2, and 3], and selenomethionine. In addition, the corresponding isotope-labeled spike compounds need to be commercially available or need to be synthesized in-house. Unfortunately, only a few certified spike compounds are commercially available at present. ^{202}Hg-labeled MMM can be obtained from the IRMM, and a mixture of MBT, DBT, and TBT labeled with ^{119}Sn can be purchased from ISC Science, Gijón, Spain. All other isotopically enriched species need to be synthesized in-house. The synthesis of low-molecular weight inorganic elemental species is usually not very complicated. For example, ^{129}I-labeled iodate can be obtained by converting commercially available Na^{129}I into the corresponding sodium iodate by treatment with a mixture of concentrated HNO_3, $HClO_3$, and $HClO_4$, followed by purification of the iodate thus obtained by anion-exchange chromatography [70]. For the determination of TML, the corresponding spike can be synthesized by dissolving commercially available ^{206}Pb-enriched metallic lead in HBr. The corresponding lead bromide is then converted by methyllithium into tetramethyllead in nonaqueous solution and, subsequently, into ^{206}Pb-labeled TML by adding elemental iodine to this intermediate [71].

As an important example of a biochemically relevant species, ^{77}Se-labeled selenomethionine can be prepared, for example, by isolation from selenized yeast, grown in a ^{77}Se-labeled selenite solution. After digestion of the yeast with protease/lipase solution, ^{77}Se-labeled selenomethionine is isolated by size-exclusion chromatography (SEC) [72, 73]. Another important example is the synthesis of ^{57}Fe-labeled transferrin [74]. The synthesis of other biologically relevant molecules, especially those of higher molecular weight, is usually too complicated, and the species-unspecific spiking mode is more convenient in these cases.

Applications involving elemental speciation relying on species-specific or species-unspecific ICP-IDMS are listed in some of the textbooks on elemental speciation [75, 76]. However, the most comprehensive and detailed compendium of elemental speciation with IDMS can be found in two reviews by Rodriguez-González *et al.* [77, 78], covering all investigations on this subject up to 2004 and between 2005 and 2009. respectively. These reviews also critically discuss new concepts, methodologies, and trends. Another useful review, by Meija and Mester [79], summarizes the fundamental prerequisites behind IDMS for elemental speciation and discusses their practical limits of validity and effects on the accuracy of the corresponding results.

Speciation Analysis: Simultaneous Determination of Multiple Species of the Same Element At the beginning of the development of species-specific isotope dilution, only a single spike compound was used [77]. Later, species-specific ICP-IDMS has also been used for the determination of more than one species of the same element, using different spike species labeled with the same isotope. Examples include the determination of mono- and dimethylmercury (MMM and DMM) and of the three butyltin compounds TBT, DBT, and MBT [80–83]. Various spike isotopes, usually of medium natural abundance, such as ^{198}Hg or ^{201}Hg and ^{117}Sn, ^{118}Sn, or ^{119}Sn, have been applied. The reason for selecting this type of spike isotope lies in cost optimization, on the one hand, but also allows one to keep the error multiplication factor of the isotope ratio measurement in the isotope-diluted sample low (see Section 8.2.3). The elemental species mentioned above were preferably determined via GC–ICP-MS after ethylation of the corresponding organometallic ions for conversion into volatile compounds amenable to GC separation. The determination of MMM and the butyltin compounds is especially relevant for samples such as seafood and sediments, due to possible accumulation of these compounds in these types of samples. An interesting example in atmospheric chemistry is the accurate determination of iodide and iodate species in aerosol samples by means of gel electrophoresis coupled with ICP-IDMS using ^{129}I-labeled spike compounds [84]. The results thus obtained can contribute to a more profound insight into the biogeochemical cycle of iodine in the environment. Another environmentally relevant analysis is the determination of trimethyllead (TML) and triethyllead (TEL) in urban dust, using the corresponding ^{206}Pb-labeled spike compounds [85]. The organolead species were extracted from the dust with an acetic acid–methanol mixture, followed by alkylation with tetrabutylammonium tetrabutyloborate and GC separation.

An application of technical interest is the determination of different thiophene derivatives, the major sulfur impurities in petroleum products. ^{34}S-labeled thiophene, dibenzothiophene, and 4-methyldibenzothiophene spikes in hexane were used to determine these sulfur compounds in different gasoline samples, naphtha, diesel fuel, gas oil, and heating oil by species-specific GC–ICP-IDMS [55]. The accuracy of the results thus obtained was demonstrated by successful analysis of the reference material NIST SRM 2296. The LOD was limited to 7 ng of sulfur per gram of sample by the background noise in the isotope ratio chromatograms.

Speciation Analysis: Simultaneous Determination of Species of Different Elements
In cases where elemental species of different elements occur in the same sample, it is desirable to be able to determine these elemental species simultaneously in one analytical run. A prominent example is the possible occurrence of MMM, TML, MBT, DBT, and TBT in the marine environment. Whereas the first two compounds can be produced in the ocean by biogenic methylation, the last three are of anthropogenic origin only (TBT is used as biocide in boat paints). In addition, TML can also be of anthropogenic origin as a degradation product of TML, previously used extensively as an anti-knock additive in petrol. The accurate determination of these highly toxic organometallic compounds is of special importance in seafood, where they can easily accumulate. A simultaneous approach for the determination of these elemental species via species-specific GC–ICP-IDMS was therefore developed [86].

After spiking, the organic matrix (e.g., seafood) is best dissolved using tetramethylammonium hydroxide (TMAH). The isotope-diluted organometallic ions are then converted into volatile species by ethylation or propylation using $NaBEt_4$ or $NaBPr_4$, respectively. After *in situ* extraction of the neutral alkylated compounds thus formed into hexane, about 1 µl of the isotope-diluted extract is injected into a gas chromatograph for separation on a capillary column. The signal intensities of

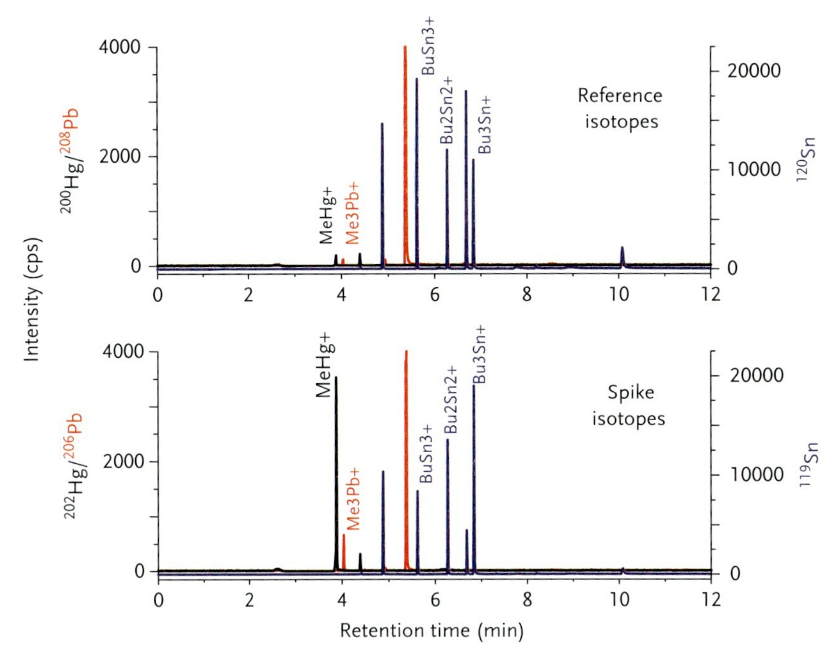

Figure 8.16 GC–ICP-MS chromatogram of the spike and reference isotopes of mercury, lead, and tin in different elemental species fractions of an isotope-diluted mussel tissue sample. Reproduced with permission from [86].

the spike and reference isotopes are measured for each of the elements of interest (Sn, Hg, and Pb). From the signal intensities, the isotope ratios of the isotope-diluted elemental species are obtained and used for calculation of the corresponding concentrations. Figure 8.16 shows the GC–ICP-MS results for the spike and reference isotopes for mercury, lead, and tin obtained for a species-specific isotope-diluted mussel tissue sample [86]. The isotope chromatograms of the spike and reference isotope are registered simultaneously, but are plotted separately in Figure 8.16 for the sake of clarity.

The accuracy of this type of simultaneous species-specific multi-element GC–ICP-IDMS determination of five elemental species of high relevance for toxicological assessment of marine food samples was demonstrated by the excellent agreement of the results obtained with the corresponding certified values for different reference materials. LODs of 1.4 ng g^{-1} for MMM, 0.06 ng g^{-1} for TML, 1.2 ng g^{-1} for DBT, and 0.3 ng g^{-1} for MBT and TBT were obtained, which are much lower than the limits required for adequate control of seafood. The tolerable weekly intake of MMM for humans as recommended by the US Environmental Protection Agency (EPA) is <0.7 μg kg^{-1} body weight and the tolerable MMM content in fish as recommended by the EU is <1 μg g^{-1}. With respect to the relative simple sample preparation and the commercial availability of certified spike solutions of MMM and the butyltin compounds, multi-species species-specific GC–ICP-IDMS has a high potential for use on a routine basis for the accurate and sensitive determination of these four toxic elemental species in seafood.

Species-Specific ICP-IDMS for Identification of Species Transformations Possible species transformations during sample pretreatment steps are an important problem in elemental speciation. Significant doubts concerning the accuracy of results for MMM in sediments obtained after using water vapor distillation during the sample pretreatment first arose when an isotope-enriched inorganic Hg^{2+} spike was added to the sample, because a substantial amount of the distilled MMM was found to be also enriched in the spike isotope [87]. Later, it could be shown via species-specific isotope dilution that significant species transformations can also occur during other types of sample pretreatment steps. For example, by using a ^{201}Hg-labeled MMM spike, it was demonstrated that ethylation of mercury species in water samples by $NaBEt_4$, followed by distillation of the neutral ethylation product and GC separation, leads to a substantial transformation of MMM into elemental Hg^0, the extent of which was seen to be dependent on the chloride concentration in the sample [88]. In Figure 8.17a, such a transformation of MMM can be seen, whereas corresponding propylation by $NaBPr_4$ under otherwise identical conditions does not show any transformation of this mercury species (Figure 8.17b) [7, 88]. Despite this difference between ethylation and propylation, the analytical results calculated from the corresponding $^{202}Hg/^{201}Hg$ isotope ratios of both MMM peaks are identical within the given standard deviations. This is because loss of the isotope-diluted substance by transformation into Hg^0 did not change the isotope ratio of the remaining MMM, such that, in this case, the transformation had no effect on the analytical result. This example demonstrates that accurate analytical results can even be obtained via species-specific isotope

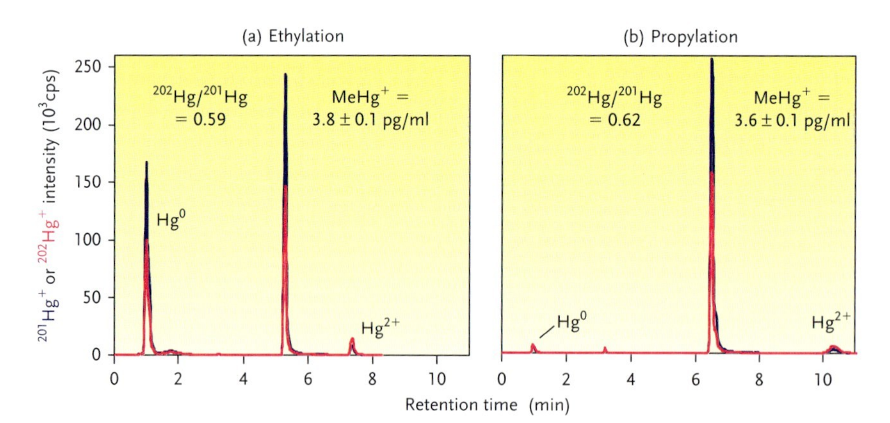

Figure 8.17 Comparison of GC–ICP-MS chromatograms for mercury species in a river water sample after species-specific isotope dilution with a ^{201}Hg-labeled MMM spike and derivatization with (a) $NaBEt_4$ and (b) $NaBPr_4$. Reproduced with permission from [88]. Only chromatogram (a) indicates conversion of MMM into Hg^0.

dilution if transformation of elemental species takes place after the isotope dilution step has been carried out. On the other hand, this technique can reveal and identify species transformations, usually not an easy task with other analytical methods.

Even though stable isotope tracer studies are not real isotope dilution techniques, precise isotope ratio measurements are also necessary in this context, and they can be combined successfully with species-specific ICP-IDMS for quantification. Such tracer studies can be applied to investigate the behavior of elements in the environment. For example, methylation of inorganic mercury into MMM in sediments, and also the reverse process of demethylation, were investigated by Hintelmann *et al.* using enriched stable mercury isotopes [89, 90]. The change in the isotope ratio of different species reflects the environmental behavior of the tracer. Using more than one isotope of the same element in different species allows one to separate the influence of several natural processes from one another or to follow the dynamics of exchange reactions within the same or between different compartments.

Multiple Spiking for Identification and Quantification of Species Transformations
Application of more than one spike compound of the same element labeled with different isotopes can be used either to follow the formation and degradation processes of species or to correct for interconversions between different species during sample pretreatment steps. A multiple spike solution is added to the sample and the variation of the isotope ratios is measured as a function of time for all different species. In the case of double spiking, two independent isotope ratios are measured in each species. Huo *et al.* used a ^{50}Cr-labeled Cr(III) spike and a ^{53}Cr-enriched Cr(VI) spike and measured the ^{53}Cr/^{52}Cr and ^{52}Cr/^{50}Cr isotope ratios to investigate interconversion reactions between Cr(III) and Cr(VI) [91]. Comparable double-spiking investigations were carried out to study the methylation and demethylation processes of mercury [90]. In case of possible interconversion reactions between more than two species, additional isotope-enriched spike compounds need to be applied. This is the case with the investigation of possible interconversions between inorganic tin Sn(IV) and the butyltin species TBT, DBT, and MBT, for which a triple-spiking technique was applied by Rodriguez-González *et al.* [92]. Spike compounds of TBT, DBT, and MBT, labeled with ^{117}Sn-, ^{118}Sn-, and ^{119}Sn-enriched isotopes, respectively, were used and the ^{120}Sn/^{117}Sn, ^{120}Sn/^{118}Sn, and ^{120}Sn/^{119}Sn isotope ratios were measured in each fraction of the different butyltin species. As can be seen from Figure 8.18, theoretically, 12 interconversion reactions need to be considered. However, some of these reactions are not very likely (dashed arrows in Figure 8.18), such that these reactions can be neglected. The extensive mathematical approach for quantifying the different interconversion reactions and thus providing the corrected concentrations of the three butyltin species is relatively complex and can be found in the literature [77, 92]. Other mathematical approaches have also been developed for multiple spiking and their merits for butyltin determination in sediments were compared [93, 94].

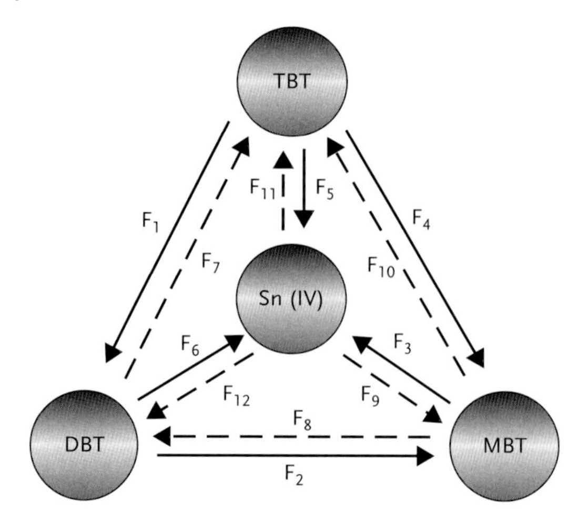

Figure 8.18 Possible interconversion reactions involving inorganic Sn(IV) and the butyltin species TBT, DBT, and MBT. Reproduced with permission from [92].

Multiple species-specific spiking is, in principle, a powerful method to quantify possible interconversion reactions and to determine simultaneously corrected species concentrations, but there are also limitations to this technique. A major limitation in accuracy and precision was observed for real samples in which the concentrations of the different species are extremely different. Such limitations were found, for example, in sediment samples where the content of inorganic mercury is usually several orders of magnitude higher than that of MMM [95]. In addition, it was also demonstrated that any corrections for species interconversion are performed at the expense of precision of the analytical result because, under these conditions, typically less effort is devoted to minimizing species interconversions during sample treatment steps [96]. Less effort towards avoiding species interconversion to the largest possible extent results in larger uncertainties for the species concentrations determined. Therefore, the species-specific multiple-spiking technique is not an analytical tool for routine analysis, but it permits the identification and quantification of interconversion reactions between different species of the same element, which is especially important in the context of validation of speciation methods.

An important improvement in the calculation procedure for species-specific multiple spiking was obtained by introducing isotope pattern deconvolution (IPD). The difference from the conventional multiple-spiking technique is the use of isotopic abundances instead of isotope ratios for calculation. The measurement of additional isotopes not used for calculation of interconversion reactions provides internal correction for mass bias effects without the application of an external isotopic standard, but also correction for spectral interferences. The basic equations applied in IPD can be found in the literature [78]. This new method is predominantly applied in speciation analysis, but also in metabolism studies using stable isotopes [97, 98]. For quantification of elemental species, including correction for interconversion reactions by double-spiking IPD, both isotope-enriched spike compounds are added to the sample at the same time, for example, ^{201}Hg-enriched inorganic mercury and ^{199}Hg-enriched MMM to determine the quantity of and possible interconversion between these mercury species. On the other hand, one of the spike compounds can also be added later to the sample. In this case, the spike added first is used as a tracer, for example, in metabolic studies, and the second for quantification of the elemental species. Meija *et al.* [99] investigated the redox interconversion reaction between Cr(III) and Cr(VI) in yeast using ^{50}Cr-enriched Cr(III) and ^{53}Cr-enriched Cr(VI) by least-squares ion intensity-based IPD. They found that the results were in exact agreement with conventional isotope dilution calculations. An example in which IPD was applied for metabolism studies is the determination of the metabolism of butyltin compounds in rats [100]. ^{119}Sn-labeled MBT, ^{118}Sn-labeled DBT, and ^{117}Sn-labeled TBT dissolved in tap water were administered orally to the animals. Degradation of the butyltin compounds was studied in different organs by least-squares IPD and their concentrations were determined via reverse isotope dilution analysis using the corresponding compounds of natural isotopic composition as spikes.

8.3.4.3 Species-Unspecific ICP-IDMS

General Remarks For elemental species the exact composition and structure are not known or for those for which isotope-enriched spike compounds are not available, only the species-unspecific spiking mode can be used. Prominent examples for the first group of elemental species mentioned are metal complexes of humic substances and high molecular weight biomolecules, such as metalloproteins. For these substances, the separation techniques used in the hyphenated ICP-MS setups are usually not able to separate distinct elemental species from one another, and only fractions of molecules with similar behavior are obtained. Therefore, it would be more correct to use the term fractionation instead of speciation in this context. The disadvantage of species-unspecific compared with species-specific spiking is the prerequisite that no loss of the elemental species may occur during separation in the case of "post-column" mixing and that certain preconditions need to be fulfilled in the case of "pre-column" mixing (see Section 8.3.4.1). On the other hand, it is a great advantage that only one spike compound need be used for different elemental species. As the measured isotope ratio varies within the chromatographic peak, the corresponding data handling is slightly more complicated than in species-specific spiking, but it also results in real-time concentrations. Some representative applications of species-unspecific ICP-IDMS are discussed in the following sections.

Sulfur Speciation in Petroleum Products Owing to the environmental problems linked to the emission of SO_2 as a result of the combustion of sulfur-containing fuels, the legal limit for sulfur in gasoline and diesel fuel has been continuously reduced in recent years so that, for example, the sulfur content in "sulfur-free" gasoline in the EU may currently not exceed 10 $\mu g\ g^{-1}$. For optimum reduction of the sulfur level during the refining of crude oil, the different sulfur species present

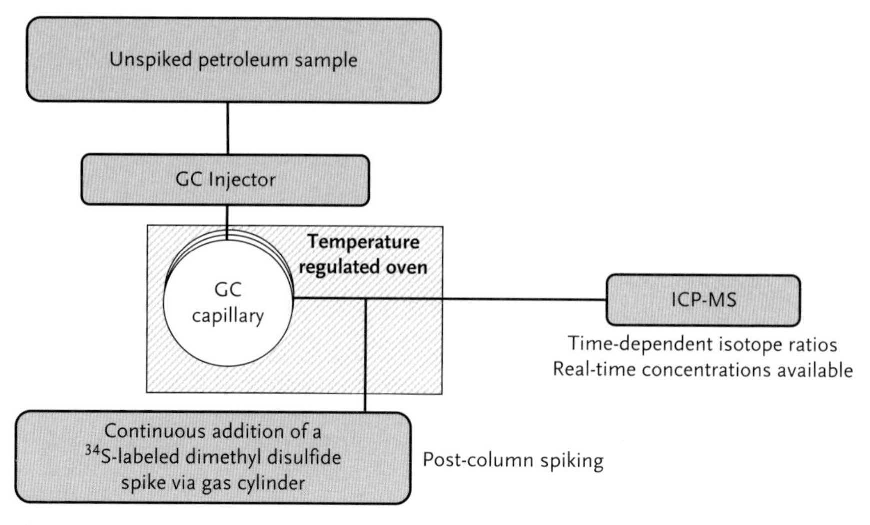

Figure 8.19 Species-unspecific GC–ICP-IDMS set-up for the determination of volatile sulfur species in petroleum products [101].

need to be known, hence speciation analysis of sulfur in crude oil and petroleum products is an important analytical task in the petroleum industry. In this context, a GC–ICP-IDMS system was developed by Heilmann and Heumann for accurate determination of sulfur species in petroleum products (Figure 8.19) [101].

The principle of the GC–ICP-IDMS system used for the determination of gaseous elemental species is similar to the HPLC–ICP-IDMS system shown in Figure 8.14 for dissolved species. The addition of a constant flow of gaseous spike and calibration of the latter, are, however, much more complicated. Volatile ^{34}S-labeled dimethyl disulfide, usually not present in petroleum products, was used as a spike compound. It was diluted with Ar and stored under pressure in a gas cylinder. Calibration of the spike gas flow rate was carried out by adding a sulfur-containing internal standard of natural isotopic composition to the sample. To avoid coelution, a sulfur standard with a retention time different from those of the different compounds present in the sample was selected for this purpose. Thus, for determination of the sulfur species in low-boiling petroleum products, the high-boiling dibenzothiophene was selected, whereas the low-boiling 2-ethylthiophene was used for high-boiling samples. The use of such an internal standard has the additional advantage that the exact value of the spike gas flow rate does not have to be known, because of the direct correlation between the integrated peak area for the known amount of internal standard and the corresponding peak areas of sulfur species in the sample. However, the precondition for such a calibration procedure is that the internal standard and the sulfur species of interest are completely eluted. Figure 8.20 shows the sulfur mass flow chromatogram of the gas oil reference material BCR CRM 107, which is certified for the total sulfur content (10.4 ± 0.15 mg g^{-1}) only. The mass flow chromatogram in Figure 8.20 was obtained by conversion of

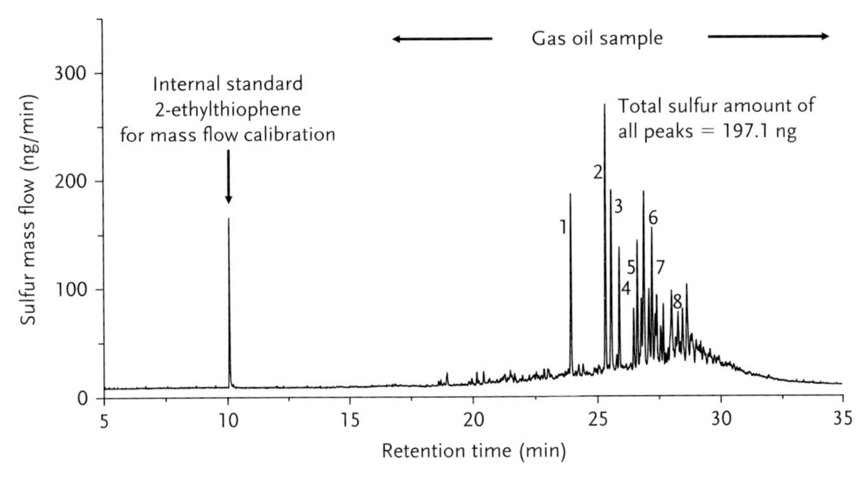

Figure 8.20 Sulfur mass flow chromatogram of gas oil reference material BCR CRM 107 obtained by species-unspecific GC–ICP-IDMS. Major peaks identified by EI-MS–MS: 1 = dibenzothiophene; 2 = 4-methyldibenzothiophene; 3 = 2-methyldibenzothiophene; 4 = 1-methyldibenzothiophene; 5 = 4,6-dimethyldibenzothiophene; 6 = 2,6-dibenzothiophene; 7 = 1,4-dibenzothiophene; 8 = 2,4,6-trimethyldibenzothiophene. Reproduced with permission from [101].

the isotope ratio chromatogram in the same way as shown in Figure 8.15. A complex chromatogram, demonstrating the presence of many different sulfur compounds, was obtained. Twelve major peaks could be identified on the basis of their retention times in combination with structural analysis via electron ionization tandem mass spectrometry (EI-MS–MS) and eight of them are indicated in Figure 8.20. However, the content of all separated sulfur compounds, even of those not identified, could be determined via this species-unspecific GC–ICP-IDMS method. The total sulfur amount obtained by integration of all peaks in the chromatogram correlated with a sulfur content in the sample of $10.5 \pm 0.07\,\mathrm{mg\ g^{-1}}$ (standard deviation from three independent chromatograms), which agrees very well with the certified value. This also demonstrates that all sulfur species in the sample had been covered by the analytical method used. Checking full recovery is an important aspect of the validation for all speciation analyses and, as mentioned above, a necessary precondition for the use of an internal standard for spike flow calibration.

Fractionation of Humic Substance Metal Complexes SEC coupled with species-unspecific ICP-IDMS was used for the first time by Rottmann and Heumann (using the setup shown in Figure 8.14) for the quantitative fractionation analysis of different heavy metal complexes of humic substances [102]. The combination of a UV flow-through detector and ICP-MS allowed the simultaneous detection of humic substances and metals, permitting correlations between the organic matter and the metals to be obtained. Figure 8.21 shows the SEC–ICP-IDMS mass flow chromatograms for molybdenum, zinc, and copper for sewage water from two different sewage plants [103, 104]. As can be seen, the distribution of the metal complexes is similar in both waste water samples, but different organic fractions

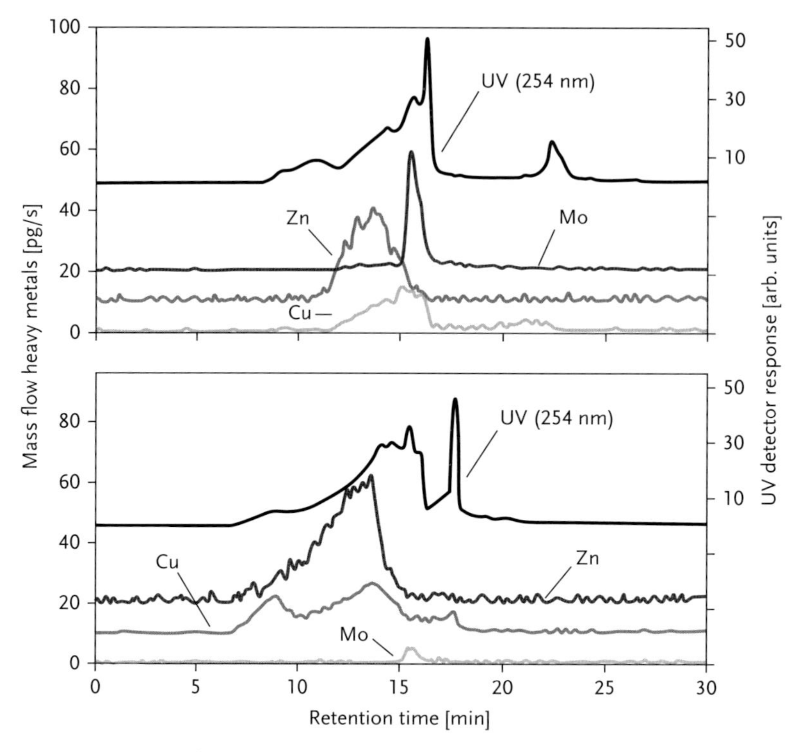

Figure 8.21 Mass flow chromatograms of molybdenum, zinc, and copper complexes with humic substances obtained by SEC–ICP-IDMS for sewage water from two different sewage plants. Reproduced with permission from [104].

interact with the various metals. Whereas zinc prefers a high molecular weight fraction of humic substances (low retention time), copper interacts with several fractions of the organic matter. Molybdenum, on the other hand, only interacts with a small fraction of medium molecular weight. Also for halogens, characteristic distributions in humic substances of specific molecular weight were found [104, 105]. For metal complexes of other origins, characteristic fingerprints of metal complexes were also observed [102–104].

Protein Analysis Quantification of proteins is required in many scientific fields, such as clinical chemistry and biochemistry. Various isotope dilution strategies, relying on the use of compounds enriched in ^{2}H, ^{13}C, ^{15}N, or ^{18}O, have been used for relative (difference in protein content between two different biochemical stages) or absolute quantification of proteins using mass spectrometric (MS) detection following soft ionization, with methods such as MALDI (matrix-assisted laser desorption ionization) or ESI (electrospray ionization) [106]. In addition to the problem of synthesizing the corresponding isotope-labeled compounds, extensive signal corrections are necessary in isotope dilution via MALDI-MS and ESI-MS to take into account the effects caused by the occurrence of several isotopes of the non-metallic elements (^{1}H and ^{2}H, ^{12}C and ^{13}C, ^{14}N and ^{15}N, ^{16}O, ^{17}O, and ^{18}O) in the molecule. Therefore, the use of ICP-MS detection with its much simpler isotope patterns of atoms and a sensitivity independent of the chemical compound in which the element is present has significant advantages for quantitative proteomics. Because ICP-MS is not adequate to detect isotopes of the nonmetals mentioned above and because phosphorus is monoisotopic, other heteroatoms in proteins need to be relied upon for isotope dilution. Sulfur, selenium, and metals, mostly or often present in proteins, have therefore been selected for isotope dilution purposes using ICP-MS. As synthesizing distinct isotope-labeled biomolecules is a great challenge, species-specific isotope dilution has been applied only very rarely in the past, even though it is the best option for the accurate quantification of proteins. Determinations of transferrin isoforms in human serum using ^{57}Fe-labeled compounds and of γ-glutamyl-Se-methylselenocysteine, a selenium compound relevant in cancer research, labeled with ^{77}Se are two especially interesting exceptions [74, 107]. As a result, species-unspecific ICP-IDMS is usually applied in proteomics. The use of sulfur and of other elements in isotopic analysis in the context of quantitative proteomics has been summarized in two reviews published in 2008 and 2010 [108, 109]. Because sample pretreatment steps and separation procedures in proteomics are typically greatly affected by loss of the analyte and because in the species-unspecific approach the isotope dilution step is at the end of such a procedure, it is of particular importance that the recovery of the analyte is well known for the absolute quantification of proteins. Otherwise, this will be the limiting factor for the accurate quantification of proteins.

The important role of sulfur in protein quantification via hyphenated ICP-IDMS techniques is exemplified in Figure 8.22, which illustrates the key role of sulfur in protein quantification by ICP-MS [108]. Cysteine and methionine are sulfur-containing amino acids that are present in most proteins. The enzymatic digestion

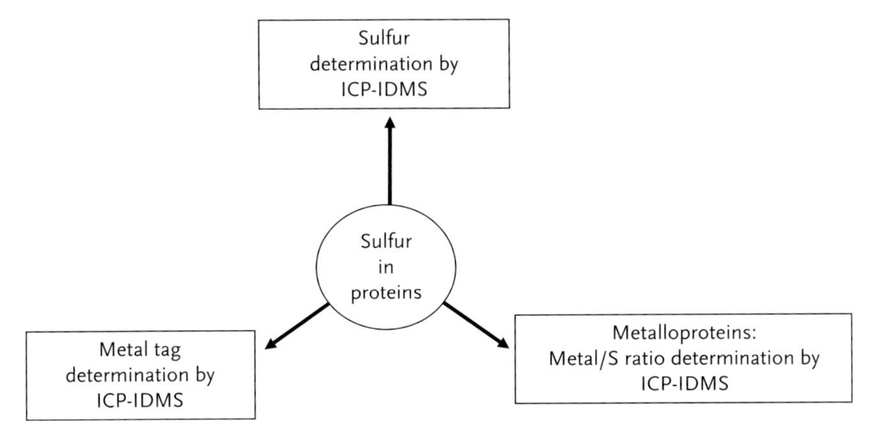

Figure 8.22 Hyphenated ICP-IDMS techniques in quantitative proteomics: applications based on sulfur.

of proteins should therefore lead to sulfur-containing peptides. Sulfur in proteins or their digests can be directly determined via species-unspecific ICP-IDMS using a ^{34}S-labeled spike compound, for example, sulfate. From the $^{32}S/^{34}S$ isotope ratio measurement, the amount of protein can be calculated if the number of sulfur atoms in the molecule is known. Metals such as zinc, cadmium, and copper in metalloproteins can be determined together with sulfur via ICP-IDMS. By means of such an analysis, the stoichiometry of individual metalloproteins can be determined. In addition, the reactivity of sulfhydryl (−SH) groups of cysteine allows reactions with other molecules, for example, metal-containing substances. This reactivity can be relied upon for tagging. Because metals are measured with much higher sensitivity than sulfur by ICP-MS, a substantial improvement in the LODs can be obtained for proteins in this way. By using more than one metal per tag molecule, an additional improvement in the detection power is achieved. This effect can be seen from Figure 8.23, in which the concentration ranges in which human plasma proteins are present are compared with the LODs attainable with ICP-MS based on sulfur and on metal tag determination [109].

The first protein quantification based on sulfur monitoring and species-unspecific isotope dilution was carried out by Schaumlöffel *et al.* using a CE unit coupled to a sector field mass spectrometer operated at a mass resolution of 3000 and aimed at analyzing metallothionein fractions of a rabbit liver [110].

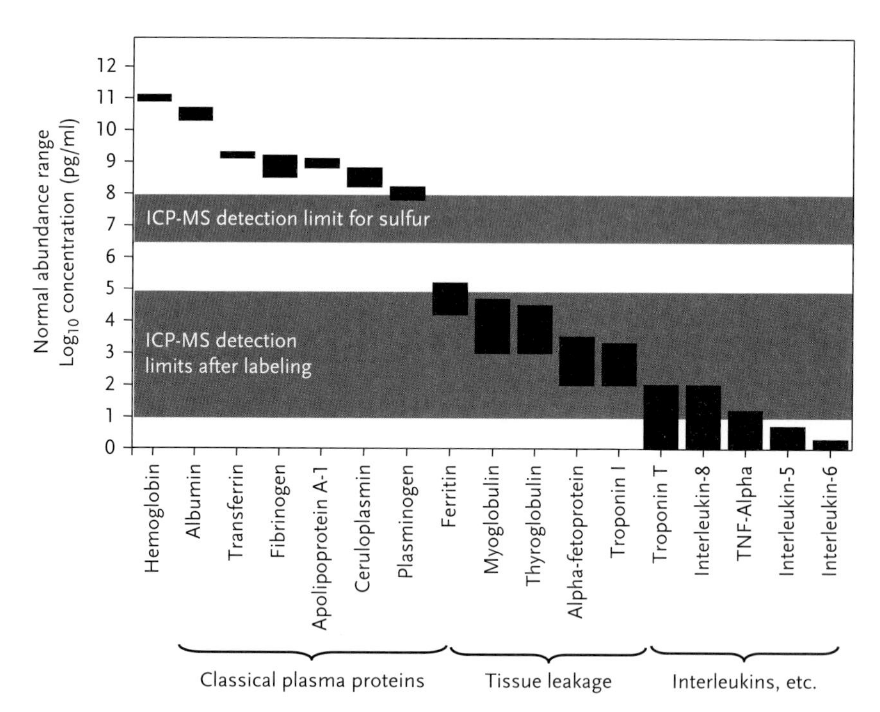

Figure 8.23 ICP-MS LODs for proteins based on monitoring of sulfur or of a metal tag compared with the concentrations of different proteins in human plasma. Reproduced with permission from [109].

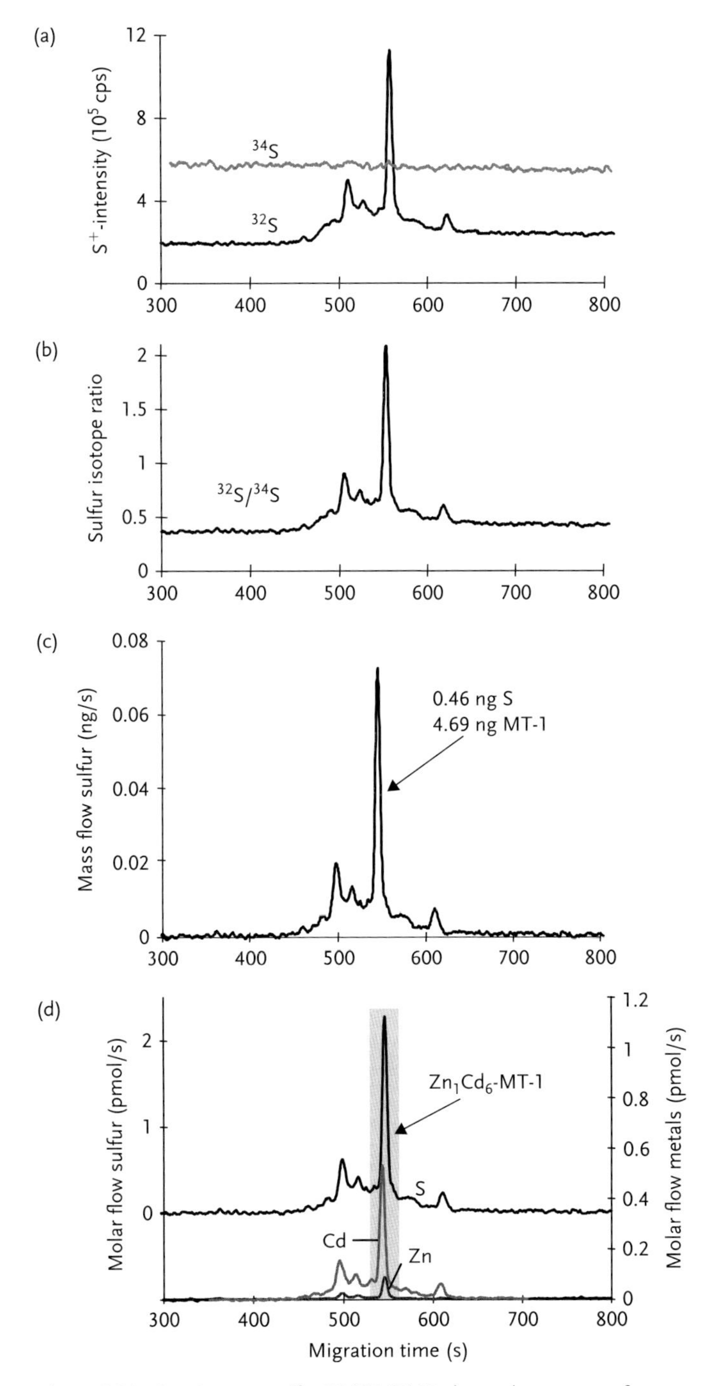

Figure 8.24 Species-unspecific CE-ICP-IDMS electropherograms of measured sulfur isotope intensities (a) calculated sulfur isotope ratios (b) and mass flows (c) of metallothioneins from a rabbit liver as well as the corresponding mass flow electropherograms of Cd and Zn (d). Reproduced with permission from [108].

Figure 8.24a represents the measurement of the sulfur isotopes ^{32}S and ^{34}S in the metallothionein MT-1 fraction. As expected for the species-unspecific spiking mode, the intensity of ^{34}S is nearly constant over the entire electropherogram because the contribution from the metallothionein is extremely small as a result of the low natural isotopic abundance of ^{34}S. From the measurements shown in Figure 8.24a, the isotope ratio electropherogram in Figure 8.24b was calculated. Conversion of these isotope ratios into the corresponding sulfur amounts (identical to the corresponding explanation for Figure 8.15) results in the mass flow of sulfur, and also the total sulfur amount of the separated protein (Figure 8.24c). From the known structure and known number of sulfur atoms (21) in the protein MT-1, the protein amount was calculated to be 4.69 ng of MT-1. Via simultaneous quantification of the metal ions Zn^{2+} and Cd^{2+} bound in MT-1, the stoichiometry of this metallothionein could be determined as Zn_1Cd_6MT-1.

By analyzing a solution of three standard proteins (lysozyme, ovalbumin, and bovine serum albumin), and also two certified protein reference materials (BCR CRM 393 human apolipoprotein A-1 and BCR CRM 486 α-fetoprotein), the accuracy but also possible limitations of absolute protein quantification via species-unspecific isotope dilution were demonstrated. A micro-HPLC system coupled to a sector field mass spectrometer, operated at a mass resolution of

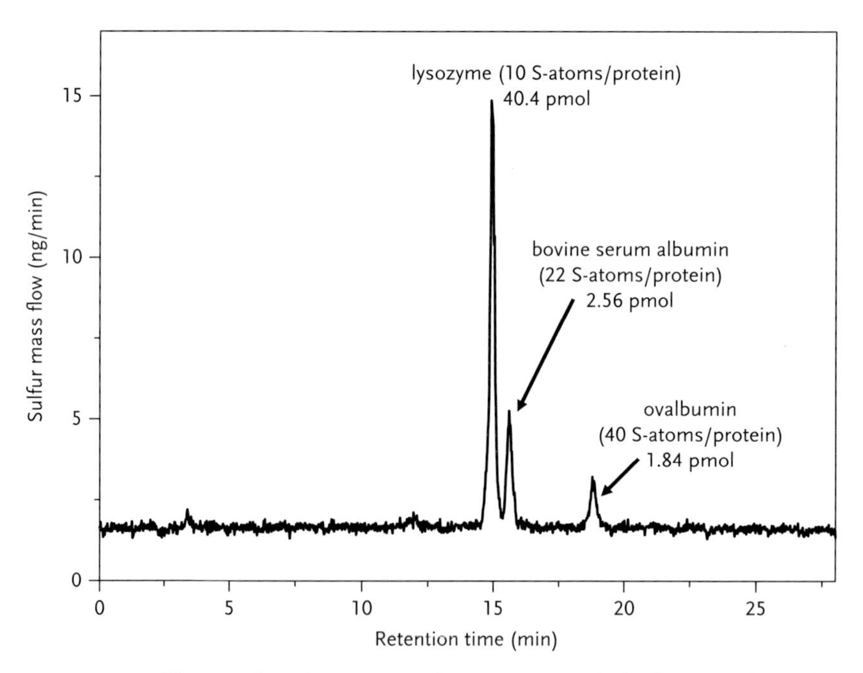

Figure 8.25 Sulfur mass flow chromatogram for a protein standard solution as obtained by species-unspecific micro-HPLC–ICP-IDMS and conversion into the amounts of the separated proteins based on the known number of sulfur atoms per protein molecule. Reproduced with permission from [111].

4000, was used, and ^{34}S-enriched sulfate was deployed as a species-unspecific spike [111]. The results for the solution of standard proteins are presented in Figure 8.25, where the sulfur mass flow is plotted as a function of the retention time.

The three proteins eluted as sharp separated peaks, thus allowing the amount of sulfur in each fraction to be calculated. Based on the known number of sulfur atoms in the different proteins, the corresponding amounts of protein (in the picomole range) could be determined. A similar chromatogram for the reference material BCR CRM 393 resulted in an apolipoprotein A-1 concentration of $37.0 \pm 1.4 \pm \text{mol l}^{-1}$, which agrees very well with the certified value of $37.7 \pm 1.8\ \mu\text{mol l}^{-1}$. In contrast, when analyzing the second reference material, only 60% of the α-fetoprotein applied was recovered via species-unspecific isotope dilution using the same micro-HPLC–ICP-IDMS system. A high percentage of this protein remained on the column and species-unspecific isotope dilution does not correct for on-column losses.

As illustrated by Figure 8.23, the LODs for proteins can be dramatically improved, by three to five orders of magnitude, via metal tagging. Rappel and Schaumlöffel studied the species-unspecific isotope dilution analysis of metal-tagged proteins [112]. They derivatized peptides with diethylenetriaminepentaacetic (DTPA) anhydride, followed by lutetium chelation, a well-known tool for modification of peptides and proteins, already used for their relative quantification via ESI-MS and MALDI-MS, respectively. Via nano-HPLC–ICP-IDMS with a species-unspecific ^{176}Lu-labeled spike, they were able to determine peptides precisely and accurately with a sensitivity about four orders of magnitude higher compared with that obtained via measurement of sulfur in the corresponding cysteine or methionine residues. The LOD for LuDTPA-tagged peptides was about 180 amol.

Derivatization of the free sulfhydryl groups of proteins by metal tags is an interesting alternative. A fundamental study on labeling proteins with mercury tags using derivatization of ovalbumin with *p*-hydroxymercurybenzoic acid (pHMB) was reported by Kutscher *et al.* [113]. A relatively strong direct S–Hg bond is one of the advantages of derivatization with pHMB. For accurate protein quantification based on such mercury-tagged bioconjugates, the number of free sulfhydryl groups needs to be known and complete derivatization of all these groups or of a constant known fraction of them must be guaranteed under the experimental conditions used. For ovalbumin, a pHMB-label/protein ratio of ~3:1 was found via MALDI-MS, whereas ESI time-of-flight MS resulted in the theoretical ratio of 4:1. The ICP-MS LOD of 1 fmol ovalbumin via the pHMB-labeled compound is about 500 times better than that based on the measurement of sulfur in unlabeled ovalbumin. The first results for protein quantification via species-specific ICP-IDMS analysis using ^{199}Hg-labeled pHMB were obtained by the same group [114]. These preliminary results demonstrate the high potential of metal tagging for highly sensitive quantification of proteins, and further investigations on this topic will be one of the future trends in the field of protein quantification via species-unspecific and species-specific ICP-IDMS.

References

1 Heumann, K.G. (1988) Isotope dilution mass spectrometry, in *Inorganic Mass Spectrometry* (eds. F. Adams, R. Gijbels, and R. van Grieken), John Wiley & Sons, Inc., New York, p. 301.
2 De Bièvre, P. (1994) *Fresenius' J. Anal. Chem.*, **350**, 277.
3 De Bièvre, P. (2010) *Accred. Qual. Assur.*, **15**, 321.
4 Heumann, K.G., Rottmann, L., and Vogl, J. (1994) *J. Anal. At. Spectrom.*, **9**, 1351.
5 Brown, A.A., Ebdon, L., and Hill, S.J. (1994) *Anal. Chim. Acta*, **286**, 391.
6 Krystek, P. and Heumann, K.G. (1999) *J. Anal. At. Spectrom.*, **14**, 1443.
7 Heumann, K.G. (2004) *Anal. Bioanal. Chem.*, **378**, 318.
8 Smith, D.H. (2000) Isotope dilution mass spectrometry, in *Inorganic Mass Spectrometry – Fundamentals and Applications* (eds. C.M. Barshick, D.C. Duckworth, and D.H. Smith), Marcel Dekker, New York, p. 223.
9 Sargent, M., Harte, R., and Harrington, C. (2002) *Guidelines for Achieving High Accuracy in Isotope Dilution Mass Spectrometry (IDMS)*, Royal Society of Chemistry, Cambridge.
10 Yip, Y. and Sham, W. (2007) *Trends Anal. Chem.*, **26**, 727.
11 Klemens, P. and Heumann, K.G. (2001) *Fresenius' J. Anal. Chem.*, **371**, 758.
12 Kipphardt, H., De Bièvre, P., and Taylor, P.D.P. (2004) *Anal. Bioanal. Chem.*, **378**, 330.
13 De Laeter, J.R., Böhlke, J.K., De Bièvre, P., Hidaka, H., Peiser, H.S., Rosman, K.J.R., and Taylor, P.D.P. (2003) *Pure Appl. Chem.*, **75**, 683.
14 Coplen, T.B., Hopple, J.A., Böhlke, J.K., Peiser, H.S., Krouse, H.R., Rosman, K.J.R., Ding, T., Vocke, R.D., Révész, K.M., Lamberty, A., Taylor, P.D.P., and De Bièvre, P. (2001) Compilation of minimum and maximum isotope ratios of selected elements in naturally occurring terrestrial materials and reagents. US Geological Survey Water-Resources Investigations, Report 01–4222, US Geological Survey, Reston, VA.
15 Begley, I.S. and Sharp, B.L. (1997) *J. Anal. At. Spectrom.*, **12**, 395.
16 Heumann, K.G., Gallus, S.M., Rädlinger, G., and Vogl, J. (1998) *J. Anal. At. Spectrom.*, **13**, 1001.
17 Watters, R.L. Jr., Eberhardt, K.R., Beary, E.S., and Fasett, J.D. (1997) *Metrologia*, **34**, 87.
18 Institute for Reference Materials and Measurements, Isotopic Reference Materials – Catalogue 2008, European Commission Joint Research Centre, Geel (2008).
19 BAM: Bundesanstalt für Materialforschung und -prüfung (Federal Institute for Materials Research and Testing) (2011) http://www.rm-certificates.bam.de/de/certificates/isotopic_materials/index.htm (last accessed February 2012).
20 Riepe, W. and Kaiser, H. (1966) *Fresenius' Z. Anal. Chem.*, **223**, 321.
21 Moser, J., Wegscheider, W., Meisel, T., and Fellner, N. (2003) *Anal. Bioanal. Chem.*, **377**, 97.
22 Fortunato, G. and Wunderli, S. (2003) *Anal. Bioanal. Chem.*, **377**, 111.
23 Kessel, W. (1993) ISO/BIPM Guideline: Uncertainty of Measurement, http://www.metrodata.de/papers/resistor_en.pdf. (last accessed 24 December 2011).
24 Ellison, S.L.R., Rosslein, M., and Williams, A. (eds.) (2000) Eurachem/CITAC Guide CG 4, Quantifying Uncertainty in Analytical Measurement, 2nd edn., http://www.measurementuncertainty.org (last accessed 24 December 2011).
25 Boulyga, S.F. and Heumann, K.G. (2005) *Anal. Bioanal. Chem.*, **383**, 442.
26 Boulyga, S.F., Tibi, M., and Heumann, K.G. (2004) *Anal. Bioanal. Chem.*, **378**, 342.
27 Vogl, J., Liesegang, D., Ostermann, M., Diemer, J., Berglund, M., Quétel, C.R., Taylor, P.D.P., and Heumann, K.G. (2000) *Accred. Qual. Assur.*, **5**, 314.

28 Diemer, J., Quétel, C., and Taylor, P.D.P. (2002) *Anal. Bioanal. Chem.*, **374**, 220.
29 Vogl, J. (2006) *J. Anal. At. Spectrom.*, **22**, 475.
30 Makishima, A. and Nakamura, E. (2006) *Geostand. Geoanal. Res.*, **30**, 245.
31 Ohtsuka, Y., Takaku, Y., Kimura, J., Hisamatsu, S., and Inaba, J. (2007) *Anal. Sci.*, **21**, 205.
32 Enzweiler, J., Potts, P.J., and Jarvis, K.E. (1995) *Analyst*, **120**, 1391.
33 Boulyga, S.F. and Heumann, K.G. (2005) *Int. J. Mass Spectrom.*, **242**, 291.
34 Boulyga, S.F., Heilmann, J., Prohaska, T., and Heumann, K.G. (2007) *Anal. Bioanal. Chem.*, **389**, 697.
35 Evans, P., Wolff-Birche, C., and Fairman, B. (2001) *J. Anal. At. Spectrom.*, **16**, 964.
36 Heilmann, J. and Heumann, K.G. (2009) *Anal. Bioanal. Chem.*, **393**, 393.
37 Günther, D., Jackson, S.E., and Longerich, H.P. (1999) *Spectrochim. Acta B*, **54**, 381.
38 Russo, R.E., Mao, X., Liu, H., Gonzales, J., and Mao, S.S. (2002) *Talanta*, **57**, 425.
39 Craig, C.-A., Jarvis, K.E., and Clarke, L.J. (2000) *J. Anal. At. Spectrom.*, **15**, 1001.
40 Kanicky, V., Otruba, V., and Mermet, J.-M. (2001) *Fresenius' J. Anal. Chem.*, **371**, 934.
41 Pickhardt, C. and Becker, J.S. (2001) *Fresenius' J. Anal. Chem.*, **370**, 534.
42 Guillong, M. and Günther, D. (2001) *Spectrochim. Acta B*, **56**, 1219.
43 Leach, J.J., Alien, L.A., Aeschliman, D.B., and Houk, R.S. (1999) *Anal. Chem.*, **71**, 440.
44 Günther, D., Frischknecht, R., Muschenborn, H.J., and Heinrich, C.A. (1997) *Fresenius' J. Anal. Chem.*, **359**, 390.
45 Reid, E.J., Horn, I., Longerich, H.P., Forsythe, L., and Jenner, G.A. (1999) *Geostand. Newslett.*, **23**, 149.
46 Tibi, M. and Heumann, K.G. (2003) *J. Anal. At. Spectrom.*, **18**, 1076.
47 Boulyga, S.F., Heilmann, J., and Heumann, K.G. (2005) *Anal. Bioanal. Chem.*, **382**, 1808.
48 Heilmann, J., Boulyga, S.F., and Heumann, K.G. (2009) *J. Anal. At. Spectrom.*, **24**, 385.
49 Tibi, M. and Heumann, K.G. (2003) *Anal. Bioanal. Chem.*, **377**, 126.
50 Fernández, B., Claverie, F., Pécheyran, C., and Donard, O.F.X. (2008) *J. Anal. At. Spectrom.*, **23**, 367.
51 Pickhardt, C., Izmer, A.V., Zoriy, M.V., Schaumlöffel, D., and Becker, J.S. (2006) *Int. J. Mass Spectrom.*, **248**, 136.
52 Yang, C.-K., Chi, P.-H., Lin, Y.-C., Sun, Y.-C., and Yang, M.-H. (2010) *Talanta*, **80**, 1222.
53 Resano, M., Aramendia, M., and Vanhaecke, F. (2006) *J. Anal. At. Spectrom.*, **21**, 1036.
54 Yu, L.L., Kelly, W.R., Fassett, J.D., and Vocke, R.D. (2001) *J. Anal. At. Spectrom.*, **16**, 140.
55 Heilmann, J. and Heumann, K.G. (2008) *Anal. Bioanal. Chem.*, **390**, 643.
56 D'Ulivio, A. (2004) *Spectrochim. Acta B*, **59**, 793.
57 Zheng, C.B., Sturgeon, R.E., Brophy, C.S., He, S.P., and Hou, X.D. (2010) *Anal. Chem.*, **82**, 2996.
58 Guo, X.M., Sturgeon, R.E., Mester, Z., and Gardner, G.J. (2004) *Anal. Chem.*, **76**, 2401.
59 Zheng, C., Yang, L., Sturgeon, R.E., and Hou, X. (2010) *Anal. Chem.*, **82**, 3899.
60 Müller, M. and Heumann, K.G. (2000) *Fresenius' J. Anal. Chem.*, **368**, 109.
61 Paliulionyte, V., Meisel, T., Ramminger, P., and Kettisch, P. (2006) *Geostand. Newslett.*, **30**, 87.
62 Shinotsuka, K. and Suzuki, K. (2007) *Anal. Chim. Acta*, **603**, 129.
63 Palesskii, S.V., Nikolaeva, I.V., Koz'menko, O.A., and Anoshin, G.N. (2009) *J. Anal. Chem.*, **64**, 272.
64 Perelygin, V.P. and Chuburkov, Y.T. (1997) *Radiat. Meas.*, **28**, 385.
65 Zheng, J. and Yamada, M. (2008) *J. Oceanogr.*, **64**, 541.
66 Ohtsuka, Y., Takaku, Y., Kimura, J., Hisamatsu, S., and Inaba, J. (2005) *Anal. Sci.*, **21**, 205.
67 Cornelis, R., Caruso, J., Crews, H., and Heumann, K.G. (2003) *Handbook of Elemental Speciation. Vol. 1, Techniques*

and Methodology, John Wiley & Sons, Ltd., Chichester.

68 Rottmann, L. and Heumann, K.G. (1994) *Fresenius' J. Anal. Chem.*, **350**, 221.

69 Schaumlöffel, D., Giusti, P., Preud'Homme, H., Szpunar, J., and Lobinski, R. (2007) *Anal. Chem.*, **79**, 2859.

70 Reifenhäuser, C. and Heumann, K.G. (1990) *Fresenius' J. Anal. Chem.*, **336**, 559.

71 Poperechna, N. and Heumann, K.G. (2005) *Anal. Chem.*, **77**, 511.

72 Larsen, E.H., Sloth, J., Hansen, M., and Moesgaard, S. (2003) *J. Anal. At. Spectrom.*, **18**, 310.

73 Encinar, J.R., Schaumlöffel, D., Ogra, Y., and Lobinski, R. (2004) *Anal. Chem.*, **76**, 6635.

74 Del Castillo Busto, M.E., Montes-Bayón, M., and Sanz-Medel, A. (2006) *Anal. Chem.*, **78**, 8218.

75 Cornelis, C., Caruso, J., Crews, H., and Heumann, K.G. (2005) *Handbook of Elemental Speciation. Vol. 2, Species in the Environment, Food, Medicine and Occupational Health*, John Wiley & Sons, Ltd., Chichester.

76 Quevauviller, P. (1998) *Method Performance Studies for Speciation Analysis*, Royal Society of Chemistry, Cambridge.

77 Rodriguez-González, P., Marchante-Gayón, J.M., Garcia Alonso, J.I., and Sanz-Medel, A. (2005) *Spectrochim. Acta B*, **60**, 151.

78 Rodriguez-González, P. and Garcia Alonso, J.I. (2010) *J. Anal. At. Spectrom.*, **25**, 239.

79 Meija, J. and Mester, Z. (2008) *Anal. Chim. Acta*, **607**, 115.

80 Wilken, R.D. and Falter, R. (1998) *Appl. Organomet. Chem.*, **12**, 551.

81 Snell, J.P., Stewart, I.I., Sturgeon, R.E., and Frech, W. (2000) *J. Anal. At. Spectrom.*, **15**, 1540.

82 Ruiz Encinar, J., Monterde Villar, M.I., Gotor Santamaria, V., Garcia Alonso, J.I., and Sanz-Medel, A. (2001) *Anal. Chem.*, **73**, 3174.

83 Inagaki, K., Takatsu, A., Watanabe, T., Kuroiwa, T., Aoyagi, Y., and Okamoto, K. (2004) *Anal. Bioanal. Chem.*, **378**, 1265.

84 Brüchert, W., Helfrich, A., Zinn, N., Klimach, T., Breckheimer, M., Chen, H., Lai, S., Hoffmann, T., and Bettmer, J. (2007) *Anal. Chem.*, **79**, 1714.

85 Yabutani, T., Motonaka, J., Inagaki, K., Takatsu, A., Marita, T., and Chiba, K. (2008) *Anal. Sci.*, **24**, 791.

86 Poperechna, N. and Heumann, K.G. (2005) *Anal. Bioanal. Chem.*, **383**, 153.

87 Hintelmann, H., Falter, R., Ilgen, G., and Evans, R.D. (1997) *Fresenius' J. Anal. Chem.*, **358**, 363.

88 Demuth, N. and Heumann, K.G. (2001) *Anal. Chem.*, **73**, 4020.

89 Hintelmann, H., Evans, R.D., and Villeneuve, J. (1995) *J. Anal. At. Spectrom.*, **10**, 619.

90 Hintelmann, H. and Evans, R.D. (1997) *Fresenius' J. Anal. Chem.*, **358**, 378.

91 Hou, D., Lu, Y., and Kingston, H.M. (1998) *Environ. Sci. Technol.*, **32**, 3418.

92 Rodriguez-González, P., Ruiz Encinar, J., Garcia Alonso, J.I., and Sanz-Medel, A. (2004) *J. Anal. At. Spectrom.*, **19**, 685.

93 Rodriguez-González, P., Monperrus, M., Garcia Alonso, J.I., Amouroux, D., and Donard, O.F.X. (2007) *J. Anal. At. Spectrom.*, **22**, 1373.

94 Ouerdane, L., Mester, Z., and Meija, J. (2009) *Anal. Chem.*, **81**, 5075.

95 Monperrus, M., Rodriguez-González, P., Amouroux, D., Garcia Alonso, J.I., and Donard, O.F.X. (2008) *Anal. Bioanal. Chem.*, **390**, 655.

96 Meija, J., Ouerdane, L., and Mester, Z. (2009) *Anal. Bioanal. Chem.*, **394**, 199.

97 Rodriguez-Castrillón, J.A., Moldovan, M., Garcia Alonso, J.I., Lucena, J.J., Garcia-Tomé, M.L., and Hernández-Apaolaza, L. (2008) *Anal. Bioanal. Chem.*, **390**, 579.

98 González Iglesias, H., Fernández Sánchez, M.L., Rodríguez-Castrillón, J.A., García-Alonso, J.I., López Sastre, J., and Sanz-Medel, A. (2009) *J. Anal. At. Spectrom.*, **24**, 460.

99 Meija, J., Yang, L., Caruso, J.A., and Mester, Z. (2006) *J. Anal. At. Spectrom.*, **21**, 1294.

100 Rodriguez-González, P., Rodriguez-Cea, A., Garcia Alonso, J.I., and Sanz-Medel, A. (2005) *Anal. Chem.*, **77**, 7724.

101 Heilmann, J. and Heumann, K.G. (2008) *Anal. Chem.*, **80**, 1952.

102 Rottmann, L. and Heumann, K.G. (1994) *Anal. Chem.*, **66**, 3709.

103 Vogl, J. and Heumann, K.G. (1997) *Fresenius' J. Anal. Chem.*, **359**, 438.

104 Heumann, K.G., Marx, G., Rädlinger, G., and Vogl, J. (2002) Heavy metal and halogen interactions with fractions of refractory organic substances separated by size exclusion chromatography, in *Refractory Organic Substances in the Environment* (eds. F.H. Frimmel, G. Abbt-Braun, K.G. Heumann, B. Hock, H.-D. Lüdemann, and M. Spiteller), Wiley-VCH Verlag GmbH, Weinheim, p. 55.

105 Rädlinger, G. and Heumann, K.G., *Fresenius' J.* (1997) *Anal. Chem.*, **359**, 430.

106 Bantscheff, M., Schirle, M., Sweetmann, G., Rick, J., and Küster, B. (2007) *Anal. Bioanal. Chem.*, **389**, 1017.

107 Goenaga-Infante, H., del Carmen Ovejero Bendito, M., Cámara, C., Evans, L., Hearn, R., and Moesgaard, S. (2008) *Anal. Bioanal. Chem.*, **390**, 2099.

108 Rappel, C. and Schaumlöffel, D. (2008) *Anal. Bioanal. Chem.*, **390**, 605.

109 Bettmer, J. (2010) *Anal. Bioanal. Chem.*, **397**, 3495.

110 Schaumlöffel, D., Prange, A., Marx, G., Heumann, K.G., and Brätter, P. (2002) *Anal. Bioanal. Chem.*, **372**, 155.

111 Zinn, N., Krüger, R., Leonhard, P., and Bettmer, J. (2008) *Anal. Bioanal. Chem.*, **391**, 537.

112 Rappel, C. and Schaumlöffel, D. (2009) *Anal. Chem.*, **81**, 385.

113 Kutscher, D.J., del Castillo Busto, M.E., Zinn, N., Sanz-Medel, A., and Bettmer, J. (2008) *J. Anal. At. Spectrom.*, **23**, 1359.

114 Kutscher, D. and Bettmer, J. (2009) *Anal. Chem.*, **81**, 9172.

9
Geochronological Dating

Marlina A. Elburg

9.1
Geochronology: Principles

Geochronology is concerned with determining the absolute age of rocks and minerals, and thereby geological events that shaped these materials. This is generally done using the natural decay of unstable (radioactive) isotopes ("parent" or "mother" nuclide) to daughter nuclides, which may either be radioactive themselves, or stable. Each radioactive isotope has a characteristic half-life ($T_{½}$), which is the time in which half of the parent nuclides decay to their daughter nuclide. The rate of radioactive decay can also be expressed as the decay constant λ, which is related to the half-life by $T_{½} = \ln 2/\lambda$.

In this chapter, dating methods applicable to the past 4 billion to 2 million years are discussed.

9.1.1
Single Phase and Isochron Dating

If a material contains only the radioactive parent at the time of its formation, then the age of the material (or "system") can be determined by measuring the parent/daughter ratio, provided that the system remained closed since its formation. If both parent and daughter are present in the system at the time of its formation, we must know the initial amount of daughter present. The derivation of the pertinent equations can be found in several textbooks on geochronology [1–3], with the essential one for geochronological purposes being

$$D_{tot} = P(e^{\lambda t} - 1) + D_0$$

where D_{tot} is the total amount of daughter isotope present, P the amount of parent isotope, λ the decay constant, t the time elapsed since closure of the system (both t and λ are typically expressed in years), and D_0 the amount of daughter isotope already present at closure of the system.

Isotopic Analysis: Fundamentals and Applications Using ICP-MS,
First Edition. Edited by Frank Vanhaecke and Patrick Degryse.

Published 2012 by WILEY-VCH Verlag GmbH & Co. KGaA

As it is easier to determine precisely the ratio of two isotopes relative to each other than the absolute concentration of an individual isotope, both sides of the equation above are generally divided by a stable isotope of the element to which the daughter belongs. In the case of the radioactive decay of ^{87}Rb to ^{87}Sr, for instance, everything is normalized to ^{86}Sr, so the equation becomes

$$\left(\frac{^{87}\mathrm{Sr}}{^{86}\mathrm{Sr}}\right)_t = \frac{^{87}\mathrm{Rb}}{^{86}\mathrm{Sr}}(e^{\lambda t} - 1) + \left(\frac{^{87}\mathrm{Sr}}{^{86}\mathrm{Sr}}\right)_0$$

The subscript t refers to the time at which we perform the measurement, to distinguish it from time zero, at which the system became closed.

The equation above obviously describes a straight line in a diagram of $(^{87}\mathrm{Sr}/^{86}\mathrm{Sr})_t$ versus $^{87}\mathrm{Rb}/^{86}\mathrm{Sr}$, with the slope of the line dependent on the age of the sample, and the intercept with the y-axis the value for $(^{87}\mathrm{Sr}/^{86}\mathrm{Sr})_0$ – sometimes also called $(^{87}\mathrm{Sr}/^{86}\mathrm{Sr})_i$ where i refers to the *i*nitial ratio. Hence if we have two or more materials, with different $^{87}\mathrm{Rb}/^{86}\mathrm{Sr}$ ratios, which formed at the same time and in equilibrium with each other, so they both have the same $(^{87}\mathrm{Sr}/^{86}\mathrm{Sr})_0$, we can determine the age at which they formed, and $(^{87}\mathrm{Sr}/^{86}\mathrm{Sr})_0$. This is called the isochron method of dating (Figure 9.1a).

The assumption of initial isotopic equilibrium between the different samples used for the isochron method is an important one, especially for materials that are relatively young and/or only show limited enrichments of the parent over the daughter isotope. The assumption of initial equilibrium is likely to hold for different minerals that have grown from a melt within a limited time, such as is assumed to be the case for extrusive magmatic rocks. It becomes slightly more doubtful when whole rocks, instead of mineral separates, are used to construct isochrons. Initial isotopic equilibrium may still hold for different magmatic rocks from a single system (e.g., basalts and rhyolites; granites and related pegmatites), but differentiation of magmas may occur concomitantly with crustal contamination, which might change the initial isotopic composition. Even less likely is initial isotopic equilibrium between different sedimentary rocks that have undergone metamorphism. If the assumption of initial isotopic equilibrium is not met, the samples will show more scatter on an isochron diagram than can be accounted for by the uncertainty of the isotope ratio measurements. Although we can still draw a straight line through the points, which may have some sort of age significance, the resulting straight line is called an errorchron instead of isochron (Figure 9.1b).

The minimum and maximum ages of materials that can be dated depend on the decay constant/half-life of the parent isotope (Table 9.1) and the relative enrichment of parent over daughter isotope, apart from the precision of the isotope ratio measurements. The materials most suitable for dating purposes are those that only incorporate the parent isotope at the time of formation, as a single sample (mineral) suffices to obtain an age, whereas at least two, but preferably three or more, samples are necessary to obtain an age by the isochron method.

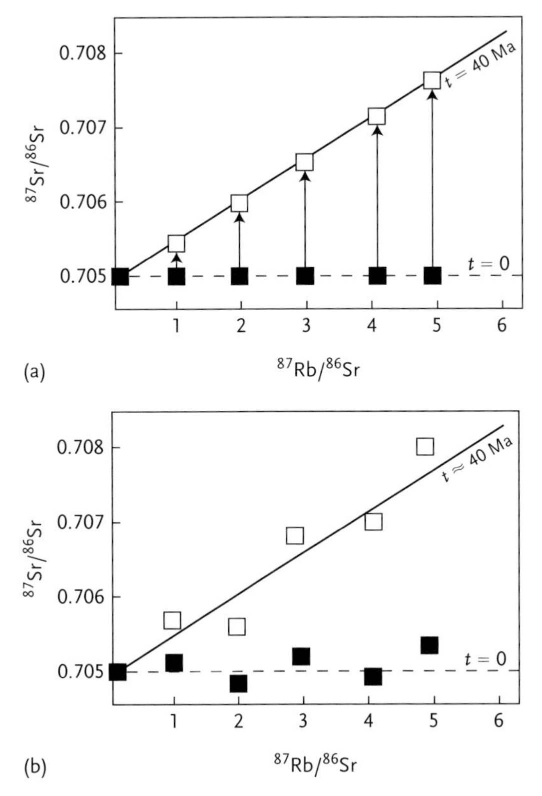

Figure 9.1 (a) Illustration of the Rb–Sr isochron method of dating. At time $t = 0$, all six materials are in equilibrium with each other and therefore have the same $^{87}Sr/^{86}Sr$ ratio, but a different $^{87}Rb/^{86}Sr$ ratio, for example, because they represent different minerals, of which some have a greater affinity for Rb and others for Sr. Over time, ^{87}Rb decays into ^{87}Sr, and materials with higher $^{87}Rb/^{86}Sr$ ratios develop higher $^{87}Sr/^{86}Sr$ ratios. The slope of the arrows is very slightly to the left of the vertical, but this is not visible on the scale of this diagram. The slope of the line connecting the open boxes depends on the time elapsed since the six minerals last equilibrated with each other or, in other words, since closure of the Rb–Sr system for these minerals. In this example, 40 million years (Ma) elapsed since closure. (b) Illustration of an errorchron. If the assumption of initial isotopic equilibrium is not met, the samples will show a greater scatter than can be accounted for by the analytical uncertainty (which we assume to be represented by the size of the symbols in this example). If the scatter at $t = 0$ was random (as is the case in this example), the best-fit line through the open symbols will still have age significance (but with a reduced precision), but is called an errorchron instead of isochron.

9.1.2
Closure Temperature

The age that is obtained by the methods described previously is the time elapsed since the mineral(s) stopped equilibrating with each other and the rest of the outside world. This represents the moment when diffusion rates for the isotopes

Table 9.1 Isotopic systems commonly used for geochronology that can be analyzed by ICP-MS.

Parent (isotopic abundance, %)	Daughter	λ (year^{-1})	Half life (year)	Normalizing isotope (isotopic abundance, %)	Mass bias correction ratio for daughter	Other isotopes of daughter (isotopic abundance, %)	Other isotopes of parent (isotopic abundance, %)	Ref.	CHUR[a]
^{87}Rb (27.83)	^{87}Sr	1.42×10^{-11}	48.8×10^{9}	^{86}Sr (9.86)	$^{88}Sr/^{86}Sr = 8.375$	84 (0.56)	85 (72.17)	[4]	$^{87}Sr/^{86}Sr = 0.7047$; $^{87}Rb/^{86}Sr = 0.031$
^{147}Sm (14.99)	^{143}Nd	6.54×10^{-12}	1.06×10^{11}	^{144}Nd (23.8)	$^{146}Nd/^{144}Nd = 0.7219$	142 (27.2), 145 (8.3), 148 (5.7), 150 (5.6)	144 (3.07), 148 (11.24), 149 (13.82), 150 (7.38), 152 (26.75), 154 (22.75)	[5]	$^{143}Nd/^{144}Nd = 0.512630$; $^{147}Sm/^{144}Nd = 0.1960$
^{176}Lu (2.59)	^{176}Hf	1.867×10^{-11}	37.1×10^{9}	^{177}Hf (18.6)	$^{179}Hf/^{177}Hf = 0.7325$	175 (0.1), 178 (27.28), 180 (35.08)	175 (97.41)	[5]	$^{176}Hf/^{177}Hf = 0.282785$; $^{176}Lu/^{177}Hf = 0.0336$
^{187}Re (62.60)	^{187}Os	1.667×10^{-11}	4.16×10^{10}	^{188}Os (13.24)	$^{192}Os/^{188}Os = 3.083$ $^{189}Os/^{188}Os = 1.21978$	184 (0.02), 186 (radiogenic), 190 (26.26)	185 (37.4)	[6]	$^{187}Os/^{188}Os = 0.127$ $^{187}Re/^{188}Os = 0.40186$
^{190}Pt (0.013)	^{186}Os	1.477×10^{-12}	4.69×10^{11}	^{188}Os (13.24)	$^{192}Os/^{188}Os = 3.083$ $^{189}Os/^{188}Os = 1.21978$	184 (0.02), 187 (radiogenic), 190 (26.26)	192 (0.782), 194 (32.97), 195 (33.832), 196 (25.242), 198 (7.163)	[7]	

^{232}Th (100)	^{208}Pb	4.9475×10^{-11}	1.401×10^{10}	^{204}Pb (only non-radiogenic isotope)	None ($^{205}Tl/^{203}Tl = 2.3871$ or standard-sample bracketing)	204, 206, 207	230 (radioactive, half-life 75 kyr)	[8,4]	$^{208}Pb/^{204}Pb = 39.081$ $^{232}Th/^{204}Pb = 4.25$
^{235}U (0.73)	^{207}Pb	9.8485×10^{-10}	0.704×10^{9}	^{204}Pb (only non-radiogenic isotope)	None ($^{205}Tl/^{203}Tl = 2.3871$ or standard-sample bracketing)	204, 206, 208	238 (99.27)	[4]	$^{207}Pb/^{204}Pb = 15.518$ $^{235}U/^{204}Pb = 0.0676$
^{238}U (99.27)	^{206}Pb	1.55125×10^{-10}	4.468×10^{9}	^{204}Pb (only non-radiogenic isotope)	None ($^{205}Tl/^{203}Tl = 2.3871$ or standard-sample bracketing)	204, 207, 208	235 (0.73)	[4]	$^{206}Pb/^{204}Pb = 18.426$ $^{238}U/^{204}Pb = 9.2$

[a] CHUR = chondritic uniform reservoir.

of interest within the mineral slowed to such an extent that the mother and daughter elements became frozen in. For a given element and mineral, diffusion rates mainly depend on temperature, so the age represents the moment that the minerals cooled through their closure temperature. For some minerals and isotopic systems, such as the U–Pb and Th–Pb system in zircon ($ZrSiO_4$) this temperature is generally indistinguishable from the mineral's crystallization temperature (>900 °C), but it can be much lower, such as 400 °C for the Rb–Sr system in biotite [9]. Apart from temperature, the size of the minerals also plays a role [10]. Recrystallization of the minerals as a result of deformation and/or fluid flow will, of course, also disturb or even reset the isotopic systems [11], as will new growth due to dissolution–precipitation.

9.2 Practicalities

9.2.1 Isobaric Overlap

The equation for the Rb–Sr isotopic system and the illustration in Figure 9.1 show that the precision with which we can determine the age of a material depends on how well we know the slope of the isochron. This is dependent upon, among other things, the age and the enrichment of the parent isotope relative to the daughter, which is largely determined by the nature of the minerals that we analyze and the chemical system in which they equilibrated. These are geological constraints that we cannot alter. In the case of young samples (relative to the half life of the isotopic system with which we are dealing) that do not show a large enrichment in parent over daughter isotope, the precision with which we can determine the age is critically dependent upon, in the case of the Rb–Sr system, the precision of the measured $^{87}Sr/^{86}Sr$ and $^{87}Rb/^{86}Sr$ ratios. For several isotopic systems (Rb–Sr, Lu–Hf, Re–Os), the parent isotope is an isobar of the daughter isotope, and the two cannot be resolved from each other by present-day mass spectrometers. For the Sm–Nd system, the parent isotope (^{147}Sm) has a different mass from the daughter (^{143}Nd), but the normalizing isotope (^{144}Nd) shows an isobaric overlap with ^{144}Sm. One could try to correct mathematically for these cases of isobaric overlap by also measuring other isotopes of the elements of interest (e.g., ^{85}Rb) that do not suffer from isobaric overlap and using the known relative abundances of the isotopes ($^{85}Rb/^{87}Rb = 72.17/27.83$). However, one generally strives to date materials that are enriched in the parent isotope relative to the daughter, as this will give the best precision on the age obtained; $^{87}Rb/^{86}Sr$ ratios of 100 are not atypical for micas, for instance. Therefore, a small error in the concentration of the parent element results in a far greater error in the corrected radiogenic isotope ratio $^{87}Sr/^{86}Sr$, which for young samples is of the order of 0.71. The precision attainable by modern multi-collector (MC) mass spectrometry are of the order of 0.001 relative % (i.e., an uncertainty in the sixth decimal). An equal precision on the measured

$^{85}Rb/^{86}Sr$ ratio of 300 would translate to an error of the ^{87}Rb-corrected $^{87}Sr/^{86}Sr$ ratio that lies within the third decimal place. Although this may be acceptable for old materials (hundreds of millions of years), it is not if one wishes to date younger materials with this method. Therefore, the best results will be obtained by dissolution of the minerals and subsequent separation of Sr from Rb by column chemistry, so the Sr fraction can be analyzed separately from the Rb fraction. To obtain precise $^{87}Rb/^{86}Sr$ data (which are a function of the ratio of Rb/Sr elemental concentrations), it will be necessary to spike the sample and determine the concentrations by isotope dilution (ID) (see also Chapter 8). Ideally, the spike should be added to the sample prior to dissolution, to ensure complete equilibration between spike and sample. In the case of Rb–Sr, the spike will generally consist of Sr strongly enriched in the isotope 84 and Rb enriched in either 87 (the least abundant isotope) or 85 (the isotope that does not interfere with the crucial ^{87}Sr signal). However, there will always be some of the other isotopes of Sr present in the spike, and the measured $^{87}Sr/^{86}Sr$ ratio will have to be corrected ("spike stripping") for the presence of ^{87}Sr, ^{86}Sr, and ^{88}Sr (which is the isotope typically involved in mass bias correction) from the spike.

9.2.2 ICP-MS versus TIMS for Geochronology

The isotopic systems used for geochronological purposes that can be analyzed by inductively coupled plasma mass spectrometry (ICP-MS) are those that were traditionally done using thermal ionization mass spectrometry (TIMS), and are given in Table 9.1. Several of the advantages of ICP-MS over TIMS for geochronological purposes are similar to those mentioned for other applications. The speed of the actual analysis is an important factor, with an Nd isotopic analysis by TIMS taking $\sim$2 h whereas the same analysis can be done by ICP-MS in less than 20 min. Another important aspect is the improved ionization for some elements, notably Hf and U, by ICP-MS. The purity of the isolations is in some respects less important for ICP-MS analysis than for TIMS. The presence of Fe in the Pb fraction and Ca in the Sr fraction was known to have a deleterious effect on ionization during TIMS analysis, for Pb often to such an extent that the sample could not be analyzed. This problem does not occur with ICP-MS analysis, although the presence of (non-isobaric) impurities may affect mass bias [12], and is therefore still undesirable (see also Chapter 5). In general, instrument-induced mass bias is more reproducible for ICP-MS than for TIMS, as variations in filament or sample position and shape, which can influence running parameters for TIMS, do no apply to ICP-MS analysis. Finally, the ability to use laser ablation (LA) as a sample introduction system has had a great impact on several aspects of geochronology, especially for the U–Pb isotopic system.

Unfortunately, some of the "general" disadvantages of ICP-MS also apply to the field of geochronology. One of them is the higher probability of memory effects from material in and adsorbed on surfaces of the pump, tubing, torch, or cones, and contamination from impurities in the argon gas. The improved ionization for some elements is sometimes offset by the enormous loss of material during aerosol formation when carrying out solution analysis. Stability of mass bias is a positive aspect, but the amount of mass bias is far greater, complicating the process of simultaneous spike stripping and mass bias correction, as would be desirable in certain geochronological applications. Oxide and hydride formation and the presence of doubly charged ions are other potential problems, which will be discussed under the heading of the isotopic system for which they cause complications.

9.3 Various Isotopic Systems

In this section, the various isotopic systems used for geochronology will be reviewed, focusing on the matters that set ICP-MS analysis apart from the traditional TIMS approach.

9.3.1 U/Th-Pb

The most important development in geochronology of the past 15 years is the use of LA–ICP-MS for U–Pb dating of (parts of) individual U-rich crystals, most notably zircon. Zircon has been the geochronologist's mineral of choice for a long time, as it accommodates virtually no Pb in its crystal lattice at the time of formation (called "common Pb," to distinguish it from the radiogenic Pb formed by the *in situ* decay of U or Th), so it can be used for single phase, rather than isochron dating. Thereby, U–Pb provides us with two geochronometers, so we can have heightened confidence in the age obtained if the $^{206}Pb/^{238}U$ and $^{207}Pb/^{235}U$ ages are the same, or concordant. A third age can be obtained from the $^{208}Pb/^{232}Th$ system, as Th is generally also incorporated in zircon and other U-bearing phases (e.g., monazite).

When no initial Pb is present, the relevant equations for the U–Pb system are

$$^{206}\mathrm{Pb} = {}^{238}\mathrm{U}(e^{\lambda_{238}t} - 1)$$

$$^{207}\mathrm{Pb} = {}^{235}\mathrm{U}(e^{\lambda_{235}t} - 1)$$

where λ_{238} is the decay constant for ^{238}U and λ_{235} that for ^{235}U. All materials that give concordant ages fall on a single curve in a $^{206}Pb/^{238}U$ versus $^{207}Pb/^{235}U$ diagram; this curve is called the concordia (Figure 9.2).

These two equations can be combined to give

$$\frac{^{207}\mathrm{Pb}}{^{206}\mathrm{Pb}} = \frac{^{235}\mathrm{U}}{^{238}\mathrm{U}}\left(\frac{e^{\lambda_{235}t} - 1}{e^{\lambda_{238}t} - 1}\right)$$

The advantage of this equation is that the present-day $^{235}U/^{238}U$ ratio is constant at 1/137.88. Although this equation cannot be solved analytically (it is a

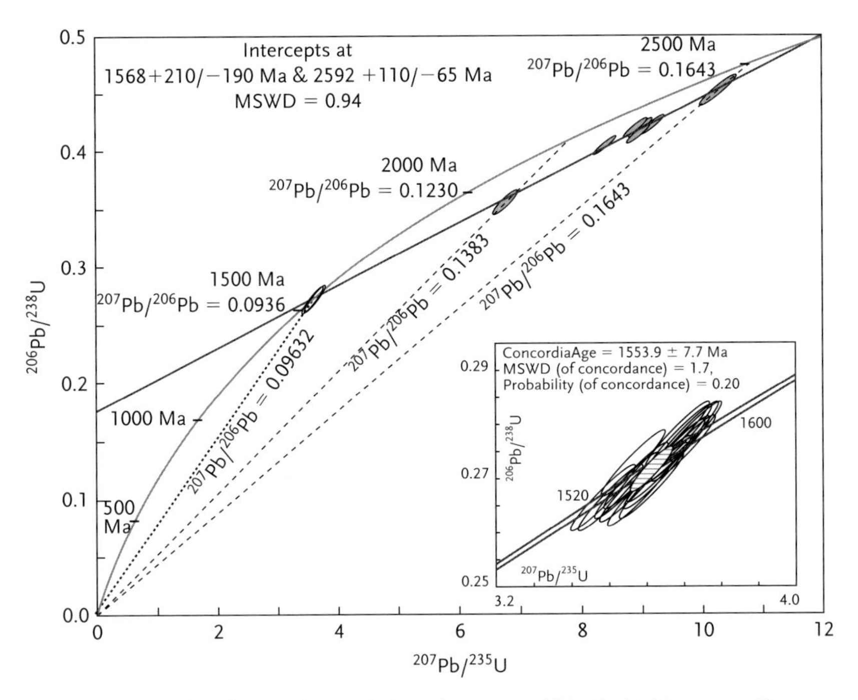

Figure 9.2 Example of concordant and discordant ages within the U–Pb system. The diagram was made with the ISOPLOT add-in [13] (downloadable from http://sourceforge.net/projects/isoplot/) for Microsoft Excel 2003. The curve, with age labels and corresponding $^{207}Pb/^{206}Pb$ ratios, represents the concordia, that is, the curve on which all closed-system U–Pb analyses fall. The seven analysis results indicated in grey are from individual zircons within a metasediment from South Australia; the cluster of analyses around 1550 Ma (inset on lower right) are zircons from a granite in the same area. The results indicated in grey for the metasedimentary zircons define a discordia, of which the upper intercept (~2600 Ma) is interpreted as the age at which the zircons formed; the lower intercept (~1570 Ma) coincides within error with the age of granitic magmatism. The U–Pb system of the metasedimentary zircons has been thermally disturbed by the granites, whereby the zircons lost part of their radiogenic Pb. The ages for the metasedimentary zircons calculated from the $^{207}Pb/^{206}Pb$ ratios vary between ~2200 and 2500 Ma, and can be visualized as the intercept between the thin dashed lines (essentially iso-$^{207}Pb/^{206}Pb$ lines, running from the origin through the individual zircon analyses), and the concordia. If the ~1550Ma zircons were to lose part of their Pb (for instance due to a thermal event) today, their analyses would plot on a similar straight line (bold dash) through the origin. In this case, the $^{207}Pb/^{206}Pb$ age would still be 1550 Ma.

transcendental equation), we can easily determine the age from a measured $^{207}Pb/^{206}Pb$ ratio by successive approximations. Therefore, even if we were not to analyze U, an age could be obtained from the $^{207}Pb/^{206}Pb$ ratio of the zircon [14]. A disadvantage of this method is that we do not have an independent check on whether the age is concordant. The $^{207}Pb/^{206}Pb$ age is even correct if any Pb or U loss or gain occurred in a geologically very recent past, as it will shift the analysis on an "iso-$^{207}Pb/^{206}Pb$" line towards the origin. Zircons that suffered Pb loss (or rarely U gain) and thereby lie below the concordia are said to be "discordant," whereas those that lie above the concordia are said to be "reversely discordant." The $^{207}Pb/^{206}Pb$ age will not have any relevance if the U–Pb isotopic system was disturbed at a time significantly before the present. In the case of a two-stage history of zircon formation (at t_1) and subsequent partial Pb loss (at t_2), the zircon analyses, when plotted as a $^{206}Pb/^{238}U$ versus $^{207}Pb/^{235}U$ diagram, will lie on a straight line. This line, called a "discordia," intersects the concordia at t_1 and t_2, and the combined $^{206}Pb/^{238}U$ and $^{207}Pb/^{235}U$ ratios can be used to unravel the sequence of events (Figure 9.2). The $^{207}Pb/^{206}Pb$ ratios, on the other hand, will only yield meaningless ages intermediate between these two events.

U–Pb zircon analyses used to be done by analyzing different fractions of zircons from a single rock sample, based on their color, shape, size, and/or magnetic characteristics. These zircons were then dissolved, necessitating the use of high-pressure bombs and concentrated HF and HNO_3 at high temperatures, as zircon is one of the most resistant minerals to chemical dissolution. After splitting the sample, spiking one fraction and separation of U and Pb by column chemistry, the actual analyses were performed using TIMS. Only in selected clean laboratories are blank levels low enough that (fragments of) individual zircons can be analyzed [15] after physical and/or chemical abrasion techniques to remove metamict domains [16]. This is necessary when radiation damage has disturbed the crystal lattice ("metamictization"), which might lead to loss of radiogenic Pb, and thereby discordant ages. These concordant single zircon analyses have not yet been surpassed in terms of analytical precision by any other method, but few laboratories can provide this (also very time-consuming) analytical service.

Since the middle of the 1980s, the traditional method of zircon geochronology was complemented by the use of the sensitive high-resolution ion microprobe (SHRIMP) [17], with which 15 μm diameter spots on zircons can be dated, and metamict or mixed-age domains avoided. Since minimal amounts of material are used, this method is still preferred in cases where the material to be analyzed is scarce, such as zircons from extraterrestrial samples, or the early Archean ($\sim$4.4 Ga) [18]. Unfortunately, the number of SHRIMP instruments is limited, and in cases where scarcity of material is not a problem, and a precision on the level of a few million years is acceptable, SHRIMP and TIMS techniques for zircon dating have been largely replaced by LA-ICP-MS analyses, pioneered in the early 1990s [19,20]. Here a laser beam is used to excavate a small (15–90 μm diameter) pit in the mineral, from which the ablated material is transferred to the plasma of an ICP-MS instrument, where ionization and analysis take place (see also Chapters 2 and 4).

9.3.1.1 **LA–ICP-MS U–Pb Dating of Zircon**

There are several different approaches to LA–ICP-MS in terms of equipment used and standardization, but most will start with the separation of zircons from a rock sample by crushing and sieving, followed by separation of the non-magnetic heavy fraction with a Frantz isomagnetic separator and heavy liquids such as sodium or lithium polytungstate, bromoform, tetrabromoethane, and/or methylene iodide.

However, in cases where spatial information is crucial, zircons (or other minerals) can also be analyzed in polished slab or thin sections of rock [21]. Separated zircons are mounted in epoxy and polished to expose the parts that will be targeted for dating. The targets for LA–ICP-MS analysis are generally selected based on cathodoluminescence imaging [22] (Figure 9.3). Prior to ablation, the sample is generally cleaned with nitric acid to remove any surface contamination, most notably Pb.

The main challenge of LA–ICP-MS analyses for U–Pb geochronological purposes is to constrain mass bias within the ICP-MS (see also Chapter 5) and elemental fractionation during laser ablation (see also Chapter 4), while simultaneously obtaining the maximum spatial resolution to resolve different age zones, as illustrated in Figure 9.3, and an adequate signal intensity to obtain the desired precision on the analysis. These are demands that unfortunately conflict with each other, so a compromise has to be found.

As with any analysis, the precision with which the isotope ratios of interest can be determined is related to the intensity of signal of the least abundant isotope. During laser ablation, this intensity is strongly related to the volume of material ablated: smaller laser pits will yield smaller amounts of material and therefore

Figure 9.3 Example of a cathodoluminescence (CL) image of a zircon from a meta-igneous lithology, subsequently used for LA–ICP-MS analysis. The inner core (within the dashed outline) gave a discordant age, which, together with three other cores, constitute a discordia with an upper intercept age of 1700 Ma. The smudged appearance of this core is rather typical for discordant metamict zircons. The oscillatory-zoned part, typical for igneous zircons, gave a near-concordant age around 1580 Ma. The outer rim was not analyzed for this zircon, but similar-looking rims on other zircons gave concordant ages around 480 Ma, similar to the lower intercept of the discordia for the inner core. The differences in brightness in CL are related to varying levels of rare earth elements, mainly dysprosium.

lower intensity signals and less precise analyses. On the other hand, small laser pits allow the user to sample thin age zones within zircons, which can be very important, for instance if one wants to date metamorphic events such as migmatite formation [23]. However, the pit diameter is only one aspect of spatial resolution, as the laser beam drills progressively down into the sample and may in this way enter a different age zone. Therefore, one can also decide to use fairly wide beams, but extremely shallow (10–30 shot) craters [24] on unmounted zircons.

Apart from a reduced intensity, small-diameter ablation craters generally display another disadvantage, namely that they cause the elements to fractionate from each other with depth. A clear correlation exists between the diameter of the crater and the amount of fractionation with depth [25], which can also be expressed as a function of the depth/width ratio of the crater [26]. This laser-induced fractionation effect is seen in glass (such as the widely used NIST 610 borate glass) in addition to minerals such as zircon.

Laser-induced elemental fractionation is the norm for the most widely used types of lasers, namely 266–193 nm lasers with pulse lengths in the nanosecond range; the effect is absent in a new generation of lasers with shorter pulse lengths, the femtosecond lasers [27]. Until these have replaced the current generation of lasers, strategies need to be adopted to correct for laser-induced elemental fractionation. This is especially important for zircon age dating, where the age is calculated on the basis of $^{206}Pb/^{238}U$ and $^{207}Pb/^{235}U$ ratios, that is, ratios of two different elements, which are fractionated from each other during laser ablation (Figure 9.4). This laser-induced fractionation is additional to the ICP-MS-induced mass discrimination as a result of which heavy isotopes appear to be overabundant relative to their lower-mass counterparts (see also Chapter 5). Correcting for these fractionation effects is more difficult for the Pb isotopic system than for most other radiogenic ratios, as Pb has only one isotope that is non-radiogenic, and an invariant ratio to which to normalize therefore does not exist.

A solution to the problem of laser-induced elemental fractionation is to use very wide laser beams (>90 μm) or raster patterns [28]. Marginal downhole fractionation generally still occurs, even with wide laser beams, but it is assumed that the $^{206}Pb/^{238}U$ ratio at the start of the analysis (the "intercept ratio") is only influenced by ICP-MS-induced mass bias. This can be corrected for by simultaneous aspiration of tracer solutions enriched in Tl and ^{233}U. Here it is assumed that the fractionation of Pb isotopes from each other can be corrected for on the basis of the measured versus known (natural) $^{205}Tl/^{203}Tl$ ratio (of 2.3871 [8]), while the measured versus known $^{205}Tl/^{233}U$ ratio is used to correct the $^{206}Pb/^{238}U$ ratio of the unknown [28].

Determination of the intercept ratio generally assumes a linear relationship between time ($\approx$ downhole depth) and elemental fractionation, but other types of curves can also be fitted to the data [29], as there is no *a priori* reason to assume linearity. Ideally, fitting a curve through the time-resolved analysis and determining the intercept should give the same result as using the very first analytical point only. However, analytical uncertainty on the single point is much greater

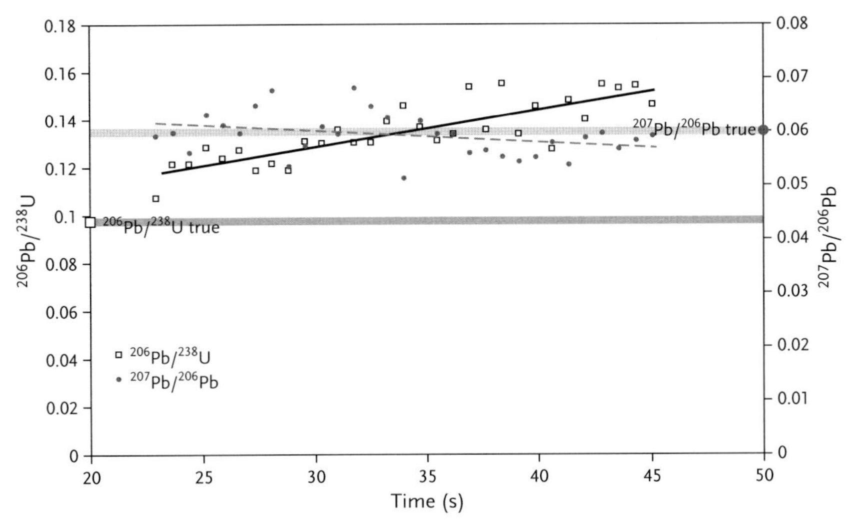

Figure 9.4 Example of time-resolved analysis of standard zircon GJ-1 by LA–sector field ICP-MS (213 nm Nd:YAG system combined with a sector field ICP-MS instrument; the spot size was 40 μm). Only a blank correction has been applied, based on the average signal for the first 20 s of the analysis, during which the laser was fired with the shutter closed. $^{206}Pb/^{238}U$ ratios (open squares) are significantly offset from their true value, and increase with time (downhole; thick drawn line is best fit through data points). This is the result of elemental fractionation during laser ablation. $^{207}Pb/^{206}Pb$ ratios (filled circles) are close to their true value, and their time-dependent change (dashed line) is insignificant. Only ICP-MS induced mass discrimination, far smaller than laser-induced elemental fractionation, is responsible for the offset from their true values. One could just use the $^{207}Pb/^{206}Pb$ ratios to calculate the age of the zircon, which would come out as 570 $\pm$ 150 (1σ) Ma for this analysis, well within error of the accepted $^{207}Pb/^{206}Pb$ age of 608 $\pm$ 0.2 Ma, but very imprecise. The $^{206}Pb/^{238}U$ age calculated from the average and standard deviation of the uncorrected data would be 816 $\pm$ 70 Ma, compared with the accepted $^{206}Pb/^{238}U$ age of 600 Ma. The far smaller error on this age, despite the downhole fractionation, demonstrates why it is preferred to try to correct for laser-induced elemental fractionation and use the U–Pb age rather than the Pb–Pb age. In a similar experiment using MC-ICP-MS, the errors on the ages derived from the blank-only corrected $^{207}Pb/^{206}Pb$ and $^{206}Pb/^{238}U$ ratios were reduced to 27 and 4 Ma, demonstrating the advantage of MC instruments for isotopic analysis. The accuracy of the ages calculated from the uncorrected data, however, was not better.

than that on the fitted curve, and curve fitting will therefore result in ages that are more precise and accurate.

If the assumption that the intercept ratio is not affected by laser-induced elemental fractionation does not hold, the approach described above does not yield the correct age. Whether it does or not can be checked by analysis of a standard zircon of known age. Some well documented standards exist, and their relevant ratios are given in Table 9.2.

An alternative correction method is based on, essentially, standard–sample bracketing, where analyses of unknowns are interspersed with analyses of standards. The basic assumption in this case is that the amount of elemental and

Table 9.2 Relevant isotope ratios and ages (Ma) for standard zircons commonly used for LA–ICP-MS Geochronology.

Name	U ($mg\ kg^{-1}$)	Pb ($mg\ kg^{-1}$)	Th/U	$^{206}Pb/^{204}Pb$ average	$^{206}Pb/^{238}U$ (2σ)[a]	$^{206}Pb/^{238}U$ age (2σ)[a]	$^{207}Pb/^{235}U$ (2σ)[a]	$^{207}Pb/^{235}U$ age (2σ)[a]	$^{207}Pb/^{206}Pb$ (2σ)[a]	$^{207}Pb/^{206}Pb$ age (2σ)[a]	$^{208}Pb/^{232}Th$	$^{208}Pb/^{232}Th$ age (2σ)[a]	Ref.
GJ-1	212–422	19–37	0.02–0.06[b]	436,000	0.09762 (11)	600 (2)[c]	0.8093 (9)	602 (1.6)[c]	0.06014 (1)	608.4 (0.4)	0.03011[b]		[28]
91500	55–82	13–17	0.3444 (18)	24,300	0.17917 (16)	1062.4 (0.8)	1.8502 (16)		0.07488 (2)	1065.4 (0.6)	0.05404 (32); radiogenic: 0.05374 (30)	1058.1 (5.6)	[31]
Temora-1	71–490	14 ± 8	0.21–0.63	72,000	0.06678 (8)	416.8 (0.5)	0.5077 (8)	416.9 (0.5)	0.05515 (6)	418.3 (2.5)	0.02092 (13)	418.4 (2.6)	[32]
Temora-2	94–271	7–20	0.34–0.43		0.06667 (19)	416.1 (1.2)	0.5072 (23)	416.6 (1.6)	0.05518 (16)	419.5 (6.4)			[33]
Mud Tank	6–36	1–4				732 (10)							[34]
Plesovice	465–3000[d]	39–116[d]	0.04–0.15[b]	51,9000	0.053682 (13)	337.1 (0.1)	0.39398 (15)	337.3 (0.1)	0.053244 (9)	339.3 (0.4)			[35]
(CD)QGNG	35–315	14–1151	0.38–1.08[b]	45,000	0.3203 (80)[c]	1796 (38)[c]	5.04 (12)[c]	1840 (11)[c]	0.11311 (10)[c]	1850 (2)[c]			[36]

[a] Weighted average, corrected for common Pb.
[b] Calculated ratio based on age and $^{208}Pb/^{206}Pb$ ratio.
[c] 95% confidence limits.
[d] High values in metamict, discordant, high-REE domains that should be avoided.

isotopic fractionation in the unknowns is exactly the same as that in the standards. This means that the ablation characteristics (energy, repetition frequency, beam diameter) need to be kept constant, which can be a disadvantage, although most workers would be happy to keep the laser at a constant small diameter, rather than a variable wide diameter. With this approach, either the intercept method could be used, or average ratios – if the latter method is applied, it is important that exactly the same time slice of the standard as of the unknown is used, since ratios vary with time. Using only part of the ablation run of an unknown may be necessary when an inclusion or a different age zone is encountered.

This method only works if standard and sample behave identically, a condition that is most likely to be met if they resemble each other in chemistry and structure (e.g., crystallinity), and are therefore "matrix matched." Considering that well-characterized zircons are not that easy to obtain, considerable effort has been spent on determining how closely matched standards and unknowns need to be, mainly with a view to using the widely available and well-known glass reference materials NIST 610 and 612 [37] as a standard for U–Pb geochronology [38]. The overwhelming conclusion is that matrix matching is essential for good-quality analyses using nanosecond lasers; combinations of certain ablation parameters (spot size, energy, frequency) may lead to fortuitous similarities between disparate materials [39], which explains why some workers report good results with using non-matrix-matched standards [38]. Pb and U show very disparate elemental fractionation during LA, with Pb/U ratios increasing downhole, both in NIST glass [26] and in zircon, but with depth-related changes more pronounced in NIST glass [39]. The contrasting fractionation is likely to be related to the fundamentally different behavior of zircon and NIST glass during ablation with nanosecond lasers, whereby at least part of the zircon incongruently melts to crystalline ZrO_2 (baddeleyite) and SiO_2 (cristobalite) [40], with U (together with Zr and Hf) selectively retained in the newly formed refractory baddeleyite [39]. Normalizing U and Pb intensities to Zr and Si, respectively, is therefore likely to give improved precision on calculated ages [41], although a constant correction factor still needs to be applied to account for the preferential loss of Pb compared with Si [39]. This kind of correction scheme could also compensate for offsets of LA–ICP-MS U–Pb ages compared with their accepted TIMS ages [33], which has been linked to the presence of certain trace elements. However, as some workers report that $^{207}Pb/^{206}Pb$ ages also vary, depending on the normalizing zircon standard used [42], perhaps not all discrepancies can be blamed on laser-induced fractionation. This would set a limit to the accuracy of the analyses, meaning that analyses from different laboratories cannot be meaningfully interpreted beyond the 4% level [42]. One medium-term solution would be to use more than one well-characterized zircon (Table 9.2) as a bracketing standard, whereby the offsets roughly cancel out.

Laser Ablation System From the discussion above, it is clear that the actual ablation and analysis can be carried out using a variety of instruments, with results from 266 nm Nd:YAG, 213 nm Nd:YAG [30], 193 nm Nd:YAG [42] and 193 nm excimer ArF* [25] lasers being reported on a routine basis. Results from femtosecond lasers used for geochronological purposes [27] are not yet common. In a comparison between 266 and 213 nm systems [30], it was found that the shorter wavelength reduced matrix-dependent elemental fractionation. Femtosecond lasers are even better for this purpose [27], and are likely to become the laser of choice in the future. The 213 nm quintupled Nd:YAG laser appears to be suitable for many geochronological purposes, with the additional benefit of lower running costs compared with the excimer ArF* system. Most workers are now using He rather than Ar as a carrier gas, as this improves analyte transport efficiency and, thus, detection limits [26], especially for the shorter wavelength lasers [43]. Beam diameter (generally between 15 and 90 μm), repetition rate (5–10 Hz), and fluence (0.1–10 J cm^{-2}) are adapted to suit the material being analyzed, with important variables being the age of the material analyzed (younger zircons contain less radiogenic Pb, and therefore need larger beams to obtain a good signal) and the spatial resolution needed. Apart from the laser wavelength and carrier gas, the design of the ablation cell can have an important influence on the resulting signal. The ablation cells delivered with off-the-shelf laser systems are often replaced by custom-designed cells. The actual design depends on the importance that the user attaches to signal stability versus response and wash-out time. Small-volume cells have a short response and wash-out time, but thereby give off signals that are not as stable as those from larger cells. Post-ablation mixing devices [44,45] can also be introduced to stabilize the signal. For time-resolved analyses, where one drills from one age zone into the next, a smaller cell with a faster response time should be preferred. Another advantage of a smaller cell is the lack of signal build-up, where maximum intensity is only reached after a number of scans, that is seen in larger cells [46]. The disadvantage of a small cell is the limited size of the sample that can be accommodated (typically of the order of a single 2.5 cm grain mount), but this problem is being circumvented by the use of a two-volume cell [47]. The surface area of the fitted sample(s) is 50 × 50 mm, but the volume in which the ablation takes place is less than 2 cm^3, thereby ensuring a quick response and rapid wash-out between samples.

As discussed, fractionation of the elements from each other during laser ablation is always a problem, especially for U/Pb ratios, and worsens as the ablation pits get deeper [26,30], so most workers avoid making the depth of the pits greater than their diameter.

ICP-MS Equipment The U/Th-Pb system is one of the few geochronological systems where the use of single-collector ICP-MS (see also Chapter 2) gives adequate results. The use of quadrupole-based and sector field single- and multi-collector ICP-MS equipment of different makes is reported in the literature.

The use of an MC instrument has the obvious advantage of the simultaneous measurement of the signal intensities for all isotopes of interest, thereby providing a better precision on the measured ratios. The main disadvantage is that only a limited number of isotopes can be measured, so the Th–Pb isotopic system cannot be used simultaneously with the U–Pb system on all MC instruments [48,49]. Also, normalization of U and Pb intensities to Zr and Si to account for second-order variations in matrix-related effects can only be deployed with single-collector equipment. As the less abundant isotopes (generally all of the Pb isotopes) need to be monitored via ion-counting electron multipliers rather than the Faraday cups with which U (and Th) ion beam intensities are measured, the different detectors need to be cross-calibrated carefully [50]. To correct for drift in the Faraday–ion counter cross-calibration, most laboratories use a procedure of external normalization by standard–sample bracketing [50,51], sometimes with an additional check on fractionation behavior by simultaneous aspiration of a U–Tl solution [52].

The use of single-collector equipment has the advantage that it permits the analysis of all isotopes and that all isotopes are analyzed on the same detector, albeit not simultaneously. Cross-calibration problems are avoided, but the precision is generally lower owing to the sequential analysis of the elements. Dwell times on the different peaks can be adjusted to suit the required precision, with longer acquisition times for the smaller peaks. Many modern instruments can switch between pulse counting mode for low-abundance isotopes to analog detection (see also Chapter 2) for higher abundance isotopes such as ^{238}U and ^{232}Th [53]. This can be especially advantageous for younger zircons, for which the concentration of radiogenic isotopes will be low, so the ablation parameters can be adjusted to increase their count rate, while the measurement of U or Th in analog mode avoids overloading ("tripping") of the detector.

Quadrupole-based instruments have the advantage that their scanning speed is significantly higher than for sector field instruments. Disadvantages include the peak shape, which is Gaussian rather than flat-topped, so small amounts of drift in peak position will influence the accuracy of the measured ratios. Sample–standard bracketing will counteract this potential problem to some extent. The slower scanning speed of *magnetic sector field* ICP-MS instruments is related to the settling time for the magnet after each mass jump. However, this can be reduced by performing part of the scanning (from ^{202}Hg to ^{238}U) by varying the acceleration potential rather than the magnet setting [44]. The flat-topped peak shape, improved isotope ratio precision, and capability of measuring at higher mass resolution are the main advantages of the use of sector field devices.

Isobaric Overlap, Common Pb and Oxide Formation The U/Th-Pb system in zircon is used because zircon incorporates very little "common" (non-radiogenic) Pb in its crystal lattice – unfortunately, it cannot be assumed that no common Pb has been incorporated. The presence of common Pb can be monitored by measuring the ^{204}Pb isotope, which is not produced by radioactive decay. However, here the enhanced blank levels of ICP-MS equipment compared with TIMS come into play, as most ICP-MS equipment displays a (gas) blank of Hg, of which the ^{204}Hg isotope overlaps irresolvably with that of ^{204}Pb. This can be corrected for by measuring a gas blank before each analysis, and/or monitoring ^{202}Hg or ^{201}Hg during analysis and mathematically subtracting the amount of the signal at a mass-to-charge ratio of 204 that is derived from ^{204}Hg, assuming natural ratios and a certain (ICP-MS-induced) mass bias. Because of uncertainty on the appropriate Hg isotope ratios, the correction is generally based on measurement of the gas blank prior to analysis. However, if the Hg blank is high, the signal coming from ^{204}Pb may be swamped. It can therefore be advantageous to invest in a mercury trap, consisting of gold coated glass fragments or an activated charcoal filter [54]. If ^{204}Pb can be detected, the isotopic composition of the common Pb incorporated in the zircon is supposed to be related to the first approximation of the $^{206}Pb/^{238}U$ zircon age, following the Stacey–Kramers [55] crustal growth curve [56]. If ^{204}Pb cannot be detected, it is often assumed that the zircons are concordant, and that the presence of common Pb is the only reason for discordance. This can then be corrected for based on measured ^{208}Pb intensities, assuming that the Th/U ratio of the zircon has not been disturbed [57]. Alternatively, the correction can be based on ^{207}Pb, using a Tera–Wasserburg diagram [58] (Figure 9.5), again assuming concordance.

Unfortunately, zircons that have high common Pb contents often also display Pb loss. This can be checked in a Tera–Wasserburg diagram, provided that the zircons belong to a single population, since common-Pb correction should reduce the data to a single point on the concordia. This does not hold, of course, when detrital zircons are analyzed, which are unlikely to belong to a single population. A more robust mathematical solution to correct for common Pb when ^{204}Pb analyses are unavailable, using both the U/Pb and Th/Pb ratios, and which does not work on the assumption of concordance, was proposed by Andersen [59]. This method, however, uses the $^{208}Pb/^{232}Th$ ratio, and one of the commonly used reference zircons, GJ-1, has very low ^{208}Pb and ^{232}Th contents and a poorly defined

Figure 9.5 The principle of a Tera–Wasserburg diagram. This 3D diagram shows the Tera–Wasserburg diagram in the *x–y* view, where $^{207}Pb/^{206}Pb$ is plotted against $^{238}U/^{206}Pb$. The curve, with tick marks for different ages, incorporates all analyses with concordant U–Pb ages and no common Pb. Zero-age materials cannot be represented on this diagram, as ^{206}Pb concentrations will be zero and the *x*-coordinate therefore at infinity. The *z*-coordinate gives the $^{204}Pb/^{206}Pb$ ratio. If this ratio can be measured, then the zircon analysis will plot somewhere within the 3D box, generally close to the *x–y* plane (as zircons typically contain low levels of common Pb and therefore contain very little ^{204}Pb). The *y–z* plane shows the evolution of common Pb through time [55]. However, the common Pb composition is determined from the actual zircon analyses and does not need to lie on this curve. The spheres within the 3D box represent the case of 1600 Ma zircons with measured $^{204}Pb/^{206}Pb$ ratios. Correcting for the presence of common Pb (which in this example lies on the curve in the *y–z* plane at 1600 Ma) is projecting the analysis from the 1600 Ma point in the y-z plane onto the *x–y* plane. In this idealized scenario, the projection falls exactly on the $^{207}Pb/^{206}Pb$ -$^{238}U/^{206}Pb$ point on the concordia at 1600 Ma. If $^{204}Pb/^{206}Pb$ ratios are unavailable, the analysis data can only be plotted on the *x–y* plane, and the result will then be a projection of the point within the 3D box along the *z*-axis. Thereby, the projected points (grey circles in inset) will plot slightly above and to the left of the 1600 Ma concordia point. Higher percentages of common Pb will offset the analyses further from the point on the concordia. Therefore, a population of zircons of the same age and containing varying amounts of common Pb with the same composition will plot on a Tera–Wasserburg diagram along a line from the $^{207}Pb/^{206}Pb$ ratio of the common Pb component (in this case 0.96) to the concordia age of the population (in this case 1600 Ma).

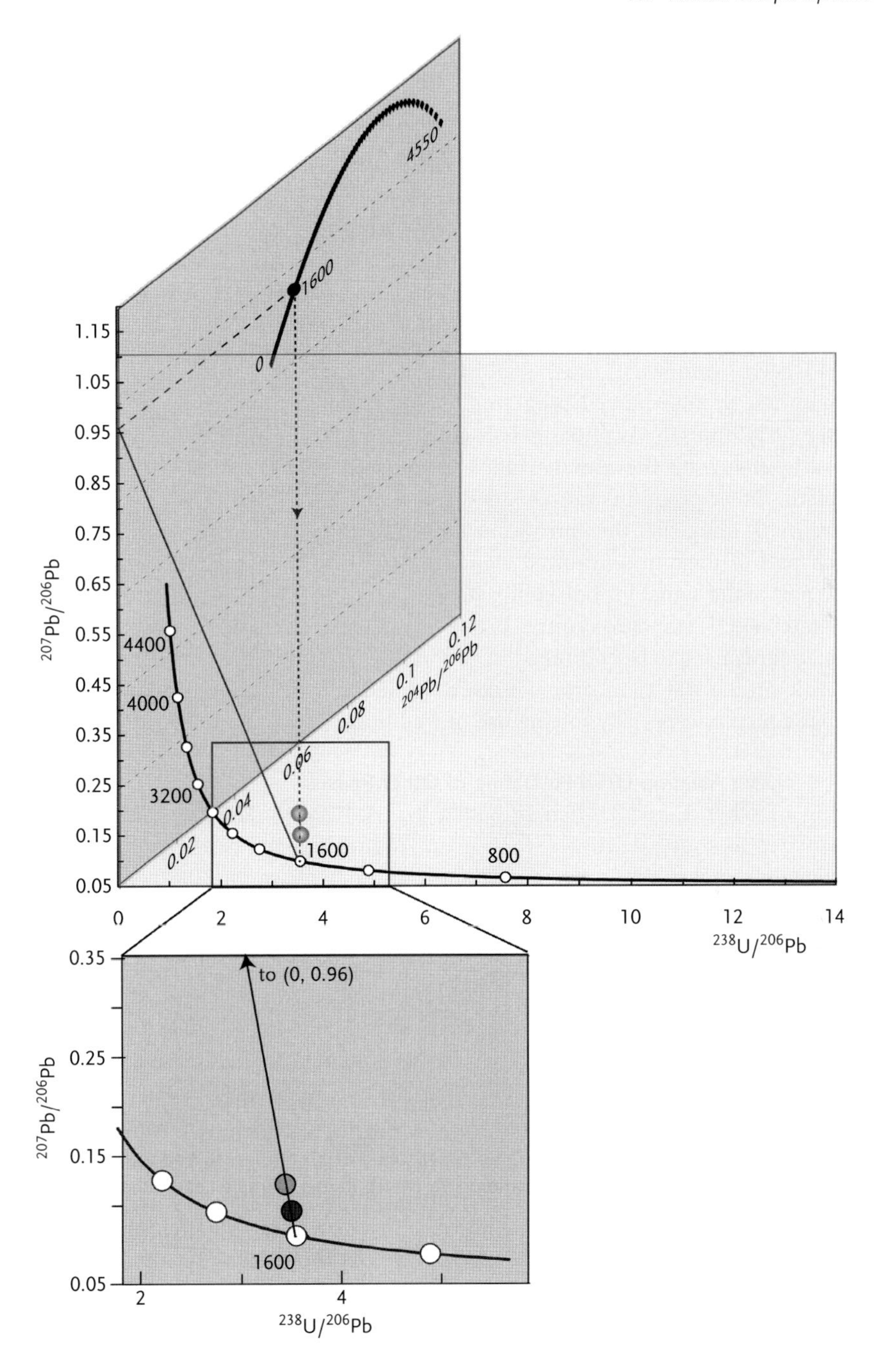

4550
1600
0
1.15
1.05
0.95
0.85
0.75
0.65
0.55
0.45
0.35
0.25
0.15
0.05
$^{207}Pb/^{206}Pb$
4400
4000
3200
1600
800
0.02
0.04
0.06
0.08
0.1
0.12
$^{204}Pb/^{206}Pb$
0
2
4
6
8
10
12
14
$^{238}U/^{206}Pb$
to (0, 0.96)
0.35
0.25
0.15
0.05
$^{207}Pb/^{206}Pb$
1600
2
4
$^{238}U/^{206}Pb$

$^{208}Pb/^{232}Th$ ratio, so analyses done with this zircon as a bracketing standard are likely to be less suited for this approach [30].

If the intercept ratio is used without standard–sample bracketing, there should be a check on the percentage of oxides formed, as this is likely to be different for U than for Pb, and may affect the U/Pb ratios. This can easily be done on single-collector instruments by monitoring oxides for co-aspirated solutions [60]. In the case of standard–sample bracketing, it is assumed that the level of oxide formation is similar for sample and standard.

Data Processing Most data processing is done on- or off-line in home-made Excel spreadsheets, somewhat more sophisticated spreadsheets provided by certain laboratories, or commercial programs. An example of commercial programs is Glitter ([61] http://www.glitter-gemoc.com/). Shareware is PEPITA [62], which works by matching the average ratios of the selected time slice to that for the standard. Other programs are LAMDATE [63] and LAMTRACE [64].

The main thing that the spreadsheets need to do is to allow the user to decide which part of their time-resolved signal is used, and then reference that to the same part of the signal for bracketing standards, or use it for determining the zero-time intercept. If standard–sample bracketing is used, errors on the unknown and the standards need to be combined in quadrature [56], and an assessment needs to be made on the way to incorporate errors related to drift. Error correlations are a further concern, especially for the use of traditional concordia diagrams [13].

9.3.1.2 Laser Ablation U/Th-Pb Dating of Other Phases

Dating of other U- or Th-rich minerals by LA–ICP-MS provides similar (dis) advantages to those discussed for zircon. Matrix matching of standards and unknowns remains the greatest challenge, as each phase has its own ablation characteristics when it comes to the downhole fractionation of Pb from U or Th [65]. Therefore, it is necessary to use the same mineral as a standard when performing standard–sample bracketing. As none of these materials are commercially available, it remains a challenge to procure the necessary homogeneous, well-characterized U–Th-rich phases, preferably within the same age bracket as the material that one wants to date.

Apart from zircon, other common accessory minerals that can be used for U/Th-Pb dating are monazite, rutile, titanite, apatite and allanite; results on rarer minerals such as columbite–tantalite, perovskite, baddeleyite and the uranium ore minerals uraninite and davidite have also been reported.

Monazite [(Ce,La,Nd,Th)PO_4] is probably the next most popular phase after zircon used for U/Th-Pb dating by LA–ICP-MS. As the mineral formula implies, Th rather than U is generally the main radioactive element present. Like zircon, common Pb contents tend to be very low, permitting single spot dating. Geochronologically, monazite behaves very differently from zircon. Zircon is an extremely resistant phase, and isotopic resetting is mainly a result of dissolution and new growth during partial melting. Monazite dissolves far more easily in acidic solutions (concentrated nitric acid is generally used in the laboratory) and can therefore be affected by hydrothermal alteration at already fairly low temperatures (200–300°C) [66]. Therefore, monazite often records the age of later metamorphic events, while zircon (cores) still record protolith or detrital ages [23], although the opposite can also be true [67], probably depending on the exact chemical environment. The closure temperature of monazite is also lower than that for zircon, around 725–800°C [68,69].

The instrumentation and analytical protocol for U/Th-Pb dating of monazite is very similar to that used for zircon [70,71]. Yet unexplained discrepancies between LA–ICP-MS and TIMS ages also exist for monazite, limiting the accuracy to 1 and 5% for the Pb–Pb and U/Th-Pb dates, respectively [71]. One potential problem with monazite U–Pb dating is the very high level of Th present in the crystal. During formation of the crystal, not only ^{232}Th but also ^{230}Th, an intermediate decay product of ^{238}U with a relatively short half-life of 7.5×10^4 years, will be incorporated in the crystal. The greater affinity of monazite for Th than for U means that this ^{230}Th will be incorporated to a greater extent than its parent, ^{238}U. This will lead over time to higher ^{206}Pb/^{238}U ratios than for systems in which this "unsupported" ^{230}Th is not present, and thereby to ^{206}Pb/^{238}U ages that are too old [72], resulting in reverse discordance on ^{207}Pb/^{235}U versus ^{206}Pb/^{238}U concordia diagrams. This problem will be more pronounced in relatively young crystals.

An interesting aspect of monazite is its excellent resistance to radiation damage. Zircons with high U contents become "metamict" over time, whereby radiation damage to the crystal lattice makes the crystal more susceptible to Pb loss, resulting in discordant ages. Monazite, on the other hand, shows a kind of "self-healing" effect, whereby the crystal lattice becomes distorted but remains intact [73], so that even old monazites with high levels of radioactive elements can yield concordant ages.

Rutile (TiO_2), **titanite** (or sphene, $CaTiSiO_5$), and **apatite** [$Ca_5(PO_4)_3(OH,F,Cl)$] commonly occur as metamorphic or primary igneous minerals, and have high to intermediate closure temperatures (titanite 680°C [74]; rutile 600°C [75]; apatite 500°C [74]). Although all three minerals typically contain appreciable amounts of U, they also contain common Pb, making them less suitable for single spot mineral dating. As the relative contribution of common Pb to the total Pb budget is not constant throughout most single crystals, or between different crystals, isochron dating is possible with multiple spot analyses. With relatively high levels of common Pb, ^{204}Pb levels easily rise above the background ^{204}Hg levels and ^{206}Pb/^{204}Pb and ^{207}Pb/^{204}Pb ratios can be measured reliably on MC instruments [75,76], albeit on the electron multiplier. The added advantage of relying on Pb–Pb isochrons is that laser-induced elemental fractionation is of no concern; instrumental mass discrimination can be corrected for by bracketing with the well-known NIST 610 glass or simultaneous aspiration of a Tl solution.

A point to note is that within-run correction for ^{204}Hg is probably best done by monitoring ^{201}Hg rather than ^{200}Hg or ^{202}Hg for rutile [75], as a result of non-zero levels of tungsten, of which the WO^+ oxide ions yield spectral overlap at masses 200 and 202. Monitoring ^{201}Hg combined with simultaneous Tl aspiration could, however, be a problem with the achievable cup configurations of some MC instruments.

As the precision attainable by this type of isochron dating improves with the incorporation of points with high ratios of radiogenic to common Pb, this approach is likely to be most successful on Precambrian crystals.

Some reported titanite U/Th-Pb LA–quadrupole (Q)-based ICP-MS ages were obtained with zircon as a calibration standard, without demonstrating the accuracy of this approach [77,78]. Considering that the phase change of zircon to baddeleyite is unlikely to take place in titanite, the downhole fractionation of U from Pb may be rather different, and acceptable results can only be expected with the intercept method for calculating U/Pb isotope ratios.

With the use of well-characterized titanite as an external standard [21], some interesting results have been obtained in dating putative microbial ichnofossils in Archean rocks, now mineralized with minute titanite crystals, giving an age of around 2.9 ± 0.1 billion years [79].

The most thorough study of **allanite** [$(Ca,REE,Th)_2(Al,Fe^{2+})_3Si_3O_{12}(OH)$] LA–Q-ICP-MS Th–Pb dating was performed by Gregory *et al.* [65], convincingly showing the need for matrix-matched standards. Allanite occurs in a range of igneous and metamorphic rocks, and can therefore be used to constrain the timing of either magmatism or metamorphism. As allanite contains common Pb, the isochron approach to dating was applied. In this case, common ^{206}Pb was used as the normalizing isotope, as the ^{204}Pb signal was swamped by the gas blank. The achievable accuracy and precision on the ages for (igneous) allanite lies within the same range as for LA–ICP-MS analyses of zircon. Small sizes and higher common Pb contents for metamorphic allanite restrict the precision attainable.

The **columbite-tantalite** [$(Fe,Mn)(Ta,Nb)_2O_6$] solid solution series of minerals mainly occurs within specific pegmatites, so their usefulness for geochronological purposes is limited. Also, columbite–tantalite U–Pb standards do not exist, and in the one study pertaining to LA–MC-ICP-MS dating of this mineral [80], a monazite standard was used. The old age of the samples (>2.5 Ga) meant that most reliance was placed on the $^{207}Pb/^{206}Pb$ ages, which should have suffered least from non-matrix-matched standardization. Material analyzed by both TIMS and LA–MC-ICP-MS gave the same U–Pb intercept age within error (2649 ± 11 versus 2664 ± 6) Ma, suggesting that even these ages suffered little from the use of a monazite instead of matrix-matched standard. Some finer details of the comparison between the two techniques may have been lost in the reverse discordance of the samples.

Perovskite ($CaTiO_3$) is found within silica-undersaturated igneous rocks, and its dating can be useful to obtain ages for kimberlite pipes, of interest for diamond exploration. Again, well-characterized standards are unavailable, but it has been claimed that accurate dating can be performed using zircon as bracketing standard [81]. Rastering combined with simultaneous aspiration of a tracer solution and the intercept method, without any external normalization, has been the other way to perform LA–Q-ICP-MS analyses [82]. Common Pb is also a problem for perovskite, especially since it cannot be safely assumed that the common Pb will conform to the crustal growth curve. Correction by regression through the data points in a Tera–Wasserburg diagram was performed, but a restricted spread of the sample points and/or relatively large scatter around the regression line limited the precision of the intercept ages in both studies, at $\sim$10% relative at the 1 σ level.

Baddeleyite (ZrO_2) occurs in silica-poor rocks in which zircon may be absent. It generally has low common Pb and high U contents and is therefore very well suited to dating. It may be the one mineral for which external standardization with zircon could work, considering that zircon itself breaks down to baddeleyite with most commonly used lasers [40]. It has been analyzed by LA–Q-ICP-MS with zircon standardization [83], giving an $^{207}Pb/^{206}Pb$ age indistinguishable from that for zircons from the same mafic dyke sample (1353 $\pm$ 14 versus 1345 $\pm$ 12 Ma). As these zircons were discordant and baddeleyite reversely discordant, it is unfortunately not possible to assess whether the zircon normalization gave the correct U–Pb ages for baddeleyite.

The one study on **uraninite** (UO_{2+x}) and **davidite** [$(La,Ce)(Y,U,Fe^{2+})(Ti,Fe^{3+})_{20}(O,OH)_{38}$] [84] is slightly unusual that it claimed the need to use a sector field ICP-MS instrument operated at medium mass resolution to avoid overlap of the signals of the polyatomic ion $^{204}(Pb,Hg)^{31}P^{+}$ and $^{235}U^{+}$ in a mineral without essential phosphorus. Another uncommon point is that no corrections for elemental fractionation and instrumental mass discrimination were applied for U/Pb age calculation from 10 to 50 μm diameter ablation spots. Although the reported highly discordant ages are likely to represent true geological processes, data treatment may not have helped to obtain concordant ages.

9.3.1.3 **Solution Pb–Pb Dating**

LA–ICP-MS dating seems to have largely replaced other methods of radiogenic Pb-based dating by ICP-MS, but its very stable mass bias compared with TIMS could also have advantages for solution-based dating. Analysis of different phases from a single rock to obtain a Pb–Pb isochron by solution analysis has the advantage that, prior to full dissolution, leaching techniques can be applied to remove any extraneous (common) Pb. Isolation and purification of the Pb fraction from the dissolved sample by column chemistry have the advantage of avoiding matrix-induced changes in mass bias. In this way, potentially very accurate and precise Pb–Pb isochron ages can be obtained, provided that there is a large enough spread in initial U/Pb ratios and/or old ages. This approach has been used successfully on chondritic meteorites [85], but not on much else – probably because the extra precision and accuracy seldom weigh up against the far greater investment of time and effort compared with LA-based U/Th-Pb dating.

9.3.2
Lu–Hf System

Hf isotopic analysis by TIMS is plagued by the need for extreme purification of the Hf fraction and, even then, difficulties with ionization. The superior ionization efficiency in the ICP has made Hf isotopic analysis a far more attainable option. On the other hand, the ease with which U–Pb zircon analyses can be done by LA–ICP-MS has caused a considerable lessening of interest in other geochronological techniques that date approximately the same temperature event. Nevertheless, Lu–Hf dating still has its use for dating relatively low-temperature, high-pressure events, during which little new zircon grew, but where metamorphic garnet formed, as in eclogite metamorphism, or for growth of phosphates in either sedimentary or metamorphic rocks. Another point of interest in the Lu–Hf system is perhaps not strictly geochronological (although model ages can be obtained), but pertains to Hf isotopic analysis in zircon using LA for sample introduction.

Uncertainty surrounded the exact decay constant of ^{176}Lu, necessary to obtain accurate ages, for several years, with values obtained by the analysis of meteorites different from those of terrestrial rocks [86]. It seems most likely that the extra-terrestrial materials used may have been subject to isotopic disturbances [5], and the "terrestrial" value of 1.867×10^{-11} (Table 9.1) is now generally accepted.

Unlike the situation for the U–Pb systems, where the enrichment of U over Pb is generally so high that single-collector ICP-MS analysis may suffice, Hf isotopic compositions can only be determined using MC instrumentation with sufficient precision to be of any use.

9.3.2.1 **Lu–Hf Isochrons with Garnet**

The Lu–Hf system is based on the radioactive decay of ^{176}Lu, with lutetium being one of the heavy rare earth elements (HREEs), to ^{176}Hf, with Hf being a high field strength element with a close chemical affinity to Zr. Materials that are most useful for Hf isochron dating are those rich in Lu and relatively poor in Hf. The mineral garnet ($X_3^{2+}Y_2^{3+}Si_3O_{12}$) is well known for its affinity for the HREEs, and although its exact Lu/Hf ratio depends on the whole rock composition and the other minerals present, garnet is generally the mineral with the highest Lu/Hf ratios, typically 1–100, whereas whole rocks have values <0.1. It is not possible to define a single closure temperature for the Lu–Hf system in garnet, as closure temperatures are grain size dependent, and garnet grain sizes vary much more than for, for example, zircon or other accessory phases. It is deemed to be slightly higher than that for the Sm–Nd isotopic system [87], at around 600–750 °C.

The variable composition of garnet means that it can form in a range of rock types, from pegmatites to mantle rocks and most metamorphosed crustal lithologies. As said, its main importance lies in dating metamorphic events during which there was no new growth of U/Th-Pb phases; another point of interest is that inclusion patterns in garnet can help to establish the timing of its formation relative to deformation of the rock [88]; therefore, dating garnet growth can also yield a (minimum or maximum) age for specific deformation events [89].

As the signals of ^{176}Lu and ^{176}Hf overlap irresolvably during mass spectrometric analyses, Lu–Hf dating can most reliably be done after dissolution of the sample and isolation of the Hf fraction [90], while a split will also need to be analyzed for Lu/Hf ratios, which is most accurately done by adding a spike enriched in one of the less abundant isotopes of each of the elements [91]. Complete equilibration between spike and sample is critical, and necessitates high-pressure (Parr bomb) digestion for garnet-bearing samples [92]. The simultaneous correction for mass bias and spike addition to permit the determination of isotope ratios and Hf concentrations from a single mass spectrometric run can be done mathematically, for instance by bracketing unknowns by standards [93].

The isochron method relies on isotopic equilibrium between the points that define the line, which can be a problem. One can either use minerals that were formed during the same metamorphic event (e.g., garnet and omphacite during eclogite facies metamorphism [94]), or assume that garnet is in equilibrium with the whole rock composition. However, this almost certainly does not hold for the zircon fraction that might be present in eclogitic LT–HP (low temperature–high pressure) metamorphic rocks, as zircon is likely to have retained the Hf isotopic composition of its (igneous) crystallization, and it also is the phase that tends to be the richest in Hf, as a result of the geochemical similarities between Hf and Zr. It can therefore be a good idea to dissolve the whole rock in such a way that zircon remains as a solid phase. Considering the very resistant nature of zircon, this is fairly easily achieved by HF/HNO_3 dissolution in capped vials, rather than high-pressure bombs [90]. A similar zircon-related problem may arise if this accessory mineral occurs as inclusions in garnet [87], which can be avoided by careful handpicking of the material to be analyzed.

9.3.2.2 Lu–Hf on Phosphates

A further application of Lu–Hf dating that yields information unavailable by the fast and simple method of U–Pb LA–ICP-MS geochronology is dating of (biogenic) phosphates. The advantage of the use of phosphates for Lu–Hf analysis is that they contain virtually no Ti, an element that was reported to hinder physically Hf isotopic analysis due to Ti oxide plating of the cones during MC-ICP-MS analysis [90], so the purification procedure can be simplified [95]. Moreover, phosphates are easy to dissolve and have relatively high $^{176}Lu/^{177}Hf$ ratios (1–100).

Lu–Hf isochrons on phosphorites or fossils give ages that are close to or younger than depositional ages of the sediments in which they are found [95,96]. Enamel or crystalline phosphorites give relatively good results, but bone material and dentine can be affected by post-depositional processes. In addition to the uncertainty of the accuracy of the age, the precision also is not always very good, with better results for Pb–Pb isochrons on the same (Neoproterozoic) samples [96]. Therefore, the method may be of most use for relatively young phosphates lacking significant U.

9.3.2.3 Zircon Hf Isotopic Model Ages

The mineral zircon is not only used for dating by U–Pb, but can also be targeted for Hf isotopic analysis. Its Lu–Hf geochemistry is more or less the opposite of that of garnet as it has a strong preference for Hf rather than Lu. By virtue of this Lu/Hf ratio close to zero, the Hf isotopic signature of the zircon is virtually identical with that of the magma from which it crystallized at the time the crystallization took place. The near absence of Lu (and Yb) in zircon permits the determination of the Hf isotopic composition by LA–MC-ICP-MS [97] (see also Chapter 4), with only very small corrections necessary for the isobaric overlap of the signals of ^{176}Yb, ^{176}Lu, and ^{176}Hf. Techniques have even been developed for the simultaneous determination of $^{207}Pb/^{206}Pb$ and the Hf isotopic composition during LA–MC-ICP-MS [98,99].

The geochronological information that can be obtained from such Hf isotopic analysis is a so-called model age (Figure 9.6), representing the moment at which the zircon would have formed from a reservoir with an Hf isotopic composition similar to the depleted mantle (DM) or chondritic uniform reservoir (CHUR), the latter being a model for the whole Earth.

These model mantle extraction ages can be interpreted either as a minimum age at which the material separated from the depleted mantle or chondrite uniform

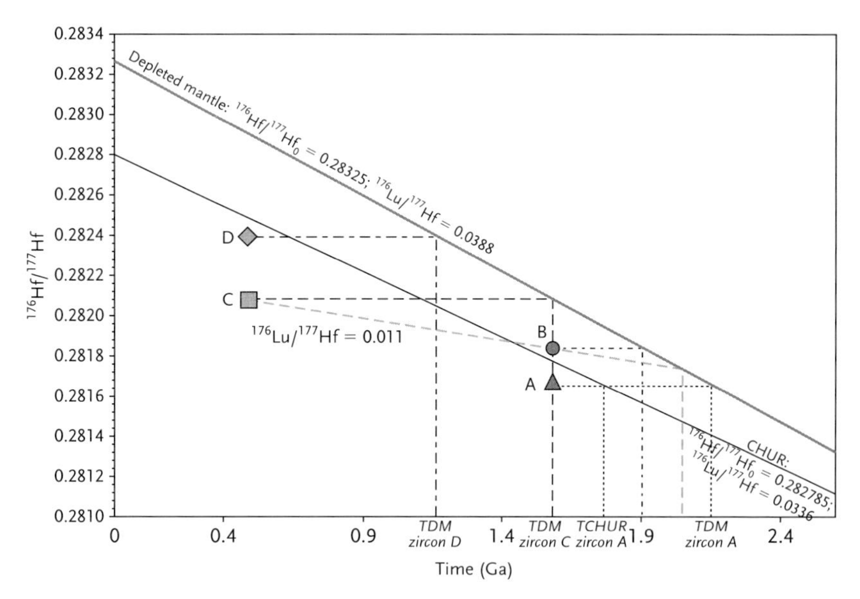

Figure 9.6 Concept of model ages for the Lu–Hf isotopic system in zircon. The labeled solid curves give the development of the $^{176}Hf/^{177}Hf$ ratio over time for depleted mantle and the CHUR. As Lu is more compatible than Hf during mantle melting, the crust has lower Lu/Hf ratios than CHUR, and CHUR again lower than depleted mantle, explaining the different slopes of the evolution lines. Zircons A and B have crystallization ages of 1.6 Ga (determined by U–Pb dating), and zircons C and D 0.45 Ga. All zircons have $^{176}Lu/^{177}Hf$ ratios close to zero, so their Hf isotopic composition will remain the same when projected back in time. For zircon A, we can define both a T_{CHUR} (age at which material with this particular $^{176}Hf/^{177}Hf$ and $^{176}Lu/^{177}Hf$ ratio would have separated from CHUR) and a T_{DM} (the same, but then for separation from the depleted mantle). As zircon B has a higher $^{176}Hf/^{177}Hf$ ratio than CHUR at the time of its formation, we can only define a T_{DM} (~1.9 Ga). Zircon C, which crystallized at 0.45 Ga, has a T_{DM} of 1.6 Ga. However, material with the Hf isotopic composition of zircon C can also be produced by starting out with material that has a $^{176}Hf/^{177}Hf$ ratio similar to zircon B and a $^{176}Lu/^{177}Hf$ ratio of 0.011 (typical of the granitic rocks in which zircon B resides) and allowing this to evolve between 1600 and 450 Ma (thick dashed line). The whole rock composition represented by this thick dashed line would have a T_{DM} of ~2 Ga. This illustrates that the depleted mantle extraction ages of zircons typically yield a minimum age, as a result of the zircons' Lu/Hf ratios being close to zero. Zircon D has a T_{DM} of ~1.1, but could also represent crystallization at 450 Ma with a contribution from material with an Hf isotopic composition similar to zircon C and material from the depleted mantle. If this is the case, the T_{DM} represents a mixed age, with contributions from older (crustal) material and zero age (at 450 Ma) mantle melts.

reservoir, or as a mixed age, indicating the relative contributions of older crustal materials and mantle melts [100,101]. The rather wide range of possible interpretations limits their usefulness to demonstrating what ages are not permitted by the data. Alternatively, they can be used for broad, continent-scale studies of crust-forming events, using large populations of zircons, for which ages are plotted on probability density plots, and for which peaks are interpreted to reflect magmatic crust-building events [102].

9.3.3 Re(-Pt)–Os System

The most abundant isotope of rhenium, ^{187}Re, decays to stable ^{187}Os, one of the less abundant (<2%) Os isotopes. The decay constant for ^{187}Re is 1.666×10^{-11} year^{-1} [6]. Despite this rather small decay constant, the contrast in relative abundances between ^{187}Re and ^{187}Os in terrestrial reservoirs means that observed variations in $^{187}Os/^{188}Os$ are large. The other radiogenic isotope of Os, ^{186}Os, is produced by the decay of ^{190}Pt, with a decay constant that is an order of magnitude smaller than that of ^{187}Re. This, combined with the fact that ^{190}Pt is the least abundant (0.0129%) of the Pt isotopes, means that observed variations in $^{186}Os/^{188}Os$ are rather small, unless dealing with Pt alloys.

In terms of geochemical behavior, the Re–Os system is completely different from the U/Th-Pb, Lu–Hf, Rb–Sr, and Sm–Nd systems, as Os is strongly compatible (and thereby remains in the residue) during mantle melting, whereas all the other elements mentioned are incompatible (albeit to varying extents). Hence the abundance of Os in the continental crust is exceedingly low (0.041 μg kg^{-1} [103]), and Re is also not very abundant (0.19 μg kg^{-1}). The Re–Os system is therefore generally applied to mantle rocks, mostly to obtain modeled Re depletion ages; and to selected crustal materials, such as molybdenites or black shales, that can yield isochron ages or even single mineral ages.

Osmium isotopic analysis has typically been performed by negative (N)-TIMS, whereby osmium trioxide ions are formed, but the accuracy of these analyses can potentially be compromised by interferences from other polyatomic species and uncertainty concerning the exact isotopic composition of the oxygen involved in oxide formation [104]. However, it remains the method of choice for small samples [105], as MC-ICP-MS analysis using Faraday collectors necessitates the use of >50 ng of sample, although far smaller amounts can be used if ion counters are used instead [106]. Whether N-TIMS or ICP-MS analysis is planned, solution analysis suffers from the added disadvantage of the need to dissolve the samples, equilibrate the materials with a spike, and isolate the rhenium and osmium fractions. Digestion generally involves the use of a quartz Carius tube with aqua regia or CrO_3 and H_2SO_4 at 200–300 °C, or fusion (see [107,108] for an overview of techniques), while isolation of osmium involves solvent extraction [109] and/or distillation [110]. These technical complications are overcome by using LA as a sample introduction technique, although the isobaric overlap of ^{187}Re on ^{187}Os restricts its usefulness to low Re/Os samples or very old materials.

9.3.3.1 **Re–Os Molybdenite Dating**

Molybdenite (MoS_2) is a mineral that occurs in various ore deposits and can thereby date (hydrothermal) mineralization, as opposed to igneous or high-grade metamorphic events dated by most other isotope systems. Molybdenite has a great affinity for Re, with concentrations generally in the mg kg^{-1} rather than the μg kg^{-1} range [111]. Os, on the other hand, is virtually excluded from the crystal lattice at the time of formation, so in this respect molybdenites are for the Re–Os system what zircon is for the U–Pb system. It is therefore possible to obtain single mineral ages from molybdenites, monitoring ^{192}Os to correct for initial (or, in the terminology corresponding to the U–Pb system, "common") Os, for which a $^{187}Os/^{188}Os$ ratio then needs to be assumed. If initial Os contents are suitably low, the exact composition that is assumed makes very little difference; for higher initial Os contents, the composition will need to be established by other means (e.g., isochron dating).

The use of LA as a means of sample introduction has been tested in molybdenite dating [112], but the inhomogeneity of most molybdenites, with decoupling of Re from Os, causes problems in this approach. An additional problem may be dimer formation, with $^{94}Mo^{94}Mo^+$ and $^{92}Mo^{95}Mo^+$ signals causing spectral interference with the measurement of $^{188}Os^+$ on $^{187}Os^+$, respectively. Therefore, molybdenites are analyzed after separation from their silicate matrix, homogenization, dissolution, and isolation of Re and Os fractions. Despite this heterogeneity, molybdenites are fairly stable, and molybdenites of varying ages have even been used to refine the decay constant for ^{187}Re [6].

As nearly all Os is radiogenic, the concentration of Os in molybdenites depends on its initial Re content and its age; it typically lies in the range from 10 μg kg^{-1} to 5 mg kg^{-1} [6,111,113,114]. The low concentrations would be difficult to measure by MC-ICP-MS using Faraday cups for attainable sample sizes, and much of the published molybdenite geochronological literature therefore consists of N-TIMS data. However, the extreme enrichment of ^{187}Os compared with the other osmium isotopes means that the precision of the analyses does not need to be very high to obtain reliable ages, and it is therefore possible to use single-collector ICP-MS instruments to achieve this goal, using spiked purified fractions of Os and Re [113,115]. Comparison of ages with those obtained by U–Pb dating of zircons from associated intrusives illustrates the accuracy of this dating method [115].

9.3.3.2 **Re–Os Dating of Black Shales**

Black organic-rich shales also have a tendency to concentrate Re over Os, albeit not to the same extent as molybdenite, so isochron rather than single sample dating needs to be performed. The ages obtained are interpreted as the depositional age of the sediments and the initial $^{187}Os/^{188}Os$ ratio as that of seawater at the time of deposition [116].

Re–Os dating of black shales needs to be done after digestion, spiking, and isolation of the spiked Re and Os fractions. For these isochrons, it is important to avoid inclusion of detrital Os-bearing material as much as possible, as their composition may vary from one sample to the next, giving scatter around the isochron. For this purpose, the aqua regia dissolution technique appears to be less suitable than the CrO_3–H_2SO_4 method [117].

The Re and Os concentrations of the shales are variable, between 0.3–1500 ng kg^{-1} for Re and 0.04–1.2 ng kg^{-1} for Os, with $^{187}Re/^{188}Os$ ratios of 50–1000 [117–120]. Again, the small amounts of Os makes it more difficult to carry out the analysis by MC-ICP-MS using Faraday cups, but the rather high $^{187}Os/^{188}Os$ ratios make it possible to obtain meaningful ages with single-collector ICP-MS [119].

9.3.3.3 Pt-Re–Os on Mantle Peridotites

As Os is compatible in the mantle, whereas Re has an incompatibility similar to that of Al or Lu [121], melting of the mantle will lead to a very strong decrease in the Re/Os ratio. Although this could be used to obtain isochrons from variably melt-depleted mantle rocks (peridotites), this is generally unsuccessful, as a result of the relatively high mobility of Re, resulting in Re addition ("refertilization") after melt depletion.

Most osmium in mantle peridotites is contained within small sulfide and alloy grains, typically with grain sizes between 20 and 150 μm. The work by Alard *et al.* [122] showed that sulfide grains from a single sample can have contrasting relative abundances of platinum group elements (PGEs: Os, Ir, Re, Rh, Pt, Pd), with one group of sulfides occurring enclosed within silicate minerals and the other between the silicate minerals. The former were interpreted to be the residues of partial melting, the latter later metasomatic additions. This gave the impetus to set up LA–MC-ICP-MS (Pt–)Re–Os dating of these different types of sulfides [123] and alloys [124]. One of the main challenges with Re–Os analysis is the isobaric overlap of the signals from ^{187}Re and ^{187}Os, which has to be corrected for by monitoring ^{185}Re. No Re–Os sulfide reference materials are available, so induced mass bias for Re has to be determined prior to LA by solution analysis [123,124]. Whether the mass bias thus determined also holds for the heavier plasma loading during LA analysis is a matter of contention. Worse still, the potential of elemental fractionation of Re and Os during laser ablation, as seen in the U–Pb system, is basically unstudied, but it is believed to be responsible for at least some discrepancies between Re–Os and Pt–Os ages [124]. To what upper limit of $^{187}Re/^{188}Os$ ratios the solution-based overlap corrections can still be applied to LA analysis is a matter of debate, with the most conservative cut-off set at 0.5 [124], whereas other laboratories are more optimistic [123]. This, together with the uncertainty concerning elemental fractionation, plus the relatively high mobility of Re, limits the reliability of Re–Os isochron dating of sulfides and alloys by LA–MC-ICP-MS at present.

Encouraging results for LA–MC-ICP-MS Pt–Os isochron dating of Pt-rich alloys were reported by Nowell *et al.* [124]. As Pt–Os follows an α-decay scheme, there is at least no isobaric overlap of the parent isotope on the daughter, but in this case, the signal of $^{186}W^+$ overlaps with that of the daughter and that of $^{190}Os^+$ with that of the parent isotope. This can be corrected for by monitoring ^{182}W and ^{188}Os. The Pt–Os method is probably slightly more robust than the Re–Os method, as Pt and Os are both PGEs and are therefore more likely to show similar behavior during LA than Re and Os. Nevertheless, overlap corrections are still based on solution analysis, which may not be completely accurate, but this will affect the position, rather than the slope, of the isochron [124], preserving age information, although initial ratios may be affected. Another advantage of Pt–Os over Re–Os dating is that the initial $^{186}Os/^{188}Os$ ratio of the mantle is likely to be more homogeneous than the $^{187}Os/^{188}Os$ ratio, so scatter around the isochron is limited, even for materials that were not necessarily in initial isotopic equilibrium. The very long half-life of ^{190}Pt makes this scheme most suited for geologically rather old materials, such as the ~2 Ga alloys that were dated in the study by Nowell *et al.* [124].

Like Hf isotope ratios, $^{187}Os/^{188}Os$ ratios can be used to derive model ages (Figure 9.7), in this case for residues of mantle melting, such as some sulfides or alloys are believed to be. The mantle extraction age (T_{MA}) is defined as

$$T_{MA} = \frac{1}{\lambda_{187}}\left[1 + \frac{(^{187}Os/^{188}Os)_{sample} - (^{187}Os/^{188}Os)_{CHUR}}{(^{187}Re/^{188}Os)_{sample} - (^{187}Re/^{188}Os)_{CHUR}}\right]$$

where λ_{187} is the ^{187}Re decay constant of 1.667×10^{-11} [6] and CHUR refers to the chondritic uniform reservoir (mantle), with $^{187}Re/^{188}Os = 0.40186$ and present-day $^{187}Os/^{188}Os = 0.127$ [125]. This model age uses the measured $^{187}Re/^{188}Os$ and $^{187}Os/^{188}Os$ ratios of a sample to back-calculate when it had the same $^{187}Os/^{188}Os$

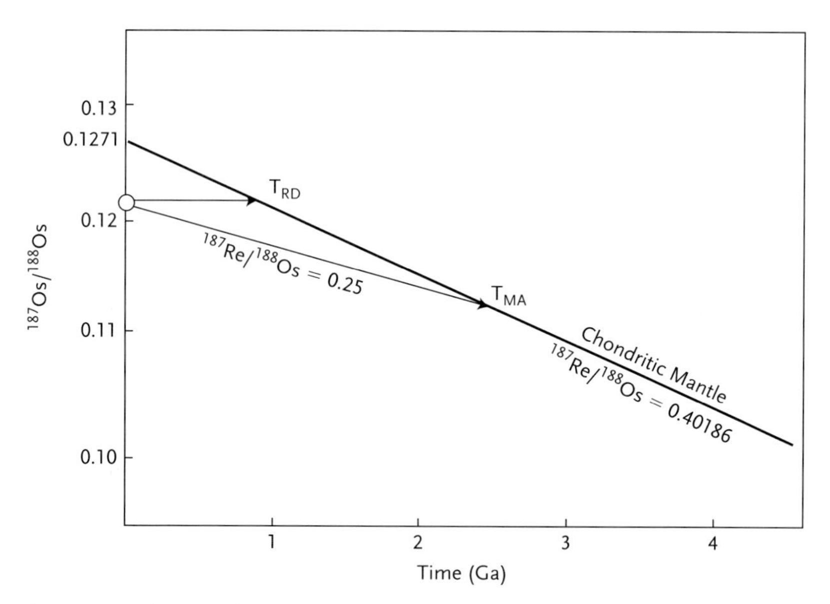

Figure 9.7 Schematic illustration of the concept of mantle extraction ages (T_{MA}) and rhenium depletion ages (T_{RD}) for a residue of partial melting, of which the present day $^{187}Os/^{188}Os$ ratio is indicated by the circle on the y-axis. The mantle extraction age uses the measured $^{187}Re/^{188}Os$ ratio to calculate the age at which the $^{187}Os/^{188}Os$ ratio of the sample was the same as that of the chondritic mantle. The rhenium depletion age assumes that all rhenium was removed during the partial melting event, and that any rhenium present in the sample now was added recently.

ratio as the chondrite uniform reservoir, believed to be a good approximation of that of the mantle. This age can be interpreted as reflecting a partial melting event, of which the analyzed material is a residue (in case the Re/Os ratio is lower than that of CHUR). As Re is relatively mobile, and can be added to or lost from the residue at a time post-dating the partial melting event, it is often preferred to calculate a rhenium depletion age (T_{RD}; Figure 9.7), defined as

$$T_{RD} = \frac{1}{\lambda_{187}}\left[1 + \frac{(^{187}Os/^{188}Os)_{sample} - (^{187}Os/^{188}Os)_{CHUR}}{0 - (^{187}Re/^{188}Os)_{CHUR}}\right]$$

In this case, it is assumed that the residue lost all of its rhenium during the partial melting event, and that any rhenium present was added recently, and therefore did not contribute any radiogenic ^{187}Os. The rhenium depletion age is always younger than the mantle extraction age.

These model ages permit age information to be obtained from materials with, ideally, very low Re/Os ratios, so that corrections for isobaric overlap of ^{187}Re and ^{187}Os signals during LA–MC-ICP-MS analyses are minor. LA–MC-ICP-MS Os isotopic analysis of large numbers of sulfide and alloy grains from mantle xenoliths or xenocrysts [126] and ophiolites [127] can thus be used to obtain probability density diagrams of rhenium depletion ages to date depletion events in the lithosphere, thought to give information on the age of crust formation.

9.4 Systems for Which ICP-MS Analysis Brings Fewer Advantages

Rb–Sr dating by the isochron method can be performed by MC-ICP-MS, but yields few advantages over the use of TIMS. Similar purification procedures are necessary as for TIMS, because of the overlap of $^{87}Rb^+$ and $^{87}Sr^+$ signals. The advantage of TIMS is that any remaining Rb in the Sr fraction can be burned off at lower temperatures, before starting the actual analysis of the Sr isotopes. On the other hand, the more reproducible mass bias of ICP-MS means that a correction can be applied by monitoring ^{85}Rb. The presence of Ca in the Sr fraction does not inhibit ionization, as is a problem with TIMS analyses. However, the Ar gas used for the plasma generally contains measurable amounts of Kr, which has isotopes with masses of 84 and 86, interfering with the isobaric Sr isotopes. This can be corrected for by analyzing a blank prior to analysis. For small sample sizes, TIMS remains the analytical method of choice.

Overlap of ^{144}Sm on ^{144}Nd means that **Sm–Nd** isochron dating is also best done after isolation of the Nd fraction, as for TIMS, although interferences can be corrected for reasonably successfully [128]. Speed of analysis is a great advantage compared with TIMS, but there is some doubt whether MC-ICP-MS analyses are as accurate as TIMS analyses [129]. This problem is likely to be compounded when spiked analyses are attempted. Fairly successful analyses of Nd isotope ratios by LA–MC-ICP-MS have been reported [130,131], but these have been geared towards isotopic fingerprinting rather than geochronological use. Considering that the magnitude of the Sm interference correction increases with increasing Sm/Nd ratio, and that high-Sm/Nd materials are most useful for dating purposes, the LA approach may not be ideally suited for Sm–Nd geochronology.

Acknowledgments

Comments on this chapter were provided by John Foden, Tom Andersen, Knut-Endre Sjåstad, Ingrid Smet, and Stijn Glorie, whose efforts are much appreciated.

References

1 Faure, G. (1986) *Principles of Isotope Geology*, 2nd edn., John Wiley & Sons, Imc., New York, p. 589.

2 Dickin, A.P. (2005) *Radiogenic Isotope Geology*, 2nd edn., Cambridge University Press, Cambridge, p. 508.

3 Allègre, C.J. (2008) *Isotope Geology*, Cambridge University Press, Cambridge, p. 507.

4 Steiger, R.H. and Jäger, E. (1977) Subcommission on geochronology: convention on the use of decay constants in geo- and cosmochronology. *Earth Planet. Sci. Lett.*, **36**, 359–362.

5 Bouvier, A., Vervoort, J.D., and Patchett, P.J. (2008) The Lu–Hf and Sm–Nd isotopic composition of CHUR: constraints from unequilibrated chondrites and implications for the bulk composition of terrestrial planets. *Earth Planet. Sci. Lett.*, **273** (1–2), 48–57.

6 Selby, D., Creaser, R.A., Stein, H.J., Markey, R.J., and Hannah, J.L. (2007) Assessment of the Re-187 decay constant by cross calibration of Re–Os molybdenite and U–Pb zircon chronometers in magmatic ore systems. *Geochim. Cosmochim. Acta*, **71** (8), 1999–2013.

7 Begemann, F., Ludwig, K.R., Lugmair, G.W., Min, K., Nyquist, L.E., Patchett, P.J., Renne, P.R., Shih, C.Y., Villa, I.M., and Walker, R.J. (2001) Call for an improved set of decay constants for geochronological use. *Geochim. Cosmochim. Acta*, **65** (1), 111–121.

8 Dunstan, L.P., Gramlich, J.W., Barnes, I.L., and Purdy, W.C. (1980) Absolute isotopic abundance and the atomic weight of a reference sample of thallium. *J. Res. Natl. Bur. Stand.*, **85** (1), 1–10.

9 Verschure, R.H., Andriessen, P.A.M., Boelrijk, N., Hebeda, E.H., Maijer, C., Priem, H.N.A., and Verdurmen, E.A.T. (1980) On the thermal stability of Rb–Sr and K–Ar biotite systems – evidence from coexisting Sveconorwegian (Ca 870 Ma) and Caledonian (Ca 400 Ma) biotites in SW Norway. *Contrib. Mineral. Petrol.*, **74** (3), 245–252.

10 Ehlers, K., Powell, R., and Stüwe, K. (1994) Cooling rate histories from garnet plus biotite equilibrium. *Am. Mineral.*, **79** (7–8), 737–744.

11 Villa, I.M. (1998) Isotopic closure. *Terra Nova*, **10** (1), 42–47.

12 Barling, J. and Weis, D. (2008) Influence of non-spectral matrix effects on the accuracy of Pb isotope ratio measurement by MC-ICP-MS: implications for the external normalization method of instrumental mass bias correction. *J. Anal. At. Spectrom.*, **23** (7), 1017–1025.

13 Ludwig, K. (2008) *User's Manual for Isoplot 3.6, a Geochronological Toolkit for Microsoft Excel*, Berkeley Geochronology Center Special Publication 4, Berkeley Geochronology Center, Berkeley, CA, p. 78.

14 Elburg, M.A., Bons, P.D., Dougherty-Page, J., Janka, C.E., Neumann, N., and Schaefer, B. (2001) Age and metasomatic alteration of the Mt Neill Granite at Nooldoonooldoona Waterhole, Mt Painter Inlier, South Australia. *Aust. J. Earth Sci.*, **48**, 721–730.

15 Schoene, B., Crowley, J.L., Condon, D. J., Schmitz, M.D., and Bowring, S.A. (2006) Reassessing the uranium decay constants for geochronology using ID-TIMS U–Pb data. *Geochim. Cosmochim. Acta*, **70** (2), 426–445.

16 Krogh, T.E. (1982) Improved accuracy of U–Pb zircon ages by the creation of more concordant systems using an air abrasion technique. *Geochim. Cosmochim. Acta*, **46** (4), 637–649.

17 Compston, W., Williams, I.S., and Meyer, C. (1984) U–Pb geochronology of zircons from Lunar Breccia 73217 using a sensitive high mass-resolution ion microprobe. *J. Geophys. Res.*, **89** (Suppl. S2), B525–B534.

18 Wilde, S.A., Valley, J.W., Peck, W.H., and Graham, C.M. (2001) Evidence from detrital zircons for the existence of continental crust and oceans on the Earth 4.4 Gyr ago. *Nature*, **409** (6817), 175–178.

19 Feng, R., Machado, N., and Ludden, J. (1993) Lead geochronology of zircon by laser probe–inductively coupled plasma-mass spectrometry (LP–ICPMS). *Geochim. Cosmochim. Acta*, **57** (14), 3479–3486.

20 Fryer, B.J., Jackson, S.E., and Longerich, H.P. (1993) The application of laser-ablation microprobe–inductively coupled plasma-mass spectrometry (LAM–ICP-MS) to *in-situ* (U)–Pb geochronology. *Chem. Geol.*, **109** (1–4), 1–8.

21 Simonetti, A., Heaman, L.M., Chacko, T., and Banerjee, N.R. (2006) *In situ* petrographic thin section U–Pb dating of zircon, monazite, and titanite using laser ablation–MC-ICP-MS. *Int. J. Mass Spectrom.*, **253** (1–2), 87–97.

22 Hanchar, J.M. and Miller, C.F. (1993) Zircon zonation patterns as revealed by cathodeluminesence and backscattered electron images: implications for interpretation of complex crustal histories. *Chem. Geol.*, **110**, 1–13.

23 Williams, I.S. (2001) Response of detrital zircon and monazite, and their U–Pb isotopic systems, to regional metamorphism and host-rock partial melting, Cooma Complex, southeastern Australia. *Aust. J. Earth Sci.*, **48** (4), 557–580.

24 Cottle, J.M., Horstwood, M.S.A., and Parrish, R.R. (2009) A new approach to single shot laser ablation analysis and its application to *in situ* Pb/U geochronology. *J. Anal. At. Spectrom.*, **24** (10), 1355–1363.

25 Horn, I., Rudnick, R.L., and McDonough, W.F. (2000) Precise elemental and isotope ratio determination by simultaneous solution nebulization and laser ablation–ICP-MS: application to U–Pb geochronology. *Chem. Geol.*, **164** (3–4), 281–301.

26 Mank, A.J.G. and Mason, P.R.D. (1999) A critical assessment of LA–ICP-MS as an analytical tool for depth analysis in silica-based glass samples. *J. Anal. At. Spectrom.*, **14** (8), 1143–1153.

27 Freydier, R., Candaudap, F., Poitrasson, F., Arbouet, A., Chatel, B., and Dupré, B. (2008) Evaluation of infrared femtosecond laser ablation for the analysis of geomaterials by ICP-MS. *J. Anal. At. Spectrom.*, **23** (5), 702–710.

28 Kosler, J., Fonneland, H., Sylvester, P., Tubrett, M., and Pedersen, R.B. (2002) U–Pb dating of detrital zircons for sediment provenance studies – a comparison of LA–ICPMS and SIMS techniques. *Chem. Geol.*, **182** (2–4), 605–618.

29 Paton, C., Woodhead, J.D., Hellstrom, J.C., Hergt, J.M., Greig, A., and Maas, R. (2010) Improved laser ablation U–Pb zircon geochronology through robust downhole fractionation correction. *Geochem. Geophys. Geosyst.*, **11**, Q0AA06.

30 Jackson, S.E., Pearson, N.J., Griffin, W.L., and Belousova, E.A. (2004) The application of laser ablation–inductively coupled plasma-mass spectrometry to *in situ* U–Pb zircon geochronology. *Chem. Geol.*, **211** (1–2), 47–69.

31 Wiedenbeck, M., Alle, P., Corfu, F., Griffin, W.L., Meier, M., Oberli, F., Vonquadt, A., Roddick, J.C., and Speigel, W. (1995) 3 natural zircon standards for U/Th-Pb, Lu–Hf, trace-element and REE analyses. *Geostand. Newsl.*, **19** (1), 1–23.

32 Black, L.P., Kamo, S.L., Allen, C.M., Aleinikoff, J.N., Davis, D.W., Korsch, R.J., and Foudoulis, C. (2003) TEMORA 1: a new zircon standard for Phanerozoic U–Pb geochronology. *Chem. Geol.*, **200** (1–2), 155–170.

33 Black, L.P., Kamo, S.L., Allen, C.M., Davis, D.W., Aleinikoff, J.N., Valley, J.W., Mundil, R., Campbell, I.H., Korsch, R.J., Williams, I.S., and Foudoulis, C. (2004) Improved $^{206}Pb/^{238}U$ microprobe geochronology by the monitoring of a trace-element-related matrix effect; SHRIMP, ID-TIMS, ELA–ICP-MS and oxygen isotope documentation for a series of zircon standards. *Chem. Geol.*, **205**, 115–140.

34 Black, L.P. and Gulson, B.L. (1978) The age of the Mud Tank carbonatite, Strangways Range, Northern Territory. *BMR J. Aust. Geol. Geophys.*, **3**, 227–232.

35 Slama, J., Kosler, J., Condon, D.J., Crowley, J.L., Gerdes, A., Hanchar, J.M., Horstwood, M.S.A., Morris, G.A., Nasdala, L., Norberg, N., Schaltegger, U., Schoene, B., Tubrett, M.N., and Whitehouse, M.J. (2008) Plesovice zircon – a new natural reference material for U–Pb and Hf isotopic microanalysis. *Chem. Geol.*, **249** (1–2), 1–35.

36 Black, L.P., Kamo, S.L., Williams, I.S., Mundil, R., Davis, D.W., Korsch, R.J., and Foudoulis, C. (2003) The application of SHRIMP to Phanerozoic geochronology, a critical appraisal of four zircon standards. *Chem. Geol.*, **200**, 171–188.

37 Pearce, N.J.G., Perkins, W.T., Westgate, J.A., Gorton, M.P., Jackson, S.E., Neal, C.R., and Chenery, S.P. (1997) A compilation of new and published major and trace element data for NIST SRM 610 and NIST SRM 612 glass reference materials. *Geostand. Newsl. J. Geostand. Geoanal.*, **21** (1), 115–144.

38 Orihashi, Y., Nakai, S., and Hirata, T. (2008) U–Pb age determination for seven standard zircons using inductively coupled plasma-mass spectrometry coupled with frequency quintupled Nd-YAG (lambda = 213 nm) laser ablation system: comparison with LA–ICP-MS zircon analyses with a NIST glass reference material. *Resour. Geol.*, **58** (2), 101–123.

39 Kuhn, B.K., Birbaum, K., Luo, Y., and Gunther, D. (2010) Fundamental studies on the ablation behaviour of Pb/U in NIST 610 and zircon 91500 using laser ablation inductively coupled plasma mass spectrometry with respect to geochronology. *J. Anal. At. Spectrom.*, **25** (1), 21–27.

40 Kosler, J., Wiedenbeck, M., Wirth, R., Hovorka, J., Sylvester, P., and Mikova, J. (2005) Chemical and phase composition of particles produced by laser ablation of silicate glass and zircon – implications for elemental fractionation during ICP-MS analysis. *J. Anal. At. Spectrom.*, **20** (5), 402–409.

41 Jackson, S.E. (2009) Correction of fractionation in LA–ICP-MS elemental and U–Pb analysis. *Geochim. Cosmochim. Acta*, **73** (13), A579.

42 Klotzli, U., Klotzli, E., Gunes, Z., and Koslar, J. (2009) Accuracy of laser ablation U–Pb zircon dating: results from a test using five different reference zircons. *Geostand. Geoanal. Res.*, **33** (1), 5–15.

43 Horn, I. and Gunther, D. (2003) The influence of ablation carrier gases Ar, He and Ne on the particle size distribution and transport efficiencies of laser ablation-induced aerosols: implications for LA–ICP-MS. *Appl. Surf. Sci.*, **207** (1–4), 144–157.

44 Tiepolo, M. (2003) *In situ* Pb geochronology of zircon with laser ablation-inductively coupled plasma-sector field mass spectrometry. *Chem. Geol.*, **199** (1–2), 159–177.

45 Tunheng, A. and Hirata, T. (2004) Development of signal smoothing device for precise elemental analysis using laser ablation–ICP-mass spectrometry. *J. Anal. At. Spectrom.*, **19** (7), 932–934.

46 Frei, D. and Gerdes, A. (2009) Precise and accurate *in situ* U–Pb dating of zircon with high sample throughput by automated LA–SF-ICPMS. *Chem. Geol.*, **261**, 261–270.

47 Muller, W., Shelley, M., Miller, P., and Broude, S. (2009) Initial performance metrics of a new custom-designed ArF excimer LA–ICPMS system coupled to a two-volume laser-ablation cell. *J. Anal. At. Spectrom.*, **24** (2), 209–214.

48 Andersen, T., Andersson, U.B., Graham, S., Aberg, G., and Simonsen, S.L. (2009) Granitic magmatism by melting of juvenile continental crust: new constraints on the source of Palaeoproterozoic granitoids in Fennoscandia from Hf isotopes in zircon. *J. Geol. Soc.*, **166**, 233–247.

49 Buhn, B., Pimentel, M.M., Matteini, M., and Dantas, E.L. (2009) High spatial resolution analysis of Pb and U isotopes for geochronology by laser ablation multi-collector inductively coupled plasma mass spectrometry (LA–MC-ICP-MS). *An. Acad. Brasil. Cienc.*, **81** (1), 99–114.

50 Gehrels, G.E., Valencia, V.A., and Ruiz, J. (2008) Enhanced precision, accuracy, efficiency, and spatial resolution of U–Pb ages by laser ablation–multicollector-inductively coupled plasma-mass spectrometry. *Geochem. Geophys. Geosyst.*, **9**, Q03017.

51 Cocherie, A. and Robert, M. (2008) Laser ablation coupled with ICP-MS applied to U–Pb zircon geochronology: a review of recent advances. *Gondwana Res.*, **14** (4), 597–608.

52 Simonetti, A., Heaman, L.M., Hartlaub, R.P., Creaser, R.A., MacHattie, T.G., and Bohm, C. (2005) U–Pb zircon dating by laser ablation–MC-ICP-MS using a new multiple ion counting Faraday collector array. *J. Anal. At. Spectrom.*, **20** (8), 677–686.

53 Chang, Z.S., Vervoort, J.D., McClelland, W.C., and Knaack, C. (2006) U–Pb dating of zircon by LA–ICP-MS. *Geochem. Geophys. Geosyst.*, **7** Q05009.

54 Hirata, T., Iizuka, T., and Orihashi, Y. (2005) Reduction of mercury background on ICP-mass spectrometry for *in situ* U–Pb age determinations of zircon samples. *J. Anal. At. Spectrom.*, **20** (8), 696–701.

55 Stacey, J.S. and Kramers, J.D. (1975) Approximation of terrestrial lead isotope evolution by a two-stage model. *Earth Planet. Sci. Lett.*, **26**, 207–221.

56 Horstwood, M.S.A., Foster, G.L., Parrish, R.R., Noble, S.R., and Nowell, G.M. (2003) Common-Pb corrected *in situ* U–Pb accessory mineral geochronology by LA–MC-ICP-MS. *J. Anal. At. Spectrom.*, **18** (8), 837–846.

57 Williams, I. (1998) U/Th-Pb geochronology by ion microprobe, in *Application of Microanalytical Techniques to Understanding Mineralizing Processes*, Reviews in Economic Geology, vol. 7 (eds. M.A. McKibben, W.C. Shanks, and W.I. Ridley), Society of Economic Geologists, Littleton, CO, pp. 1–35.

58 Tera, F. and Wasserburg, G.J. (1972) U/Th-Pb systematics in 3 Apollo 14 basalts and problem of initial Pb in lunar rocks. *Earth Planet. Sci. Lett.*, **14** (3), 281.

59 Andersen, T. (2002) Correction of common lead in U–Pb analyses that do not report Pb-204. *Chem. Geol.*, **192** (1–2), 59–79.

60 Cox, R.A., Wilton, D.H.C., and Kosler, J. (2003) Laser-ablation U/Th-Pb *in situ* dating of zircon and allanite: an example from the October Harbour granite, central coastal Labrador, Canada. *Can. Mineral.*, **41**, 273–291.

61 Griffin, W.L., Powell, W.J., Pearson, N.J., and O'Reilly, S.Y. (2008) *Glitter: data reduction software for LA–ICP-MS, in Laser Ablation ICP–MS in the Earth Sciences: Current Practices and Outstanding Issues, Short Course Series,* vol. **40** (ed. P. Sylvester), Mineralogical Association of Canada, Quebec, pp. 308–311.

62 Dunkl, I., Mikes, T., Simon, K., and von Eynatten, H. (2007) Data handling, outlier rejection and calculation of isotope concentrations from laser ICP-MS analyses by PEPITA software. *Geochim. Cosmochim. Acta*, **71** (15), A243.

63 Kosler, J., Forst, L., and Slama, J. (2008) LAMDATE and LAMTOOL: spreadsheet-based data reduction for LA–ICP-MS, in Laser Ablation ICP–MS in the Earth Sciences: Current Practices and Outstanding Issues, Short Course Series, vol. 40 (ed. P. Sylvester), Mineralogical Association of Canada, Quebec, pp. 315–317.

64 Jackson, S.E. (2008) LAMTRACE data reduction software for LA–ICP-MS, in Laser Ablation ICP–MS in the Earth Sciences: Current Practices and Outstanding Issues, Short Course Series, vol. 40 (ed. P. Sylvester), Mineralogical Association of Canada, Quebec, pp. 305–307.

65 Gregory, C.J., Rubatto, D., Allen, C.M., Williams, I.S., Hermann, J., and Ireland, T. (2007) Allanite micro-geochronology: a LA–ICP-MS and SHRIMP U/Th-Pb study. *Chem. Geol.*, **245** (3–4), 162–182.

66 Poitrasson, F., Chenery, S., and Bland, D.J. (1996) Contrasted monazite hydrothermal alteration mechanisms and their geochemical implications. *Earth Planet. Sci. Lett.*, **145** (1–4), 79–96.

67 Schaltegger, U., Fanning, C.M., Gunther, D., Maurin, J.C., Schulmann, K., and Gebauer, D. (1999) Growth, annealing and recrystallization of zircon and preservation of monazite in high-grade metamorphism: conventional and *in-situ* U–Pb isotope, cathodoluminescence and microchemical evidence. *Contrib. Mineral. Petrol.*, **134** (2–3), 186–201.

68 Parrish, R.R. (1990) U–Pb dating of monazite and its application to geological problems. *Can. J. Earth Sci.*, **27** (11), 1431–1450.

69 Rubatto, D., Williams, I.S., and Buick, I.S. (2001) Zircon and monazite response to prograde metamorphism in the Reynolds Range, central Australia. *Contrib. Mineral. Petrol.*, **140** (4), 458–468.

70 Paquette, J.L. and Tiepolo, M. (2007) High resolution (5 μm) U/Th-Pb isotope dating of monazite with excimer laser ablation (ELA)–ICPMS. *Chem. Geol.*, **240** (3–4), 222–237.

71 Kohn, M.J. and Vervoort, J.D. (2008) U/Th-Pb dating of monazite by single-collector ICP-MS: pitfalls and potential. *Geochem. Geophys. Geosyst.*, **9**, Q04031.

72 Schärer, U. (1984) The effect of initial Th-230 disequilibrium on young U–Pb ages – the Makalu case, Himalaya. *Earth Planet. Sci. Lett.*, **67** (2), 191–204.

73 Seydoux-Guillaume, A.M., Wirth, R., Deutsch, A., and Schärer, U. (2004) Microstructure of 24–928Ma concordant monazites; implications for geochronology and nuclear waste deposits. *Geochim. Cosmochim. Acta*, **68** (11), 2517–2527.

74 Dahl, P.S. (1997) A crystal-chemical basis for Pb retention and fission-track annealing systematics in U-bearing minerals, with implications for geochronology. *Earth Planet. Sci. Lett.*, **150** (3–4), 277–290.

75 Vry, J.K. and Baker, J.A. (2006) LA–MC-ICPMS Pb–Pb dating of rutile from slowly cooled granulites: confirmation of the high closure temperature for Pb diffusion in rutile. *Geochim. Cosmochim. Acta*, **70** (7), 1807–1820.

76 Willigers, B.J.A., Baker, J.A., Krogstad, E.J., and Peate, D.W. (2002) Precise and accurate *in situ* Pb–Pb dating of apatite, monazite, and sphene by laser ablation multiple-collector ICP-MS. *Geochim. Cosmochim. Acta*, **66** (6), 1051–1066.

77 Storey, C.D., Smith, M.P., and Jeffries, T.E. (2007) *In situ* LA–ICP-MS U–Pb dating of metavolcanics of Norrbotten, Sweden: records of extended geological histories in complex titanite grains. *Chem. Geol.*, **240** (1–2), 163–181.

78 Li, J.W., Deng, X.D., Zhou, M.F., Lin, Y.S., Zhao, X.F., and Guo, J.L. (2010) Laser ablation ICP-MS titanite U/Th-Pb dating of hydrothermal ore deposits:

a case study of the Tonglushan Cu–Fe–Au skarn deposit, SE Hubei Province, China. *Chem. Geol.*, **270** (1–4), 56–67.

79 Banerjee, N.R., Simonetti, A., Furnes, H., Muehlenbachs, K., Staudigel, H., Heaman, L., and Van Kranendonk, M.J. (2007) Direct dating of Archean microbial ichnofossils. *Geology*, **35** (6), 487–490.

80 Smith, S.R., Foster, G.L., Romer, R.L., Tindle, A.G., Kelley, S.P., Noble, S.R., Horstwood, M., and Breaks, F.W. (2004) U–Pb columbite–tantalite chronology of rare-element pegmatites using TIMS and laser ablation–multi collector-ICP-MS. *Contrib. Mineral. Petrol.*, **147** (5), 549–564.

81 Batumike, J.M., Griffin, W.L., Belousova, E.A., Pearson, N.J., O'Reilly, S.Y., and Shee, S.R. (2008) LAM–ICPMS U–Pb dating of kimberlitic perovskite: Eocene–Oligocene kimberlites from the Kundelungu Plateau, DR Congo. *Earth Planet. Sci. Lett.*, **267** (3–4), 609–619.

82 Cox, R.A. and Wilton, D.H.C. (2006) U–Pb dating of perovskite by LA–ICP-MS: an example from the Oka carbonatite, Quebec, Canada. *Chem. Geol.*, **235** (1–2), 21–32.

83 Zhang, S.H., Zhao, Y., Yang, Z.Y., He, Z.F., and Wu, H. (2009) The 1.35Ga diabase sills from the northern North China Craton: implications for breakup of the Columbia (Nuna) supercontinent. *Earth Planet. Sci. Lett.*, **288** (3–4), 588–600.

84 Chipley, D., Polito, P.A., and Kyser, T.K. (2007) Measurement of U–Pb ages of uraninite and davidite by laser ablation–HR-ICP-MS. *Am. Mineral.*, **92** (11–12), 1925–1935.

85 Bouvier, A., Blichert-Toft, J., Moynier, F., Vervoort, J.D., and Albarède, F. (2007) Pb–Pb dating constraints on the accretion and cooling history of chondrites. *Geochim. Cosmochim. Acta*, **71** (6), 1583–1604.

86 Soderlund, U., Patchett, J.P., Vervoort, J.D., and Isachsen, C.E. (2004) The Lu-176 decay constant determined by Lu–Hf and U–Pb isotope systematics of Precambrian mafic intrusions. *Earth Planet. Sci. Lett.*, **219** (3–4), 311–324.

87 Scherer, E.E., Cameron, K.L., and Blichert-Toft, J. (2000) Lu–Hf garnet geochronology: closure temperature relative to the Sm–Nd system and the effects of trace mineral inclusions. *Geochim. Cosmochim. Acta*, **64** (19), 3413–3432.

88 Passchier, C.W. and Trouw, R.A.J. (2005) *Microtectonics*, 2nd edn., Springer, Berlin.

89 Wolf, D.E., Andronicos, C.L., Vervoort, J.D., Mansfield, M.R., and Chardon, D. (2010) Application of Lu–Hf garnet dating to unravel the relationships between deformation, metamorphism and plutonism: an example from the Prince Rupert area, British Columbia. *Tectonophysics*, **485** (1–4), 62–77.

90 Cheng, H., King, R.L., Nakamura, E., Vervoort, J.D., and Zhou, Z. (2008) Coupled Lu–Hf and Sm–Nd geochronology constrains garnet growth in ultra-high-pressure eclogites from the Dabie orogen. *J. Metamorph. Geol.*, **26** (7), 741–758.

91 Vervoort, J.D., Patchett, P.J., Soderlund, U., and Baker, M. (2004) Isotopic composition of Yb and the determination of Lu concentrations and Lu/Hf ratios by isotope dilution using MC-ICPMS. *Geochem. Geophys. Geosyst.*, **5**, Q11002.

92 Mahlen, N.J., Beard, B.L., Johnson, C.M., and Lapen, T.J. (2008) An investigation of dissolution methods for Lu–Hf and Sm–Nd isotope studies in zircon- and garnet-bearing whole-rock samples. *Geochem. Geophys. Geosyst.*, **9**, Q01002.

93 Lapen, T.J., Mahlen, N.J., Johnson, C.M., and Beard, B.L. (2004) High precision Lu and Hf isotope analyses of both spiked and unspiked samples: a new approach. *Geochem. Geophys. Geosyst.*, **5**, Q01010.

94 Thoni, M., Miller, C., Blichert-Toft, J., Whitehouse, M.J., Konzett, J., and Zanetti, A. (2008) Timing of high-pressure metamorphism and exhumation of the eclogite type-locality (Kupplerbrunn-Prickler Halt, Saualpe, south-eastern Austria): constraints from correlations of the Sm–Nd, Lu–Hf, U–Pb and Rb–Sr isotopic systems. *J. Metamorph. Geol.*, **26** (5), 561–581.

95 Barfod, G.H., Otero, O., and Albarède, F. (2003) Phosphate Lu–Hf geochronology. *Chem. Geol.*, **200** (3–4), 241–253.

96 Barfod, G.H., Albarède, F., Knoll, A.H., Xiao, S.H., Telouk, P., Frei, R., and Baker, J. (2002) New Lu–Hf and Pb–Pb age constraints on the earliest animal fossils. *Earth Planet. Sci. Lett.*, **201** (1), 203–212.

97 Griffin, W.L., Pearson, N.J., Belousova, E., Jackson, S.E., van Achterbergh, E., O'Reilly, S.Y., and Shee, S.R. (2000) The Hf isotope composition of cratonic mantle: LAM–MC-ICPMS analysis of zircon megacrysts in kimberlites. *Geochim. Cosmochim. Acta*, **64** (1), 133–147.

98 Woodhead, J., Hergt, J., Shelley, M., Eggins, S., and Kemp, R. (2004) Zircon Hf-isotope analysis with an excimer laser, depth profiling, ablation of complex geometries, and concomitant age estimation. *Chem. Geol.*, **209** (1–2), 121–135.

99 Kemp, A.I.S., Foster, G.L., Scherstén, A., Whitehouse, M.J., Darling, J., and Storey, C. (2009) Concurrent Pb–Hf isotope analysis of zircon by laser ablation multi-collector ICP-MS, with implications for the crustal evolution of Greenland and the Himalayas. *Chem. Geol.*, **261** (3–4), 244–260.

100 Scherer, E.E., Whitehouse, M.J., and Münker, C. (2007) Zircon as a monitor of crustal growth. *Elements*, **3** (1), 19–24.

101 Andersen, T., Griffin, W.L., and Sylvester, A.G. (2007) Sveconorwegian crustal underplating in southwestern Fennoscandia: LAM–ICPMS U–Pb and Lu–Hf isotope evidence from granites and gneisses in Telemark, southern Norway. *Lithos*, **93** (3–4), 273–287.

102 Kemp, A.I.S., Hawkesworth, C.J., Paterson, B.A., and Kinny, P.D. (2006) Episodic growth of the Gondwana supercontinent from hafnium and oxygen isotopes in zircon. *Nature*, **439** (7076), 580–583.

103 Rudnick, R.L. and Gao, S. (2003) Composition of the continental crust, in *The Crust* (ed. R.L. Rudnick) Elsevier, Oxford, pp. 1–64.

104 Luguet, A., Nowell, G.M., and Pearson, D.G. (2008) Os-184/Os-188 and Os-186/Os-188 measurements by negative thermal ionisation mass spectrometry (N-TIMS): effects of interfering element and mass fractionation corrections on data accuracy and precision. *Chem. Geol.*, **248** (3–4), 342–362.

105 Nowell, G.M., Luguet, A., Pearson, D.G., and Horstwood, M.S.A. (2008) Precise and accurate Os-186/Os-188 and Os-187/Os-188 measurements by multi-collector plasma ionisation mass spectrometry (MC-ICP-MS). Part I: solution analyses. *Chem. Geol.*, **248** (3–4), 363–393.

106 Schoenberg, R., Nagler, T.F., and Kramers, J.D. (2000) Precise Os isotope ratio and Re–Os isotope dilution measurements down to the picogram level using multicollector inductively coupled plasma mass spectrometry. *Int. J. Mass Spectrom.*, **197**, 85–94.

107 Qi, L. and Zhou, M.F. (2008) Determination of platinum-group elements in OPY-1: comparison of results using different digestion techniques. *Geostand. Geoanal. Res.*, **32** (3), 377–387.

108 Qi, L., Zhou, M.F., Gao, J.F., and Zhao, Z. (2010) An improved Carius tube technique for determination of low concentrations of Re and Os in pyrites. *J. Anal. At. Spectrom.*, **25** (4), 585–589.

109 Cohen, A.S. and Waters, F.G. (1996) Separation of osmium from geological materials by solvent extraction for analysis by thermal ionisation mass spectrometry. *Anal. Chim. Acta*, **332** (2–3), 269–275.

110 Shirey, S.B. and Walker, R.J. (1995) Carius tube digestion for low-blank rhenium–osmium analysis. *Anal. Chem.*, **67** (13), 2136–2141.

111 Stein, H.J., Markey, R.J., Morgan, J.W., Hannah, J.L., and Scherstén, A. (2001) The remarkable Re–Os chronometer in molybdenite: how and why it works. *Terra Nova*, **13** (6), 479–486.

112 Kosler, J., Simonetti, A., Sylvester, P.J., Cox, R.A., Tubrett, M.N., and Wilton, D.H.C. (2003) Laser-ablation ICP-MS measurements of Re/Os in molybdenite and implications for Re–Os geochronology. *Can. Mineral.*, **41**, 307–320.

113 Liu, J.M., Zhao, Y., Sun, Y.L., Li, D.P., Liu, J., Chen, B.L., Zhang, S.H., and Sun, W.D. (2010) Recognition of the latest Permian to early Triassic Cu–Mo mineralization on the northern margin of the North China block and its geological significance. *Gondwana Res.*, **17** (1), 125–134.

114 Conliffe, J., Selby, D., Porter, S.J., and Feely, M. (2010) Re–Os molybdenite dates from the Ballachulish and Kilmelford Igneous Complexes (Scottish Highlands): age constraints for late Caledonian magmatism. *J. Geol. Soc.*, **167** (2), 297–302.

115 Su, Y., Tang, H., Sylvester, P.J., Liu, C., Qu, W., Hou, G., and Cong, F. (2007) Petrogenesis of Karamaili alkaline A-type granites from East Junggar, Xinjiang (NW China) and their relationship with tin mineralization. *Geochem. J.*, **41** (5), 341–357.

116 Ravizza, G. and Turekian, K.K. (1989) Application of the Re-187–Os-187 system to black shale geochronometry. *Geochim. Cosmochim. Acta*, **53** (12), 3257–3262.

117 Kendall, B.S., Creaser, R.A., Ross, G.M., and Selby, D. (2004) Constraints on the timing of Marinoan "Snowball Earth" glaciation by Re-187–Os-187 dating of a Neoproterozoic, post-glacial black shale in Western Canada. *Earth Planet. Sci. Lett.*, **222** (3–4), 729–740.

118 Kendall, B., Creaser, R.A., and Selby, D. (2006) Re–Os geochronology of postglacial black shales in Australia: constraints on the timing of "Sturtian" glaciation. *Geology*, **34** (9), 729–732.

119 Pasava, J., Oszczepalski, S., and Du, A.D. (2010) Re–Os age of non-mineralized black shale from the Kupferschiefer, Poland, and implications for metal enrichment. *Miner. Deposita*, **45** (2), 189–199.

120 Singh, S.K., Trivedi, J.R., and Krishnaswami, S. (1999) Re–Os isotope systematics in black shales from the Lesser Himalaya: their chronology and role in the Os-187/Os-188 evolution of seawater. *Geochim. Cosmochim. Acta*, **63** (16), 2381–2392.

121 Rudnick, R.L. and Walker, R.J. (2009) Interpreting ages from Re–Os isotopes in peridotites. *Lithos*, **112**, 1083–1095.

122 Alard, O., Griffin, W.L., Lorand, J.P., Jackson, S.E., and O'Reilly, S.Y. (2000) Non-chondritic distribution of the highly siderophile elements in mantle sulphides. *Nature*, **407** (6806), 891–894.

123 Pearson, N.J., Alard, O., Griffin, W.L., Jackson, S.E., and O'Reilly, S.Y. (2002) *In situ* measurement of Re–Os isotopes in mantle sulfides by laser ablation multicollector-inductively coupled plasma mass spectrometry: analytical methods and preliminary results. *Geochim. Cosmochim. Acta*, **66** (6), 1037–1050.

124 Nowell, G.M., Pearson, D.G., Parman, S.W., Luguet, A., and Hanski, E. (2008) Precise and accurate Os-186/Os-188 and Os-187/Os-188 measurements by multi-collector plasma ionisation mass spectrometry. Part II: laser ablation and its application to single-grain Pt–Os and Re–Os geochronology. *Chem. Geol.*, **248** (3–4), 394–426.

125 Walker, R.J. and Morgan, J.W. (1989) Rhenium–osmium isotope systematics of carbonaceous chondrites. *Science*, **243** (4890), 519–522.

126 Griffin, W.L., Spetsius, Z.V., Pearson, N.J., and O'Reilly, S.Y. (2002) *In situ* Re–Os analysis of sulfide inclusions in kimberlitic olivine: new constraints on depletion events in the Siberian lithospheric mantle. *Geochem. Geophys. Geosyst.*, **3**, 1069.

127 Pearson, D.G., Parman, S.W., and Nowell, G.M. (2007) A link between large mantle melting events and continent growth seen in osmium isotopes. *Nature*, **449** (7159), 202–205.

128 Luais, B., Telouk, P., and Albarède, F. (1997) Precise and accurate neodymium isotopic measurements by plasma-source mass spectrometry. *Geochim. Cosmochim. Acta*, **61** (22), 4847–4854.

129 Debaille, V., Brandon, A.D., and Mattielli, N. (2009) Comparison of Nd isotope ratio measurements between MC-ICPMS and TIMS. 8th International Sector Field ICP-MS Conference: Ghent, Belgium, p. 61.

130 Foster, G.L. and Vance, D. (2006) *In situ* Nd isotopic analysis of geological materials by laser ablation MC-ICP-MS. *J. Anal. At. Spectrom.*, **21** (3), 288–296.

131 Gregory, C.J., McFarlane, C.R.M., Hermann, J., and Rubatto, D. (2009) Tracing the evolution of calc-alkaline magmas: *in-situ* Sm–Nd isotope studies of accessory minerals in the Bergell Igneous Complex, Italy. *Chem. Geol.*, **260**, 73–86.

10
Application of Multiple-Collector Inductively Coupled Plasma Mass Spectrometry to Isotopic Analysis in Cosmochemistry

Mark Rehkämper, Maria Schönbächler, and Rasmus Andreasen

10.1
Introduction

Cosmochemistry is a close relative of geochemistry in that both research fields share the analytical and data-based approach, which involves the acquisition and interpretation of chemical and isotopic data for relevant samples. But whereas geochemistry focuses on terrestrial samples to investigate the Earth and its internal reservoirs, research in cosmochemistry explores the evolution of the Universe, the solar system and its diverse planetary bodies. As such, cosmochemistry is also loosely related to the research fields of meteoritics and planetary science [1, 2].

Analysis of extraterrestrial materials, and in particular meteorites, is an important focus of cosmochemical research, as such samples preserve chemical and isotopic records of early solar system conditions and processes. The first studies of meteorites, which recognized that such samples have an extraterrestrial origin, date back to the late eighteenth century [3], but modern research in cosmochemistry has a much more recent origin. This is traced back by many to the founder of contemporary geochemistry, V.M. Goldschmidt, as he produced early, but well-founded, compilations of cosmic element abundances, based on data acquired for meteorites [4, 5]. Goldschmidt's work was later continued and extended by Suess in collaboration with Urey and their study on the abundances of the elements [5] is still an important milestone in cosmochemistry.

The development of modern sector field mass spectrometry, primarily driven by Nier [6], overlapped in time with these early advances in cosmochemistry and this enabled isotopic investigations to play a key role in cosmochemistry from the inception of the research field. Of particular initial importance was the age dating that could be achieved, based on the abundances of isotopes produced by long-lived radionuclides. The application of the U—Pb chronometer to establish the age of the Earth was pioneered by Houtermans and Holmes, using isotopic data originally acquired by Nier [7, 8]. More then a decade after these initial investigations. Patterson used a similar approach, but with refined methods and more suitable

Isotopic Analysis: Fundamentals and Applications Using ICP-MS,
First Edition. Edited by Frank Vanhaecke and Patrick Degryse.

Published 2012 by WILEY-VCH Verlag GmbH & Co. KGaA

samples, to carry out a landmark study which yielded the first accurate Pb–Pb isochron age for the solar system and the Earth at 4.55 ± 0.07Ga [9].

Long-lived radioactive decay systems still provide our only means for dating past events and processes on geological time scales, hence such methods continue to be a mainstay of cosmochemical research. Since the 1950s, it was also recognized, however, that extraterrestrial samples host an unusual wealth of isotopic variations and anomalies, which reflect sources and processes other than long-lived radioactive nuclides [2]. This includes anomalies (i) produced by now extinct radioactive isotopes that were "alive" only in the early solar system, (ii) resulting from reactions induced by high-energy cosmic rays, or (iii) caused by mass-dependent isotope fractionation. This isotopic diversity has been exploited in numerous studies, to further our understanding on questions ranging from the formation of the elements to the conditions that prevailed in the early solar nebula and the timing of planet formation and differentiation.

Given the importance of isotopic analysis to cosmochemical research, it is unsurprising that improvements in mass spectrometric methods and instrumentation have played a significant role in advancing the scientific field. The advent of multiple-collector (MC) inductively coupled plasma mass spectrometry (ICP-MS) in the early 1990s [10–13] is no exception in this regard. The success of this instrumental technique, which is now routinely used in well over 100 laboratories world-wide, has had an enormous effect on cosmochemistry, with many studies published over the last two decades that showcase novel methods and applications.

This chapter provides an overview of the impact that MC-ICP-MS has had on cosmochemical research. To this end, it encompasses (i) an introduction to the extraterrestrial samples, and in particular meteorites, which are analyzed; (ii) a brief explanation of the origin and significance of the most important types of isotopic anomalies that are measured; (iii) a brief introduction to the particular advantages that MC-ICP-MS provides for isotopic analysis in cosmochemistry and to common analytical procedures; and (iv) an overview of important applications of MC-ICP-MS in cosmochemistry. This last section highlights selected novel findings and their scientific significance, while also discussing particular analytical procedures and potential pitfalls.

10.2
Extraterrestrial Samples

10.2.1
Introduction

Although chemical and isotopic analysis by remote sensing using either Earth- or satellite-based instruments play an important role in some cosmochemical studies, laboratory measurements on "real" samples remain the predominant means of data acquisition. In cosmochemistry, extraterrestrial materials are the most common samples analyzed, while terrestrial samples are used mainly (i) as reference materials that provide a baseline to which results obtained for extraterrestrial materials are compared or (ii) for understanding the evolution of the early Earth as a planetary body.

Broadly, two types of extraterrestrial samples are available for laboratory analysis: the meteorites that we can find on Earth and materials that are collected on targeted sample return missions. Not all available samples are suitable for isotopic analysis by MC-ICP-MS, however, as such measurements are destructive and typically require between several milligrams to grams of material. In practice, this requirement rules out MC-ICP-MS analysis of particularly small or unique and precious samples, such as the tiny amounts of cometary and solar wind material that were collected during the NASA Stardust and Genesis Missions, respectively, or the sub-millimeter-sized inter-planetary dust particles (IDPs) which are sampled in the Earth's upper atmosphere using airplanes. Further current and planned space missions are aimed at returning more material from a range of extraterrestrial bodies, but such samples will continue to be available in only small quantities that can support only a limited number of destructive analyses.

In contrast, several thousand different meteorites have been identified and many of these have masses from several hundred grams to kilograms. The NASA Apollo missions furthermore brought almost 300 kg of lunar rocks and soils to Earth. These materials will continue to be the main focus of cosmochemical research in the coming decades, not only because they are available in sufficient quantities for destructive chemical and isotopic analysis, but also due to their scientific value. The chemically diverse lunar rocks and soils from seven Apollo landing sites in Maria (low-lying plains) on the Moon's near side continue to play a key role in establishing a better understanding of the Moon's formation and geochemical evolution. The meteorites are of even greater importance in many respects, as they provide samples from a large number of distinct extraterrestrial bodies and their analysis can hence shed light on a wide range of cosmochemical issues.

10.2.2
Classification of Meteorites

Meteorites are solid extraterrestrial materials that have survived the descent through the Earth's atmosphere. They are derived from *meteoroids*, which are natural objects with diameters of up to 10–100 m that orbit in space, while a *meteor* is the visible path of a meteoroid that enters the Earth's (or another body's) atmosphere (i.e., a shooting star). The meteoroids (and hence all meteorites) are ultimately derived from a parent body, from which they were separated by impact events that launched larger debris into new orbits. Most meteorite parent bodies appear to be in the asteroid belt, as can be inferred from orbits that were calculated for meteorites that were observed to fall [3]. Some meteorites have even been linked to specific asteroids, and there is good evidence that others are derived from Mars and the Moon (Table 10.1). As such, meteorites provide us with direct chemical and isotopic information on a large range of planetary bodies. Many meteorites furthermore have very old ages and petrological, mineralogical, and chemical characteristics that imply derivation from primitive bodies, which have experienced only limited secondary processing. The analysis of such meteorites can hence provide information on the conditions and processes that shaped the early solar system.

There are numerous ways to classify meteorites [14, 15]. Meteorites *finds*, which cannot be directly linked to a specific observed fall, are much more common than *falls*. The distinction between stones (silicate meteorites), iron meteorites, and stony-irons is straightforward but ignores the significant genetic differences amongst stony meteorites. More recent nomenclature systems make a primary distinction between chondrites and non-chondrites, whereby the latter group includes all types of igneously differentiated meteorites (Table 10.1).

Table 10.1 Meteorite groups, important examples, and additional information[a].

Categories/classes	Groups (and important examples)	Comments
Chondrites		
Carbonaceous chondrites	CI1 (Ivuna, Orgueil)	Volatiles-rich composition, best approximates solar composition (= bulk solar system)
	CM (Murchison), CO, CR, CB, CV (Allende, Efremovka), CK	Petrological types 1–3 are most common. CV3 chondrites rich in CAIs, the oldest dated objects in the solar system
Ordinary chondrites	H, L, LL	Petrological types 3–6; most common class of meteorites
Enstatite chondrites	EH, EL	Petrological types 3–6
Primitive non-chondrites	Acapulcoites, lodranites, winonaites	Record incipient melting of parent bodies.
Differentiated non-chondrites		Derived from molten and igneously differentiated parent bodies with metallic core and silicate mantle
Achondrites	HED meteorites: howardites, eucrites (Juvinas), diogenites	Vesta is thought to be the HED parent body
	SNC meteorites	From Mars
	Lunar meteorites	From the Moon; provide material from lunar regions (highlands) not sampled by Apollo missions
	Other groups: angrites, aubrites, ureilites	Rare achondrite groups
Stony irons	Pallasites, mesosiderites	Pallasites derived from core-mantle boundary of asteroid
Irons	IAB (Canyon Diablo, Toluca), IIICD	Non-magmatic irons
	IIAB, IIC, IID IIIAB, IIIE, IVA, IVB	Main groups of magmatic irons; from cores of asteroids

[a]The meteorite classification used here follows [14, 15].

10.2.3 Chondritic Meteorites

Chondrites represent the most primitive type of meteorite (Figure 10.1, Table 10.1) and they are derived from unmolten parent bodies that have not differentiated to produce a metallic core and silicate mantle. As such, they are still recognizable as direct products of condensation from the solar nebula [16, 17].

The main constituent of most chondrites are chondrules (Figure 10.1), which are nearly spherical igneous particles, typically of millimeter size, that were probably formed by rapid heating, melting, and quenching of silicates [16, 17]. In chondrites, chondrules are embedded in dark, fine-grained matrix material. Two further important constituents of chondrites are metallic Fe, Ni (and rarer sulfide) grains and refractory inclusions. The refractory Ca–Al-rich inclusions (CAIs) (Figure 10.1) are products of very early high-temperature processes and they are generally recognized as the oldest dateable objects in the solar system. Hence the ~4568 Ma formation age of CAIs [18, 19] is commonly used as an absolute age marker for the formation of the solar system. In addition, chondrites also contain very small, but apparently almost ubiquitous, presolar grains, most commonly in the form of diamonds, graphite, or SiC [20]. Such grains are thought to have

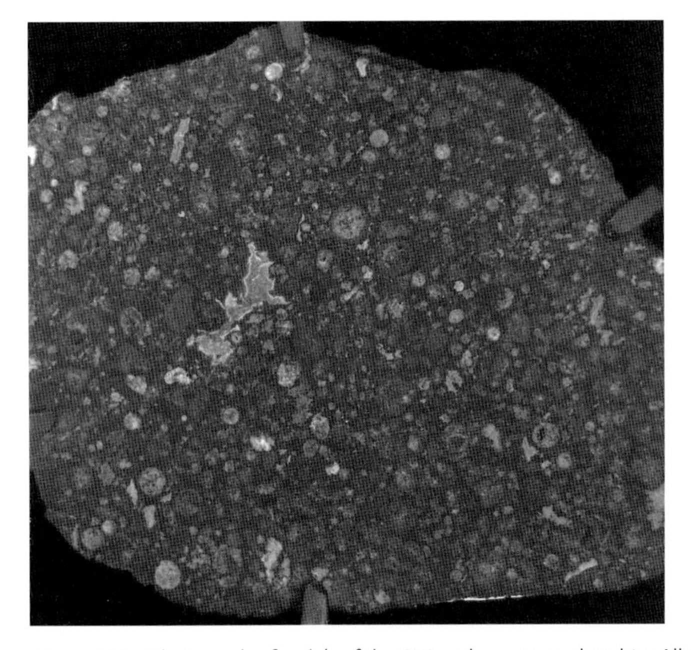

Figure 10.1 Photograph of a slab of the CV3 carbonaceous chondrite Allende at the American Museum of Natural History. The spherical chondrules and a large elongated calcium–aluminum-rich inclusion (CAI; just left of center) are embedded in a dark, fine-grained matrix. The chondrules typically have diameters of 1–2 mm.
Source: posted on Flickr by *Shiny Things* and used in accordance with a CC BY 2.0 license.

formed in the stars that contributed material to the molecular cloud from which the solar system formed. Presolar grains hence provide a direct window to processes that preceded the formation of the solar nebula, such as the nucleosynthetic (element-forming) reactions that occur in stars. Given that the place and mode of origin of chondrite components are still only poorly understood, it is not surprising that the formation of these meteorites remains a very active field of research today.

There are three main classes of chondrites (Table 10.1) that are most readily distinguished by whole-rock chemical and O isotopic compositions [14, 17]. The ordinary chondrites are the most abundant type of meteorite, as they account for about 80% of all meteorite falls. Enstatite chondrites have very reduced compositions and they are rich in reduced Fe and Fe-poor silicate phases, such as enstatite ($MgSiO_3$). Most carbonaceous chondrites have high contents of carbon and other volatile elements that only condensed into solids at very low temperatures in the solar nebula.

The three chondrite classes are subdivided into groups that are denoted by letters, for example, CI, CM, CO, H, L, or EH (Table 10.1), and each group is thought to originate from a different parent body. Numbers are added to the letters (e.g., CM2, CV3) to distinguish between different *petrological types* that denote the mode and extent of secondary processing on the asteroidal parent bodies [14, 15]. Type 3 chondrites display the least modification by secondary processes, while types 1 and 2, which are only found in carbonaceous chondrites, denote aqueous alteration from weak (2) to stronger (1). In contrast, the petrological types 4, 5, and 6, which are observed primarily in ordinary and enstatite chondrites, denote increasingly stronger thermal metamorphism. Chondrites of type 3 are commonly called *unequilibrated* because their mineralogy best reflects the original materials that presumably formed in different parts of the solar nebula and which were mixed during the accretion processes that built up small primitive asteroids. In contrast, chondritic meteorites with higher petrological grades 4–6 are denoted *equilibrated,* because they record increasing extents of thermal metamorphism, which induced petrological equilibration and recrystallization of minerals. The widely accepted onion-shell model states that the different petrological types of ordinary and enstatite chondrites correspond to the depth of burial on the meteorite parent body [17]. The unequilibrated (type 3) samples, which show little or no sign of thermal metamorphism, are derived from near the surface. The equilibrated samples of types 4, 5, and 6 are derived from increasing depths of the parent body. The peak metamorphic conditions of type 6 chondrites correspond to $T \approx 950°C$.

The carbonaceous chondrites are a particularly important meteorite group. Petrographically they record the lowest temperatures of thermal metamorphism ($\leq 150°C$) and the highest degree of aqueous alteration of all meteorites, particularly in CI1 chondrites. Hence almost all carbonaceous chondrites are of petrological types 1, 2, and 3 [15]. The chemical compositions render the carbonaceous chondrites especially noteworthy, as they have the highest abundances of volatile elements and are hence considered to feature the most pristine chemical compositions that most closely approximate the bulk composition of the solar system. This match is particularly striking for the CI carbonaceous chondrites, which have element abundances that closely match those determined for the Sun [17].

10.2.4 Non-Chondritic Meteorites

There are four main types of non-chondritic meteorites (Table 10.1). Primitive achondrites, such as the acapulcoites and lodranites, are thought to be from asteroids that experienced only incipient or limited melting (Table 10.1). In contrast, achondrites, iron meteorites, and stony-irons are considered to represent parent bodies that featured widespread melting processes, which ultimately led to planetary differentiation and the formation of a metallic core and a silicate-rich mantle and crust [14, 15].

Achondrites (Table 10.1) are silicate-rich mafic (basaltic) and ultramafic meteorites of igneous origin that are derived from the crust and mantle of differentiated asteroids and planets [14, 21, 22]. Amongst the most abundant achondrites are the so-called HED meteorites (howardites, eucrites, diogenites) that are all thought to be derived from a single parent asteroid, which is speculated to be the asteroid 4 Vesta, based on spectroscopic studies (Table 10.1). Two other prominent types of achondrites are (i) lunar meteorites, which are of particular interest as they provide access to rock types and localities (lunar highlands) not covered by the Apollo samples [15], and (ii) SNC meteorites, named after the Shergotty, Nakhla, and Chassigny finds, which are thought to come from Mars (Table 10.1).

Iron meteorites consist mainly of Fe–Ni metal with some troilite (FeS) inclusions. There are more than 10 different groups of irons, for example, IAB, IIAB, and IIIAB (Table 10.1), that are defined based on chemical differences, and each may be derived from a distinct parent body [22, 23]. The *magmatic iron meteorites* (Table 10.1), which encompass most groups of irons, are thought to represent the cores of differentiated asteroids. In contrast, the *non-magmatic* irons (IAB, IIICD) are probably derived from Fe-rich melts that formed on a chondritic parent body. The stony-iron meteorites are well described by their name, as they consist of Fe–Ni metal plus silicate minerals. Best known amongst this type are the *pallasites*, which are an assemblage of olivine and Fe–Ni metal, and they are thought to represent samples from the core–mantle boundary of differentiated asteroids (Table 10.1).

10.3 Origin of Cosmochemical Isotopic Variations

The extraterrestrial materials that are analyzed by cosmochemists commonly feature isotope anomalies for a much broader range of elements than terrestrial samples and this variability is generally ascribed to five principle sources, which are introduced below (see also Chapter 1).

10.3.1
Radiogenic Isotope Variations from the Decay of Long-Lived Radioactive Nuclides

Long-lived radioactive decay systems are based on unstable parent nuclides with half-lives that are sufficiently long, such that only part of the inventory that was present at the start of the solar system has since decayed away (Figure 10.2). Consequently, most long-lived radioactive parents feature half-lives that are within one to two orders of magnitude of the Earth's age, with values of about $T_{1/2} \approx 1$–100 Ga (Table 10.2). Such decay systems are ideally suited for studying events that occurred several million to billions of years ago and their application in geochronology provides the only means of obtaining absolute age information on geological time scales (see also Chapter 9).

10.3.2
Radiogenic Isotope Variations from the Decay of Extinct Radioactive Nuclides

Extinct radioactive decay systems are based on short-lived parent isotopes that were alive during the early history of the solar system. These radioactive parents are no longer present today, because they have essentially decayed away completely, as they feature short half-lives of only about 0.1–100 Ma (Figure 10.2, Table 10.2). Their former presence can be inferred from meteorites, based on abundance variations for the daughter isotopes that we can measure today. Differences in the isotopic compositions of the daughter elements can then provide very precise temporal constraints on processes that occurred in the earliest solar system and information on the kinds of stars that last added material to the molecular cloud from which the solar system formed.

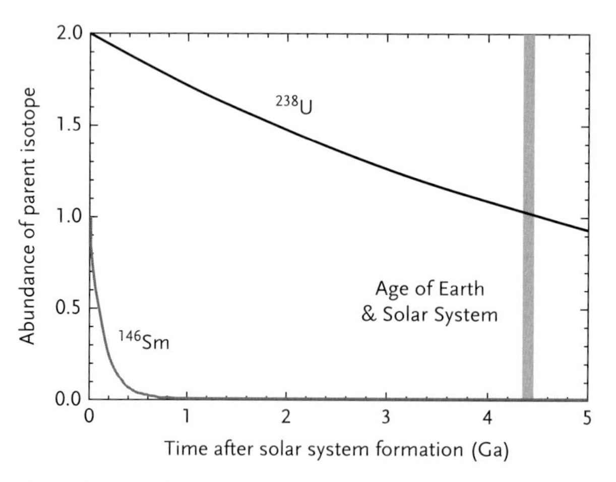

Figure 10.2 Radioactive decay diagram to illustrate the difference between long-lived and extinct radionuclides. Shown are the decay of long-lived ^{238}U ($T_{1/2} = 4.47$ Ga) and extinct ^{146}Sm ($T_{1/2} = 103$ Ma), which are assigned arbitrary initial abundances of 2 and 1, respectively. The age of the Earth (~4.47 Ga) and the solar system (4.57 Ga) are marked by the shaded field. Today, about 50% of the ^{238}U that was present at the start of the solar system (at $t = 0$) is still "alive," whereas essentially all ^{146}Sm has decayed to ^{142}Nd.

Table 10.2 Selected extinct and long-lived radionuclide systems that are used in cosmochemistry.

Parent	Daughter	Half-life
Extinct decay systems		
^{26}Al	^{26}Mg	0.73 Ma
^{60}Fe	^{60}Ni	2.6 Ma
^{53}Mn	^{53}Cr	3.7 Ma
^{107}Pd	^{107}Ag	6.5 Ma
^{182}Hf	^{182}W	9.0 Ma
^{205}Pb	^{205}Tl	15 Ma
^{129}I	^{129}Xe	16 Ma
^{92}Nb	^{92}Zr	36 Ma
^{146}Sm	^{142}Nd	103 Ma
Long-lived decay systems		
^{235}U	^{207}Pb	707 Ma
^{40}K	^{40}Ar, ^{40}Ca	1.28 Ga
^{238}U	^{206}Pb	4.47 Ga
^{232}Th	^{208}Pb	14 Ga
^{176}Lu	^{176}Hf	37 Ga
^{187}Re	^{187}Os	42 Ga
^{87}Rb	^{87}Sr	48 Ga
^{147}Sm	^{143}Nd	106 Ga

10.3.3 Nucleosynthetic Isotope Anomalies

Nucleosynthetic isotopic variations are caused by heterogeneities in the distribution of stable, non-radiogenic isotopes within the solar nebula, which reflect the incomplete mixing of material that is derived from distinct sources of element production.

Most elements of the periodic table, with the exception of H, He, and traces of a few other light elements (e.g., Li, B), are formed either within stars or during highly energetic supernova events. This occurs primarily by nuclear fusion in stellar cores for elements up to Fe and neutron capture and additional reactions for the heavier elements [24, 25]. The direct signatures of such nucleosynthetic processes are captured in presolar grains, which are found in tiny quantities in chondritic meteorites. Such grains can be isolated for analysis, by preferential dissolution of other meteorite components [20]. However, the same nucleosynthetic fingerprints, albeit in more dilute form, can also be detected in solutions that are obtained by partial dissolution (*leaching*) of meteorite phases or bulk meteorite samples.

The study of nucleosynthetic isotopic variations is of interest because such anomalies provide direct evidence of the sites and mechanisms of element production in stars. Furthermore, the detection of presolar grains places constraints on the thermal history of the solar nebula. Also, the survival of anomalous distributions of such grains, as demonstrated by the prevalence of nucleosynthetic isotope anomalies in individual meteorites or even asteroids, provides clues to the efficiency of the physical mixing processes that acted to erase such variations in the solar nebula.

10.3.4
Mass-Dependent Isotope Fractionation

Variations in isotopic compositions that are generated by isotope fractionation associated with chemical, physical, or (on Earth) biological processes are generally of *mass-dependent* nature. This implies that the magnitude of an isotope effect is proportional to the mass difference of the respective isotopes. Such mass-dependent isotope effects are hence generally most prevalent for lighter elements, which feature the largest relative differences in isotopic masses, and classic stable isotope studies were therefore focused on the elements H, C, N, O, and S. However, more recent studies, often conducted by MC-ICP-MS, have shown that natural isotope fractionation is also common for many heavier elements in both terrestrial rocks and meteorites [26, 27].

There are two principal pathways for creating mass-dependent variations in isotopic compositions. *Equilibrium* isotope effects arise from reversible isotope exchange reactions that establish thermodynamic equilibrium. In cosmochemistry, such effects may be relevant during condensation of solid matter at high gas pressures in the solar nebula or the segregation of liquid metal and silicate phases during planetary differentiation. In contrast, *kinetic* isotope effects are associated with irreversible, incomplete, or unidirectional processes, such as evaporation into vacuum, solid-state diffusion during cooling or (in terrestrial environments) biologically mediated reactions. Kinetically controlled processes often generate large isotope effects, whereby the reaction products are typically enriched in the light isotopes [26].

Many common processes and reactions generate isotope effects, hence there are also many possible applications. For example, isotope ratio data can be applied to obtain constraints on environmental conditions (e.g., temperatures, gas pressures) and reaction pathways. In addition, they are commonly used both as a material tracer and for the quantification of the mass fluxes [26].

10.3.5
Cosmogenic Isotope Anomalies

Cosmogenic isotope effects are generated by the interaction of matter with high-energy cosmic rays that are produced primarily by the Sun. Such reactions are induced not only by the cosmic rays, which consist largely of protons and α-particles (He nuclei), but also by the secondary particles, particularly neutrons, which are generated in primary reactions. In many cases, such reactions produce short-lived radioactive nuclides, such as ^{14}C and ^{10}Be. The abundance of such nuclides can be measured to determine the *cosmic ray exposure ages* of meteorites, which provide information on the time that meteorites were in space following the breakup of the parent body and until they arrived on Earth [28, 29]. In addition, such nuclides can also be used to infer the terrestrial ages of meteorite samples, which marks the time between the fall and discovery of a sample [30].

However, cosmogenic reactions can also produce small variations in the abundances of stable isotopes. Such anomalies can be difficult to resolve from radiogenic, nucleosynthetic, and stable isotope effects and, if undiscovered, can lead to erroneous data interpretations. Targeted measurements of isotopes that are generated at particularly high levels by such cosmogenic reactions (e.g., ^{150}Sm, $^{156-158}Gd$) have therefore been used as a monitor of possible cosmogenic isotope anomalies [28].

10.4 Use of MC-ICP-MS in Cosmochemistry

Most elements of the periodic table have been investigated for the presence of isotopic anomalies as part of various cosmochemical studies (Figure 10.3) and many have even been studied at great detail. For light elements that are either present as gases in samples or which are readily volatilized, such analyses are typically conducted by gas source mass spectrometry. In the past, isotopic analysis of metallic and metalloid elements were primarily carried out using either thermal ionization mass spectrometry (TIMS) or, for in situ measurements, secondary

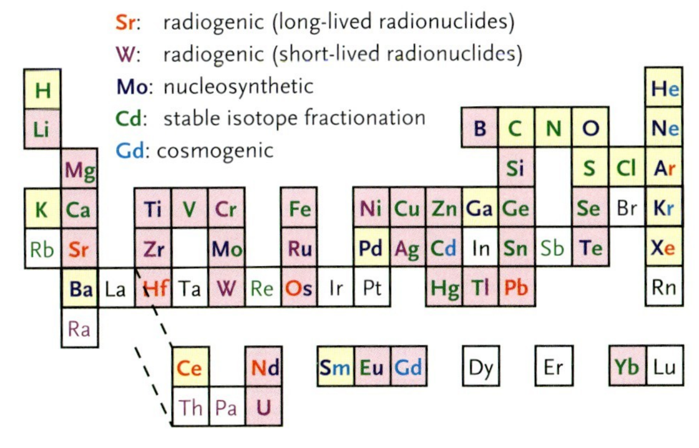

Figure 10.3 Periodic table of multi-isotopic elements. Colored elemental symbols are used for all elements for which radiogenic, nucleosynthetic, and/or mass-dependent isotopic variations have been investigated in natural samples. Different colors denote the origin of any measured (or sought) isotope anomalies. Fields with colored shading denote isotopic analyses in extraterrestrial materials whilst the red shading denotes cosmochemical isotopic analyses that were conducted by MC-ICP-MS. Only very few elements have not been studied to date and most elements that were studied displayed at least some natural isotopic variability.

ion mass spectrometry (SIMS). The advent of MC-ICP-MS has, however, had an enormous impact on isotopic studies of such elements in the last 20 years (Figure 10.3). This impact reflects that MC-ICP-MS is used alongside TIMS for isotopic analysis of some elements (e.g., Nd, Pb), whilst it has completely replaced TIMS for others (Hf). In addition, MC-ICP-MS has further allowed routine investigations of previously unstudied isotope systems, such as Mg and Fe.

In the following, we provide (i) a short overview of the particular advantages that the MC-ICP-MS technique features for isotopic analysis in cosmochemistry and (ii) a brief outline of the analytical techniques that are utilized in such measurements.

10.4.1 Specific Advantages of MC-ICP-MS

The ion optical and detector systems of MC-ICP-MS instruments are optimized for precise isotopic analysis via static multiple collection and, as in TIMS, they are suitable for the resolution of small isotopic anomalies at the 10–100 ppm level [31, 32]. In addition, the ICP ion source of MC-ICP-MS provides excellent ionization efficiencies for most elements of the periodic table. For example, MC-ICP-MS can achieve ion yields of up to about 0.5, 1, and 2%, when sample solutions of Zn, Cd, and Pb, respectively, are analyzed using a desolvating nebulizer. As such, the technique of MC-ICP-MS is extremely versatile, allowing precise isotopic measurements for a wide range of elements (Figure 10.3).

For some elements, TIMS can achieve ion yields that are superior to MC-ICP-MS, but in many cases, this requires the application of carefully optimized and time-consuming filament loading procedures and analytical conditions. The development and reproducible implementation of such methods is furthermore often extremely challenging and may require several steps of incremental optimization. In contrast, the solution nebulization methods that are most commonly used for sample introduction in MC-ICP-MS are straightforward and support rapid routine measurements. In addition, they are also undemanding, with respect to the development of new methods of analysis for novel isotope systems. The latter advantage is of particular importance for cosmochemical research, as this often investigates isotopic variations for a much wider range of elements than are utilized in geochemistry (Figure 10.3).

The high ionization efficiency that characterizes plasma source mass spectrometers also has drawbacks, however. In particular, isotopic analysis by MC-ICP-MS is more susceptible than TIMS measurements to spectral interferences from isobars, molecular ions, and doubly charged species. Such interferences are furthermore much more problematic for isotope ratio than for element concentration measurements, as isotopic compositions typically need to be measured with an accuracy and precision of better than $\sim$100 ppm. In order to reduce spectral interferences to tolerable levels, isotopic analyses of complex natural samples in geo- and cosmochemistry are generally carried out following separation of the analyte elements from the sample matrix. In most cases, such separations are achieved using chemical procedures, which involve ion-exchange or extraction chromatography [32].

The technique of *internal normalization* is commonly applied in both MC–ICP–MS and TIMS for the precise correction of the instrumental mass bias (see also Chapter 5) that is encountered during the analysis of radiogenic isotopic compositions [33, 34]. The ICP ion source of MC-ICP-MS, however, also features two characteristics that play an important role for isotopic analysis, where internal normalization cannot be applied. First, an ICP source operates at steady state and therefore mass fractionation is not primarily a time-dependent process, as in TIMS where the measured isotopic compositions change with time due to the progressive evaporation of a sample from the filament. The steady-state operation of an ICP ion source is beneficial for the correction of instrumental mass bias by *external standardization*, where the isotope ratio data obtained for a sample are referenced to the values obtained for bracketing analyses of an isotopic standard [27, 35]. Hence, this procedure is commonly termed *standard-sample bracketing*.

Second, the mass discrimination associated with plasma source mass spectrometry is, to first order, a relatively simple function of mass, such that elements with overlapping mass ranges display a nearly identical mass bias [27, 32, 35]. Using a solution containing a mixture of two elements with overlapping mass ranges, the mass discrimination observed for an element of known isotopic composition can be utilized to determine the unknown isotopic composition of the second element. This procedure of *external normalization* was first suggested by Longerich *et al.* [36] to improve the precision of Pb isotope ratio measurements via quadrupole-based ICP-MS, using Tl as the reference element. It has since been applied with success in numerous stable isotope analyses by MC-ICP-MS for elements such as Cu, Zn, Cd and others [37, 38]. Furthermore, it has also been used for radiogenic isotope measurements of elements that feature only two isotopes, such as Ag and Tl [39, 40], where internal normalization cannot be applied.

In addition to these analytical advantages, the technique of MC-ICP-MS features further general benefits, particularly ease of operation, high sample throughput, and reasonable capital and operating costs, which have made such instruments very popular for isotopic research in many geo- and cosmochemical laboratories.

10.4.2 Analytical Procedures

There are two principal means of sample introduction for isotopic analysis by MC-ICP-MS in geo- and cosmochemistry. Most common is the technique of solution nebulization, whilst laser ablation (LA) systems have been applied for *in situ* isotope ratio measurements [32]. As such, MC-ICP-MS complements both the TIMS and SIMS methodologies, which are the most common alterative methods for bulk sample and *in situ* isotopic analysis, respectively, in cosmochemical research.

The application of LA–MC-ICP-MS for isotopic measurements (see also Chapters 2 and 4) was first pioneered in the 1990s [41–43], but although the technique continues to be applied in several laboratories [44–46], it has not evolved to become a routine analytical method, despite significant improvements in LA instrumentation. The somewhat limited success of such LA techniques is a consequence of both the rigorous demands of isotopic analysis and the particular analytical characteristics of MC-ICP-MS. In many cases, particularly in cosmochemistry, isotopic measurements must feature an accuracy and precision of better than 100 ppm to be useful, but the acquisition of such data by LA–MC-ICP-MS are limited to elements that have very high concentrations (more than about 1000 ppm) in the relevant samples. Of additional concern are the abundant spectral interferences that are produced by plasma ionization and the matrix effects, which may have an impact on the instrumental mass bias. As a result, isotopic analyses via LA–ICP-MS are comparatively rare and its application in cosmochemistry has been limited to studies of elements that are major constituents of the phases analyzed. Examples of such applications are isotopic analysis of (i) Mg carried out on individual chondrules and CAIs from carbonaceous chondrites [47] and (ii) Hf in meteoritic zircon grains [48]. As such, SIMS remains the most commonly applied method for *in situ* isotopic analysis in cosmochemistry. The SIMS technique, however, suffers from some of the same disadvantages as LA—ICP-MS and its routine application is therefore also limited to isotopic measurements of lighter elements up to Si, which occur at relatively high concentrations in meteorite phases [20].

In contrast, the application of solution nebulization in conjunction with MC-ICP-MS now readily rivals or even surpasses the importance of TIMS for isotopic analysis of metallic and metalloid elements in bulk meteorite samples. Invariably, such analyses encompass (i) sample dissolution, most often by acid digestion, and (ii) a highly selective separation of the analyte element from the bulk sample matrix, typically by one or several stages of column chromatography. Such separations are carried out prior to isotopic analysis of both major elements (e.g., Mg, Si, Fe) and trace constituents, primarily to ensure that spectral interferences and matrix effects are reduced to either insignificant or at least tolerable levels. In addition, the chemical isolation procedure also acts as a preconcentration method, which allows the isotopic analysis to be carried using solutions that feature element concentrations that are optimized for precise data acquisition.

In practice, the purified elemental separates are dissolved in a small volume of dilute acid (in many cases 0.1–1M HNO_3) to obtain analyte concentrations of about 10 ppb to 1 ppm. Using standard instrumental configurations, such concentrations are typically sufficient for most elements to produce ion beam intensities of $>10^{-11}$ A for the isotopes of interest and to permit isotopic analyses that achieve a precision (2σ) of about ± 100 ppm or better. Such analyses typically utilize high-efficiency desolvating nebulizer systems, which produce *dry plasma* conditions, as this allows significantly higher ion yields (by a factor of $\sim$3–10) than conventional nebulizers that generate a *wet plasma*.

Isotopic analysis in cosmochemistry has further profited immensely from instrumental improvements that were developed over the last decade, including new skimmer cone geometries, improved desolvation nebulizer systems, and the use of higher capacity vacuum pumps for the expansion chamber. These developments, which have led to significantly better ion yields for essentially all elements, now allow routine isotopic analyses of trace elements, even in samples of limited size (a problem commonly encountered in cosmochemistry). Recent Ni isotope ratio measurements have furthermore shown that MC-ICP-MS can even achieve precisions (2σ) of ± 10 ppm or better, which rival the best reproducibilities attained by TIMS [49]. Such analyses require analytical strategies that are optimized for MC-ICP-MS and which involve multiple alternating standard and sample measurements that are carried out at high ion beam intensities.

10.5 Applications of MC-ICP-MS in Cosmochemistry

The following sections provide an overview of the use of MC-ICP-MS in cosmochemistry. To this end, the text highlights selected novel findings and their scientific significance, while also discussing any particular analytical procedures and potential problems.

10.5.1 Nucleosynthetic Isotope Anomalies

The solar nebula, from which our solar system formed, consisted of material produced in many different stellar sources such as supernovae, red giants, and novae. These materials carried very distinct isotopic signatures, which are characteristic fingerprints of their production sites. Mixing of these components in the molecular cloud and in the protosolar disk before and during solar system formation produced the abundance pattern of elements and isotopes that is currently observed in the solar system. The general homogeneity in isotopic compositions determined for bulk solar system material, when effects generated by radioactive decay, cosmic rays, and mass-dependent fractionation are excluded, reflects the efficiency of the mixing processes. This homogeneity is documented, for example, by the high-precision Te and Zr isotope ratio data that have been obtained by MC-ICP-MS for a number of primitive and differentiated meteorites [50–52].

However, evidence for preserved isotopic heterogeneities in our solar system at the ppm level due to the incomplete mixing of different nucleosynthetic materials has become more and more evident during the past decade [49, 53–55]. The evidence is based on new high-precision data available as the result of the development of improved analytical techniques for MC-ICP-MS and TIMS analysis. The challenge of identifying nucleosynthetic isotope anomalies is that there is no independent control mechanism to distinguish true nucleosynthetic isotope anomalies from spectral interferences or matrix effects. Hence, analytical artifacts are not easily identified and, for this reason, such anomalies are often controversially discussed in the literature, for example, results obtained for Zr [51, 56, 57] and Mo isotopes [53, 58, 59]. Spectral interferences and matrix effects therefore need to be carefully assessed for the isotopic analysis of each element of interest. A good separation of interfering elements, most commonly achieved by ion exchange chemistry prior to

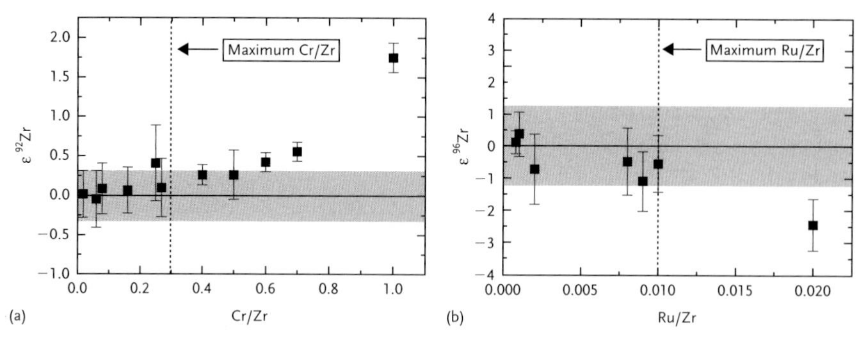

Figure 10.4 Zr isotope compositions obtained for Alfa Aesar Zr standard solutions doped with variable amounts of (a) Cr and (b) Ru, adapted from [60]. The gray bars mark the external reproducibility of the standard (2σ). $\varepsilon^{9X}Zr$ denotes variations in the $^{9X}Zr/^{90}Zr$ ratio, which are quoted in parts per 10 000. (a) Cr/Zr ratios of up to 0.3 are acceptable, but higher amounts of Cr may significantly influence $\varepsilon^{92}Zr$. The Cr/Zr ratios therefore need to be checked prior to each measurement. (b) The Ru interference on ^{96}Zr can be adequately corrected for Ru/Zr up to about 1×10^{-2}. This correction is usually negligible for terrestrial samples, whereas it is crucial for most chondrites, due to the high Ru contents.

analysis, is crucial to avoid analytical artifacts. This is illustrated in Figure 10.4 [60] for the determination of Zr isotope compositions. Larger amounts of Ru (Ru/Zr >0.01) and Cr (Cr/Zr >0.3) present in the sample solutions analyzed lead to erroneous Zr isotope ratio data. The amount of Ru and Cr left behind after ion-exchange chemistry strongly depends on the original sample matrix. The Ru contents of terrestrial and lunar bulk rock samples and evolved stony meteorites (e.g., basaltic eucrites) are low, such that the chemical separation of Ru from Zr is not problematic. However, for primitive meteorites (chondrites) and iron meteorites with Ru/Zr >0.1, a clean chemical separation, and also correction for residual Ru in the sample solution, become vital to ascertain the accuracy of the data [60].

Based on optimized chemical separation and interference correction procedures, Schönbächler *et al.* [60] demonstrated that, in contrast to previous reports [56, 57], different meteorite classes (carbonaceous, enstatite, and ordinary chondrites, eucrites, and mesosiderites; Table 10.1) as well as lunar and terrestrial samples have identical Zr isotope compositions [51]. Further improvements in the Zr isotope measurement techniques were made in a recent study [61], which took advantage of the larger dynamic range of the collectors deployed to measure higher signal intensities. This recent work showed that small differences in Zr isotope compositions (<100 ppm) exist between carbonaceous chondrites and lunar/terrestrial samples, which could not be resolved previously [61]. These variations, which are limited to the $^{96}Zr/^{90}Zr$ ratios, are attributed to the inhomogeneous distribution of ^{96}Zr in the solar system. The isotope ^{96}Zr requires a higher neutron flux [62] to be produced in stellar environments than the other Zr isotopes, which are predominately formed by the *s*-process (through capture of *slow* neutrons). Thus, variations in ^{96}Zr can be explained by a heterogeneous distribution of a nucleosynthetic component that was produced in a stellar environment with high neutron fluxes.

Another complicating factor for studies involving nucleosynthetic anomalies is the mass bias introduced by the MC-ICP-MS instrument, which is much larger ($\sim$2% for Zr) than the actual measurement precision (sub-permil). The mass bias is usually corrected utilizing the exponential mass fractionation law and a second isotope ratio of the target element, which is assumed to be invariant (see also Chapter 5). Zirconium has five isotopes, hence, it is possible to use different ratios to study the effect of the mass bias correction. Notably, choosing any isotope ratio involving ^{90}Zr, ^{91}Zr, ^{92}Zr, and ^{94}Zr yields variations in the ^{96}Zr/^{90}Zr ratio only and this corroborates the conclusion that ^{96}Zr is the isotope that is heterogeneously distributed. If ^{96}Zr is used for mass bias correction, the variations in ^{96}Zr propagate through the calculations and generate apparent effects in other Zr isotopes. Therefore, the normalization ratio needs to be carefully chosen according to nucleosynthetic theory. The discussion of the correct normalization ratio has been particularly important for the study of nucleosynthetic Mo isotope effects [53, 58, 59]. For Mo isotopes, the nucleosynthetic anomalies are best resolved when the mass bias correction uses isotopes produced by the *s*-process or mixtures of the *s*- and *r*- (rapide neutron capture during conditions of higher neutron fluxes) processes (i.e., ^{96}Mo, ^{98}Mo).

Isotopic studies involving MC-ICP-MS revealed evidence for a heterogeneous distribution of Zr, Ti, Mo, and Ni isotopes in the solar system [49, 53–55, 59, 61]. Variations in relative isotopic abundances of ^{96}Zr and ^{50}Ti between carbonaceous chondrites and terrestrial samples provide evidence for a heterogeneous distribution of CAIs (Table 10.1) as carriers of these nucleosynthetic anomalies. This conclusion has implications for mixing processes in the solar nebula and accretion models of planets.

Potential carrier phases of nucleosynthetic anomalies can be investigated by stepwise dissolution of carbonaceous chondrites. These meteorites contain presolar phases (e.g., SiC and graphite grains) with characteristic isotopic compositions that formed around distant stars and have survived the formation of the solar system [63]. These phases exhibit different resistances to acid digestion. Stepwise dissolutions of the carbonaceous chondrites Orgueil (CI), Murchison (CM), and Allende (CV) revealed large nucleosynthetic anomalies for Zr isotopes that contrast with the almost uniform compositions reported for bulk meteorites [64]. Two complementary nucleosynthetic components were observed, which were either enriched or depleted in *s*-process nuclides relative to the composition of bulk materials (Figure 10.5). The *s*-process depletion (characterized by excess ^{96}Zr) is most distinctive in the acetic acid leachate, with ε^{96}Zr values of up to about +50 (where ε denotes isotopic variation in parts per 10000). The ^{96}Zr excess decreases with increasing acid strength and the final leaching steps of the experiment are strongly depleted in ^{96}Zr. Presolar SiC grains are likely host phases for part of the anomalous *s*-process Zr released during these later stages because, on average, they exhibit large depletions in ^{96}Zr, which are documented by single grain analysis performed by resonance ionization spectrometry [65].

However, mass balance indicates that presolar SiC grains cannot account for the ^{96}Zr excesses observed in the early leaching steps and this therefore hints at the

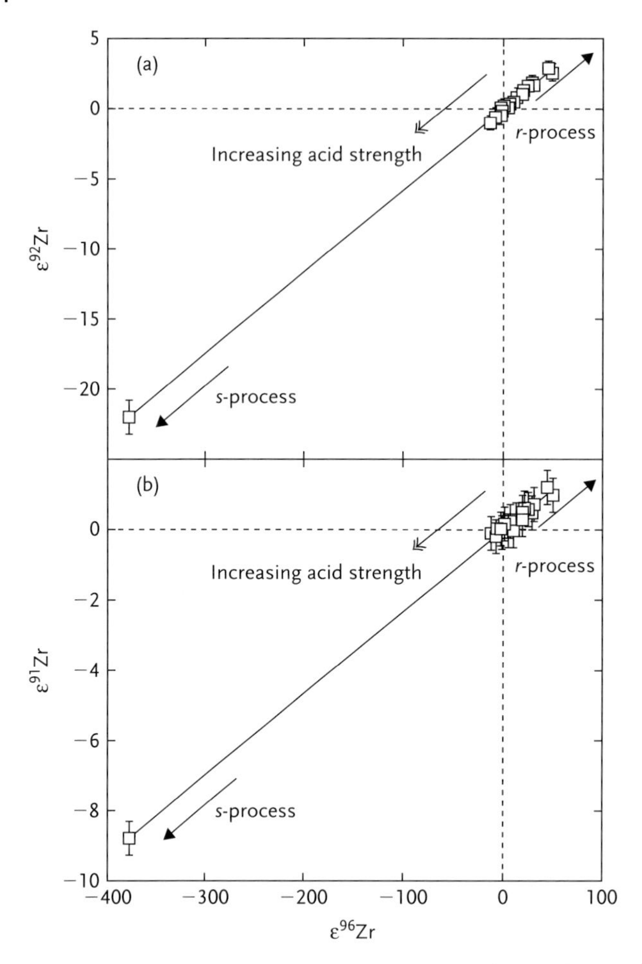

Figure 10.5 Variations in Zr isotope compositions illustrated in (a) ε^{91}Zr and (b) ε^{92}Zr versus ε^{96}Zr diagrams for leachates of the carbonaceous chondrites Orgueil (CI), Murchison (CM), and Allende (CV) (Table 10.1). ε^{9X}Zr denotes variations in the ^{9X}Zr/^{90}Zr ratio, given in parts per 10 000. The leachates were obtained by treatment of bulk rock samples with sequentially stronger mineral acids: 50% HAc, 4M HNO_3, 6M HCl, 13.5M HF + 3M HCl, followed by dissolution in a Parr bomb with HNO_3–HF. The dashed line denotes the Zr isotopic composition of average solar system material. Adapted from [64].

presence of at least one additional carrier phase with significant amounts of anomalous Zr. This phase either (i) carries an *r*-process signature and readily dissolves in weak acid or (ii) it is an acid-resistant *s*-process carrier phase. Notably, Te isotope data obtained by MC-ICP-MS on the same sample aliquots as the Zr isotope results [66] also show hints of a small excess of the *r*-process nuclide ^{130}Te in the early leach fractions, which is consistent with an overabundance of an *r*-process component or the depletion of s-process material relative to the average solar system. The Zr isotope data (Figure 10.5) strikingly illustrate that average solar system material consists of a homogenized mixture of different nucleosynthetic components, which can be partly resolved by leaching experiments on carbonaceous chondrites. The precursor material included *r*- and *s*-process components that were distributed nearly homogeneously at a bulk rock or larger scale.

10.5.2
Long-Lived Radioactive Decay Systems

10.5.2.1 The ^{87}Rb–^{87}Sr Decay System

The β-decay of ^{87}Rb to ^{87}Sr, with a half-life of 48.8 billion years (Table 10.2), is one of the classic radiometric dating systems and has been used in cosmo- and geochronology for almost half a century. Strontium has four stable isotopes and as more than two stable non-radiogenic isotopes are thus available, instrumental mass bias can be corrected by internal normalization (see also Chapter 5). For the correction of the radiogenic ^{87}Sr/^{86}Sr ratios, ^{88}Sr/^{86}Sr is most suitable, as this blankets the isotope of interest (^{87}Sr) and features two isotopes with relatively high abundances, which minimizes the analytical uncertainty of the measurements.

Owing to its susceptibility to resetting by metamorphism [67], the Rb–Sr system is not frequently used in cosmochronology. The high concentration of Sr in phases such as plagioclase makes it possible to use LA–MC-ICP-MS to obtain precise ^{87}Sr/^{86}Sr ratios. However, spectral interferences from the ablated and ionized matrix, mainly from Ca dimers and/or Ca argides, can be high and impossible to correct for accurately [68]. Multi-collector ICP-MS offers rapid sample analysis compared with TIMS, but in cosmochronology, where sample throughput is less of an issue, the time-limiting step is more often the chemical separation of Sr. Where MC-ICP-MS is most advantageous compared with TIMS is for the analysis of stable Sr isotopic variations, which has been given more attention in recent years in both geo- and cosmochemistry [69].

10.5.2.2 The ^{147}Sm–^{143}Nd Decay System

The α-decay of ^{147}Sm to ^{143}Nd, with a half-life of 106 Ga (Table 10.2), has been used in geo- and cosmochronology since the early 1970s. Neodymium has seven isotopes and Nd isotopic analysis can hence be corrected internally for instrumental mass bias. The large number of relatively abundant isotopes means that a number of isotope ratios can (and have been) be used for mass bias correction, but the current convention is to use ^{146}Nd/^{144}Nd for the correction of the radiogenic ^{143}Nd/^{144}Nd ratios. The elements Sm–Nd furthermore offer a second, extinct decay system, as short-lived ^{146}Sm decays to ^{142}Nd with a half-life of 103 Ma (Table 10.2). This coupling of a long- and a short-lived decay scheme renders the Sm–Nd system useful not only for absolute dating but also for the precise tracking of events that took place early in the history of the solar system.

Traditionally, TIMS has been used for Nd isotopic analysis and although this technique does not offer great ionization efficiency, the Nd concentrations of stony meteorites are generally high enough that sample limitation is not an issue. Additionally, samples with very low Nd contents can be analyzed for the isotopic composition of Nd by TIMS relying on the formation of the oxide ion NdO^+, which gives higher ionization efficiency, but introduces additional uncertainties in the correction for mass fractionation and from spectral interferences. MC-ICP-MS offers a rapid analysis time compared with TIMS, although this is rarely essential in cosmochronology. Furthermore, for $^{147}Sm-^{143}Nd$ chronology, the latest generation of TIMS instruments offers slightly better precision than MC-ICP-MS.

Nonetheless, MC-ICP-MS has been used successfully for Nd isotopic analysis in cosmochemistry. A good example is the combined use of the Sm–Nd and Lu–Hf systems by Blichert-Toft *et al.* [70]. They analyzed 18 eucrites (Table 10.1) and found an $^{147}Sm-^{143}Nd$ age of 4464 ± 75 Ma, with three cumulative eucrites defining a better constrained isochron of 4470 ± 22 Ma. This suggests that cumulative eucrites have a crystallization age that postdates the formation of the solar system by about 100 Ma and implies a fairly protracted crystallization history for the eucrite parent body, which is thought to be the asteroid 4 Vesta (Table 10.1). Combined $^{146}Sm-^{142}Nd$ and $^{147}Sm-^{143}Nd$ ages of eucrites [71, 72] give a somewhat older age of 4554 (+24/−28) Ma, with no indication of a younger age for the cumulative eucrites. It is not known whether this difference in results is due to disturbance of the long-lived $^{147}Sm-^{143}Nd$ system, sampling uncertainty in the small number of samples used for combined $^{142}Nd-^{143}Nd$ chronometry, or, less likely, uncertainty in the assumed initial abundance of ^{146}Sm.

The low initial abundance of ^{146}Sm and the challenge of accurately correcting $^{142}Nd/^{144}Nd$ (which is not bracketed in mass by $^{146}Nd/^{144}Nd$) for mass bias render the application of the extinct $^{146}Sm-^{142}Nd$ decay scheme particularly challenging. For example, SNC meteorites from Mars (Table 10.1) define a larger range in ^{142}Nd abundances than samples from any other planetary body, but the $^{142}Nd/^{144}Nd$ ratios still vary by only about 1 ε unit (1 part in 10 000). In contrast, the $^{143}Nd/^{144}Nd$ isotope ratios of Martian meteorites display a variability of about 60 ε. Although ^{142}Nd can be measured by MC-ICP-MS [73, 74], the stability of the plasma source is generally not sufficient to resolve the small differences in $^{142}Nd/^{144}Nd$ of most natural samples, with the possible exclusion of Martian meteorites. Hence ^{142}Nd analyses are generally conducted by TIMS, which can provide better precision for a comparable sample size [75].

10.5.2.3 The $^{176}Lu-^{176}Hf$ Decay System

The decay of ^{176}Lu to ^{176}Hf with a half-life of 37.1Ga (Table 10.2) has been used in geo- and cosmochronology since the early 1980s [76]. Hafnium has six stable isotopes, and $^{179}Hf/^{177}Hf$ is commonly used for internal normalization of the radiogenic $^{176}Hf/^{177}Hf$ isotope ratio. The application of the Lu–Hf decay scheme has become much more widespread since the advent of MC-ICP-MS, as TIMS achieves only a very low ionization efficiency for Hf.

Blichert-Toft and Albarède [77] carried out the first published cosmochemical Hf isotopic study. They investigated the Lu–Hf systematics of chondrites, which featured only imprecise published TIMS data, due to the relatively low Hf contents of these meteorites. Their results yielded an initial solar system ratio of $^{176}Hf/^{177}Hf = 0.27974 \pm 3$. This was later revised by Blichert-Toft *et al.* [70] to

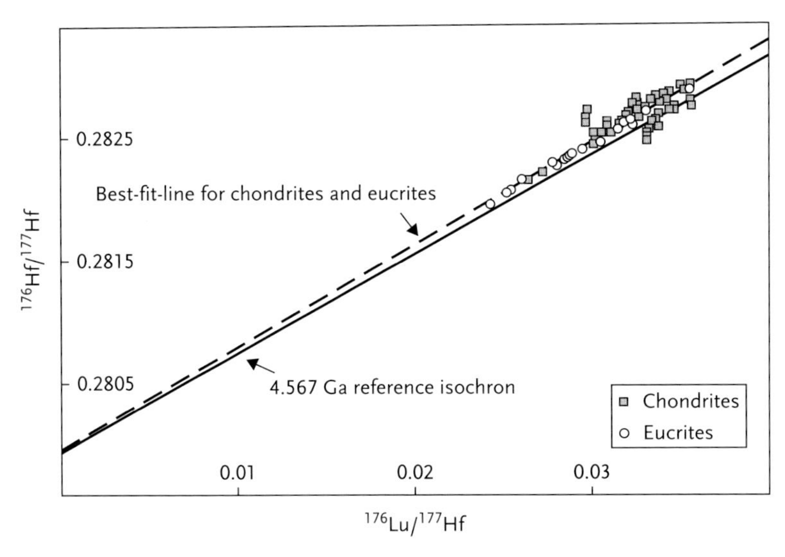

Figure 10.6 Lu-Hf isochron diagram for chondrites and eucrites, as modified from [78]. The data symbols used here are larger than the analytical uncertainties. The reference isochron is calculated using a $^{176}Hf/^{177}Hf$ initial value of 0.27968 and a decay constant of $1.867 \times 10^{-11}\ a^{-1}$, such that an age of 4.567 Ga is obtained. This isochron, however, does not provide a good fit to the data. In contrast, the best-fit line for chondrites and eucrites gives an $\sim$4.8 Ga isochron age that is clearly too old, presumably due to excess ^{176}Hf that is positively correlated with the Lu/Hf ratio. This indicates that the excess ^{176}Hf is caused by accelerated ^{176}Lu decay, rather than being of nucleosynthetic origin. Data from [70, 76, 77, 81, 82].

0.27966 $\pm$ 2, based on data for eucrites with a crystallization age of 4604 $\pm$ 39Ma and a much larger range in Lu/Hf ratios than the chondrites. This latter result is in agreement with a recent estimate of 0.279680 $\pm$ 1 that was obtained from MC-ICP-MS analysis of angrite meteorites [78].

The extensive high-precision Hf isotopic analyses that were conducted with MC-ICP-MS for terrestrial and meteorite samples [79, 80] since the late 1990s furthermore revealed that the previously used ^{176}Lu decay constant, which is based on counting experiments, may be erroneous. In particular, the Lu–Hf isochron studies of chondrites and eucrites yielded ages that were about 4% too old, in comparison with ages that were obtained for the same meteorites by Pb–Pb chronology (Figure 10.6). Attempts to correct for this offset by using a decay constant calibrated to the well-determined Pb isotope ages produced an excess of ^{176}Hf for the meteoritic samples (Figure 10.6). This excess is apparently not of nucleosynthetic origin [83], but most likely reflects accelerated decay of ^{176}Lu in early nebular condensates, due to exposure to highly energetic gamma-rays [78, 84, 85].

10.5.2.4 **The U/Th-Pb Decay Systems**

Two isotopes of U and ^{232}Th decay to three different Pb isotopes, with half-lives between $\sim$700 Ma and 14 Ga (Table 10.2). The elements U and Th partition into very different phases than Pb and this results in some minerals (such as zircons and pyroxenes) having very high parent daughter/ratios, which produce highly radiogenic Pb isotopic compositions over time. This makes it possible to obtain very precise U–Pb ages from isochron analyses of suitable minerals or the leachates and leaching residues of samples.

Lead has four stable isotopes, ^{204}Pb, ^{206}Pb, ^{207}Pb, and ^{208}Pb, but three of these are radiogenic, such that there is no invariant isotope that can be used for internal correction of instrumental mass fractionation. This poses a significant analytical challenge for precise Pb isotopic analysis. This problem is most readily overcome with MC-ICP-MS, because the Pb isotope ratio measurements can utilize the $^{205}Tl/^{203}Tl$ ratio of added Tl for external normalization of the Pb isotope ratio data (see also Chapter 5). The simplicity of this procedure is the main reason why MC-ICP-MS Pb isotopic analysis is now commonly used in geo- and cosmochemistry. This approach needs to be applied with care, however, and it has been argued that Pb isotopic analysis that utilizes the double spike methodology (either in conjunction with TIMS or MC-ICP-MS) is typically superior in accuracy and precision [86, 87].

The coupled $^{238}U-^{206}Pb$ and $^{235}U-^{207}Pb$ decay systems can provide age information based on $^{207}Pb/^{204}Pb$ and $^{206}Pb/^{204}Pb$ only. This is advantageous as Pb isotopic compositions can be determined with higher precision than the elemental parent/daughter ratio that must be applied in conventional chronological studies. This technique of Pb–Pb dating can provide the most precise absolute age data for solids formed early in the solar system and is the lynchpin for the calibration of extinct radioactive decay systems. Analytical improvements that have been achieved

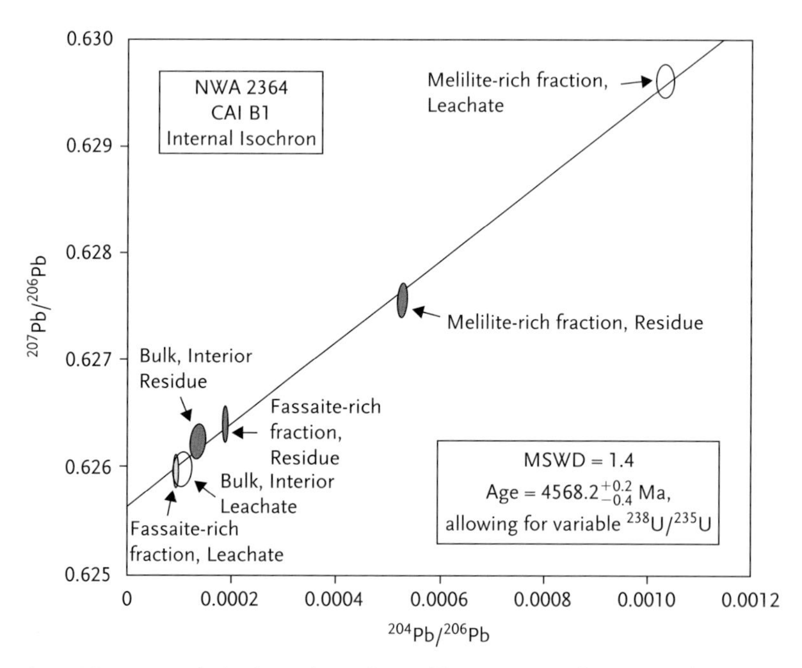

Figure 10.7 Internal Pb–Pb isochron obtained by MC-ICP-MS for a CAI inclusion in the CV3 carbonaceous chondrite NWA 2364, giving an age of 4568.2 Ma. Reproduced from [19].

in the application of the Pb–Pb chronometer have made ages with a precision of $\pm$0.1–0.2 Ma possible for meteorites and meteorite components with highly radiogenic Pb isotopic compositions, such as angrites and CAIs, respectively. For example, the oldest Pb–Pb isochron age for solar system material was recently determined by MC-ICP-MS as 4568.2 $\pm$ 0.2 Ma (Figure 10.7), for a CAI inclusion in a CV3 carbonaceous chondrite [19]. This age is about 0.5–1.0 Ma older than previously analyzed CAIs from two other primitive CV3 chondrites (Allende and Efremovka) and identical with the $^{26}Al-^{26}Mg$ model age obtained in the same study [19].

A basic premise of Pb–Pb chronology is that all extraterrestrial materials have the same $^{238}U/^{235}U$ ratio, but this has been shown to be incorrect by some recent MC-ICP-MS studies [88]. A correction for variations of up to 0.1% in $^{238}U/^{235}U$ was applied by Amelin *et al.* [89] in the calculation of Pb–Pb isochron ages for early solar system materials and this approach was able to resolve previous inconsistencies in the relative Pb–Pb, $^{26}Al-^{26}Mg$, and $^{182}Hf-^{182}W$ ages of CAIs and chondrules. Importantly, all isotope systems now indicate that CAIs formed about 2–3 Ma before chondrules [89].

10.5.3 Extinct Radioactive Decay Systems

Some short-lived radionuclides were sufficiently abundant at the start of the solar system to produce variations in the abundance of their daughter isotopes in early-formed objects (Table 10.2). The half-lives of these nuclides are between about 0.1 and 100 Ma (Table 10.2). Hence, the parent isotopes are no longer present today, but they were synthesized in stars shortly before solar system formation and therefore they were present in the early solar nebula. The isotopic record of these nuclides provides information about stellar nucleosynthetic sites active shortly before the birth of the solar system and the time scales over which the early solar system formed and first differentiated. Depending on half-life and chemical affinities of parent and daughter isotopes, extinct radionuclide systems can be used to date processes as diverse as the formation of CAIs and chondrules, volatile element depletion and planetary differentiation (e.g., core segregation and differentiation of early silicate reservoirs). In particular, they are powerful tools to study the Earth's accretion and core formation [90–92].

In contrast to long-lived decay systems, extinct radionuclides provide relative ages and they need to be calibrated relative to a long-lived chronometer (usually the Pb–Pb chronometer) to obtain absolute ages. The most reliable evidence for the former existence of an extinct radionuclide in the solar system is provided by internal isochrons. The determination of such an isochron requires a sample of old age with a co-genetic suite of mineral phases that have remained undisturbed since their formation. Furthermore, the phases must display variations in the parent/daughter elemental ratios. If this is the case, the mineral phases will define a single correlation line or isochron in the diagram of isotope versus elemental ratio (Figure 10.8). In the following, selected examples are discussed to illustrate how extinct radionuclides can be used to study the formation history of small planetary bodies.

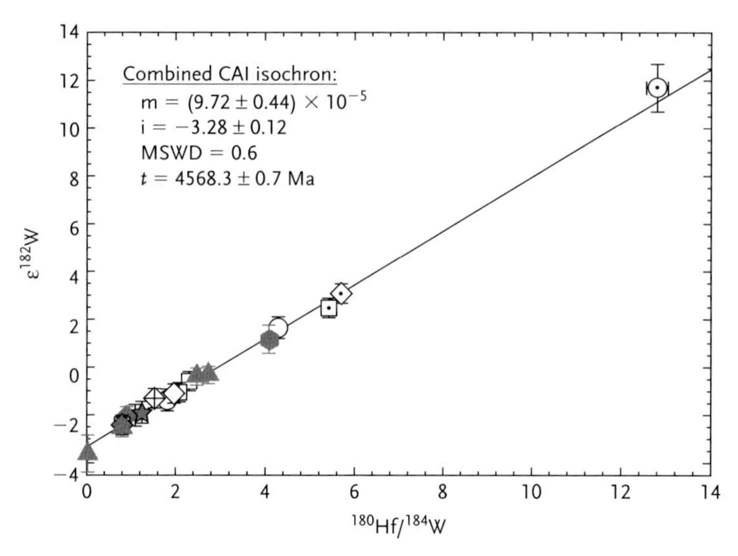

Figure 10.8 Combined Hf–W isochron for various CAIs. Symbols indicate different fractions separated from CAIs, which all yield identical ages within uncertainties. The gray triangles show data for different separates from a single Allende CAI (ALL-MS-1) and thus define an internal isochron for this particular inclusion. The absolute age (4568.3 ± 0.7 Ma) of this sample was obtained relative to absolute Pb–Pb ages for angrites (see [93] and references therein for more details). m: initial $^{182}W/^{184}W$ ratio of CAIs; i: initial $\varepsilon^{182}W$ of CAIs. Adapted from [93].

All rocky planets and many asteroids of our solar system underwent differentiation to segregate a metallic core. The only samples of such cores that are available for laboratory studies are iron meteorites [94]. These meteorites are grouped as (i) magmatic irons, believed to represent fragments of cores from asteroids, and (ii) non-magmatic irons (Table 10.1). The latter show evidence of a more complex formation history and contain abundant silicate inclusions. Chronological studies of iron meteorites can provide the time scales of asteroid accretion, core segregation, and the time that was necessary for the originally molten metal to cool and crystallize. Due to the unique composition of iron meteorites (mainly Fe–Ni metal), only a few radiogenic systems are able to date these processes [95]. The systems include Hf–W, Pd–Ag, and Pb–Tl (Table 10.2), which are all primarily analysed using MC-ICP-MS. Ag and Tl possess two isotopes only and this hinders the instrumental mass fractionation correction in TIMS analysis. During MC-ICP-MS analysis the instrumental mass bias can be monitored by external normalization and this leads to superior analytical uncertainties of about ±0.5–1 ε compared to the best TIMS data with precisions of ±10–20 ε [39, 96].

Iron meteorites exhibit variations in the $^{182}W/^{184}W$ ratio, which were originally thought to result from the decay of ^{182}Hf [97, 98]. Since the short-lived nuclide ^{182}Hf is lithophile and virtually absent in metal, iron meteorites preserve the $^{182}W/^{184}W$ ratio at the time of metal–silicate differentiation on their parent bodies. Newer W MC-ICP-MS isotope data indicate, however, that some of the variations reflect burnout and production of W isotopes by interaction with thermal neutrons that are generated by cosmic rays [99–102]. A comprehensive suite of iron meteorites were analyzed in recent studies [99, 100, 102] that also attempted corrections for such cosmogenic effects. These studies concluded that, following correction, all magmatic iron meteorite groups and the non-magmatic IIICDs have initial W isotope compositions identical with those of CAIs. This provides evidence that the parent bodies of these iron meteorites experienced early silicate–metal segregation within about 1Ma of CAI formation.

The further core crystallization history of iron meteorites can be studied using the Pd–Ag and Pb–Tl decay systems. For example, the IIIAB iron Grant shows an initial $^{107}Pd/^{108}Pd$ ratio of 1.55 $(\pm 0.11) \times 10^{-5}$ [40, 103] (Figure 10.9a). This

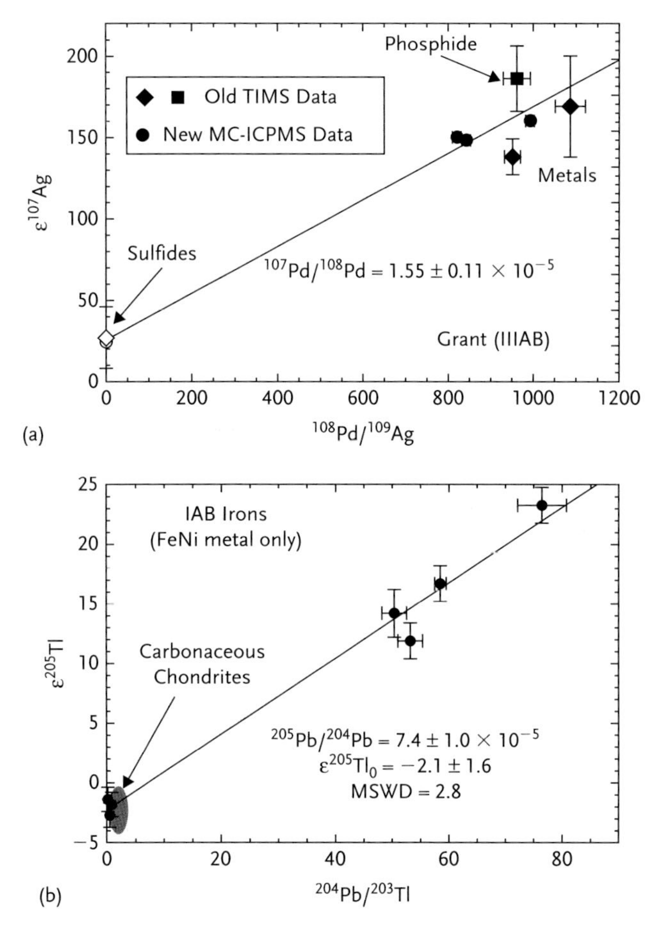

Figure 10.9 (a) Pd–Ag and (b) Pb–Tl isochron data for iron meteorites and mineral inclusions. (a) Shown are Pd–Ag results for metal fragments, troilite, and phosphide inclusions from the IIIAB iron meteorite Grant. Data from [40, 103, 104]. (b) Pb–Tl isochron data for metal fragments from the IAB irons Canyon Diablo and Toluca [39] define an isochron, which first established that live ^{205}Pb was present in the early solar system. The results for eight carbonaceous chondrites fall into the gray field [105].

value corresponds to an age of 13.0 (+3.5/−4.9) Ma, if the solar system initial $^{107}Pd/^{108}Pd = 5.9\ (\pm 2.2) \times 10^{-5}$ derived from carbonaceous chondrites is applied [106]. Hf–W ages, corrected for cosmogenic effects, indicate that Grant segregated early, within less than 1.2 Ma of Hf–W system closure in CAI [99]. This leaves about ~12Ma to cool from the melting temperature of the metal to 1100 K, which is the closure temperature estimate for Pd–Ag in iron meteorite [107]. This translates to a cooling rate of ~42 K Ma^{-1}, which is in good agreement with a previously determined metallographic cooling rate of ~50 K Ma^{-1} and implies a IIIAB parent body radius of 25 ± 7 km [108].

First evidence of live ^{205}Pb in the early solar system was discovered relatively recently in iron meteorites by Nielsen *et al.* [39]. This study reported Pb–Tl data for metal fragments of the non-magmatic IAB iron meteorites and found that the $^{205}Tl/^{203}Tl$ isotope ratios displayed a correlation with $^{204}Pb/^{203}Tl$ (Figure 10.9b). When interpreted as an isochron and applying the initial $^{205}Pd/^{204}Pd$ ratio of the solar system deduced from carbonaceous chondrites, the correlation defines an age of 57 (+10/−14)Ma after the start of the solar system [105]. The slightly younger Pd–Ag age for IAB metals of 19 (+24/−10)Ma [106] might be due to the different closing temperatures of the two systems [105]. Taken together, the data provide evidence that IAB iron meteorites experienced a protracted metal crystallization history, which stands in contrast to the crystallization history of magmatic IIIAB irons. This conclusion is in good agreement with other petrological and isotopic evidence from IABs, which also indicates that the IAB iron meteorite history is more complicated and potentially involved a catastrophic disruption of the parent body at a later stage [109, 110].

10.5.4
Stable Isotope Fractionation

Prior to the advent of MC-ICP-MS, geo- and cosmochemical investigations of stable isotope variations were focused almost exclusively on a few light elements particularly H, C, O, and S. For heavier elements, isotope fractionation was essentially unexplored because the effects were too small to be readily resolved by commonly accessible mass spectrometric techniques. The development of MC-ICP-MS fundamentally changed this situation, as the technique was shown to be suitable for the routine determination of small natural variations in the isotopic composition of heavier elements. As a result, numerous MC-ICP-MS studies that investigate novel *non-traditional* stable isotope systems have been published over the last decade.

A particular difficulty with stable isotope analysis follows from the necessity to correct the mass spectrometric data of a sample for changes in instrumental mass bias, without erasing the natural isotope fractionation that is also recorded in the results. The latter requirement is not fulfilled by the method of internal normalization, which is commonly applied in the correction for instrumental mass bias in radiogenic and nucleosynthetic isotope analysis. This procedure is therefore unsuitable for stable isotope measurements. The specific characteristics of the ICP ion source, however, permit precise isotopic analysis for a wide range of elements, using three different techniques to correct for instrumental mass bias effects (see also Chapter 5). Each of these methods has distinct advantages and drawbacks [27, 32, 35].

The steady-state operation of the ICP source is beneficial for the correction of instrumental mass bias by standard-sample bracketing, where the *raw* (measured and uncorrected) isotope ratio data of a sample are referenced to the results obtained for an isotopic standard, which is preferentially analyzed before and after each sample [27, 35]. This technique is similar to the standardization method commonly used in gas source isotope ratio mass spectrometry. To account best for drifts in instrumental mass bias, which can be particularly severe for light elements such as Li and B, data collection often utilizes multiple but short analytical measurements for samples that are each bracketed by standard analyses. Switching between samples and standards can be very rapid, if long washout protocols are not required, and mass spectrometric measurements of about 5 min or less have been used to optimize the precision of Li and Mg isotope ratio measurements by MC-ICP-MS [111, 112].

The principal advantage of the standard–sample bracketing approach is that its application is very straightforward and only two interference-free isotopes are needed, in principle, for an analysis. The method, however, is also particularly susceptible to the generation of analytical artifacts from variations in instrumental mass bias that are induced by residual matrix elements, which are present in the sample solutions but absent in the pure reference standards (Figure 10.10). Hence this approach is typically used only for elements (i) that are present at high concentrations in samples (e.g., the major elements Mg, Si, and Fe of silicate meteorites or Fe from iron meteorites) or (ii) for which the technique of external normalization is not readily applicable (e.g., Li, B; see below).

External normalization is the most commonly used procedure for correction for instrumental mass bias in isotopic analysis by MC-ICP-MS [27, 35]. This method makes use of the observation that the mass discrimination associated with plasma source mass spectrometry is, to first order, primarily a function of mass, such that elements with similar masses display a nearly identical mass bias. Using a solution containing a mixture of two such elements, the mass discrimination observed for an admixed element of known isotopic composition can be used to determine the unknown isotopic composition of the second (analyte) element [27, 35].

The main disadvantage of external normalization is that four isotopes of two elements need to be measured to determine two isotope ratios, which must all be free of spectral interferences. External normalization, however, is much less susceptible to the generation of analytical artifacts by matrix effects than standard sample bracketing (Figure 10.10). The approach furthermore appears to work best for element pairs that have essentially overlapping mass ranges, presumably because such elements show the highest degree of correlation in their responses to matrix-induced changes in instrumental mass bias. Hence external normalization has been most successful for isotopic analysis of intermediate- or high-mass elements, such as Cu, Zn, Cd, and Tl, using added Ni or Zn, Cu, Ag, and Pb, respectively, for the mass bias correction [27, 35]. In contrast, the method has been

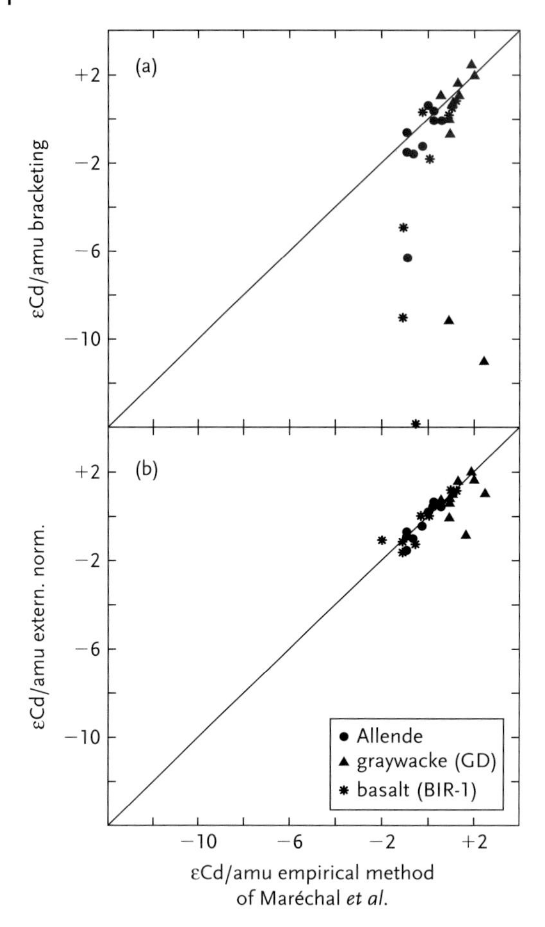

Figure 10.10 Comparison of Cd isotope ratio data obtained using different normalization procedures for repeated measurements of two terrestrial rocks samples and the Allende CV3 chondrite (from [38]). (a) Comparison of results obtained by standard-sample bracketing and by external normalization (based on added Ag) using the empirical method of Maréchal *et al.* [37]. The bracketing approach yields inaccurate results for some samples, due to matrix-induced changes in instrumental mass bias. (b) Comparison of results obtained by external normalization using Ag using either the exponential law in combination with standard-sample bracketing or the empirical technique of Maréchal *et al.* [37]. Both methods yield similar results for all samples. The isotopic data are shown as εCd/amu values, which denote the variation in Cd isotopic composition relative to a terrestrial standard and normalized to a mass difference of 1u [38, 113].

less successful for lighter elements, such as Li and Mg, and these are therefore more commonly analyzed relying on standard sample bracketing, albeit after careful chemical isolation to avoid matrix effects.

The third technique that is applied to correct stable isotope ratio measurements for instrumental mass bias involves the use of a double spike [114, 115]. In this case, a spike solution, which is highly enriched in two isotopes of the target element, is added to the samples. The natural isotope fractionation shown by the target element in the sample is then determined from mass spectrometric data, which are collected for three isotope ratios (with a common denominator) of the sample–spike mixture. This approach was originally developed for TIMS isotope ratio measurements [114], but has more recently been applied to MC-ICP-MS for a number of elements, including Zn [116, 117], Mo [118], and Cd [119]. The double-spike data reduction technique that was published for Mo [118] is particularly suited for MC-ICP-MS results, as its iterative approach permits correction for the instrumental mass bias with the exponential or even general power law [118, 119].

The primary disadvantages of the double-spike technique are that (i) the preparation and calibration of a new double spike require significant effort and (ii) four interference-free isotope signals are needed for accurate data reduction, and this also rules out double-spike analysis of elements that feature only two or three isotopes. In many cases, however, these factors will be outweighed by the advantages of the method: (i) it offers an instrumental mass bias correction that is similar in application and reliability to internal normalization and hence is even more robust towards matrix effects than external normalization; (ii) the approach can correct for laboratory-induced mass fractionation effects, if the spike is added to the samples prior to the chemical processing; and (iii) precise elemental concentration data are obtained as a byproduct of the double-spike method. Hence the double-spike method has recently found increasing popularity in MC-ICP-MS stable isotope analysis of non-traditional elements.

Using the approaches outlined above, mass-dependent isotope fractionation has been investigated via MC-ICP-MS in extraterrestrial materials for numerous elements during the last 10 years (Figure 10.3). Particularly comprehensive data sets were collected for the elements Mg, Fe, Ni, Zn, Cu, and Cd. These studies have shown that the most prominent isotope fractionation is recorded in meteorites as a result of (i) solid–vapor partitioning during either nebular and planetary processes and/or (ii) selective partitioning between solid and liquid phases (e.g., liquid silicate–liquid metal) during planetary melting and differentiation. A few examples of such investigations are outlined below.

A detailed Fe isotopic study that was carried out on more than 30 magmatic and non-magmatic iron meteorites established clear evidence for Fe isotope fractionation between the metal and troilite (FeS) phases of these samples [120]. In particular, it was found that the troilite inclusions generally feature lighter Fe isotopic compositions than coexisting metal, but no systematic relationship between the metal–sulfide fractionation factor $\Delta^{57/54}Fe_{M-FeS}$ ($\Delta^{57/54}Fe_{M-FeS} = \delta^{57/54}Fe_{metal} - \delta^{57/54}Fe_{FeS}$, where $\delta^{57/54}Fe$ is the deviation of the $^{57}Fe/^{54}Fe$ isotope ratio of a sample from the isotopic reference material in permil) and bulk compositional parameters or meteorite group could be identified. However, the $\Delta^{57/54}Fe_{M-FeS}$ values obtained were found to display a positive correlation with the metallographic cooling rates, whereby slowly cooled samples feature the larger Fe isotope fractionation. This indicates that the isotopic offset between metallic Fe and FeS most likely reflects equilibrium Fe isotope partitioning and fractionation, which was established by solid-state diffusion during cooling. The largest

$\Delta^{57/54}Fe_{M-FeS}$ values of about 0.8 ± 0.1‰ are hence thought to represent best the true equilibrium metal–sulfide Fe isotope fractionation factor [120]. Mass balance calculations that were carried out based on this result suggest that the particularly heavy Fe isotopic compositions of metal samples from IIAB iron meteorites are the inherited fingerprint of an S-rich core, a conclusion that is in accord with the independent results of previous studies.

A particularly contentious issue of isotope cosmochemistry is the question of whether there exists a significant difference in the Fe isotopic compositions of the silicate portions of Mars, asteroid 4 Vesta (Table 10.1), the Moon and Earth. Some studies have concluded that the Earth and Moon are isotopically lighter, by about 0.1–0.2‰ in $\delta^{57/54}Fe$, than Mars and Vesta, potentially as a result of Fe loss from the Earth and particularly the Moon by partial evaporation during the Moon-forming Giant Impact [121, 122]. This conclusion, however, is disputed based on the observation that terrestrial peridotites are identical in their Fe isotopic composition with samples from Mars and 4 Vesta [123]. The peridotite data were furthermore thought to provide the best estimate for the Fe isotopic composition of the bulk silicate Earth, as the slightly higher $\delta^{57/54}Fe$ values of basalts were interpreted to reflect Fe isotope fractionation during partial melting and melt differentiation. In addition, most lunar magmatic rocks were also found to be indistinguishable from the terrestrial basalts in $\delta^{57/54}Fe$, which argues for a solar system with a very homogeneous Fe isotopic composition throughout [123, 124].

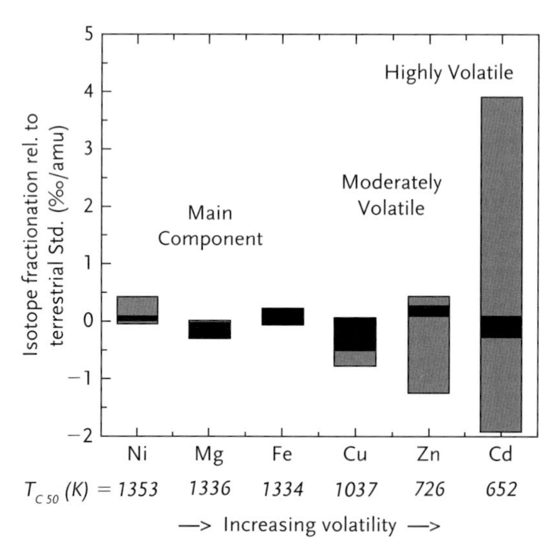

Figure 10.11 Diagram that compares the isotopic variability measured for five elements with different volatilities, as expressed by half-mass condensation temperatures $T_{C\,50}$ [125], in chondritic (unmolten and undifferentiated) meteorites. The light and dark fields denote the variability measured in ordinary/enstatite chondrites and carbonaceous chondrites, respectively. The isotopic variability is expressed as δ/amu, which denotes relative variations in parts per 1000 (permil or ‰), normalized to an isotopic mass difference of 1 u. Data from [38, 105, 126–135].

The largest isotope fractionations thus far identified in extraterrestrial rocks appear to be linked to partial evaporation and condensation. This interpretation is strongly supported by the observation that, in chondritic meteorites that are derived from undifferentiated asteroidal parent bodies (Table 10.1), the more volatile elements display the largest extent of isotopic variability (Figure 10.11). Of particular significance is that for the moderately volatile element Zn and highly volatile Cd, most of the isotopic variability is found in ordinary chondrites (Figures 10.11 and 10.12). In contrast, these two elements show approximately the same extent of isotope fractionation in carbonaceous chondrites as the more refractory metals Ni, Mg, and Fe (Figure 10.11). This suggests that the large depletions that have been observed for the abundances of moderately volatile elements in CM, CV, CO, and other carbonaceous chondrites (Table 10.1) relative to the CI chondrites, which are thought to best approximate the composition of the bulk solar system, were not caused by processes involving partial evaporation, as this would generate larger differences in isotopic compositions due to kinetic isotope fractionation [136, 136a]. The limited variability seen in Zn and Cd isotopic compositions rather indicates that the variable volatile depletion of carbonaceous chondrites (and potentially other planetary bodies) is most readily explained by partial condensation at near-equilibrium conditions or two-component mixing between volatile-rich and volatile-poor material [38, 127, 134].

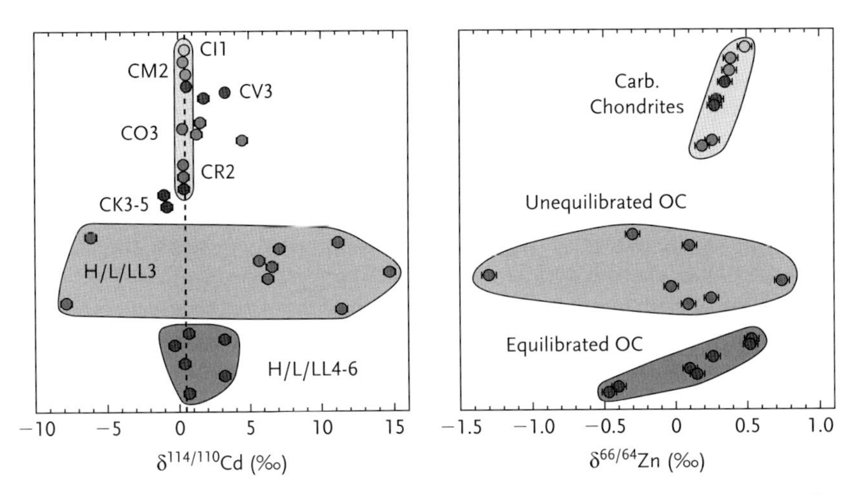

Figure 10.12 Diagram comparing Cd and Zn isotope compositions for different classes of carbonaceous and ordinary chondrites. The Cd and Zn isotope ratio data are shown in δ-notation to denote isotopic variation in $^{114}Cd/^{110}Cd$ and $^{66}Zn/^{64}Zn$, respectively, relative to a terrestrial standard reference material in permil. For both Cd and Zn, pronounced isotope fractionation is apparent for the ordinary chondrites (and particularly unequilibrated ordinary chondrites of petrological type 3), whereas such effects are much more limited for the carbonaceous chondrites. This suggests that nebular processes only generate small Zn and Cd isotope fractionations, whereas large isotope effects are produced as a result of partial evaporation/condensation during parent body metamorphism [134]. Data from [38, 105, 127, 134].

The large extent of Cd isotope fractionation seen in ordinary chondrites is unlikely to reflect nebular processes, as it is not seen in the carbonaceous materials (Figure 10.12). Hence it is most likely caused by mobilization and associated isotope fractionation of Cd during thermal metamorphism on the parent asteroids [38, 134]. This conclusion is also in accord with the Zn isotope ratio data, which show clearly resolvable but smaller isotope fractionations in ordinary chondrites (Figure 10.12) [127, 129]. The more limited variability is expected, given that Zn is much less volatile than Cd (Figure 10.11) and hence will be mobilized to a lesser extent. Importantly, the Cd and Zn isotope ratio data available for ordinary chondrites are in accord with the onion shell model for these parent bodies [17] as (i) Cd depletions and smaller isotope fractionation are typically observed in equilibrated ordinary chondrites of petrological types 4–6, which are thought to be derived from the hotter interiors of the parent asteroids, and (ii) larger isotope fractionations associated with both Cd depletions and enrichments are found in unequilibrated (type 3) ordinary chondrites, which are expected to be from the cooler parent body surface regions (Figure 10.12).

10.5.5 Cosmogenic Isotope Variations

Variations in isotopic abundances that are caused by nuclear reactions induced by cosmic rays are most commonly utilized in cosmic ray exposure dating, but this employs isotopes that are measured by either accelerator or noble gas mass spectrometry [28, 29]. In fact, there are only a very limited number of elements that are suitable for the study of cosmogenic isotopic variations, which can be readily analyzed by either TIMS or MC-ICP-MS [28]. The most important application of these techniques are studies of the secondary neutron fluxes that are generated by (primary) cosmic rays. Such measurements aim to detect anomalies in Sm, Gd, and Cd isotopic abundances that are produced by (n,γ) reactions, for example ^{114}Cd(n, γ)^{113}Cd. Many of these investigations were conducted by TIMS [137–139], but some cosmogenic Cd isotope variations of lunar rocks and soils were evaluated based on MC-ICP-MS isotope ratio data that were originally acquired as part of a stable isotope study [134].

More important to MC-ICP-MS measurements are the minor cosmogenic isotope anomalies that can interfere with studies of nucleosynthetic, radiogenic and stable isotope effects. A prominent example are the cosmochemical MC-ICP-MS W isotopic analyses that are carried out primarily to detect variations in the abundance of ^{182}W from the decay of now extinct ^{182}Hf (Table 10.2). Such variations are most commonly detected by the measurement of ^{182}W/^{184}W or ^{182}W/^{183}W, which are internally normalized for instrumental mass bias correction relative to various W isotope ratios [100, 140, 140a]. Early W isotopic analyses of extraterrestrial materials ascribed all anomalies in the abundance of ^{182}W to radiogenic ingrowth. However, subsequent work that was carried out at improved precision and in combination with additional investigations (e.g., of the abundance of cosmogenic

^{3}He) were able to show that some iron meteorites and lunar rocks also have cosmogenic W isotope anomalies that are superimposed on the radiogenic isotope effects. For the lunar rocks, these anomalies were shown to be due to the $^{181}Ta(n,\gamma)^{182}Ta$ reaction, induced by cosmic radiation, whereby the short-lived isotope ^{182}Ta ($T_{½} = 114$ days) subsequently decays to ^{182}W [141]. For the iron meteorites, cosmic ray-induced reactions lead to both the production and consumption of various W isotopes. As a result of these reactions, many samples of both magmatic and non-magmatic irons appear to have $^{182}W/^{184}W$ ratios that are too high by up to about 1 ε (0.01%) [100, 102].

Given the high precision that is now attained in the isotopic analysis of various elements by MC-ICP-MS, it is important that cosmochemical studies carefully consider and evaluate possible cosmogenic effects. Without such an evaluation, it is possible that cosmogenic isotopic variations are erroneously interpreted as nucleosynthetic, radiogenic, or mass-dependent stable isotope effects.

10.6 Conclusion

The application of MC-ICP-MS has had a profound impact on isotopic research in cosmochemistry over the last two decades. This immense impact primarily reflects two factors. First, MC-ICP-MS instruments are comparatively affordable and straightforward to use. As a result, there are now many laboratories world-wide in which MC-ICP-MS instruments are in routine use on a daily basis. The second factor is the performance characteristics of the instrumental technique, which is both versatile and suitable for high-precision isotopic analysis. As such, MC-ICP-MS can been applied to resolve small natural isotopic variations for a wide range of metallic and metalloid elements. Furthermore, it is equally suitable for the analysis of radiogenic and nucleosynthetic isotope anomalies and also mass-dependent isotope fractionations. As such, the technique of MC-ICP-MS is ideally suited for exploring the wealth of isotopic variations that are present in extra-terrestrial materials and many successful investigations, which have yielded novel and important results, have been carried out in the recent past.

In particular, the technique of MC-ICP-MS has significantly furthered the routine application of a number of extinct radionuclide systems, such as $^{182}Hf-^{182}W$, $^{26}Al-^{26}Mg$, and $^{107}Pd-^{107}Ag$, for a wide range of extraterrestrial samples, enabled routine analysis of the long-lived $^{176}Lu-^{176}Hf$ chronometer, and provided a straightforward means for the acquisition of precise Pb isotope ratio data. In addition, it has permitted precise studies of nucleosynthetic isotope anomalies to be carried out on a much wider range of elements and samples than was previously feasible. Taken together, such investigations have significantly improved our understanding of the nature and timing of the processes which characterized the dynamic early solar system and which were completed with the formation of the planetary system that we know today. Furthermore, MC-ICP-MS also facilitated the first systematic studies of "non-traditional" stable isotope systems in extraterrestrial materials. Such investigations are still at an early stage at present for many elements but are likely to lead to important conclusions, for example, regarding the origin of the volatile-depleted nature of many solar system bodies, in the near future.

Acknowledgments

The contents of this chapter were shaped by discussions that the authors had with many colleagues over the years, including Gretchen Benedix, Phil Bland, Manuela Fehr, Alex Halliday, Sara Russell, and Frank Wombacher. Frank Vanhaecke and Patrick Degryse are thanked for giving us the opportunity to prepare this chapter and their patient editorial handling.

References

1 Cowley, C.R. (1995) *An Introduction to Cosmochemistry*, Cambridge University Press, Cambridge.

2 McSween, H.Y. and Huss, G.R. (2010) *Cosmochemistry*, Cambridge University Press, Cambridge.

3 McSween, H.Y. (1999) *Meteorites and Their Parent Bodies*, Cambridge University Press, Cambridge.

4 Suess, H.E. (1988) V.M. Goldschmidt and the origin of the elements. *Appl. Geochem.*, **3**, 385–391.

5 Suess, H.E. and Urey, H.C. (1956) Abundances of the elements. *Rev. Mod. Phys.*, **28**, 53–74.

6 Nier, A.O. (1989) Some reminiscences of mass spectrometry and the Manhattan Project. *J. Chem. Educ.*, **66**, 385–388.

7 Holmes, A., (1946) An estimate of the age of the Earth. *Nature*, **157**, 680–684.

8 Houtermans, F.G. (1946) Die Isotopenhäufigkeiten im natürlichen Blei und das Alter des Urans. *Naturwissenschaften*, **33**, 185–186.

9 Patterson, C.E. (1956) Age of meteorites and the Earth. *Geochim. Cosmochim. Acta*, **10**, 230–237.

10 Halliday, A.N., Lee, D.-C., Christensen, J.N., Walder, A.J., Freedman, P.A., Jones, C.E., Hall, C.M., Yi, W., and Teagle, D. (1995) Recent developments in inductively coupled plasma magnetic sector multiple collector mass spectrometry. *Int. J. Mass Spectrom. Ion Processes*, **146/147**, 21–33.

11 Walder, A.J. and Freedman, P.A. (1992) Isotopic ratio measurement using a double focusing magnetic sector mass analyser with an inductively coupled plasma as an ion source. *J. Anal. At. Spectrom.*, **7**, 571–575.

12 Walder, A.J. and Furuta, N. (1993) High-precision lead isotope ratio measurement by inductively coupled plasma multiple collector mass spectrometry. *Anal. Sci.*, **9**, 675–680.

13 Walder, A.J., Platzner, I., and Freedman, P.A. (1993) Isotope ratio measurement of lead, neodymium and neodymium–samarium mixtures, hafnium and hafnium–lutetium mixtures with a double focusing multiple collector inductively coupled plasma mass spectrometer. *J. Anal. At. Spectrom.*, **8**, 19–23.

14 Krot, A.N., Keil, K., Goodrich, C.A., Scott, E.R.D., and Weisberg, M.K. (2005) Classification of meteorites, in *Meteorites, Comets and Planets* (ed. A.M. Davis), Elesevier, Amsterdam, pp. 83–128.

15 Weisberg, M.K., McCoy, T.J., and Krot, A.N. (2006) Systematics and evaluation of meteorite classification, in *Meteorites and the Early Solar System II* (eds. D.S. Lauretta and H.Y. McSween), University of Arizona Press, Tucson, AZ, pp. 19–52.

16 Brearly, A.J. and Jones, R.H. (1998) Chondritic meteorites, in *Planetary Materials* (ed. Papike, J.J.), Mineralogical Society of America, Washington, DC, pp. 3-1 – 3-198

17 Scott, E.R.D. and Krot, A.N. (2005) Chondrites and their components, in *Metorites, Comets, and Planets* (ed. A.M. Davis), Elsevier, Amsterdam, pp. 143–200.

18 Amelin, Y., Krot, A.N., Hutcheon, I.D., and Ulyanov, A.A. (2002) Lead isotopic ages of chondrules and calcium–aluminum-rich inclusions. *Science*, **297**, 1678–1683.

19 Bouvier, A. and Wadhwa, M. (2010) The age of the solar system redefined by the oldest Pb–Pb age of a meteoritic inclusion. *Nat. Geosci.*, **3**, 637–641.

20 Zinner, E.K. (2005) Presolar grains, in *Meteorites, Comets and Planets* (ed. A.M. Davis), Elsevier, Amsterdam, pp. 17–39.

21 Mittlefehldt, D.W. (2005) Achondrites, in *Meteorites, Comets, and Planets* (ed. A.M. Davis), Elsevier, Amsterdam, pp. 291–324.

22 Mittlefehldt, D.W., McCoy, T.J., Goodrich, C.A., and Kracher, A., (1998). Non-chondritic meteorites from asteroidal bodies, in *Planetary Materials* (ed. J.J. Papike), Mineralogical Society of America, Washington, DC, pp. 4-01 – 4-195.

23 Haack, H. and McKoy, T.J. (2005) Iron and stony-iron meteorites, in *Meteorites, Comets, and Planets* (ed. A.M. Davis), Elsevier, Amsterdam, pp. 325–345.

24 Burbidge, E.M., Burbidge, G.R., Fowler, W.A., and Hoyle, F. (1957) Synthesis of elements in stars. *Rev. Mod. Phys.*, **29**, 547–650.

25 Truran, J. and Heger, A. (2005) Origin of the elements, in *Meteorites, Comets and Planets* (ed. A.M. Davis), Elsevier, Amsterdam, pp. 1–15.

26 Hoefs, J. (2004) *Stable Isotope Geochemistry*, Springer, Berlin.

27 Rehkämper, M., Wombacher, F., and Aggarwal, J.K. (2004) Stable isotope analysis by multiple collector ICP-MS, in *Handbook of Stable Isotope Analytical Techniques*, vol. **I** (ed. P. de Groot), Elsevier, Amsterdam, pp. 652–691.

28 Eugster, O., Herzog, G.F., Marti, K., and Caffee, M.W. (2006) Irridation records, cosmic-ray exposure ages, and transfer times of meteorites, in *Meteorites and the Early Solar System II* (eds. D.S. Lauretta and H.Y. McSween), University of Arizona Press, Tucson, AZ, pp. 829–867.

29 Herzog, G.F. (2005) Cosmic-ray exposure ages of meteorites, in *Meteorites, Comets and Planets* (ed. A.M. Davis), Elsevier, Amsterdam, pp. 347–380.

30 Jull, A.J.T. (2006) Terrestrial ages of meteorites, in *Meteorites and the Early Solar System II*, (eds. D.S. Lauretta and H.Y. McSween), University of Arizona Press, Tucson, AZ, pp. 889–905.

31 Halliday, A.N., Christensen, J.N., Lee, D.-C., Rehkämper, M., Hall, C.M., and Luo, X., (2000) Multiple-collector inductively coupled plasma mass spectrometry, in *Inorganic Mass Spectrometry* (eds. C.M. Barslik, D.C. Duckworth, and D.H. Smith), Marcel Dekker, New York, pp. 291–328.

32 Rehkämper, M., Schönbächler, M., and Stirling, C.H. (2001) Multiple collector ICP-MS: introduction to instrumentation, measurement techniques and analytical capabilities. *Geostand. Newsl.*, **25**, 23–40.

33 Wasserburg, G.J., Jacobsen, S.B., DePaolo, D.J., McCulloch, M.T., and Wen, T. (1981) Precise determination of Sm/Nd ratios, Sm and Nd isotopic abundances in standard solutions. *Geochim. Cosmochim. Acta*, **45**, 2311–2323.

34 Wombacher, F. and Rehkämper, M. (2003). Investigation of the mass discrimination of multiple collector ICP-MS using neodymium isotopes and the generalised power law. *J. Anal. At. Spectrom.*, **18**, 1371–1375.

35 Albarède, F. and Beard, B. (2004) Analytical methods for non-traditional isotopes. *Rev. Mineral. Geochem.*, **55**, 113–152.

36 Longerich, H.P., Fryer, B.J., and Strong, D.F. (1987) Determination of lead isotopic ratios by inductively coupled plasma mass spectrometry

(ICP-MS). *Spectrochim. Acta B*, **42**, 39–48.

37 Maréchal, C.N., Télouk, P., and Albarède, F. (1999) Precise analysis of copper and zinc isotopic compositions by plasma-source mass spectrometry. *Chem. Geol.*, **156**, 251–273.

38 Wombacher, F., Rehkämper, M., Mezger, K., and Münker, C. (2003) Stable isotope compositions of cadmium in geological materials and meteorites determined by multiple collector-ICPMS. *Geochim. Cosmochim. Acta*, **67**, 4639–4654.

39 Nielsen, S.G., Rehkämper, M., and Halliday, A.N. (2006) Thallium isotopic variations in iron meteorites and evidence for live lead-205 in the early solar system. *Geochim. Cosmochim. Acta*, **70**, 2643–2657.

40 Woodland, S.J., Rehkämper, M., Halliday, A.N., Lee, D.-C., Hattendorf, B., and Günther, D. (2005) Accurate measurement of silver isotopic compositions in geological materials including low Pd/Ag meteorites. *Geochim. Cosmochim. Acta*, **69**, 2153–2163.

41 Christensen, J.N., Halliday, A.N., Godfrey, L.V., Hein, J.R., and Rea, D.K. (1997) Climate and ocean dynamics and the lead isotopic records in Pacific ferromanganese crusts. *Science*, **277**, 913–918.

42 Christensen, J.N., Halliday, A.N., Lee, D.-C., and Hall, C.M. (1995) *In situ* Sr isotopic analyses by laser ablation. *Earth Planet. Sci. Lett.*, **136**, 79–85.

43 Thirlwall, M.F. and Walder, A.J. (1995) *In situ* hafnium isotope ratio analysis of zircon by inductively coupled plasma multiple collector mass spectrometry. *Chem. Geol.*, **122**, 241–247.

44 Horn, I., von Blanckenburg, F., Schoenberg, R., Steinhoefel, G., and Markl, G. (2006) *In situ* iron isotope ratio determination using UV–femtosecond laser ablation with application to hydrothermal ore formation processes. *Geochim. Cosmochim. Acta*, **70**, 3677–3688.

45 Simonetti, A., Heaman, L.M., Hartlaub, R.P., Creaser, R.A., MacHattie, T.G., and Böhm, C., (2005) U–Pb zircon dating by laser ablation–MC-ICP-MS using a new multiple ion counting Faraday collector array. *J. Anal. At. Spectrom.*, **20**, 677–686.

46 Woodhead, J., Swearer, S., Hergt, J., and Maas, R. (2005) *In situ* Sr-isotope analysis of carbonates by LA–MC-ICP-MS: interference corrections, high spatial resolution and an example from otolith studies. *J. Anal. At. Spectrom.*, **20**, 22–27.

47 Young, E.D., Ash, R.D., Galy, A., and Belshaw, N.S. (2002) Mg isotope heterogeneity in the Allende meteorite measured by UV laser ablation–MC-ICPMS and comparisons with O isotopes. *Geochim. Cosmochim. Acta*, **66**, 683–698.

48 Iizuka, T. and Hirata, T. (2005) Improvements of precision and accuracy in *in situ* Hf isotope microanalysis of zircon using the laser ablation–MC-ICPMS technique. *Chem. Geol.*, **220**, 121–137.

49 Regelous, M., Elliott, T., and Coath, C.D. (2008) Nickel isotope heterogeneity in the early solar system. *Earth Planet. Sci. Lett.*, **272**, 330–338.

50 Fehr, M.A., Rehkämper, M., Halliday, A.N., Wiechert, U., Hattendorf, B., Günther, D., Ono, S., Eigenbrode, J.L., and Rumble, D. III (2005) Tellurium isotopic composition in the early solar system – a search for effects resulting from stellar nucleosynthesis, ^{126}Sn decay, and mass-independent fractionation. *Geochim. Cosmochim. Acta*, **69**, 5099–5112.

51 Schönbächler, M., Lee, D.C., Rehkämper, M., Halliday, A.N., Fehr, M.A., Hattendorf, B., and Günther, D. (2003) Zirconium isotope evidence for incomplete admixing of *r*-process components in the solar nebula. *Earth Planet. Sci. Lett.*, **216**, 467–481.

52 Schönbächler, M., Rehkämper, M., Halliday, A.N., Lee, D.-C., Bourot-Denise, M., Zanda, B., Hattendorf, B., and Günther, D. (2002). Niobium–zirconium chronometry and early solar system development. *Science*, **295**, 1705–1708.

53 Dauphas, N., Marty, B., and Reisberg, L. (2002) Molybdenum evidence for inherited planetary scale isotope heterogeneity of the protosolar nebula. *Astrophys. J.*, **565**, 640–644.

54 Leya, I., Schönbächler, M., Wiechert, U., Krähenbühl, U., and Halliday, A.N. (2008) Titanium isotopes and the radial heterogeneity of the solar system. *Earth Planet. Sci. Lett.*, **266**, 233–244.

55 Trinquier, A., Elliott, T., Ulfbeck, D., Coath, C., Krot, A.N., and Bizzarro, M. (2009) Origin of nucleosynthetic isotope heterogeneity in the solar protoplanetary disk. *Science*, **324**, 374–376.

56 Sanloup, C., Blichert-Toft, J., Trélouk, P., Gillet, P., and Albarède, F. (2000) Zr isotope anomalies in chondrites and the presence of ^{92}Nb in the early solar system. *Earth Planet. Sci. Lett.*, **184**, 75–81.

57 Yin, Q.Z., Jacobsen, S.B., Blichert-Toft, J., Télouk, P., and Albarède, F. (2001) Nb–Zr and Hf–W isotope systematics: applications to early solar system chronology and planetary differentiation, 32nd Lunar and Planetary Science Conference, 12–16 March 2001, Houston, TX, Abstract 2128.

58 Becker, H. and Walker, R.J. (2003) Efficient mixing of the solar nebula from uniform Mo isotopic composition of meteorites. *Nature*, **425**, 152–155.

59 Yin, Q.Z., Jacobsen, S.B., and Yamashita, K., (2002) Diverse supernova sources of pre-solar material inferred from molybdenum isotopes in meteorites. *Nature*, **415**, 881–883.

60 Schönbächler, M., Rehkämper, M., Lee, D.-C., and Halliday, A.N. (2004) Ion exchange chromatography and high precision isotopic measurements of zirconium by MC-ICP-MS. *Analyst*, **129**, 32–37.

61 Akram, W.M., Schönbächler, M., Williams, H., and Halliday, A.N. (2011) The origin of nucleosynthetic zirconium-96 heterogeneities in the inner solar system, 42nd Lunar and Planetary Science Conference, 7–11 March 2011, Houston, TX, Abstract 1908.

62 Lugaro, M., Davis, A.M., Gallino, R., Pellin, M.J., Straniero, O., and Käppeler, F. (2003) Isotopic composition of strontium, zirconium, molybdenum and barium in single presolar SiC grains and asymptotic giant branch stars. *Astrophys. J.*, **593**, 486–508.

63 Nittler, L.R. (2003) Presolar dust in meteorites: recent advances and scientific frontiers. *Earth Planet. Sci. Lett.*, **209**, 259–273.

64 Schönbächler, M., Rehkämper, M., Fehr, M.A., Halliday, A.N., Hattendorf, B., and Günther, D. (2005) Nucleosynthetic zirconium isotope anomalies in acid leachates of carbonaceous chondrites. *Geochim. Cosmochim. Acta*, **69**, 5113–5122.

65 Nicolussi, G.K., Davis, A.M., Pellin, M.J., Lewis, R.S., Clayton, R.N., and Amari, S. (1997) *s*-Process zirconium in presolar silicon carbide grains. *Science*, **277**, 1281–1283.

66 Fehr, M.A., Rehkämper, M., Halliday, A.N., Schönbächler, M., Hattendorf, B., and Günther, D. (2006) Search for nucleosynthetic and radiogenic tellurium isotope anomalies in carbonaceous chondrites. *Geochim. Cosmochim. Acta*, **70**, 3436–3448.

67 Birck, J.L. and Allegre, C.J. (1978) Chronology and chemical history of parent body of basaltic achondrites studied by Rb-87–Sr-87 method. *Earth Planet. Sci. Lett.*, **39**, 37–51.

68 Waight, T., Baker, J., and Peate, D. (2002). Sr isotope ratio measurements by double-focusing MC-ICPMS: techniques, observations and pitfalls. *Int. J. Mass Spectrom.*, **221**, 229–244.

69 Moynier, F., Agranier, A., Hezel, D.C., and Bouvier, A. (2010) Sr stable isotope composition of Earth, the Moon, Mars, Vesta and meteorites. *Earth Planet. Sci. Lett.*, **300**, 359–366.

70 Blichert-Toft, J., Boyet, M., Telouk, P., and Albarède, F. (2002). Sm-147–Nd-143 and Lu-176–Hf-176 in eucrites and the differentiation of the HED parent body. *Earth Planet. Sci. Lett.*, **204**, 167–181.

71 Andreasen, R. and Sharma, M., (2006) Solar nebula heterogeneity in *p*-process samarium and neodymium isotopes. *Science*, **314**, 806–809.

72 Boyet, M. and Carlson, R.W. (2005) Nd-142 evidence for early (>4.53 Ga) global differentiation of the silicate Earth. *Science*, **309**, 576–581.

73 Boyet, M., Blichert-Toft, J., Rosing, M., Storey, M., Telouk, P., and Albarède, F. (2003) Nd-142 evidence for early Earth differentiation. *Earth Planet. Sci. Lett.*, **214**, 427–442.

74 Foley, C.N., Wadhwa, M., Borg, L.E., Janney, P.E., Hines, R., and Grove, T.L. (2005) The early differentiation history of Mars from W-182–Nd-142 isotope systematics in the SNC meteorites. *Geochim. Cosmochim. Acta*, **69**, 4557–4571.

75 Andreasen, R. and Sharma, M. (2009) Fractionation and mixing in a thermal ionization mass spectrometer source: implications and limitations for high-precision Nd isotope analyses. *Int. J. Mass Spectrom.*, **285**, 49–57.

76 Patchett, P.J. and Tatsumoto, M. (1980) Lu–Hf total-rock isochron for the eucrite meteorites. *Nature*, **288**, 571–574.

77 Blichert-Toft, J. and Albarède, F. (1997) The Lu–Hf isotope geochemistry of chondrites and the evolution of the mantle–crust system. *Earth Planet. Sci. Lett.*, **148**, 243–258.

78 Thrane, K., Connelly, J.N., Bizzarro, M., Meyer, B.S., and The, L.S. (2010) Origin of excess Hf-176 in meteorites. *Astrophys. J.*, **717**, 861–867.

79 Amelin, Y., 2005. Meteorite phosphates show constant ^{176}Lu decay rate since 4557 million years ago. *Science*, **310**, 839–841.

80 Söderlund, U., Patchett, J.P., Vervoort, J.D., and Isachsen, C.E. (2004) The Lu-176 decay constant determined by Lu–Hf and U–Pb isotope systematics of precambrian mafic intrusions. *Earth Planet. Sci. Lett.*, **219**, 311–324.

81 Bizzarro, M., Baker, J.A., Haack, H., Ulfbeck, D., and Rosing, M. (2003). Early history of Earth's crust–mantle system inferred from hafnium isotopes in chondrites. *Nature*, **421**, 931–933.

82 Patchett, P.J., Vervoort, J.D., Söderlund, U., and Salters, V.J.M. (2004) Lu–Hf and Sm–Nd isotopic systematics in chondrites and their constraints on the Lu–Hf properties of the Earth. *Earth Planet. Sci. Lett.*, **222**, 29–41.

83 Sprung, P., Scherer, E.E., Upadhyay, D., Leya, I., and Mezger, K. (2010) Non-nucleosynthetic heterogeneity in non-radiogenic stable Hf isotopes: implications for early solar system chronology. *Earth Planet. Sci. Lett.*, **295**, 1–11.

84 Albarède, F., Scherer, E.E., Blichert-Toft, J., Rosing, M., Simionovici, A., and Bizzarro, M. (2006) Gamma-ray irradiation in the early solar system and the conundrum of the Lu-176 decay constant. *Geochim. Cosmochim. Acta*, **70**, 1261–1270.

85 Amelin, Y. and Davis, W.J. (2005) Geochemical test for branching decay of Lu-176. *Geochim. Cosmochim. Acta*, **69**, 465–473.

86 Baker, J., Peate, D., Waight, T., and Meyzen, C., 2004. Pb isotopic analysis of standards and samples using a Pb-207–Pb-204 double spike and thallium to correct for mass bias with a double-focusing MC-ICP-MS. *Chem. Geol.*, **211**, 275–303.

87 Thirlwall, M.F. (2002) Multicollector ICP-MS analysis of Pb isotopes using a $^{207}Pb/^{204}Pb$ double spike demonstrates up to 400 ppm/amu systematic errors in Tl-normalization. *Chem. Geol.*, **184**, 255–279.

88 Brennecka, G.A., Weyer, S., Wadhwa, M., Janney, P.E., Zipfel, J., and Anbar, A.D. (2010) U-238/U-235 variations in meteorites: extant Cm-247 and implications for Pb–Pb dating. *Science*, **327**, 449–451.

89 Amelin, Y., Kaltenbach, A., Iizuka, T., Stirling, C.H., Ireland, T.R., Petaev, M., and Jacobsen, S.B. (2010) U–Pb chronology of the solar system's oldest solids with variable U-238/U-235. *Earth Planet. Sci. Lett.*, **300**, 343–350.

90 Halliday, A.N. and Kleine, T. (2006) Meteorites and the timing, mechanisms, and conditions of terrestrial planet accretion and early differentiation, in *Meteorites and the Early Solar System II* (eds. D.S. Lauretta and H.Y. McSween

Jr), University of Arizona Press, Tucson, AZ, pp. 775–801.

91 Schönbächler, M., Carlson, R.W., Horan, M., Mock, T.D., and Hauri, E.H. (2010) Heterogeneous accretion and the moderately volatile element budget of Earth. *Science*, **328**, 884–887.

92 Wood, B.J., Nielsen, S.G., Rehkämper, M., and Halliday, A.N. (2008). The effects of core formation on the Pb- and Tl-isotopic composition of the silicate Earth. *Earth Planet. Sci. Lett.*, **269**, 326–336.

93 Burkhardt, C., Kleine, T., Bourdon, B., Palme, H., Zipfel, J., Friedrich, J.M., and Ebel, D.S. (2008) Hf–W mineral isochron for Ca,Al-rich inclusions: age of the solar system and the timing of core formation in planetesimals. *Geochim. Cosmochim. Acta*, **72**, 6177–6197.

94 Chabot, N.L. and Haack, H. (2006) Evolution of asteroidal cores, in *Meteorites and the Early Solar System II* (eds. D.S. Lauretta, and H.Y. McSween Jr), University of Arizona Press, Tucson, AZ, pp. 747–771.

95 Wadhwa, M., Srinivasan, G., and Carlson, R.W. (2006). Timescales of planetesimal differentiation in the early solar system, in *Meteorites and the Early Solar System II* (eds. D.S. Lauretta and H.Y. McSween Jr), University of Arizona, Tucson, AZ, pp. 715–731.

96 Schönbächler, M., Carlson, R.W., Horan, M.F., Mock, T.D., and Hauri, E.H. (2007) High precision Ag measurements in geologic materials by multiple collector ICPMS: an evaluation of dry- versus wet-plasma. *Int. J. Mass Spectrom.*, **261**, 183–191.

97 Horan, M.F., Smoliar, M.I., and Walker, R.J. (1998) ^{182}W and ^{187}Re–^{187}Os systemantics of iron meteorites: chronology of melting, differentiation, and crystallization in asteroids. *Geochim. Cosmochim. Acta*, **62**, 545–554.

98 Lee, D.-C. (2005) Protracted core formation in asteroids: evidence from high-precision W isotopic data. *Earth Planet. Sci. Lett.*, **237**, 21–32.

99 Markowski, A., Leya, I., Quitté, G., Ammon, K., Halliday, A.N., and Wieler, R. (2006) Correlated helium-3 and tungsten isotopes in iron meteorites: quantitative cosmogenic corrections and planetesimal formation times. *Earth Planet. Sci. Lett.*, **250**, 104–115.

100 Markowski, A., Quitté, G., Halliday, A.N., and Kleine, T. (2006). Tungsten isotopic compositions of iron meteorites: chronological constraints vs. cosmogenic effects. *Earth Planet. Sci. Lett.*, **242**, 1–15.

101 Qin, L., Dauphas, N., Wadhwa, M., Masarik, J., and Janney, P.E. (2008) Rapid accretion and differentiation of iron meteorite parent bodies inferred from ^{182}Hf–^{182}W chronometry and thermal modeling. *Earth Planet. Sci. Lett.*, **273**, 94–104.

102 Scherstén, A., Elliott, T., Hawkesworth, C., Russell, S., and Masarik, J. (2006) Hf–W evidence for rapid differentiation of iron meteorite parent bodies. *Earth Planet. Sci. Lett.*, **241**, 530–542.

103 Carlson, R.W. and Hauri, E.H. (2001) Extending the ^{107}Pd–^{107}Ag chronometer to low Pd/Ag meteorites with multicollector plasma-ionization mass spectrometry. *Geochim. Cosmochim. Acta*, **65**, 1839–1848.

104 Chen, J.H. and Wasserburg, G.J. (1996) Live ^{107}Pd in the early solar system and implications for planetary evolution, in *Earth Processes – Reading the Isotopic Code* (eds. A. Basu and S. Hart), American Geophysical Union, Washington, DC, pp. 1–20.

105 Baker, R.G.A., Schönbächler, M., Rehkämper, M., Williams, H., and Halliday, A.N. (2010) The thallium isotope composition of carbonaceous chondrites – new evidence for live ^{205}Pb in the early solar system. *Earth Planet. Sci. Lett.*, **291**, 39–47.

106 Schönbächler, M., Carlson, R.W., Horan, M.F., Mock, T.D., and Hauri, E. H. (2008) Silver isotope variations in chondrites: volatile depletion and the intitial ^{107}Pd abundance of the solar system. *Geochim. Cosmochim. Acta*, **72**, 5330–5341.

107 Sugiura, N. and Hoshino, H. (2003) Mn–Cr chronology of five IIIAB iron meteorites. *Meteorit. Planet. Sci.*, **38**, 117–143.

108 Rasmussen, K.L. (1989). Cooling rates of IIIAB iron meteorites. *Icarus*, **80**, 315–325.

109 Benedix, G.K., McCoy, T.J., Keil, K., and Love, S.G. (2000) A petrographic study of the IAB iron meteorites: constraints on the formation of the IAB–Winonaite parent body. *Meteorit. Planet. Sci.*, **35**, 1127–1141.

110 Vogel, N. and Renne, P.R. (2008) ^{40}Ar–^{39}Ar dating of plagioclase grain size separates from silicate inclusions in IAB iron meteorites and implications for the thermochronological evolution of the IAB parent body. *Geochim. Cosmochim. Acta*, **72**, 1231–1255.

111 Galy, A., Belshaw, N.S., Halicz, L., and O'Nions, R.K. (2001) High-precision measurement of magnesium isotopes by multiple-collector inductively coupled plasma mass spectrometry. *Int. J. Mass Spectrom.*, **208**, 89–98.

112 Jeffcoate, A.B., Elliott, T., Thomas, A., and Bouman, C. (2004) Precise, small sample size determinations of lithium isotopic compositions of geological reference materials and modern seawater by MC-ICP-MS. *Geostand. Geoanal. Res.*, **28**, 161–172.

113 Wombacher, F. and Rehkämper, M. (2004) Problems and suggestions concerning the notation of Cd stable isotope compositions and the use of reference materials. *Geostand. Geoanal. Res.*, **28**, 173–178.

114 Dodson, M.H. (1963) A theoretical study of the use of internal standards for precise isotopic analysis by the surface ionization technique. Part I: general first-order algebraic solutions. *J. Sci. Instrum.*, **40**, 289–295.

115 Rudge, J.F., Reynolds, B.C., and Bourdon, B. (2009) The double spike toolbox. *Chem. Geol.*, **265**, 420–431.

116 Arnold, T., Schönbächler, M., Rehkämper, M., Dong, S., Zhao, F.-J., Kirk, G.J.D., Coles, B.J., and Weiss, D.J. (2010) Determination of zinc stable isotope compositions in geological and biological samples by double spike MC-ICPMS. *Anal. Bioanal. Chem.*, **398**, 3115–3125.

117 Bermin, J., Vance, D., Archer, C., and Statham, P.J. (2006) The determination of the isotopic composition of Cu and Zn in seawater. *Chem. Geol.*, **226**, 280–297.

118 Siebert, C., Nägler, T.F., and Kramers, J.D. (2001) Determination of molybdenum isotope fractionation by double-spike multicollector inductively coupled plasma mass spectrometry. *Geochem. Geophys. Geosyst.*, **2** (7), 1032, doi: 10.1029/2000GC000124.

119 Ripperger, S. and Rehkämper, M. (2007) Precise determination of cadmium isotope fractionation in seawater by double-spike MC-ICPMS. *Geochim. Cosmochim. Acta*, **71**, 631–642.

120 Williams, H., Markowski, A., Quitté, G., Halliday, A.N., Teutsch, N., and Levasseur, S. (2006) Fe isotope fractionation in iron meteorites: new insights into metal–sulphide segregation and planetary accretion. *Earth Planet. Sci. Lett.*, **250**, 486–500.

121 Poitrasson, F. (2007) Does planetary differentiation really fractionate iron isotopes?. *Earth Planet. Sci. Lett.*, **256**, 484–492.

122 Poitrasson, F., Halliday, A.N., Lee, D.-C., Levasseur, S., and Teutsch, N. (2004) Iron isotope differences between Earth, Moon, Mars and Vesta as possible records of contrasted accretion mechanisms. *Earth Planet. Sci. Lett.*, **223**, 253–266.

123 Weyer, S., Anbar, A.D., Brey, G.P., Münker, C., Mezger, K., and Woodland, A.B. (2005) Iron isotope fractionation during planetary differentiation. *Earth Planet. Sci. Lett.*, **240**, 251–264.

124 Weyer, S., Anbar, A.D., Brey, G.P., Münker, C., Mezger, K., and Woodland, A.B. (2007) Fe-isotope fractionation during partial melting on Earth and the current view on the Fe-isotope budgets of the planets. *Earth Planet. Sci. Lett.*, **256**, 638–646.

125 Lodders, K. (2003) Solar system abundances and condensation temperatures of the elements. *Astrophys. J.*, **591**, 1220–1247.

126 Kehm, K., Hauri, E.H., Alexander, C.M.O.D., and Carlsen, R.W. (2003) High precision iron isotope measurements of meteoritic material by cold plasma ICP-MS. *Geochim. Cosmochim. Acta*, **67**, 2879–2891.

127 Luck, J.-M., Ben Othman, D., and Albarède, F. (2005) Zn and Cu isotopic variations in chondrites and iron meteorites: early solar nebula reservoirs and parent-body processes. *Geochim. Cosmochim. Acta*, **69**, 5351–5363.

128 Luck, J.M., Ben Othman, D., Barrat, J.A., and Albarède, F. (2003) Coupled ^{63}Cu and ^{16}O excesses in chondrites. *Geochim. Cosmochim. Acta*, **67**, 143–151.

129 Moynier, F., Blichert-Toft, J., Telouk, P., Luck, J.-M., and Albarède, F. (2007) Comparative stable isotope geochemistry of Ni, Cu, Zn, and Fe in chondrites and iron meteorites. *Geochim. Cosmochim. Acta*, **71**, 4365–4379.

130 Poitrasson, F., Levasseur, S., and Teutsch, N. (2005) Significance of iron isotope mineral fractionation in pallasites and iron meteorites for the core–mantle differentiation of terrestrial planets. *Earth Planet. Sci. Lett.*, **234**, 151–164.

131 Schiller, M., Handler, M.R., and Baker, J.A. (2010) High-precision Mg isotopic systematics of bulk chondrites. *Earth Planet. Sci. Lett.*, **297**, 165–173.

132 Teng, F.-Z., Li, W.-Y., Ke, S., Marty, B., Dauphas, N., Hunag, S., Wu, F.-Y., and Pourmand, A. (2010) Magnesium isotopic composition of the Earth and chondrites. *Geochim. Cosmochim. Acta*, **74**, 4150–4166.

133 Wiechert, U. and Halliday, A.N. (2007) Non-chondritic magnesium and the origins of the inner terrestrial planets. *Earth Planet. Sci. Lett.*, **256**, 360–371.

134 Wombacher, F., Rehkämper, M., Mezger, K., Bischoff, A., and Münker, C. (2008) Cadmium stable isotope cosmochemistry. *Geochim. Cosmochim. Acta*, **72**, 646–667.

135 Zhu, X.K., Guo, Y., O'Nions, R.K., Young, E.D., and Ash, R.D. (2001) Isotopic homogeneity of iron in the early solar system. *Nature*, **412**, 311–313.

136 Humayun, M. and Cassen, P. (2000) Processes determining the volatile abundances of the meteorites and terrestrial planets, in *Origin of the Earth and Moon* (eds. R.M. Canup and K. Righter), University of Arizona Press, Tucson, AZ, pp. 3–23.

136a Davis, A.M. (2006) Volatile evolution and loss, in *Meteorites and the Early Solar System II* (eds. D.S. Lauretta and H.Y. McSween), University of Arizona Press, Tucson, AZ, pp. 295–307.

137 Bogard, D.D., Nyquis, L.E., Bansal, B.M., Garrison, D.H., Wiesmann, H., Herzog, G.F., Albrecht, A.A., Vogt, S., and Klein, J. (1995) Neutron-capture ^{36}Cl, ^{41}Ca, ^{36}Ar, and ^{150}Sm in large chondrites: evidence for high fluences of thermalized neutrons. *J. Geophys. Res.*, **100**, 9401–9416.

138 Eugster, O., Tera, F., Burnett, D.S., and Wasserburg, G.J. (1970) The isotopic composition of Gd and the neutron capture effects in samples from Apollo 11. *Earth Planet. Sci. Lett.*, **8**, 20–30.

139 Sands, D.G., de Laeter, J.R., and Rosman, K.J.R. (2001) Measurements of neutron capture effects on Cd, Sm and Gd in lunar samples with implications for the neutron energy spectrum. *Earth Planet. Sci. Lett.*, **186**, 335–346.

140 Lee, D.-C. and Halliday, A.N. (1995) Hafnium–tungsten chronometry and the timing of terrestrial core formation. *Nature*, **378**, 771–774.

140a Schoenberg, R., Kamber, B.S., Collerson, K.D., and Eugster, O. (2002) New W-isotope evidence for rapid terrestrial accretion and very early core formation. *Geochim. Cosmochim. Acta*, **66**, 3151–3160.

141 Lee, D.-C., Halliday, A.N., Leya, I., Wieler, R., and Wichert, U. (2002) Cosmogenic tungsten and the origin and earliest differentiation of the Moon. *Earth Planet. Sci. Lett.*, **198**, 267–274.

11
Establishing the Basis for Using Stable Isotope Ratios of Metals as Paleoredox Proxies

Laura E. Wasylenki

11.1
Introduction

In terms of the numbers of dedicated instruments, scientists, and recent publications, one of the applications of multi-collector (MC) inductively coupled plasma mass spectrometry (ICP-MS) attracting the most attention is the measurement of stable isotope ratios of transition and post-transition metals. Much of the excitement is inspired by the immeasurable impact that stable isotope geochemistry of "light" elements (H, C, O, N, and S) has had in so many areas of earth science, biology, and forensic science. The surprising turn-of-the-millennium discovery that stable isotope ratios of many heavier elements, including even Tl and U, also vary significantly in nature has raised the prospect that stable isotope geochemistry of H, C, O, N, and S may be only the proverbial tip of the iceberg. Consequently, the past few years have witnessed a flood of activity aimed at characterizing and interpreting metal isotope variability in nature.

Much of this activity has been undertaken by earth scientists interested in finding records of the climatic, biological, and geological processes of the Earth's past in ancient rocks. All that remains from the Earth's distant past are the rocks, and the quest to reconstruct a complete history requires resourceful use of as many geochemical signals as possible. The hope is that isotopic signatures of metals will provide a powerful new set of paleoproxy tools that can reveal the magnitudes, senses, and rates of change of ancient processes and variables that have thus far been very poorly constrained. Hence the first decade of work included many efforts to interpret isotopic signatures of metals as records of changes in metal fluxes to the oceans, marine and atmospheric redox status, seawater temperature, seawater pH, nutrient levels, and activities of particular biological metabolisms.

A limitation on the usefulness of metal isotope ratio paleoproxies thus far is a lack of certainty that patterns in isotopic compositions observed in natural samples truly reflect the environmental variables of interest, rather than other variables or

Isotopic Analysis: Fundamentals and Applications Using ICP-MS,
First Edition. Edited by Frank Vanhaecke and Patrick Degryse.

Published 2012 by WILEY-VCH Verlag GmbH & Co. KGaA

diagenetic processes or some combination of these. The initial wave of exploration of isotope ratios of metals in nature has taught us that real data sets are complicated and that we rarely see the simple patterns of variation that we might expect. Most publications include long discussions of multiple possible explanations for the patterns observed, and it is clear that we do not yet have a sufficient grasp of all the variables that govern isotopic signatures. Many papers advise caution, and metal stable isotope proxies have yet to yield many unambiguous answers, even in modern settings, where characterization of the systems should be more straightforward. Rather than let enthusiasm wane, we must find ways to overcome complexities and develop more nuanced approaches to reading the signals recorded by metal isotopes in natural samples.

Part of the way forward is to build a better fundamental understanding of the processes and mechanisms governing isotope ratios in natural materials. This is best done through experimentation with simple systems, sometimes complemented by theoretical work. If they are well designed, experimental and theoretical investigations limit the number of variables at play and control conditions such that relationships between intended variables and isotopic behavior can be systematically quantified. These investigations thus serve as a critical grounding for interpretation of the more complex data found in natural samples. Despite the importance of such work to application of metal isotope geochemistry to the geological record, only a few processes relevant to metal isotope ratio paleoproxies have yet been studied with this fundamental approach.

The aim of this chapter is to review contributions to the use of metal isotopes as paleoredox proxies from fundamental investigations of the mechanisms responsible for fractionation. The chapter focuses on ancient redox conditions, because this has been an area of widespread interest in the geochemistry community and because a significant amount of fundamental work has already been published that bears on the interpretation of existing and future data. Readers seeking comprehensive reviews of natural data sets from recent or ancient settings, proxies for other environmental parameters, or proxies for particular biological processes are referred to other helpful reviews for such information, including Volume 55 of the *Reviews in Mineralogy and Geochemistry* series Johnson *et al.*, [1], Anbar and Rouxel [2], Johnson *et al.* [3], Montero-Serrano *et al.* [4], Lyons and co-workers [5, 6], Severmann and Anbar [7], and Bullen [8].

Throughout this chapter, variations in isotopic compositions will be described using conventional stable isotope notation. Fractionation factors are defined in terms of alpha, with

$$\alpha_{\text{sample 1/sample 2}} = \frac{{}^{heavy}E_{\text{sample 1}}/{}^{light}E_{\text{sample 1}}}{{}^{heavy}E_{\text{sample 2}}/{}^{light}E_{\text{sample 2}}}$$

where *heavy* and *light* are integers referring to the atomic masses of the heavier and lighter isotopes of element *E*, and sample 1 and sample 2 refer to the two pools of element *E* being compared. The isotopic compositions of single samples relative to a standard are described in terms of the relative difference in the isotope ratios:

$$\delta^{\text{heavy/light}}E_{\text{sample 1/standard}} = \left(\frac{{}^{heavy}E_{\text{sample 1}}/{}^{light}E_{\text{sample 1}} - {}^{heavy}E_{\text{standard}}/{}^{light}E_{\text{standard}}}{{}^{\text{heavy}}E_{\text{standard}}/{}^{light}E_{\text{standard}}}\right) \times 1000‰$$

Differences in the isotopic compositions of two samples are denoted by

$$\Delta^{heavy/light}E_{\text{sample 1/sample 2}} = \delta^{heavy/light}E_{\text{sample 1/standard}} - \delta^{heavy/light}E_{\text{sample 2/standard}}$$

Note that Δ is also used in another context elsewhere and does not refer here to mass-independent fractionation (see also Chapter 1).

It is important to note that the instruments used for measuring isotope ratios of heavier elements have multiple collectors, whether the ions are produced in an inductively coupled plasma or thermal ionization source. This setup is required because the magnitudes of variation in isotope ratios for heavier elements are far smaller than for H, C, O, N, and S, largely because the relative mass difference between isotopes is typically only 1–5%. Isotope ratios in nature typically vary by a few permil or few tenths of permil only, such that resolving real differences between samples requires high analytical precision. In addition, efficient ionization, which permits the analysis of very small samples (from a few nanograms to a few micrograms of metal) and fast sample throughput are also important factors driving the recent explosion in research on stable isotope geochemistry of metals using MC-ICP-MS.

11.2 Isotope Ratios of Metals as Paleoredox Proxies

To date, reconstruction of redox conditions in the deep past is the most common objective of publications on the stable isotope geochemistry of metals. This is partly because some of the earliest work on metal isotopes was undertaken by paleoceanographers and partly because the chemical behavior of several transition and post-transition metals changes profoundly as a function of redox conditions. Such differences in chemical behavior drive isotope fractionation effects, and therefore shifts in redox conditions over time should be recorded in the isotopic signatures of rocks deposited during those shifts. Much attention has been directed to reconstructing the history of the two most significant increases in oxygenation of the oceans and atmosphere at ~2.3 Ga and ~550 Ma, because of the profound consequences that those changes had for all of life on Earth, but other studies have addressed more recent episodes of anoxia in the Phanerozoic Eon.

A particularly appealing aspect of paleoredox proxies based on isotope ratios of metals is the possibility of interpreting records of *global* redox conditions in the oceans or atmosphere, as opposed to the local snapshots in time and space provided by the occurrence of particular rock types, such as banded iron formations. Those elements that are reasonably abundant in seawater, have long residence times in the ocean relative to the ~1600 year mixing time [9], and have changes in valence state and/or speciation in response to environmental redox shifts show the highest potential as global proxies. The isotopic systems proposed as paleoredox proxies in the literature thus far include Mo and U (for the oceans) and Cr (for the atmosphere), and these three are reviewed below.

Especially for Mo and Cr, significant experimental and theoretical work has already been conducted to constrain the fractionation mechanisms behind the isotope ratio variations observed and to lay quantitative groundwork for the accurate interpretation of natural samples. As will be clear from the following pages, the conceptual models for paleoredox proxies are simple, with few reservoirs and processes involved, but we are quickly learning from natural data sets and from experimental and theoretical efforts that the isotope systematics may in fact be less straightforward. Further experiments and theory will therefore play a critical role in the development of our ability to model the isotope budgets of redox-sensitive metals and to interpret the signatures of redox evolution in the rock record.

11.2.1
Molybdenum Isotope Ratios and Global Ocean Paleoredox

One of the first isotope ratio variations to be documented for a metal in nature was the wide range of molybdenum isotope ratios among samples of seawater, ferromanganese nodules, Black Sea sediment, Devonian Ohio Shale, and molybdenite [10]. Relative to an in-house standard, the isotopic composition of a single seawater sample was measured as $\delta^{97/95}$Mo = +1.48‰, whereas ferromanganese nodules were much lighter, ranging from −0.26 to −0.75‰, with a mean value of −0.51 ± 0.14‰. In contrast to ferromanganese sediments, recent euxinic sediments from the Black Sea ranged from +1.11 to +1.60‰, averaging +1.33 ± 0.20‰, and Devonian black shales were also enriched in the heavier isotopes ($\delta^{97/95}$Mo from +0.82 to +1.16; mean = +1.02 ± 0.11‰). Especially in 2001, when the field of stable isotope geochemistry of metals was very new, and even now, such a large range of isotopic compositions was surprising, especially since Mo is present in the same redox state, Mo(VI), in nearly all of the samples analyzed. Figure 11.1, from Arnold *et al.* [11], shows a compilation of data from continental and marine samples collected by Barling *et al.* [10], Siebert *et al.* [12], and Arnold *et al.* [11].

Based on the observation that the oxic and euxinic marine sinks for Mo display such contrasting isotopic signatures, Barling *et al.* [10] proposed the isotopic analysis of Mo in black shales as a potential tool for reconstructing paleoredox conditions in the global ocean through geological time. The proxy concept rests on the assumption that the Mo flux balance into and out of the global ocean reservoir is at a steady state, and therefore the inventory of Mo isotopes coming into the ocean should balance what is removed. Most Mo comes in from rivers and is ultimately derived from oxidative weathering of continental rocks, and early analyses suggested that

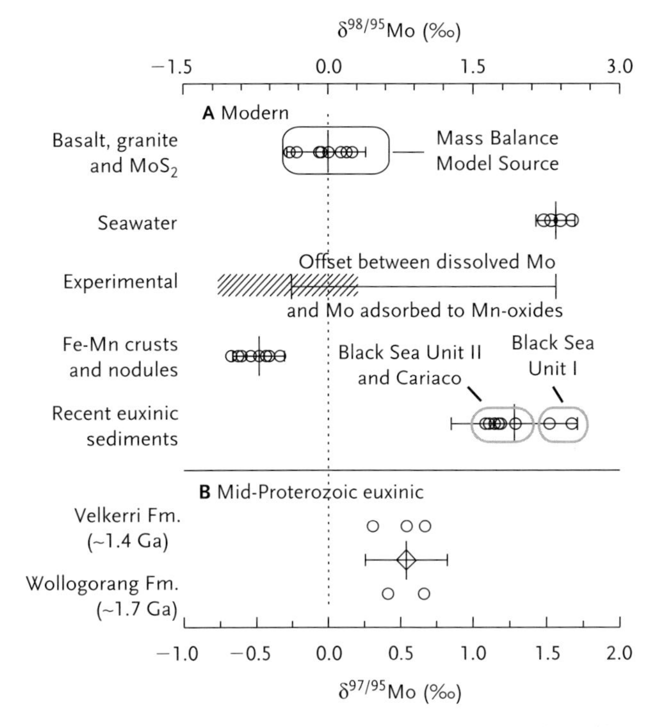

Figure 11.1 Data for continental and marine samples collected by Barling *et al.* [10], Siebert *et al.* [12], and Arnold *et al.* [11], shown in terms of both $\delta^{97/95}$Mo and $\delta^{98/95}$Mo. The "experimental" data come from the experimental study of Barling and Anbar [13], in which Mo was partially adsorbed on synthetic manganese oxyhydroxide particles. Reproduced from [11].

continental rocks are isotopically homogeneous [10] Siebert *et al.*, [12]; Arnold *et al.*, [11]. Two dominant sinks for Mo from the ocean were identified, one oxic (ferromanganese sediments) and the other euxinic (organic-rich sediments that will become black shales during diagenesis). Removal to the oxic sink is accompanied by a large isotope fractionation of $\Delta^{97/95}\text{Mo}_{\text{seawater–ferromanganese}} = +1.8$ to 2.0‰, with lighter isotopes preferentially incorporated into ferromanganese sediments. Because recent euxinic sediments and modern seawater have similar values of $\delta^{97/95}$Mo and because Mo is severely depleted from the water column in the Black Sea, a modern euxinic environment, Barling *et al.* [10] hypothesized that Mo is completely or almost completely removed from the water column in euxinic conditions, resulting in little or no isotopic difference between seawater and sediments. Given these systematics, shifts over time in the proportion of Mo removed via the oxic sink should shift the isotopic composition of Mo for the global ocean, and this shift should be faithfully recorded in rocks laid down in euxinic environments. When oxic deposition is widespread, the Mo in the ocean should be isotopically relatively heavy, due to removal of more light Mo in oxic depositional zones and vice versa. This simple mass balance model is illustrated in Figure 11.2, adapted from Barling *et al.* [10] and

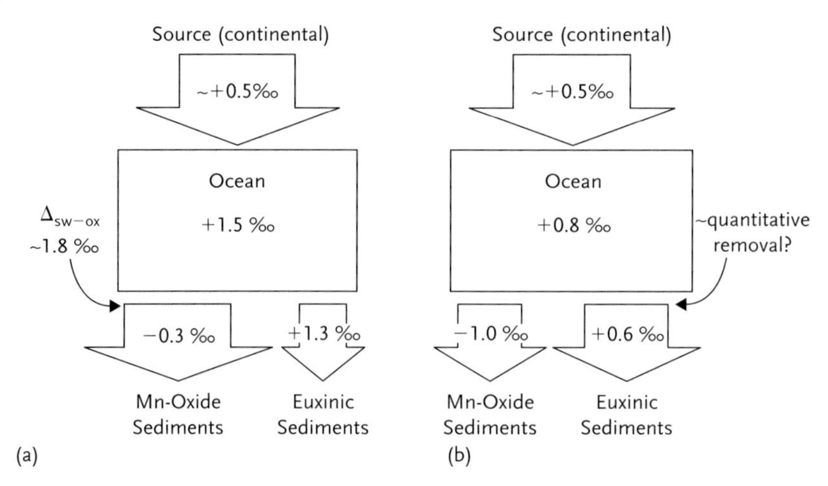

Figure 11.2 Simple two-sink box models for the Mo isotope budget of the oceans in (a) well-oxygenated conditions and (b) oxygen-depleted conditions. The value for the continental source is from [14]. Other values are based on those compiled by Arnold *et al.* [11].

Arnold *et al.* [11]. The isotopic composition of the input to the oceans from continental weathering in this figure is based on the analyses of river water by Archer and Vance [14]. Suboxic and low-sulfide, anoxic sinks were not included in this earliest model of the marine Mo isotope budget.

Since the first paper in 2001, the Mo isotope paleoproxy has become a popular study topic. At the time this chapter was written, there were 60 peer-reviewed publications on stable isotope ratios of Mo, and 48 of these directly address the use of Mo isotope ratios to reconstruct past redox shifts in oceans, estuaries, or lakes. Many of these papers are discussed in review articles by Severmann and Anbar [7], Montero-Serrano *et al.* [4], Lyons and co-workers [5, 6], Anbar and Rouxel [2], and Anbar [15]. Most of the published investigations belong to one of four groups: (i) those presenting analyses of Mo isotopic signatures in ancient samples that formed at times about which we know little about the timing and extent of fluctuations in redox conditions; (ii) analyses of natural samples formed before, during, and after large shifts in marine redox status already identified from other paleoredox proxies, for example, the Great Oxidation Event and the Ocean Anoxic Events of the Late Cretaceous Period; (iii) studies of modern waters and sediments, aimed at characterizing Mo isotopic behavior in modern or Recent systems in order to validate, refine, or refute the Mo paleoredox proxy concept; and (iv) experimental or theoretical studies designed to constrain the fundamental mechanisms of Mo isotope fractionation relevant to the paleoproxy application.

Those studies in the second and third categories have clearly demonstrated that robust application of the Mo isotope paleoredox proxy may not be perfectly straightforward and requires careful consideration of at least three complications. First, the isotopic composition of Mo in riverine waters flowing into the ocean is apparently not constant, but varies spatially and possibly temporally [14, 16]. This may be due in part to weathering of different bedrock types, but whether there is fractionation during weathering or during processes that partially remove dissolved Mo from riverine waters could obviously also be important. Second, the mechanism by which Mo is removed from euxinic waters and the conditions under which it is quantitatively removed are vigorously debated [5, 6, 17]. A third complication is the identification of a significant, intermediate sink in the form of suboxic or anoxic continental margin sediments, with Mo showing an isotopic composition intermediate between those of the oxic and euxinic sinks (e.g., [18–20]); this introduces more variables to the mass balance equations underlying the model. Given these complications, it is currently not possible to state with certainty whether the Mo isotopic composition of seawater is, in fact, systematically correlated with the global redox status of the ocean, or whether any particular rock type (let alone any particular sample set) reliably and permanently reflects the ocean's Mo isotopic signature. Answers to these questions will depend, in part, on careful investigation of the chemical mechanisms by which Mo moves from one model reservoir to another and of the isotope fractionation effects associated with those chemical pathways. There have been few such investigations thus far in the fourth category above, but nonetheless they serve as the beginnings of a fundamental understanding of the relationships between Mo isotope ratio systematics and marine paleoredox conditions. The purpose of the following paragraphs is to review those papers and their implications.

Most of the experimental and theoretical work published thus far concerning the Mo isotope paleoredox proxy relates to isotope fractionation during removal of Mo from seawater to the oxic sink, represented by ferromanganese sediments ($\Delta^{97/95}Mo_{seawater-ferromanganese} \sim 1.8‰$, or $\Delta^{98/95}Mo_{seawater-ferromanganese} \sim 2.7‰$). Barling and Anbar [13] performed the first experiments to determine whether adsorption of Mo to synthetic particles of birnessite, the most common Mn oxyhydroxide mineral in cold water ferromanganese chemical precipitates, would result in the same fractionation in the laboratory as Barling *et al.* [10] had observed between seawater and natural ferromanganese nodules. In these experiments, a solution of dissolved Mo of known isotopic composition was mixed with a suspension of birnessite particles, and adsorption was allowed to occur for up to 96 h. Following filtration of the birnessite particles with adsorbed Mo from the solution, the Mo remaining in solution was purified by ion-exchange chromatography and analyzed for its isotopic composition. Dissolved Mo was always enriched in the heavier Mo isotopes relative to the starting solution, hence lighter isotopes preferentially adsorbed on birnessite. The fraction of the total Mo in the experiment that remained in solution varied as a function of experimental duration and pH (6.5–8.5), but mass balance calculations indicated that the fractionation between dissolved and adsorbed pools of Mo was constant at $\Delta^{97/95}Mo_{dissolved-adsorbed} \sim 1.8 \pm 0.5‰$, strongly suggesting an equilibrium fractionation process. Excellent agreement with the fractionation observed in nature implies that the same fractionation process may well be the one operating in nature, despite differences in Mo concentrations and ionic strength between experiment and nature ([Mo] = 25 μM versus 0.105 μM Mo in seawater and an ionic strength of 0.1 M versus ~0.7 M in seawater).

Wasylenki *et al.* [20] conducted similar experiments, but directly determined the isotopic composition of both the dissolved and adsorbed pools of Mo for every experiment, with improved analytical precision, and extended the experimental conditions to include high ionic strength (0.7 M) and temperatures ranging from 1 to 50°C. Higher ionic strength resulted in less adsorption, but was found to have no significant effect on isotope fractionation, as shown in Figure 11.3, and the temperature dependence of fractionation was very small, with $\Delta^{97/95}Mo_{dissolved-adsorbed}$ ~1.88‰ at 1°C and 1.63‰ at 50°C. These results strengthened the argument that adsorption on birnessite is the process responsible for the fractionation observed in nature and that hydrogenetic ferromanganese sediments have likely always been offset by ~1.8‰ from seawater.

Siebert *et al.* [12] and Barling and Anbar [13] suggested that the equilibrium isotope fractionation effect during adsorption of Mo on birnessite could result either from equilibration of two chemical species of Mo in solution, followed by adsorption of just one of them, or from fractionation during the adsorption reaction itself, between one aqueous species and one adsorbed species not present in solution. In order to address this, Tossell [21] used quantum mechanical calculations to explore which trace species of aqueous Mo might fractionate from the

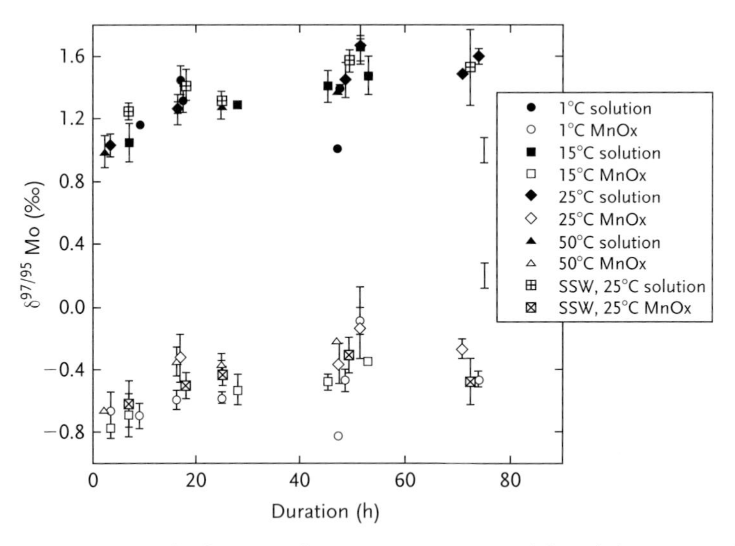

Figure 11.3 Results from Mo adsorption experiments. Solid symbols represent dissolved Mo and open symbols Mo adsorbed on synthetic birnessite (MnOx). Symbol shapes vary with experimental temperature according to the legend. SSW denotes experiments run at an ionic strength of 0.7 M. The two bold error bars indicate the typical magnitude of comprehensive experimental uncertainty, based on a set of identical replicates. Reproduced from [20].

predominant species, MoO_4^{2-}, with the appropriate magnitude. Tossell [21] computed isotope exchange equilibrium constants for MoO_4^{2-} versus each of a long list of possible trace species, including oxide, hydroxide, oxyhydroxide, and thiomolybdate ions. The equilibrium constants were calculated from computations of the vibrational, rotational, and translational contributions to free energies of the molecules in the gas phase. The species with the best match to the observed fractionation at 273 K (0°C) was MoO_3, and therefore equilibration of MoO_4^{2-} and MoO_3, followed by adsorption of MoO_3, was proposed as the mechanism driving the isotope effect. Although the fractionation at 273 K was in excellent agreement with nature and the experiments of Barling and Anbar [13], Wasylenki *et al.* [20] later showed that the temperature dependence of the MoO_3 results of Tossell [21] was much steeper than that obtained from experiments (see Figure 11.4).

A second theoretical study was published by Weeks *et al.* [22], who used slightly different computational methods (see also the corrigendum in Weeks *et al.*, [23]). Weeks *et al.* also calculated equilibrium fractionations between MoO_4^{2-} and various other candidate trace species, but used density functional theory (DFT) to compute vibrational frequencies and then reduced partition function ratios for different isotopomers (isotopic isomers) of the species considered. Weeks *et al.* found that all of the calculated equilibrium fractionation factors were too small in magnitude to match the effect observed. The closest match was MoO_3, the preferred species in Tossell [21], but this species was rejected on the grounds that Raman spectroscopic evidence from Oyerinde *et al.* [24] indicates that $MoO_3(H_2O)_3$, rather than MoO_3, is

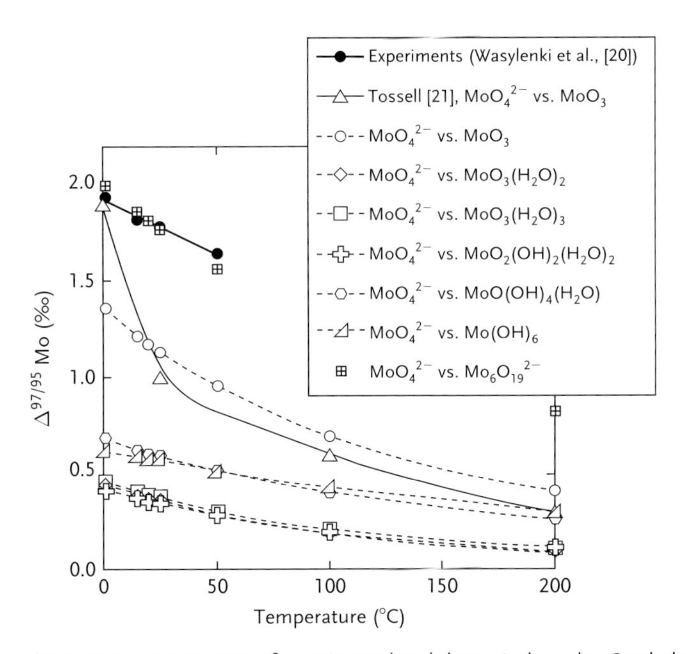

Figure 11.4 Comparison of experimental and theoretical results. Symbols as denoted in the legend. Reproduced from [20].

the stable form of molybdic acid in solution. Wasylenki *et al.* [20] demonstrated that the slopes of the temperature dependences computed by Weeks *et al.* [22] matched the experimental results well, but the lack of any calculated fractionation matching nature or experiment led Wasylenki *et al.* [20] to state that the mechanism remained unknown and may not be due to equilibration of two solution species, but rather due to equilibration of MoO_4^{2-} in solution with an adsorbed species yet to be identified.

Liu [25] presented computational results for which the aqueous solvent was explicitly modeled, in contrast to the *in vacuo* modeling of Tossell [21] and Weeks *et al.* [24]. Liu found that modeling the coordinating water molecules had a significant impact on the calculated electron distributions, bond vibrational frequencies, and fractionation factors. Liu [25] found that $Mo(OH)_6$ fractionated from MoO_4^{2-} with the right magnitude, but Tossell [21] had rejected $Mo(OH)_6$ as a valid species in solution because his computations imply a high energy of formation for this species.

Wasylenki *et al.* [26] directly investigated the coordination environment of Mo when adsorbed on birnessite by collecting extended X-ray absorption fine structure (EXAFS) spectra for an experimental sample and for a natural, hydrogenetic ferromanganese nodule. The results indicated that Mo is certainly in distorted octahedral coordination in both cases and is likely adsorbed in the form of a polymerized molybdate, for example, $Mo_6O_{19}^{2-}$. This coordination is in stark contrast to the near-perfect tetrahedral arrangement of oxygen atoms in MoO_4^{2-}, so this coordination difference between the predominant aqueous Mo species and the predominant adsorbed species likely explains the isotope effect. The polymolybdate apparently assembles on the birnessite surface, as equilibrium speciation models indicate that this species has a vanishingly small concentration in the experimental solution (25 μM). The isotopic composition of the natural ferromanganese sample was 1.8‰ lighter than that of seawater in terms of $\Delta^{97/95}Mo$, as expected. DFT calculations were then performed to compute the equilibrium fractionation factor between MoO_4^{2-} and the newly identified sorbed species, $Mo_6O_{19}^{2-}$. The results were in excellent agreement with experimental results, in terms of both fractionation magnitude and temperature dependence of that fractionation, as shown in Figure 11.4. Wasylenki *et al.* [26] concluded from the agreement of isotopic analyses on natural and experimental samples, EXAFS spectra for natural and experimental samples, and the agreement of theoretical and experimental results that the formation of a polymolybdate adsorption complex on the surface of birnessite is the molecular mechanism behind the observed fractionation when Mo is removed to oxic, ferromanganese sediments from seawater. Kashiwabara et al. [27] recently published an independent EXAFS spectrum of Mo on synthetic birnessite and interpreted it as Mo monomers with distorted octahedral coordination, but agreed that coordination change likely drives the isotope effect in this system.

Natural ferromanganese sediments are, of course, mixtures of several mineral phases, including a significant proportion of Fe oxyhydroxides. Therefore, it is possible that adsorption to manganese oxides is not the only way in which Mo is incorporated in oxic sediments. Koschinsky and Hein [28] conducted sequential leaching experiments to determine whether various metals are primarily associated with Mn or with Fe oxyhydroxide phases in ferromanganese crusts and nodules, and they found that less Mo was removed from the sediment during the step designed to dissolve Mn oxyhydroxides reductively than in the step designed to dissolve Fe oxyhydroxides, implying that most Mo is associated with Fe oxyhydroxides. This result is seemingly inconsistent with the observation that adsorption on pure birnessite in the laboratory results in the same isotopic fractionation and nearly the same EXAFS spectrum as natural, mixed-oxide samples. Hence it is important to know how Mo isotopes fractionate during adsorption to Fe oxyhydroxides. Goldberg *et al.* [29] conducted Mo adsorption experiments with four Fe oxyhydroxide phases, including magnetite, ferrihydrite, goethite, and hematite. As with birnessite, lighter isotopes were preferentially adsorbed, and the fractionations, in terms of $\Delta^{98/95}Mo_{dissolved-adsorbed}$, were 0.83 ± 0.60, 1.11 ± 0.15, 1.40 ± 0.48, and 2.19 ± 0.54‰, respectively. Different fractionation factors for these four different ferric minerals suggest that the bonding geometry of adsorbed Mo must vary from one phase to the next, but thus far, only Mo adsorbed on goethite has been studied by EXAFS. Arai [30] inferred a polymeric Mo complex adsorbed on goethite, with Mo in distorted octahedral coordination at circumneutral pH, but fitted a combination of this complex and tetrahedrally coordinated Mo at lower pH. Perhaps the intermediate isotope fractionations for Mo adsorbed on the various ferric oxyhydroxide phases can be explained in terms of different proportions of octahedrally coordinated, polymeric Mo (strongly fractionated) and tetrahedrally coordinated Mo (not fractionated) on the mineral surfaces. However, the near constant Mo isotope ratio values in all modern ferromanganese sediments analyzed so far strongly indicate that adsorption on Mn oxyhydroxides, and not Fe oxyhydroxides, governs the isotopic signature of ferromanganese sediments.

The experiments just discussed may also help explain the Mo isotopic composition of river waters. Archer and Vance [14] analyzed waters from several large rivers and found a wide range of $\delta^{98/95}$Mo values of +0.2 to +2.3‰. Riverine water is apparently variably enriched in heavier isotopes, although earlier results from Barling *et al.* [10] and Siebert *et al.* [12] indicated that $\delta^{98/95}$Mo in most continental rock types is near ~0‰. The Mo flux-weighted average of the samples analyzed by Archer and Vance [14] was +0.7‰.

There are three possible causes of the positive value of riverine waters relative to average continental rock. The first is that continental lithologies actually vary significantly in Mo isotopic composition and that the earliest studies failed to sample those lithologies that are enriched in the heavier isotopes. Neubert *et al.* [16] measured riverine waters from small catchments and found values ranging from 0 to +1.9‰. The isotope ratio data in this study were strongly correlated with bedrock lithology, and Neubert *et al.* [16] concluded that bedrock lithology is the primary control on riverine Mo isotope ratios. Positive values were found especially in those catchments where rock and soil contained Mo associated with sulfates and sulfides. A second possible explanation for heavy riverine waters is that weathering of continental rocks results in preferential release of heavier isotopes of Mo. A straightforward way to test this is to simulate weathering of Mo-bearing minerals and rocks in the laboratory. Very few experimental data have been published thus far. Siebert *et al.* [12] presented a single data point for Mo leached from a powdered granite. About 63% of the Mo was released during 24 h of leaching in very dilute HCl, and the isotopic composition of Mo in the leachate was identical to that of the bulk granite, implying no fractionation. Liermann *et al.* [31] leached a black shale at pH ~6 in several batch experiments, with some replicates containing metal-chelating ligands or bacteria and some only containing bacterial growth medium (potassium phosphate, pH buffer, glucose). In all experiments, regardless of the presence of ligands or bacteria, Mo released from the black shale was enriched in the heavier isotopes of Mo, with $\Delta^{98/95}Mo_{leachate\text{–}bulk\ shale}$ ranging from +0.3 to +0.5‰. It is not known whether the fractionation was driven by preferential release of heavier Mo isotopes from mineral grains, as might occur if isotopically heavy mineral phases dissolve at faster rates than isotopically light phases, or whether no fractionation accompanies the leaching, but the overall effect results from a small extent of preferential adsorption of lighter Mo isotopes from the leachate fluid on mineral surfaces.

Preferential adsorption of lighter Mo isotopes is the third possible cause of enrichment of riverine waters in heavier Mo isotopes compared with most continental lithologies. Perhaps no fractionation occurs during weathering, but instead adsorption on solid minerals, perhaps including Fe or Mn oxyhydroxides, causes riverine waters to lose isotopically light Mo during transport to the oceans. Given the paucity of natural and experimental data, it remains poorly understood what role each of these three possibilities plays in governing the isotopic composition of Mo flowing into the global oceans. Future work must address this question, especially with the aim of determining whether the Mo flux into the oceans is likely to have remained constant over geological time.

Far less experimental or theoretical work has addressed fractionation during removal of Mo into euxinic sinks, although knowledge about this process is absolutely critical to successful application of the paleoredox proxy. If one can conclusively demonstrate for a given set of ancient samples that Mo was, in fact, quantitatively removed from the water column, then the isotopic composition of Mo in the water column at the time of deposition was faithfully recorded, and any fractionation occurring in this environment is of no consequence. If Mo is not completely removed, it is essential to know whether the process of partial removal of Mo from euxinic water results in isotope fractionation. At this time, it is not even known what phase hosts Mo in the organic-rich, sulfide-rich sediments that form in euxinic depositional areas. In these areas, MoO_4^{2-} is believed to convert to thiomolybdate ions (MoO_3S^{2-}, $MoO_2S_2^{2-}$, $MoOS_3^{2-}$, MoS_4^{2-}), and these thiomolybdates are "particle reactive" and believed to adhere to organic matter sinking to the sediment–water interface. The various thiomolybdates are thought to fractionate from each other [32–34], but so far, there is no direct quantification or mechanistic investigation of this effect. The quantum mechanical calculations of Tossell [21] predict very large fractionations equivalent to $\Delta^{98/95}Mo = 3.3‰$ for MoO_4^{2-} versus $MoO_2S_2^{2-}$ (MoO_4^{2-} heavier) and $\Delta^{98/95}Mo = 6.8‰$ for $MoO_2S_2^{2-}$ versus MoS_4^{2-}. If this is indeed the case, then robust application of the paleoredox proxy will rest critically on demonstrations that Mo was completely removed from seawater when those shales were deposited.

Recent observations of Mo concentration profiles in a seasonally stratified lake (Rogoznica in Croatia) caused Helz *et al.* [17] to cast doubt on the hypothesis that

Mo is removed from sulfidic waters by sorption of thiomolybdate ions on sinking particles. Rather, they postulated that Mo is incorporated into an Fe–Mo sulfide mineral phase that precipitates in euxinic waters below the depth where FeS becomes supersaturated and that solubility of this phase controls the Mo concentration in the sulfidic water column. This may be the same Fe–Mo phase that was experimentally precipitated by Helz *et al.* [35] and was shown by EXAFS analysis to contain Mo tetrahedrally coordinated by S, as appears to be the case in some samples of black shale. To understand better the conditions in which Mo might precipitate as an Fe–Mo sulfide, Helz *et al.* [17] used thermodynamic methods to calculate the solubility of the putative sulfide phase as a function of pH, [Mo], and [H_2S]. Their calculations predicted that only a narrow range of pH–[Mo]–[H_2S] conditions will result in complete insolubility of Mo. Their model succeeded in predicting the Mo concentrations in deep waters in several other modern anoxic and euxinic areas. Helz *et al.* [17] concluded that the complete removal of Mo from the water column in the Black Sea is not representative of the general behavior of Mo in euxinic basins over time, but is instead the exception to the rule.

Helz *et al.* [17] could eventually prove to be correct in postulating that the solubility of an Fe–Mo sulfide phase controls the extent to which Mo is removed from euxinic waters. The actual mechanism of removal, however, must also explain the robust and widespread (spatially and temporally) observation that Mo concentrations in black shales and organic-rich sediments are very strongly correlated with total organic content (TOC) ([36] and references therein). At the time of this writing, multiple groups are working on experiments to quantify isotope fractionation between the various thiomolybdate species, measure fractionation during precipitation of Mo-bearing sulfides and during formation of Mo–organic complexes, and on the application of EXAFS analysis to Mo in black shales. Perhaps the next few years will see some important advances for these most critical components of the Mo isotope paleoredox proxy.

11.2.2 Cr Isotope Ratios and Paleoredox Conditions of the Atmosphere

Also motivated by the desire to constrain further the history of oxidation of the atmosphere and oceans, Frei *et al.* [37] determined the isotopic compositions of Cr in banded iron formation (BIF) samples, representing the Archean (>2.5 Ga ago) and Proterozoic (2.5 Ga to 542 Ma ago) Eons. They anticipated that isotopic variations in those rocks would correlate with changes in global redox conditions over time because of previous work reporting isotope ratio variations correlated with the Cr oxidation state in natural samples [38, 39], experiments [40, 41], and theoretical calculations [42].

Frei *et al.* [37] hypothesized that a paleoredox proxy based on Cr isotope ratios would likely reflect atmospheric oxygenation, rather than marine oxygenation. When oxygen is abundant in the atmosphere, Cr(III) in exposed rocks on land is converted during oxidative weathering to Cr(VI), which is highly soluble in water and transported by rivers and groundwater to the ocean. Since all of the previous studies mentioned above reported heavier values of $\delta^{53/52}$Cr for Cr(VI) than for Cr(III), Frei *et al.* proposed that the pool of Cr(VI) delivered to the oceans should be isotopically heavy, with an isotope fractionation from the Cr(III) remaining in continental rocks possibly as large as 6‰, since that is the equilibrium fractionation between Cr(VI) and Cr(III) in aqueous solution, as calculated theoretically by Schauble *et al.* [42]. The higher the extent of oxidative weathering, the more heavy Cr could be delivered to the ocean. Upon encountering dissolved Fe(II) in seawater, isotopically heavy Cr(VI) should be rapidly reduced to Cr(III) and coprecipitated with Fe oxyhydroxides (sediments that will become BIF). Hence the $\delta^{53/52}$Cr values of BIF should correlate positively with the intensity of oxidative weathering at the time of deposition.

Frei *et al.* [37] compared the ranges of isotopic compositions they observed for Cr in BIF samples of different ages. In accordance with expectations, only relatively light Cr isotopic compositions in BIF older than 2.8 Ga ($\delta^{53/52}$Cr from −0.25 to −0.09‰; see Figure 11.5) were observed. These values all fall within the range defined by analyses of magmatic rocks [39]. From 2.8 to 2.61 Ga, most samples also have compositions overlapping those of magmatic rocks, but a few are significantly heavier, up to +0.29‰. Frei *et al.* [37] interpreted such values as resulting from transient pulses of elevated oxygen in the atmosphere, beginning up to 300 million years prior to the marine Great Oxidation Event, as recorded by other geochemical proxies [43–45]. Samples that are 2.48–2.45 billion years old,

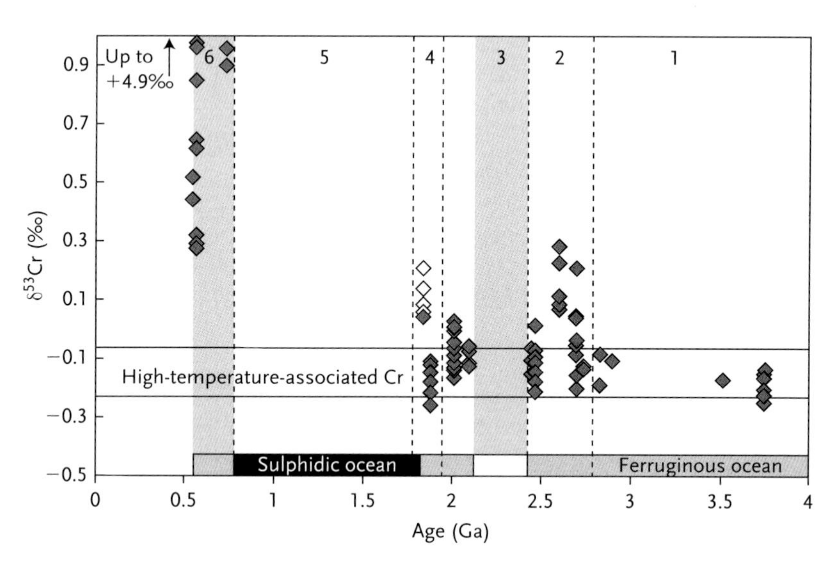

Figure 11.5 Cr isotopic signatures for banded iron formation samples as a function of age. The shaded bar across the bottom delineates six stages in which samples with variable degrees of enrichment in the heavier Cr isotopes were found. The maximum degrees of enrichment in each stage were used to infer six stages in the evolution of atmospheric chemistry, as described in the text. Reproduced from [37].

just prior to the marine Great Oxidation Event, all show negative $\delta^{53/52}$Cr values, except one at +0.01‰. From 2.4 to 2.1 Ga, BIF are absent from the rock record, presumably because oxygenation of seawater and rapid precipitation of ferric oxyhydroxides drastically drew down the concentration of Fe in seawater, making further BIF deposition impossible. BIF reappear between 2.1 and 1.88 billion years ago, with Cr isotopic compositions ranging from −0.26 to +0.03‰. From 1.878 billion year old samples of the Gunflint Formation in Ontario, Canada ($\delta^{53/52}$Cr from −0.26 to −0.11‰), Frei *et al.* [37] inferred a decrease in atmospheric oxygen following the Great Oxidation Event. This idea is supported by the presence of BIF that were deposited in high-energy, near-shore environments [46], which requires that even shallow waters near the coastlines were low enough in oxygen to allow high concentrations of dissolved Fe(II) to be transported there before being oxidized and precipitated. The range of isotope ratio values reported by Frei *et al.* [37] suggest that oxygen levels at 1.878 Ga could possibly have been even lower than between 2.8 and 2.61 Ga, although mass-independent sulfur isotope fractionation effects are absent at 1.878 Ga ([47] and references therein), suggesting that oxygen levels were higher at this time than during the Archean Eon. The youngest Gunflint Formation samples at 1.84 Ga show a marked increase in $\delta^{53/52}$Cr up to +0.21‰, suggesting an increase in atmospheric oxygen at this time. Increased oxidative weathering likely led to increased sulfate input to the oceans, increased bacterial sulfate reduction, and hence high sulfide concentrations in seawater throughout the Mesoproterozoic [48]. BIF disappear from the rock record after 1.84 Ga because Fe was deposited in sulfide phases rather than as Fe oxyhydroxides. BIF make a brief reappearance in the Late Neoproterozoic, during periods of low oxygen, driven by episodes of extreme glaciation [49]. Cr isotopic compositions of samples between 0.75 and 0.55 billion years old are all significantly heavier than magmatic rocks and range from +0.31 to +5.00‰, which is consistent with much greater extents of oxidative weathering once the atmosphere reached modern oxygen levels in the Late Neoproterozoic.

If the interpretation of Cr isotope ratio variation by Frei *et al.* [37] is valid, Cr isotope ratios provide an extremely valuable complement to other geochemical proxies for paleoredox conditions, since they reflect atmospheric, rather than marine, oxygenation and can be measured in a rock type that is not suitable for applying other paleoredox proxies, such as Mo or S isotope ratios. Some critical assumptions in Frei *et al.*'s study [37] must be verified, namely (i) that oxidative weathering of Cr(III) from rocks indeed results in an isotopically heavy pool of dissolved Cr and (ii) that reduction and coprecipitation of Cr with Fe oxyhydroxides either does not fractionate Cr isotopes or results in complete removal of Cr from the water column in continental shelf environments where BIF are deposited. Although Cr isotope ratio behavior is not yet fully understood at a fundamental level, several studies have been published that provide important insights into this system.

Schauble *et al.* [42] calculated equilibrium fractionation factors between various Cr species, using both published vibrational spectra and *ab initio* force field models. The findings were consistent with similar work on other elements, including Fe and Zn, in that heavier isotopes preferentially partition into species with higher valence states [50–53]. Calculations of reduced partition function ratios for equilibrium isotope exchange between CrO_4^{2-} and Cr^{3+} in aqueous solution predicted an equilibrium fractionation of +6–7‰ in $\Delta^{53/52}Cr_{VI-III}$.

In order for this large equilibrium fractionation to be expressed in the products of oxidative weathering, isotopic equilibration of Cr remaining as Cr(III) in chromite or other Cr-bearing minerals and Cr that has been oxidized and dissolved into weathering fluids must occur. (A kinetic isotope fractionation effect during leaching of Cr and/or oxidation would lead to an isotopically *light* product, since reaction rates are faster for lighter isotopes; see also Chapter 1.) As Frei *et al.* [37] pointed out, oxidation of Cr(III) in nature often results from interaction with Mn(IV) in manganese oxides, with reduction of Mn(IV) to Mn(II) or Mn(III). The exact mechanism of oxidative dissolution of Cr is not known, hence the likelihood of isotopic exchange between the dissolved, oxidized Cr pool and the Cr pool remaining as Cr(III) in mineral lattices is also unknown.

No one has yet published an experimental study of Cr isotope behavior during weathering of Cr(III)-bearing mineral phases, but Zink *et al.* [54] conducted experiments in which aqueous Cr(III) was partially oxidized into aqueous Cr(VI). In those experiments, Cr(III) was kept in solution at high pH (10.5) and high ionic strength, and different amounts of H_2O_2 were added to several aliquots of this solution to oxidize various fractions of the Cr(III). Oxidation occurred within seconds, as evidenced by color changes in the solutions. Cr(VI) (present as CrO_4^{2-}) and Cr(III) were separated from one another by anion-exchange chromatography. When results were plotted as $\delta^{53/52}Cr$ versus the fraction of Cr oxidized, the trends for both the Cr(III) and Cr(VI) pools took the shape of Rayleigh fractionation trends (see also Chapter 1), with the Cr(VI) reaction product consistently enriched in the heavier isotope. This result indicates either an equilibrium fractionation process that is irreversible or an equilibrium process with a reverse reaction with much slower kinetics. Again, kinetic isotope fractionation is ruled out because the reaction product is enriched in heavy isotopes. The puzzling aspect of the results is that the best-fit fractionations for the two trends defined by Cr(III) and Cr(VI) are not equal; the Cr(III) trend fits a fractionation of $\Delta^{53/52}Cr_{VI\text{-}III} = +0.6$‰, while Cr(VI) values fit a fractionation of only +0.2‰. Zink *et al.* [54] inferred that oxidation of Cr does not have one simple reaction mechanism. Possible complications include formation of unstable intermediates with Cr(IV) and Cr(V), disproportionation of these intermediates to Cr(III) and Cr(VI), and a reductive reverse reaction of Cr(VI) to lower oxidation states. The kinetics and isotope fractionation effects associated with these complications are unknown and cannot be constrained from any experiments published so far. A complex oxidation mechanism is also implied by the results of Bain and Bullen [55], who oxidized aqueous Cr(III) with birnessite, a common Mn oxide phase. In those experiments, the isotopic composition of the Cr(VI) pool varied from −2.5 to +0.7‰, and isotopic compositions continued to change even after net Cr(VI) production ceased, implying continued reaction and/or reverse reaction of intermediate valence states.

Zink *et al.* [54] also attempted to equilibrate isotopically spiked aqueous Cr(III) ($\delta^{53/52}Cr = +800‰$) with aqueous Cr(VI) of natural isotopic composition to determine whether exchange would occur on short time scales. They reported no exchange up to 120 h after the two species were mixed. Whether dissolved Cr(VI) and Cr(III) can equilibrate on time scales appropriate for weathering reactions is unknown.

None of the experiments published so far resulted in expression of the anticipated equilibrium fractionation, nor did they result in any fractionations as large as the observed differences in $\delta^{53/52}Cr$ between magmatic rocks and the isotopically heaviest samples analyzed by Frei *et al.* [37], and it therefore remains unknown whether oxidative weathering governs the Cr isotopic composition of banded iron formation sediments. Both Zink *et al.* [54] and Bain and Bullen [55] mentioned an alternative way for Cr delivered to the oceans to become enriched in heavier isotopes, that is, partial reduction of dissolved Cr(VI) during transport in groundwater, soil pore waters, or rivers. The reduced Cr would be enriched in lighter isotopes and would be left behind during transport because of low solubility. Open-system Rayleigh fractionation during transport and redox cycling of Cr could generate especially large fractionations between source rocks and marine Cr, such as those observed by Frei *et al.* [37] for Late Neoproterozoic samples. This does not necessarily mean that the isotopic compositions of Cr in ancient sediments cannot be used to infer paleoredox conditions, but further research will be needed to resolve the relationships between the extent of oxidative weathering of terrestrial rocks, post-weathering redox reactions, and isotopic composition of marine sediments.

An important part of this endeavor will be to understand Cr isotope fractionation during reduction. Several studies have addressed this question experimentally, using various reductants, including magnetite and natural reducing sediments [40], microbes [41], H_2O_2 [54], and aqueous Fe(II) [56]. Some of these studies were undertaken with an environmental application in mind: since Cr(VI) is carcinogenic and mobile, whereas Cr(III) is relatively nontoxic and immobile, researchers are interested in whether isotope ratio systematics can help track and predict the transport and fate of Cr where pollution or oxidation of naturally occurring Cr(III) has created health hazards. Nonetheless, the results of these experiments can also be applied to the use of Cr isotope ratios as a paleoredox proxy.

Using thermal ionization mass spectrometry (TIMS) for isotopic analysis, Ellis *et al.* [40] conducted three sets of experiments in which aqueous Cr(VI) was reduced by magnetite particles, by sediment from an intertidal mud flat in the San Francisco Estuary, or by sediment from a microbially rich pond in Urbana, IL, USA. Samples of Cr in solution were taken over time from all three experiments, and a gradual decrease in Cr(VI) and progressive depletion of lighter isotopes were observed. Trends in $\delta^{53/52}Cr$ versus fraction of Cr(VI) reduced fit Rayleigh fractionation curves, with best-fit fractionation factors of $\alpha^{53/52}Cr_{III-VI} = 0.9965$, 0.9967, and 0.9965 for the three reductants, respectively. These correspond to fractionations of $\Delta^{53/52}Cr_{VI-III} = +3.5$, +3.7, and +3.5‰. Ellis *et al.* [40] attributed these results to kinetic isotope fractionation during the reduction reactions, whereby the rate of breakage of Cr(VI)–oxygen bonds is faster for ^{52}Cr than for ^{53}Cr. When the two sediments were autoclaved first, the results were the same, so Ellis *et al.* [40] inferred a common, abiotic reduction mechanism for all three reducing agents.

Sikora *et al.* [41] investigated fractionation during biologically mediated reduction, using cultures of *Shewanella oneidensis* bacteria, strain MR-1. In eight batch cultures, growth was stimulated with 3–100 μM lactate or formate solution, and the data for all eight cultures fell fairly well on a Rayleigh fractionation trend with a best-fit fractionation factor of $\Delta^{53/52}Cr_{VI-III} = +4.1‰$. This is a slightly larger instantaneous fractionation factor than Ellis *et al.* [38] observed during abiotic reduction, suggesting a slightly different reduction mechanism when reduction is mediated by microbes.

In one culture experiment, *S. oneidensis* fed on 10 mM lactate solution, and the reduction rates were severalfold faster than in the more scantly amended cultures. The best-fit fractionation factor for this experiment was only +1.8‰. As Sikora *et al.* discussed, decreases in the magnitude of fractionation with increases in bacterial oxyanion reduction rates have previously been observed for sulfate, nitrate, and selenate (see the references in [41]). Rees [57] and Canfield [58] developed a model to explain this phenomenon for multi-step pathways of sulfate metabolism, but the concept may well apply to Cr isotope behavior also. According to this model, if different steps in a metabolic pathway impart different fractionations, then the overall fractionation for the metabolism will be the sum of the fractionations occurring during the steps up to and including the rate-limiting step. If availability of an oxyanion substrate is limited, then the first step in the metabolism of that oxyanion may be rate limiting and therefore will govern the overall fractionation. An excess of that substrate may result in some other step becoming rate limiting, and thus a different isotope fractionation may become apparent. Cr uptake and reduction have been shown to be a multi-step process, possibly involving intermediate oxidation states [59–60], so perhaps the smaller fractionation factor in the lactate-rich experiment of Sikora *et al.* [41] results from a shift in which step is rate limiting in this system, a hypothesis that could be tested with further experimentation.

Zink *et al.* [54] conducted Cr reduction experiments using H_2O_2 as reductant. In the first set of experiments, certain details of the experimental procedure were varied among replicates to test their influence on the results, but in each of these experiments, enough H_2O_2 was added to a solution of Cr(VI) in HCl (pH ~0.7) to reduce 70% of the Cr(VI). Despite the minor variations in procedure, all replicates resulted in a fractionation of $\Delta^{53/52}Cr_{VI-III} = +7.11$ to 7.32‰. Since the fraction of Cr reduced was constant, it is not possible to deduce whether an equilibrium or Rayleigh fractionation trend is more appropriate, but the data can be fitted both ways, given the initial Cr(VI) isotopic composition, the fraction reduced, and the final Cr(VI) and Cr(III) isotope ratio values. If the isotope fractionation effect is thermodynamically governed (equilibrium), then $\alpha^{53/52}Cr_{VI-III} \approx 0.9928$, since $\Delta^{53/52}Cr_{VI-III} = +7.11–7.32‰$. The best-fit Rayleigh trend suggests that if the reaction is irreversible and/or accompanied by kinetic isotope fractionation, the instantaneous fractionation factor is $\alpha^{53/52}Cr_{VI-III} \approx 0.9956$ ($\Delta^{53/52}Cr_{VI-III} = 4.2‰$). In the next set of experiments, various fractions of Cr were reduced at pH < 1. The results indicate an equilibrium isotope fractionation effect with fractionation factor $\alpha^{53/52}Cr_{VI-III} \approx 0.9965$ ($\Delta^{53/52}Cr_{VI-III} = 3.54 \pm 0.53‰$; see Figure 11.6). Interestingly, when repeated at pH 7, the experiments yielded data that clearly fit a Rayleigh trend rather than an equilibrium trend, with a best-fit fractionation factor of $\alpha^{53/52}Cr_{VI-III} \approx 0.9950$ ($\Delta^{53/52}Cr_{VI-III} = 5.0‰$), for those experiments with less than 85% of Cr reduced. Smaller fractionations were observed between Cr(III) and the small amount of Cr(VI) remaining in the experiments with >85% of Cr reduced, and Zink *et al.* [54] attributed this to imperfect separation of Cr(VI) and Cr(III) as the

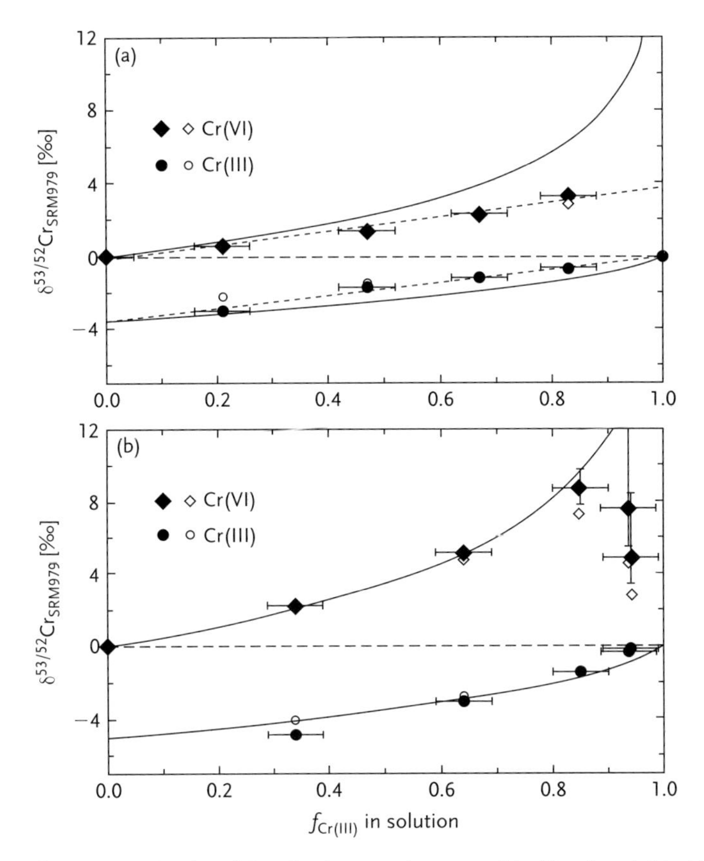

Figure 11.6 Results of Cr reduction experiments at (a) pH $\ll 1$ and (b) pH ~ 7. The *x*-axis in both plots is the fraction of original Cr(VI) reduced in each experiment. Black diamonds represent Cr(VI) remaining in solution and black circles denote reduced Cr(III). These isotopic compositions have been corrected for a small degree of cross-contamination during separation of the species; uncorrected values are shown as open symbols. The short-dashed lines in (a) represent the best-fit equilibrium fractionation. Solid curves in both plots represent best-fit Rayleigh trends. Uncertainties on $\delta^{53/52}Cr$ values are smaller than the symbols. Reproduced from [54].

experimental products were processed. Imperfect separation may have affected the other replicates to some extent also, but it is in the replicates with the smallest remaining pool of Cr(VI) that this artifact can be most apparent, since a smaller amount of Cr(III) contamination already suffices to shift the isotopic composition of the small Cr(VI) pool, and the isotopic difference between the Cr(VI) and Cr(III) pools is largest at high fractions of Cr reduced.

Zink *et al.* [54] pointed out that the magnitude of isotope fractionation in their series of low-pH experiments is similar to that observed by Ellis *et al.* [40] for Cr reduction by magnetite at circumneutral pH, but the systematics observed by Ellis *et al.* [40] suggest a kinetic isotope fractionation effect, whereas the data of Zink *et al.* [54] clearly reflect an equilibrium isotope fractionation effect. Because Cr speciation shifts as a function of pH and because the reducing agent differs between the two studies, it is likely that different reaction mechanisms are the reason for the contrasting behavior in these two studies. The predominant species at low pH are $HCrO_4^-$ and Cr^{3+}, but at circumneutral pH, CrO_4^{2-} and $Cr(OH)_2^+$ are more abundant, so different changes in bonding environments for Cr and thus different isotope behavior can be expected. At pH 7, Zink *et al.* [51] did observe fractionation fitting a Rayleigh trend, but their best-fit instantaneous fractionation was +5.0‰, which differs from the +3.3 to 3.5‰ observed by Ellis *et al.* [40]. Since the reducing agents differ and because the reaction mechanisms are unknown, including the possible involvement of intermediate Cr(V) or Cr(IV) species, it is not possible to explain this discrepancy without further research.

Døssing *et al.* [55] recently conducted Cr(VI) reduction experiments with aqueous ferrous iron as reductant, inside an N_2–H_2-purged glovebox. In batch experiments in which 3–6% of initial aqueous Cr(VI) was reduced over the course of 1 h (pH = 7), $\Delta^{53/52}Cr_{VI-III}$ ranged from +3.1 to +4.6‰ (measured by MC-TIMS). Døssing *et al.* pointed out that this result is comparable to the kinetic fractionation observed during reduction of Cr(VI) by solid Fe(II)-bearing phases by Ellis *et al.* [40]. Døssing *et al.* also conducted two constant addition experiments at pH 6.8 and 8.1, in which dissolved $FeCl_2$ (2000 mgl^{-1}) was pumped continuously at 0.8 ml min^{-1} into a beaker with 200 ml of 200 mgl^{-1} CrO_4^{2-}. The pH was held constant by titration with NaOH. The authors expected a 3:1 ratio of Fe(II) to Cr(VI) for complete reaction, but instead, more complex stoichiometry was observed. Early in the experiments (first 30 min), the ratio of Fe(II) required to reduce Cr(VI) was higher than expected. Between 30 and 70 min, both experiments exhibited approximately the 3:1 ratio expected, but after 70 min, the ratio was again higher than expected. Døssing *et al.* inferred that multiple stages occurred in their constant addition experiments: in the first stage, they speculated that much of the Fe(II) added to the beaker precipitated as mixed valence state solid phases, such as green rust [$Fe(II)_3Fe(III)(OH)_8Cl$], whereas reduced Cr formed solid CrOOH. The X-ray diffraction pattern for the recovered solids indicated a poorly crystalline material with a number of *d*-spacings matching lepidocrocite or goethite, which are Fe(III) phases, but the presence of some green rust may not be entirely ruled out on this basis. In the second stage, green rust may have reacted with CrO_4^{2-} to form

CrOOH and FeOOH or a solid solution of these two components. Døssing *et al.* suggested that in the third stage, CrO_4^{2-} previously adsorbed on solids was reduced to form CrOOH. The approach taken here to explain complex experimental results differs from the studies discussed above; rather than looking to intermediate valence states of Cr for explanation, Døssing *et al.* called on the distinction between homogeneous reduction of Cr(VI) by aqueous Fe(II) and inhomogeneous reduction of Cr(VI) intercalated in green rust.

Given that multiple, poorly constrained reaction mechanisms are at play, interpretation of the isotopic data of Døssing *et al.* [56] is challenging. Values of $\Delta^{53/52}Cr_{III-VI}$ are mostly negative (reduced Cr is lighter), ranging from −2.62 to +0.75‰. Those data points for which the fraction of Cr that has been reduced is less than 0.7 are all negative and define a nearly linear trend in $\Delta^{53/52}Cr_{III-VI}$ versus fraction of Cr reduced. Døssing *et al.* modeled these data with a Rayleigh fractionation curve with $\alpha^{53/52}Cr_{III-VI} = 0.9985$, which is considerably smaller than the fractionation factors observed by Ellis *et al.* [40], Bain and Bullen [55], Sikora *et al.* [41], and Zink *et al.* [54]. Døssing *et al.* explained that homogeneous reduction of Cr(VI) by Fe(II) may have the fractionation factor of 0.9965 reported by Ellis *et al.* [40], but some Cr in their experiments is reduced inhomogeneously by green rust, and this process may not measurably fractionate Cr isotopes. Hence a weighted sum of two fractionation factors, for homogeneous and inhomogeneous reduction, may help explain the smaller isotope fractionation observed overall. Reconciliation of this explanation with the three stages of Cr reduction in the experiments remains problematic.

What do all of the experimental results obtained so far mean for the potential of Cr isotope ratios to serve as a paleoredox proxy? Certainly it is discouraging that the attempts thus far to quantify isotope systematics during Cr oxidation did not generate pools of significantly heavy Cr(VI) and that Zink *et al.* [54] did not observe any shift towards equilibrium in their short-duration, spiked Cr(VI)–Cr(III) exchange experiment. Nonetheless, because the speciation of Cr and the time scales of published experiments do not match the conditions under which weathering occurs in nature, these experiments do not necessarily rule out the possibility that the Cr isotope ratio could reflect the extent of continental weathering somehow. For the model of Frei *et al.* [37] to work, it is also important that no strong kinetic isotope fractionation occurs during reduction and removal of Cr from seawater during BIF formation, as BIF must either faithfully record the isotopic composition of Cr in seawater or be offset from seawater by a constant amount that does not vary with small shifts in pH, temperature, and so on. The very small number of experiments published so far obviously disagree strongly on the nature of Cr isotope fractionation during reduction, probably because of complex reaction mechanisms that have yet to be well constrained. If we are to understand Cr isotope systematics during redox reactions better, future experiments will require clever design so that the molecular-scale mechanisms of Cr reaction can be directly monitored. Until such work is done, the meaning of the Cr isotope ratio variations observed by Frei *et al.* [37] in ancient BIF remains difficult to interpret.

11.2.3
Uranium Isotope Ratios and Marine Paleoredox

Uranium has no stable isotopes, but ^{235}U and ^{238}U have extremely long half-lives, and any variations in $\delta^{238/235}U$ that might be observed in rocks of similar type and age, at least within a few millions or tens of millions of years, can be considered to be due to "stable" isotope fractionation, rather than radiogenic decay (except for the Oklo reactor; see Chapter 1). The relative mass difference between ^{235}U and ^{238}U is only 0.42%, so the lack of any measurable variation in nature in $\delta^{238/235}U$ would not have been surprising. Nonetheless, the first two studies looking for such variations found plenty, including a span of more than 1‰ in $\delta^{238/235}U$ in sedimentary samples from the marine environment [61, 62].

Uranium is the heaviest naturally occurring element, and it happens that the large size of the nucleus plays a significant role in governing "stable" isotope fractionation for this element, and also for thallium, mercury, and several other heavy metals [63–65]. For these heaviest elements, measurable fractionation is not driven only by mass-dependent differences in bond vibrational frequencies as it is for lighter elements. Instead, two other effects may influence or even dominate the measured fractionation effects. First, differences in size between the highly positively charged nuclei of, for example, ^{238}U and ^{235}U affect the electron density distribution around those nuclei, thus influencing isotopic behavior. The result for uranium is that, at equilibrium, the heavier isotope (larger nucleus) preferentially partitions into sites where U is tetravalent rather than hexavalent. This mass-independent fractionation effect is referred to as the nuclear field shift effect or the nuclear volume effect and was thoroughly discussed by Bigeleisen [63] and Schauble [64]. Second, Epov *et al.* [65] and Malinovsky and Vanhaecke [66] have documented isotope fractionation due to hyperfine coupling between the nuclear spin of odd-numbered isotopes and the electrons and, at least for Hg, this effect is larger than the nuclear volume effect.

Compilation and direct comparison of the natural data from Stirling *et al.* [61] and Weyer *et al.* [62] require care, since each group used a different reference material as standard. Stirling *et al.* [61] reported their data in terms of epsilon units, with $\varepsilon^{235/238}U = (^{235}U/^{238}U_{sample} - {}^{235}U/^{238}U_{CRM\text{-}145})/(^{235}U/^{238}U_{CRM\text{-}145}) \times$ 10000. Weyer *et al.* [62], on the other hand, used standard delta notation, with $\delta^{238/235}U_{SRM\text{-}950a}$ defined as zero. The epsilon units of Stirling *et al.* [61] can be converted to delta units by dividing each value by 10 and changing the sign, but the use of different standards means that the resulting delta values are not necessarily directly comparable to the values in the Weyer *et al.* [62] dataset. Each group analyzed a small number of seawater samples, modern corals, and ferromanganese sediments. The offsets between the Stirling *et al.* averages and the Weyer *et al.* averages for these three types of samples were 0.07, 0.07, and 0.03‰, respectively. For the purpose of compiling data from both studies into a single figure, it was assumed here that CRM-145 has a value of +0.06‰ relative to SRM-950a. In Figure 11.7, each data point in the Stirling *et al.* [61] dataset has been divided by 10, multiplied by −1, and added to 0.06‰. Although this exercise involves a small

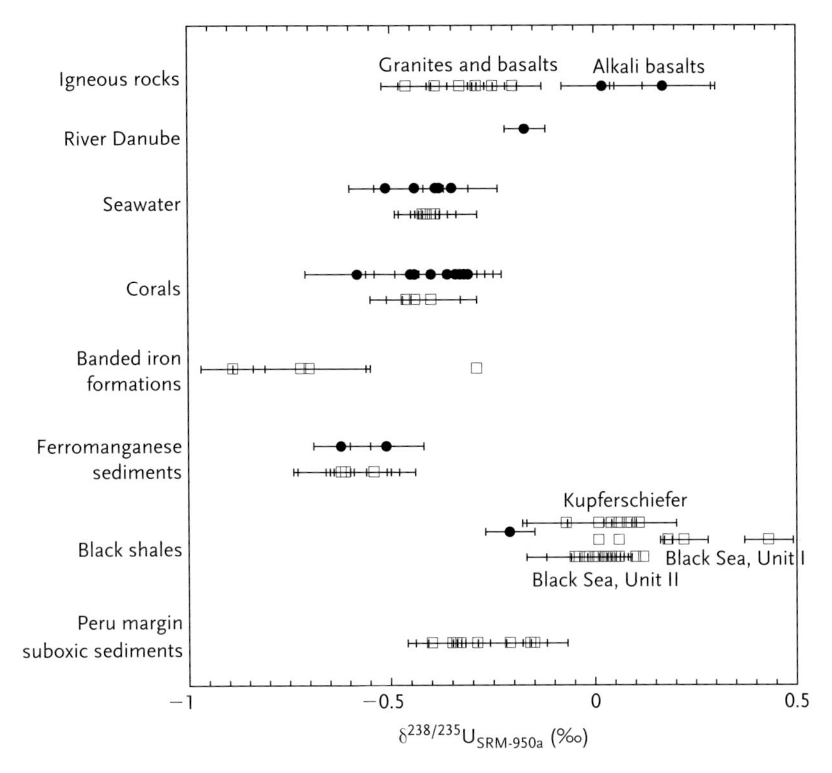

Figure 11.7 A compilation of data from [61] (filled circles) and [62] (open squares). The original data were reported with different notations and standards, but here they are all shown in terms of $\delta^{238/235}$U relative to the SRM-950a standard. $\varepsilon^{235/238}U_{CRM\text{-}145}$ values from [61] were divided by 10, multiplied by −1, and added to +0.06‰. The offset of +0.06 was chosen by considering the differences between the data reported by both groups for seawater, modern corals, and ferromanganese sediments.

degree of assumption and approximation, it is helpful for seeing what is thus far known about U isotope ratio systematics in the marine environment.

The compiled data show that modern seawater samples define a narrow range of isotopic compositions, from −0.51 to −0.35‰, and that modern and fossil corals (up to 330 000 years old) all fall in this range within analytical uncertainties. Relative to this range, ferromanganese sediments are slightly enriched in ^{235}U, suggesting that removal of U from seawater into such oxidizing sediments is associated with a small fractionation of $\Delta^{238/235}U_{\text{seawater–oxic sink}} \approx +0.2$‰. All but one of the banded iron formation samples analyzed by Weyer *et al.* [62] are considerably lighter than seawater, ranging from −0.89 to −0.70‰, implying either that a larger isotope fractionation occurred when U was removed from Precambrian seawater into BIF than occurs now between seawater and ferromanganese sediments or that the same fractionation occurs, but Precambrian seawater was lighter in $\delta^{238/235}$U than it is now. One outlying BIF sample was slightly heavier than seawater, at −0.29‰. Meanwhile, both studies found black shales to be considerably enriched in the heavier isotope compared with seawater. This implies a fractionation during removal of U to sediments in reducing, sulfidic conditions with a magnitude $\Delta^{238/235}U_{seawater-black\ shale} \approx -0.45‰$. This is presumably due to preferential reduction and burial of the heavier U isotope. Recent suboxic sediments from the Peru margin, chosen to represent a major sink for U from the modern ocean, were similar to or slightly heavier than seawater, ranging from −0.40 to −0.15‰ [62]. Water from the River Danube, presumably representing U input to the oceans from weathering of continental rocks, had a value of $\delta^{238/235}U = -0.15‰$ [61], although a set of granites and basalts analyzed by Weyer *et al.* [62] had values ranging from −0.20 to −0.46‰, and alkali basalts analyzed by Stirling *et al.* [61] showed values of +0.02 to 0.17‰.

The concept for a marine paleoredox proxy based on variations in the U isotope ratio in sediments is very similar to the Mo isotope paleoredox proxy described earlier in this chapter. Since incorporation of U in oxidizing sediments such as ferromanganese crusts results in preferential removal of isotopically light U, and incorporation of U in reducing sediments results in preferential removal of isotopically heavy U, changes in the relative proportions of these two sinks over time should drive systematic shifts in the isotopic composition of U in seawater. In turn, these temporal shifts in the isotopic composition of seawater should be manifested as vertical fluctuations in $\delta^{238/235}U$ in the sedimentary record. Excursions in the sedimentary record towards lighter values of $\delta^{238/235}U$ (in any of these rock types) could signify less removal of isotopically light U to oxic sediments and/or increased removal of isotopically heavy U to reducing sinks, and vice versa, and should therefore track the degree of oxygenation of the atmosphere and hydrosphere. As with Mo, the hope is that due to a long residence time in the oceans (~500 000 years for U [67]) relative to the ocean mixing time (~1600 years), the U isotope ratio record may yield a measure of oxygenation of the global ocean reservoir through time.

Based on this idea, Montoya-Piño *et al.* [68] analyzed black shales deposited before, during, and after the mid-Cretaceous Ocean Anoxic Event 2 (OAE2), thought to be a globally extensive interval of marine anoxia at ~93 Ma [69, 70]. They also measured modern samples of organic-rich sediment from the Black Sea. The range of values of $\delta^{238/235}U$ was similar for recent euxinic sediment and for samples above and below the OAE2 horizon (averages of $+0.10 \pm 0.07$ and $+0.06 \pm 0.06‰$, respectively; see Figure 11.8), although a few of the Black Sea samples were as heavy as +0.2 and +0.4‰. As expected, an excursion towards lighter values of $\delta^{238/235}U$ was observed in black shales in the OAE2 interval, with an average of $-0.07 \pm 0.09‰$. Montoya-Piño *et al.* [68] used the measured ~0.15‰ shift during OAE2 to estimate the factor by which the removal of U to anoxic sinks increased during the anoxic event. Assuming that input to the oceans was constant with $\delta^{238/235}U = -0.3‰$ and that Cretaceous seawater and black shales differed by 0.5‰, as they do in the modern Black Sea, a mass balance computation suggests that $40 \pm 20\%$ of U was removed to anoxic sediments during OAE2, as compared with ~10% now.

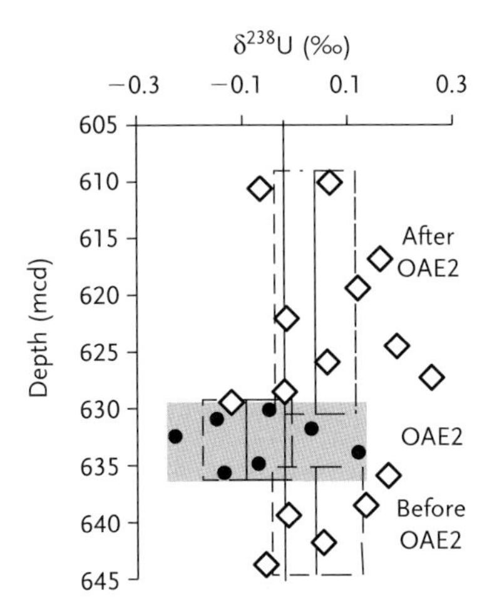

Figure 11.8 Isotopic composition of U in black shales from before, during, and after Ocean Anoxic Event 2 in the Late Cretaceous. Reproduced with permission from [68].

Other recent efforts aimed at using the isotopic signature of U in carbonates to interpret paleoredox shifts, including that at the Permian–Triassic boundary [71, 72]. Assuming that the U isotope ratios recorded in carbonates are not affected by diagenetic processes, carbonates may be the best rock type in which to look for global marine redox-driven U isotope ratio variations because of their ubiquity in the rock record, regardless of global redox conditions. The similarity of $\delta^{238/235}$U values for seawater and the modern coral samples of Stirling *et al.* [61] and Weyer *et al.* [62] suggest that biogenic aragonite may directly record the isotopic composition of seawater, but Brennecka *et al.* [72] reported U isotopic compositions of modern ooids (calcite) that are ~0.1‰ heavier than seawater, so the isotope systematics of U removal to carbonates may differ according to the mineral phase and have yet to be well established for either phase.

A small amount of theoretical work has been published on the fractionation of uranium isotopes. As mentioned above, Schauble [64] demonstrated that equilibrium isotope fractionation between species of the heaviest elements is not dominated by differences in bond vibrational frequencies, as they are for lighter elements, but by the nuclear field shift effect. This effect is due to interactions between electron shells, especially s shells, that have high electron density near the very large nuclei of heavy atoms. The heavier isotopes partition into those species with fewer s electrons or in which s electrons are shielded by more p, d, or f electrons. Schauble [64] presented calculations for various oxidation states and species of Tl and Hg and the same general conclusions apply to U. Calculations for uranium species were presented at a conference by Schauble [73]. The largest fractionations are predicted to occur when U(IV) and U(VI) species equilibrate, with values of $\alpha_{U(IV)-U(VI)}$ as large as 0.0012 at 273 K [$\Delta^{238/235}_{U(IV)-U(VI)} \approx 1.2‰$ at 0°C]. U(IV) has two 5f electrons that apparently shield s electrons from the isotopically heavy, positively charged uranium nucleus. Schauble's theoretically calculated fractionation is consistent with the difference between the isotopic composition of seawater and those of the heaviest black shales measured by Weyer *et al.* [62]. This similarity suggests that fractionation between seawater and organic-rich sediment may be governed by equilibrium fractionation of U(IV) and U(VI). However, the average fractionation between seawater and black shales is somewhat smaller, at $\Delta^{238/235}U_{shale-seawater} \approx 0.5‰$, so perhaps this does not simply reflect an equilibrium isotope effect between reduced and oxidized forms of uranium.

More recently, Epov *et al.* [65] reviewed what is known about mechanisms of fractionation not due to mass-dependent differences in bond vibrational frequencies. These mechanisms include the nuclear field shift effect, but also the nuclear spin effect, which results from interaction between the magnetic field associated with a nucleus with nonzero spin (such as ^{235}U) and the magnetic fields associated with electron spin angular momentum. They pointed out that so far, no data unambiguously reflect this effect, but it is possible that fractionation of U in the contexts described here results from the nuclear spin effect, rather than the nuclear field shift effect.

Fujii *et al.* [74] conducted a clever experiment in which they reduced U on an ion-exchange resin to determine U(IV)–U(VI) isotope fractionation factors at temperatures ranging from 87 to 160°C. In this experiment, U(VI) was loaded into a series of four long titanium columns filled with a specialized, high-temperature anion-exchange resin. After the U(VI) had been loaded, a solution containing V(III) was loaded to reduce U(VI) partially into U(IV) and elute U from the resin. Chromatographic separation of oxidized and reduced U occurred along the length of the column, since U(IV) has a lower binding affinity for the resin than U(VI). Successive elution cuts were collected, and the proportions of U(IV) and U(VI), and also the $^{238/235}U$ isotope ratios, were measured in this series of samples. The resulting data were used to compute the distribution coefficients of U(IV) and U(VI) between resin and solution and the isotope fractionation between the two species. Reduced U (reaction product) was found to be enriched in the heavier isotope, indicating that equilibration of U(IV) and U(VI) occurred in the beds of exchange resin. Fractionation factors ranged from 1.4 to 1.1‰, and the magnitude of fractionation generally decreased with increase in temperature, as expected for an equilibrium isotope effect. This magnitude of fractionation is very consistent with the theoretical results of Schauble [73] and also with the observed fractionation between the heaviest black shales and seawater analyzed by Weyer *et al.* [62].

Rademacher *et al.* [75] also undertook experimental work to determine how uranium isotopes fractionate during reduction, but they conducted batch experiments and chose bacteria (*Geobacter sulfurreducens* and *Anaeromyxobacter dehalogenans*) and zerovalent iron as reducing agents. The biological experiments are particularly important, since reduction of many metals, possibly including uranium, is mediated by bacteria in natural waters. Culture media under anaerobic conditions were amended with 14 mgl^{-1} U in the form of UO_2Cl_2, inoculated with bacteria, incubated, and sampled at intervals of up to 96 h. Aliquots of the media were analyzed for U concentration and isotopic composition. Except for some scatter in the early hours of the experiment, U concentrations were observed to decrease steadily, and U isotope ratios in the media became progressively enriched in the heavier isotope. The complement of the U remaining in the media must therefore have been enriched in the lighter isotopes and was presumably the pool of U that was reduced by the bacteria. The isotopic compositions of the media were modeled with the Rayleigh fractionation equation, and the best-fit fractionations were computed as $\Delta^{238/235}U = -0.31‰$ for *G. sulfurreducens* and $-0.34‰$ for *A. dehalogenans*. Note that the isotopically light, reduced reaction product and the fit to an open-system Rayleigh trend indicate a kinetic isotope fractionation effect and that the sense and magnitude of fractionation in these experiments are in contrast to the fractionation observed in nature between seawater and reducing sediments and to the theoretical equilibrium fractionation. This implies that either (i) different taxa with a different mechanism of fractionating isotopes are responsible for U reduction in euxinic, marine settings than used in this bacterial experiment; (ii) the same bacterial reduction and fractionation mechanism is at play in the marine setting, but on longer time scales, other reactions reverse the kinetic fractionation and drive an approach to isotopic equilibration between U(VI) and U(IV); or (iii) reduction of U in the marine setting is abiotic and, therefore, isotope systematics in euxinic sediments bear no relationship to the experiments of Rademacher *et al.* [75]. Further investigation of reduction mechanisms in the natural setting will likely be necessary to distinguish between these possibilities. When Fe^0 was used as reducing agent, Rademacher *et al.* [75] observed decreasing U concentrations in solution as with the bacteria, with the complement presumably precipitating as U(IV) oxide particles, but in this case the U isotopic compositions of aliquots of the experimental solution did not evolve in terms of $\delta^{238/235}U$. This suggests that the mechanism of U reduction by Fe^0 does not drive any kinetic or equilibrium isotope fractionation and therefore is not likely to be the reaction mechanism governing isotope fractionation between seawater and marine sediments. Despite this negative result for Fe^0, other abiotic reduction reactions may yet prove to drive isotope fractionation for U.

To explore possible mechanisms of fractionation between seawater and oxic sediments, such as ferromanganese sediments and banded iron formations, Brennecka *et al.* [76] conducted experiments in which dissolved U(VI) was partially adsorbed on the manganese oxyhydroxide birnessite. These experiments were similar to those of Barling and Anbar [13] and Wasylenki and co-workers [20, 26] and were designed to determine whether a coordination change for U between the dominant aqueous species and the dominant adsorbed species is responsible for the fractionation observed in nature between seawater and ferromanganese sediments. In each experiment, an aliquot of a diluted, commercially available uranium standard solution was added to a suspension of synthetic birnessite particles that had been adjusted to pH ~5 and allowed to equilibrate with air. The U-amended suspensions were agitated for durations ranging from 10 min to 48 h and then filtered to separate dissolved from adsorbed U. Following purification by ion-exchange chemistry, all samples were analyzed for $\delta^{238/235}U$. On average, U remaining in solution was 0.22 ± 0.09‰ (2σ) heavier than adsorbed U, in excellent agreement with the fractionation measured between seawater and ferromanganese sediments by Weyer *et al.* [62].

The agreement between the experimental result and nature indicates that adsorption may well be the fractionating process, but in the absence of a redox change for U, a change in coordination chemistry must be driving the isotope effect. In order to constrain the fractionation mechanism, Brennecka *et al.* [76] collected an EXAFS spectrum from an experimental sample and compared it with spectra for aqueous $UO_2(CO_3)_3^{4-}$ and UO_2^{2+}, which are the dominant solution species at the pH of seawater and at pH ~5, respectively. None of the EXAFS spectra showed any evidence for U(IV), only for U(VI). All three spectra displayed strong scattering at 0.18 nm (1.8 Å), indicating two multiply bonded, collinear, axial oxygen atoms, but differences were observed among all three patterns for coordinating equatorial atoms. The spectra for $UO_2(CO_3)_3^{4-}$ and UO_2^{2+} were best fitted with arrangements of equatorial carbonate groups and water molecules that are symmetrically distributed around the central U atom with respect to the collinear axial oxygens, but the fit for the adsorbed U spectrum was improved significantly by modeling the coordination geometry with an asymmetric arrangement of equatorial oxygen atoms, with some at a distance of 0.236 nm (2.36 Å) and some at 0.249 nm (2.49 Å) from the central U atom. This result suggests a loss of symmetry when U forms an adsorbed complex on birnessite. Brennecka *et al.* [76] hypothesized that this loss of symmetry is the molecular-scale mechanism driving fractionation in their experiments and likely in nature between seawater and ferromanganese sediments. The same loss of symmetry and pair of U–O distances was observed previously for U adsorbed on ferrihydrite [77], hematite [78], and montmorillonite [79]. As Brennecka *et al.* [76] pointed out, the possibility that uranium isotopes fractionate by 0.2‰ during adsorption on all of these phases may have important implications for future work on uranium isotopes in marine settings, whether for marine paleoredox or for environmental applications.

11.3
Diagenesis: a Critical Area for Further Work

Taken together, the studies discussed above represent a remarkable amount of scientific progress in little more than a decade. Many specific challenges remain for the application of stable isotope geochemistry of metals to the ancient rock record, but the rate at which these challenges are being identified and addressed bodes well for the future. One area of uncertainty, however, has scarcely been mentioned in the literature so far, namely whether isotopic signatures of metals are altered by post-depositional, diagenetic processes.

All geochemical paleoproxy tools are subject to this particular challenge. Even when a particular environmental or biological variable has been recorded directly and straightforwardly in sediment, preservation of that clear chemical or isotopic signal through time is never guaranteed. Burial, shifts in redox conditions, elevated temperatures, and especially flux of fluids with variable dissolved contents may all act to erase precious information and confound our efforts to interpret conditions at the time of deposition. Accurate reconstruction of ancient conditions and processes requires demonstration that the chemical signal of interest has not been disturbed or reset since the geochemical signature of interest was recorded. Almost no work has yet been done to investigate how diagenesis affects isotopic signatures of metals in ancient sediments, nor to set clear criteria for determining whether a given sample set has been disturbed.

Probably the strongest approach to avoiding problems related to diagenetic resetting of a specific geochemical paleoproxy record is to demonstrate, with the same sample set, that another geochemical system for which clear criteria exist for recognizing diagenetic disturbance has not been reset. In an effort to do this, Anbar *et al.* [43] published a highly concordant Re–Os isochron for the same samples for which Mo concentrations [43] and Mo isotopic signatures had been determined [80]. The lack of discordance in the Re–Os system, which is highly sensitive to post-depositional disturbances, provides confidence that the Mo isotope ratios were never significantly altered, or at least not altered since the Re–Os isochron was locked in for those 2.5 billion year old black shales. A similar, undisturbed Re–Os isochron was published alongside Mo isotope ratio data by Kendall *et al.* [81] for the 1.4 Ga Velkerri Formation. The nearby Wollogorang Formation, however, showed more scattered Re–Os data, with an age considerably younger than the U–Pb zircon age of 1.7 Ga for the same rocks; hence Kendall *et al.* [81] inferred that the Wollogorang Mo isotope ratio data have likely been overprinted by interaction with hydrothermal fluids.

Even the concordant Re–Os isochrons do not provide absolute certainty that Mo isotope ratio records in ancient rocks have been properly preserved. McManus *et al.* [82] demonstrated that Mo isotope ratios in pore waters in Santa Monica Basin sediment are fractionated by $\sim$0.7‰ from the Mo isotope ratios in the sediment itself, due to redox shifts during burial of organic-rich material, suggesting that the very earliest stages of diagenesis can shift isotopic signatures of Mo, well before an Re–Os isochron is set. Poulson Brucker *et al.* [19] showed that Mn cycling during redox changes associated with early diagenesis can affect Mo isotope ratios, presumably because isotopically light Mo is released to pore waters during reductive dissolution of Mn oxides. Neubert *et al.* [34] contended that there is probably fractionation between the various thiomolybdate species, and this implies that sulfur cycling just below the water–sediment interface could also confound Mo isotopic signatures.

Given these observations in natural samples, systematic investigations of which processes can reset metal isotope ratios, how those processes shift isotope records, and what signs can be used to detect, and possibly correct for, the effects of diagenesis are absolutely essential. Some of the experimental and theoretical studies discussed above have yielded results that will help build a fundamental understanding of how diagenesis affects stable isotopic signatures of metals, but to date no studies have explicitly and systematically addressed this issue, neither for the Mo, Cr, or U paleoredox proxies, nor for other metal isotope proxy records in sedimentary rocks. The metal stable isotope field is poised to make many advances in the next few years with analysis and interpretation of the signals in ancient samples, but to do so with confidence and reliability will require much better constraint of the role of diagenesis in creating the geochemical signals that we most wish to understand.

References

1 Johnson C.M., Beard B.L., Roden E.E., Newman D.K., Nealson K.H. (2004) Isotopic constraints on biogeochemical cycling of Fe. *Rev. Mineral. Geochem.*, **55**, 359–408.

2 Anbar, A.D. and Rouxel, O. (2007) Metal stable isotopes in paleoceanography. *Annu. Rev. Earth Planet. Sci.*, **35**, 717–746.

3 Johnson C.M., Beard B.L., and Roden E.E. (2008) The iron isotope fingerprints of redox and biogeochemical cycling in modern and ancient earth. *Annu. Rev. Earth Planet. Sci.*, **36**, 457–493.

4 Montero-Serrano, J.-C., Martinez-Santana, M., Tribovillard, N., Riboulleau, A., and Garban, G. (2009) Geochemical behavior of molybdenum and its isotopes in the sedimentary environment – a bibliographic review. *Rev. Biol. Mar. Oceanogr.*, **44** (2), 263–275 (in Spanish).

5 Lyons, T.W., Reinhard, C.T., and Scott, C. (2009) Redox redux. *Geobiology*, **7** (5), 489–494.

6 Lyons, T.W., Anbar, A.D., Severmann, S., Scott, C., and Gill, B.C. (2009) Tracking euxinia in the ancient ocean: a multiproxy perspective and proterozoic case study. *Annu. Rev. Earth Planet. Sci.*, **37**, 507–534.

7 Severmann S. and Anbar A.D. (2009) Reconstructing paleoredox conditions through a multitracer approach: the key to the past is the present. *Elements.*, **5**(6), 359–364.

8 Bullen, T.D. (2011) Stable isotopes of transition and post-transition metals as tracers in environmental studies. *In Handbook of Environmental Isotope Geochemistry, Advances in Isotope Geochemistry* (ed. M. Baskaran), Springer-Verlag, pp. 177–203.

9 Primeau, F. (2005) Characterizing transport between the surface mixed layer and the ocean interior with a forward and adjoint global ocean transport model. *J. Phys. Oceanogr.*, **35**, 545–564.

10 Barling, J., Arnold, G.L, and Anbar, A.D. (2001) Natural mass-dependent variations in the isotopic composition of molybdenum. *Earth Planet. Sci. Lett.*, **193** (3–4), 447–457.

11 Arnold G.L., Anbar A.D., Barling J. and Lyons T.W. (2004) Molybdenum isotope evidence for widespread anoxia in mid-proterozoic oceans. *Science.*, **304**, 87–90.

12 Siebert C., Nägler T. F., von Blanckenburg F., and Kramers J. D. (2003) Molybdenum isotope records as a potential new proxy for paleoceanography. *Earth Planet. Sci. Lett.*, **211**, 159–171.

13 Barling, J. and Anbar, A.D. (2004) Molybdenum isotope fractionation during adsorption by manganese oxide. *Earth Planet. Sci. Lett.*, **217** (3–4), 315–329.

14 Archer, C. and Vance, D. (2008) The isotopic signature of the global riverine molybdenum flux and anoxia in the ancient oceans. *Nat. Geosci.*, **1**, 597–600.

15 Anbar A.D. (2004) Molybdenum stable isotopes: observations, interpretations and directions. *Rev. Mineral. Geochem.*, **55**, 429–454.

16 Neubert, N., Heri, A.R., Voegelin, A.R., Nägler, T.F., Schlunegger, F., and Villa,

I.M. (2011) The molybdenum isotopic composition in river water: constraints from small catchments. *Earth Planet. Sci. Lett.*, **304** (1–2), 180–190.
17 Helz G.R., Bura-Nakic E., Mikac N., and Ciglenecki I. (2011) New model for molybdenum behavior in euxinic waters *Chemical Geology.*, **284**, 323–332.
18 Poulson R.L., Siebert C., McManus J., Berelson W.M. (2006) Authigenic molybdenum isotope signatures in marine sediments. *Geology.*, **34**, 617–620.
19 Poulson Brucker, R.L., McManus, J., Severmann, S., and Berelson, W.M. (2009) Molybdenum behavior during early diagenesis: insights from Mo isotopes. *Geochem. Geophys. Geosyst.*, **10** (6), Q06010, doi: 10.1029/2008GC002180.
20 Wasylenki, L.E., Rolfe, B.A., Weeks, C. L., Spiro, T.G., and Anbar, A.D. (2008) Experimental investigation of the effects of temperature and ionic strength on Mo isotopic fractionation during adsorption to manganese oxide. *Geochim. Cosmochim. Acta*, **72** (24), 5997–6005.
21 Tossell, J.A. (2005) Calculating the partitioning of the isotopes of Mo between oxidic and sulphidic species in aqueous solution. *Geochim. Cosmochim. Acta*, **69**, 2981–2993.
22 Weeks C.L., Anbar A.D., Wasylenki L.E., and Spiro T.G. (2007) Density functional theory analysis of molybdenum isotopefractionation. *J. Phys. Chem.*, A **111**, 12434–12438.
23 Weeks C.L., Anbar A.D., Wasylenki L.E. and Spiro T. (2008) Density functional theory analysis of molybdenum isotope fractionation (correction to v111A, 12434, 2007). *J. Phys. Chem.*, A **112**, 10703.
24 Oyerinde O.F., Weeks C.L., Anbar A.D. and Spiro T.G. (2008) Solution structure of molybdic acid from Raman spectroscopyand DFT analysis. *Inorg. Chim. Acta.*, **361**, 1000–1007.
25 Liu, Y. (2008) Theoretical study on the mechanism of the removal of Mo from seawater in oxic environment. *Geochim. Cosmochim. Acta*, **72** (12) (Suppl. 1), A564.
26 Wasylenki, L.E., Weeks, C.L., Bargar, J.R., Spiro, T.G., Hein, J.R., and Anbar, A.D. (2011) The molecular mechanism of Mo isotope fractionation during adsorption to birnessite. *Geochim. Cosmochim. Acta*, **75**, 5019–5031.
27 Kashiwabara T., Takahashi Y., TanimizuM., and Usui A. (2011) Molecular-scale mechanisms of distribution and isotopic fractionation of molybdenum between seawater and ferromanganese oxides. *Geochim. Cosmochim. Acta.*, **75**, 5762–5784.
28 Koschinsky, A. and Hein, J.R. (2003) Uptake of elements from seawater by ferromanganese crusts: solid-phase association and seawater speciation. *Mar. Geol.*, **198**, 331–351.
29 Goldberg T., Archer C., Vance D. and Poulton S. (2009) Mo isotope fractionation during adsorption to Fe (oxyhydr)oxides. *Geochim. Cosmochim. Acta.*, **73**, 6502–6516.
30 Arai, Y. (2010) X-ray spectroscopic investigation of molybdenum multinuclear sorption mechanism at the goethite–water interface. *Environ. Sci. Technol.*, **44**, 8491–8496.
31 Liermann, L.J., Mathur, R., Wasylenki, L.E., Nuester, J., Anbar, A.D., and Brantley, S.L. (2011) Extent and isotopic composition of Fe and Mo release from two Pennsylvania shales in the presence of organic ligands and bacteria. *Chem. Geol.*, **281** (3–4), 167–180.
32 Nägler, T.F., Siebert, C., Lüshcen, H., and Böttcher, M.E. (2005) Sedimentary Mo isotope record across the Holocene fresh–brackish water transition of the Black Sea. *Chem. Geol.*, **219**, 283–295.
33 Siebert C., McManus J., Bice A., Poulson R., and Berelson W.M. (2006) Molybdenum isotope signatures in continental margin marine sediments. *Earth Planet. Sci. Lett.*, **241**, 723–733.
34 Neubert N., Nägler T.F., Böttcher M.E. (2008) Sulfidity controls molybdenum isotope fractionation into euxinic sediments: Evidence from the modern Black Sea. *Geology* **36**, 775–778.
35 Helz, G.R., Miller, C.V., Charnock, J.M., Mosselmans, J.F.W., Pattrick, R.A.D., Garner, C.D., Vaughan, D.J., (1996).

Mechanism of molybdenum removal from the sea and its concentration in black shales. *Geochim. Cosmochim. Acta.*, **60**, 3631–3642.

36 Algeo, T.J. and Lyons, T.W. (2006) Mo–TOC covariation in modern anoxic marine environments: implications for analysis of paleoredox and hydrographic conditions. *Paleoceanography*, **21**, PA1016.

37 Frei, R., Gaucher, C., Poulton, S.W., and Canfield, D.E. (2009) Fluctuations in Precambrian atmospheric oxygenation recorded by chromium isotopes. *Nature*, **461**, 250–253.

38 Izbicki J.A., Kulp T.R., Bullen T.D., Ball J.W. and O'Leary D.R. (2008) Chromium mobilization from the unsaturated zone. *Geochim. Cosmochim.* **72**(12) *Suppl.* 1, A417.

39 Schoenberg, R., Zink, S., Staubwasser, M., and von Blanckenburg, F. (2008) The stable Cr isotope inventory of solid Earth reservoirs determined by double spike MC-ICP-MS. *Chem. Geol.*, **249** (3–4), 294–306.

40 Ellis, A.S., Johnson, T.M., and Bullen, T. D. (2002) Chromium isotopes and the fate of hexavalent chromium in the environment. *Science*, **295**, 2060–2062.

41 Sikora, E.R., Johnson, R.M., and Bullen, T.D. (2008) Microbial mass-dependent fractionation of chromium isotopes. *Geochim. Cosmochim. Acta*, **72**, 3631–3641.

42 Schauble E.A., Rossman G.R., and Taylor H.P. (2004) Theoretical estimates of equilibrium chromium-isotope fractionations. *Chemical Geology.*, **205**, 99–114.

43 Anbar, A.D., Duan, Y., Lyons, T.W., Arnold, G.L., Kendall, B., Creaser, R.A., Kaufman, A.J., Gordon, G.W., Scott, C., Garvin, J., and Buick, R. (2007) A whiff of oxygen before the Great Oxidation Event? *Science*, **317**, 1903–1906.

44 Kaufman, A.J., Johnston, D.T., Farquhar, J., Masterson, A.L., Lyons, T. W., Bates, S., Anbar, A.D., Arnold, G.L., Garvin, J., and Buick, R. (2007) Late Archean biosheric oxygenation and atmospheric evolution. *Science*, **317**, 1900–1903.

45 Wille, M., Kramers, J.D., Nägler, T.F., Voegelin, A.R., Beukes, N.J., Schröder, S., Lacassie, J.P., and Meisel, T. (2007) Evidence for a gradual rise of oxygen between 2.6 and 2.5 Ga from Mo isotopes and Re-PGE signatures in shales. *Geochim. Cosmochim. Acta*, **71**, 2417–2435.

46 Canfield, D.E. (2005) The early history of atmospheric oxygen: homage to Robert M. Garrels. *Annu. Rev. Earth Planet. Sci.*, **33**, 1–36.

47 Johnston, D.T., Poulton, S.W., Fralick, P.W., Wing, B.A., Canfield, D.E., and Farquhar, J. (2006) Evolution of the oceanic sulfur cycle at the end of the Paleoproterozoic. *Geochim. Cosmochim. Acta*, **70**, 5723–5739.

48 Canfield D.E. (2008) A new model for Proterozoic ocean chemistry. *Nature.*, **396**, 450–453.

49 Kirschvink, J. (1992) Late Proterozoic low-latitude global glaciation: the snowball Earth, in *The Proterozoic Biosphere: a Multidisciplinary Study* (eds. J.W. Schopf and C. Klein), Cambridge University Press, Cambridge, pp. 51–52.

50 Schauble, E.A., Rossman, G.R., and Taylor, H.P. Jr (2001) Theoretical estimates of equilibrium Fe-isotope fractionations from vibrational spectroscopy. *Geochim. Cosmochim. Acta*, **65**, 2487–2497.

51 Schauble, E.A. (2004) Applying stable isotope fractionation theory to new systems. *Rev. Mineral. Geochem.*, **55**, 65–111.

52 Hill P.S., Schauble E.A., and Young E. D. (2010) Effects of changing solution chemistry on Fe(3+)/Fe(2+) isotope fractionation in aqueous Fe-Cl solutions. *Geochim. Cosmochim. Acta* **74**, 6669–6689.

53 Black J.R., Kavner A., and Schauble E.A. (2011) Calculation of equilibrium stable isotope partition function ratios for aqueous zinc complexes and metallic zinc. *Geochim. Cosmochim. Acta.*, **75**, 769–783.

54 Zink, S., Schoenberg, R., and Staubwasser, M. (2010) Isotopic fractionation and reaction kinetics between Cr(III) and Cr(VI) in aqueous media. *Geochim. Cosmochim. Acta*, **74**, 5729–5745.

55 Bain D.J. and Bullen T.D. (2005) Chromium isotope fractionation during oxidation of Cr(III) by manganese oxides. *Geochim. Cosmochim. Acta.*, **69**, *Suppl. 1*, A212.

56 Døssing, L.N., Dideriksen, K., Stipp, S.L. S., and Frei, R. (2011) Reduction of hexavalent chromium by ferrous iron: a process of chromium isotope fractionation and its relevance to natural environments. *Chem. Geol.*, **285**, 157–166.

57 Rees, C.E. (1973) A steady-state model for sulphur isotope fractionation in bacterial reduction processes. *Geochim. Cosmochim. Acta*, **37**, 1141–1162.

58 Canfield, D.E. (2001) Biogeochemistry of sulfur isotopes, in *Stable Isotope Geochemistry* (eds. J.W. Valley and D.R. Cole), Mineralogical Society of America, Washington, DC, pp. 607–636.

59 Myers, C., Carstens, B., Antholine, W., and Myers, J. (2000) Chromium(VI) reductase activity is associated with the cytoplasmic membrane of anaerobically grown *Shewanella putrefaciens* MR-1. *J. Appl. Microbiol.*, **88**, 98–106.

60 Kalabegishvili, T., Tsibakhashvili, N., and Holman, H. (2003) Electron spin resonance study of chromium(V) formation and decomposition by basalt-inhabiting bacteria. *Environ. Sci. Technol.*, **37**, 4678–4684.

61 Stirling, C.H., Andersen, M.B., Potter, E., and Halliday, A.N. (2007) Low-temperature isotopic fractionation of uranium. *Earth Planet. Sci. Lett.*, **264**, 208–225.

62 Weyer, S., Anbar, A.D., Gerdes, A., Gordon, G.W., Algeo, T.J., and Boyle, E.A. (2008) Natural fractionation of $^{238}U/^{235}U$. *Geochim. Cosmochim. Acta*, **72**, 345–359.

63 Bigeleisen J. (1996) Nuclear size and shape effects in chemical reactions. Isotope chemistry of the heavy elements. *J. Amer. Chem. Soc.*, **118**, 3676–3680.

64 Schauble, E.A. (2007) Role of nuclear volume in driving equilibrium stable isotope fractionation of mercury, thallium and other heavy elements. *Geochim. Cosmochim. Acta*, **71**, 2170–2189.

65 Epov, V.N., Malinovsky, D., Vanhaecke, F., Bégué, D., and Donard, O.F.X. (2011) Modern mass spectrometry for studying mass-independent fractionation of heavy stable isotopes in environmental and biological sciences. *J. Anal. At. Spectrom.*, **26**, 1142–1156.

66 Malinovsky, D. and Vanhaecke, F. (2011) Mass-independent isotope fractionation of heavy elements measured by MC-ICPMS: a unique probe in environmental sciences. *Anal. Bioanal. Chem.*, **400** (6), 1619–1624.

67 Colodner, D., Edmond, J., and Boyle, E. (1995) Rhenium in the Black Sea: comparison with molybdenum and uranium. *Earth Planet. Sci. Lett.*, **131**, 1–15.

68 Montoya-Piño, C., Weyer, S., Anbar, A. D., Pross, J., Oschmann, W., Van de Schootbrugge, B., and Arz, H.W. (2010) Global enhancement of ocean anoxia during Oceanic Anoxic Event 2: a quantitative approach using U isotopes. *Geology*, **38** (4), 315–318.

69 Arthur, M.A., Dean, W.E., and Pratt, L.M. (1988) Geochemical and climatic effects of increased marine organic carbon burial at the Cenomanian/Turonian boundary. *Nature*, **335**, 714–717.

70 Erbacher, J., Friedrich, O., Wilson, P.A., Birch, H., and Mutterlose, J. (2005) Stable carbon isotope stratigraphy across Oceanic Anoxic Event 2 of Demerara Rise, western tropical Atlantic. *Geochem. Geophys. Geosyst.*, **6**, Q06010, doi: 10.1029/2004GC000850.

71 Herrmann, A.D., Wasylenki, L., and Anbar, A.D. (2010) Uranium isotopic composition of carbonate sediments as a potential redox proxy. *Geol. Soc. Am. Abstr. Programs*, **42** (5), 514.

72 Brennecka, G.A., Herrmann, A.D., Saltzman, M., and Anbar, A.D. (2009) Using $^{238}U/^{235}U$ ratios in carbonates as a paleoredox indicator: variations across the Permian–Triassic boundary. *Geol. Soc. Am. Abstr. Programs*, **41** (7), 566.

73 Schauble, E.A. (2006) Equilibrium uranium isotope fractionation by nuclear volume and mass-dependent processes, American Geophysical Union, Fall Meeting, abstract V21B-0570.

74 Fujii, Y., Higuchi, N., Haruno, Y., Nomura, M., and Suzuki, T. (2006) Temperature dependence of isotope effects in uranium chemical exchange reactions. *J. Nucl. Sci. Technol.*, **43** (4), 400–406.

75 Rademacher, L.K., Lundstrom, C.C., Johnson, T.M., Sanford, R.A., Zhao, J., and Zhang, Z. (2006) Experimentally determined uranium isotope fractionation during reduction of hexavalent U by bacteria and zero valent iron. *Environ. Sci. Technol.*, **40**, 6943–6948.

76 Brennecka, G.A., Wasylenki, L.E., Bargar, J.R., Weyer, S., and Anbar, A.D. (2011) Uranium isotope fractionation during adsorption to Mn-oxyhydroxides. *Environ. Sci. Technol.*, **45**, 1370–1375.

77 Waite, T.D., Davis, J.A., Payne, T.E., Waychunas, G.A., and Xu, N. (1994) Uranium(VI) adsorption to ferrihydrite: application of a surface complexation model. *Geochim. Cosmochim. Acta*, **58**, 5465–5478.

78 Bargar, J.R., Reitmeyer, R., Lenhart, J.J., and Davis, J.A. (2000) Characterization of U(VI)–carbonato ternary complexes on hematite: EXAFS and electrophoretic mobility measurements. *Geochim. Cosmochim. Acta*, **64**, 2737–2749.

79 Catalano, J.G. and Brown, G.E. Jr (2005) Uranyl adsorption on montmorillonite: evaluation of binding sites and carbonate complexation. *Geochim. Cosmochim. Acta*, **69**, 2995–3005.

80 Duan, Y., Anbar, A.D., Arnold, G.L., Lyons, T.W., Gordon, G.W., and Kendall, B. (2010) Molybdenum isotope evidence for mild environmental oxygenation before the Great Oxidation Event. *Geochim. Cosmochim. Acta*, **74** (23), 6655–6668.

81 Kendall, B., Creaser, R.A., Gordon, G. W., and Anbar, A.D. (2009) Re–Os and Mo isotope systematics of black shales from the Middle Proterozoic Velkerri and Wollogorang Formations, McArthur Basin, northern Australia. *Geochim. Cosmochim. Acta*, **73**, 2534–2558.

82 McManus, J., Nägler, T.F., Siebert, C., Wheat, C.G., and Hammond, D.E. (2002) Oceanic molybdenum isotope fractionation: diagenesis and hydrothermal ridge-flank alteration. *Geochem. Geophys. Geosyst.*, **3** (12), 1078, doi: 10.1029/2002GC000356.

12
Isotopes as Tracers of Elements Across the Geosphere–Biosphere Interface

Kurt Kyser

12.1
Description of the Geosphere–Biosphere Interface

The distribution of elements in the near-surface environment of the Earth is a function of both abiogenic and biogenic processes. Characterizing these elemental distributions is difficult, because they vary over both space and time and they are controlled largely by the interface between the geosphere (the non-living part of the Earth) and the biosphere (the living part of the Earth). The distribution of elements across the geosphere–biosphere interface is fundamental for discovering buried mineral and energy resources, managing our environment, understanding human effects on animal ecology and climate, and determining the bioaccessibility of toxic elements in the environment. All of these aspects of the geosphere–biosphere interface are critical for maintaining and improving our standard of living globally, and also for protecting ecosystems [1].

Understanding which processes control the cycle of elements across the geosphere–biosphere interface is the recent focus of many research efforts that integrate both concentrations of elements and their isotopic compositions to characterize the source, transport, and depositional characteristics of natural and anthropogenic compounds in the near-surface environment. Isotopic tracing of elements has been applied to a variety of materials, including rocks, minerals, soils, tissues, fluids, and foodstuffs, with most of the effort using stable isotopes of the "light elements" H, C, N, O, and S, the building blocks of the biosphere. The goal of these research efforts is to understand element cycling at scales ranging from the upper 2 km of the crust for recognizing ore deposit signatures down to the nanoscale for characterizing ingested dust particles or food. Our ability to track elements in the near-surface environment depends on understanding complete element cycles in both the geosphere and biosphere and how these are controlled by the ability of fluids and microorganisms to mobilize elements to specific depositional sites.

Recent advances in isotope ratio mass spectrometry, particularly the advent of multi-collector (MC) inductively coupled plasma mass spectrometry (ICP-MS),

Isotopic Analysis: Fundamentals and Applications Using ICP-MS, First Edition. Edited by Frank Vanhaecke and Patrick Degryse.

Published 2012 by WILEY-VCH Verlag GmbH & Co. KGaA

using both solution nebulization and laser ablation (LA) as means of sample introduction, have afforded us the ability to analyze new isotopic systems at low concentrations in a myriad of both geological and biological materials [2, 3]. The purpose of this chapter is to discuss some of the results and advances that have been made in tracing elements across the geosphere–biosphere interface using MC-ICP-MS and to elucidate possible future directions of this technology. This chapter is meant to demonstrate that element tracing via isotopic analysis of metals is a viable technique that adds value to our understanding of the biosphere and its relation with the geosphere and to discuss selected recent results. The health of our natural environment and our citizens, the survival of our wildlife, the development of mineral resources, and the fostering of high-technology industries are anticipated benefits of these results.

Here, the geosphere refers to that part of the solid Earth that is not living, but certainly is not devoid of life. It is where humans and other organisms live and exploit for food, fuels, and resources. The geosphere encompasses the deep mantle, the upper part of the continental and oceanic crusts of the Earth, sediments deposited and preserved during Earth history, suspended particles in the ocean, soil, and the uppermost weathered layer of rock (Figure 12.1). The geosphere is interfaced with the atmosphere, hydrosphere, anthrosphere, and biosphere, the last denoting the environment inhabited by living organisms, of which the anthrosphere is part. The geosphere, atmosphere, hydrosphere, and biosphere are in physical contact with each other and their boundaries overlap

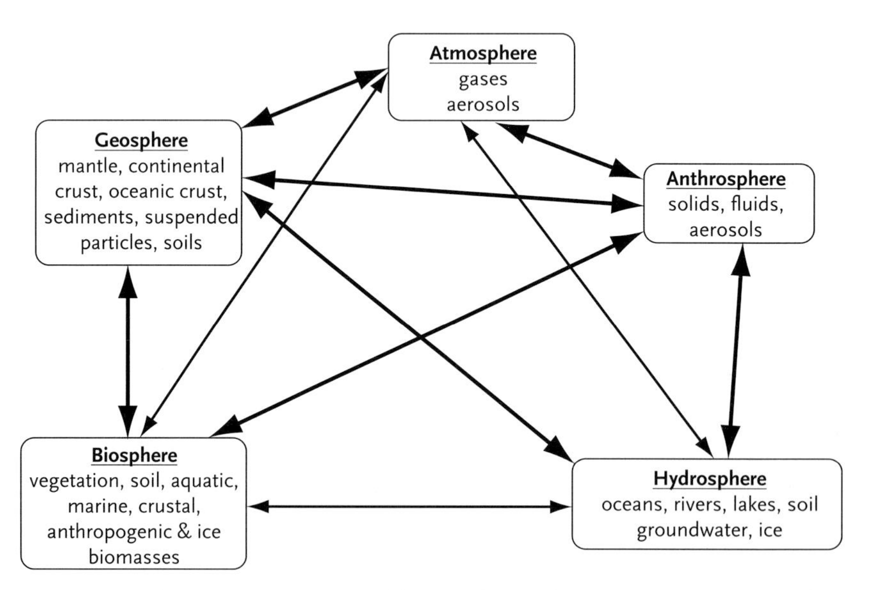

Figure 12.1 Schematic representation of the geosphere and its interaction with other spheres of the Earth's surface. Thick arrows indicate more substantial interactions. Transport of material and energy from the geosphere affects the content of the other spheres and the two-way flows between them.

significantly because there is significant interaction among them and the anthrosphere (Figure 12.1). The geosphere provides physical habitat, nutrients, and other elements to the biosphere and receives these back primarily in organic matter that is subsequently altered. Similarly, the geosphere and biosphere exchange gases and fluids with the atmosphere, hydrosphere, and anthrosphere and react with them chemically, during which they all relinquish and receive elements across their interfaces. The nature and magnitude of the exchanges between the geosphere and the biosphere have changed considerably throughout the Earth's history [4] and will likely continue to do so in the future.

The mass of the geosphere and the diversity of its chemical and mineral composition are the most substantial of all reservoirs [5] and it is the ultimate source of all elements. The mass of the entire geosphere down to 2 km depth is about 10^{22} kg. The hydrosphere holds 1.4×10^{21} kg of water, with 97.5% of it in the oceans. The hydrosphere amounts to 15% of the mass of the geosphere and hosts only about 5×10^{19} kg of dissolved solids. The mass of the atmosphere is only 5×10^{18} kg, which is a small fraction of the geosphere. The mass of dry organic matter in the biosphere is about 1.5×10^{15} kg, contained mostly in land plants, with animals accounting for only 1% of the total and marine life for an even smaller fraction [6].

Human activities occur primarily on land, mankind derives most of its needed fuels and raw materials from the Earth's sediments and crust, and its agricultural activities affect primarily the land vegetation and soils. Activities originating in the geosphere perturb the hydrosphere, biosphere, and atmosphere through the exchange of elements (Figure 12.1). The biosphere consists mainly of C, N, P, S, O, and H, but also contains critical trace elements, such as B, Cu, Cr, Fe, Zn, Cd, Sr, Se, Nd, Pb, and U, which are the focus of studies using MC-ICP-MS.

The geosphere is far from immutable, with changes caused by volcanic and seismic activity, erosion of the continents, waning ice volume, the area of the land and vegetation cover. Most of these natural changes are far slower than changes to the geosphere attributable to anthropogenic activities. The present industrial age is characterized by intensive extraction of organic and inorganic materials from the geosphere and extensive alteration of both the terrestrial and marine environments, including the biosphere [7]. Agricultural activity has resulted in deforestation, replanting, accelerated soil erosion, and changes in the residence time of many elements on land [6]. Although the cultivated area of the world is about 15% of the land surface, deforestation, particularly of the tropical forests, is the second most important source of carbon dioxide emissions from land to the atmosphere after burning of fossil fuels [8]. Agricultural and other land-use activities are responsible for a greater transport of elements from land to the oceans and have led to faster recycling of organic matter and the elements therein.

The International Geosphere–Biosphere Programme (IGBP) [9] is a research program that aims to understand how the physical, chemical, and biological processes regulate the Earth system and how humans are influencing global elemental cycles. The IGBP published a synthesis in 2005, stating that because of anthropogenic influences, the Earth is in a state well outside the range of natural variability during at least the last several hundred years [10]. This is reflected in land use, wherein half of the land surface has been domesticated for human use, in the world's fisheries, which are over-exploited, in alterations to the composition of the atmosphere relative to a century ago, and in enhanced extinctions. Although some of these are moot points, there is a strong coupling between present-day perturbations of the geosphere and the density of human populations. As the geosphere does not absorb or store all the products of human activities, these are passed on to the hydrosphere, the biosphere, and the atmosphere in progressively greater amounts and with uncertain consequences for the future direction of impact on the global environment. Given this, any study dealing with the interaction between the geosphere and the biosphere, whether for exploration of deep ore deposits or the bioavailability of elements in the environment, must consider anthropogenic influences [11].

12.2
Elements That Typify the Geosphere–Biosphere Interface

MC-ICP-MS techniques have capitalized on both thermal ionization mass spectrometry (TIMS) and ICP-MS protocols to enhance research using non-traditional stable isotopes, with special attention to transition metals. Almost 20 elements, namely Li, B, Mg, Si, Ca, Cd, Cr, Zn, Cu, Fe, Se, Tl, Sr, Mo, Nd, Hf, Hg, Pb, and U, in addition to the traditional light elements H, C, N, O, and S, have isotope systems that allow one to track elements across the geosphere and biosphere. Lithium is useful because the mass difference between its two isotopes, ^{6}Li and ^{7}Li, is substantial, so it is a tracer of the alkali metals and how they partition between solids and fluids. The same is true for Mg and Ca, which show mass-dependent isotopic fractionation as the primary process involved in variations in their isotopic compositions. Boron isotopes fractionate between the two common forms of B, $B(OH)_3$ and $B(OH)_4^-$, and the distribution of B over these two compounds is a function of the prevailing pH. Those elements that are redox-sensitive and hence display isotopic fractionation because of significant difference in bonding environment, include Cr, Tl, Cu, Se, Mo, Cd, Hg, and U. Those that have isotope ratios involving an isotope that is radiogenic include Sr, Nd, Hf, and Pb. The uses of these isotope systems are varied, ranging from tracing elements through the environment for remediation purposes, to exploiting the origin of biogenic minerals for engineering structures, to tracing elements in media at the surface that have migrated from ore deposits at depth. Using concentrations alone does not always track the origin of an element: often, isotopic analysis is the only definitive way of tracing its origin and history [12].

Among the benefits from using isotopic analysis to study trace element movement across the geosphere and biosphere is the realization that Hg shows mass-independent fractionation that enhances our ability to define the sources of this toxic element [13]. Isotopes of Ca and Mg in biogenic carbonates indicate that there are both equilibrium mineralogical controls and kinetic controls on how these elements are incorporated by organisms from the geosphere and

Pb isotopes reveal that elements migrate both from below and also from above the surface as they are incorporated into vegetation and soils. Redox reactions promote fractionation of the isotopes of several elements, with the heavier isotope generally favored in the more oxidized phase. Organisms, which rely on redox reactions for their survival, fractionate metal isotopes, even in ore deposits, which tend to be havens for microbial activity that promotes element mobility into the surrounding environment.

12.3 Microbes at the Interface

The geosphere is intimately tied to its biosphere [14], a major link between the two being the microorganisms that grow in and on rocks and minerals, relying on their substrate for critical compounds, particularly those which need redox-sensitive elements to produce energy [15]. The presence of a metabolizing cell on a mineral substrate has a significant effect on the mineral texture and geochemistry of the surrounding microenvironment. In Nature, microorganisms exist in microbial communities as mats or biofilms, growing on a solid substrate. As such, they cover a vast volume of the Earth's surface [16] and are a major target for research.

Iron is a redox-active element that is extremely abundant in the geosphere and critical for organisms in the biosphere. Variations in the isotopic composition of Fe in Nature often reflect the equilibrium fractionation of 3‰ in the $^{56/54}Fe$ ratios between dissolved Fe^{3+} complexes and dissolved Fe^{2+} (Figure 12.2), the latter having the lower ratio and being the most mobile form of Fe in solution [18]. Chemical sediments that form from Fe^{2+}, such as organic-rich black shales or hydrothermal ores, normally have lower $^{56/54}Fe$ ratios than typical igneous rocks (Figure 12.2). Isotopic fractionation of Fe during redox reactions mediated by biological processes can be more extreme than during abiotic redox reactions [19], so that the isotopic composition of Fe may identify biological mediation of Fe in the rock record [20]. Microbial Fe redox cycling during repetitive oxidation and reduction of Fe in surface sediments may be an important source of Fe^{2+} having low $^{56/54}Fe$ ratios compared with the oceans [21]. For example, Archean sedimentary rocks, which record anoxia in the oceans, normally have variable $^{56/54}Fe$ ratios (Figure 12.2) that may reflect an enhanced supply of Fe^{2+} from processes such as microbial Fe^{3+} reduction. High $^{56/54}Fe$ ratios are ascribed to oxic removal processes where Fe^{3+} oxides are formed in the ocean [22], perhaps associated with bacteria that can photo-oxidize Fe^{2+} in the absence of dissolved O_2 [20].

Isotopic compositions of Mo and Fe measured by MC-ICP-MS indicate that during assimilation of Mo and Fe via nitrogen fixation by *Azotobacter vinelandii*, the lighter isotope of Mo (^{95}Mo versus ^{97}Mo) is preferentially assimilated by the bacteria with a fractionation factor between assimilated and dissolved Mo of 0.9997 [2]. In contrast, the heavier isotope of Fe (^{56}Fe versus ^{54}Fe) was removed from solution by the bacteria with a fractionation factor of 1.0011. Fractionation of Mo isotopes results from kinetic effects involving metal-binding ligands, whereas Fe

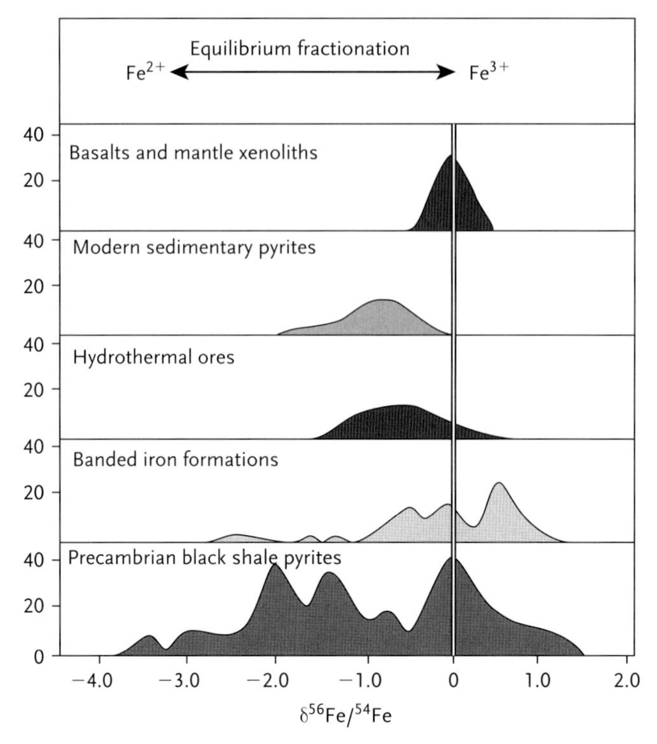

Figure 12.2 Variations in $\delta^{56}Fe/^{54}Fe$ values for various geological materials. The vertical line marks the isotopic composition of igneous rocks at 0‰. The bias towards negative $\delta^{56/54}Fe$ values reflects Fe from Fe^{2+}, which is isotopically light.
Modified from [17].

isotopic fractionation reflects equilibrium between different Fe complexes or adsorption on cell surfaces. Therefore, isotope fractionation studies using MC-ICP-MS may provide new constraints on the processes by which microbes extract metals from their surroundings.

12.4
Element Tracing in Environmental Science and Exploration of Metal Deposits

The isotopic compositions of Zn and Cu are used separately and in tandem to study geosphere–biosphere interactions. Zinc has five stable isotopes, ^{64}Zn, ^{66}Zn, ^{67}Zn, ^{68}Zn, and ^{70}Zn, with average natural isotopic abundances of 48.63, 27.90, 4.10, 18.75, and 0.62%, respectively. Zinc is vital for most living organisms and plays an important role in various biological processes as an essential element, with its numerous isotopes potentially offering a decisive tool for studying geosphere–biosphere interactions. In environmental sciences, Zn isotopic compositions have been investigated in aerosols [23], lichens [24], peat cores [25], river water [26], and stream water affected by mining activities and deposition around smelters. Viers *et al.* [27] determined the cycle of Zn in a nonimpacted watershed in Cameroon using isotopes in soils and trees.

In contrast to Zn, Cu has only two stable isotopes, ^{63}Cu and ^{65}Cu, with average natural isotopic abundances of 69.17 and 30.83%, respectively. Copper is extremely toxic for all aquatic photosynthetic microorganisms, but it is extensively used in biological processes by other organisms. The two common oxidation states and the ability of Cu to engage in various complexes, particularly as hydrates and halides, result in significant isotopic fractionations for Cu. Copper isotopic compositions may provide direct evidence of the source of metals and fluid pathways in mineralizing systems [28], but also information on isotopic fractionation during adsorption of aqueous Cu on metal oxide, hydroxide, and bacterial surfaces [29]. Fractionation accompanying redox-driven reactions is significant [30, 31], in addition to that during dissolution and precipitation processes at low temperatures [30, 32]. Copper normally occurs at a concentration level one to two orders of magnitude lower than that of Zn, and has two natural oxidation states, Cu^{+} and Cu^{2+}, whereas Zn has only one, Zn^{2+}.

How Cu and Zn isotopes can be used to differentiate the impact of the geosphere and anthrosphere on the biosphere is exemplified by the work of Thapalia *et al.* [12], who used combined isotopic analysis of Zn and Cu to identify sources and timing of deposition of these metals in a sediment core from Lake Ballinger near Seattle, WA, USA. The lake is 53 km north of a Pb smelter that became operational in 1890 and then switched to a Cu ore smelter, used from 1905 until 1985. Isotopic changes with time are preserved, with the pre-smelter period in the core having $\delta^{66}Zn$ of $+0.39‰$ and $\delta^{65}Cu$ of $+0.77‰$, whereas later than 1900 when the smelter was operating, $\delta^{66}Zn$ is lower at $+0.14‰$ and $\delta^{65}Cu$ is higher at $+0.94$. For the post-smelting period after 1985, characterized by urban land use, an even lower value of $\delta^{66}Zn$ near 0.00‰ and $\delta^{65}Cu$ near $+0.82‰$ were established. Early urbanization during the post-World War II era increased metal loading to the lake, but the $\delta^{66}Zn$ and $\delta^{65}Cu$ values suggest that increased metal loads during this time were derived mainly from mobilization of historically contaminated soils. Urban sources of Cu and Zn after 1980 are consistent with pollution from tire wear as the likely source of Zn. However, the authors also suggested that $\delta^{66}Zn$ values of atmospheric emissions from smelting sources may be homogeneous over much larger areal extents than previously suspected. The $\delta^{66}Zn$ of smelter emissions should become significantly lighter as particle sizes decrease with increasing distance from a smelter, but this is not what was observed at Lake Ballinger. Although Zn and Cu isotopes may be useful for assessing the impact of urban metal sources, the ranges of Zn and Cu isotopic compositions in natural samples and those from anthropogenic sources overlap significantly.

The combination of the Zn and Cu isotopic systems has been applied in exploration geochemistry [33], wherein soils and vegetation over a volcanic massive sulfide (VMS) ore deposit at 100 m depth were used to understand the transport pathways of these elements from the ore body to the surface (Figure 12.3). Microbes within the deposit help mobilize Cu and Zn complexes that migrate to the surface and into the overlying geosphere and biosphere. Although isotopic fractionations

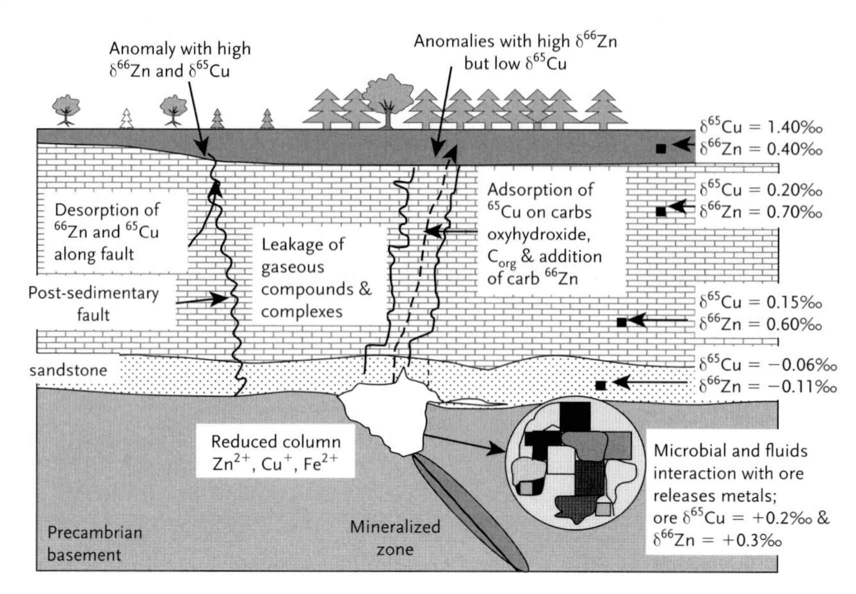

Figure 12.3 Simplified model for the various processes governing the Cu and Zn isotopic composition collected in the Talbot area of Canada [33]. The ore zone is redox-active, with interactions between microbes and fluids on the sulfide ores mobilizing Cu and Zn as aqueous and gaseous complexes into the overlying sandstone and carbonate, which are about 100 m thick in total. Migration of these components occurs through micro- and macro-fractures and then they are trapped in the near-surface environment on clay minerals in the till or incorporated into the biosphere, including trees, moss, and organic matter in the soils.

occur as these complexes migrate through the overlying carbonates, there are distinct isotopic anomalies in various surface media over the deposit and over a fault system that transects the deposit at depth. Relative to background values, high δ^{66}Zn and low δ^{65}Cu values were found in soil, moss, black spruce tree bark, and tree cores over the VMS deposit, whereas over a post-sedimentary fault, soils showed both high δ^{66}Zn values and high δ^{65}Cu values (Figure 12.3). Variations of Cu and Zn isotopic compositions for samples collected at the surface can be traced to the VMS deposit buried at 100 m depth and also reflect different processes that occur during metal migration to the surface. Despite the complexities, Cu and Zn isotopic compositions may be definitive tracers of mineralization at depth.

The isotopes of Pb have a demonstrated utility in tracing the impact of the use of leaded gasoline on the environment. A chronology of the concentrations and isotopic compositions of oceanic Pb in the North Atlantic Ocean for the past 220 years was determined by Boyle *et al.* [34] in annual bands of corals that grew in coastal seawaters near Bermuda and from seawater samples collected from the North Atlantic since 1980. Anthropogenic Pb emissions in this area reflect the industrialization of North America since 1840 and the introduction of leaded gasoline in 1920 until its cessation by the late 1990s. Coral Pb concentrations are constant at 5 nmol Pb per mole Ca from late 1700s to 1850, increase slightly to a plateau at 25 nmol Pb per mole Ca in the 1930s and increase rapidly in late 1940s to 60 nmol Pb per mole Ca, but Pb concentrations began to decline in 1975 to the same levels as in the early twentieth century. Lead isotope ratios determined using MC-ICP-MS are the least radiogenic before 1900 due to the early dominance of Upper Mississippi Valley Pb ore in the USA. After 1900, $^{207}Pb/^{206}Pb$ ratios increase slightly until 1970, when sources of common Pb were phased out in the USA. Leaded gasoline utilization continued in Europe where the Pb additive was more radiogenic, and this is recorded as an increase in $^{207}Pb/^{206}Pb$ ratios in the North Atlantic compared with their most radiogenic ratios in the past two centuries.

Rock, soil, and plant samples collected at 3 km intervals along a 120 km long transect through the city of Oslo, Norway, record a northward increase in both $^{207}Pb/^{206}Pb$ and $^{208}Pb/^{206}Pb$ ratios measured with single-collector sector field ICP-MS and an increase in Pb concentrations in urban areas [35]. In addition, low variability in the concentrations and Pb isotopic compositions in all plant samples and O-horizon soil profiles compared with those in rocks and mineral soils is observed along the transect. Reimann *et al.* [35] suggested that (i) there are two different sources for the Pb, the lithosphere and the biosphere, and (ii) in the latter Pb undergoes mass fractionation as a result of biological processes, therefore resulting in less radiogenic and more constant Pb isotopic compositions that differ from the underlying soils. However, this study does not integrate results from numerous previous studies on local, regional, or global scales, that indicate anthropogenic Pb in the atmosphere from leaded gasoline, coal burning, and metallurgy has been dispersed worldwide [36]. Pre-anthropogenic Pb isotopic compositions in continental archives of atmospheric deposition such as peat bogs [37, 38] indicate that (i) these Pb isotopic signatures are geogenic and regionally derived and (ii) fluxes of Pb were considerably lower before deposition became dominated by human activities [39]. The Pb isotopic signatures recorded in modern biogenic samples more likely reflect mixing of twentieth century anthropogenic Pb and local geogenic background Pb rather than effects of mass-dependent fractionation, which is exceedingly small for heavy elements, or mass-independent fractionation. The relationship between Pb concentrations and isotopic compositions of the biogenic media indicates mixing between two end-members, one of which has the highest Pb contents and highest $^{207}Pb/^{206}Pb$ and is geogenic and the other having more common Pb isotopic compositions and is anthropogenic (see also Chapter 1).

Lahd Geagea *et al.* [40] measured the Pb, Sr, and Nd isotopic compositions of lichen, moss, bark, and soil litter from different regions in the Rhine Valley and particles in ice of the Rhône and Oberaar glaciers and lichens from the Swiss Central Alps. These regions record isotopic compositions of the corresponding basement rocks or soils at the same sites, but their isotopic compositions are very different from those of atmospheric baseline samples from the Vosges Mountains and the Rhine Valley. This is because atmospheric Pb and Sr isotopic compositions are affected by traffic, industrial, and urban emissions even in remote areas such as the Vosges Mountains and the Rhine Valley. Lichen samples from below the Rhône and Oberaar glaciers reflect the geogenic baseline isotopic compositions. The Nd isotope ratios are highly variable and consistent with wet deposition and aerosols originating from regional natural and industrial environments and from distant sources such as the Sahara in North Africa.

12.5 Isotopes as Indicators of Paleoenvironments

The study of paleoenvironments using isotopic analysis has traditionally involved the light elements, H, C, N, O, and S. However, recent advances in MC-ICP-MS have expanded the field of elements that can be used to evaluate paleoceanography. The cycle of nutrients in the oceans is one of the more obvious examples of the flow of material and energy between the geosphere and biosphere. The Si cycle is coupled to both the carbon cycle and global climatic change in that biological precipitation of amorphous silica (opal) by diatoms removes $Si(OH)_4$ from seawater, transporting silica and organic C to the seafloor. The $\delta^{30}Si$ values of sponge spicules are a potential proxy for changes in Si cycling, as revealed by Hendry *et al.* [41], who used modern sponges and seawater from the Southern Ocean. They reported that the concentration of $Si(OH)_4$ varied with the $\delta^{30}Si$ values and that deep waters of the Southern Ocean were not enriched in $Si(OH)_4$ during the last glacial maximum, despite recording low $\delta^{13}C$ values.

Molybdenum isotopes are fractionated during adsorption to ferromanganese oxides [42] and precipitation of Fe–Mo–S solids [43]. In the open oceans, Mo is a conservative element, the concentration and isotopic composition of which do not change with depth. The residence time of Mo in the modern oceans is about 10^6 years. The isotopic fractionation of Mo sequestered into sediments has an impact on the entire ocean basins. Uptake of Mo onto marine particles by oxic adsorption processes preferentially removes light isotopes from solution, so that the $^{98}Mo/^{95}Mo$ isotope ratio for seawater is higher than riverine input by $>1.6‰$ [44]. Similarly, Mo isotopic compositions are lower during times of reduced ocean oxygenation and can be monitored using black shales deposited during periods of anoxia, as there is minimal isotopic fractionation of Mo since all of the dissolved Mo is sequestered into the sediments (see also chapter 11).

12.6 Tracing the Geosphere Effect on Vegetation and Animals

Plant tissues show a range of $\delta^{66}Zn$ values from 0.6 to 1.4‰, exceeding the variability known for geological terrestrial materials [45]. Such a large variation in $\delta^{66}Zn$ values results from mass-dependent fractionation from the growth medium and among plant tissues. There are minimal data available for $\delta^{66}Zn$ values in animal tissues, although plankton and lobster organs have higher $\delta^{66}Zn$ values than seawater and human hair has higher $\delta^{66}Zn$ values than blood [23]. Integrated studies of Zn isotope fractionations in various organs of sheep raised on a controlled diet showed that δ^{66}Zn variability exceeds 10‰, with bone, muscle, serum, and urine being enriched in the heavy isotopes, whereas feces, red blood cells, kidney, and liver are enriched in light isotopes relative to the diet value [46]. Enrichment in ^{66}Zn of serum likely occurs in the digestive tract via preferential binding of lighter isotopes with phytic acid, which controls the uptake of metallic elements. Maximum δ^{66}Zn values relative to diet are reached after about 10 years, thus relating Zn isotopic compositions with maturity of animals.

The isotope ^{87}Sr is a decay product of ^{87}Rb, so that relative to the stable isotope ^{86}Sr, the amount of ^{87}Sr is a function of the age of the rock and the amount of Rb within it. The older or more Rb-rich a rock, the higher is its ^{87}Sr/^{86}Sr ratio. This Sr isotopic composition will be preserved in soils that form *in situ* on the bedrock. Plants take up Sr from the soil and transfer the ^{87}Sr/^{86}Sr ratios of the soils and underlying rock into the biosphere, where its natural variations can be used for comparative and provenance studies. As such, Sr isotopic analysis has been used in a number of studies to trace the origins of migratory fish [47], mammals [48], and humans [49] and as tracers for migratory birds [50]. However, the isotopic composition of Sr in plants may not always directly reflect that in the bedrock. For example, Degryse *et al.* [51] found that the Sr isotopic composition of the water consumed by the plant, which would be dependent on both plant species and local-scale hydrology, can also greatly affect the Sr isotopic composition of the plant. They caution that a definite relation between the isotopic composition of bedrock and the plants growing on this bedrock is not always evident.

An example of how isotopes moving from the geosphere into the biosphere can provide unique solutions to fundamental problems in the ecology and evolution of migratory organisms was provided by Sellick *et al.* [50], who used a multi-proxy approach involving the determination of both δD values (via gas source isotope ratio MS) and ^{87}Sr/^{86}Sr ratios (via MC-ICP-MS) in the feathers of a migratory songbird, the tree swallow (*Tachycineta bicolor*), for samples coming from 18 sampling sites across North America (Figure 12.4). They found that δD values were correlated with both latitude and the isotopic composition of meteoric water of the sampling site where molting occurred, whereas ^{87}Sr/^{86}Sr ratios of the same feather were correlated with longitude, and hence the geology (Figure 12.4). Using simulation models, the number of individuals correctly assigned to their site of origin increased from less than 40% using either δD values or ^{87}Sr/^{86}Sr ratios alone to 74% using both isotopes. Their results suggest that combining isotope systems influenced by vastly different global-scale processes links the population dynamics of migratory animals across large geographic ranges.

Strontium isotope ratios measured with LA–ICP-MS in otoliths of fall-run Chinook salmon (*Oncorhynchus tshawytscha*) from all major natural and hatchery spawning sites in the California Central Valley have been applied to identifying the origin of adult Pacific salmon from individual rivers and hatcheries in the ocean [52]. The isotopic composition of Sr is a natural tag to identify the natal origin with high accuracy using juvenile portions of otoliths accreted in natal streams and hatcheries. Adults of known origin were used to verify the early life-histories using Sr isotopic compositions, so that isotopes can be used to track the natal origin and

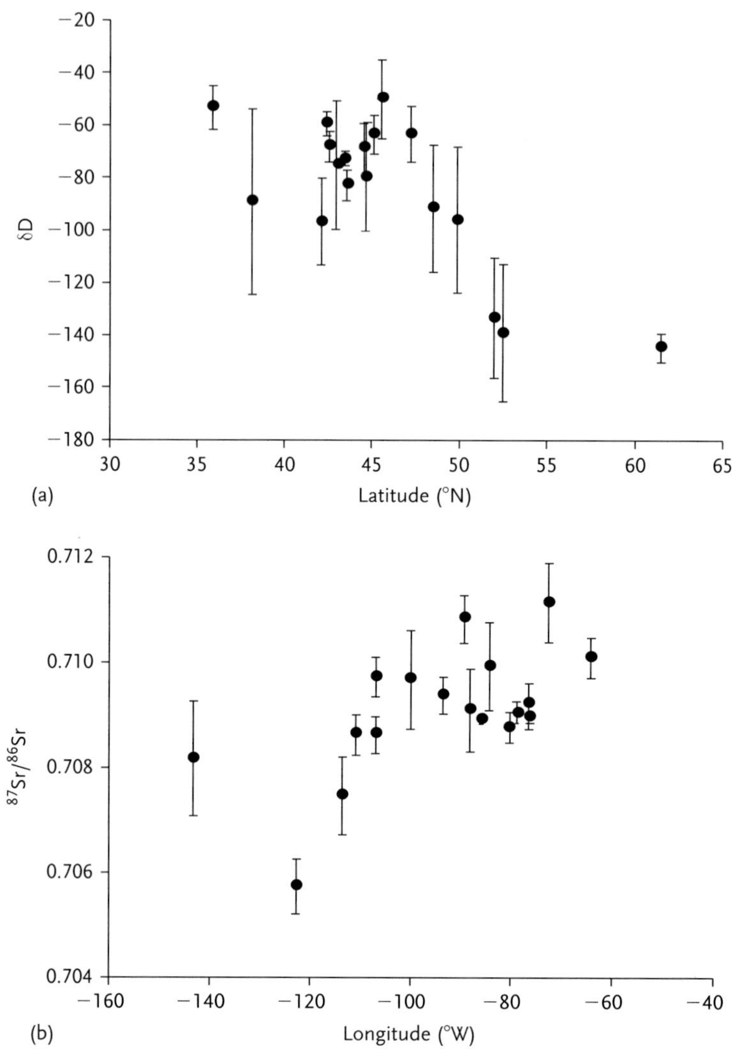

Figure 12.4 Relationship between geographic coordinates and mean (± SD) isotopic compositions of H and Sr in feathers of tree swallows (*Tachycineta bicolor*) across 18 North American breeding sites. (a) Latitude versus δD ($r_s = -0.47$, $p = 0.05$); (b) longitude versus $^{87}Sr/^{86}Sr$ ($r_s = 0.62$, $p = 0.009$). Reproduced from [50].

movement of salmonids from freshwater to marine environments and identify habitats that promote success of the salmon.

MC-ICP-MS has facilitated the use of Ca isotopes in soils and vegetation [53–55], marine precipitates [56], and animal tissues [57]. Studies of trees indicate that $\delta^{44/42}$Ca values are highest in the leaves and lower in roots and stemwood, although all tree tissues have lower values than the extractable Ca in their soils

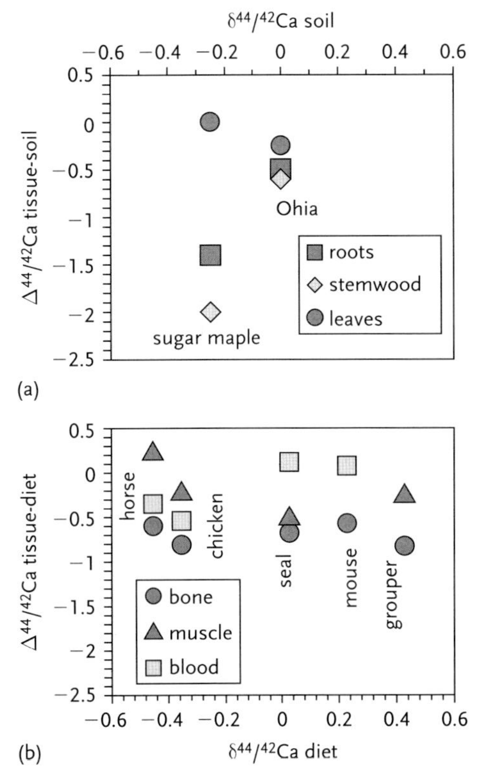

Figure 12.5 (a) Relationship between the $\delta^{44/42}$Ca values in soils and the difference between the values in various tree tissues and their soils ($\delta^{44/42}$Ca) for trees from hardwood forests in New York (sugar maple) and from tropical forests in Hawaii (Ohia). (b) Relationship between the $\delta^{44/42}$Ca values in diet and the difference between the values in various tissues and diet ($\delta^{44/42}$Ca) for animals, showing the general decrease in $\delta^{44/42}$Ca values with trophic level and preferential sequestration of the light isotope of Ca into bone.
Modified from [58] and references therein.

(Figure 12.5). Preferential enrichment in the heavy isotope in the leaves is attributed to a chromatographic effect wherein the light isotope of Ca is sequestered in plant tissues during transport of the Ca from the roots to the leaves [55]. The opposite effect is seen for the isotopes of Zn and Fe, wherein the heavy isotopes are enriched in the roots relative to the leaves [45].

Some studies imply a Ca trophic level effect in both animals and marine invertebrates (Figure 12.5), where $\delta^{44/42}$Ca values decrease with trophic level [59]. Lower $\delta^{44/42}$Ca values in vertebrates reflect fractionation during bone formation, but vertebrates at higher trophic levels have low $\delta^{44/42}$Ca values that reflect eating of bone or cartilage (Figure 12.5). Bone has lower $\delta^{44/42}$Ca values than diet, compared with soft tissue, which is similar to diet [60]. Female sheep have higher δ^{44}Ca values than males because of lactation by the ewes [60], but dairy consumption results in lower and variable human $\delta^{44/42}$Ca values because human

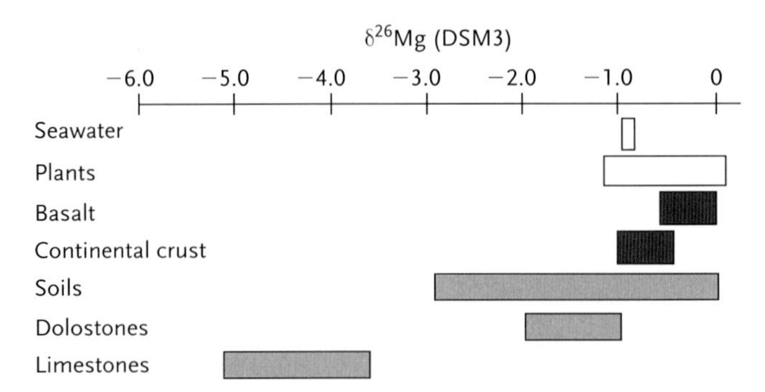

Figure 12.6 Ranges in the δ^{26}Mg values relative to the DSM3 standard of various media in the geosphere and biosphere.
Modified from [53].

bone predating animal domestication and dairy availability do not show any difference in their $\delta^{44/42}$Ca values.

During mineral weathering, Mg^{2+} is released from minerals to soil solutions, where it can be taken up by plants as an essential nutrient and included as a central ion in chlorophyll molecules [61]. Magnesium has three stable isotopes ^{24}Mg, ^{25}Mg, and ^{26}Mg, with average natural isotopic abundances of 78.992, 10.003, and 11.005%, respectively. Mg isotopic fractionation results in a terrestrial range of about 5‰ for $\delta^{26/24}$Mg values (Figure 12.6). Siliceous soils have higher δ^{26}Mg values by 0.4‰ relative to the crustal value of −0.8‰ [61], whereas surface waters draining silicates have much lower δ^{26}Mg values of −7 to −3.9‰ [62]. No systematic difference between the Mg isotopic composition of soils, waters, and unweathered rocks is observed in basaltic catchments.

Incorporation of Mg into the chlorophyll molecule induces Mg isotope fractionation, including a range of nearly 1‰ in δ^{26}Mg values of different chlorophylls for phytoplankton from the Northwest Pacific, suggesting biological mass-dependent fractionation of Mg isotopes among photosynthetic organisms and chlorophyll forms [63]. For example, rye grass and clover have higher δ^{26}Mg values relative to their nutrient source, with enrichment in the heavy isotopes of Mg related to organic acids in the roots. Both Fe and Zn are mobile in the phloem in plants and are enriched in heavy isotopes in their roots relative to the solution and their leaves, with the effect for Fe isotopes dependent on whether the Fe is taken up in oxidized form or is reduced in the root zone. However, the heavy isotopes of Ca and Si accumulate in leaves (Figure 12.5), the opposite to Fe and Zn.

12.7 Tracing in the Marine Environment

Most skeletal carbonates have variably lower δ^{26}Mg values than seawater, depending on taxon-specific processes, including the mineralogy precipitated (Figure 12.7). Unlike Ca isotopes, which show a small temperature-dependent

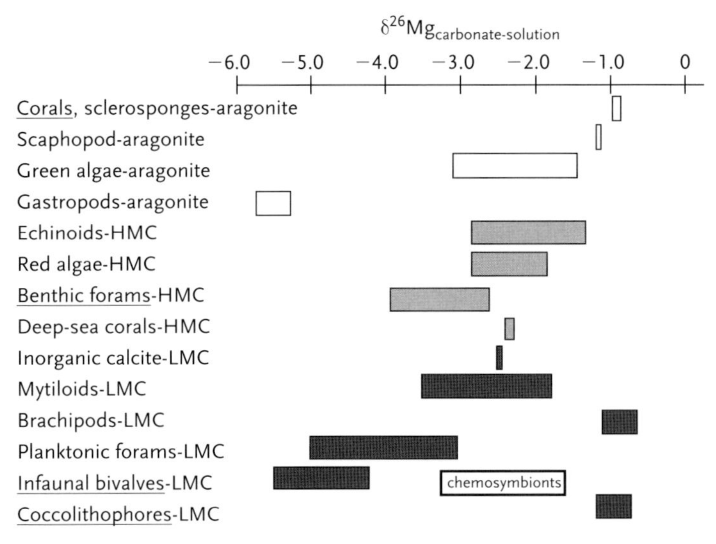

Figure 12.7 Variation in the $\delta^{26/24}$Mg values of carbonates, with the organism and mineralogy indicated. Although there is a general relationship between mineralogy and δ^{26}Mg value, there are significant exceptions for some organisms. HMC = high-Mg calcite, LMC = low-Mg calcite, and underlined organisms indicate those with symbionts. Adapted from [64], with additional data from this study.

fractionation of isotopes between the carbonate precipitated and dissolved Ca, δ^{26}Mg values of inorganic precipitates [65], foraminifera [66], endobenthic echinoids, and coralline red algae show neither consistent salinity nor temperature effects, consistent with its use as a proxy of past seawater Mg isotopic composition. However, bivalves do appear to record some salinity, temperature, and growth rate effects in their δ^{26}Mg values.

The most pronounced fractionations of Mg isotopes between carbonate minerals and dissolved Mg can be attributed to mineralogy and rate of precipitation, although there are some significant exceptions (Figure 12.7). In contrast to Ca isotopic compositions wherein $\delta^{44/42}$Ca values tend to be higher in calcite and closer to seawater values, relative to biogenic aragonite, aragonitic skeletons of corals and sclerosponges have the highest δ^{26}Mg values and are closer to seawater than high-magnesium calcite (HMC), whereas low-magnesium calcite (LMC) of planktonic foraminifera has the lowest δ^{26}Mg values. Coccolithophores, which precipitate LMC, and benthic foraminifera, which precipitate HMC, have higher δ^{26}Mg values than do LMC planktonic foraminifera. The low δ^{26}Mg values of down to −4.5‰ in some calcifying organisms may reflect Mg transport across channels and pumps, embedded in their cell membranes [67]. Inorganically precipitated calcite produced both naturally and in the laboratory has δ^{26}Mg values controlled by equilibrium fractionation, whereas the low δ^{26}Mg values of LMC in planktonic foraminifera is controlled by kinetic fractionation related to the rate of LMC precipitation [68]. The aberrantly high δ^{26}Mg values of benthic foraminifera and of coccolithophores are more consistent with multi-stage rather than single-stage mass-fractionation pathways. Given that coccolithophores represent the major carbonate-precipitating organisms in the ocean today, their effect on the Ca, Mg, and C cycles is substantial, but as yet not well understood.

The use of Sr isotopic compositions in carbonates to reflect the evolution of Sr isotopes in seawater throughout geological time is well established and useful because of the long residence time of Sr in the marine system [69]. In contrast, other radiogenic isotopes, such as Nd and Pb, have much shorter residence times in seawater and therefore will reflect more immediate effects from the environment on element cycles [70]. Nd has oceanic residence times of 400–1000 years, similar to the global mixing time of the ocean. The $^{143}Nd/^{144}Nd$ ratio of seawater is affected by dissolution of the continents and exchange of seawater with particulate phases, which varies with depth in the oceans. For example, aeolian inputs can influence the $^{143}Nd/^{144}Nd$ ratios of surface waters by partial dissolution of dust particles. These water masses have distinct $^{143}Nd/^{144}Nd$ ratios that can serve as tracers of water mass mixing and ocean circulation once away from any inputs of additional Nd.

Nd, and also Pb, Hf, and Os, are extracted in the deep ocean by authigenic ferromanganese crusts and nodules [71], which provide age resolution. In the surface and deep waters, these elements and their isotopes are sequestered in carbonate shells, but Fe oxides that are scavenged by the shells must be removed to avoid any elements that have been extracted and may not reflect the isotopic composition of the waters in which the carbonates were precipitated. Solitary deep-water corals, which can be dated by U–Th methods, have recently been explored as a new archive for modern and past seawater $^{143}Nd/^{144}Nd$ ratios, particularly at intermediate water depths [72]. For example, Colin *et al.* [73] examined Holocene intermediate-depth water $^{143}Nd/^{144}Nd$ ratios extracted from deep-sea corals from the Northeast Atlantic. Significant changes in the $^{143}Nd/^{144}Nd$ ratios are interpreted as changes in the relative proportions of upper-intermediate waters originating in the sub-tropical and sub-polar Atlantic, which are effected by convection and overturning in the North Atlantic. This is yet another example of the geosphere interacting with the biosphere that can be exploited to understand better the subtleties and histories of ocean circulation.

Another target for the reconstruction of intermediate- and deep-water $^{143}Nd/^{144}Nd$ ratios has been ferromanganese coatings that form at the sediment–seawater interface on fossil fish debris and fish teeth [74]. Mediterranean outflow water (MOW) is characterized by higher temperatures and salinities than other ambient water masses and has been a source of salinity for the Atlantic meridional overturning circulation in the North Atlantic. High-resolution Nd and Pb isotope ratio records of past ambient seawater were obtained from authigenic ferromanganese coatings of biogenic sediments in three cores along the Portuguese margin complemented by a selection of surface sediments to reconstruct the extent and pathways of MOW over the past 23 000 years [75]. The $^{143}Nd/^{144}Nd$ ratios throughout the cores are near the present-day composition of MOW, but do not reflect the present-day Nd isotopic stratification of the water column. In contrast, Pb isotope ratio records show significant and systematic variations, which provide evidence for a significantly different pattern of the MOW pathways between 20 000 and 12 000 years ago. In this study, the biosphere plays a passive, yet critical, role in the isotope tracing process.

12.8 Future Directions

The past decade has revealed unexpected effects in isotopic compositions, even for heavy metals such as Sr, Hg, Tl, and U [13, 76–78]. More conventional systems were explored through field observations, theoretical studies and experimental results that provided a framework for interpreting natural studies. After techniques had been developed to make measuring these isotope systems routine, emphasis shifted to reveal the processes behind variations in the isotopic compositions. This should be the same path that metal isotope research follows, but as yet it is more of a survey of what variations there are.

The field of metal stable isotope biogeochemistry emerged as an analytical pursuit driven by advances in mass spectrometry that were applied to interesting geological materials, associated with the geosphere–biosphere interface. Theoretical and experimental studies within this pursuit tend to be rare and, as this new field develops further, it will be necessary to balance technical and analytical innovations to explore more isotope systems and identify existing proxies. Among the unexplored isotope systems, V and Re have potential as geosphere–biosphere tracers because of their chemical similarities to Mo and U [79]. However, Re and V have only two stable isotopes (^{185}Re and ^{187}Re and ^{50}V and ^{51}V), so their utility will be limited for the effort required. Future applications of ICP-MS will focus on additional tracer experiments and the development of isotope dilution techniques, together with nanoflow injection for the analysis of small volumes of biological samples.

The application of metal stable isotopes to examining the flux of elements between the geosphere and biosphere should benefit from increasing analytical capabilities. However, progress requires that we develop multi-proxy approaches [17], using a combination of metal isotopes and more conventional and biosphere-related isotope systems, such as those involving H, C, N, O, and S. We also need to relate the processes traced by isotope systems to elemental concentrations that are more easily measured and are currently part of the repertoire of exploration and environmental strategies. Unless the isotopic results add value to these fields, it will be difficult to justify their use in applied research. We need to probe more deeply into modern analog environments and pursue experimental and theoretical studies that can lead to increased understanding of the processes driving isotopic fractionation, thereby enabling predictive power about the consequences of the anthrosphere on the geosphere–biosphere interface.

References

1 Walker, G. (2005) Sociological theory and the natural environment. *Hist. Hum. Sci.*, **18** (1), 77–106.
2 Wasylenki, L.E., Anbar, A.D., Liermann, L.J., Mathur, R., Gordon, G.W., and Brantley, S.L. (2007) Isotope fractionation during microbial metal uptake measured by MC-ICP-MS. *J. Anal. At. Spectrom.*, **22** (8), 905–910.
3 Siebert, C. (2008) Double spiking and the analysis of non-traditional stable isotopes by MC-ICPMS; an update. *Geochim. Cosmochim. Acta*, **72** (12S), A867.
4 Dobretsov, N.L., Kolchanov, N.A., and Suslov, V.V. (2006) On the early stages of the evolution of the geosphere and biosphere. *Paleontol. J.*, **40** (4), S407–S424.
5 Berner, E.K. and Berner, R.A. (eds.) (1996) *Global Environment; Water, Air, and Geochemical Cycles*, Prentice Hall, Upper Saddle River, NJ, p. 376.
6 Mackenzie, F.T. and Lerman, A. (eds.) (2006) *Carbon in the Geobiosphere; Earth's Outer Shell*, Springer, New York, p. 402.
7 Lucht, W. (2006) Earth system analysis and the future of the biosphere. *J. Phys. I*, **139**, 143–155.
8 Andreae, M.O. and Merlet, P. (2001) Emission of trace gases and aerosols from biomass burning. *Global Biogeochem. Cycles*, **15** (4), 955–966.
9 Abelson, P.H. (1986) The International Geosphere–Biosphere Program. *Science*, **234** (4777), 657.
10 Steffen, W., Sanderson, A., Tyson, P.D., Jager, J., Matson, P.A., Moore, B., Oldfield, F., Richardson, K., Schellnhuber, H.J., Turner, B.L., and Wasson, R.J. (2004) *Global Change and the Earth System: a Planet Under Pressure*. Global Change – The IGBP Book Series, Springer, Berlin, p. 366.
11 Steffen, W., Andreae, M.O., Bolin, B., Cox, P.M., Crutzen, P.J., Cubasch, U., Held, H., Nakicenovic, N., Scholes, R.J., Talaue-McManus, L., and Turner, B.L. II (2004) Abrupt changes in Achilles' heels of the Earth system. *Environment*, **46** (3), 8–20.
12 Thapalia, A., Borrok, D.M., and Van Metre, P.C., Musgrove, M., and Landa, E.R. (2010) Zn and Cu isotopes as tracers of anthropogenic contamination in a sediment core from an urban lake. *Environ. Sci. Technol.*, **44**, 1544–1550.
13 Kritee, K., Blum, J.D., Johnson, M.W., Bergquist, B.A., and Barkay, T. (2007) Mercury stable isotope fractionation during reduction of Hg(II) to Hg(0) by mercury resistant microorganisms. *Environ. Sci. Technol.*, **41** (6), 1889–1895.
14 Dahlgren, R.A. (2006) Biogeochemical processes in soils and ecosystems: from landscape to molecular scale. *J. Geochem. Explor.*, **88** (1–3), 186–189.
15 Dittrich, M. and Luttge, A. (2008) Microorganisms, mineral surfaces, and aquatic environments: learning from the past for future progress. *Geobiology*, **6** (3), 201–213.
16 Fredrickson, J.K. and Balkwill, D.L. (2006) Geomicrobial processes and biodiversity in the deep terrestrial subsurface. *Geomicrobiol. J.*, **23** (6), 345–356.
17 Severmann, S. and Anbar, A.D. (2009) Reconstructing paleoredox conditions through a multitracer approach: the key to the past is the present. *Elements*, **5** (6), 359–364.
18 Johnson, C.M., Beard, B.L., Roden, E.E., Newman, D.K., and Nealson, K.H. (2004) Isotopic constraints on biogeochemical cycling of Fe. *Rev. Mineral. Geochem.*, **55** (1), 359–408.
19 Crosby, H.A., Roden, E.E., Johnson, C. M., and Beard, B.L. (2007) The mechanisms of iron isotope fractionation produced during dissimilatory Fe(III) reduction by *Shewanella putrefaciens* and *Geobacter sulfurreducens*. *Geobiology*, **5** (2), 169–189.
20 Johnson, C.M., Beard, B.L., Klein, C., Beukes, N.J., and Roden, E.E. (2008) Iron isotopes constrain biologic and abiologic processes in banded iron formation genesis. *Geochim. Cosmochim. Acta*, **72** (1), 151–169.

21 Severmann, S., Johnson, C.M., Beard, B.L., and McManus, J. (2006) The effect of early diagenesis on the Fe isotope compositions of pore waters and authigenic minerals in continental margin sediments. *Geochim. Cosmochim. Acta*, **70** (8), 2006–2022.

22 Rouxel, O.J., Bekker, A., and Edwards, K.J. (2005) Iron isotope constraints on the Archean and Paleoproterozoic ocean redox state. *Science*, **307** (5712), 1088–1091.

23 Cloquet, C., Carignan, J., Lehmann, M., and Vanhaecke, F. (2008) Variation in the isotopic composition of zinc in the natural environment and the use of zinc isotopes in biogeosciences: a review. *Anal. Bioanal. Chem.*, **390** (2), 451–463.

24 Dolgopolova, A., Weiss, D.J., Seltmann, R., Kober, B., Mason, T.F.D., Coles, B., and Stanley, C.J. (2006) Use of isotope ratios to assess sources of Pb and Zn dispersed in the environment during mining and ore processing within the Orlovka–Spokoinoe mining site (Russia). *Appl. Geochem.*, **21** (4), 563–579.

25 Weiss, D.J., Rausch, N., Mason, T.F.D., Coles, B.J., Wilkinson, J.J., Ukonmaanaho, L., Arnold, T., and Nieminen, T.M. (2007) Atmospheric deposition and isotope biogeochemistry of zinc in ombrotrophic peat. *Geochim. Cosmochim. Acta*, **71** (14), 3498–3517.

26 Chen, J., Gaillardet, J.R.M., and Louvat, P. (2008) Zinc isotopes in the Seine river waters, France: a probe of anthropogenic contamination. *Environ. Sci. Technol.*, **42** (17), 6494–6501.

27 Viers, J., Oliva, P., Nonell, A., Gélabert, A., Sonke, J.E., Freydier, R., Gainville, R., and Dupré, B. (2007) Evidence of Zn isotopic fractionation in a soil–plant system of a pristine tropical watershed (Nsimi, Cameroon). *Chem. Geol.*, **239** (1–2), 124–137.

28 Graham, S., Pearson, N., Jackson, S., Griffin, W., and O'Reilly, S.Y. (2004) Tracing Cu and Fe from source to porphyry: *in situ* determination of Cu and Fe isotope ratios in sulfides from the Grasberg Cu–Au deposit. *Chem. Geol.*, **207** (3–4), 147–169.

29 Pokrovsky, O.S., Viers, J., Emnova, E.E., Kompantseva, E.I., and Freydier, R. (2008) Copper isotope fractionation during its interaction with soil and aquatic microorganisms and metal oxy (hydr)oxides: possible structural control. *Geochim. Cosmochim. Acta*, **72** (7), 1742–1757.

30 Ehrlich, S., Butler, I., Halicz, L., Rickard, D., Oldroyd, A., and Matthews, A. (2004) Experimental study of the copper isotope fractionation between aqueous Cu(II) and covellite, CuS. *Chem. Geol.*, **209** (3–4), 259–269.

31 Mathur, R., Ruiz, J., Titley, S., Liermann, L., Buss, H., and Brantley, S. (2005) Cu isotopic fractionation in the supergene environment with and without bacteria. *Geochim. Cosmochim. Acta*, **69** (22), 5233–5246.

32 Mathur, R., Titley, S., Barra, F., Brantley, S., Wilson, M., Phillips, A., Munizaga, F., Maksaev, V., Vervoort, J., and Hart, G. (2009) Exploration potential of Cu isotope fractionation in porphyry copper deposits. *J. Geochem. Explor.*, **102** (1), 1–6.

33 Lahd Geagea, M., Kyser, K., and Chipley, D. (2012) Characterization of metal mobilization in soils and vegetation over a VMS ore deposit using Zn and Cu isotopic systems. *J. Geochem. Explor.*, in press.

34 Boyle, E.A., Kelly, A.E., Reuer, M.K., and Goodkin, N.F. (2009) Lead concentrations and isotopes in corals and water near Bermuda, 1780–2000. *Earth Planet. Sci. Lett.*, **283** (1–4), 93–100.

35 Reimann, C., Flem, B., Arnoldussen, A., Englmaier, P., Finne, T.E., Koller, F., and Nordgulen, O. (2008) The biosphere: a homogeniser of Pb-isotope signals. *Appl. Geochem.*, **23** (4), 705–722.

36 Bindler, R., Renberg, I., Klaminder, J., and Emteryd, O. (2004) Tree rings as Pb pollution archives? A comparison of Pb-206/Pb-207 isotope ratios in pine and other environmental media. *Sci. Total Environ.*, **319** (1–3), 173–183.

37 Weiss, D., Shotyk, W., Appleby, P.G., Kramers, I.D., and Cheburkin, A.K. (1999) Atmospheric Pb deposition since the industrial revolution recorded by five Swiss peat profiles: enrichment factors, fluxes, isotopic composition, and sources. *Environ. Sci. Technol.*, **33** (9), 1340–1352.

38 Klaminder, J., Renberg, I., Bindler, R., and Emteryd, O. (2003) Isotopic trends and background fluxes of atmospheric lead in northern Europe: analyses of three ombrotrophic bogs from south Sweden. *Global Biogeochem. Cycles*, **17** (1), 1019, doi: 10.1029/2002GB001921.

39 Shotyk, W. (2008) Comment on "The biosphere: a homogeniser of Pb-isotope signals" by C. Reimann, B. Flem, A. Arnoldussen, P. Englmaier, T.E. Finne, F. Kolle, and O. Nordgulen. *Appl. Geochem.*, **23** (8), 2514–2518.

40 Lahd Geagea, M., Stille, P., Gauthier-Lafaye, F., and Millet, M. (2008) Tracing of industrial aerosol sources in an urban environment using Pb, Sr, and Nd isotopes. *Environ. Sci. Technol.*, **42** (3), 692–698.

41 Hendry, K.R., Georg, R.B., Rickaby, R.E.M., Robinson, L.F., and Halliday, A.N. (2010) Deep ocean nutrients during the last glacial maximum deduced from sponge silicon isotopic compositions. *Earth Planet. Sci. Lett.*, **292** (3–4), 290–300.

42 Wasylenki, L.E., Rolfe, B.A., Weeks, C.L., Spiro, T.G., and Anbar, A.D. (2008) Experimental investigation of the effects of temperature and ionic strength on Mo isotope fractionation during adsorption to manganese oxides. *Geochim. Cosmochim. Acta*, **72** (24), 5997–6005.

43 Neubert, N., Naegler, T.F., and Boettcher, M.E. (2008) Sulfidity controls molybdenum isotope fractionation into euxinic sediments; evidence from the modern Black Sea. *Geology*, **36** (10), 775–778.

44 Archer, C. and Vance, D. (2008) Transition metal isotopes in rivers and the weathering environment; Abstracts of the 18th annual V. M. Goldschmidt conference. *Geochim. Cosmochim. Acta*, **72** (12S), A31.

45 Moynier, F., Pichat, S., Pons, M.-L., Fike, D., Balter, V., and Albarède, F. (2009) Isotopic fractionation and transport mechanisms of Zn in plants. *Chem. Geol.*, **267** (3–4), 125–130.

46 Balter, V., Zazzo, A., Moloney, A.P., Moynier, F., Schmidt, O., Monahan, F.J., and Albarede, F. (2010) Bodily variability of zinc natural isotope abundances in sheep. *Rapid Commun. Mass Spectrom.*, **24**, 605–612.

47 Outridge, P., Chenery, S., Babaluk, J., and Reist, J. (2002) Analysis of geological Sr isotopes markers in fish otoliths with subannual resolution using laser ablation–multicollector-ICP-mass spectrometry. *Environ. Geol.*, **42**, 891–899.

48 Pellegrini, M., Donahue, R.E., Chenery, C., Evans, J., Lee-Thorp, J., Montgomery, J., and Mussi, M. (2008) Faunal migration in late-glacial central Italy: implications for human resource exploitation. *Rapid Commun. Mass Spectrom.*, **22** (11), 1714–1726.

49 Montgomery, J. (2010) Passports from the past: investigating human dispersals using strontium isotope analysis of tooth enamel. *Ann. Human Biol.*, **37** (3), 325–346.

50 Sellick, M.J., Kyser, T.K., Wunder, M.B., Chipley, D., and Norris, D.R. (2009) Geographic variation of strontium and hydrogen isotopes in avian tissue: implications for tracking migration and dispersal. *Plos One*, **4** (3), e4735.

51 Degryse, P., Shortland, A., de Muynck, D., Van Heghe, L., Scott, R., Neyt, B., and Vanhaecke, F. (2010) Considerations on the provenance determination of plant ash glasses using strontium isotopes. *J. Archaeol. Sci.*, **37** (12), 3129–3135.

52 Barnett-Johnson, R., Pearson, T.E., Ramos, F.C., Grimes, C.B., and MacFarlane, R.B. (2008) Tracking natal origins of salmon using isotopes, otoliths, and landscape geology. *Limnol. Oceanogr.*, **53** (4), 1633–1642.

53 Tipper, E.T., Louvat, P., Capmas, F., Galy, A., and Gaillardet, J. (2008) Accuracy of stable Mg and Ca isotope data obtained by MC-ICP-MS using the standard addition method. *Chem. Geol.*, **257**, 65–75.

54 Ewing, S.A., Yang, W., DePaolo, D.J., Michalski, G., Kendall, C., Stewart, B.W., Thiemens, M., and Amundson, R. (2008) Non-biological fractionation of stable Ca isotopes in soils of the Atacama Desert, Chile. *Geochim. Cosmochim. Acta*, **72**, 1096–1110.

55 Chabaux, F., Cenki-Tok, B., Lemarchand, D., Schmitt, A.D., Pierret, M.C., Viville, D., Bagard, M.L., and Stille, P. (2009) The impact of water–rock interaction and vegetation on calcium isotope fractionation in soil and stream waters of a small, forested catchment (the Strengbach case). *Geochim. Cosmochim. Acta*, **73** (8), 2215–2228.

56 Teichert, B.M.A., Gussone, N., and Torres, M.E. (2009) Controls on calcium isotope fractionation in sedimentary porewaters. *Earth Planet. Sci. Lett.*, **279**, 373–382.

57 Reynard, L.M., Hedges, R.E.M., and Henderson, G.M. (2008) Stable calcium isotope ratios ($\delta^{44/42}$Ca) in bones and teeth for the detection of dairying by ancient humans. *Geochim. Cosmochim. Acta*, **72** (12), A790.

58 von Blanckenburg, F., von Wiren, N., Guelke, M., Weiss, D.J., and Bullen, T.D. (2009) Fractionation of metal stable isotopes by higher plants. *Elements*, **5**, 375–380.

59 Clementz, M.T., Holden, P., and Koch, P.L. (2003) Are calcium isotopes a reliable monitor of trophic level in marine settings? *Int. J. Osteoarchaeol.*, **13** (1–2), 29–36.

60 Reynard, L.M., Henderson, G.M., and Hedges, R.E.M. (2010) Calcium isotope ratios in animal and human bone. *Geochim. Cosmochim. Acta*, **74** (13), 3735–3750.

61 Bolou-Bi, E.B., Poszwa, A., Leyval, C., and Vigier, N. (2010) Experimental determination of magnesium isotope fractionation during higher plant growth. *Geochim. Cosmochim. Acta*, **74** (9), 2523–2537.

62 Tipper, E.T., Galy, A., Gaillardet, J., Bickle, M.J., Elderfield, H., and Carder, E.A. (2006) The magnesium isotope budget of the modern ocean: constraints from riverine magnesium isotope ratios. *Earth Planet. Sci. Lett.*, **250**, 241–253.

63 Ra, K. and Kitagawa, H. (2007) Magnesium isotope analysis of different chlorophyll forms in marine phytoplankton using multi-collector ICP-MS. *J. Anal. At. Spectrom.*, **22**, 817–821.

64 Hippler, D., Buhl, D., Witbaard, R., Richter, D.K., and Immenhauser, A. (2009) Towards a better understanding of magnesium-isotope ratios from marine skeletal carbonates. *Geochim. Cosmochim. Acta*, **73** (20), 6134–6146.

65 Buhl, D., Immenhauser, A., Smeulders, G., Kabiri, L., and Richter, D.K. (2007) Time series ^{26}Mg analysis in speleothem calcite: kinetic versus equilibrium fractionation, comparison with other proxies and implications for palaeoclimate research. *Chem. Geol.*, **244**, 715–729.

66 Chang, V.T.C., Williams, R.J.P., Makishima, A., Belshawl, N.S., and O'Nions, R.K. (2004) Mg and Ca isotope fractionation during $CaCO_3$ biomineralisation. *Biochem. Biophys. Res. Commun.*, **323** (1), 79–85.

67 Zeebe, R.E. and Sanyal, A. (2002) Comparison of two potential strategies of planktonic foraminifera for house building: Mg^{2+} or H^+ removal? *Geochim. Cosmochim. Acta*, **66** (7), 1159–1169.

68 Kisakurek, B., Niedermayr, A., Muller, M.N., Taubner, I., Eisenhauer, A., Dietzel, M., Buhl, D., Fietzke, J., and Erez, J. (2009) Magnesium isotope fractionation in inorganic and biogenic calcite. *Geochim. Cosmochim. Acta*, **73** (13), A663.

69 McArthur, J.M., Howarth, R.J., and Bailey, T.R. (2001) Strontium isotope stratigraphy: LOWESS version 3: best fit to the marine Sr-isotope curve for 0–509 Ma and accompanying look-up table for deriving numerical age. *J. Geol.*, **109** (2), 155–170.

70 O'Nions, R.K., Frank, M., von Blanckenburg, F., and Ling, H.F. (1998) Secular variation of Nd and Pb isotopes in ferromanganese crusts from the

Atlantic, Indian and Pacific Oceans. *Earth Planet. Sci. Lett.*, **155** (1–2), 15–28.

71 Frank, M., Whiteley, N., Kasten, S., Hein, J.R., and O'Nions, K. (2002) North Atlantic deep water export to the Southern Ocean over the past 14 Myr: evidence from Nd and Pb isotopes in ferromanganese crusts. *Paleoceanography*, **17** (2), 1022, doi: 10.1029/2000PA000606.

72 Van de Flierdt, T., Robinson, L.R., and Adkins, J.F. (2009) Atlantic Ocean water mass distribution over the past 32,000 yrs from Nd isotopes in deep-sea corals. *Geochim. Cosmochim. Acta*, **73** (13), A1367.

73 Colin, C., Frank, N., Copard, K., and Douville, E. (2010) Neodymium isotopic composition of deep-sea corals from the NE Atlantic: implications for past hydrological changes during the Holocene. *Quat. Sci. Rev.*, **29**, 2509–2517.

74 Martin, E. and Scher, H. (2004) Preservation of seawater Sr and Nd isotopes in fossil fish teeth: bad news and good news. *Earth Planet. Sci. Lett.*, **220**, 25–39.

75 Stumpf, R., Frank, M., Schonfeld, J., and Haley, B.A. (2010) Late Quaternary variability of Mediterranean outflow water from radiogenic Nd and Pb isotopes. *Quat. Sci. Rev.*, **29**, 2462–2472.

76 Fietzke, J., Liebetrau, V., Gunther, D., Gurs, K., Hametner, K., Zumholz, K., Hansteen, T.H., and Eisenhauer, A. (2008) An alternative data acquisition and evaluation strategy for improved isotope ratio precision using LA–MC-ICP-MS applied to stable and radiogenic strontium isotopes in carbonates. *J. Anal. At. Spectrom.*, **23**, 955–961.

77 Weyer, S., Anbar, A.D., Gerdes, A., Gordon, G.W., Algeo, T.J., and Boyle, E.A. (2008) Natural fractionation of $^{238}U/^{235}U$. *Geochim. Cosmochim. Acta*, **72** (2), 345–359.

78 Nielsen, S.G., Mar-Gerrison, S., Gannoun, A., LaRowe, D., Klemm, V., Halliday, A.N., Burton, K.W., and Hein, J.R. (2009) Thallium isotope evidence for a permanent increase in marine organic carbon export in the early Eocene. *Earth Planet. Sci. Lett.*, **278**, 297–307.

79 Morford, J.L., Emerson, S.R., Breckel, E.J., and Suk Hyun, K. (2005) Diagenesis of oxyanions (V, U, Re, and Mo) in pore waters and sediments from a continental margin. *Geochim. Cosmochim. Acta*, **69**, 5021–5032.

13
Archeometric Applications

Patrick Degryse

13.1
Introduction

In archeology, ancient artifacts are not simply materials intended for classification, but are direct proof of how ancient production and technology evolved. A particular purpose in the investigation of such artifactual evidence is the study of the provenance and trade of objects and raw materials, on which this chapter will focus. Such research must be based on a combination of archeological and archeometric research, where the former offers insight into the typology (form) and chronology (date) of artifacts, and the latter provides a (unique) scientific characterization. In this way, not only can questions on mining and quarrying, arts and crafts, or trade be answered, but also social stratigraphy (e.g., the presence of exotic goods), subsistence strategies (e.g., provenance of food based on the provenance of their containers), and chronological changes in the standard of living (e.g., changing proportions of import versus local production) can be documented. All these topics are extremely important to the evaluation of the ancient socio-economy of a site or region.

Within the field of archeometry, the scientific examination of archeological and historical artifacts relies on the assumption that there is a scientifically measurable property that will link an artifact to a particular source or production site (the provenance postulate [1]). The provenance determination of stone and ceramics is often accomplished through chemical and mineralogical—petrographic analysis. Combined thin-section studies and other techniques such as (but not exclusive to) elemental and light (carbon and oxygen) isotopic analysis have proven very useful. In general, many techniques from the geo- and other sciences are explored for provenancing purposes applied to archeological materials. The story of ASMOSIA (Association for the Study of Marble and Other Stones Used in Antiquity), founded in 1988 to bring together archeologists, geologists, and geochemists to promote a better interpretation of data concerning stone provenancing, is a good example in this respect. Marble and other stone, especially from the Mediterranean, has received much attention since then [2–8] and many new methods and databases have been developed. It is recognized, however, that a single method of

Isotopic Analysis: Fundamentals and Applications Using ICP-MS,
First Edition. Edited by Frank Vanhaecke and Patrick Degryse.

Published 2012 by WILEY-VCH Verlag GmbH & Co. KGaA

analysis alone is not sufficient for provenance studies of, for example, classical white marbles, as overlaps are common when potential sources are compared [9]. Since the introduction of petrography and geochemistry in the 1950s and 1960s [10], also in provenance studies of (especially Mediterranean) ceramics, these techniques have been routinely used over the years. In such studies, the analytical data from the ceramics are compared with the geological characteristics of the potential raw material sources. As with marble, a combination of techniques offers good results.

As for other material categories, some diagnostic chemical compositions of ancient glasses also provide a characterization which suggests a specific source [11–13]. However, although attempts to provide a provenance for glass by elemental analysis continue, a direct relationship between mineral raw materials and the artifacts made from them can be altered by the high temperature used during its manufacturing. The precise nature of the raw materials used in glass production and the geographic location where their transformation into prefabricated or completely finished artifacts occurred often remain unclear. Archeological scientists have become increasingly aware that smelting or melting may have had a dramatic effect on the concentrations of many minor and trace elements [14, 15]. Consequently, whereas the study of the provenance of and trade in stone and ceramics is already well advanced, this is not necessarily the case for vitreous materials and metals [16]. Such issues have been addressed using heavy (often radiogenic) isotopes. The realization that transformations such as melting have little effect on heavy isotope ratios has provided the basis for the use of this technique.

As a first approximation, it can be stated that the isotopic composition of the elements is constant in Nature. Nevertheless, small variations in their isotopic composition occur as a consequence of a number of processes (see also Chapter 1). Certain elements (e.g., Sr, Nd, Pb) display variations in their isotopic composition because one or more of their isotopes is the end product of the decay of naturally occurring and long-lived radionuclides. The isotopic composition of elements with such radiogenic isotopes varies, depending on the initial parent/daughter ratio and the time that these nuclides have spent together. In this way, the isotopic composition of these elements can not only be used for geochronological dating, but can also be very useful in tracing the sources of detrital matter in the sedimentary and biogeochemical cycles [17]. Moreover, it has been known for a considerable time that light elements (H, Li, B, C, N, O, S) also display natural variations in their isotopic composition as a consequence of kinetic and thermodynamic isotope fractionation effects. The introduction of multi-collector (MC) inductively coupled plasma mass spectrometry (ICP-MS) allowed it to be demonstrated that such isotope fractionation effects also exist for an increasing number of heavier elements, such as Cu and Zn [18], Cd [19], and even Tl [20]. Although the variations in isotopic composition induced by fractionation effects are often (very) small, they contain valuable information that could also be used for provenancing and/or source-tracing studies, as variations in many lighter stable isotope ratios reflect different geological origins, due to different formation processes. The isotopic composition of a raw material is thus largely dependent on the geological age and origin of that material. Conversely, especially the heavy isotopes, owing to their relatively high masses and low internal relative mass differences [21, 22], are not fractionated during technical processing. The isotopic composition of the artifact will hence be identical, within analytical errors, with that of the raw material from which it was derived, whereas the signatures of different raw materials used, and hence the resulting artifacts, may differ [23]. Isotope ratios may therefore be used in tracing raw materials in craft production.

In summary, MC-ICP-MS, as a relatively new and powerful technique, is a reliable tool for highly precise isotope ratio determination of a wide range of target elements [24]. Except for H, C, N, O, and the noble gases, MC-ICP-MS can be used for the study of the isotopic composition of nearly all of the elements of the periodic table, in all matrices [25].Therefore, this technique, although relatively new in the application to archeological raw material research, has great potential for use in provenance studies.

13.2 Current Applications

It is beyond the scope of this chapter to review the chemical principles and analytical methodology of isotopic analysis aimed at the provenancing of archeological artifacts. In summary, these analytical procedures utilize well-established chemical methods of acid dissolution and ion-exchange chromatography for the isolation and purification of microgram quantities of target element from the sample. By using laser ablation, the sample consumption can be strongly reduced and artifacts can be analyzed in a quasi-nondestructive way, a development with great prospects in archeological research, where sampling of an object may be problematic. The chemical preparation of the material sampled is done in a clean laboratory environment where necessary, to avoid and exclude contamination. As for application to different material categories, the use of isotopic analysis in archeometallurgical studies was reviewed by Rehren and Pernicka [26] and in archeological vitreous materials by Degryse *et al.* [27]. Aberg [28] and Bentley [29] reviewed the use of isotopic analysis in environmental studies and skeletal remains, respectively. Upon reviewing the literature, it becomes clear that of those isotope systems that can be studied by MC-ICP-MS, only a limited number of elements have been exploited so far.

13.2.1 Lead

Lead isotopes were the first (radiogenic) isotopes to be used to investigate ancient materials. Lead isotopic analysis has been extensively used in archeometry to trace the provenance of lead, silver, and bronze metal (mainly from the Mediterranean Bronze Age period) since the 1960s [23, 30–47]. Recently, the provenancing of ancient iron has also been attempted via Pb isotopic analysis [34, 48, 49]. These efforts, but also the disadvantages and controversy around the use of lead isotopes, have been extensively reviewed (for recent reviews, see [50, 51]). The potentially large chemical and Pb isotopic variation within metal ores is a general problem in provenance studies based on lead isotopes. Lead isotopic signatures can be "broad" and lead isotope ratios of distinct ore districts may overlap considerably or may be difficult to define statistically, making it impossible to identify specific ore types used for metal production. Because of these pitfalls in the use of lead isotopes as a technique, very little new or innovative archeological interpretation had been published since the end of the 1990s. Issues on accuracy and precision of instrumentation, fractionation, and human influence on isotope ratios, the relevance of sampling strategies, and the definition of ore fields have demanded much effort. By now however, many of these questions have been resolved or are much better understood, either by extensive testing or by the technological developments in new instrumentation, more specifically MC-ICP-MS. Hence the use of lead isotopes is taking off again in archeological research on metals [52–55], and also on vitreous materials and ceramics.

From the beginning, lead isotopes have also been used in a range of archeological vitreous materials, in particular by Brill and co-workers [56–59]. In these studies, it was possible to distinguish between vitreous materials from different regions; however, some overlap was noted between "fields of origin," as was/is not unusual with lead isotopes. In further studies, lead isotopes were used with varying success to link raw material mixtures or regional signatures to glass production units [60–69]. The low lead concentrations in glass, in the 10–100 ppm range, are likely to originate from heavy or non-quartz mineral constituents in the silica source [63], although low levels of lead have also been ascribed to recycling [70] or to impurities in plant ashes used as a flux [71]. The discovery of a homogenous lead isotopic composition in a range of glass samples is likely to suggest that a homogeneous silica raw material was used for the manufacture of those glass samples, and perhaps that that glass was produced at a single location [65]. Heterogeneous lead isotopic compositions of a range of glass samples, particularly with high lead contents, is an indication of recycling of glass, whereas a linear trend in lead isotopic composition may be indicative of the mixing of two or more end members to produce a glass type [65, 67]. High lead concentrations may be caused by the addition of a mineral compound to the glass batch as a (de)colorant, possibly strongly influencing or obscuring the original signal of the sand raw material. Moreover, the recycling of glass has typical chemical consequences. The incorporation of old colored and opaque glass into a batch, with each cycle of re-use, will lead to a progressive increase in the concentration of colorant elements such as copper and antimony, but also of lead [60, 72–75]. Recycled glass may have up to a few thousand ppm of lead, which evidently will entirely alter the lead isotopic composition of the glass, and delete its original raw material signature. Also, a lead compound may be used for the production of a glaze [76, 77]. Although lead isotopes may help to trace silica sources used in glass and glaze production, caution is clearly needed in interpreting the results.

So far, lead isotopes have hardly ever been applied to characterize the source materials of ceramics [78–80].

13.2.2
Strontium

Sr isotopes have been used to trace the provenance of a number of organic and inorganic modern and archeological materials. The principle is that the Sr isotopic composition, or the $^{87}Sr/^{86}Sr$ isotope ratio, of the product investigated is identical, within analytical errors, with that of the geological raw material or bedrock from which it was derived [23]. Sr isotopes were especially used in the study of the provenance of glass [63, 81, 82], iron [49], ceramics [83], and stone [84], in addition to organic materials such as ivory [85, 86], wine and cider [87–89], ginseng [90], asparagus [91], cheese [92], and caviar [93], and also human remains [94–96]. Most studies so far were performed using thermal ionization mass spectrometry (TIMS), but the use of MC-ICP-MS is growing rapidly [97]. The use of Sr to trace inorganic artifacts in the form of glass and iron and organic material in the form of skeletal matter is discussed in detail below.

13.2.2.1 Inorganics: Glass and Iron

The application of isotopic analysis on ancient glass relies primarily on the assumption that the Sr in the glass is incorporated with the lime-bearing constituents [63]. It has been assumed that the contribution of natron to the Sr balance of glass is negligible [81], and minor contributions may be attributed to feldspars or heavy minerals in the silica raw material [81, 82]. Where the lime in a natron glass was derived from Holocene beach shell, the Sr isotopic composition of the glass is that of modern sea water [63, 81, 98]. Where the lime was derived from "geologically aged" limestone, the Sr signature in the glass is a reflection of seawater at the time this limestone was deposited, possibly modified by diagenetic alteration [81]. Such signatures can be read from the sea water evolution curve, as defined by Burke *et al.* [99, 100]. It has been argued that plant ash glasses are likely to have been produced from low-lime sand or (pure) quartz pebbles [66]. In this case, the Sr in the glass is derived from the plant ash, the $^{87}Sr/^{86}Sr$ isotope ratio thus reflecting the bioavailable strontium of the soil on which the plants grew. This, in turn, reflects the geological origin of the soil [81]. Both the $^{87}Sr/^{86}Sr$ isotope ratio and Sr concentrations are useful indicators. Aragonite in shell may contain a few thousand ppm of Sr. However, conversion of aragonite to calcite during diagenesis or chemical precipitation of calcite or limestone will incorporate only a few hundred ppm of Sr [81]. Plant ash glasses can have high Sr contents, sometimes of the same order of magnitude as or higher than those in glasses made from natron and sand with shell [81]. Sr concentrations in plant ash glass often show considerable variations, indicative of the varied or complex source of the strontium and the concentrating effect of ashing the plant [66, 71]. Henderson *et al.* [101] illustrated how important it is to establish variations in Sr isotopic signatures in the landscape as part of trying to provide a provenance for plant ash glasses. The effects of recycling on the strontium isotopic composition of glass were studied by Degryse *et al.* [82]. With mixing lines, a plot of isotopic signatures versus concentrations, it was demonstrated that the Sr in a locally, secondarily produced glass was a mixture of the signatures in two imported end members. A first survey of the Sr isotopic composition of glass throughout the ancient world has indicated the promising nature of the technique in classifying glasses according to their origin [102].

Degryse *et al.* [49] demonstrated that the provenance determination of iron artifacts through Sr isotopes is possible and decisively overcomes the limitations of Pb isotope ratio data. Whereas lead isotopic signatures of artifacts and ores may be ambiguous owing to the often highly variable chemistry of the raw materials, $^{87}Sr/^{86}Sr$ ratios may represent a more straightforward and powerful tool to exclude or corroborate specific point sources of ores used for metal production. As such, $^{87}Sr/^{86}Sr$ ratios can also help to discriminate between overlapping lead isotope populations.

13.2.2.2 **Organics: Skeletal Matter**

Strontium isotopic analysis used in archeological science to trace organic materials back to their geological origin is based on the fact that Sr in the geological bedrock moves into soil and groundwater and, hence, into the food chain [103, 104] without measurable fractionation. Sr concentrations and isotope ratios in rock, groundwater, soil, plants, and animals thus depend (at least partly) on the local geology [105–107]. The Sr isotopic compositions of bones and teeth match the average diet of an individual, which in turn is assumed to reflect the composition of the geology on which the foodstuff consumed grew or lived [94, 108]. The isotopic composition of teeth and bone can thus serve as a tracer of the geology in which the individuals were born and grew up on the one hand, and in which they died on the other, because consumed Sr is incorporated into the skeleton during bone formation and remodeling. Bone undergoes continual replacement of its inorganic phase [109], so that measurements of bone Sr reflect the later years of the life of the individual. Tooth enamel, on the other hand, forms during childhood and undergoes relatively little change [110]. Differences in strontium isotopic composition between bone and tooth enamel in a single individual thus reflect changes in the residence history of that person [94, 108]. However, the $^{87}Sr/^{86}Sr$ isotope ratio of Sr can differ substantially between bedrock and other environments. Differences between the Sr isotopic composition of plants and the bioavailable Sr signature in general on the one hand, and the bulk soil Sr signature on the other, suggests that potential applications of $^{87}Sr/^{86}Sr$ relationships should use biologically available Sr as a starting point, rather than the geological substrate as such [111], although the latter can still be used for comparison. Also, the water used or ingested by all organisms in the food chain should be taken into account [112–114]. The question then becomes how best to measure this biologically available Sr value. This is solved in archeology through the study of populations of small animals [108]. The results of studies on small animal remains (e.g., mice, guinea pigs, rabbits, squirrels, snails) demonstrate that they provide a homogeneous signal of local Sr and a strong correspondence to indigenous larger animals and humans [108].

However, the $^{87}Sr/^{86}Sr$ isotope ratio in modern animals may differ from that of local sources of Sr for several reasons, such as the consumption of imported foods or pollution from fertilizer or airborne sources of Sr. Moreover, bioavailable Sr and Sr forensic maps are rare [113]. Also, Sr isotopic analysis applied to humans needs to take into account the fact that "foreign" Sr may be ingested through foodstuff imported over long distances. Aggarwal *et al.* [115] suggested that modern populations with their diverse and diversely sourced diet no longer relate in terms of isotopic composition to the geographical—geological area where they live or were born. The Sr technique has been presented in a more theoretical way for forensic work [94, 116], but also applied in practice to murder victims [117].

13.2.3 Neodymium

The introduction of Nd isotopic analysis to archeological science, in particular glass studies, is very recent. The Nd in glass is likely to have originated from the heavy or non-quartz mineral content of the sand raw material used. For example, the strong correlation between the contents in Nd, FeO, and TiO_2 for high iron, manganese, titanium (HIMT) glass is an indication of this [118]. Nd isotope ratios are used as an indicator of the provenance of siliciclastic sediments in a range of sedimentary basin types [17]. Moreover, the effect of recycling on the Nd isotopic composition of a glass batch does not seem to be significant. Also, the rare earth element (REE) pattern of a glass batch is not expected to change significantly with the addition of colorants or opacifiers [67]. This offers great potential in tracing the origins of primary glass production. Freestone *et al.* [118], Degryse and co-workers [119–122], and Henderson *et al.* [101, 123] applied this technique successfully to trace the silica sources of early Bronze Age to Roman and Byzantine glass production around the Mediterranean. As with Sr, the first studies on Nd were performed using TIMS, but now MC-ICP-MS is used routinely.

13.2.4 Osmium

The elemental composition (particularly of platinum group metals) in gold artifacts does not provide reliable information on the provenance of the gold source used [124]. An alternative to chemical analysis of the inclusions in gold, in the form of the determination of their Os isotope ratios, was presented by Junk and Pernicka [125]. Similarly to Pb, Os exhibits a variable isotopic composition due to the radioactive decay of ^{187}Re and ^{190}Pt to ^{187}Os and ^{186}Os, respectively. It was hoped that the Os isotopic signature of an artifact could be characteristic of a gold deposit, but it was shown that the compositional variation within a single object can be extremely large due to fractionation of the relatively volatile Os during high-temperature processes. It particular circumstances, it was deemed feasible by Junk and Pernicka to distinguish between different types of gold sources, but the method has not been used in published archeometric studies since then.

13.3 New Applications

13.3.1 Copper

Lead isotopes are used to distinguish different metal sources, in particular ancient copper alloys from one another (see above). However, significant overlap may exist between different ore fields and areas of origin. It was not until the advent of MC-ICP-MS instruments that precise isotopic measurements of Cu isotopes for bronze provenancing became possible. Maréchal *et al.* [126] published the first measurements of Cu isotope compositions in a variety of minerals and biological materials. The application of isotopic analysis of Cu in archeometry to distinguish different groups of artifacts and metal sources from one another offers new perspectives, especially in combination with that of Pb [127]. Differences in the Cu isotopic composition between ore deposits have been reported and as such differences between sulfidic and oxidized ores exist, these can be used to distinguish the type of ore used in ancient technology [128, 129].

13.3.2 Tin

Tin was a vital commodity in ancient times, but the provenance of prehistoric Sn, used for the production of bronze from the third millennium BC onwards, is still unsolved. As Pb in ancient bronze comes mostly from the copper ore used, Pb isotopic analyses cannot be applied to determine the provenance of Sn [130]. The first suggestions to use Sn isotope ratios to solve archeological provenance issues were made by Gale [131], not finding isotopic fractionation using TIMS. Clayton *et al.* [132], using MC-ICP-MS, showed variations in the isotopic composition of cassiterite ores from different locations. Other investigations in the field of tin isotopes with an archeological background were reported by Gillis and Clayton [133], Nowell *et al.* [134], Gillis *et al.* [135], and Yi *et al.* [136]. However, it was not until Haustein *et al.* [137] showed that large numbers of ore deposit samples investigated were homogeneous regarding their Sn isotope composition, but significantly different between deposits, that a possible basis for provenancing ancient Sn was laid.

13.3.3 Antimony

The most striking feature of ancient glass is often its color, obtained naturally or intentionally via the use of (de)colorizers and careful control of furnace parameters, resulting in a wide variety of hues [138]. The largest proportion of ancient glass has a blue—green color, owing to Fe impurities introduced into the glass via the raw materials [138]. The addition of metal ions to the glass melt may result in different colors, for example, purple—blue glass is obtained by the addition of Co, and the addition of Cu results in glass with a dark-blue, green, brown, or orange color [139]. The provenance of Cu-based colorants through archeological time can be evaluated via Cu isotopic analysis. This technique provides comparative data to identify the Cu ores used and, more importantly, allows a diachronic comparison of Cu-colored glass from different periods and origins. In addition to colored glass, an increasing popularity of colorless glass can be observed from the early Roman period (first century AD) onwards [138]. Colorless glass can be produced either by selecting raw materials that are low in Fe minerals (high-purity sands) or, more commonly, by the addition of a decolorizing agent. Such agents were based on either Mn or Sb, where both elements decolorize a glass by converting Fe to its colorless oxidation state [138, 140–142]. Sb is a more efficient decolorizer than Mn and results in a more brilliant glass, likely by removing dissolved gases [138]. Despite the fact that many analytical studies have been devoted to decolorizing agents [138, 140–142], compositional analysis has not been able to reveal the provenance of the materials used. A most likely origin is an addition in the form of Mn- or Sb-bearing minerals, such as pyrolusite (MnO_2) and stibnite (Sb_2S_3). Although unlikely in the case of Sb, it has also been suggested that decolorizers were added accidentally as contaminants in the raw materials. The use of isotopic analysis via MC-ICP-MS can also shed new light on this debate. As Mn is monoisotopic, it cannot be used in isotope ratio studies, but isotopic analysis of Sb present in ancient glass, combined with Sb isotope ratio analysis of Sb ores, can help to identify the nature of the Sb decolorizing agent. The naturally occurring Sb isotopes display a very small relative mass difference, so that minimal natural fractionation effects are expected. So far, only one study has been devoted to this isotope system, in which the occurrence of natural Sb isotopic fractionation was demonstrated [143].

13.3.4 **Boron**

In addition to sand as a raw material for glass production, a flux is used to lower the melting temperature of pure quartz. After using ash from plants as a flux for more than a 1000 years to produce plant ash glass, mineral soda (*natrun*, mainly sodium carbonates) gradually took over this function in the early first millennium BC. To the west of the Euphrates, it became the principle source of alkali by the fifth century BC, although to the east, plant ash continued to be used as flux. Sodium carbonates in their pure form do not occur frequently in Nature. An overview of the known sources of natron in antiquity was given by Shortland *et al.* [144], of which the Wadi Natrun and al-Barnuj in Egypt are presumed to be the main suppliers in the antique world. So far, however, there is no direct evidence that soda was exported (e.g., for glass production) outside Egypt in Roman times. Also, enormous amounts of natron are used in glass production, and it is uncertain whether the Egyptian lakes could on their own supply such quantities. No earlier studies were performed to identify different origins of flux materials in glass. The element B is inherited in natron glass through the flux and the B isotopic composition of evaporites (such as natron) has been used successfully in geological studies to distinguish different environments [145, 146]. Therefore, ancient flux and glass can be analyzed to detect differences in B isotopic composition and, hence, different sources of the raw materials.

13.4 Conclusion

With the specific information provided by the strategies for isotopic analysis described here, archeologists may understand the origin (and in some cases the nature) of the raw materials used in artifact production and can begin to contribute to models of trade, (socio-)economic models involving manufacture, and eventually models of the ancient economy. Ideally, the interpretation should combine isotopic results with evidence from archeological excavations, historical sources and typo-chronological study. The main issue in these archeological studies is the fact that the access to analytical facilities for isotope studies may be restricted and/or expensive. This is a critical aspect for the future of what could be termed *isotope archeology* in material studies. A critical number of laboratories and research groups investigating isotopes in artifacts should be established to allow for a broad base of research, enough input of new materials and new applications of different isotope systems, and in particular new ideas and inter-laboratory comparisons. In particular, a synergy with the disciplines of chemistry and earth sciences would be effective in this respect, and would facilitate the development of a revived field in isotope archeology.

References

1 Weigand, P.C., Harbottle, G., and Sayre, E.V. (1977) Turquoise sources and source analysis: mesoamerica and the southwestern U.S.A., in *Exchange Systems in Prehistory* (eds. T.K. Earle and J.E. Ericson), Academic Press, New York, pp. 15–34.

2 Herz, N. and Waelkens, M. (eds.) (1988) *Classical Marble: Geochemistry, Technology, Trade,* NATO ASI Series E, Applied Sciences, vol. **153**, Kluwer, Dordrecht.

3 Waelkens, M., Herz, N., and Moens, L. (eds.) (1992) *Ancient Stones: Quarrying, Trade and Provenance – Interdisciplinary Studies on Stones and Stone Technology in Europe and the Near East from the Prehistoric to the Early Christian period.* Acta Archaeologica Lovaniensia, Monographiae 4, Leuven University Press, Leuven.

4 Schvoerer, M. (ed.) (1999) *Archéomatériaux – Marbres et Autres Roches. Actes de la IVème Conférence Internationale de l'Association pour l'Étude des Marbres et Autres Roches Utilisés dans le Passé,* Presses Universitaires de Bordeaux, Bordeaux.

5 Herrmann, J., Herz, N., and Newman, R. (eds.) (2002) *ASMOSIA 5, Interdisciplinary Studies on Ancient Stone. Proceedings of the Fifth International Conference of the Association for the Study of Marble and Other Stones in Antiquity,* Archetype Publications, London.

6 Lazzarini, L. (ed.) (2002) *Interdisciplinary Studies on Ancient Stone – ASMOSIA VI. Proceedings of the Sixth International Conference of the Association for the Study of Marble and Other Stones in Antiquity*, Bottega d'Erasmo Aldo Ausilio Editore, Padua.

7 Maniatis, Y. (ed.) (2009) *Proceedings of the 7th International Conference of the Association for the Study of Marble and Other Stones in Antiquity, Thassos, 15–20 September, 2003*, Bulletin de Correspondance Hellenique, Paris, vol. **51**.

8 Jockey, P. (ed.) (2009) *Leukos Lithos: Marbres et Autres Roches de la Méditerranée Antique: Études Interdisciplinaires. Actes du 8e Colloque ASMOSIA, Aix-en-Provence, 2006*, Maisonneuve et Larose, Maison Méditerranéenne des Sciences de l'Homme, l'Atelier Méditerranéen, Aix-en-Provence.

9 Kempe, D.R.C. and Harvey, A.P. (1983) *The Petrology of Archaeological Artifacts*, Clarendon Press, Oxford.

10 Peacock, D.P.S. (1970) The scientific analysis of ancient ceramics: a review. *World Archaeol.*, **1**, 375–389.

11 Henderson, J. (1988) Electron probe microanalysis of mixed-alkali glasses. *Archaeometry*, **30**, 77–91.

12 Brill, R.H. (1989) Thoughts on the glass of Central Asia with analyses of some glasses from Afghanistan, in *Proceedings of the XVth International Congress on Glass*, International Commission on Glass, pp. 19–24.

13 Dussubieux, L., Kusimba, C.M., Gogte, V., Kusimba, S.B., Gratuze, B., and Oka, R. (2008) The trading of ancient glass beads: new analytical data from south Asian and east African soda—alumina glass beads. *Archaeometry*, **50**, 797–821.

14 Wilson, L. and Pollard, A.M. (2001) The provenance hypothesis, in *Handbook of Archaeological Sciences* (eds. D.R. Brothwell and A.M. Pollard), John Wiley & Sons, Ltd., Chichester.

15 Rehren, T. (2008) A review of factors affecting the composition of early Egyptian glasses and faience: alkali and alkali earth oxides. *J. Archaeol. Sci.*, **35**, 1345–1354.

16 Henderson, J. (2009) The provenance of archaeological plant ash glasses, in *From Mine to Microscope: Advances in the Study of Ancient Technology* (eds. A.J. Shortland, I.C. Freestone, and T. Rehren), Oxbow Books, Oxford, pp. 129–138.

17 Banner, J.L. (2004) Radiogenic isotopes: systematics and applications to Earth surface processes and chemical stratigraphy. *Earth Sci. Rev.*, **65**, 141–194.

18 Maréchal, C. and Albarède, F. (2002) Ion-exchange fractionation of copper and zinc isotopes. *Geochim. Cosmochim. Acta*, **66**, 1499–1509.

19 Cloquet, C., Rouxel, O., Carignan, J., and Libourel, G. (2005) Natural cadmium isotopic variations in eight geological reference materials (NIST SRM 2711, BCR 176, GSS-1, GXR-1, GXR-2, GSD-12, Nod-P-1, Nod-A-1) and anthropogenic samples, measured by MC-ICP-MS. *J. Geostand. Geoanal. Res.*, **29**, 95–106.

20 Rehkämper, M., Frank, M., Hein, J.R., Porcelli, D., Halliday, A., Ingri, J., and Liebetrau, V. (2002) Thallium isotope variations in seawater and hydrogenetic, diagenetic, and hydrothermal ferromanganese deposits. *Earth Planet. Sci. Lett.*, **197**, 65–81.

21 Faure, G. (1986) *Principles of Isotope Geology*, 2nd edn., John Wiley & Sons, Inc., New York.

22 Faure, G. (2001) *Origin of Igneous Rocks, the Isotopic Evidence*, Springer, Berlin.

23 Brill, R.H. and Wampler, J.M. (1965) Isotope studies of ancient lead. *Am. J. Archaeol.*, **71**, 63–77.

24 Rehkämper, M., Schönbächler, M., and Stirling, C.H. (2001) Multiple collector ICP-MS: introduction to instrumentation, measurement technique and analytical capabilities. *Geostand. Newsl.*, **25**, 23–40.

25 Halliday, A.N., Lee, D.-C., Christensen, J.N., Rehkomper, M., Yi, W., Luo, X., Hall, C.M., Ballentine, C.J., Pettke, T., and Stirling, C. (1998) Applications of

multiple collector-ICPMS to cosmochemistry, geochemistry and palaeoceanography. *Geochim. Cosmochim. Acta*, **62**, 919–940.

26 Rehren, T. and Pernicka, E. (2008) Coins, artifacts and isotopes. *Archaeometal. Archaeomet. Archaeom.*, **50**, 232–248.

27 Degryse, P., Henderson, J., and Hodgins, G. (eds.) (2009) *Isotopes in Vitreous Materials*, Studies Archaeological Sciences, vol. **1**, Leuven University Press, Leuven.

28 Aberg, G. (1995) The use of natural strontium isotopes as tracers in environmental studies. *Water Air Soil Pollut.*, **79**, 309–322.

29 Bentley, R.A. (2006) Strontium isotopes from the Earth to the archaeological skeleton: a review. *J. Archaeol. Method Theory*, **13**, 135–187.

30 Grögler, N., Geiss, J., Grünenfelder, M., and Houtermans, F.G. (1966) Isotopenuntersuchungen zur Bestimmung der Herkunft römischer Bleirohre und Bleibarren. *Z. Naturforsch.*, **21a**, 1167–1172.

31 Budd, P., Gale, D., Pollard, A.M., Thomas, R.G., and Williams, P.A. (1993) Evaluating lead isotope data: further observations. *Archaeometry*, **35**, 241–247.

32 Gale, N.H. and Stos-Gale, Z.A. (1981) Cycladic lead and silver metallurgy. *Ann. Br. School Athens*, **76**, 169–224.

33 Gale, N.H. and Stos-Gale, Z. (1982) Bronze Age copper sources in the Mediterranean: a new approach. *Science*, **216**, 11–19.

34 Gale, N.H., Bachmann, H.G., Rothenberg, B., Stos-Gale, Z.A., and Tylecote, R.F. (1990) The adventitious production of iron in the smelting of copper, in *Researches in the Arabah 1959–1984* (eds. B. Rothenberg and H.G. Bachmann), University College London, London.

35 Gebel, A. and Schmidt, K.H. (2000) Analyse der Pb-Isotopie römischer Silbermünzen mit Hilfe der Laserablation-ICP-MS, in *Metallanalytische Untersuchungen an Münzen der Römischen Republik*, Berliner Numismatische Forschungen, Neue Folge 6 (ed. W. Hollstein), Gebr. Mann Verlag, Berlin, pp. 55–70.

36 Niederschlag, E., Pernicka, E., Seifert, T., and Bartelheim, M. (2003) The determination of lead isotope ratios by multiple collector ICP-MS: a case study of early Bronze Age artifacts and their possible relation with ore deposits of the Erzgebirge. *Archaeometry*, **45**, 61–100.

37 Pernicka, E. (1992) Evaluating lead isotope data: comments on Sayre *et al.*, "Statistical evaluation of the presently accumulated lead isotope data from Anatolia and surrounding regions," Comments III. *Archaeometry*, **34**, 322–326.

38 Pernicka, E. (1993) Comments on Budd *et al.*, "Evaluating lead isotope data: further observations," Comments III. *Archaeometry*, **35**, 259–262.

39 Pernicka, E., Begemann, F., Schmitt-Strecker, S., Todorova, H., and Kuleff, I. (1997) Prehistoric copper in Bulgaria: its composition and provenance. *Eurasia Antiqua*, **3**, 41–180.

40 Sayre, E.V., Yener, K.A., Joel, E.C., and Barnes, I.L. (1992) Statistical evaluation of the presently accumulated lead isotope data from Anatolia and surrounding regions. *Archaeometry*, **34**, 73–105.

41 Sayre, E.V., Joel, E.C., Blackman, M.J., Yener, K.A., and Özbal, H. (2001) Stable lead isotope studies of Black Sea Anatolian ore sources and related Bronze Age and Phrygian artifacts from nearby archaeological sites. Appendix: new central Taurus ore data. *Archaeometry*, **43**, 77–115.

42 Stos-Gale, Z., Gale, N.H., Houghton, J., and Speakman, R. (1995) Lead isotope data from the Isotrace Laboratory, Oxford: *Archaeometry* data base 1, ores from the western Mediterranean. *Archaeometry*, **37**, 407–415.

43 Stos-Gale, Z.A., Maliotis, G., Gale, N. H., and Annetts, N. (1997) Lead isotope characteristics of the Cyprus copper ore deposits applied to provenance studies of copper oxide ingots. *Archaeometry*, **39**, 83–124.

44 Wagner, G.A. and Weisgerber, G. (eds.) (1985) *Silber, Blei und Gold auf Sifnos, prähistorische und antike Metallproduktion*, Der Anschnitt, Beiheft 3, Deutsches Bergbau-Museum, Bochum.

45 Wagner, G.A. and Weisgerber, G. (eds.) (1988) *Antike Edel- und Buntmetallgewinnung auf Thasos*, Der Anschnitt, Beiheft 6, Deutsches Bergbau-Museum, Bochum.

46 Wagner, G.A., Wagner, I., Öztunali, Ö., Schmitt-Strecker, S., and Begemann, F. (2003) Archäometallurgischer Bericht über Feldforschung in anatolien und bleiisotopische Studien an Erzen und Schlacken, in *Man and Mining – Mensch und Bergbau: Studies in Honour of Gerd Weisgereber on the Occasion of His 65th Birthday*, Der Anschnitt, Beiheft 16 (eds. T. Söllner, G. Körlin, G. Steffens, and J. Cierny), Deutsches Bergbau-Museum, Bochum, pp. 475–494.

47 Yener, K.A., Sayre, E.V., Joel, E.C., Özbal, H., Barnes, I.L., and Brill, R.H. (1991) Stable lead isotope studies of Central Taurus ore sources and related artifacts from eastern Mediterranean chalcolothic and Bronze Age sites. *J. Archaeol. Sci.*, **18**, 541–577.

48 Schwab, R., Höppner, B., and Pernicka, E. (2003) Studies in technology and provenance of iron artifacts from the Celtic Oppidum of Manching (Bavaria), in *Proceedings of the International Conference on Archaeometallurgy in Europe, 24–26 September 2003, Milan*, Associazione Italiana di Metallurgia, Milan, pp. 545–554.

49 Degryse, P., Schneider, J., Kellens, N., Muchez, Ph., Haack, U., Loots, L., and Waelkens, M. (2007) Tracing the resources of iron working at ancient Sagalassos: a combined lead and strontium isotope study on iron artifacts and ores. *Archaeometry*, **49** (1), 75–86.

50 Pollard, A.M. and Heron, C. (2008) Lead isotope geochemistry and the trade in metals, in *Archaeological Chemistry*, 2nd edn., Royal Society of Chemistry, Cambridge, ch. 9, pp. 302–345.

51 Pollard, A.M. (2009) What a long strange trip it's been: lead isotopes in archaeology, in From Mine to Microscope: Advances in the Study of Ancient Technology (eds. A.J. Shortland, I.C. Freestone, and T. Rehren), Oxbow Books, Oxford, pp. 181–189.

52 Srinivasan, S. (1999) Lead isotope and trace element analysis in the study of over a hundred South Indian metal icons. *Archaeometry*, **41**, 91–116.

53 Al-Saad, Z. (2000) Technology and provenance of a collection of Islamic copper-based objects as found by chemical and lead isotope analysis. *Archaeometry*, **42**, 385–397.

54 Attanasio, D., Bultrini, G., and Ingo, G.M. (2001) The possibility of provenancing a series of bronze punic coins found at Tharros (western Sardinia), using the literature lead isotope database. *Archaeometry*, **43**, 529–547.

55 Ponting, M., Evans, J.A., and Pashley, V. (2003) Fingerprinting of Roman mints using laser-ablation MC-ICP-MS lead isotope analysis. *Archaeometry*, **45**, 591–597.

56 Brill, R.H., Barnes, I.L., and Adams, B. (1974) Lead isotopes in some Egyptian objects, in *Recent Advances in the Science and Technology of Materials* (ed. A. Bishay), Plenum Press, New York. pp. 9–27.

57 Brill, R.H., Yamazaki, K., Barnes, I.L., Rosman, K.J.R., and Diaz, M. (1979) Lead isotopes in some Japanese and Chinese glasses. *Ars Orientalis*, **11**, 87–109.

58 Brill, R.H., Barnes, I.L., and Jeol, E.C. (1991) Lead isotopes studies of early Chinese glasses, in *Scientific Research in Early Chinese Glass: Proceedings of the Archaeometry of Glass Sessions of the 1984 International Symposium on Glass, Beijing* (ed. J.H. Martin), Corning Museum of Glass, Corning, NY, pp. 65–83.

59 Barncs, I.L., Brill, R.H., and Deal, E.C. (1986) Lead isotopes studies of the finds of the Serçe Limani shipwreck, in *Proceedings of the 24th International Archaeology Symposium* (eds. J.S. Olin and J.M. Blackmann), Smithsonian Institution Press, Washington, DC, pp. 1–12.

60 Brill, R.H. (1988) Scientific investigations of the Jalame glass and related finds, in *Excavations at Jalame, Site of a Glass Factory in Late Roman Palestine* (ed. G.D. Weinberg), University of Missouri Press, Columbia, MO, pp. 257–294.

61 Lilyquist, C. and Brill, R. (eds.) (1993) *Studies in Early Egyptian Glass*, Metropolitan Museum of Art, New York.

62 Wedepohl, K.H., Krueger, I., and Hartmann, G. (1995) Medieval lead glass from north Western Europe. *J. Glass Stud.*, **37**, 65–82.

63 Wedepohl, K.H. and Baumann, A. (2000) The use of marine molluscan shells in the Roman glass and local raw glass production in the Eifel area (Western Germany). *Naturwissenschaften*, **87**, 129–132.

64 Schultheis, G., Prohaska, T., Stingeder, G., Dietrich, K., Jembrih-Simbürger, D., and Schreiner, M. (2004) Characterisation of ancient and art nouveau glass samples by Pb isotopic analyses using laser ablation coupled to a magnetic sector field inductively coupled plasma mass spectrometer. *J. Anal. At. Spectrom.*, **19**, 838–843.

65 Degryse, P., Schneider, J., Poblome, J., Muchez, P., Haack, U., and Waelkens, M. (2005) Geochemical study of Roman to Byzantine glass from Sagalassos, southwest Turkey. *J. Archaeol. Sci.*, **32**, 287–299.

66 Henderson, J., Evans, J.A., Sloane, H.J., Leng, M.J., and Doherty, C. (2005) The use of oxygen, strontium and lead isotopes to provenance ancient glasses in the Middle East. *J. Archaeol. Sci.*, **32**, 665–673.

67 Freestone, I.C., Wolf, S., and Thirlwall, M. (2005) The production of HIMT glass: elemental and isotopic evidence, in Proceedings of the 16th Congress of the Association Internationale pour l'Histoire du Verre, London, pp. 153–157.

68 El-Goresy, A., Tera, F., Schlick-Nolte, B., and Pernicka, E. (1998) Chemistry and lead isotopic compositions of glass from a Ramesside workshop at Lisht and Egyptian lead ores: a test for a genetic link and for the source of glass, in *Proceedings of the Seventh International Congress of Egyptologists*, Orientalia Lovaniensia Analecta 82 (ed. C.J. Eyre), Peeters, Leuven, pp. 471–481.

69 Shortland, A.J. (2006) The application of lead isotopes to a wide range of late Bronze Age Egyptian materials. *Archaeometry*, **48**, 657–671.

70 Henderson, J. (2000) *The Science and Archaeology of Materials*, Routledge, London.

71 Barkoudah, Y. and Henderson, J. (2006) Plant ashes from Syria and the manufacture of ancient glass: ethnographic and scientific aspects. *J. Glass Stud.*, **48**, 297–321.

72 Henderson, J. and Holand, I. (1992) The glass from Borg, an early medieval chieftain's farm in northern Norway. *Medieval Archaeol.*, **36**, 29–58.

73 Henderson, J. (1993) Aspects of early medieval glass production in Britain, in *Proceedings of the 12th Congress of the International Association for the History of Glass*, International Association for the History of Glass, Amsterdam, pp. 247–259.

74 Jackson, C.M. (1996) From Roman to early medieval glasses. Many happy returns or a new birth?, in *Proceedings of the 13th Congress of the International Association for the History of Glass*, International Association for the History of Glass, Lochem, pp. 289–302.

75 Freestone, I.C., Ponting, M., and Hughes, J. (2002) The origins of Byzantine glass from Maroni Petrera, Cyprus. *Archaeometry*, **44**, 257–272.

76 Wolf, S., Stos, S., Mason, R., and Tite, M.S. (2003) Lead isotope analyses of Islamic pottery glazes from Fustat, Egypt. *Archaeometry*, **45**, 405–420.

77 Walton, M. (2005) A materials chemistry investigation of archaeological lead glazes, Doctoral Thesis, University of Oxford, Linacre College.

78 Knacke-Loy, O., Satir, M., and Pernicka, E. (1995) Zur Herkunftsbestimmung der bronzezeitlichen Keramik von Troia: Chemische und Isotopengeochemische (Nd, Sr, Pb) Untersuchungen. *Studia Troica*, **5**, 145–175.

79 Renson, V., Coenaerts, J., Nys, K., Mattielli, N., Åström, P., and Claeys, P. (2007) Provenance determination of pottery from Hala Sultan Tekke using lead isotopic analysis: preliminary results, in *Hala Sultan Tekke 12. Tomb 24, Stone Anchors, Faunal Remains and Pottery Provenance*, Studies in Mediterranean Archaeology, Vol **45** (12) (eds. P. Åström and K. Nys), Paul Åströms Förlag, Sävedalen, pp. 53–60.

80 Renson, V., Coenaerts, J., Nys, K., Mattielli, N., Vanhaecke, F., Fagel, N., and Claeys, P. (2011) Lead isotopic analysis for the identification of Late Bronze pottery from Hala Sultan Tekke (Cyprus). *Archaeometry*, **53**, 37–57.

81 Freestone, I.C., Leslie, K.A., Thirlwall, M., and Gorin-Rosen, Y. (2003) Strontium isotopes in the investigation of early glass production: Byzantine and early Islamic glass from the Near East. *Archaeometry*, **45**, 19–32.

82 Degryse, P., Schneider, J., Haack, U., Lauwers, V., Poblome, J., Waelkens, M., and Muchez, P. (2006) Evidence for glass "recycling" using Pb and Sr isotopic ratios and Sr-mixing lines: the case of early Byzantine Sagalassos. *J. Archaeol. Sci.*, **33**, 494–501.

83 Li, B.P., Zhao, J.X., Greig, A., Collerson, K.D., Feng, Y.X., Sun, X.M., Guo, M.S., and Zhuo, Z.X. (2006) Characterisation of Chinese Tang sancai from Gongxian and Yaozhou kilns using ICP-MS trace element and TIMS Sr—Nd isotopic analysis. *J. Archaeol. Sci.*, **33**, 56–62.

84 Brilli, M., Cavazzini, G., and Turi, B. (2005) New data of $^{87}Sr/^{86}Sr$ ratio in classical marble: an initial database for marble provenance determination. *J. Archaeol. Sci.*, **32**, 1543–1551.

85 Van der Merwe, N.J., Lee-Thorp, J.A., Thackeray, J.F., Hall-Martin, A., Kruger, F.J., Coetzee, H., Bell, R.H., and Lindeque, M. (1990) Source-area determination of elephant ivory by isotopic analysis. *Nature*, **346**, 744–746.

86 Singh, R.R., Goyal, S.P., Khanna, P.P., Mukherjee, P.K., and Sukumar, R. (2006) Using morphometric and analytical techniques to characterize elephant ivory. *Forensic Sci. Int.*, **162**, 144–151.

87 Almeida, C.M. and Vasconselos, M.T. (2001) ICP-MS determination of strontium isotope ratio in wine in order to be used as a fingerprint of its regional origin. *J. Anal. At. Spectrom.*, **16**, 607–611.

88 Barbaste, M., Robinson, K., Guilfoyle, S., Medina, B., and Lobinski, R. (2002) Precise determination of the strontium isotope ratios in wine by inductively coupled plasma sector field multicollector mass spectrometry (ICP-SF-MC-MS). *J. Anal. At. Spectrom.*, **17**, 135–137.

89 García-Ruiz, S., Moldovan, M., Fortunato, G., Wunderli, S., and García Alonso, I.J. (2007) Evaluation of strontium isotope abundance ratios in combination with multi-elemental analysis as a possible tool to study the geographical origin of ciders. *Anal. Chim. Acta*, **590**, 55–66.

90 Choi, H.K., Lim, Y.S., Kim, Y.S., Park, S.Y., Lee, C.H., Hwang, K.W., and Kwon, D.Y. (2008) Free-radical-scavenging and tyrosinase-inhibition activities of Cheonggukjang samples fermented for various times. *Food Chem.*, **106**, 564–568.

91 Swoboda, S., Brunner, M., Boulyga, S.F., and Galler, P. (2008) Identification of Marchfeld asparagus using Sr isotope ratio measurements by MC-ICP-MS. *Anal. Bioanal. Chem.*, **390**, 487–494.

92 Fortunato, G., Mumic, K., Wunderli, S., Pillonel, L., Bosset, J.O., and Gremaud, G. (2004) Application of strontium abundance ratios measured by MC-ICP-MS for food authentication. *J. Anal. At. Spectrom.*, **19**, 227–234.

93 Rodushkin, I., Bergman, T., Douglas, G., Engström, E., Sörlin, D., and Baxter,

D.C. (2007) Authentication of Kalix (N.E. Sweden) vendace caviar using inductively coupled plasma-based analytical techniques: evaluation of different approaches. *Anal. Chim. Acta,* **583**, 310–318.

94 Beard, B.L. and Johnson, C.M. (2000) Strontium isotope composition of skeletal material can determine the birth place and geographic mobility of humans and animals. *J. Forensic Sci.*, **45**, 1049–1061.

95 Müller, W., Fricke, H., Halliday, A.N., McCulloch, M.T., and Wartho, J.A. (2003) Origin and migration of the Alpine Iceman. *Science,* **302**, 862–866.

96 Montgomery, J., Evans, J.A., and Cooper, R.E. (2007) Resolving archaeological populations with Sr-isotope mixing models. *Appl. Geochem.*, **22**, 1502–1514.

97 De Muynck, D., Huelga-Suarez, G., Van Heghe, L., Degryse, P., and Vanhaecke, F. (2009) Systematic evaluation of a strontium-specific extraction chromatographic resin for obtaining a purified Sr fraction with quantitative recovery from complex and Ca-rich matrices. *J. Anal. At. Spectrom.*, **24**, 1498–1510.

98 Huisman, D.J., De Groot, T., Pols, S., Van Os, B.J.H., and Degryse, P. (2009) Compositional variation in Roman colourless glass objects from the Bocholtz burial (The Netherlands). *Archaeometry*, **51**, 413–439.

99 Burke, W.H., Denison, R.E., Hetherington, E.A., Koepnick, R.B., Nelson, H.F., and Otto, J.B. (1982) Variation of seawater $^{87}Sr/^{86}Sr$ throughout Phanerozoic time. *Geology*, **10**, 516–519.

100 Dungworth, D., Degryse, P., and Schneider, J. (2009) Kelp in historic glass: the application of Sr isotope analysis, in *Isotopes in Vitreous Materials* (eds. P. Degryse, J. Henderson, and G. Hodgins), Leuven University Press, Leuven, ch. 6, pp. 113–130.

101 Henderson, J., Evans, J., and Barkoudah, Y. (2009) The provenance of Syrian plant ash glass: an isotopic approach, in *Isotopes in Vitreous Materials* (eds. P. Degryse, J. Henderson, and G. Hodgins), Leuven University Press, Leuven, ch. 4, pp. 73–98.

102 Brill, R.H. and Fullagar, P.D. (2006) Strontium isotope analysis of historical glasses and some related materials: a progress report. Presented at the 17th International Conference of the Association Internationale de l'Histoire de Verre, 4–8 September 2006, Antwerp.

103 Sillen, A. and Kavanagh, M. (1982) Strontium and paleodietary research: a review. *Yearbook Phys. Anthropol.*, **25**, 67–90.

104 Price, T.D. (1989) *The Chemistry of Prehistoric Human Bone,* Cambridge University Press, Cambridge.

105 Dasch, E.J. (1969) Strontium isotopes in weathering profiles, deep sea sediments and sedimentary rocks. *Geochim. Cosmochim. Acta*, **33**, 1521–1522.

106 Hurst, R.W. and Davis, T.E. (1981) Strontium isotopes as traces of air borne fly ash from coal-fired plants. *Environ. Geol.*, **3**, 363–397.

107 Graustein, W.C. (1989) $^{87}Sr/^{86}Sr$ ratios measure the sources and flow of strontium in terrestrial ecosystems, in *Stable Isotopes in Ecological Research* (eds. P.W. Rundel, J.R. Ehleringer, and K.A. Nagy), Springer, New York, pp. 491–512.

108 Price, T.D., Burton, J.H., and Bentley, R.A. (2002) The characterisation of biologically available strontium isotope ratios for the study of prehistoric migration. *Archaeometry*, **44**, 117–135.

109 Jowsey, J. (1971) The internal remodeling of bones, in *The Biochemistry and Physiology of Bone III: Development and Growth* (ed. G.H. Bourne), Academic Press, New York, pp. 201–238.

110 Hillson, S. (1997) *Dental Anthropology*, Cambridge University Press, Cambridge.

111 Sillen, A., Hall, G., and Armstrong, R. (1998) $^{87}Sr/^{86}Sr$ ratios in modern and fossil food-webs of the Sterkfontein Valley: implications for early hominid habitat preference. *Geochim. Cosmochim. Acta*, **62**, 2463–2478.

112 Montgomery, J., Evans, J.A., and Wildman, G. (2006) $^{87}Sr/^{86}Sr$ isotope composition of bottled British mineral waters for environmental and forensic purposes. *Appl. Geochem.*, **21**, 1526–1534.

113 Evans, J.E., Montgomery, J., Wildman, G., and Boulton, N. (2010) Spatial variations in biosphere $^{87}Sr/^{86}Sr$ in Britain. *J. Geol. Soc.*, **167**, 1–4.

114 Degryse, P., Freestone, I.C., Schneider, J., and Jennings, S. (2010) Technology and provenance study of Levantine plant ash glass using Sr—Nd isotope analysis, in *Glas in Byzanz – Produktion, Verwendung, Analysen* (eds J. Drauschke and D. Keller), Romisch-Germanischen Zentralmuseums, Mainz, pp. 83–91.

115 Aggarwal, J., Habicht-Mauche, J., and Juarez, C. (2008) Application of heavy stable isotopes in forensic isotope geochemistry. *Rev. Appl. Geochem.*, **23**, 2658–2666.

116 Juarez, C.A. (2008) Strontium and geolocation, the pathway to identification for deceased undocumented Mexican border-crossers: a preliminary report. *J. Forensic Sci.*, **53**, 46–49.

117 Pye, K. (2004) Isotopic and trace element analysis of human teeth and bones for forensic purposes, in *Forensic Geoscience: Principles, Techniques and Applications*, Special Publication 232 (eds. K. Pye and D.J. Croft), Geological Society of London, London, pp. 15–236.

118 Freestone, I.C., Degryse, P., Shepherd, J., Gorin-Rosen, Y., and Schneider, J. (in press) Neodymium and strontium isotopes indicate a Near Eastern origin for late Roman glass in London. *J. Archaeol. Sci.*

119 Degryse, P. and Schneider, J. (2008) Pliny the Elder and Sr—Nd radiogenic isotopes: provenance determination of the mineral raw materials for Roman glass production. *J. Archaeol. Sci.*, **35**, 1993–2000.

120 Degryse, P., Schneider, J., Lauwers, V., Henderson, J., Van Daele, B., Martens, M., Huisman, H., De Muynck, D., and Muchez, P. (2009) in *Isotopes in Vitreous Materials* (eds. P. Degryse, J. Henderson, and G. Hodgins), Leuven University Press, Leuven, ch. 3, pp. 53–72.

121 Degryse, P., Boyce, A., Erb-Satullo, N., Eremin, K., Kirk, S., Scott, R., Shortland, A.J., Schneider, J., and Walton, M. (2010) Isotopic discriminants between late Bronze Age glasses from Egypt and the Near East. *Archaeometry*, **52**, 380–388.

122 Degryse, P., Shortland, A., De Muynck, D., Van Heghe, L., Scott, R., Neyt, B., and Vanhaecke, F. (2010) Considerations on the provenance determination of plant ash glasses using strontium isotopes. *J. Archaeol. Sci.*, **37** (12), 3129–3135.

123 Henderson, J., Evans, J., and Nikita, K. (2010) Isotopic evidence for the primary production, provenance and trade of late Bronze Age glass in the Mediterranean. *Mediterranean Archaeol. Archaeom.*, **10** (1), 1–24.

124 Meeks, N.D. and Tite, M.S. (1980) The analysis of platinum-group element inclusions in gold antiquities. *J. Archaeol. Sci.*, **7**, 267–275.

125 Junk, A.S. and Pernicka, E. (2003) An assessment of osmium isotope ratios as a new tool to determine the provenance of gold with platinum-group metal inclusions. *Archaeometry*, **45**, 313–331.

126 Maréchal, C.N., Télouk, P., and Albarède, F. (1999) Precise analysis of copper and zinc isotope compositions by plasma-source mass spectroscopy. Chemical Geology, v. 156, pp. 251–273.

127 Gale, N.H., Woodhead, A.P., Stos-Gale, Z.A., Walder, A., and Bowen, I. (1999) Natural variations detected in the isotopic composition of copper: possible applications to archaeology and geochemistry. *International Journal of Mass Spectrometry* 184, 1–9.

128 Klein, S., Domergue, C., Lahaye, Y., Brey, G.P. and von Kaenel, H.-M. (2009) The lead and copper isotopic composition of copper ores from the Sierra Morena (Spain). *Journal of Iberian Geology* 35(1):59–68

129 Klein, S., Brey, G.P., Durali-Müller, S., and Lahaye, Y. (2010) Characterisation of the raw metal sources used for the production of copper and copper-based objects with copper isotopes. *Archaeological and Anthropological Sciences*, 2: 45–56.

130 Begemann, F., Kallas, K., Schmitt-Strecker, S., and Pernicka, E. (1999) Tracing ancient tin via isotope analysis, in *The Beginnings of Metallurgy*, Der Anschnitt, Beiheft 9 (eds. A. Hauptmann, E. Pernicka, T. Rehren, and Ü. Yalcin), Deutsches Bergbau-Museum, Bochum, pp. 277–284.

131 Gale, N.H. (1997) The isotopic composition of tin in some ancient metals and the recycling problem in metal provenancing. *Archaeometry*, **39**, 71–82.

132 Clayton, R.E., Andersson, P., Gale, N.H., Gillis, C., and Whitehouse, M. (2002) Precise determination of the isotopic composition of tin using MC-ICP-MS. *J. Anal. At. Spectrom.*, **17**, 1248–1256.

133 Gillis, C. and Clayton, R. (2008) Tin and the Aegean in the Bronze Age, in *Aegean Metallurgy in the Bronze Age* (ed. I. Tzachili), Ta Pragmata Publications, Athens, pp. 133–142.

134 Nowell, G., Clayton, R.E., Gale, N.H., and Stos-Gale, Z.A. (2002) Sources of tin – is isotopic evidence likely to help?, in *Die Anfänge der Metallurgie in der alten Welt* (eds. E. Pernicka and M. Bartelheim), Marie Leidorf, Rahden/Westf, pp. 291–302.

135 Gillis, C., Clayton, R.E., Pernicka, E., and Gale, N.H. (2003) Tin in the Aegean Bronze Age, in *METRON: Measuring the Aegean Bronze Age. Proceedings of the 9th International Aegean Conference, Yale University, New Haven,* Aegaeum 24 (eds. K. Polinger Foster and R. Laffineur), Program in Aegean Scripts and Prehistory (PASP), University of Texas at Austin, Austin, TX, pp. 103–110.

136 Yi, W., Budd, P., McGill, R.A.R.S., Young, M.M., Halliday, A.N., Haggerty, R., Scaife, B., and Pollard, A.M. (1999) Tin isotope studies of experimental and prehistoric bronzes, in *The Beginnings of Metallurgy*, Der Anschnitt, Beiheft 9 (eds. A. Hauptmann, E. Pernicka, T. Rehren, and Ü. Yalcin), Deutsches Bergbau-Museum, Bochum, pp. 285–290.

137 Haustein, M., Gillis, C., and Pernicka, E. (2010) Tin Isotopy–A New Method for Solving Old Questions. *Archaeometry*, 52, 816–832.

138 Jackson, C.M. (2005) Making colourless glass in the Roman period. *Archaeometry*, **47**, 763–780.

139 Heck, M. and Hoffmann, P. (2002) Analysis of early medieval glass beads: the raw materials to produce green, orange and brown colours. *Mikrochim. Acta*, **139**, 71–76

140 Baxter, M.J., Cool, H.E.M., and Jackson, C.M. (2005) Further studies in the compositional variability of colourless Romano-British vessel glass. *Archaeometry*, **47**, 47–68.

141 Paynter, S. (2006) Analyses of colourless Roman glass from Binchester, County Durham. *J. Archaeol. Sci.*, **33**, 1037–1057.

142 Silvestri, A., Molin, G., and Salviulo, G. (2008) The colourless glass of Iulia Felix. *J. Archaeol. Sci.*, **35**, 331–341.

143 Rouxel, O., Ludden, J., and Fouquet, Y. (2003) Antimony isotope variations in natural systems and implications for their use as geochemical tracers. *Chem. Geol.*, **200**, 25–40.

144 Shortland, A.J., Schachner, L., Freestone, I.C., and Tite, M. (2006) Natron as a flux in the early vitreous materials industry-sources, beginnings and reasons for decline. *Journal of Archaeological Science* 33, 521–530.

145 Swihart, G.H., Moore, P.B., and Callis, E.L. (1986) Boron isotopic composition of marine and nonmarine evaporite borates. *Geochim. Cosmochim. Acta*, **50**, 1297–1301.

146 Xiao, Yingkai., Sun, Dapeng., Wang, Yunhui., Qi, Hairing., and Jin, Lin. (1992) Boron isotopic compositions of brine, sediments, and source water in Da Qaidam Lake, Qinghai, China. *Geochim. Cosmochim. Acta*, **56**, 1561–1568

14
Forensic Applications

Martín Resano and Frank Vanhaecke

14.1
Introduction

14.1.1
What is Forensics?

In the strictest sense, forensic sciences, a term most often shortened to forensics, comprise the application of different scientific tools to answer questions of interest in relation to a crime or a civil action. The word forensic, like so many legal terms, has a Latin root, the adjective *forensis*, which means "of the Forum," because it was at the Forum where many types of businesses and public affairs, such as debates and actions by courts of law, were conducted in Roman times.

The term forensics has been extended to other circumstances where it is interesting to establish the source of a material, not only for legal reasons but also for academic purposes (e.g., in archaeological studies). This chapter focuses on those cases with a clear relation with legal circumstances, since other applications are discussed in other chapters throughout the book. The applications covered here fall into four major categories: crime scene investigation, nuclear forensics, food authentication, and environmental monitoring.

14.1.2
The Role of ICP-MS in Forensics

Obviously, when attempting to link a piece of evidence to a particular person or object, chemical analysis can play a decisive role. There are numerous analytical techniques that can offer this type of information, but when it comes to elemental analysis, very few of them can compete with inductively coupled plasma mass spectrometry (ICP-MS) in terms of detection power, selectivity, and multi-element capabilities.

However, elemental analysis is often not discriminative enough. It has been demonstrated that, on some occasions, it is possible to differentiate between

Isotopic Analysis: Fundamentals and Applications Using ICP-MS,
First Edition. Edited by Frank Vanhaecke and Patrick Degryse.

Published 2012 by WILEY-VCH Verlag GmbH & Co. KGaA

various objects based on the information provided by major, minor, and trace elements [1–3], but on other occasions the elemental fingerprint of an object may depend on numerous factors different from those subject to investigation. For instance, the chemical composition of a bullet will largely depend on the composition of the ores used for its production, but can also be affected by modifications occurring during the production process and can always be disturbed by impurities arising in different steps of the manufacturing process, and also during or after use (e.g., material evidence buried for a prolonged period may have incorporated elements from the soil). Therefore, samples coming from the same production lot and being fired by the same gun may not show exactly the same elemental composition upon analysis [4]. Therefore, rather than relying on elemental information only, the determination of the isotopic composition of specific elements present in the sample may provide a more robust means to link a sample selectively with a source, since it is more likely (although never 100% certain) that the isotopic fingerprint of an ore remains practically unaffected through production and usage into the evidence found [5].

There is also another important point to consider. When preparing a fake, it is possible that criminals may manage to imitate very well the chemical composition of an original object, if they have enough information on the manufacturing process and the components typically used. It is much more difficult, however, for them to also manage to obtain a product containing the elements with exactly the same isotopic composition as that in the original one, since this requires a very high degree of sophistication (unless the same raw material can be used or if old material without economical value can be remelted). For instance, let us assume that a sculptor wants to copy an art object made of bronze by an old master. He may know very well how to imitate the original style and how alloys were prepared in that particular period, such that he may prepare a very similar material, but the Pb and Cu isotopic compositions of the original bronze may be characteristic of ores coming from a remote mine or from a mine that is nowadays depleted. In order to copy it, he would need to determine the isotopic compositions first, which requires knowledge and access to advanced instrumentation, and then, possibly, mix ores from various sources in the right proportions, such that the final isotopic compositions would match that of the original piece.

Therefore, isotopic analysis is a powerful tool that is often used in forensic investigations, although it is not always simple to achieve unambiguous conclusions, as discussed in Section 14.2 [6, 7].

One of the advantages of using ICP-MS is that it can provide both multi-elemental and isotopic information, making its use very appealing in forensics. However, it must be noted that the methodologies used for multi-element and isotopic analysis are not even similar, so that different procedures have to be used, even if the same instrument is utilized. Moreover, the best results of isotopic analysis are obtained when dedicated ICP-MS instrumentation is used [e.g., multi-collector (MC) ICP-MS], so that is very likely that different ICP-MS instruments will be utilized in a laboratory if both elemental and isotopic analysis are targeted.

14.2 Forensic Applications Based on ICP-MS Isotopic Analysis

14.2.1 Crime Scene Investigation

This is probably the first type of application that comes to mind in connection with forensics, particularly considering the influence of recent TV shows. The most conclusive way to link a bullet that has been fired to a certain weapon is still by identification of the unique markings from the barrel on the surface of the bullet. However, on occasions, the bullet may be so deformed and damaged that the quality of the marks is not sufficient to carry out a reliable comparison. It is in those circumstances that chemical analysis can play a pivotal role in helping to identify the source of the bullet [8].

Although ICP-MS is not the only technique that can be used for this type of investigation and results obtained with ICP-optical emission spectrometry (multi-elemental analysis only) [9] or secondary ion mass spectrometry (SIMS) [10] have been reported, it is clear that ICP-MS is becoming more and more popular for this type of application. Table 14.1 shows the main applications published on this topic based on the use of ICP-MS.

Probably the first work based on ICP-MS analysis in this context was published by Dufosse and Touron [11]. They investigated the potential of quadrupole-based ICP-MS for investigating the elemental fingerprint and the Pb isotopic composition of different bullets or bullet fragments found at crime scenes, after their

Table 14.1 Selected applications of crime scene investigation based on isotopic analysis by means of ICP-MS.

Sample	Analytes	Technique[a]	Ref.
Bullets and bullet fragments found at crime scenes	Pb isotope ratios, Pb/Sb ratio, trace elements	Q-ICP-MS	[11]
Bullets and bullet fragments found at crime scenes	Pb isotope ratios, Pb/Sb ratio, trace elements	Q-ICP-MS SF-ICP-MS	[4]
Bullets from various geographical origins	Pb isotope ratios	MC-ICP-MS	[12]
Projectiles, cartridge cases, firearms residues in barrels of firearms and in gunshot entries	Pb isotope ratios	MC-ICP-MS	[13]
Bullets from various manufacturers	Pb isotope ratios	Q-ICP-MS	[14]
Bullets of various types (unjacketed, semi-jacketed, and full-jacketed) and skin samples from gunshot entry wounds	Pb isotope ratios	Q-ICP-MS	[15]
Glass samples	Pb isotope ratios	MC-ICP-MS	[16]

[a]Q, quadrupole.

dissolution. They reported that Pb isotope ratios may act as a signature of bullet origin and give reliable results, while the chemical composition may be random. In addition, they also found that the Pb/Sb inter-elemental ratio also shows a high discriminant power, as Sb is added by manufacturers in different proportions to harden the bullets. Moreover, they stressed that trace element analysis can provide further confirmation because, even though the concentrations may differ within a batch of bullets, the elemental fingerprint should be similar from a qualitative point of view.

Although the conclusions of this study are sound, it is also clear that it is a pioneering work in the field. From an analytical point of view, the precision [better than 1% relative standard deviation (RSD)] and accuracy (better than 2% deviation) reported for Pb isotopic analysis are rather poor and would hardly be acceptable nowadays. In part, this is due to the sequential nature of the process of detection and quantification of ions when quadrupole-based ICP-MS is deployed. Still, it is possible to improve these results with an updated data acquisition methodology and a better correction for instrumental mass bias, as is discussed later.

Subsequent work proved this point, while also profiting from new instrumental developments. For instance, Ulrich *et al.* [4] compared the performances of quadrupole-based ICP-MS and sector field (SF)-ICP-MS for the isotopic analysis of Pb in the context of two real crime scene investigations. They concluded that the latter showed better performance in terms of precision. This aspect is well known and is due to the flat-topped (trapezoidal) peaks obtained at low mass resolution with a double-focusing SF mass spectrometer (see also Chapter 2). The authors also stressed that, although qualitative comparison of isotope ratios could be achieved without proper instrument calibration, it is better practice to measure absolute ratios by normalizing the measured raw ratios to an isotope ratio standard, preferably an isotopic reference material (IRM).

The latter aspect marks a clear difference compared with the previous work. It is related to the fact that the raw ratios obtained after measurement with an ICP-MS instrument will not reflect the real isotopic composition of the sample, because different effects occurring during the measurement process will affect the response of heavier and lighter isotopes in different ways. Moreover, this instrumental bias may not be the same in different working sessions or during a long measurement session. Hence monitoring a standard of known isotopic composition between samples permits the calculation of a correction factor that will lead to more accurate values and, therefore, a higher discriminant power (see Chapter 5 for more details on different procedures for mass bias correction). In this way, precision values in the range 0.13–0.3% RSD were reported for $^{208}Pb/^{206}Pb$ and $^{207}Pb/^{206}Pb$ ratios, which are substantially better than those reported previously. In addition, Ulrich *et al.* [4] also indicated the usefulness of the Pb/Sb ratio, and stressed that, for some elements, the content may vary very significantly, even for bullets originating from the same package (e.g., Cu or Zn). In any case, even when using isotopic analysis, it seems difficult to achieve unambiguous conclusions, so that it is not recommended to use a unique indicator, unless legislation forces manufacturers to use distinctive fingerprints in the future.

Further work on this topic already made use of MC-ICP-MS, with the corresponding improvements in terms of precision and accuracy. Buttigieg *et al.* reported a similar performance to that attainable with thermal ionization mass spectrometry (TIMS) in this context, and at least one order of magnitude better than that provided by quadrupole-based ICP-MS devices [12]. In this work, the potential of Pb isotopic analysis for establishing the origin of bullets was evaluated. It was concluded that those bullets coming from economically isolated regions (e.g., the former Soviet Union and some surrounding countries, and also South Africa) retain a significant regional Pb isotopic character. However, in countries wherein large amounts of Pb may be imported and recycled, regional comparisons are not meaningful [12].

Also using MC-ICP-MS, Zeichner *et al.* investigated possible memory effects derived from shooting bullets [13]. It was concluded that no mechanical or chemical method completely removes lead deposits from the barrels of firearms. That obviously makes the identification of bullets much more difficult, since the isotopic composition of a bullet fragment could be affected by those from bullets shot earlier from the same weapon. Still, the authors indicated that if the ammunition brands involved (bullets and primers) in a shoot-out incident differ significantly in their original isotopic composition, it should be possible to find out which ammunition or firearm was responsible for a particular gunshot entry.

More recent work on this topic has been published in Japanese [14], with the abstract, tables and figures in English. The authors used a quadrupole-based ICP-MS unit and reported satisfactory results for the discrimination of bullets.

Wunnapuk *et al.* investigated a possible relation between the type of bullet used (unjacketed, semi-jacketed, or jacketed) and the Pb isotopic composition of the skin surrounding a gunshot wound [15]. They found higher $^{208}Pb/^{206}Pb$ and $^{208}Pb/^{207}Pb$ ratios for wounds caused by semi-jacketed bullets. However, the poor correlation found between the isotopic composition of the bullets and that of the skin wounds makes these results hardly reliable. It is also not clear why such a difference could be observed, since the origin of the ores used (for Pb and also for the other metals used in the jackets, which could include Pb traces), and not the type of bullet, should be the critical factor giving rise to major differences. It is interesting, however, that this study was carried out with a quadrupole-based ICP-MS instrument designed to work at significantly lower Ar flows ($\sim$9 l min^{-1} total Ar consumption, as opposed to a typical value of $\sim$15–20 l min^{-1}), which is a trend that may be seen in future instrumentation.

On a different topic, Sjåstad *et al.* evaluated the potential of Pb isotopic analysis in glass by means of MC-ICP-MS [16]. They explained that glass is a typical form of material evidence in criminal cases as it breaks easily, and forensic scientists often try to establish a link between glass fragments collected at the crime scene and those collected from a suspect. The authors demonstrated that Pb isotopic analysis can help significantly in this context.

14.2.2 Nuclear Forensics

This is a field that has been growing in importance owing to different social and political factors. Obviously, the phenomenon of smuggling and illicit trafficking of nuclear materials has resulted in increased interest in the development of methods that can help in determining their origin. Since nuclear materials are of anthropogenic origin, the composition of the ores from which they were originally extracted, and also the mode of production, should have an effect on the elemental and isotopic composition of the final materials. ICP-MS can play an important role in establishing nuclear fingerprints, in combination with other techniques, for instance, for examination of the macroscopic and microscopic appearance of the samples [17].

However, establishing the origin of nuclear materials (e.g., uranium ore concentrates) is not the only subject of investigation in this field. It is necessary to develop methodologies to reveal undeclared clandestine nuclear activities, and also to test possible effects thereof on animals and humans. As can be seen in Table 14.2, numerous applications have tackled these issues during the past 5 years, and the most relevant are briefly discussed below.

Table 14.2 Selected applications of nuclear forensics based on isotopic analysis by means of ICP-MS.

Sample	Analytes	Technique	Ref.
Single uranium particles	U isotope ratios	LA–SF-ICP-MS	[18]
Environmental microsamples	U, Nd, and Ru/Tc isotope ratios	LA–MC-ICP-MS	[19]
Various types of solid samples	U and Pb isotope ratios plus multi-elemental analysis	LA–ICP-TOF-MS	[20]
Microscopic uranium oxide grains	U isotope ratios	LA–MC-ICP-MS	[21]
Various environmental samples	Pu isotope ratios	SF-ICP-MS	[22]
Swipe samples	Pu and U isotope ratios	SF-ICP-MS	[23]
No samples, only standards	Pu isotope ratios	Electrochemical separation coupled to Q-ICP-MS	[24]
Lake sediments	U isotope ratios	Q-ICP-MS	[25]
Uranium pellets and powder	U isotope ratios, trace elements, particle size and shapes	MC-ICP-MS (for isotopic analysis)	[26]
Pine needles as bioindicators	U isotope ratios	SF-ICP-MS	[27]
Human blood	U isotope ratios	SF-ICP-MS	[28]
Human hair	U isotope ratios	Q-ICP-MS	[29]
Uranium ore concentrates	Pb and Sr isotope ratios	MC-ICP-MS	[30]
Uranium ores	Pb isotope ratios	MC-ICP-MS	[31]

The use of laser ablation (LA)–ICP-MS is of particular interest for this type of application, because the analysis of small particles that have been transported in aerosol form provides a powerful method for tracking down nuclear activities. Therefore, the analysis of so-called hot particles (originating from nuclear weapons tests, accidental releases, or unauthorized activities) is of great importance.

Although LA can provide sufficient spatial resolution for the analysis of individual particles down to a few micrometers in size, achieving reliable analytical data, particularly when isotopic analysis is aimed at, is not so straightforward in this context. There are a number of reasons for this. When single-collector ICP-MS instruments are used for analysis, the isotopes of interest are actually monitored in a sequential manner. Hence all the isotopes are not monitored simultaneously and the corresponding signals may be affected by noise sources in different ways, degrading the overall precision attained in the determination of the isotope ratios. This problem can be somewhat mitigated by increasing the total measuring time when solution analysis is carried out (the average signal for a long period of time for every isotope should provide a fairly constant value), but when LA of small particles is required, the signals last for only a few seconds. Furthermore, the total number of counts is an important factor to consider in order to obtain the necessary precision. In solution analysis, by increasing the measuring time it is possible to achieve a higher number of counts, as the sample volume is often not so limited. For LA–ICP-MS analysis of small particles, the signal is ultimately limited by the tiny amount of sample available. A more detailed discussion of the problems associated with isotopic analysis by LA–single-collector-ICP-MS, and means to minimize them, can be found elsewhere [32], and the performance of LA–MC-ICP-MS is discussed in Chapter 4.

For these reasons, even if an MC-ICP-MS instrument is deployed for the task, the precision achieved in this kind of isotopic analysis can hardly match that attainable with solution-based methods. Fortunately, in this particular context, this may not be necessary anyway. In fact, differences in the isotopic signatures resulting from nuclear activities are expected to be very high in comparison with the typical natural variations. For instance, the abundance of ^{236}U is almost negligible in natural uranium samples. The detection of this nuclide in a sample is therefore a strong indication of an anthropogenic nuclear activity. Thus, even if the precision finally obtained is only a few per cent RSD, it could be more than enough for achieving clear evidence.

Researchers in this field have not ignored this issue, as they have obviously tried to achieve the best possible precision considering the circumstances. For instance, Cottle *et al.* [33] examined different ways to process the transient signals when using LA–MC-ICP-MS, and came to the conclusion that it is best to integrate the whole signal for each isotope separately and then calculate the ratios, since otherwise the time offset between the signal intensities of the Faraday collectors and the electron multipliers may affect the result. Varga [18] proposed the use of a lower energy setting for the laser device, in order to expand the signal obtained in time and thus to achieve a better precision when monitoring individual uranium oxide particles by means of LA–SF-ICP-MS. He proposed the use of a medium resolution setting ($R = 4000$) to remove interferences from polyatomic ions affecting the monitoring of uranium isotopes. Eventually, he reported precision values ranging between 0.9 and 5.0% RSD for the measurement of $^{234}U/^{238}U$, $^{235}U/^{238}U$, and $^{236}U/^{238}U$, with lateral resolution down to 10 μm, and stressed that better values could be expected when using an MC-ICP-MS instrument.

Boulyga and Prohaska evaluated the use of LA–MC-ICP-MS to analyze micro-samples collected in the vicinity of Chernobyl [19]. The authors observed different isotopic patterns. While Nd isotopic abundances in the samples was not distinguishable from the natural isotopic composition, the $^{101}Ru/(^{99}Ru + ^{99}Tc)$ and $^{102}Ru/(^{99}Ru + ^{99}Tc)$ ratios were found to be significantly lower than natural ratios. That is because, while it is not possible to differentiate between ^{99}Ru and ^{99}Tc with ICP-MS, ^{99}Tc levels in natural samples are negligible, while in these exposed samples the contribution of the ^{99}Tc signal to that ratio was not insignificant. Monitoring of $^{235}U/^{238}U$ and $^{236}U/^{238}U$ also provided very sensitive indicators of nuclear activities in the area.

Bürger and Riciputi investigated the potential of a different kind of single-collector device for multi-nuclide screening of nuclear and non-nuclear solid samples, a time-of-flight (TOF) ICP-MS instrument [20]. Although this type of instrumentation is currently not widely used in the analytical community for a number of reasons, the most noteworthy being its typically lower sensitivity (5–10-fold lower than that of a quadrupole-based unit), it is better suited to deal with transient signals because of the way in which it operates, and it is still considerably more cost-effective than an MC-ICP-MS unit. The basic principle of an ICP-TOF-MS instrument is to inject an isokinetic package of ions electrostatically into a field-free flight tube, which in this case is orthogonally positioned relative to the original ion path. Their subsequent drift permits the ions to be separated according to their mass-to-charge ratio as a result of the corresponding different velocities and arrival times at the detector (see also Chapter 2). The instrument detector is capable of processing up to 29 400 complete spectra per second. Even more relevant is the fact that each of these packages of ions is sampled from the plasma simultaneously, such that it represents the exact composition of ions in the plasma at that very moment. Thus, the level of correspondence achieved for different nuclides is very high, permitting most sources of noise to be filtered out. A more detailed evaluation of the typical performance of an ICP-orthogonal TOF-MS device can be found elsewhere [34, 35].

This instrument, in combination with LA, permits the screening of suspect uranium samples. providing information on their enrichment (U isotopic analysis), and also on the possible presence or absence of other actinide elements, such as thorium or plutonium. The authors reported that it is feasible to achieve a precision and accuracy of 4% for U and Pb isotopic analysis, without performing any mass bias correction, which, in this context, can be considered fit-for-purpose, as discussed earlier.

Regarding solution analysis, a number of papers have explored the potential of ICP-MS for the isotopic analysis of Pu or U in environmental or clinical samples. Generally, owing to the low levels at which these elements are expected to be present (ng g^{-1} and lower), it is necessary to develop methodologies that minimize contamination and matrix effects and can, if possible, preconcentrate the analyte. For instance, Agarande *et al.* discussed the potential of SF-ICP-MS, after sample dissolution and chromatographic isolation of the analyte to avoid spectral overlap ($^{239}Pu^+$ and $^{238}U^1H^+$), for the determination of ^{241}Pu, ^{240}Pu, and ^{239}Pu in field samples. It was shown that the $^{240}Pu/^{239}Pu$ ratio varies from 0.194 to 0.260, depending on the source of contamination. In isotopic analysis, this is a vast difference and it may permit the identification of the source of a release [22].

Széles *et al.* also developed a procedure, based on microwave-assisted digestion followed by extraction chromatography and subsequent SF-ICP-MS analysis, that permits the determination of the isotope ratios and isotope concentrations for U and Pu (^{234}U, ^{235}U, ^{236}U, ^{238}U, ^{239}Pu, ^{240}Pu, and ^{241}Pu) in swipe samples [23]. This procedure provides precision values of 7–10% RSD for $^{240}Pu/^{239}Pu$ and between 0.23–3% RSD for $^{235}U/^{238}U$, $^{234}U/^{238}U$, and $^{236}U/^{238}U$. These values comply well with the requirements of the International Atomic Energy Agency (IAEA) Network of Analytical Laboratories (NWAL). A different example of separation was published by Liezers *et al.*, who used electrochemically modulated separations for extracting and concentrating Pu [24]. Other groups have investigated the presence of depleted uranium in humans, after developing methods for the analysis of blood (using SF-ICP-MS for determining the $^{235}U/^{238}U$ ratio [28]) or hair (by means of quadrupole-based ICP-MS [29]).

Finally, Pb and Sr isotopic analysis of uranium ore concentrates has been investigated as a means to reveal illicit traffic of these samples. *A priori*, monitoring Pb isotopic variations in these samples seems sensible. Since ^{206}Pb, ^{207}Pb, and ^{208}Pb are produced by radiogenic decay of Th or U isotopes, while the remaining stable Pb isotope, ^{204}Pb, is not radiogenic, large variations in the Pb isotopic signature are expected for these samples. However, the authors concluded that owing to the high variation of the Pb isotopic compositions within mine sites and the high contribution of natural lead during production, the determination of the $^{87}Sr/^{86}Sr$ ratio is a more meaningful indicator of the origin of the ore [30]. However, in some cases, it is possible to correct mathematically for the natural Pb contribution (primordial Pb found as a trace element in the rock), estimated from the signal of ^{204}Pb, thus enhancing the provenancing power of Pb isotopic analysis [31].

14.2.3 Food Authentication

Determination of food authenticity is a topic of increasing importance in food quality control laboratories, for a number of reasons. On the one hand, the appearance of diseases linked to a particular foodstuff, originating from a particular region, makes it necessary to ensure that it is possible to control this aspect in order to ensure consumer safety. On the other hand, there are certain agricultural products and foodstuffs that are produced, processed, and/or prepared in a particular geographical region with a recognized know-how. This has led to the creation of different quality labels (for instance, Protected Designation of Origin within the European Union). Obviously, the products distinguished with these labels receive greater appreciation from customers, which is often reflected in a significantly higher price compared with other products that may otherwise be similar. In this situation, it becomes relevant to ensure that any mislabeling or misinformation about the origin of a particular product, or any adulteration resulting in the incorporation of substances of inferior or unrecognized quality, can be detected and notified to the relevant authorities.

Again, chemical analysis plays a crucial role in this type of forensic work. A wide variety of techniques can be used to achieve information on a product that can be linked to its origin [36, 37], but the use of ICP-MS to obtain trace element and isotopic information is becoming more popular.

As can be seen in Table 14.3, the isotopic systems that have been investigated most often for this purpose, excluding those of light elements not suitable for ICP-MS monitoring, are mainly Pb and Sr. This makes sense because these elements have isotopes that are radiogenic. Thus, the soil and rocks of different geographical areas where agricultural products are cultivated may show significant variations in their isotopic signatures. If these differences are preserved throughout the production chain and incorporated into the final foodstuff, it may be possible to differentiate among these and even establish whether the ingredients truly originate from a certain area. Other elements that have been used for this purpose include B, which, unlike other light non-metals, can be monitored by ICP-MS, and, more recently, S, for which the arrival of MC-ICP-MS has opened up new possibilities.

The sample that has attracted most attention so far, and that was studied first, is wine [62]. This is logical, because it is a product of great social, economic, and cultural relevance, and its quality and price are linked to the area of grape collection and wine production. The evolution of the instrumentation selected for analysis can be also perceived in Table 14.3, with a tendency to use MC-ICP-MS preferentially in the last 5 years. Moreover, during this period, other food products have also received attention.

Already in 1997, Augagneur *et al.* compared the performance of quadrupole-based ICP-MS with that of TIMS for Pb isotopic analysis of wine samples. They concluded that there was good agreement between the results provided by the two techniques, and reported a precision for quadrupole-based ICP-MS ranging between 0.1 and 0.3% RSD for ratios involving ^{208}Pb, ^{207}Pb, and ^{206}Pb [38].

One of the problems to be taken into account when analyzing wine samples, and samples with a high organic content in general, is the effect that the matrix may have on the instrument, which may lead to instabilities, signal enhancement or suppression effects, and overall higher imprecision values. Almeida and Vasconcelos evaluated different sample pretreatment procedures for wine analysis and came to the conclusion that a relatively simple protocol, based on attack with HNO_3 and H_2O_2 and UV irradiation, followed by filtering and dilution, was sufficient to minimize matrix effects greatly [39, 40]. A precision of 0.3% for the ratios of the main Pb isotopes was achieved and proved sufficient to observe differences among Port wines.

Table 14.3 Selected applications of food authentication based on isotopic analysis by means of ICP-MS.

Sample	Analytes	Technique	Ref.
Different types of wines	Pb isotope ratios	Q-ICP-MS	[38]
Port wine samples	Pb isotope ratios	Q-ICP-MS	[39]
Port wine samples	Pb isotope ratios	Q-ICP-MS	[40]
Wine samples from different countries	Pb isotope ratios	Q-ICP-MS ICP-TOF-MS MC-ICP-MS	[41]
French wines	Pb isotope ratios	Q-ICP-MS	[42]
Aerosol, soils, vine leaves, grapes, and wine samples	Pb isotope ratios	Q-ICP-MS	[43]
Vineyard soil and wines	Pb isotope ratios	Q-ICP-MS	[44]
Italian wines	Pb isotope ratios	Q-ICP-MS	[45]
Soil, grape juice, and wines	Sr isotope ratios	Q-ICP-MS	[46]
Wines from Portugal and France	Sr isotope ratios	Q-ICP-MS	[47]
Wine samples from different countries	Sr isotope ratios	MC-ICP-MS	[48]
Wine samples from different countries	B isotope ratios	Q-ICP-MS	[49]
Cheese samples from different countries	Sr isotope ratios	MC-ICP-MS	[50]
Rice samples from different regions	Sr isotope ratios	MC-ICP-MS	[51]
Cider samples from different regions	Sr isotope ratios Multi-elemental analysis	MC-ICP-MS	[52]
Caviar samples from different regions and different processing methods	Sr and Os isotope ratios Sr/Ca, Sr/Mg, and Sr/Ba ratios	MC-ICP-MS SF-ICP-MS	[53]
Asparagus from different countries	Sr isotope ratios	MC-ICP-MS	[54]
Natural mineral waters all across Europe, plus wheat and honey	Sr isotope ratios	MC-ICP-MS	[55]
French mineral water samples	Sr isotope ratios	MC-ICP-MS	[56]
Processed spices produced from paprika and soils	Sr isotope ratios Multi-elemental analysis	MC-ICP-MS SF-ICP-MS	[57]
Poultry meat and dried beef samples from different countries	Sr isotope ratios (O isotope ratios monitored by IRMS)	MC-ICP-MS SF-ICP-MS	[58]
Chinese and Korean ginsengs	Sr isotope ratios	Q-ICP-MS	[59]
Cane and beet sugars from different regions	Pb and Sr isotope ratios Multi-elemental analysis (C isotope ratios monitored by IRMS)	MC-ICP-MS SF-ICP-MS and Q-ICP-MS	[60]
Beer from different brands	S isotope ratios	MC-ICP-MS	[61]

Barbaste *et al.* compared the performance of different types of instruments, including quadrupole-based ICP-MS, ICP-TOF-MS, and MC-ICP-MS for Pb isotopic analysis of wine [41]. They concluded that ICP-TOF-MS provides the worst sensitivity, as already mentioned in the previous section, an observation further aggravated by matrix suppression due to the presence of ethanol. Therefore, Pb separation from the matrix and preconcentration were necessary, steps that are also typical if the highest accuracy is aimed for when deploying MC-ICP-MS instrumentation. Under these conditions, ICP-TOF-MS provided a precision very similar to that offered by MC-ICP-MS, and significantly better than that realized by quadrupole-based ICP-MS: quadrupole ICP-MS: 0.14–2.7% RSD, ICP-TOF-MS: 0.04–0.17% RSD, and MC-ICP-MS: 0.01–0.12% RSD for $^{208}Pb/^{206}Pb$ and $^{207}Pb/^{206}Pb$ ratios.

However, when Pb is the target element, it has to be taken into account that not all the Pb present in the wine may come from rock and soil weathering, and that some contribution from anthropogenic sources (e.g., fertilizers, pesticides, airborne pollution, or wine production) is likely to occur [42–44]. This can affect the efficiency of Pb as a proxy for the determination of wine origin, even though it is possible that some of these anthropogenic sources are also origin-dependent [42]. In this regard, Almeida and Vasconcelos [43] compared the results obtained for two different wines produced by two different wineries and reported that major sources of Pb were found in the vinification system, the more traditional method introducing significantly higher amounts than the modern method. In the end, only one-quarter to one-third of the total Pb actually came from the soil and atmospheric depositions. This situation is expected to improve as modern wineries remove Pb sources in tubes and containers. For similar reasons (contamination), and also the complexity of viticultural orography, Larcher *et al.* concluded in a study of Italian wines that, although some differences can be appreciated, Pb isotope ratios are not a very effective tool for the authentication of the origin of Italian wines [45].

Researchers have also turned their attention to other isotopic systems, such as $^{87}Sr/^{86}Sr$. Almeida and Vasconcelos evaluated the manufacturing process of two different wines, and concluded that the Sr level increased during vinification, but this was probably because Sr was liberated from the grape seeds and skins, richer in Sr than the pulp, and, therefore, this fact did not change the isotopic composition of the final wine [46], making $^{87}Sr/^{86}Sr$ ratios a suitable indicator of the origin.

In order to obtain accurate results for this ratio, regardless of the type of ICP-MS instrumentation used, it is necessary to perform a chromatographic separation to avoid overlap between ^{87}Rb and ^{87}Sr, or at least to decrease the ^{87}Rb signal intensity to a point where its contribution can be mathematically corrected for [47, 48]. Under these conditions and when using MC-ICP-MS, it is possible to obtain a level of precision of 0.002–0.003% RSD. Coetzee and Vanhaecke demonstrated that monitoring $^{11}B/^{10}B$ ratios can also be used to characterize wines from different geographical origins [49].

In addition to wine, other food products have also been investigated in recent years and, in most of cases, the usefulness of the $^{87}Sr/^{86}Sr$ ratio has been proven. The procedures used are very similar, most of them based on using a suitable digestion approach, followed by chromatographic isolation of Sr, use of an appropriate mass bias correction strategy (for Sr, it is possible to use the so-called internal method of correction, since the $^{88}Sr/^{86}Sr$ ratio is practically constant in Nature, and thus the extent of mass bias calculated for that ratio can be propagated to predict and correct for the mass bias when measuring $^{87}Sr/^{86}Sr$), and mathematical correction of some spectral overlaps (e.g., remaining ^{87}Rb, and Kr isotopes, which are present as impurities in Ar), as described in detail for cheese analysis [50]. Other samples investigated include rice [51], cider [52], caviar [53], asparagus [54], mineral water [55, 56], paprika [57], poultry meat and dried beef [58], ginseng [59], and cane and beet sugars [60].

In some of these cases, successful results have only been achieved when combining Sr isotopic information with trace element composition, after suitable statistical treatment [52, 57, 60]. In the case of Kalix vendace caviar, the results suggested that Sr is superior to Os ($^{187}Os/^{188}Os$) as a tracer, and some inter-element ratios (Sr/Ca, Sr/Mg, and Sr/Ba) also showed discriminant power.

The work by Swoboda *et al.* has shown that, in order to link the product to a specific region, the Sr isotope ratio of NH_4NO_3 soil extracts (representing the mobile, bioavailable Sr phase) corresponds better to the Sr ratio found for the asparagus samples than the Sr isotope ratio derived from analysis of the total amount of Sr in thc soils (obtained upon complete digestion of the soils) [54]. In this context, the work of Voerkelius *et al.* represents a major step forward, since they carried out a large-scale analysis of hundreds of mineral waters from all across Europe. This permitted the publication of a geological map that permits the prediction of the Sr isotopic composition of ground water, which may represent the composition of bioavailable Sr that is accessible for plants and can be transferred into the food chain [55]. An recent study, however, questioned that establishing a link between the soil and the vegetation/animal life would be straightforward, as the $^{87}Sr/^{86}Sr$ ratio found in plants may depend on numerous aspects (e.g., local geology, balance between water intake from rain and groundwater), and could even be influenced by the plant species [63]. In any case, the authors showed that results obtained for honey and wheat from certain regions correspond well with the map predictions.

A case in which it was shown that the $^{87}Sr/^{86}Sr$ ratio did not help in determining geographical origin was in the investigation of poultry meat and dried beef [58]. The authors explained these results by indicating that, nowadays, animals (and particularly chickens) are housed in such a way that they have relatively little contact with the environment, and that feed components and mineral supplements are traded globally, thus contaminating the fingerprint of local drinking water.

It should also be noted that it is important to be rigorous when developing a new analytical methodology, because many factors (e.g., mass bias, spectral overlap, contamination) can significantly affect the results found otherwise. In this regard, the values published by Choi *et al.* on the analysis of Chinese ginseng [59] have been strongly contested [64, 65], suggesting that the ratios reported in the study are non-existent in natural materials on Earth.

Finally, an example of the new possibilities of MC-ICP-MS has recently been reported by Giner Martínez-Sierra *et al.*, showing the capabilities of this technique for monitoring S isotope ratios in beer. Since the presence of sulfur in beer nowadays is mostly related to the addition of sulfites as antioxidants, the study was not really conceived as a way to link beer to a certain area, but more as a first step towards evaluating sulfur isotopic variations in foodstuffs [61].

14.2.4
Monitoring Environmental Pollution

In this particular field, the analyte most investigated is Pb. This is not surprising since Pb is a global pollutant and Pb isotopic analysis is a very efficient tool for tracing the sources of this Pb pollution. Thus, Pb isotopic analysis of a large variety of environmental samples, including soils, sediments, peat deposits, aerosols, dust, sewage, algae, and lichens, has been investigated in this context. Several reviews have already focused on this topic [66, 67], where more detailed information can be found. The recent availability of MC-ICP-MS has made it possible to investigate other isotopic systems (in addition to Pb) that can be of interest in provenancing pollution. The elements evaluated in the last 5 years include Ag, Cd, Cu, Hg, Sb, and Zn. The main applications reported are presented in Table 14.4.

Focusing on Pb analysis, Townsend's group has been particularly active in the field, investigating anthropogenic sources of Pb in Antarctica. The obvious interest of this research is to establish responsibility for the contamination found in some sites, in order to apportion the cleaning costs adequately. This group always made use of a single-collector SF-ICP-MS instrument for all the research carried out, proving that, for Pb analysis, the information attainable with such an instrument may be sufficient.

Townsend and Snape first investigated marine sediments near Casey Station, Antarctica, and clearly demonstrated the usefulness of isotopic analysis as a tracer of pollution, as even samples that contain low concentrations of Pb (e.g., $20\,\mu g\,g^{-1}$) show a significant deviation from the natural ratio [68]. Thus, Pb isotopic analysis provides a very sensitive method for assessing contamination in originally pristine environments. The source of anthropogenic Pb could be identified and discarded Pb batteries of Australian origin were postulated to be the main cause of this contamination. In subsequent work, batteries were further demonstrated to be the dominant source of contamination (in comparison with fuel spills and paint samples) [69]. Moreover, the authors also investigated samples originating from a different area (Wilkes), where both US (1957–1959) and Australian (1959–1969) expeditions were once located. Sediments from this area displayed isotopic signatures consistent with contamination from multiple Pb sources. This aspect is illustrated in Figure 14.1. Overall, the sediments analyzed from Wilkes display signatures that fit a triangular field, with natural/geogenic (represented by the ellipse), Australian (Broken Hill and Mount Isa)/Idaho and Missouri Pb as end members. This work shows the potential of isotopic analysis by means of ICP-MS to track the sources of environmental pollution even in complex situations.

Table 14.4 Selected applications of environmental forensics based on isotopic analysis by means of ICP-MS.

Sample[a]	Analytes	Technique	Ref.
Marine sediments near East Antarctica	Pb isotope ratios	SF-ICP-MS	[68]
Marine sediments near East Antarctica and Pb sources	Pb isotope ratios	SF-ICP-MS	[69]
Marine sediments near East Antarctica and Pb sources	Pb isotope ratios	SF-ICP-MS	[70]
Antarctic macroalga	Pb isotope ratios	SF-ICP-MS	[71]
Peat core from Okefenokee swamp	Pb isotope ratios	Q-ICP-MS	[72]
Blood from condors, diet sources and ammunition	Pb isotope ratios	SF-ICP-MS	[73]
Feathers and blood from condors	Pb isotope ratios	SF-ICP-MS	[74]
Peat and lichens	Pb isotope ratios	LA–Q-ICP-MS LA–MC-ICP-MS	[75]
Layered paint chip	Pb isotope ratios	LA–Q-ICP-MS LA–MC-ICP-MS	[76]
Lichens and bus air-filter aerosols	Pb isotope ratios Multi-elemental analysis	MC-ICP-MS	[77]
Sediment monolith from a marsh in Iceland	Pb isotope ratios Pb/Ti and Pb/Li ratios	MC-ICP-MS Q-ICP-MS	[78]
Suspended particulate matter from Nagoya City	Pb isotope ratios	SF-ICP-MS	[79]
Different types (sediments, soils, dust, sewage, etc.)	Pb isotope ratios	SF-ICP-MS	[80]
Gas samples from fermenters	Sb isotope ratios	HPLC–MC-ICP-MS	[81]
Environmental CRMs	Ag isotope ratios	MC-ICP-MS	[82]
Ag metal, Ag nanomaterials, Silversoft socks and dietary supplement	Ag isotope ratios	MC-ICP-MS	[83]
Sediments, soils, and Hg ores	Hg isotope ratios	MC-ICP-MS	[84]
Chinese river sediments	Cd isotope ratios (Pb isotope ratios monitored by TIMS)	MC-ICP-MS	[85]
Lichens, ambient particulate matter, and bus air-filter aerosols	Zn and Pb isotope ratios	MC-ICP-MS	[86]
Particulate matter sampled at different points inside and outside a smelter plant	Zn isotope ratios	MC-ICP-MS	[87]
Sediments and tires	Cu and Zn isotope ratios	MC-ICP-MS	[88]
Ore concentrates, fumes, effluents, and final metals	Cd, Pb, and Zn	MC-ICP-MS	[89]

[a]CRM, Certified Reference Material.

Also, the possibility of using partial extraction pretreatment of the sediments with 1 M HCl, instead of complete dissolution using HF, was evaluated and shown to be a more powerful, cost-effective, and rapid approach. The advantage of this methodology is that only anthropogenic Pb is leached out,

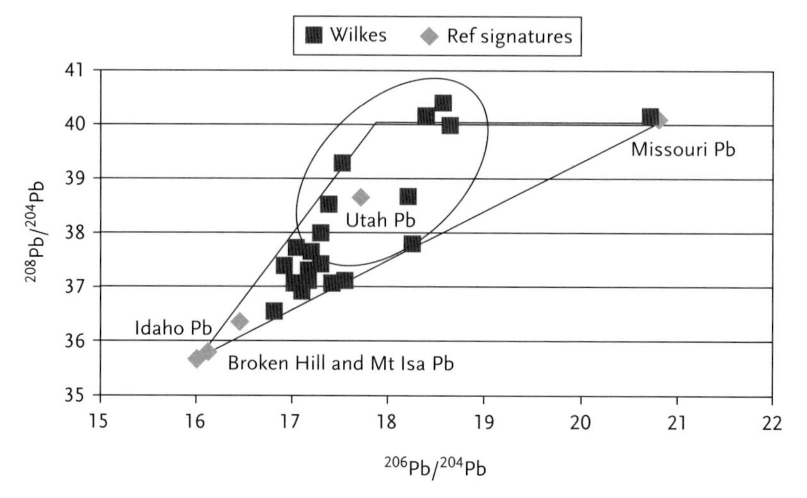

Figure 14.1 Lead isotopic signatures (displayed as $^{208}Pb/^{204}Pb$ versus $^{206}Pb/^{204}Pb$) of 24 sediments collected from the Wilkes area (Antarctica). The ellipse shows background sediment ratios. Isotopic signatures of major Australian (Broken Hill and Mount Isa) and US (Idaho, Missouri, and Utah) deposits are included. Error bars representing instrumental precision (1 SD) are comparable in size to the symbols used.
Reproduced with permission from Elsevier from [69].

whereas complete dissolution not only sets anthropogenic free Pb but also geogenic Pb [70].

This group has also focused on investigating marine macroalgae as a tracer to investigate the origin and the extent of metal flux in the marine environment [71]. As discussed earlier, the analysis of organic samples is not too simple for ICP-MS. The authors proposed to spike all samples and standard with 1% ethanol as a matrix-matching strategy, to compensate for the possible presence of undigested organic matter in the samples, which could otherwise lead to matrix effects and inaccurate results.

Pb sources have also been investigated in another supposedly pristine environment, Okefenokee Swamp (Georgia, USA) [72]. The authors collected peat core and monitored the temporal evolution of its Pb isotopic composition, coming to the conclusion that coal combustion emission was the major source of Pb in the area. The positive outcome of the research is that Pb restrictions in gasoline and the designation of the swamp as a Class 1 Air Quality Area seemed to have effectively reduced Pb concentrations in the area. From an analytical point of view, this work was carried out with a quadrupole-based ICP-MS instrument equipped with a collision/reaction cell, which was pressurized with a nonreactive gas in order to achieve collisional focusing and thermalization of ions, leading to improved precision (better than 0.1% when ^{206}Pb and ^{207}Pb signals were at a level of 200 000 cps or higher). More information on this topic can be found elsewhere [90, 91].

Another interesting application was reported by Smith and co-workers, who demonstrated the usefulness of monitoring the isotopic evolution of Pb found in feathers of Californian condors [73, 74]. As opposed to blood analysis, spatially resolved analysis of feathers permits the long-term monitoring of Pb poisoning. The authors identified accidental ingestion of ammunition in carcasses of animals that had been killed by hunters as the main source of Pb that threatens the recovery of this endangered species.

The use of LA–quadrupole-based ICP-MS has also proved advantageous in this context. Kylander *et al.* evaluated the potential benefit of using this technique in terms of sample throughput when screening peat and lichens for routine biomonitoring. The rationale is that this technique may allow one to quickly determine which samples should be further analyzed by means of more precise solution-based approaches. Alternatively, for more precise results, the coupling of LA to MC-ICP-MS was evaluated, allowing the bias versus the reference method to be reduced from up to 3.1% to less than 1% [75].

Ghazi and Millette demonstrated the possibilities of LA–ICP-MS when depth profiling analysis of small samples is attempted [76]. They separately analyzed the six different layers present in a paint chip sample. They applied ablation in drilling mode (the laser beam, of 20 μm diameter, was always located at the same sample spot), penetrating the sample down to a depth of 500 μm, and observed the different signal profiles as a function of depth. The last four layers contained significant amounts of Pb. Based on the Pb isotopic signature of these layers, different probable sources for each layer could be identified.

In solution analysis, the use of MC-ICP-MS has allowed the precision to be significantly increased, as discussed in previous sections. This has been also demonstrated in this particular context for the analysis of lichens, used as biotracers for mapping the dispersion of atmospheric pollutants in urban areas [77], and for monitoring the evolution of sources of anthropogenic Pb in Iceland, after analysis of a sediment monolith [78]. However, in addition, the major advantage of this technique is that it opens doors for measuring other isotopic systems that are also of environmental interest.

For instance, Wehmeier *et al.* demonstrated improved precision for the monitoring of transient signals obtained when using gas chromatography coupled to MC-ICP-MS for the isotopic analysis of different Sb species [81]. The precision attained (0.02% RSD) permitted isotope fractionation to be observed when Sb is methylated by anaerobic bacteria, opening up ways to investigate environmental pathways of Sb.

Yang and co-workers investigated Ag by means of MC-ICP-MS and observed significant variations in environmental samples and in commercially available products, proving that it may be possible to fingerprint sources of Ag in the environment [82, 83]. In a similar way, isotopic data may permit the pollution history of Hg [84, 92], Cd [85], Zn [86–88, 93], and Cu [88] to be traced.

However, in order to identify the potential sources of these elements, it has to be taken into account that, in contrast to the situation with Pb, metallurgical processing may lead to isotopic fractionation. In a recent study, Shiel *et al.* showed for Zn that, whereas the fume and effluent were significantly fractionated isotopically, the Zn ore concentrate and the roasting product showed practically the same Zn isotopic composition, owing to a recovery in the process of $>98\%$ [89]. However, the situation is different for Cd, as for that element only 72–92% of the initial metal is recovered. In that case, the refined metal is isotopically heavier than the starting material. Thus, the original material, the fumes emitted, and the final product all have different isotopic fingerprints for this element.

14.2.5 Other Applications

There are a few interesting forensic applications that do not exactly fall into any of the previous categories and that will be discussed in this section. Table 14.5 shows their main characteristics.

For instance, a typical situation that demands the development of fingerprinting methodologies is the detection of counterfeit drugs, a problem that has escalated over the last decade. A counterfeit product is a generic or branded pharmaceutical drug that is deliberately mislabeled with respect to its identity and/or source. These counterfeit products need to be traced, especially as they might not contain the active pharmaceutical ingredients (APIs) at all or contain them in a wrong amount.

Santamaria-Fernandez *et al.* investigated different approaches to identify such counterfeit products. First, they compared the discriminant power of approaches relying on S isotopic analysis for a large-scale investigation of a certain antiviral product [94]. They evaluated three different approaches: dissolution of the samples followed by analysis by means of MC-ICP-MS, isolation of the API (active pharmaceutical ingredient) by high-performance liquid chromatography (HPLC)

Table 14.5 Various types of forensic applications based on isotopic analysis by means of ICP-MS.

Sample	Analytes	Technique	Ref.
Antiviral drugs	S isotope ratios	(LA and HPLC)–MC-ICP-MS	[94]
Antiviral drugs	S and Mg isotopes ratios (C and N isotope ratios monitored by IRMS)	MC-ICP-MS	[95]
Pharmaceutical packages	Ca and Pb isotope ratios	LA–MC-ICP-MS	[96]
Black and colored toners	Cu isotope ratios plus multi-elemental analysis	LA–ICP-TOF-MS	[97]
Single human hair strands	S isotope ratios	LA–MC-ICP-MS	[98]
Cu-bearing andesines	Cu isotope ratios Multi-elemental analysis Ar isotope ratios	LA–MC-ICP-MS LA–Q-ICP-MS	[99]

and measurement of S isotopic ratios by MC-ICP-MS, and direct analysis of the tablets using LA–MC-ICP-MS. Figure 14.2a shows a typical HPLC–MC-ICP-MS signal. A Si spike was used to correct for mass bias (continuous signal). The encouraging conclusion of the work was that all of the approaches tested permit genuine tablets to be clearly differentiated from counterfeit tablets, as shown in

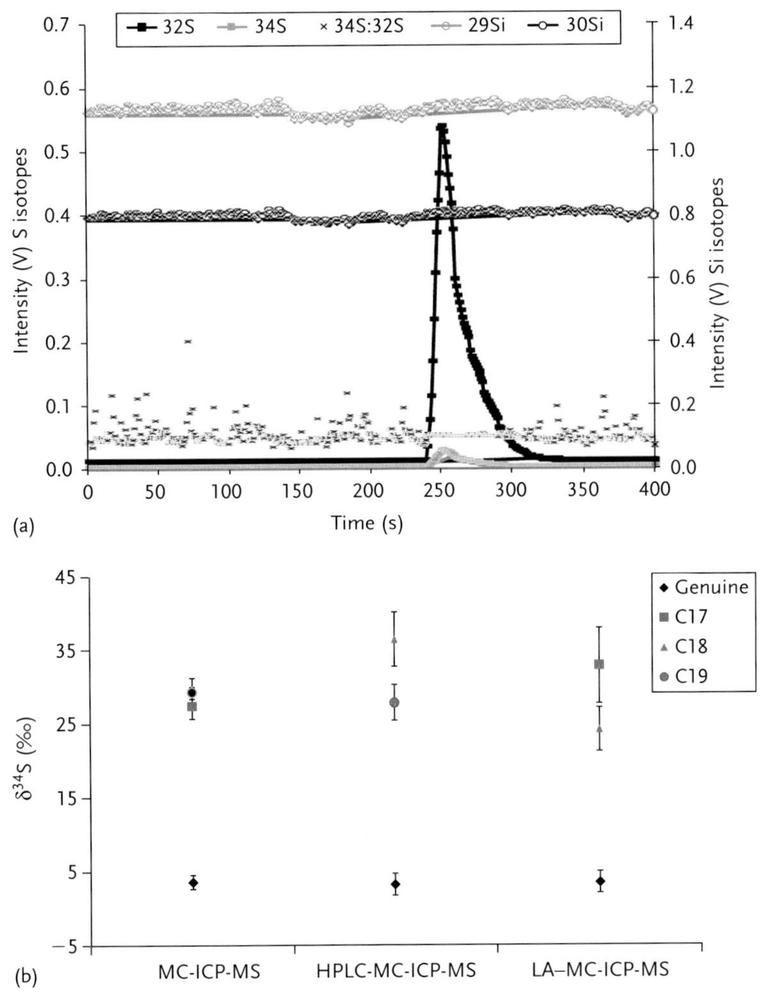

Figure 14.2 (a) Transient signal profile for ^{34}S and ^{32}S obtained upon HPLC separation of the active pharmaceutical ingredient from antiviral tablets and ICP-MS measurement. Silicon is admixed and its isotopes ^{29}Si and ^{30}Si are continuously monitored. $^{34}S/^{32}S$ ratios are also plotted. (b) $\delta^{34}S$ authenticity ranges obtained when using the different approaches and results for counterfeit tablets: counterfeit 17, 18, and 19. Error bars represent combined expanded uncertainties ($k = 2$; $n = 3$) for each tablet; measurements were performed in triplicate.
Reproduced with permission of the Royal Society of Chemistry from [94].

Figure 14.2b, and, in fact, even to reveal different groups of counterfeit tablets. Obviously, the use of LA is beneficial in this context, as it would allow the analysis of a larger number of samples per day.

A second study by this group demonstrated that it is often best to combine the information provided by various techniques [95]. They aimed at the detection of counterfeit varieties of the antiviral drug Heptodin (lamivudine) and compared the results obtained by S and Mg isotopic analysis using MC-ICP-MS and by C and N isotopic analysis using a carbon and nitrogen elemental analyzer coupled to a mass spectrometer [isotope ratio mass spectrometry (IRMS)]. They concluded that best results were achieved by using the information attained when combining at least three isotopic systems (C, S, and N) and applying principal component analysis (PCA) to the results. This approach permitted genuine drugs to be distinguished from counterfeit drugs and to link some suspect counterfeit samples to each other. This example further demonstrates that, in forensic analysis, ICP-MS should not be seen as a replacement of traditional IRMS approaches, but as a complementary technique.

Finally, the same group also evaluated the possibility of investigating the packaging, instead of the product, to detect counterfeit goods [96]. For this purpose, LA–MC-ICP-MS was used for the isotopic analysis of Ca and Pb in the cardboard/ink of different packages. The advantages of using LA in this context are clear, providing lateral resolution at the micrometer level to focus on the areas of interest only. The use of Ca demonstrated a higher discriminant power in comparison with that of Pb, with expanded uncertainties of 0.06% for $^{43}Ca/^{44}Ca$ and $^{42}Ca/^{44}Ca$ ratios. This precision permitted genuine and counterfeit pharmaceutical packages to be clearly separated from one another.

A similar type of application was developed by Szynkowska *et al.*, who investigated the potential of LA–ICP-TOF-MS for the analysis of different toners from different brands, with the aim of tracing illegal activities that make use of photocopy machines and printers [97]. Their conclusion was that black toners can be differentiated based on multi-elemental information. However, for prints containing cyan toner, Cu isotopic analysis is recommended, because cyan toner is typically produced using an organic pigment (Phthalocyanine Blue BN), which is a complex of Cu and phthalocyanine.

Another study by Santamaria-Fernandez *et al.* focused on investigating the potential of monitoring longitudinal variations in the sulfur isotopic composition of hair as an indicator of recent geographical movements, which can, for example, be significant in counter-terrorism investigations [98]. Again, the coupling of LA to MC-ICP-MS combines the spatial resolution necessary for this kind of application (a laser beam with a diameter of 12 μm was used) with the sensitivity and precision required, such that spatially (and hence temporally) resolved variations can be established for a frequent traveler.

A very different type of application based on the use of LA–MC-ICP-MS, among other techniques, was reported recently by Fontaine *et al.*, who investigated whether colored andesines (silicate gemstones) originating from Tibet were authentic or were produced from colorless andesines by Cu diffusion under heat treatment [99]. The authors analyzed laboratory-diffused samples and proved that, owing to this diffusion process, these samples showed large variations of the $^{65}Cu/^{63}Cu$ ratio within the same mineral (up to 3‰). The authors concluded that the large variability also observed in the $^{65}Cu/^{63}Cu$ ratios of the samples from Tibet under investigation, together with other indications (e.g., low concentration of radiogenic ^{40}Ar) seemed to indicate that colored andesine samples labeled as coming from Tibet are most probably Cu-diffused stones, using initially colorless Mongolian andesines as starting material.

It should be noted that the successful application of LA–ICP-MS for isotopic analysis depends on a number of issues already discussed earlier (such as the advantage offered by instrumentation with truly simultaneous capabilities; see Chapters 3 and 4 for more details), but also on the development of a suitable strategy to correct for instrumental mass bias. In the case of LA, the situation is more complicated than for solution-based work, in which the analyte is often first isolated from the matrix. This is not feasible when using LA and the matrix is expected to affect the raw data finally obtained, thus necessitating proper correction. This effect can be compensated for by monitoring matrix-matched standards of known isotopic composition. These are not always to be found, but it may be possible to use one sample or a material very similar to it and determine its true isotopic composition by means of a complementary technique, thus producing an in-house reference. For instance, Santamaria-Fernandez *et al.* [98] used a horse hair sample as an external standard to correct for mass bias effects during the analysis of the human hair samples. Alternatively, together with the dry aerosol originating from the ablation of the sample, it is possible to aspirate a solution containing an element of known isotopic signature that is similar in mass to the analyte of interest (e.g., Tl for Pb analysis, Pd for Ag analysis, or Si for S analysis). The mass bias observed in the isotopic signature of this reference element can be calculated and subsequent corrections for the analyte can be performed, according to a suitable propagation law (see Chapter 5 for more details). In this case, a desolvation membrane can be used in order to keep the plasma dry [94]. However, not using such a membrane and using a wet plasma instead may also show some advantages. The plasma conditions are then governed by the presence of water, making them less prone to matrix effects [18, 100, 101].

14.3 Future Outlook

Forensics is a field in which the use of ICP-MS is expanding. This technique can offer multi-elemental and isotopic information that, in combination with the information provided by other techniques (e.g., IRMS), can significantly improve the possibilities for fingerprinting and provenancing.

A clear trend towards the use of MC-ICP-MS can also be observed, as this technique provides the precision needed to obtain relevant information for new isotopic systems. When the target of the investigation is a solid sample, the use of LA is becoming more and more popular, owing to the high speed of analysis and the spatial resolution thus obtained. LA–ICP-MS offers the best results when MC-ICP-MS, or another technique that provides simultaneous or quasi-simultaneous signal monitoring (e.g., ICP-TOF-MS), is used. In this regard, the introduction of new types of instruments offering this characteristic (e.g., ICP-MS instrumentation based on the Mattauch–Herzog design and new semiconductor detectors) at a reduced cost may even further increase the popularity of this methodology in forensic laboratories.

Acknowledgments

The authors acknowledge the Government of Aragon (ARAID Foundation and Ibercaja) and the Spanish Ministry of Science and Innovation (Project **CTQ2009-08606**) for financial support.

References

1 Resano, M., García-Ruiz, E., Alloza, R., Marzo, M.P., Vandenabeele, P., and Vanhaecke, F. (2007) Laser ablation–inductively coupled plasma mass spectrometry for the characterization of pigments in prehistoric rock art. *Anal. Chem.*, **79**, 8947–8955.

2 Resano, M., Pérez-Arantegui, J., García-Ruiz, E., and Vanhaecke, F. (2005) Laser ablation–inductively coupled plasma mass spectrometry for the fast and direct characterization of antique glazed ceramics. *J. Anal. At. Spectrom.*, **20**, 508–514.

3 Pérez-Arantegui, J., Resano, M., García-Ruiz, E., Vanhaecke, F., Roldán, C., Ferrero, J., and Coll, J. (2008) Characterization of cobalt pigments found in traditional Valencian ceramics by means of laser ablation–inductively coupled plasma mass spectrometry and portable X-ray fluorescence spectrometry. *Talanta*, **74**, 1271–1280.

4 Ulrich, A., Moor, C., Vonmont, H., Jordi, H.R., and Lory, M. (2004) ICP-MS trace-element analysis as a forensic tool. *Anal. Bioanal. Chem.*, **378**, 1059–1068.

5 Resano, M., García-Ruiz, E., and Vanhaecke, F. (2010) Laser ablation–inductively coupled plasma mass spectrometry in archaeometric research. *Mass Spectrom. Rev.*, **29**, 55–78.

6 Balcaen, L., Moens, L., and Vanhaecke, F. (2010) Determination of isotope ratios of metals (and metalloids) by means of inductively coupled plasma-mass spectrometry for provenancing purposes – a review. *Spectrochim. Acta B*, **65**, 769–786.

7 Aggarwal, J., Habicht-Mauche, J., and Juarez, C. (2008) Application of heavy stable isotopes in forensic isotope geochemistry: a review. *Appl. Geochem.*, **23**, 2658–2666.

8 Zeichner, A. (2003) Recent developments in methods of chemical analysis in investigations of firearm-related events. *Anal. Bioanal. Chem.*, **376**, 1178–1191.

9 Koons, R.D. and Buscaglia, J. (2005) Forensic significance of bullet lead compositions. *J. Forensic Sci.*, **50**, 341–351.

10 Stupian, G.W., Ives, N.A., Marquez, N., and Morgan, B.A. (2011) The application of lead isotope analysis to bullet individualization in two homicides. *J. Forensic Sci.*, **46**, 1342–1351.

11 Dufosse, T. and Touron, P. (1998) Comparison of bullet alloys by chemical analysis: use of ICP-MS method. *Forensic Sci. Int.*, **91**, 197–206.

12 Buttigieg, G.A., Baker, M.E., Ruiz, J., and Bonner Denton, M. (2003) Lead

isotope ratio determination for the forensic analysis of military small arms projectiles. *Anal. Chem.*, **75**, 5022–5029.

13 Zeichner, A., Ehrlich, S., Shoshani, E., and Halicz, L. (2006) Application of lead isotope analysis in shooting incident investigations. *Forensic Sci. Int.*, **158**, 52–64.

14 Tamura, S., Hokura, A., Oishi, M., and Nakai, I. (2006) Lead isotope ratio analysis of bullet samples by using quadrupole ICP-MS. *Bunseki Kagaku*, **55**, 827–834.

15 Wunnapuk, K., Minami, T., Durongkadech, P., Tohno, S., Ruangyuttikarn, W., Moriwake, Y., Vichairat, K., Sribanditmongkol, P., and Tohno, Y. (2009) Discrimination of bullet types using analysis of lead isotopes deposited in gunshot entry wounds. *Biol. Trace Elem. Res.*, **129**, 278–289.

16 Sjåstad, K.-E., Simonsen, S.L., and Andersen, T. (2011) Use of lead isotopic ratios to discriminate glass samples in forensic science. *J. Anal. At. Spectrom.*, **26**, 325–333.

17 Mayer, K., Wallenius, M., and Ray, I. (2005) Nuclear forensics – a methodology providing clues on the origin of illicitly trafficked nuclear materials. *Analyst*, **130**, 433–441.

18 Varga, Z. (2008) Application of laser ablation inductively coupled plasma mass spectrometry for the isotopic analysis of single uranium particles. *Anal. Chim. Acta*, **625**, 1–7.

19 Boulyga, S.F. and Prohaska, T. (2008) Determining the isotopic compositions of uranium and fission products in radioactive environmental microsamples using laser ablation ICP-MS with multiple ion counters. *Anal. Bioanal. Chem.*, **390**, 531–539.

20 Bürger, S. and Riciputi, L.R. (2009) A rapid isotope ratio analysis protocol for nuclear solid materials using nanosecond laser-ablation time-of-flight ICP-MS. *J. Environ. Radioact.*, **100**, 970–976.

21 Lloyd, N.S., Parrish, R.R., Horstwood, M.S.A., and Chenery, S.R.N. (2009) Precise and accurate isotopic analysis of microscopic uranium-oxide grains using LA–MC-ICP-MS. *J. Anal. At. Spectrom.*, **24**, 752–758.

22 Agarande, M., Benzoubir, S., Neiva-Marques, A.M., and Bouisset, P. (2004) Sector field inductively coupled plasma mass spectrometry, another tool for plutonium isotopes and plutonium isotope ratios determination in environmental matrices. *J. Environ. Radioact.*, **72**, 169–176.

23 Széles, E., Varga, Z., and Stefánka, Z. (2010) Sample preparation method for analysis of swipe samples by inductively coupled plasma mass spectrometry. *J. Anal. At. Spectrom.*, **25**, 1014–1018.

24 Liezers, M., Lehn, S.A., Olsen, K.B., Farmer, O.T. III, and Duckworth, D.C. (2009) Determination of plutonium isotope ratios at very low levels by ICP-MS using on-line electrochemically modulated separations. *J. Radioanal. Nucl. Chem.*, **282**, 299–304.

25 Zheng, J. and Yamada, M. (2006) Determination of U isotope ratios in sediments using ICP-QMS after sample cleanup with anion-exchange and extraction chromatography. *Talanta*, **68**, 932–939.

26 Wallenius, M., Mayer, K., and Ray, I. (2006) Nuclear forensic investigations: two case studies. *Forensic Sci. Int.*, **156**, 55–62.

27 Buchmann, J.H., Sarkis, J.E.S., Kakazu, M.H., and Rodrigues, C. (2006) Environmental monitoring as an important tool for safeguards of nuclear material and nuclear forensics. *J. Radioanal. Nucl. Chem.*, **270**, 291–298.

28 Todorov, T.I., Xu, H., Ejnik, J.W., Mullick, F.G., Squibb, K., McDiarmid, M.A., and Centeno, J.A. (2009) Depleted uranium analysis in blood by inductively coupled plasma mass spectrometry. *J. Anal. At. Spectrom.*, **24**, 189–193.

29 D'Ilio, S., Violante, N., Senofonte, O., Majorani, C., and Petrucci, F. (2010) Determination of depleted uranium in human hair by quadrupole inductively coupled plasma mass spectrometry: method development and validation. *Anal. Methods*, **2**, 1184–1190.

30 Varga, Z., Wallenius, M., Mayer, K., Keegan, E., and Millet, S. (2009) Application of lead and strontium isotope ratio measurements for the origin assessment of uranium ore concentrates. *Anal. Chem.*, **81**, 8327–8334.

31 Svedkauskaite-LeGore, J., Mayer, K., Millet, S., Nicholl, A., Rasmussen, G., and Baltrunas, D. (2007) Investigation of the isotopic composition of lead and of trace elements concentrations in natural uranium materials as a signature in nuclear forensics. *Radiochim. Acta*, **95**, 601–605.

32 Aramendía, M., Resano, M., and Vanhaecke, F. (2010) Isotope ratio determination by laser ablation–single collector-inductively coupled plasma-mass spectrometry. General capabilities and possibilities for improvement. *J. Anal. At. Spectrom.*, **25**, 390–404.

33 Cottle, J.M., Horstwood, M.S.A., and Parrish, R.R. (2009) A new approach to single shot laser ablation analysis and its application to *in situ* Pb/U geochronology. *J. Anal. At. Spectrom.*, **24**, 1355–1363.

34 Willie, S., Mester, Z., and Sturgeon, R.E. (2005) Isotope ratio precision with transient sample introduction using ICP orthogonal acceleration time-of-flight mass spectrometry. *J. Anal. At. Spectrom.*, **20**, 1358–1364.

35 Sturgeon, R.E., Lam, J.W.H., and Saint, A. (2000) Analytical characteristics of a commercial ICP orthogonal acceleration time-of-flight mass spectrometer (ICP-TOFMS). *J. Anal. At. Spectrom.*, **15**, 607–616.

36 Gonzalvez, A., Armenta, S., and de la Guardia, M. (2009) Trace-element composition and stable-isotope ratio for discrimination of foods with protected designation of origin. *Trends Anal. Chem.*, **28**, 1295–1311.

37 Kelly, S., Heaton, K., and Hoogewerff, J. (2005) Tracing the geographical origin of food: the application of multi-element and multi-isotope analysis. *Trends Food Sci. Technol.*, **16**, 555–567.

38 Augagneur, S., Medina, B., and Grousset, F. (1997) Measurement of lead isotope ratios in wine by ICP-MS and its applications to the determination of lead concentration by isotope dilution. *Fresenius' J. Anal. Chem.*, **357**, 1149–1152.

39 Almeida, C.M.R. and Vasconcelos, M.T. S.D. (1999) Determination of lead isotope ratios in Port wine by inductively coupled plasma mass spectrometry after pre-treatment by UV-irradiation. *Anal. Chim. Acta*, **396**, 45–53.

40 Almeida, C.M.R. and Vasconcelos, M.T. S.D. (1999) UV-irradiation and MW-digestion pre-treatment of Port wine suitable for the determination of lead isotope ratios by inductively coupled plasma mass spectrometry. *J. Anal. At. Spectrom.*, **14**, 1815–1821.

41 Barbaste, M., Halicz, L., Galy, A., Medina, B., Emteborg, H., Adams, F.C., and Lobinski, R. (2001) Evaluation of the accuracy of the determination of lead isotope ratios in wine by ICP MS using quadrupole, multicollector magnetic sector and time-of-flight analyzers. *Talanta*, **54**, 307–317.

42 Medina, B., Augagneur, S., Barbaste, M., Grouset, F.E., and Buat-Meard, P. (2000) Influence of atmospheric pollution on the lead content of wines. *Food Addit. Contam.*, **17**, 435–445.

43 Almeida, C.M.R. and Vasconcelos, M.T. S.D. (2003) Lead contamination in Portuguese red wines from the Douro region: from the vineyard to the final product. *J. Agric. Food Chem.*, **51**, 3012–3023.

44 Mihaljevic, M., Ettler, V., Sebek, O., Strnad, L., and Chrastny, V. (2006) Lead isotopic signatures of wine and vineyard soils – tracers of lead origin. *J. Geochem. Explor.*, **88**, 130–133.

45 Larcher, R., Nicolini, G., and Pangrazzi, P. (2003) Isotope ratios of lead in Italian wines by inductively coupled plasma mass spectrometry. *J. Agric. Food Chem.*, **51**, 5956–5961.

46 Almeida, C.M.R. and Vasconcelos, M.T. S.D. (2004) Does the winemaking process influence the wine $^{87}Sr/^{86}Sr$? A case study. *Food Chem.*, **85**, 7–12.

47 Almeida, C.M.R. and Vasconcelos, M.T. S.D. (2001) ICP-MS determination of strontium isotope ratio in wine in order to be used as a fingerprint of its regional origin. *J. Anal. At. Spectrom.*, **16**, 607–611.

48 Barbaste, M., Robinson, K., Guilfoyle, S., Medina, B., and Lobinski, R. (2002) Precise determination of the strontium isotope ratios in wine by inductively coupled plasma sector field multicollector mass spectrometry (ICP-SF-MC-MS). *J. Anal. At. Spectrom.*, **17**, 135–137.

49 Coetzee, P.P. and Vanhaecke, F. (2005) Classifying wine according to geographical origin via quadrupole-based ICP-mass spectrometry measurements of boron isotope ratios. *Anal. Bioanal. Chem.*, **383**, 977–984.

50 Fortunato, G., Mumic, K., Wunderli, S., Pillonel, L., Bosset, J. O., and Gremaud, G. (2004) Application of strontium isotope abundance ratios measured by MC-ICP-MS for food authentication. *J. Anal. At. Spectrom.*, **19**, 227–234.

51 Kawasaki, A., Oda, H., and Hirata, T. (2002) Determination of strontium isotope ratio of brown rice for estimating its provenance. *Soil Sci. Plant Nutr.*, **48**, 635–640.

52 García-Ruiz, S., Moldovan, M., Fortunato, G., Wunderli, S., and García Alonso, J.L. (2007) Evaluation of strontium isotope abundance ratios in combination with multi-elemental analysis as a possible tool to study the geographical origin of ciders. *Anal. Chim. Acta*, **590**, 55–66.

53 Rodushkin, I., Bergman, T., Douglas, G., Engström, E., Sörlin, D., and Baxter, D.C. (2007) Authentication of Kalix (NE Sweden) vendace caviar using inductively coupled plasma-based analytical techniques: evaluation of different approaches. *Anal. Chim. Acta*, **583**, 310–318.

54 Swoboda, S., Brunner, M., Boulyga, S. F., Galler, P., Horacek, M., and Prohaska, T. (2008) Identification of Marchfeld asparagus using Sr isotope ratio measurements by MC-ICP-MS. *Anal. Bioanal. Chem.*, **390**, 487–494.

55 Voerkelius, S., Lorenz, G.D., Rummel, S., Quétel, C.R., Heiss, G., Baxter, M., Brach-Papa, C., Deters-Itzelsberger, P., Hoelzl, S., Hoogewerff, J., Ponzevera, E., Van Bocxstaele, M., and Ueckermann, H. (2010) Strontium isotopic signatures of natural mineral waters, the reference to a simple geological map and its potential for authentication of food. *Food Chem.*, **118**, 933–940.

56 Brach-Papa, C., Van Bocxstaele, M., Ponzevera, E., and Quétel, C.R. (2009) Fit for purpose validated method for the determination of the strontium isotopic signature in mineral water samples by multi-collector inductively coupled plasma mass spectrometry. *Spectrochim. Acta B*, **64**, 229–234.

57 Brunner, M., Katona, R., Stefánka, Z., and Prohaska, T. (2010) Determination of the geographical origin of processed spice using multielement and isotopic pattern on the example of Szegedi paprika. *Eur. Food Res. Technol.*, **231**, 623–634.

58 Franke, B.M., Koslitz, S., Micaux, F., Piantini, U., Maury, V., Pfammatter, E., Wunderli, S., Gremaud, G., Bosset, J.O., Hadorn, R., and Kreuzer, M. (2008) Tracing the geographic origin of poultry meat and dried beef with oxygen and strontium isotope ratios. *Eur. Food Res. Technol.*, **226**, 761–769.

59 Choi, S.M., Lee, H.S., Lee, G.H., and Han, J.K. (2008) Determination of the strontium isotope ratio by ICP-MS ginseng as a tracer of regional origin. *Food Chem.*, **108**, 1149–1154.

60 Rodushkin, I., Baxter, D.C., Engström, E., Hoogewerff, J., Horn, P., Papesch, W., Watling, J., Latkoczy, C., van der Peijl, G., Berends-Montero, S., Ehleringer, J., and Zdanowicz, V. (2011) Elemental and isotopic characterization of cane and beet sugars. *J. Food Comp. Anal.*, **24**, 70–78.

61 Giner Martínez-Sierra, J., Santamaria-Fernandez, R., Hearn, R., Marchante-Gayón, J.M., and García-Alonso, J.I. (2010) Development of a direct

procedure for the measurement of sulfur isotope variability in beers by MC-ICP-MS. *J. Agric. Food Chem.*, **58**, 4043–4050.

62 Suhaj, M. and Korenovska, M. (2005) Application of elemental analysis for identification of wine origin. A review. *Acta Aliment.*, **34**, 393–401.

63 Degryse, P., Shortland, A., De Muynck, D., Van Heghe, L., Scott, R., Neyt, B., and Vanhaecke, F. (2010) Considerations on the provenance determination of plant ash glasses using strontium isotopes. *J. Archaeol. Sci.*, **37**, 3129–3135.

64 Rosner, M. (2010) Geochemical and instrumental fundamentals for accurate and precise strontium isotope data of food samples: comment on "Determination of the strontium isotope ratio by ICP-MS ginseng as a tracer of regional origin (Choi *et al.*, 2008)". *Food Chem.*, **121**, 918–921.

65 You, C.F., Wang, B.S., Chung, C.S., and Huang, K.F. (2009) Letter to Editor on "Determination of the strontium isotope ratio by ICP-MS ginseng as a tracer of regional origin". *Food Chem.*, **115**, 387.

66 Komárek, M., Ettler, V., Chrastny, V., and Mihaljevic, M. (2008) Lead isotopes in environmental sciences: a review. *Environ. Int.*, **34**, 562–577.

67 Cheng, H. and Hu, Y. (2010) Lead (Pb) isotopic fingerprinting and its applications in lead pollution studies in China: a review. *Environ. Pollut.*, **158**, 1134–1146.

68 Townsend, A.T. and Snape, I. (2002) The use of Pb isotope ratios determined by magnetic sector ICP-MS for tracing Pb pollution in marine sediments near Casey Station, East Antarctica. *J. Anal. At. Spectrom.*, **17**, 922–928.

69 Townsend, A.T. and Snape, I. (2008) Multiple Pb sources in marine sediments near the Australian Antarctic Station, Casey. *Sci. Total Environ.*, **389**, 466–474.

70 Townsend, A.T., Snape, I., Palmer, A. S., and Seen, A.J. (2009) Lead isotopic signatures in Antarctic marine sediment cores: a comparison between 1M HCl partial extraction and HF total digestion pre-treatments for discerning anthropogenic inputs. *Sci. Total Environ.*, **408**, 382–389.

71 Runcie, J.W., Townsend, A.T., and Seen, A.J. (2009) The application of lead isotope ratios in the Antarctic macroalga *Iridaea cordata* as a contaminant monitoring tool. *Mar. Pollut. Bull.*, **58**, 961–966.

72 Jackson, B.P., Winger, P.V., and Lasier, P.J. (2004) Atmospheric lead deposition to Okefenokee Swamp, Georgia, USA. *Environ. Pollut.*, **130**, 445–451.

73 Church, M.E., Gwiazda, R., Risebrough, R.W., Sorenson, K., Chamberlain, C.P., Farry, S., Heinrich, W., Rideout, B.A., and Smith, D.R. (2006) Ammunition is the principal source of lead accumulated by California condors re-introduced to the wild. *Environ. Sci. Technol.*, **40**, 6143–6150.

74 Finkelstein, M.E., George, D., Scherbinski, S., Gwiazda, R., Johnson, M., Burnett, J., Brandt, J., Lawrey, S., Pessier, A.P., Clark, M., Wynne, J., Grantham, J., and Smith, D.R. (2010) Feather lead concentrations and $^{207}Pb/^{206}Pb$ ratios reveal lead exposure history of California condors (*Gymnogyps californianus*). *Environ. Sci. Technol.*, **44**, 2639–2647.

75 Kylander, M.E., Weiss, D.J., Jeffries, T.E., Kober, B., Dolgopolova, A., Garcia-Sanchez, R., and Coles, B.J. (2007) A rapid and reliable method for Pb isotopic analysis of peat and lichens by laser ablation–quadrupole-inductively coupled plasma-mass spectrometry for biomonitoring and sample screening. *Anal. Chim. Acta*, **582**, 116–124.

76 Ghazi, A.M. and Millette, J.R. (2004) Environmental forensic application of lead isotope ratio determination: a case study using laser ablation sector ICP-MS. *Environ. Forensics*, **5**, 97–108.

77 Cloquet, C., Carignan, J., and Libourel, G. (2006) Atmospheric pollutant dispersion around an urban area using trace metal concentrations and Pb isotopic compositions in epiphytic lichens. *Atmos. Environ.*, **40**, 574–587.

78 Marshall, W.A., Clough, R., and Gehrels, W.R. (2009) The isotopic record of atmospheric lead fall-out on an Icelandic salt marsh since AD 50. *Sci. Total Environ.*, **407**, 2734–2748.

79 Zhu, Y., Kashiwagi, K., Sakaguchi, M., Aoki, M., Fujimori, E., and Haraguchi, H. (2006) Lead isotopic compositions of atmospheric suspended particulate matter in Nagoya City as measured by HR-ICP-MS. *J. Nucl. Sci. Technol.*, **43**, 474–478.

80 Soto-Jiménez, M.F. and Flegal, A.R. (2009) Origin of lead in the Gulf of California ecoregion using stable isotope analysis. *J. Geochem. Explor.*, **101**, 209–217.

81 Wehmeier, S., Ellam, R., and Feldmann, J. (2003) Isotope ratio determination of antimony from the transient signal of trimethylstibine by GC–MC-ICP-MS and GC–ICP-TOF-MS. *J. Anal. At. Spectrom.*, **18**, 1001–1007.

82 Luo, Y., Dabek-Zlotorzynska, E., Celo, V., Muir, D.C.G., and Yang, L. (2010) Accurate and precise determination of silver isotope fractionation in environmental samples by multicollector-ICPMS. *Anal. Chem.*, **82**, 3922–3928.

83 Yang, L., Dabek-Zlotorzynska, E., and Celo, V. (2009) High precision determination of silver isotope ratios in commercial products by MC-ICP-MS. *J. Anal. At. Spectrom.*, **24**, 1564–1569.

84 Yin, R., Feng, X., and Shi, W. (2010) Application of the stable-isotope system to the study of sources and fate of Hg in the environment: a review. *Appl. Geochem.*, **25**, 1467–1477.

85 Gao, B., Liu, Y., Sun, K., Liang, X., Peng, P., Sheng, G., and Fu, J. (2008) Precise determination of cadmium and lead isotopic compositions in river sediments. *Anal. Chim. Acta*, **612**, 114–120.

86 Cloquet, C., Carignan, J., and Libourel, G. (2006) Isotopic composition of Zn and Pb atmospheric depositions in an urban/periurban area of northeastern France. *Environ. Sci. Technol.*, **40**, 6594–6600.

87 Mattielli, N., Petit, J.C.J., Deboudt, K., Flament, P., Perdrix, E., Taillez, A., Rimetz-Planchon, J., and Weis, D. (2009) Zn isotope study of atmospheric emissions and dry depositions within a 5 km radius of a Pb–Zn refinery. *Atmos. Environ.*, **43**, 1265–1272.

88 Thapalia, A., Borrok, D.M., Van Metre, P.C., Musgrove, M., and Landa, E.R. (2010) Zn and Cu isotopes as tracers of anthropogenic contamination in a sediment core from an urban lake. *Environ. Sci. Technol.*, **44**, 1544–1550.

89 Shiel, A.E., Weis, D., and Orians, K.J. (2010) Evaluation of zinc, cadmium and lead isotope fractionation during smelting and refining. *Sci. Total Environ.*, **408**, 2357–2368.

90 Bandura, D.R., Baranov, V.I., and Tanner, S.D. (2000) Effect of collisional damping and reactions in a dynamic reaction cell on the precision of isotope ratio measurements. *J. Anal. At. Spectrom.*, **15**, 921–928.

91 Resano, M., Marzo, P., Pérez-Arantegui, J., Aramendía, M., Cloquet, C., and Vanhaecke, F. (2008) Laser ablation–inductively coupled plasma-dynamic reaction cell-mass spectrometry for the determination of lead isotope ratios in ancient glazed ceramics for discriminating purposes. *J. Anal. At. Spectrom.*, **23**, 1182–1191.

92 Feng, X., Foucher, D., Hintelmann, H., Yan, H., He, T., and Qiu, G. (2010) Tracing mercury contamination sources in sediments using mercury isotope compositions. *Environ. Sci. Technol.*, **44**, 3363–3368.

93 Cloquet, C., Carignan, J., Lehmann, M. F., and Vanhaecke, F. (2008) Variation in the isotopic composition of zinc in the natural environment and the use of zinc isotopes in biogeosciences: a review. *Anal. Bioanal. Chem.*, **390**, 451–463.

94 Santamaria-Fernandez, R., Hearn, R., and Wolff, J.C. (2008) Detection of counterfeit tablets of an antiviral drug using ^{34}S measurements by MC-ICP-MS and confirmation by LA-MC-ICP-MS and HPLC–MC-ICP-MS. *J. Anal. At. Spectrom.*, **23**, 1294–1299. (http://pubs.rsc.org/en/content/articlelanding/2008/ja/b802890g)

95 Santamaria-Fernandez, R., Hearn, R., and Wolff, J.C. (2009) Detection of counterfeit antiviral drug Heptodin and classification of counterfeits using isotope amount ratio measurements by multicollector inductively coupled plasma mass spectrometry (MC-ICPMS) and isotope ratio mass spectrometry (IRMS). *Sci. Justice*, **49**, 102–106.

96 Santamaria-Fernandez, R. and Wolff, J.C. (2010) Application of laser ablation multicollector inductively coupled plasma mass spectrometry for the measurement of calcium and lead isotope ratios in packaging for discriminatory purposes. *Rapid Commun. Mass Spectrom.*, **24**, 1993–1999.

97 Szynkowska, M.I., Czerski, K., Paryjczak, T., and Parczewski, A. (2010) Ablative analysis of black and colored toners using LA–ICP-TOF-MS for the forensic discrimination of photocopy and printer toners. *Surf. Interface Anal.*, **42**, 429–437.

98 Santamaria-Fernandez, R., Giner Martínez-Sierra, J., Marchante-Gayón, J.M., García-Alonso, J.I., and Hearn, R. (2009) Measurement of longitudinal sulfur isotopic variations by laser ablation MC-ICP-MS in single human hair strands. *Anal. Bioanal. Chem.*, **394**, 225–233.

99 Fontaine, G.H., Hametner, K., Peretti, A., and Günther, D. (2010) Authenticity and provenance studies of copper-bearing andesines using Cu isotope ratios and element analysis by fs-LA–MC-ICPMS and ns-LA–ICPMS. *Anal. Bioanal. Chem.*, **398**, 2915–2928.

100 O'Connor, C., Sharp, B.L., and Evans, P. (2006) On-line additions of aqueous standards for calibration of laser ablation inductively coupled plasma mass spectrometry: theory and comparison of wet and dry plasma conditions. *J. Anal. At. Spectrom.*, **21**, 556–565.

101 Resano, M., Marzo, M.P., Alloza, R., Saénz, C., Vanhaecke, F., Yang, L., Willie, S., and Sturgeon, R.E. (2010) Laser ablation single-collector inductively coupled plasma mass spectrometry for lead isotopic analysis to investigate evolution of the Bilbilis mint. *Anal. Chim. Acta*, **677**, 55–63.

15 Nuclear Applications

Scott C. Szechenyi and Michael E. Ketterer

15.1 Introduction

Both quadrupole-based and sector field inductively coupled plasma mass spectrometry (ICP-MS) find extensive use in nuclear applications. The latter has been applied with both single- and multi-collector ion detection systems. ICP-MS is particularly advantageous in the determination of long-lived radionuclides, but also finds applications in the determination of isotope ratios such as $^{241}Pu/^{239}Pu$ that involve (an) isotope(s) having a shorter half-life ($T_{½}$). Considerable overlap exists between the application ranges of mass spectrometry (MS) and decay-counting techniques in the context of nuclear applications. ICP-MS is strongly advantageous for the determination of isotopes with a $T_{½}$ exceeding 10^6 years, but neither ICP-MS nor any other form of MS can be expected to displace decay-counting measurements for very short-lived isotopes. This chapter presents several examples of ICP-MS-based isotope ratio measurements for nuclear science applications in (i) monitoring of Pu isotopic compositions in new and spent nuclear fuel, (ii) provenancing of sources of U and Pu in the environment, (iii) using U, Pb, and Sr isotope ratios as geolocators for investigating the origin of U ores, and (iv) determining the age of Pu in nuclear materials and environmental samples.

15.2 Rationale

The determination of naturally occurring and synthetic radionuclides at low, environmentally relevant activities, and also higher, process-level activities, is most often undertaken by decay-counting radiometric techniques (e.g., alpha spectrometry, liquid scintillation counting, and gamma spectrometry). Nevertheless, MS has extensive applicability in nuclear measurements and has long been employed as a complementary approach. Even during the early development of ICP-MS, the potential applicability in the nuclear industry was recognized [1]. The

Isotopic Analysis: Fundamentals and Applications Using ICP-MS,
First Edition. Edited by Frank Vanhaecke and Patrick Degryse.

Published 2012 by WILEY-VCH Verlag GmbH & Co. KGaA

decay-counting and MS techniques are fundamentally different in one major respect, as the former detects the analyte by recording a particle or photon associated with the nuclear decay. For decay-counting measurement, the sensitivity and measurement time required are thus closely associated with the activity being measured and the relevant half-lives. In contrast, MS is an atom-counting approach that counts the atoms themselves, irrespective of their decay mode, specific activity, or half-life. The efficacy of the MS measurement is determined largely by the number of atoms, along with the relevant mass spectral background, background introduced by chemical sample preparation, the influence of neighboring masses (i.e., abundance sensitivity; see also Chapter 2), and the presence of any interfering ions.

The fundamental distinction between decay counting and atom counting implies that there are many applications where decay counting is preferred or mandatory, such as for the measurement of short-lived isotopes with high specific activities and, therefore, typically fewer atoms available for measurement. For example, in a nuclear accident scenario, ^{131}I is a relevant fission product that definitively indicates recent release of reactor or bomb debris, but it would be out of the pragmatic realm to use MS in the determination of an isotope with a half-life of only 8 days. At the other extreme, however, the measurement of a very long-lived radioisotope by decay-counting approaches is constrained by the low probability of decay on the counting time scale (minutes to weeks), and in this situation, MS is faster (a few minutes per determination), more practical, and considerably more sensitive. Although decay-counting procedures are still used in some laboratories for extremely long-lived nuclides such as ^{232}Th and ^{238}U ($T_{½} = 1.4 \times 10^{10}$ and 4.5×10^9 years, respectively), MS is much preferred in these situations.

The trade-offs between decay-counting and atom-counting methods imply considerable overlap of the two approaches, with isotopes of intermediate half-life (10^1–10^6 years) being routinely determined by either method. The trade-offs and relative merits of both approaches have been discussed elsewhere in recent reviews, for example, by Henry *et al.* [2], Becker [3, 4], Larivière *et al.* [5], Ketterer and Szechenyi [6], and Hou and Roos [7]. Several different MS techniques have been commonly applied in nuclear analysis, including thermal ionization mass spectrometry (TIMS), resonance ionization mass spectrometry (RIMS), accelerator mass spectrometry (AMS), glow discharge mass spectrometry (GDMS), and secondary ionization mass spectrometry (SIMS). The technique of ICP-MS, the focus in this book, is prominently positioned versus these other, not necessarily as a result of its superior performance, but rather because it is advantageous with respect to its pragmatic attributes. The main advantages of ICP-MS lie in its combination of relatively straightforward sample introduction requirements (most commonly, introduction of aqueous solutions by pneumatic nebulization), broad elemental coverage, high sample throughput, relatively low capital and operating costs, and less demanding operator skill requirements [6]. A list of intermediate- and long-lived radionuclides for which measurement by ICP-MS is commonplace would include both naturally occurring and synthetic isotopes. This list includes

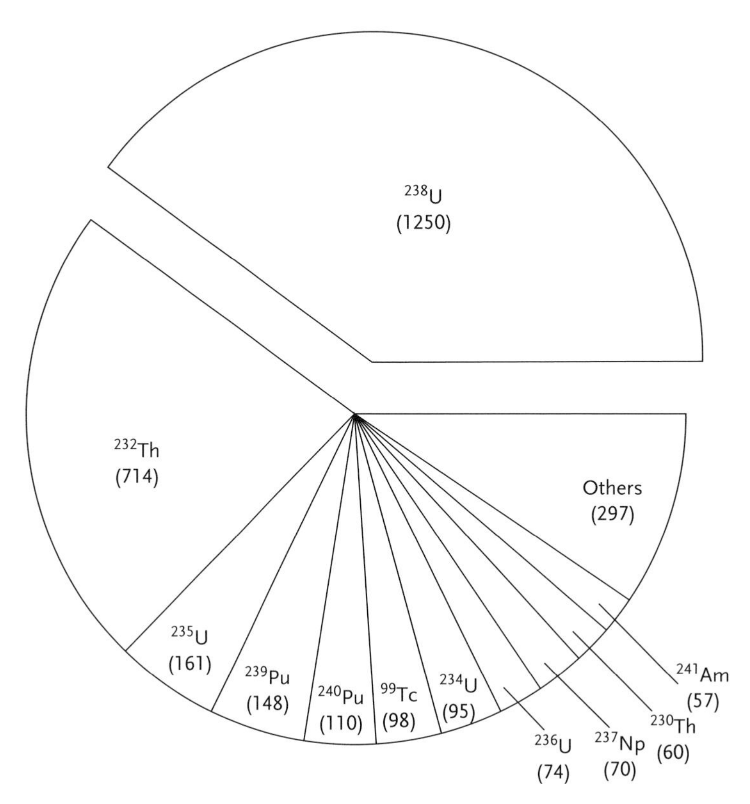

Figure 15.1 Distribution of papers published in the period 1989–2005 using ICP-MS for radionuclide determination by the isotope measured [5].

the isotopes ^{238}U, ^{235}U, ^{234}U, ^{232}Th, ^{231}Pa, ^{230}Th, and ^{226}Ra that are all naturally occurring, the isotopes ^{239}Pu, ^{237}Np, ^{236}U, ^{129}I, and ^{99}Tc that are largely synthetic, but also occur in ultra-trace quantities naturally, and the synthetic isotopes ^{242}Pu, ^{241}Pu, ^{241}Am, and ^{240}Pu. Figure 15.1, taken from Larivière *et al.* [5], depicts the distribution of published studies describing measurements of radionuclides by ICP-MS in the period 1989–2005. The publications represented in Figure 15.1 reflect both high-activity nuclear industry/process-related analyses, along with studies of radionuclides present at much lower activities for environmental monitoring applications. Between 2005 and the present, the situation is relatively similar, but has perhaps evolved somewhat more in the direction of studies focusing on low activities in environmental media.

Since the aim of this book is to describe the fundamental principles and major application areas of ICP-MS-based *isotope ratio* measurements, this specific chapter will focus on examples in which *isotope ratios* are being measured in nuclear applications, as opposed to measurements of solely *activities* or *concentrations* of a single isotope, although it is often relevant to determine both. As the literature base is very extensive and a variety of existing reviews have covered the broad subject of the use of ICP-MS in radionuclide analysis, the intent herein is to pursue a *tutorial* approach, namely, to provide descriptions of a few specific application topics that will give the reader a good sense of what has been done and what is possible. The remainder of this chapter follows this tutorial model by describing examples of published studies using ICP-MS isotope ratio measurements in the following three topical areas: process control and monitoring in the nuclear industry (Section 15.3), isotopic studies of the distribution of U and Pu in the environment (Section 15.4), and nuclear forensics and attribution of the origin of nuclear materials (Section 15.5).

15.3
Process Control and Monitoring in the Nuclear Industry

Isotope ratio measurements are of great relevance in monitoring materials and processes in the nuclear industry, in terms of both examining the isotopic compositions of input fuels and determining the isotopic characteristics of "spent" fuel post-irradiation. Knowledge of the fuel composition pre- and post-irradiation is essential for licensing, safeguards programs, and improving the industry's understanding of reactor operational processes. In order to achieve these goals, precise and accurate isotope ratio measurements, particularly of U and Pu, along with several fission product elements, are essential. Depending on the reactor type (light-water versus heavy-water), the U used in the fuel may be naturally occurring U (0.72% ^{235}U) or enriched U (>0.72% ^{235}U). Different degrees of enrichment are used in different types of reactors. In the USA, for instance, low-enriched U containing ~3% ^{235}U is used as a fuel in light-water reactors, whereas highly enriched uranium (HEU) with >90% ^{235}U fuels propulsion reactors in the US Navy submarine fleet. A solid understanding of nuclear engineering can be found in the book by Lamarsh and Baratta [8].

Plutonium can be formed in nuclear fuel as a byproduct of neutron capture by ^{238}U or it can be a pre-existing ingredient when mixed oxide fuel (MOX) is used. The most important Pu isotopes investigated in spent fuels are ^{238}Pu, ^{239}Pu, ^{240}Pu, ^{241}Pu, and ^{242}Pu. Classically, decay-counting methods such as alpha spectrometry have been applied to studies of Pu isotopic composition in spent nuclear fuel. In this context, alpha spectrometry is limited by the need for complex spectral deconvolution approaches owing to difficulties in resolving alpha particles of similar energies from ^{239}Pu and ^{240}Pu, and it provides incomplete information because one important isotope, ^{241}Pu, undergoes beta (and not alpha) decay. TIMS is well known as a routine procedure for determination of Pu isotopic compositions in MOX and spent fuels [9]. However, both sample preparation and the actual measurement for TIMS procedures are typically much more time consuming than those for ICP-MS and multi-collector (MC) ICP-MS, leading to increasing interest in application of ICP-MS.

In a seminal study, Günther-Leopold *et al.* [10] used an on-line chromatographic separation system combined with MC-ICP-MS for determination of Pu isotope ratios in nuclear fuel samples. The determination of the five relevant Pu isotopes necessitates separation and removal of U, the major mass component of both new and spent fuel. In ICP-MS measurements, the presence of ^{238}U is detrimental because of the direct isobaric interference of $^{238}U^+$ on $^{238}Pu^+$ and the formation of interfering polyatomic ions, such as $^{238}U^1H^+$ [11]. It is also necessary to eliminate ^{241}Am as a possible interference on ^{241}Pu. Günther-Leopold *et al.* [10] developed an effective, on-line separation of U and Pu based upon a Dionex CS5A mixed-bed high-resolution ion-exchange column, offering both anion- and cation-exchange capacity. The on-line separation permitted remote handling of high-level samples and emplacement of the chromatographic apparatus, and also the ICP-MS sample introduction system, within a glove-box enclosure. The use of MC-ICP-MS in this context allowed the simultaneous monitoring of all isotopes with a multiple Faraday collector array. With a dwell time of 1.049 s per data point, it was possible to obtain simultaneous chromatographic traces for each isotope. Utilizing $HClO_4$ with Pu adsorbed on a CS5A column, Pu is converted to cationic Pu(VI) in the form of $PuO_2{}^{2+}$ (aq.). Pu is eluted with 0.4 M HNO_3, whereas U is more strongly retained, ultimately being eluted from the column with 1 M HCl. Instrumental mass discrimination is a well-known source of systematic error in (MC-)ICP-MS ratio measurements (see also Chapter 5). Since no constant natural ratio can be used for corrections in Pu measurements, Günther-Leopold *et al.* [10] used a bracketing approach with injections of a Pu Certified Reference Material solution (NIST SRM 948), interspersed with analyses of unknown samples. Sample solutions contained ~100–1000 ng of total Pu, which is similar to the masses analyzed for stable element ratios by MC-ICP-MS. An example of a Pu isotopic composition in a MOX sample reported by Günther-Leopold *et al.* [10] is ^{238}Pu 3.667 ± 0.037% (one standard deviation of the results), ^{239}Pu 34.648 ± 0.039%, ^{240}Pu 35.716 ± 0.031%, ^{241}Pu 13.192 ± 0.013%, and ^{242}Pu 12.776 ± 0.018%. The authors indicated that these results were satisfactory for routine use in the nuclear industry applications aimed at, and constituted a greatly simplified approach compared with off-line separation and traditional analysis by TIMS.

A recent study by Isnard *et al.* [12] compared the use of TIMS and MC-ICP-MS for the determination of Cs isotope ratios in spent nuclear fuels. Cs isotopic information is relevant in investigations of reactor processes and in the assessment of burn-up of the fuel – slightly different Cs isotopic compositions are anticipated for the fission of ^{235}U versus ^{239}Pu or ^{241}Pu. In this isotopic analysis application, Cs has four relevant isotopes: ^{133}Cs (stable), ^{134}Cs ($T_{½} = 2.07$ years), ^{135}Cs ($T_{½} = 2 \cdot 10^6$ years) and ^{137}Cs ($T_{½} = 30.1$ years). Isnard *et al.* [12] utilized an MC-ICP-MS unit modified to handle radioactive samples that was equipped with a PFA microconcentric nebulizer–quartz spray chamber combination for sample introduction and nine Faraday collectors. This mass region suffers from spectral interference from stable Ba, hence the Cs was purified by liquid chromatographic methods prior to analysis. The authors noted that Cs is somewhat problematic with respect to memory effects in the sample introduction system and therefore care was taken to rinse adequately between samples. An interpolative method for determining mass bias correction factors based upon such factors determined for both a lighter element (Sb) and a heavier element (Eu) was used. Excellent agreement was observed between TIMS and MC-ICP-MS results. Both methods realized uncertainties of 0.2–0.5% for $^{133}Cs/^{135}Cs$ and $^{137}Cs/^{135}Cs$. The Cs isotopic data could also be used for determining fuel burn-up via measurements of $^{135}Cs/^{238}U$ atom ratios with a global uncertainty of 0.5% (with a coverage factor $k=2$).

15.4
Isotopic Studies of the Distribution of U and Pu in the Environment

A strong point of both single- and multi-collector ICP-MS is their ability to investigate the isotopic compositions of the elements U and Pu in environmental settings. Naturally, other techniques such as TIMS and AMS are also applicable to these analytical problems. However, the widespread availability of ICP-MS, its relative ease of use, high sample throughput capability, relatively low capital and per-analysis costs, and fairly straightforward sample introduction needs imply that ICP-MS is advantageous in this type of work. These points have been discussed in more detail in relevant reviews [6,7,13].

In Nature, U has a relatively fixed isotopic composition of 0.0055% ^{234}U, 0.72% ^{235}U, and 99.27% ^{238}U (see also Chapter 1). In natural systems, the $^{238}U/^{235}U$ ratio has been found to be affected by isotope fractionation to a limited extent of ~2‰ total range [14]. Small traces of ^{236}U are found in Nature, particularly in U ores [15], although the abundances of ^{236}U are $\sim 10^{-8}$% or smaller. It follows that, for environmental monitoring studies, the presence of a significant fraction of ^{236}U is specifically associated with anthropogenic nuclear contamination [16]. Natural variation in $^{234}U/^{238}U$ is commonplace, in association with disequilibria between ^{234}U and ^{238}U resulting from alpha recoil or selective leaching processes in open-system surficial environments [17]. The alpha recoil mechanism produces excess activity of ^{234}U relative to ^{238}U in aqueous phases in contact with U-bearing mineral grains. In this process, a fraction of the ^{234}Th decay product atoms are ejected into the aqueous phase, where they quickly decay further to ^{234}U. In geological settings and time frames where natural waters have sufficient residence times, aqueous phase activities of ^{234}U may also be enhanced due to preferential leaching. Preferential leaching of ^{234}U occurs at damaged crystal lattice sites as the fractured lattice exposes internal ^{234}U atoms to solution uptake in a higher proportion than would otherwise be available on the surface of the crystal. In addition, the ejected ^{234}U atoms are in the +VI oxidation state, which is a more soluble form than the ^{238}U atoms present in the +IV state in minerals such as uraninite (UO_2). Aside from these effects, deviations in $^{238}U/^{235}U$ and/or the presence of significantly elevated $^{236}U/^{238}U$ ratios represent unequivocal environmental probes for contamination from anthropogenic sources of "enriched" U ($>0.72\%$ ^{235}U) or "depleted" U ($<0.72\%$ ^{235}U). The use of isotopic analysis is generally much more sensitive than looking for a simple elevation in concentration or activity, even within the limitations of single-collector quadrupole-based ICP-MS, which can only realistically distinguish $^{238}U/^{235}U$ deviations from Nature at the level of $\sim 0.5\%$ or higher, and a very refined level of discrimination is possible.

In studies of the distribution of depleted uranium (DU) in soils at Colonie, NY, USA, Lloyd *et al.* [18] employed quadrupole-based ICP-MS and MC-ICP-MS for measurements of $^{234}U/^{238}U$, $^{235}U/^{238}U$, and $^{236}U/^{238}U$ isotope ratios. Colonie was the site of a facility that produced U metal by reduction of UF_4 and machined metal articles from DU metal from 1958 to 1984. Processes at the Colonie facility released finely divided U aerosols that accumulated in soils surrounding the site. For analysis by quadrupole-based ICP-MS, soil samples were subjected to total dissolution in a microwave digestion system with HNO_3, HF, $HClO_4$, and H_2O_2, followed by evaporation and re-dissolution in 0.16 M HNO_3. Digests were diluted to achieve $0.6–1.0 \times 10^6$ counts s^{-1} for ^{238}U. Selected samples were analyzed by MC-ICP-MS following trace/matrix separation with UTEVA chromatographic resin. In a related study, the same group also used laser ablation (LA)–ICP-MS for measurement of U isotope ratios in selected individual environmental particles from Colonie [19]. Lloyd *et al.* [18] observed excellent agreement between quadrupole-based ICP-MS and MC-ICP-MS results, with the latter [19] exhibiting ~ 2.5-fold better measurement precision. The minimum detectable $^{236}U/^{238}U$ ratio by quadrupole-based ICP-MS was estimated as 2×10^{-6} under conditions where all U signals were maintained within the pulse-counting mode of the electron multiplier. Figure 15.2 depicts results from the work of Lloyd *et al.* [18]. The soil sample ratios definitely reflect a two-component mixing process between naturally occurring U and DU contamination. The isotopic composition of the DU end-member was estimated as $^{234}U/^{238}U = (7.3 \pm 0.3) \times 10^{-6}$, $^{235}U/^{238}U = 0.00205 \pm 0.00006$ and $^{236}U/^{238}U = (3.2 \pm 0.1) \times 10^{-5}$. The $^{234}U/^{238}U$ versus

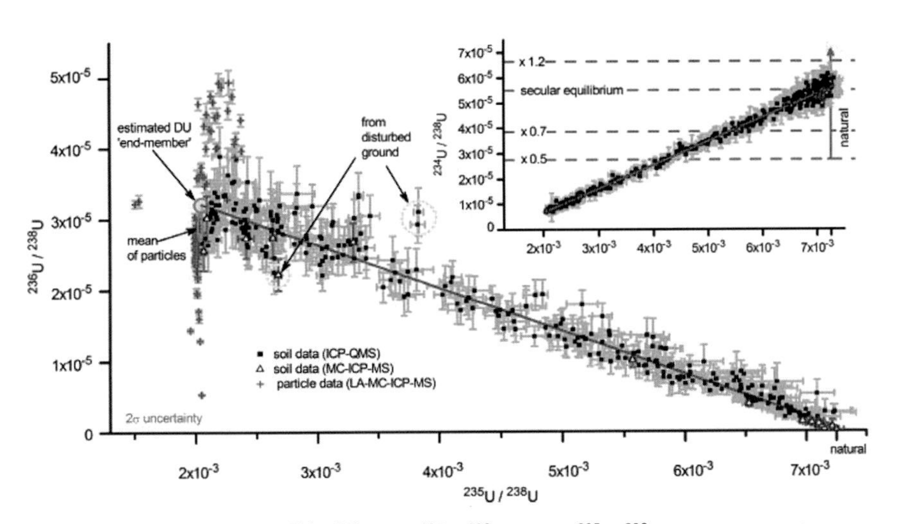

Figure 15.2 Mixing plots of $^{236}U/^{238}U$ and $^{234}U/^{238}U$ versus $^{235}U/^{238}U$ for soils from the vicinity of Colonie, NY, indicating two-component mixing between naturally occurring U and DU end-members [18].

$^{235}U/^{238}U$ mixing plot exhibits some scatter in $^{234}U/^{238}U$ for the naturally occurring U end-member, as ^{234}U and ^{238}U are not expected to be in perfect secular equilibrium ($^{234}U/^{238}U = 0.000055$) in surface soils. The presence of a single straight-line mixing relationship for $^{236}U/^{238}U$ versus $^{235}U/^{238}U$ is remarkable, considering the possibility that different DU blends could have contained widely varying levels of ^{236}U, depending on whether or not some or all of the DU had been produced from "recycled" reactor-irradiated U. The methodology was highly effective for mapping the contaminant dispersion emanating from the facility, with relatively small DU inputs of $\sim 0.05\ \mu g\ g^{-1}$ being detectable when added on top of a baseline of $1.05 \pm 0.06\ \mu g\ g^{-1}$ of naturally occurring U. The authors also underscored the efficacy of quadrupole-based ICP-MS, with which unseparated samples were analyzed in a "dilute and shoot" mode, without the need for the time-consuming U isolation via column separation as performed prior to MC-ICP-MS analysis.

Howe *et al.* [20] examined enriched U in sediments collected near a U enrichment plant at Capenhurst, UK. They utilized quadrupole-based ICP-MS to measure $^{238}U/^{235}U$ in fractions separated by anion exchange with Dowex AG1-X8 resin. Total U concentrations ranged from $24.1\ \mu g\ g^{-1}$ at a distance of 0.3 km downstream of the site to $<1\ \mu g\ g^{-1}$ U at ≥ 4 km from the enrichment facility. Sequential extractions were also used to demonstrate that the enriched U was present in all fractions, but was more pronounced in the less labile phases of the sediment, leading to the finding that the enriched U was contained in more refractory forms than the naturally occurring background U.

Another study identifying enriched U in the ambient environment near a nuclear facility was reported by Bellis *et al.* [21]. They used digested samples of tree bark as reporters of particulate U released from the Tokai-Mura nuclear facility in Japan. A quadrupole-based ICP-MS system was used to measure $^{235}U/^{238}U$ ratios in the bark samples. Controls from locations at Yakushima and Tokyo exhibited naturally occurring U with $^{235}U/^{238}U = 0.00725$, whereas samples from the Tokai-Mura vicinity exhibited $^{235}U/^{238}U$ values of up to 0.0145, signifying mixing between naturally occurring U and enriched U.

In contrast to U, Pu is almost entirely of synthetic origin, though very small amounts are also identifiable in U ores [22]. As a result, the isotopic composition of Pu varies according to its source and mode of production. Environmental Pu exhibits a relatively consistent worldwide signature in surface soils, resulting from 1950–1960s stratospheric fallout, a consequence of atmospheric nuclear weapons tests [23]. The average composition in the 31–70°N zone of the Northern Hemisphere is $^{240}Pu/^{239}Pu = 0.180 \pm 0.014$, $^{241}Pu/^{239}Pu = 0.00194 \pm 0.00028$ and $^{242}Pu/^{239}Pu = 0.00387 \pm 0.00071$, with uncertainties all expressed as 2SD errors [23]. The $^{241}Pu/^{239}Pu$ ratio is continuously decreasing with time, due to the 14.4 year half-life of ^{241}Pu with 0.00194 pertaining to the 1 January 2000 reference date quoted by Kelley *et al.* [23]. Surface inventories of Pu in soil vary fairly widely, but the global fallout background generates ^{239}Pu concentrations on the order of $\sim 1\ pg\ g^{-1}$ for mid-latitude surface soils in the Northern Hemisphere. The use of isotopic analysis, therefore, is important in distinguishing the presence of specific local or regional Pu sources that contrast with this worldwide, non-zero baseline signal.

In studies of the influence of Chernobyl-derived contamination in surface soils from Poland, Ketterer *et al.* [24] used single-collector sector field ICP-MS to determine the Pu isotope ratios $^{240}Pu/^{239}Pu$, $^{241}Pu/^{239}Pu$, and $^{242}Pu/^{239}Pu$ in NdF_3–Pu coprecipitated alpha sources. These alpha sources were dissolved in 16 M HNO_3, after which Pu was isolated using EIChrom TEVA resin. An ultrasonic nebulizer system was used to enhance the signal intensities for the low solution concentrations of Pu. The Pu isotope ratios ranged from values congruent with a global fallout-only signature (observed in southern Poland) to the elevated values of $^{240}Pu/^{239}Pu$, $^{241}Pu/^{239}Pu$, and $^{242}Pu/^{239}Pu$ found in surface soils from the Chernobyl-impacted zone in northeastern Poland. As such, environmental Pu in Poland could be construed as resulting from contributions from both the global fallout end-member [23] and an isotopically heavier Chernobyl end-member [25]. Since the original NdF_3 alpha source preparation had employed ^{236}Pu rather than the more commonly used ^{242}Pu as yield tracer, the indigenous $^{242}Pu/^{239}Pu$ ratios could be investigated, although some of the $^{242}Pu/^{239}Pu$ ratio measurements were affected by ^{242}Pu yield tracer contamination in the sample introduction system and/or unknown polyatomic ion interferences. Nevertheless, a well-defined $^{241}Pu/^{239}Pu$ versus $^{240}Pu/^{239}Pu$ mixing plot was obtained (Figure 15.3) and a mixing model generated similar apportionments as had been found previously through modeling with $^{238}Pu/^{239+240}Pu$ activity ratios from alpha spectrometry [26].

The use of ICP-MS-based Pu isotope ratios is also well grounded in studies of other settings with local or regional contributions from non-fallout Pu. One well-known example is the mixing between global fallout Pu and isotopically heavier Pu from the Pacific Proving Ground (PPG). Yamada *et al.* [27] demonstrated the

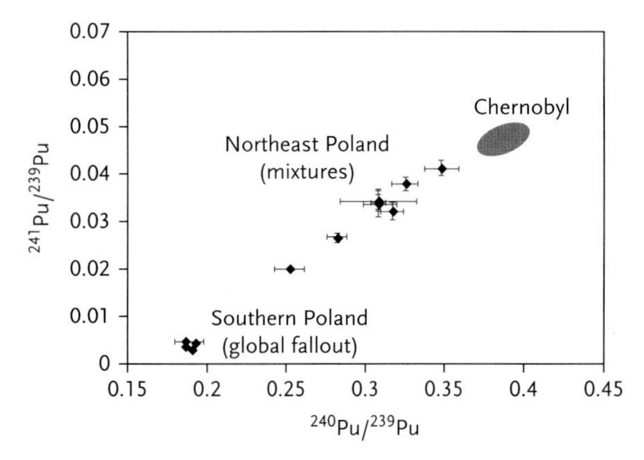

Figure 15.3 Mixing plot of $^{241}Pu/^{239}Pu$ versus $^{240}Pu/^{239}Pu$ for Polish forest soils; error bars are ± 1 standard deviation, based upon results in Ketterer *et al.* [24]. The Chernobyl end-member is based upon results in Muramatsu *et al.* [25].

widespread presence of PPG-Pu dissolved in surface waters, in the marine water column, and in sediments in widespread locations along the Pacific Margin. Another example is the study of Pu isotopes in sediment cores from the Sea of Okhotsk and Pacific Ocean locations near the Japanese coast [28]. The PPG fallout is characterized by a relatively higher $^{240}Pu/^{239}Pu$ ratio of 0.30–0.36 [29]. Zheng and Yamada [28] used a single-collector sector field ICP-MS unit equipped with a concentric glass nebulizer and a Scott-type spray chamber. Pu was isolated by anion-exchange methods and a Pu isotopic standard (NIST SRM 947) was used with sample—standard bracketing to correct externally for instrumental mass bias in the $^{240}Pu/^{239}Pu$ measurements. The results presented by Zheng and Yamada [28] demonstrate that the Pu near the Japanese coast is comprised of mixtures of global fallout Pu with PPG-Pu (Figure 15.4). All core intervals have $^{240}Pu/^{239}Pu$ systematically higher than the global fallout range of 0.180 ± 0.014 [23]. However, the PPG influence was very limited in the Sea of Okhotsk, where the Pu was essentially congruent with global fallout.

Other studies have used ICP-MS to investigate mixing between global fallout Pu and regional North American fallout from the Nevada Test Site (NTS) in western USA [30]. The NTS end-member is characterized by lower values of $^{240}Pu/^{239}Pu$ and $^{241}Pu/^{239}Pu$ relative to global fallout. LA was used with single-collector sector field ICP-MS to re-analyze previously prepared electrodeposited Pu alpha sources. Although the sources had aged 9–10 years, with significant decay of ^{241}Pu to ^{241}Am having occurred, LA–ICP-MS effectively measured a $(^{241}Pu + ^{241}Am)/^{239}Pu$ atom ratio, exhibiting a clear mixing relationship between global fallout and NTS-Pu. The $(^{241}Pu + ^{241}Am)/^{239}Pu$ atom ratios measured by LA–ICP-MS in 2007 were interpreted as representing the $^{241}Pu/^{239}Pu$ ratios present when the alpha sources were originally prepared. Cizdziel *et al.* [30] also demonstrated good

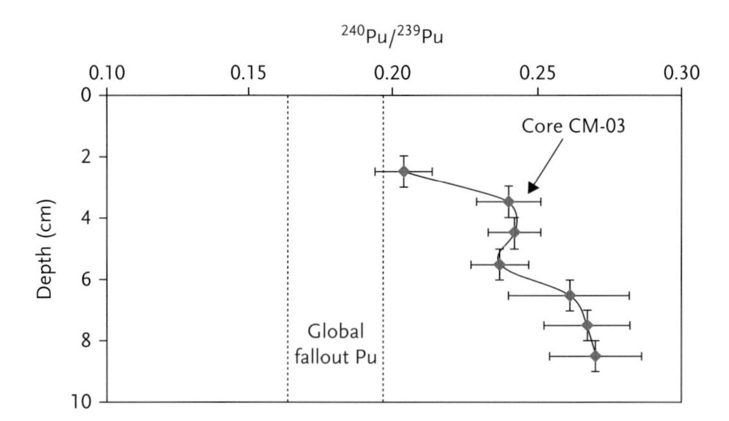

Figure 15.4 Plot of the vertical depth distributions of $^{240}Pu/^{239}Pu$ in sediment core CM-03 in the Pacific Ocean near Honsu Island, (35°59′N and 141°47.9′E, at a water depth of 2323 m). Error bars are ± 1 standard deviation of the ICP-MS measurements. Results given in Zheng and Yamada [28]. The global fallout range is from Kelley *et al.* [23].

agreement between the $^{239+240}Pu$ activities (using a ^{242}Pu spike present in the alpha source) measured by LA–ICP-MS and the alpha spectrometric $^{239+240}Pu$ activity originally obtained.

15.5 Nuclear Forensics

The field of nuclear forensics is now fairly well established [31]. The main objective of nuclear forensics is to detect illicit, undeclared activity associated with the production, transport, or deployment of nuclear materials [32]. Examples of illicit activities could consist of the clandestine operation of U isotope enrichment processes, the extraction of Pu from irradiated U fuel rods removed from power production reactors, or the construction of a "dirty bomb" device. Within these objectives, individual cases often require the interrogation of raw samples of nuclear materials or environmental samples, such as wipes, dust, soil, water, or vegetation [33]. These samples are analyzed in order to address questions about origin, age, mode of production, identities of parties responsible, and device design. Although some analytical procedures are employed immediately or within short periods of time in the field, other investigations are undertaken over the following weeks or months within laboratory settings and it is there that MS has a critical role. The evaluation of isotopic signatures of U, Pu, and associated elements is essential to addressing the critical questions in individual case studies and a repertoire of MS techniques have important application niches. In a similar vein to other areas of nuclear applications, ICP-MS-based isotope ratio measurements evolved into robust contributors to nuclear forensic analysis.

U ores represent the initial starting point for raw materials used in the nuclear fuel cycle and for the production of fissile materials, and many of the confirmed incidents of nuclear trafficking investigated by the International Atomic Energy Agency (IAEA) involve low-grade materials such as U ores. U ores vary widely with respect to their geographic origin, mineralogy, and contamination with concomitant elements [34]. Several recent published studies have reported ICP-MS-based measurements of U isotope ratios and also concentrations and/or isotope ratios of concomitant elements, with the objective of developing "geolocators" [35]. U ores exhibit (i) small variances in $^{238}U/^{235}U$, associated with natural fractionation processes [14], (ii) $^{234}U/^{238}U$ disequilibria, and (iii) varying natural signatures of $^{236}U/^{238}U$ [15]. ICP-MS is very feasible for the first two of these isotope ratio measurements, although the abundance sensitivity advantages of AMS make this technique the method of choice for measuring $^{236}U/^{238}U$ at relevant levels of $\sim 10^{-10}$ in U ores.

The use of ICP-MS for the determination of isotope ratios for concomitant elements present in U ores is appealing and has been the subject of recent work [34,36,37]. Elements such as Sr and Pb exhibit well-known variations in isotopic compositions due to radiogenic isotope ingrowth processes (see also Chapter 1), leaving potentially distinctive markers of origin, along with the U minerals.

Pb isotopic analysis is particularly advantageous because it is well established that U ores will have very different Pb isotopic compositions compared with "common" Pb. The U-series end-members ^{206}Pb and ^{207}Pb grow in rapidly from U-series decay [38]. Keegan *et al.* [36] reported quadrupole ICP-MS-based Pb isotope ratio results from three prominent Australian ore bodies, showing vastly different Pb isotopic signatures that could serve as an attribution method. It follows that more refined results with 10–20-fold better precision would be possible using MC-ICP-MS for Pb ratio measurements. A similar study by Svedkauskaite-LeGore *et al.* [37] used MC-ICP-MS to investigate $^{207}Pb/^{206}Pb$ and $^{208}Pb/^{206}Pb$ ratios in 45 samples of U ores and yellow cake concentrates from various worldwide locations, concluding that Pb isotope ratio data would greatly reduce the number of possible origins of any seized U ore materials. In Varga *et al.*'s study [34], MC-ICP-MS data were used to determine the $^{207}Pb/^{206}Pb$-based ages of U ores. $^{87}Sr/^{86}Sr$ results, also obtained by MC-ICP-MS, were used to demonstrate a relatively wide range of ratios for the dataset, ranging from 0.7068 to 0.7606. Varga *et al.* concluded that Sr isotope ratios might shed more light on the geographic origin of the host rock associated with the U deposits; for instance, the Sr isotopic signature of U ores obtained from marine phosphorite deposits would reflect the Sr signature for seawater at the time of the phosphorite precipitation [34].

Isotope ratio measurements are also essential in nuclear forensics applications centered on age determination. As an example, it is critical to determine the "age" of Pu, that is, the period of time that has passed since the material was purified from U and fission products. Age determination of Pu can be based upon several parent–daughter pairs, including ^{238}Pu–^{234}U, ^{239}Pu–^{235}U, ^{240}Pu–^{236}U, and ^{241}Pu–^{241}Am [39], proceeding in a manner analogous to methods commonplace in radiogenic isotope geology. In a study that used single-collector sector field ICP-MS, Nygren *et al.* [40] determined ages of Pu in two Institute for Reference Materials and Measurements (IRMM) reference materials, IRMM081 (^{239}Pu) and IRMM083 (^{240}Pu), in sediment from the Marshall Islands and in a vitreous mineral "trinitite" from the New Mexico Trinity site, where the first nuclear device test was conducted in July 1945. They prepared the samples using fusion with lithium borate, followed by separation using solid-phase extraction chromatography. Spikes of ^{233}U, ^{242}Pu, and ^{243}Am were added at the start of the sample preparation and a self-aspirating glass concentric nebulizer with cyclonic spray chamber was employed in the analysis. It was found that ^{241}Pu–^{241}Am was more effective than ^{240}Pu–^{236}U for age determination, since ICP-MS measurements of ^{236}U were hampered by isobaric $^{235}U^{1}H^{+}$ and/or ^{236}U present in the sample at $t = 0$. The ^{241}Pu–^{241}Am ages agreed well with the known preparation age of IRMM081 and IRMM083, and also the known detonation time frames of the Marshall Islands and Trinity devices.

In a study entitled "Nuclear archeology in a bottle," Schwantes *et al.* [41] utilized quadrupole-based ICP-MS to investigate the isotopic composition and age of a sample of several hundred milligrams of Pu excavated from a waste burial trench at Hanford (WA, USA) in 2004. Isotopic analysis of three independent parent–daughter pairs, ^{241}Pu–^{241}Am, ^{239}Pu–^{235}U, and ^{240}Pu–^{236}U, conducted on the sample in 2007, indicated that the Pu had been separated from the irradiated Hanford reactor fuel 61.6 $\pm$ 4.5 years earlier. The isotopic composition of this material, one of the oldest known samples of Pu, exhibited an extremely low $^{240}Pu/^{239}Pu$ isotope ratio of 0.000363 $\pm$ 0.000021 (2SD), which is much lower than the ratios of $\sim$0.05–0.06 present in Pu used for Cold War-era nuclear devices [23]. This clearly demonstrates the potential in the burgeoning specialty of nuclear archeology.

15.6 Prospects for Future Developments

The favorable attributes of ICP-MS have led to its now prominent role in the measurement of isotope ratios for nuclear applications. The relatively low cost of ICP-MS instrumentation makes it very competitive versus techniques such as AMS and TIMS and its widespread worldwide availability affords many opportunities in a large variety of different laboratories for investigating analytical problems, such as those that have been described here. Most of these applications can be regarded as being "well developed" or "mature," although some room exists for improvements to the measurements. One avenue for extending applicability stems from continued improvements in ion-counting efficiency, that is, the fraction of ions counted. Advancing from efficiencies of $<10^{-6}$ possible with first-generation quadrupole-based ICP-MS instruments marketed in the 1980s, state-of-art sector field ICP-MS systems equipped with high-efficiency sample introduction systems are now quoting efficiencies of up to 5%, rivaling the performance of TIMS in similar applications. ICP-MS still has significant pitfalls that the user must bear in mind, such as abundance sensitivity issues (e.g., in the determination of ^{236}U), spectral overlap (e.g., the interference of $^{238}U^{1}H^{+}$ on ^{239}Pu), relatively pronounced instrumental mass bias, and memory effects in the sample introduction system. It is certainly possible to envision specific extensions of ICP-MS in nuclear analysis if the technique offered "perfect" freedom from spectral interferences and abundance sensitivity encumbrances. Of course, ICP-MS will never displace the need for decay-counting techniques in the determination of short-lived isotopes, hence the user must consider the technique as one part of a broad-based toolbox in the nuclear analysis laboratory. The vast base of literature and the relatively few representative case studies and application areas described here emphasize the broad range of possibilities.

Acknowledgment

The authors dedicate this work to the memory of David C. Gerlach, a quiet contributor and member of the Nuclear Materials Analysis group at Pacific Northwest National Laboratory, who passed before his time and will be missed.

References

1 Blair, P.D. (1986) The application of ICPMS in the nuclear industry. *Trends Anal. Chem.*, **5**, 220–223.

2 Henry, R., Koller, D., Liezers, M., Farmer, O.T., Barinaga, C., Koppenaal, D.W., and Wacker, J. (2000) New advances in inductively coupled plasma-mass spectrometry (ICP-MS) for routine measurements in the nuclear industry. *J. Radioanal. Nucl. Chem.*, **29**, (1), 103–108.

3 Becker, J.S. (2003) Review: mass spectrometry of long-lived radionuclides. *Spectrochim. Acta B*, **58**, 1757–1784.

4 Becker, J.S. (2005) Review: ICPMS and laser ablation ICPMS for isotope analysis of long-lived radionuclides. *Int. J. Mass Spectrom.*, **242**, 183–195.

5 Larivière, D., Taylor, V.F., Evans, R.D., and Cornett, R.J. (2006) Review: radionuclide determination in environmental samples by inductively coupled plasma mass spectrometry. *Spectrochim. Acta B*, **61**, 877–904.

6 Ketterer, M.E. and Szechenyi, S.C. (2008) Review: determination of plutonium and other transuranic elements by inductively coupled plasma mass spectrometry: a historical perspective and new frontiers in the environmental sciences. *Spectrochim. Acta B*, **63**, 719–737.

7 Hou, X. and Roos, P. (2008) Review article: critical comparison of radiometric and mass spectrometric methods for the determination of radionuclides in environmental, biological, and nuclear waste samples. *Anal. Chim. Acta*, **608**, 105–139.

8 Lamarsh J.R. and Baratta, A.J. (2001) *Introduction to Nuclear Engineering*, 3rd edn., Prentice Hall, Upper Saddle River, NJ.

9 Aggarwal, S.K., Kumar, S., Saxena, M. K., Shah, P.M., and Jain, H.C. (1995) Investigations for isobaric interference of ^{238}Pu at ^{238}U during TIMS of U and Pu from the same filament loading. *Int. J. Mass Spectrom. Ion Processes*, **151**, 127–135.

10 Günther-Leopold, I., Kobler Waldis, J., Wernli, B., and Kopatjic, Z. (2005) Measurement of Pu isotope ratios in nuclear fuel samples by HPLC–MC-ICP-MS. *Int. J. Mass Spectrom.*, **242**, 197–202.

11 Ketterer, M.E., Watson, B.R., Matisoff, G., and Wilson, C.G. (2002) Rapid dating of recent aquatic sediments using Pu activities and ^{240}Pu/^{239}Pu as determined by quadrupole ICPMS. *Environ. Sci. Technol.*, **36**, 1307–1311.

12 Isnard, H., Granet, M., Caussignac, C., Ducarme, E., Nonell, A., Tran, B., and Chartier, F. (2009) Comparison of TIMS and MC-ICPMS for Cs isotope ratio measurements. *Spectrochim. Acta B*, **64**, 1280–1286.

13 Kim, C.S., Kim, C.K., Martin, P., and Sansone, U. (2007) Determination of plutonium concentrations and isotope ratio by inductively coupled plasma mass spectrometry: a review of analytical methodology. *J. Anal. At. Spectrom.*, **22**, 827–841.

14 Stirling, C.H., Andersen, M.B., Potter, E.K., and Halliday, A.N. (2007) Low-temperature isotopic fractionation of uranium. *Earth Planet. Sci. Lett.*, **264**, 208–225.

15 Berkovits, D., Feldstein, H., Ghelberg, S., Hershkowitz, A., Navon, E., and Paul, M. (2000) ^{236}U in uranium minerals and standards. *Nucl. Instrum. Methods Phys. Res. B*, **172**, 372–376.

16 Ketterer, M.E., Hafer, K.M., Link, C.L., Royden, C.S., and Hartsock, W.J. (2003) Anthropogenic ^{236}U at Rocky Flats, Ashtabula River harbor, and Mersey estuary: three case studies by sector ICPMS. *J. Environ. Radioact.*, **67**, 191–206.

17 Zielinski, R.A., Chafin, D.T., Banta, E. R., and Szabo, B.J. (1997) Use of ^{234}U and ^{238}U isotopes to evaluate contamination of near-surface groundwater with uranium-mill effluent: a case study in south-central Colorado, USA. *Environ. Geol.*, **32**, 124–136.

18 Lloyd, N.S., Chenery, S.R.N., and Parrish, R.R. (2009) The distribution of depleted uranium contamination in Colonie, NY, USA. *Sci. Total Environ.*, **408**, 397–407.

19 Lloyd, N.S., Parrish, R.R., Horstwood, M.S.A., and Chenery, S.R.N. (2009) Precise and accurate isotopic analysis of microscopic uranium-oxide grains using LA–MC-ICPMS. *J. Anal. At. Spectrom.*, 24, 752–758.

20 Howe, S.E., Davidson, C.M., and McCartney, M. (2002) Determination of U concentration and isotope composition by means of ICP-MS in sequential extracts of sediment from the vicinity of a U enrichment plant. *J. Anal. At. Spectrom.*, **17**, 497–501.

21 Bellis, D., Ma, R., Bramall, N., and McLeod, C.W. (2001) Airborne emission of enriched uranium at Tokai-Mura, Japan. *Sci. Total Environ.*, **264**, 283–286.

22 Curtis D., Fabryka-Martin J., Dixon P., and Cramer, J. (1999) Nature's uncommon elements: plutonium and technetium. *Geochim. Cosmochim. Acta*, **63**, 275–285.

23 Kelley, J.M., Bond, L.A., and Beasley, T. M. (1999) Global distribution of Pu isotopes and ^{237}Np. *Sci. Total Environ.*, **237/238**, 483–500.

24 Ketterer, M.E., Hafer, K.M., and Mietelski, J.W. (2004) Resolving Chernobyl vs. global fallout contributions in soils from Poland using plutonium atom ratios measured by ICPMS. *J. Environ. Radioact.*, **73**, 183–201.

25 Muramatsu, Y., Rühm, W., Yoshida, S., Tagami, K., Uchida, S., and Wirth, E. (2000) Concentrations of ^{239}Pu and ^{240}Pu and their isotopic ratios determined by ICP-MS in soils collected from the Chernobyl 30-km zone. *Environ. Sci. Technol.*, **34**, 2913–2917.

26 Mietelski, J.W. and Was, B. (1995) Plutonium from Chernobyl in Poland. *Appl. Radiat. Isot.*, **46**, 1203–1211.

27 Yamada, M., Zheng, J., and Wang, Z.-L. (2005) ^{137}Cs, $^{239+240}$Pu and ^{240}Pu/^{239}Pu atom ratios in the surface waters of the western North Pacific Ocean, eastern Indian Ocean and their adjacent seas. *Sci. Total Environ.*, **366**, 242–252.

28 Zheng, J. and Yamada, M. (2006) Determination of Pu isotopes in sediment cores in the Sea of Okhotsk and the NW Pacific by sector field ICP-MS. *J. Radioanal. Nucl. Chem.*, **267**, 73–83.

29 Muramatsu, Y., Hamilton, T., Uchida, S., Tagami, K., Yoshida, S., and Robison, W. (2001) Measurement of ^{240}Pu/^{239}Pu isotopic ratios in soils from the Marshall Islands using ICP-MS. *Sci. Total Environ.*, **278**, 151–159.

30 Cizdziel, J.V., Ketterer, M.E., Farmer, D., Faller, S.H., and Hodge, V.F. (2008) ^{239,240,241}Pu fingerprinting of plutonium in western US soils using ICPMS: solution and laser ablation measurements. *Anal. Bioanal. Chem.*, **390**, 521–530.

31 Mayer, K., Wallenius, M., and Fanghänel, T. (2007) Nuclear forensic science: from cradle to maturity. *Alloys Compd.*, **444–445**, 50–56.

32 Moody, K.J., Hutcheon, I.D., and Grant, P.M. (2005) *Nuclear Forensic Analysis*, CRC Press, Boca Raton, FL.

33 Donohue, D.L. (1998) Strengthening IAEA safeguards through environmental sampling and analysis. *J. Alloys Compd.*, **271–273**, 11–18.

34 Varga, Z., Wallenius, M., Mayer, K., Keegan, E., and Miller, S. (2009) Application of Pb and Sr isotope ratio measurements for the origin assessment of U ore concentrates. *Anal. Chem.*, **81**, 8327–8334.

35 Bürger, S., Riciputi, L.R., Bostick, D.A., Turgeon, S., McBay, E.H., and Lavelle, M. (2009) Isotope ratio analysis of actinides, fission products, and geolocators by high-efficiency multi-collector TIMS. *Int. J. Mass Spectrom.*, **286**, 70–82.

36 Keegan, E., Richter, S., Kelly, I., Wong, H., Gadd, P., Kuehn, H., and Alonso-Munoz, A. (2008) The provenance of Australian U ore concentrates by elemental and isotopic analysis. *Appl. Geochem.*, **23**, 765–777.

37 Svedkauskaite-LeGore, J., Rasmussen, G., Abousahl, S., and Van Belle, P. (2008) Investigation of the sample characteristics needed for determination of the origin of uranium-bearing materials. *J. Radioanal. Nucl. Chem.*, **278**, 201–209.

38 Dickson, B.L., Gulson, B.L., and Snelling, A.A. (1985) Evaluation of lead isotopic methods for uranium exploration, Koongarra area, Northern Territory, Australia. *J. Geochem. Explor.*, **24**, 81–102.

39 Wallenius, M., Peerani, P., and Koch, L. (2000) Origin determination of Pu material in nuclear forensics. *J. Radioanal. Nucl. Chem.*, **246**, 317–321.

40 Nygren, U., Rameback, H., and Nilsson, C. (2007) Age determination of plutonium using inductively coupled plasma mass spectrometry. *J. Radioanal. Nucl. Chem.*, **272**, 45–51.

41 Schwantes, J.M., Douglas, M., Bonde, S. E., Briggs, J.D., Farmer, O.T., Greenwood, L.R., Lepel, E.A., Orton, C. R., Wacker, J.F., and Luksic, A.T. (2009) Nuclear archeology in a bottle: evidence of pre-Trinity US weapons activities from a waste burial site. *Anal. Chem.*, **81**, 1297–1306.

16
The Use of Stable Isotope Techniques for Studying Mineral and Trace Element Metabolism in Humans

Thomas Walczyk

16.1
Essential Elements

Evolution of life in all its complexity is tightly linked to the chemical properties of the elements. Life emerged in the oceans, which gave water, sodium, potassium, and chlorine their functional roles. Elements such as carbon, hydrogen, oxygen, and nitrogen joined them as the building blocks of proteins, lipids, and carbohydrates, functionalized further by phosphorus and sulfur. Other, less readily available elements became part of our elemental set-up as they added new molecular functionalities at the intracellular and extracellular level. To date, more than 20 elements have been identified as being essential to the human body, that is, inevitable losses of these elements must be replenished to maintain physiological function and to prevent disorders arising from an insufficient supply. The mass elements such as hydrogen, carbon, and oxygen are available in abundance and are usually only a limiting factor for life in the form of the vitamins and those essential amino and fatty acids that the organism cannot synthesize. Other elements, such as iron, iodine, and selenium, are less readily available and an insufficient supply through the diet may result in pathological conditions.

Assessment of food intake and the element content of dietary sources alone is insufficient to determine if an individual or population is at risk of suffering from an insufficient supply of an essential element. Element absorption from the diet is always incomplete and can vary for an element such as iron from less than 1% to near complete absorption in the individual on a given day. Furthermore, what has been absorbed from the diet is not necessarily retained by the body and quantitatively incorporated into a functional biomolecule. The efficiency of element uptake and utilization for physiological function is further subject to variations both within and between individuals, and also population groups. Understanding factors that enhance or inhibit element uptake and utilization is central, not only to give reliable recommendations to the public regarding an optimal supply of a given essential element, but also for the design of effective strategies to correct or

Isotopic Analysis: Fundamentals and Applications Using ICP-MS, First Edition. Edited by Frank Vanhaecke and Patrick Degryse.

Published 2012 by WILEY-VCH Verlag GmbH & Co. KGaA

prevent such deficiencies by food fortification or supplementation programs. Here, fortification levels or dosages are determined by the gap between dietary supply and actual needs, taking element bioavailability into account. The higher the bioavailability of a nutrient, the lower the amount of nutrient that has to be supplied to the body to cover physiological needs [1].

In vitro experiments or even animal studies are unreliable sources of information owing to the complexity and uniqueness of human physiology. To obtain the necessary information, tracer studies in humans are considered the gold standard [2]. Replacement of radiotracers by stable isotopes in such studies has opened up a new field of application for isotope ratio mass spectrometry, in general, and inductively coupled plasma mass spectrometry (ICP-MS), in particular [3]. A critical appraisal of the current status of the use of ICP-MS for such applications, including the possibilities and current limitations in this field of research, is the subject of this chapter. Basic concepts on how to use stable isotope techniques for studying element metabolism will be explained to highlight current analytical challenges and restrictions to assist in the identification of the most suitable ICP-MS instrumentation for the various methodological approaches.

16.2
Stable Isotopic Labels Versus Radiotracers

Early attempts to study element absorption in humans were based on chemical balance studies. They are often referred to as "cold" balance studies as they do not involve the use of isotopic tracers. In such studies, volunteers are fed a highly controlled diet and absorption of the element is assessed by subtracting the amount recovered in feces from the amount present in the food consumed. Element retention can be assessed in a similar manner by subtracting the amount of element recovered in urine over the observational period.

Digestive physiology, however, limits data quality in chemical balance studies. First, it remains elusive what element amount recovered in feces comes from digestive juices. Second, intestinal passage of the digested food moiety usually takes several days in a healthy individual. This requires subjects to be set on a standardized diet for several days before assessment of the element balance, which gives rise to adaptive effects. Diet standardization over 1–2 weeks can be particularly demanding for study subjects when repeated feeding of the same meal over days is required to study the effect of a specific dietary component on element absorption. The value of chemical balance studies is further limited when fractional element absorption is low. In such cases, element ingestion and recovery may differ by only a few percent, which makes the result highly dependent on the accuracy and precision of the analysis.

Some of the technical difficulties associated with chemical balance studies can be overcome by measuring the increase in element concentration in plasma over several hours after test meal intake instead of analyzing the feces. In such tests, the area under the curve describing changes in element concentration is proportional to the amount of element absorbed [4]. The usefulness of such non-isotopic plasma deconvolution tests, however, can be debated. Commonly, measurable changes can only be induced with pharmacological doses, which are usually not representative of element concentrations in food. Furthermore, it is difficult to translate temporal changes in element concentrations in blood into fractional absorption data, that is, the percentage of an element that has been absorbed from a certain test meal or diet. For the above reasons, non-isotopic approaches to study element metabolism in humans are widely regarded as unreliable and secondary to tracer techniques. In contrast to "cold" balance studies, isotopic labeling permits element absorption from a single test meal to be studied. By using a second, intravenous tracer given in parallel to the oral label, unabsorbed fractions of element in excreta can be distinguished from element fractions coming from digestive juices and secretions.

The first ever application of a radiotracer in a biological experiment dates back to 1923 when George de Hevesy used ^{212}Pb to study plant uptake of lead from solution [5]. His seminal work was honored by the Nobel Prize in Chemistry in 1943 and made him the father of isotope tracing, a tool that is still indispensable in virtually any area of scientific research. The first use of a stable isotope to study mineral metabolism was reported in 1963, when Lowman and Krivit injected stable ^{58}Fe together with radioactive ^{59}Fe into a human subject to compare the plasma clearance of the two isotopes [6]. However, it was not until the 1980s that stable isotope techniques were explored systematically to study mineral and trace element metabolism in humans. This was not only due to the increasing recognition of health hazards associated with the use of radioisotopes. Mass spectrometric techniques had to be refined to measure isotope ratios of the heavier elements at a precision suitable for the exploitation of isotopically enriched elements as tracers. Stable isotopic labels are made up from the same isotopes as the natural element, from which they differ only in terms of composition, that is, in the relative abundances of their isotopes.

Today, stable isotope techniques have replaced radioisotope techniques widely in research-oriented metabolic studies. Stable isotopes do not impose any health hazards on researchers or volunteers and no special regulations have to be adhered to as in the use of radioisotopes. This comes, however, with a number of different challenges. In an ideal setting, tracer amounts should be so small that they do not alter the native element content of a test meal significantly or disturb homeostatic mechanisms of the body. Dose requirements for stable isotopes are much higher than those for radioisotopes for reasons of analytical sensitivity. This makes the use of stable isotopes as "true" tracers often difficult or impossible. Isotope ratio measurements at the highest attainable precision are required for minimizing isotope doses, but also for cost reasons (see also Chapter 8). Despite the often significant amounts of label that need to be administered, stable isotopic labels are widely referred to as tracers, a practice that will also be followed in this chapter. Second, the element to be studied must have at least two isotopes that are stable or have a sufficiently long half-life ($>50\,000$ years) to be considered stable over the course of the study. For a monoisotopic element, the abundance of its single stable isotope self-evidently cannot be altered, a necessary prerequisite to turn it into a stable isotopic tracer. This excludes elements of interest such as phosphorus or manganese from investigation simply because they are made up of a single stable isotope only. Finally, whole-body retention cannot be measured directly using stable isotope techniques. For radiotracers, element uptake and retention are directly accessible using a whole-body counter, a device made up of a radiation detector placed over the seated volunteer in a lead-shielded chamber. This disadvantage is compensated for to some extent by the possibility of tracing and analyzing stable isotopic labels over extended periods of time as they do not decay.

16.3 Quantification of Stable Isotopic Tracers

Radiotracers of an element differ physically from its stable isotopes in the number of neutrons and, as they are unstable, can be quantified directly via monitoring of the decay process through detection of emitted particles and/or gamma rays. Stable isotopic tracers, in contrast, are composed of the same isotopes as the natural element. The two differ only in the relative abundances of their isotopes with tracers being commonly enriched in the isotope(s) that is/are least abundant in nature. This means that stable isotopic tracers can only be quantified indirectly via the changes in the isotopic abundances of the natural element introduced by tracer addition, which is measurable by mass spectrometric techniques.

Stable isotopic tracers can be quantified using basic concepts originally developed for elemental analysis by isotope dilution mass spectrometry (IDMS) (see also Chapter 8) [7, 8]. In essence, the abundances of the natural isotopes are altered when the natural element is mixed with the stable isotopic tracer in proportion to the mixing ratio. To calculate the amount of element in the sample coming from a single isotopic label n_{iso} that is highly enriched in the isotope of mass number X, the abundances θ of the isotope of mass number X and the second (reference) isotope of mass number W of both the isotopic label (${}^{X}\theta_{\mathrm{iso}}$ and ${}^{W}\theta_{\mathrm{iso}}$) and the natural element (${}^{X}\theta_{\mathrm{nat}}$ and ${}^{W}\theta_{\mathrm{nat}}$) must be known. Together with the measured isotope abundance ratio (or shorter isotope ratio) ${}^{X/W}R_{\mathrm{sample}}$ of the isotope-diluted sample, the mixing ratio (or tracer to tracee ratio) can then be calculated as follows

$$\frac{n_{\mathrm{iso}}}{n_{\mathrm{nat}}} = \frac{{}^{X}\theta_{\mathrm{iso}} - {}^{X/W}R_{\mathrm{sample}} \cdot {}^{W}\theta_{\mathrm{iso}}}{{}^{X/W}R_{\mathrm{sample}} \cdot {}^{W}\theta_{\mathrm{nat}} - {}^{X}\theta_{\mathrm{nat}}} \tag{16.1}$$

where n_{nat} is the amount of natural element in the sample (tracee) and n_{iso} the amount of tracer, both expressed in moles. The isotopic abundance ${}^{W}\theta$ of the reference isotope of mass number W for both the isotopically enriched tracer and the natural element can be obtained from the sum of all measured isotope ratios

$\sum^{i} R$ with the abundance of the isotope of mass number W as the common denominator:

$$^{W}\theta = \left(1 + \sum{}^{i}R\right)^{-1} \tag{16.2}$$

The isotopic abundance of the isotope of mass number X and all other isotopes can then be calculated from the measured isotope abundance ratios:

$$^{X}\theta = {}^{W}\theta \cdot {}^{X/W}R \tag{16.3}$$

Combining Eqs. (16.1) and (16.2) yields an alternative equation, which is based entirely on isotope abundance ratios:

$$\frac{n_{\text{iso}}}{n_{\text{nat}}} = \frac{{}^{X/W}R_{\text{iso}} - {}^{X/W}R_{\text{sample}}}{{}^{X/W}R_{\text{sample}} - {}^{X/W}R_{\text{nat}}} \cdot \frac{1 + \sum^{i} R_{\text{nat}}}{1 + \sum^{i} R_{\text{iso}}} \tag{16.4}$$

where $\sum^{i} R_{\text{nat}}$ and $\sum^{i} R_{\text{iso}}$ represent the sum of all isotope abundance ratios for the natural element and isotopic label, respectively, with the abundance of the reference isotope W as the common denominator.

Conventional applications of IDMS in quantitative elemental analysis usually involve the use of a single isotopically enriched tracer/spike. This is different in metabolic studies. The use of two or more tracers of the same element in parallel permits the use of more refined study designs and methodologies, as discussed later. When two isotopic labels X and Y are used in parallel [9], the tracer to tracee ratio for label Y in the sample can be calculated using

$$\frac{n_{\text{iso},Y}}{n_{\text{nat}}} = \frac{\alpha - \beta}{\gamma - \delta} \tag{16.5}$$

and the tracer to tracee ratio for label X in the sample using

$$\frac{n_{\text{iso},X}}{n_{\text{nat}}} = \frac{\alpha\gamma - \beta\delta}{\gamma - \delta} \tag{16.6}$$

where α, β, γ, and δ are defined as

$$\alpha = \frac{{}^{X}\theta_{\text{nat}} - {}^{W}\theta_{\text{nat}} \cdot {}^{X/W}R_{\text{sample}}}{{}^{W}\theta_{\text{iso},X} \cdot {}^{X/W}R_{\text{sample}} - {}^{X}\theta_{\text{iso},X}} \tag{16.7}$$

$$\beta = \frac{{}^{Y}\theta_{\text{nat}} - {}^{W}\theta_{\text{nat}} \cdot {}^{Y/W}R_{\text{sample}}}{{}^{W}\theta_{\text{iso},X} \cdot {}^{Y/W}R_{\text{sample}} - {}^{Y}\theta_{\text{iso},X}} \tag{16.8}$$

$$\gamma = \frac{{}^{Y}\theta_{\text{iso},Y} - {}^{W}\theta_{\text{iso},Y} \cdot {}^{Y/W}R_{\text{sample}}}{{}^{W}\theta_{\text{iso},X} \cdot {}^{Y/W}R_{\text{sample}} - {}^{Y}\theta_{\text{iso},X}} \tag{16.9}$$

$$\delta = \frac{{}^{X}\theta_{\text{iso},Y} - {}^{W}\theta_{\text{iso},Y} \cdot {}^{X/W}R_{\text{sample}}}{{}^{W}\theta_{\text{iso},X} \cdot {}^{X/W}R_{\text{sample}} - {}^{X}\theta_{\text{iso},X}} \tag{16.10}$$

Alternatively, calculations can be based entirely on isotope abundance ratios with α, β, γ, and δ being then defined as

$$\alpha = \frac{{}^{X/W}R_{\text{nat}} - {}^{X/W}R_{\text{sample}}}{{}^{X/W}R_{\text{sample}} - {}^{X/W}R_{\text{iso},X}} \cdot \frac{1 + \sum^{i} R_{\text{iso},X}}{1 + \sum^{i} R_{\text{nat}}} \tag{16.11}$$

$$\beta = \frac{{}^{Y/W}R_{\text{nat}} - {}^{Y/W}R_{\text{sample}}}{{}^{Y/W}R_{\text{sample}} - {}^{Y/W}R_{\text{iso},X}} \cdot \frac{1 + \sum^{i} R_{\text{iso},X}}{1 + \sum^{i} R_{\text{nat}}} \tag{16.12}$$

$$\gamma = \frac{{}^{Y/W}R_{\text{iso},Y} - {}^{Y/W}R_{\text{sample}}}{{}^{Y/W}R_{\text{sample}} - {}^{Y/W}R_{\text{iso},X}} \cdot \frac{1 + \sum^{i} R_{\text{iso},X}}{1 + \sum^{i} R_{\text{iso},Y}} \tag{16.13}$$

$$\delta = \frac{{}^{X/W}R_{\text{iso},Y} - {}^{X/W}R_{\text{sample}}}{{}^{X/W}R_{\text{sample}} - {}^{X/W}R_{\text{iso},X}} \cdot \frac{1 + \sum^{i} R_{\text{iso},X}}{1 + \sum^{i} R_{\text{iso},Y}} \tag{16.14}$$

For experiments involving three different isotopic labels of the same element, which is only possible for elements having at least four stable isotopes (e.g., iron), it is more convenient to perform calculations by matrix techniques. In a sample spiked with three isotopic labels enriched in the isotopes of mass number X, Y, and Z, respectively, the amount of each of the isotopes ${}^{X}n$, ${}^{Y}n$, ${}^{Z}n$, and ${}^{W}n$ in the sample is determined by the isotopic abundances θ and the amounts n of each of the tracers in the blend:

$${}^{X}n_{\text{sample}} = {}^{X}\theta_{\text{nat}} \cdot n_{\text{nat}} + {}^{X}\theta_{\text{iso},X} \cdot n_{\text{iso},X} + {}^{X}\theta_{\text{iso},Y} \cdot n_{\text{iso},Y} + {}^{X}\theta_{\text{iso},Z} \cdot n_{\text{iso},Z} \tag{16.15}$$

$${}^{Y}n_{\text{sample}} = {}^{Y}\theta_{\text{nat}} \cdot n_{\text{nat}} + {}^{Y}\theta_{\text{iso},X} \cdot n_{\text{iso},X} + {}^{Y}\theta_{\text{iso},Y} \cdot n_{\text{iso},Y} + {}^{Y}\theta_{\text{iso},Z} \cdot n_{\text{iso},Z} \tag{16.16}$$

$${}^{Z}n_{\text{sample}} = {}^{Z}\theta_{\text{nat}} \cdot n_{\text{nat}} + {}^{Z}\theta_{\text{iso},X} \cdot n_{\text{iso},X} + {}^{Z}\theta_{\text{iso},Y} \cdot n_{\text{iso},Y} + {}^{Z}\theta_{\text{iso},Z} \cdot n_{\text{iso},Z} \tag{16.17}$$

$${}^{W}n_{\text{sample}} = {}^{W}\theta_{\text{nat}} \cdot n_{\text{nat}} + {}^{W}\theta_{\text{iso},X} \cdot n_{\text{iso},X} + {}^{W}\theta_{\text{iso},Y} \cdot n_{\text{iso},Y} + {}^{W}\theta_{\text{iso},Z} \cdot n_{\text{iso},Z} \tag{16.18}$$

The molar amount of each of the isotopes in the sample is related to the measured isotope ratios R_{sample} via the total element amount in the sample ${}^{\text{tot}}n_{\text{sample}}$ and the abundance ${}^{W}\theta$ of the isotope in the sample used as the reference for expression of isotope ratios:

$${}^{X}n_{\text{sample}} = {}^{X/W}R_{\text{sample}} \cdot {}^{W}\theta_{\text{sample}} \cdot {}^{\text{tot}}n_{\text{sample}} \tag{16.19}$$

$${}^{Y}n_{\text{sample}} = {}^{Y/W}R_{\text{sample}} \cdot {}^{W}\theta_{\text{sample}} \cdot {}^{\text{tot}}n_{\text{sample}} \tag{16.20}$$

$${}^{Z}n_{\text{sample}} = {}^{Z/W}R_{\text{sample}} \cdot {}^{W}\theta_{\text{sample}} \cdot {}^{\text{tot}}n_{\text{sample}} \tag{16.21}$$

$${}^{W}n_{\text{sample}} = {}^{W/W}R_{\text{sample}} \cdot {}^{W}\theta_{\text{sample}} \cdot {}^{\text{tot}}n_{\text{sample}} \tag{16.22}$$

The abundance ${}^{W}\theta_{\text{sample}}$ in Eqs. (16.19)–(16.22) can be transformed using Eq. (16.2) and Eqs. (16.15)–(16.18) can then be combined with Eqs. (16.15)–(16.18) and expressed in matrix form:

$$\begin{bmatrix} {}^{X}\theta_{\text{nat}} & {}^{X}\theta_{\text{iso},X} & {}^{X}\theta_{\text{iso},Y} & {}^{X}\theta_{\text{iso},Z} \\ {}^{Y}\theta_{\text{nat}} & {}^{Y}\theta_{\text{iso},X} & {}^{Y}\theta_{\text{iso},Y} & {}^{Y}\theta_{\text{iso},Z} \\ {}^{Z}\theta_{\text{nat}} & {}^{Z}\theta_{\text{iso},X} & {}^{Z}\theta_{\text{iso},Y} & {}^{Z}\theta_{\text{iso},Z} \\ {}^{W}\theta_{\text{nat}} & {}^{W}\theta_{\text{iso},X} & {}^{W}\theta_{\text{iso},Y} & {}^{W}\theta_{\text{iso},Z} \end{bmatrix} \begin{bmatrix} \dfrac{n_{\text{nat}}}{{}^{\text{tot}}n_{\text{sample}}} \\ \dfrac{n_{\text{iso},X}}{{}^{\text{tot}}n_{\text{sample}}} \\ \dfrac{n_{\text{iso},Y}}{{}^{\text{tot}}n_{\text{sample}}} \\ \dfrac{n_{\text{iso},Z}}{{}^{\text{tot}}n_{\text{sample}}} \end{bmatrix} = \begin{bmatrix} \dfrac{{}^{X/W}R_{\text{sample}}}{1+\sum^{i} R_{\text{sample}}} \\ \dfrac{{}^{Y/W}R_{\text{sample}}}{1+\sum^{i} R_{\text{sample}}} \\ \dfrac{{}^{Z/W}R_{\text{sample}}}{1+\sum^{i} R_{\text{sample}}} \\ \dfrac{{}^{W/W}R_{\text{sample}}}{1+\sum^{i} R_{\text{sample}}} \end{bmatrix} \tag{16.23}$$

The amount of natural element (n_{nat}) and of the isotopic tracers ($n_{\text{iso},X}$, $n_{\text{iso},Y}$, and $n_{\text{iso},Z}$), in moles, can be obtained from Eq. (16.23) by solving the matrix $\mathbf{x} = \mathrm{A}^{-1}\mathrm{b}$ with $\mathbf{x}$, A, and b being defined as

$$\mathbf{x} = \begin{bmatrix} \dfrac{n_{\text{nat}}}{{}^{\text{tot}}n_{\text{sample}}} \\ \dfrac{n_{\text{iso},X}}{{}^{\text{tot}}n_{\text{sample}}} \\ \dfrac{n_{\text{iso},Y}}{{}^{\text{tot}}n_{\text{sample}}} \\ \dfrac{n_{\text{iso},Z}}{{}^{\text{tot}}n_{\text{sample}}} \end{bmatrix} \tag{16.24}$$

$$\mathrm{A} = \begin{bmatrix} {}^{X}\theta_{\text{nat}} & {}^{X}\theta_{\text{iso},X} & {}^{X}\theta_{\text{iso},Y} & {}^{X}\theta_{\text{iso},Z} \\ {}^{Y}\theta_{\text{nat}} & {}^{Y}\theta_{\text{iso},X} & {}^{Y}\theta_{\text{iso},Y} & {}^{Y}\theta_{\text{iso},Z} \\ {}^{Z}\theta_{\text{nat}} & {}^{Z}\theta_{\text{iso},X} & {}^{Z}\theta_{\text{iso},Y} & {}^{Z}\theta_{\text{iso},Z} \\ {}^{W}\theta_{\text{nat}} & {}^{W}\theta_{\text{iso},X} & {}^{W}\theta_{\text{iso},Y} & {}^{W}\theta_{\text{iso},Z} \end{bmatrix} \tag{16.25}$$

$$\mathrm{b} = \begin{bmatrix} \dfrac{{}^{X/W}R_{\text{sample}}}{1+\sum^{i} R_{\text{sample}}} \\ \dfrac{{}^{Y/W}R_{\text{sample}}}{1+\sum^{i} R_{\text{sample}}} \\ \dfrac{{}^{Z/W}R_{\text{sample}}}{1+\sum^{i} R_{\text{sample}}} \\ \dfrac{{}^{W/W}R_{\text{sample}}}{1+\sum^{i} R_{\text{sample}}} \end{bmatrix} \tag{16.26}$$

As a general rule, IDMS calculations should be based on amounts (in moles) rather than masses (in grams) of element, because a mass spectrometer delivers ratios R of isotope abundances or isotope amounts, with both being numerically identical. Element amounts and masses can be easily converted into each other using the atomic weight of the element, A_r (in grams per mole), which is defined by the abundances θ and the atomic masses m_a of each of its isotopes W, X, Y, and Z:

$$A_r = {}^W\theta \cdot {}^W m_a + {}^X\theta \cdot {}^X m_a + {}^Y\theta \cdot {}^Y m_a + {}^Z\theta \cdot {}^Z m_a \quad (16.27)$$

The isotopic abundances of the tracee or the tracer can be obtained from measured isotope ratios using Eqs. (16.2) and (16.3), while atomic masses m_a can be taken from the literature [10]. Based on IDMS principles, measured isotope ratios can be converted into tracer to tracee ratios in the sample analyzed. The amount of tracee can then be calculated if the amount of tracer in the sample is known. The latter can be determined either by IDMS using an additional tracer or by non-IDMS techniques for quantitative analysis employing external calibration or standard addition techniques.

Although mathematical algorithms and equations for IDMS applications are not very complex, various attempts have been made in the past to simplify calculations. This includes assumptions that tracers are monoisotopic, which is never given, or that the atomic weight of the tracer equals that of the tracee. These simplifications can introduce significant bias. As an example, atomic weights differ by nearly 10% for natural zinc and a highly enriched ^{70}Zn tracer, that is, 1 g of a highly enriched ^{70}Zn tracer contains about 9% less zinc atoms than 1 g of natural zinc.

16.4 Isotope Labeling Techniques

Studies on element metabolism can only be meaningful if the tracer used for isotopic labeling of the element mimics closely the tracee in the experimental setting, that is, the element or element species/compound under investigation. Similarly to the situation in the human body, essential elements are usually present in plant- or animal-derived food products in the form of functional biomolecules with a distinct biochemical function. Upon digestion in the stomach, the element may or may not be released from the biomolecule or food matrix in ionic form. The same holds true if a specific element compound/species is studied, for example, a mineral salt used for food fortification. The compound may or may not dissolve completely in the gastric juice before leaving the stomach. Dissolution, however, is a precondition for elements to be absorbed in the intestine by common pathways.

To avoid possible artifacts due to incomplete element release in the stomach, the element in the food matrix or compound should ideally be labeled intrinsically in metabolic studies. This requires the natural element to be replaced by the isotopic label at the atomic/molecular level. This can be achieved by either synthetic or biological means. Biosynthesis entails the administration of the isotopic label usually by intravenous injection in animals or by stem injection or hydroponic culture in plants during growth [11, 12]. Timing of the isotope dosing is essential to maximize the isotope concentration in tissues. Once the label has been utilized and incorporated in plant or body tissues and equilibrated with the natural element, which may take days to weeks, plants can be harvested and meat, milk, or eggs can be obtained for test meal preparation. For specific compounds to be studied, the isotopic label has to be converted into the chemical form under investigation. For poorly acid-soluble compounds, *in vitro* experiments are possibly necessary to guarantee that the acid solubility of the isotopically labeled compound synthesized is comparable to that of the unlabeled compound. Solubility is determined not only by the chemical properties of the compound, but also by physical characteristics such as particle size and surface area [13, 14].

Intrinsic labeling can be considered the gold standard, but it can be technically challenging and costly, especially if labeling by biological means is considered necessary. Intrinsic labeling, however, is unnecessary if the element is (i) released quantitatively from the food matrix or element compound in the stomach, (ii) all of it is present in ionized form, and (iii) in the same species/oxidation state at some point in time during or after gastric digestion. If the same applies to the isotopic label, the label can simply be added extrinsically to the test meal in aqueous solution. Labels that are highly soluble in water or in dilute HCl at gastric pH (2–3) can also be added in solid form.

Extrinsic labeling requires that at some point during gastric digestion the native element and the isotopic label equilibrate quantitatively with each other, that is, they are present in the same chemical form/oxidation state. Once they have entered a common chemical pool, interaction with other food compounds or changes in pH or oxidation state affect both the native element and the isotopic label to the same extent. Hence the tracer transfer efficiency at the physiological or molecular level should be identical with that of the tracee. This can be ascertained by feeding the same test meal either intrinsically or extrinsically labeled to volunteers and by comparing the recoveries of both tracers in body fluids or excreta. Such studies have been conducted in the past using either radioactive or stable isotopic tracers for the relevant essential elements. A comprehensive element-by-element review, however, would go beyond the scope of this chapter. As a general rule, it can be stated that intrinsic labeling is required when (i) absorption is assessed from food matrices that are incompletely digested, such as nuts and seeds, (ii) absorption is measured from compounds that are poorly water or acid soluble and (iii) when the absorption efficiency differs between element species. The latter is illustrated for iron as a prominent example. Heme-bound iron in animal foods is absorbed via different pathways as compared to non-heme iron from plant foods. Iron in animal products is mostly present in the form of hemoglobin or myoglobin, from which the iron is released by the gastric juice as heme, a porphyrin ring system with Fe(II) in its center. The heme molecule survives gastric passage and is taken up intact by the intestinal mucosa by heme carrier transporter 1 (HCP1), whereas ionized iron is taken up via divalent metal transporter 1 (DMT1), either directly or after reduction by duodenal cytochrome B

(DcytB) [15]. This entails that extrinsic labeling will result in bias if iron absorption from meat or meat products as opposed to plant foods is assessed [16].

16.5 Concepts of Using Tracers in Studies of Element Metabolism in Humans

16.5.1 Overview

Stable isotope techniques for studying element metabolism *in vivo* evolved historically as a non-hazardous alternative to radiotracer techniques. Basic concepts for using stable isotopic labels follow principles originally developed for radiotracers. Stable isotopes, however, offer more than safety to volunteers and researchers. Stable isotopes do not decay, which makes it possible to detect the isotopic label over extended periods of time with analytical sensitivity as the primary limitation. Second, the natural isotopic abundances of an element can be altered slightly by chemical and physical processes as occur in Nature, including the human body. Such isotope fractionation effects are well known and have been studied for the light elements such as hydrogen, oxygen, and carbon since the 1930s. With the advent of new techniques for high-precision isotope ratio measurements, namely multi-collector (MC) thermal ionization mass spectrometry (TIMS) and MC-ICP-MS, it became possible over the past decade to measure isotope fractionation effects also for the heavier elements, including elements that are essential for the human body. This opened a new door to study element transfer processes in the human body without the use of tracers.

In the following, the various concepts available for studying human element metabolism *in vivo* will be briefly explained. Because the field is so wide, the focus of this chapter will be more on underlying principles rather than historical developments and experimental details. An in-depth discussion of the practicalities of how to design and conduct stable isotope studies in humans would go beyond the scope of this chapter and will only be highlighted when considered relevant. For technical details, the interested reader is referred to the references cited, which have been selected as representative examples for the subject matter, or available monographs [17–21].

16.5.2 Fecal Balance Studies (Single Isotopic Label)

The crudest approach to study element absorption is by taking the human body as a "black box." As for chemical balance studies, element absorption can then be assessed as the difference between dietary intake of an isotopic label and its fecal recovery. This approach, however, does not deliver the fraction of element that has truly been absorbed, that is, the fraction of element in the digested food moiety that has passed the intestinal mucosa. Digestion can only occur by action of endogenous secretions. This includes saliva for initial carbohydrate breakdown in the mouth, gastric juice for breakdown of the food matrix into chyme, pancreatic juice for chyme neutralization and enzymatic breakdown of macro-nutrients, and also bile for solubilization of lipophilic components by micelle formation. These secretions do not just represent major pathways of habitual element loss from the body. An absorbed isotopic label enters the body and may leave the body rapidly through these pathways also. This results in a higher recovery of the isotopic label in feces and thus a systemic underestimation of element absorption. As a result, the information accessible by this methodology is denoted by the term "apparent absorption."

Fecal balance studies using a single isotopic label represent a generally applicable approach to study the absorption of any element that has at least two stable or very long-lived isotopes. The labeled test meal is usually fed after an overnight fast. The diet is commonly standardized including drinks on the day of test meal administration as the different meals taken in over a day can mix in the intestine and interfere with each other. A technical difficulty is to ensure quantitative recovery of all unabsorbed isotopic label in feces, as intestinal passage, in particular through the colon, can take days. To standardize fecal collection, a dye can be fed in the evening before test meal administration and again on the evening of the seventh day following test meal administration. All feces is collected completely from the moment of appearance of the first dye until the second dye appears in the stool [22].

Collected feces samples are weighed, pooled, and homogenized either freshly or after freeze-drying and milling. A sample is finally taken to determine the concentration of the isotopic label in the stool to calculate the label recovery. Tests are required to confirm that the homogenization procedure results in homogeneous distribution of the label in the feces sample. To check for completeness of fecal collection, a known number of non-digestible radiopaque pellets can be included in the test meal [23]. As feces collection over days is rather tedious, feeding of poorly absorbable rare earth elements in aqueous solution together with the test meal has been suggested. Fecal recovery of added rare earth elements can then be used not only to confirm subject compliance, but also to correct for incomplete collection of fecal material [24–26].

16.5.3
Fecal Balance Studies (Double Isotopic Label)

"True absorption," as the element fraction that is taken up by the intestinal mucosa and released into circulation, cannot be assessed by fecal balance techniques using a single oral isotopic label only. A second label needs to be administered intravenously to determine the fraction of absorbed element that re-enters the intestine by secretory pathways. Intravenous isotopic labels in metabolic studies are usually intended to mimic freshly absorbed element from the diet. This entails that labels must be injected slowly over several minutes about 1 h after test meal administration or, ideally, as an infusion over 20–30 min in physiological saline.

Another factor to be taken into consideration is the dosage of the intravenous isotopic label. Dosages need to be kept as low as possible to ensure that they do not trigger a homeostatic response of the body, as this may result in an increased discharge of both labels into other body compartments and urine. Because it is almost impossible to mimic plasma uptake of the oral isotopic label exactly by label infusion, this is a potential source of bias. This makes the isotope with the lowest natural abundance the preferred label for intravenous administration. The lower the natural abundance, the lower is the amount of label required to induce a measurable shift in the isotope ratios (see below).

To make use of two isotopic labels in parallel, the element must have at least three stable or long-lived isotopes. This excludes, as an example, copper from such studies, as natural copper consists of only two stable isotopes. In principle, copper has two radioactive isotopes (^{64}Cu and ^{67}Cu) that can be used as tracers [27], but their half-lives are relatively short for balance studies at 12.7 and 61.8 h, respectively. Copper retention, however, has been successfully determined in the past using ^{67}Cu by whole-body counting [28]. This example illustrates that apart from safety, radioisotope and stable isotope techniques differ also in scope of use.

16.5.4
Plasma Appearance

Similarly to fecal balance studies, assessment of element absorption by plasma deconvolution is an approach adopted from non-isotopic measurements. After feeding the test meal containing the isotopic label, serum samples are taken at regular intervals over several hours. Samples are analyzed for the isotopic label to assess the appearance and disappearance of the label over time in blood. The area under the resulting curve for label concentration is then proportional to the fraction of label absorbed. The larger the area, the more of the label has passed the intestinal mucosa [29].

When a single isotopic label is used for this purpose, the basic limitation is that the absorbed amount of isotopic label cannot be assessed, as also holds true for non-isotopic deconvolution tests. In single isotope tests, comparisons between test meals can only be made in a relative manner by comparing areas under the curves obtained for different meals. This can be changed by giving a known amount of a second isotopic label in parallel by infusion. The second label enters circulation directly, which would compare to 100% absorption of the element via the intestinal mucosa. As the area under the curve for the intravenous isotopic label corresponds to the intravenous dose, the area under the curve for the oral label can be transformed into the amount of oral label that has been absorbed [30]. The ratio of the area under the curve for the oral and intravenous label equals the amount ratio of the two labels in the body.

When plasma deconvolution techniques are applied, the isotopic label should be infused slowly and doses for the intravenous label should be kept as low as possible in order not to disturb element homeostasis. Plasma deconvolution is rather invasive and more labor-intensive as it involves the analysis of a dozen or more plasma samples per subject, as opposed to a single sample in most other techniques. Plasma deconvolution techniques, however, can be a viable tool for elements that are absorbed efficiently from the gastrointestinal tract, such as calcium and magnesium. High absorption efficiency results in a relatively high enrichment of the isotopic label in serum/plasma, which facilitates isotopic analysis by mass spectrometric instrumentation, such as basic quadrupole-based ICP-MS, despite its limitations in measuring isotope ratios precisely.

16.5.5 Urinary Monitoring

The intravenous administration of a second isotopic label can also be used to assess true absorption of the oral isotopic labels for elements that are excreted via the kidneys, such as calcium and magnesium. As in plasma appearance studies, the intravenous label serves as a reference dose. When the second label is infused or injected slowly following administration of the labeled test meal, both labels appear in parallel in serum/plasma. Some of the circulating labels, however, are excreted via the kidneys, during which the amount ratio of the two labels in blood should ideally be preserved. If this is so, a single spot urine sample can be used, at least in theory, to determine the amount ratio of the two labels in serum. As the intravenous dose is known, this ratio can then be converted into the amount of oral label absorbed. As for most stable isotope techniques, this method was originally developed for radiotracers [31] and adapted for stable isotopes in the 1980s [32].

A technical difficulty associated with any technique involving the administration of an intravenous label is the matching of the rate of appearance of the intravenous label in blood with that of the oral label, which is hardly possible. As a result, the amount ratio of the two labels changes over time in blood and consequently urine, which translates into a possible bias determined by the point in time at which the urine sample is taken. In theory, this can be controlled by collecting urine over 24 h up to 1 week following isotope administration. In practical terms, the amount ratio in blood might be similar to the "true" ratio of absorbed oral label to intravenous label at some point in time. This permits, in theory, the use of a single spot urine sample for analysis [33].

16.5.6 Compartmental Modeling

Element metabolism can be described at the systemic level by transfer processes occurring between different compartments or pools, which may represent organs, physiological spaces, or chemical species of the element under investigation. Following absorption, the element circulates in plasma and is distributed to various tissues or organs for biochemical function or for storage. Provided that a compartment is not a sink for the element, the element circulates between these compartments until it is lost or excreted from the body.

To monitor fluxes between compartments, an oral and a second intravenous tracer are administered simultaneously. Both tracers are distributed over time between the different compartments according to compartment size and flux rates between compartments. Both parameters determine the time course of appearance and disappearance of the tracers in plasma, urine, feces, saliva, or accessible tissues, such as red blood cells. Time-dependent changes in label concentration can be measured and analyzed using commercially available modeling software developed for pharmacokinetic studies, such as SAAM II (University of Washington, USA) or NONMEM (University of California, USA). Provided that the test person is in a steady state, compartment sizes and flux rates can be determined by fitting model parameters, that is, compartment sizes and flux rates, to observed data [34–36].

Compartmental modeling delivers more information than any other isotopic technique, but is also technically the most demanding *in vivo* application of isotopic tracers. For each compartment or pathway of physiological loss, about a dozen samples need to be taken during up to 2 weeks and analyzed for isotopic composition of the target element. In a second step, the appropriate compartment model needs to be identified using the data generated. This entails testing of models with various numbers and arrangements of compartments (model building) finally to select the model that results in the smallest possible difference between observed and modeled data [37]. Compartmental analysis delivers absolute numbers for compartment sizes (in milligrams) and flux rates, that is, fraction of compartment size transferred per time unit from one compartment to another. It must be stressed, however, that a modeling compartment does not necessarily represent an actual organ or a single physiological space. Compartments in underlying models are defined by their kinetic parameters, not by their physiological function.

16.5.7 Tissue Retention

The bioavailability of an element in the diet is determined as the fraction of element that has been absorbed in the intestine, retained by the body, and utilized for physiological function, including deposition into storage tissues [18]. The actual element utilization of an administered label, however, is difficult to measure directly. Major functional or storage compartments, such as liver, muscle tissue, and bone, are difficult to access. They require tissue biopsies, which can be taken with minimal associated health risks, but can hardly be considered a routine sampling approach for research purposes in healthy volunteers. The only exception is iron, for which two-thirds of body iron circulates as hemoglobin in blood, which can be easily sampled. This has led to the development of two different approaches to assess iron absorption and bioavailability in the human body.

The first approach is based on the early work of Dubach *et al.* [38], who used radiotracers to determine iron bioavailability. In healthy individuals it takes about 14 days until freshly absorbed iron is incorporated into reticulocytes, that is, freshly matured red blood cells. The amount ratio of stable isotopic label to natural iron, that is, the tracer to tracee ratio, can be determined using IDMS concepts in a single venous blood sample taken 14 days after oral administration of the label. The amount of label in circulation after 14 days can then be calculated from the iron/hemoglobin concentration in blood and blood volume. The latter can be estimated from the height and weight of the subject, using gender- and age-specific empirical equations [39–41]. If the efficiency by which freshly absorbed iron is incorporated into red blood cells is known, circulating amounts of isotopic label can be translated into the amount of iron absorbed and utilized for body function. The efficiency of red blood cell incorporation of the label is about 80% in healthy adults [42]. The incorporation efficiency can be estimated in the individual by injection of a known dose of iron isotopic label into blood and calculation of its recovery in blood from the analyzed tracer to tracee ratio and the estimated amount of circulating iron. The feasibility of this approach using the stable isotope ^{58}Fe instead of radio-iron was demonstrated in the 1980s [43].

The above approach is based on estimates of blood volume and efficiency of erythrocyte incorporation of the isotopic label, unless it is determined directly in the studied subject. These potential sources of bias cancel out for relative measurements to compare iron absorption from one test meal against another in the same individual. This is possible by using two different isotopes, such as ^{57}Fe and ^{58}Fe, to label meals differently, which are then fed on two consecutive days [9, 44]. Such an approach is often used to identify the effect of a specific food component on iron absorption or to compare iron absorption from two different food fortification compounds.

In certain situations, absolute information on iron absorption is required. This may include studies to assess iron absorption during pregnancy [45–47] and infancy [48] and studies in other population groups where blood volume is either difficult to assess or the efficiency by which iron is incorporated in erythrocytes is uncertain or highly variable [49]. Following similar principles to those described for the urinary monitoring technique, a second isotopic label can be injected as a reference dose following administration of the labeled meal. After 14 days, a blood sample is taken and the amount ratio of the two isotopic labels is calculated. With the known amount of injected label, the amount of absorbed oral label can then be calculated without knowledge of blood volume or the actual incorporation efficiency of the isotope labels into red blood cells. It should be noted, however, that this approach delivers true element absorption as information and not element utilization or bioavailability, similarly to the urinary monitoring technique.

16.5.8
Element Turnover Studies

Stable isotopes do not decay, which makes them a safe alternative to radiotracers, but also offers the possibility of following administered stable isotope labels in the human body over extended periods of time. This can be made use of in two different ways. Observational periods for compartmental analysis can be extended from a few days to months or even years, which allows better assessment of slowly exchanging body compartments, such as bone, where a tracer can reside over years, until it is remobilized [37, 50]. When a tracer is remobilized from a compartment and re-enters plasma from where it is redistributed to another compartment, it is only a matter of time until the tracer to tracee ratios in all body tissues become indistinguishable. Once this condition is reached, element turnover by the human body can be measured directly, that is, the amount of element in the human body that is lost and replaced per unit of time.

Element losses together with element bioavailability determine how much of an element has to be provided through the diet to remain in element balance. For assessment of losses, a label needs to be administered once either orally or intravenously. Compartmental modeling techniques permit to calculate when the label has equilibrated with the natural element in all body compartments. When isotopic labeling has been achieved, continuous replacement of lost isotopic label with the natural element from the diet results in a continuous decline of the body's isotopic enrichment. The change in tracer to tracee ratio in blood corresponds directly to the fraction of the body's element inventory that has been lost and replaced.

Current estimates of physiological iron losses in humans are based on a study by Green *et al.* [51], who used ^{55}Fe, an iron radionuclide with a relatively long half-life of 2.94 years, for labeling body iron pools. A similar radionuclide-based approach was used recently to confirm the earlier measurements [52]. This complemented measurements of iron turnover in adolescents [53] and toddlers [54] using the stable isotope ^{58}Fe. The technical feasibility of such studies, however, is limited by the need for subject follow-up over months and possibly years and for high analytical sensitivity for detecting changes in tracer to tracee ratios over time, which is limited by the precision of the isotope ratio measurement.

16.5.9
Isotope Fractionation Effects

Except for elements with one or more radiogenic isotopes, the isotopic composition of an element in Nature can be considered constant for most applications of stable isotope techniques. However, they are not invariable (see also Chapter 1). Stable isotopes of an element differ slightly in mass, which may cause changes in the element's isotopic composition if a chemical or physical transfer process is sensitive to the isotopes' masses and if the transfer is incomplete. For reasons of mass balance, the isotopic abundances cannot be changed if all matter is transferred from a source to a target compartment.

Physical transfer processes include evaporation and precipitation, and also membrane diffusion as a basic mechanism of element transfer in the human body. A difference in diffusion rates between isotopes inevitably results in their fractionation over the course of element transport (kinetic isotope fractionation effect). When an element undergoes a chemical reaction, this can also be described as an element transfer process from one element species into another. Again, if reaction rates between isotopologs differ, this gives rise to a difference in isotopic composition between educt and product in an equilibrium reaction (equilibrium isotope fractionation effect). In either case, the transfer efficiency determines the difference between source and target compartments. This permits, in theory, the determination of the fraction of element that is present in either compartment without introducing isotopic tracers into the system.

Mass-dependent isotope fractionation is dependent not on absolute, but on *relative* mass differences between the isotopes of an element. Accordingly, isotope fractionation effects are much smaller for the mineral elements and trace elements as opposed to the light elements such as hydrogen, oxygen, and carbon. Because of technical limits in isotopic analysis by TIMS and ICP-MS, mass-dependent isotope fractionation effects for metals remained largely elusive until the end of the 1990s [55–58]. Mass-dependent isotope fractionation also occurs in the ion source of a mass spectrometer with processes following similar systematics to those in Nature. Differentiating the effects induced during isotopic analysis from genuine isotope fractionation effects in the sample is technically challenging.

The past decade has seen tremendous developments in the measurement of metal isotope ratios by ICP-MS, namely the introduction of MC-ICP-MS, at precisions similar to or better those that achievable with MC-TIMS at commonly higher mass spectrometric sensitivity [59]. Developments driven mainly by the geological community made it finally possible to study the systematics of isotope fractionation phenomena in higher organisms, including the human body. Of the essential elements, iron, calcium, and zinc have received the most attention in this context and will be discussed in more detail in this chapter.

16.6 ICP-MS in Stable Isotope-Based Metabolic Studies

16.6.1 Measurement Precision

The use of stable isotope labels in metabolic studies self-evidently involves the measurement of isotope ratios. Isotope ratios are translated into label amounts or tracer to tracee ratios following IDMS principles. IDMS *per se* is a primary method, that is, it is possible to identify all sources of uncertainty in the analysis and to state the combined uncertainty in the measurement result [60]. When IDMS is used for element quantification by means of an isotopically enriched spike, the uncertainty in the isotope ratio measurement is usually the major contributor to the uncertainty of the analysis [61]. At a precision of 0.5% for the actual isotope ratio measurement, which can be achieved comfortably, even with basic ICP-MS instruments, combined relative measurement uncertainties of 1–2% can easily be attained. This points to measurement precision not being a major obstacle in stable isotope-based metabolic studies. Most modern ICP-MS instruments are capable of measuring isotope ratios at that level of precision. However, as is often the case, reality turns out to be much more complex on second sight.

A true tracer should ideally be mass free in order not to disturb the system by its introduction. This holds especially true for the human body. For most of the elements concerned, homeostatic mechanisms are in place to maintain element concentrations in a relatively narrow physiological range in body fluids. When homeostasis is disturbed by tracer introduction, label uptake, utilization, and/or excretion may no longer mimic the normal state and may result in artifacts.

Isotope dosage is generally of minor concern for radiotracers owing to the sensitivity with which they can be detected. In stable isotope-based studies, dose requirements are much higher and can become a major hurdle in conducting meaningful experiments. Stable isotope labels cannot be detected free of background as they are made of the same isotopes as the natural element. To keep dose levels low entails isotope ratios in the sample containing the label(s) being close to the natural isotope ratios of the element. In IDMS terms, this relates to a condition of extreme "underspiking," which is commonly avoided. As a rule of thumb, in IDMS analysis both the isotopic spike and natural element should not be in excess of each other by more than an order of magnitude to minimize the measurement uncertainty of the analytical result [61]. Error propagation principles dictate that the measurement uncertainty increases in a nearly exponential manner when these limits are exceeded (see also Chapter 8).

By definition, the analytical limit of detection (LOD) for a change in an isotope ratio is three times the standard deviation of the isotope ratio measurement, and 10 times its standard deviation for the limit of quantification (LOQ). To keep the combined measurement uncertainty for the tracer to tracee ratio at a tolerable level, the relevant isotope ratio should be shifted by at least 20 times the reproducibility of the measurement including day-to-day variations. For a measurement precision of 0.5% for a given isotope ratio, this translates into a minimum shift of 10% over its natural value. To reach that enrichment using the erythrocyte incorporation technique, calculated dose requirements for a 75 kg healthy adult male would be 160 mg for a ^{57}Fe label (98% enrichment) and 50 mg for a ^{58}Fe label (95% enrichment) at an expected minimal absorption of 2% of both labels. Calculations are based on routinely achievable precisions for basic quadrupole-based instruments of 0.5% for the $^{57}Fe/^{56}Fe$ ratio and 1% for the more extreme $^{58}Fe/^{56}Fe$ ratio. For this setting, combined relative uncertainties for the amount of isotopic label in circulation would be about 25%, which is considered acceptable. Dose levels, however, would be well beyond current recommendations for dietary intake of 10–30 mg of iron per day taking differences in iron bioavailability from the habitual diet into account [62]. Placing considerations of costs aside, which would be substantial, dose requirements would be forbiddingly high and would be acceptable only, at most, to assess iron absorption from supplemental iron in a reliable manner.

The above example illustrates that a measurement precision that might be satisfactory for classical IDMS applications can be unsatisfactory for conducting metabolic studies [3, 63]. It should be noted, however, that dose requirements differ between stable isotope methods and that the erythrocyte incorporation technique in the given example has by far the highest dose requirements of all techniques available for the measurement of iron absorption. For fecal monitoring, as the alternative approach, dose requirements would be 4 mg for the ^{57}Fe label and 2 mg for the ^{58}Fe label at a combined measurement uncertainty of 1% for label recovery in feces. But how can this substantial difference in dose requirements be explained? First, iron absorption is relatively low, which results in a nearly 50-fold difference between the amount of orally administered label that enters the feces and the amount incorporated into red blood cells. Second, about 1.5 g of natural iron is present in red blood cells, as opposed to an estimated 70 mg of iron in a 7 day fecal pool. The lower the amount of natural iron in a sample, the less iron label is required to alter the iron isotope ratio to the necessary extent.

Different types of ICP-MS instruments differ significantly in the precision attainable in isotopic analysis. The evolution of ICP-MS has largely been governed by the attempt to control spectral interferences for conventional elemental analysis better and to improve the capabilities for measurements of transient signals. This led to the evolution of sector field instruments from first-generation quadrupole-based instruments, to the use of collision or reaction cells in quadrupole-based instruments and the exploration of time-of-flight (TOF) techniques for ion separation [64]. Users interested in isotope ratio measurements, who are still a minority among ICP-MS users, also benefited from these developments, but it was the advent of MC-ICP-MS that made the greatest difference. As a dedicated technique for isotopic analysis, it became competitive with MC-TIMS as the gold standard for high-precision isotopic analysis (see also Chapter 3). This permitted the exploration of isotope fractionation effects for the metals for which effects are much too small to be measurable by conventional single-collector ICP-MS instruments. When ion beams of the different isotopes/isotopologs of an element can be detected simultaneously, drift or flicker in ion currents cancel out when isotope ratios are measured. In single-collector instruments, isotope ratios are measured by measuring the signals sequentially, which makes the stability of the ion current the limiting factor for measurement precision.

The precision attainable in isotopic analysis is the decisive factor when it comes to the question of which of the different study concepts can be exploited in studies of element metabolism for a given instrumental setting. The precision in isotopic analysis can range from 1% down to 0.001%, depending on the element, isotope ratio, and choice of instrumentation. As a general rule, this precision needs to be assessed experimentally first, using the available laboratory set-up, instrumentation, and individual skills. For a given method and isotope dose, it is possible to model how much of the dose can be expected to be present in the sample. Equations (16.15)–(16.18) can then be used to model the isotope ratio in the sample and to determine the dose requirements to reach the targeted minimum enrichment in order to decide finally if the approach chosen is feasible. Because of significant differences between elements and methodologies, the possibilities and limitations will be discussed individually for each element of interest (see below).

16.6.2
Mass Spectrometric Sensitivity

In addition to measurement precision, a high mass spectrometric sensitivity can also facilitate the use of certain methodological principles. A higher mass spectrometric sensitivity may also permit a reduction in sample size. Whereas there is no problem with taking urine and feces samples, a reduction in sample volume can be particularly important when blood samples need to be obtained from infants or children. Taking intravenous blood samples is especially of ethical concern when patients are moderately or severely anemic. Here, intravenous samples of a few milliliters of blood could be replaced by a finger or heel prick of 100–200 μl [48]. In calcium kinetic studies, where frequent venous blood samples need to be taken, saliva samples can potentially be used as a surrogate for plasma samples. Tracer to tracee ratios in plasma and saliva were found to be very similar, but require a much higher sensitivity as calcium concentrations and sample volumes are significantly lower [65].

ICP-MS and TIMS remain the two mass spectrometric techniques of choice in stable isotope-based metabolic studies, each of them having distinct advantages and disadvantages. For most elements of the periodic system, a high mass spectrometric sensitivity is most certainly one of the greatest assets of ICP-MS. This is inherently associated with the ionization principle. At the high ionization temperature of the plasma of about 7500 K, the ionization efficiency of sample material introduced into the plasma of an ICP-MS system approaches 100% (see also Chapter 2). This is different for TIMS, which operates at more moderate temperatures of up to 2800 K. For elements having a first ionization energy of less than 7.5 eV, positively charged ions are formed by transfer of an electron from the atom/molecule to be ionized to the metal filament, usually made from rhenium or tungsten, on which the sample is loaded [positive thermal ionization mass spectrometry (PTIMS)]. For elements with an ionization energy higher than 7.5 eV, formation of negative ions by uptake of an electron from the filament material can be stimulated [negative thermal ionization mass spectrometry (NTIMS)]. Ion currents in TIMS can last for hours, during which 1–30% of the sample can be ionized. However, the actual measurement takes much less time, with often only 1% or less of the sample material being ionized. This translates into differences in sample requirements between ICP-MS and TIMS of one order of magnitude or even more, rendering ICP-MS the technique of choice when ultimate mass spectrometric sensitivity is needed.

16.6.3
Measurement Accuracy and Quality Control

When comparing TIMS and ICP-MS, it seems that ICP-MS is the best choice for stable isotope-based metabolic studies. It commonly surpasses in terms of mass spectrometric sensitivity and also permits the measurement of isotope ratios at comparable and sometimes better precision than TIMS, at least for MC-ICP-MS.

In addition, samples can be presented to the instrument in aqueous solution directly after digestion, which requires fewer preparative steps than in TIMS, for which the element has to be chemically separated from the sample matrix. As measurements do not require a "heating up" procedure of the sample as in TIMS, measurements require about 50% of the time, which increases sample throughput. This is a considerable advantage for studies involving compartmental modeling, taking into account that such studies involve usually 10–20 subjects who generate dozens of samples to be analyzed.

However, is the use of ICP-MS really that straightforward and are there no drawbacks to be taken into consideration? In terms of isotope ratio measurements, one of the major advantages of ICP-MS is also its major disadvantage. ICP-MS is a multi-element technique, that is, ionization is not element-specific as in TIMS. Hence elements in the sample other than the analyte element are also ionized, which potentially gives rise to spectral interference from atomic or polyatomic ions. This includes also the plasma gas, which is usually argon. The latter is of direct relevance for measurements of two of the most interesting elements in metabolic studies, iron and calcium, which are prime examples of elements subject to heavy interference in ICP-MS. The ArO^+ mass spectrum interferes with the iron mass spectrum, as a result of the overlap of the signals from $^{40}Ar^{16}O^+$ and $^{56}Fe^+$, $^{40}Ar^{17}O^+$ and $^{57}Fe^+$, and $^{40}Ar^{18}O^+$ and $^{58}Fe^+$, and the signal from the most abundant calcium isotope, ^{40}Ca, suffers from isobaric overlap with the signal from ^{40}Ar. Within the mass range 27–82 u, where most metals of biological relevance have their ion signal(s), there is virtually not a single isotope that is not potentially subject to interference. In TIMS, spectral interferences also occur, but they are much rarer and less pronounced because of chemical separation of the element from the sample matrix and ionization in vaccum.

Techniques have been developed and implemented to reduce the effect of spectral interferences, either by using a double-focusing sector field mass spectrometer to attain higher mass resolution or by removal of interfering species in collision/reaction cells by selective ion–molecule (cell gas) interactions (see also Chapter 2). Suppression of interferences can be further improved by using "cool plasma" conditions to reduce the occurrence of Ar-based ions or a desolvating nebulizer to effect water removal (see also Chapter 2). Interfering atomic ions, however, remain a problem as mass differences between interfering and analyte ions are too small to be mass resolved by a double-focusing sector field mass spectrometer. For iron, as an example, this includes the ion pairs $^{54}Fe^+$–$^{54}Cr^+$ and $^{58}Fe^+$–$^{58}Ni^+$. Such interferences need to be corrected for mathematically by measuring the chromium and nickel ion signals in addition, to calculate contributions of $^{54}Cr^+$ and $^{58}Ni^+$ to the iron mass spectrum. Potential sources of bias in this approach include first the accuracy with which the often low ion intensities of the interfering element(s) can be measured and second the fact that interfering elements, in addition to the element to be analyzed, are subject to both natural and instrumental isotope fractionation effects.

Isotope fractionation is a major source of bias in any mass spectrometric technique. For mass spectrometric analysis, the element in the sample to be analyzed needs to be ionized to permit the separation of the generated ions based on their mass-to-charge ratio. In TIMS, the mass boas, that is, the difference between the true and observed isotope ratio, is caused by a difference in the evaporation/ionization efficiency of the isotopes [66]. In ICP-MS, it is caused by differences in the transfer efficiency of the different isotopes from the atmospheric pressure in the plasma into the high vacuum of the mass spectrometer [67]. In TIMS, the mass bias is smaller, usually by 1–2 orders of magnitude compared with ICP-MS, but changes significantly with time. In ICP-MS, the mass bias, can be as high as several percent, but its drift over time is minimal. This permits external correction of mass bias effects by bracketing of the sample with a reference material of known isotopic composition, usually of the same element, an approach that is impossible with TIMS [57]. Correction for fractionation in TIMS is performed by internal normalization, which uses one measured ratio for which the true value is known and extrapolation of the observed mass bias for this ratio to the other ratios for correction [68]. A deeper insight into the algorithms and techniques of mass bias correction can be gained from Chapter 5. For isotope-diluted samples, iterative procedures can be applied [69, 70]. Internal normalization can be used both in TIMS and ICP-MS, but it is restricted to elements with at least three stable isotopes for which two isotope ratios can be measured. In addition, it cannot be used to study natural isotope fractionation effects, unless double-isotope spiking techniques are employed [71]. Regularities of isotope fractionation in Nature and in the mass spectrometer are very similar.

Because of the much higher extent with which measured isotope ratios must be corrected in ICP-MS, ICP-MS is much more susceptible to induced bias than measurements by TIMS. This gives rise to the risk of the less experienced user confusing measurement precision and accuracy. Whereas precise measurements with TIMS are usually only possible if the element has been separated effectively from the sample matrix and the ionization process is well under control, isotope ratio measurements by ICP-MS are usually precise but rather inaccurate without further refinement. Accordingly, much more attention must be paid to the validation of an ICP-MS technique than a TIMS technique [72]. As a general rule, techniques must be thoroughly validated from scratch by the individual user, irrespective of reported successful applications in the literature. It is often believed that following a published, functional procedure closely will generate reliable data. This is a serious and even dangerous misconception.

Over the past decade, awareness has grown significantly that isotopic analysis by ICP-MS is full of potential sources of bias that need to be controlled carefully to generate not only precise but also accurate data. Whereas in the early days the speed advantage of ICP-MS over TIMS in isotope ratio measurements has been emphasized has been emphasized because it permits direct analysis of digested samples, this advantage has now mostly vanished. Because of much better control of spectral and non-spectral interferences, it is now common practice in ICP-MS as in TIMS to separate the element chemically from the matrix when high analytical accuracy is of concern. This does not equalize the two techniques fully in terms of sample throughput, but makes the latter less decisive when having to choose between them.

Publications discussing sources of bias at both an introductory and an advanced level are plentiful [59, 67, 73], but so far no standard test procedures could be agreed upon that go beyond the analysis of available reference materials. Meanwhile, certified isotopic reference materials can be obtained from national and international bodies for various metals, but often they are certified at a degree of uncertainty well below the precision that can now be achieved by state-of-the-art instrumentation (see also Chapter 6). Their usefulness is further limited by the availability of a single material only that is usually representative of the natural isotopic composition. For reliable testing of measurement accuracy, however, at least a second material of non-natural isotopic composition is needed, which can be either an isotope-diluted sample for tracer studies or a sample in which the element of interest is significantly affected by isotope fractionation for studies on natural isotopic variations. Because of the general shortage of matrix-matched isotopic reference materials, it is advisable to conduct other tests, which may include gravimetric mixing experiments. As an example, a characterized whole-blood sample containing an iron isotope label can be mixed with a characterized whole-blood sample of natural iron isotopic composition. If the amount of iron coming from each of the blood samples is known, iron isotope ratios in the mixture can be calculated and compared with the measured ratios. The same holds for a sample of natural iron and a sample showing an iron isotope fractionation effect. For testing measurement accuracy without the use of any reference material, three-isotope plots can be employed for elements with at least three stable isotopes, that is, elements for which two isotopc ratios can be measured [73]. Unless mass-independent isotope fractionation effects are present (see also Chapter 1), which remain the exception, both isotope ratios must be correlated according to the underlying law describing the mass dependence of the fractionation process. When plotting measured isotope ratios against each other, deviations from the calculated, theoretical fractionation line can be used to identify offsets in either of the ratios (see also Chapter 3).

In essence, ICP-MS is indeed ideally suited for studies involving stable isotope labels as tracers or for studying natural isotope fractionation phenomena. Stringent quality control protocols, however, are needed for making use of ICP-MS in a reliable manner and to ensure data accuracy and comparability across studies. This is of particular importance in studies which may be of direct relevance to public health. This may include studies to assist national or international bodies in designing food fortification programs by providing information on functional fortification levels or studies that may result in recommendations to the public on how to improve their nutritional status. Such generally applicable quality control protocols are still lacking, which leaves the assessment of measurement accuracy largely to the experience of the instrument user – which is highly variable, as for any instrumental technique or field of research.

16.7
Element-by-Element Review

16.7.1
Calcium

Although >99% of body calcium is located in the skeleton, its physiological role as an essential nutrient goes much further than maintaining skeletal integrity. In addition to its structural role in bone in the form of hydroxyapatite, calcium is needed for nerve signal transmission and muscle contraction. Furthermore, free Ca^{2+} ions or calmodulin-bound calcium activates various enzyme systems involved in gene expression, immune function, and cell growth. Because of its vital role in cell signaling, circulating concentrations in serum are kept in a narrow range, with bone acting as the central storage tissue. Homeostatic mechanisms are supported by the kidneys, which can adjust calcium reabsorption from the primary urine and the intestine, where calcium absorption efficiency is modulated through antagonistic interaction of calcitonin with parathyroid hormone (PTH) and vitamin D [74].

Calcium consists of six stable isotopes, ^{40}Ca, ^{42}Ca, ^{43}Ca, ^{44}Ca, ^{46}Ca, and ^{48}Ca, with average natural abundances of 96.941, 0.647, 0.135, 2.086, 0.004, and 0.187%, respectively. All stable calcium isotopes can be used as tracers apart from ^{40}Ca with its high natural abundance. The least abundant isotopes ^{43}Ca, ^{46}Ca, and ^{48}Ca are most suitable, in particular for intravenous administration. The lower the natural abundance, the smaller is the required dose and, thus, its effect on calcium homeostasis in plasma. For isotopic labeling, differences in calcium speciation between the isotopic label and calcium in the meal or compound to be tested must be taken into account. Calcium absorption from intrinsically and extrinsically labeled meals was found to be equivalent for wheat [75] and dairy products [76], but not for green leafy vegetables [77, 78]. Heaney *et al.* [79] found a difference in calcium absorption from labeled calcium chloride and tricalcium phosphate when added to soy milk, but no difference for calcium carbonate and calcium citrate [80]. In the latter studies, the calcium salts were directly synthesized from the label as the standard procedure for intrinsic labeling of chemical compounds.

The kidneys play a central role in homeostasis, with excessively absorbed calcium being rapidly excreted in urine. Calcium discharge in urine amounts to about 100–500 mg per day, depending on calcium intake. This makes urinary monitoring the method of choice for studying true calcium absorption, that is, the fraction of calcium in the intestinal lumen that is transferred through the mucosa into circulation. Assessment of calcium absorption by urinary monitoring requires two different isotope labels. One isotope label is used to label the calcium in the meal, which is fed in a controlled setting, usually after an overnight fast. The second label is injected intravenously following test meal intake, usually as $CaCl_2$ in physiological saline. It can then be used as a reference as its plasma uptake is comparable to a situation in which the label is absorbed quantitatively via the oral route. Both isotopic labels are cleared from the plasma with some of it being discharged in urine. The absorbed amount of oral label can then be calculated from the amount ratio of oral to intravenous label in the urine [33].

A major obstacle associated with the urinary monitoring technique is the need for an intravenous label and extended urine collections. To control for offsets in plasma and urinary appearance between the two labels, urine should be collected over 24 h following isotope administration to minimize the effect of differences in the urinary appearance of the labels on the measurement result [81]. Urine collections over 24 h or longer permit late colonic absorption of calcium also to be covered, which was found to be 5–10% of total calcium absorption [82], and can be affected by probiotics in the diet [83]. Attempts have been made to use single serum samples instead of 24 h urine collections [84]. This is basically possible considering that at some point in time both labels are likely to occur in the "correct" amount ratio in serum and/or urine. The application of such simplified protocols, however, is strictly limited to the same population and age group and the timings, dosages, and form of isotope administration described in the original protocol which must be strictly followed. The same holds true for the radioisotope method suggested by Heaney and Recker [85], which simplified the technique further by using a single oral label and a single serum collection only, an approach adopted recently for stable isotopes [50, 86]. Here, differences in body calcium were considered by assessment of body weight and linear regression was used for calibrating the technique against the standard double-isotope technique with 24 h urine collections. The resulting empirical approach is certainly more practical to use than the more accurate physiological approach, but can never yield more than a surrogate measure, which may or may not be significantly biased.

Both the efficiency of calcium absorption from the diet and physiological losses through urine, secretions into the gastrointestinal tract, and other pathways, such as sweat or hair, determine the amount of calcium that must be provided through the diet to remain in calcium balance [87]. Additional calcium is needed during the period of growth for building up skeletal mass. Through the entire life, bone is remodeled with bone accretion dominating during growth and bone resorption when the body is aging. Variations in bone balance must be understood in order to make age-adjusted recommendations for an optimum diet to maintain skeletal integrity and physiological function [88]. Calcium absorption can be studied by the dual isotope urinary monitoring technique. Endogenous losses can be easily quantified by giving an intravenous tracer and measuring the tracer recovery in pooled fecal samples [89]. Other parameters of calcium metabolism are more difficult to assess and require compartmental modeling techniques [90]. This refers in particular to the size of the exchangeable bone calcium pool, the rate of calcium deposition or release from bone, and bone calcium balance.

Compartmental modeling involves the administration of an oral tracer, usually ^{42}Ca or ^{44}Ca, and an intravenous tracer of low natural abundance (^{43}Ca, ^{48}Ca, or, ideally, ^{46}Ca). The intravenous tracer is infused slowly to mimic the time course of uptake of the oral isotopic label. Plasma, feces, and urine samples are collected at regular time intervals over several days to determine the appearance and disappearance of the tracer with time. Tracer kinetics are modeled using dedicated software, which delivers compartment sizes and flux rates [36]. The quality of the information generated depends not only on the accuracy and number of available data points, but also on the model of calcium metabolism selected. Over the past decades, several models have been suggested and used [37]. Their application has permitted the study of the effect of interventions on calcium metabolism in controlled clinical trials, and also differences between population groups related to age, gender, ethnicity, and so on, and have even been used during space flights [91]. Observable changes in bone resorption and deposition rates were found to parallel changes in biochemical markers of bone metabolism in serum and urine in several studies [92, 93].

For technical and practical reasons, compartmental modeling techniques are confined to data collected over 1–2 weeks after isotope administration. This is a relatively short time span considering that osteoclasts need about 2 weeks for bone resorption and osteoblasts about 100 days to build up new bone matrix [94]. A longer observational period would permit more accurate information on long-term processes to be gathered. Dose requirements that would permit the detection of calcium stable isotopic labels over months and even years, however, would be massive. This gap can be closed by using ^{41}Ca, a very long-lived calcium radionuclide with a half-life of 105 000 years [95]. Because ^{41}Ca is virtually nonexistent in Nature, but detectable at ultra-trace levels using accelerator mass spectrometry (AMS), it became possible to label the calcium in the skeleton isotopically [37, 50]. Amount ratios of ^{41}Ca to natural calcium in serum equal those in urine [96], which permits the study of tracer kinetics using urine samples alone. Changes in urinary ^{41}Ca excretion have been suggested not only as a sensitive marker of changes in bone resorption and turnover, but also as a sensitive marker for cancers affecting bone turnover [97].

Measurement precision and, consequently, dose requirements are of minor concern when studying calcium metabolism using stable isotopic labels. First, dietary calcium intake is higher by 1–2 orders of magnitude compared with trace elements such as iron and zinc. This increases significantly the margins for physiologically meaningful oral doses. Second, the abundance of the minor isotopes of calcium (^{43}Ca, ^{46}Ca, and ^{48}Ca) is so low that even with a moderate precision in isotopic analysis, doses for intravenous administration can be kept low enough not to disturb plasma homeostasis. As a rule of thumb, about 10% of plasma calcium is acceptable as the maximum dose [90]. Finally, calcium amounts in urine or feces are high enough not to pose a significant challenge to mass spectrometric sensitivity. For serum samples, a technique employing acid equilibration has been developed that uses the high sensitivity of sector field ICP-MS for reducing the sample size to only 20 μl of serum [98]. In a study using TIMS, it was demonstrated that saliva samples can also be used for kinetic studies [65].

All of the above taken together may explain why calcium is among the elements most intensively studied using stable isotope techniques. A comprehensive review of the state-of-the-art of calcium isotope ratio measurements was presented recently by Boulyga [99]. Single-collector ICP-MS or TIMS permits the determination of calcium isotope ratios at a precision of the order of 0.1–0.5%, depending on the ratio, instrumentation used, and experimental conditions. Such precisions are usually sufficient for calcium tracer studies, but not satisfactory for studies of natural isotope fractionation phenomena of calcium. As an emerging field with potential applications outside the Earth and Planetary Sciences [100], such studies require precisions of the order of 0.01% for $^{42}Ca/^{44}Ca$ analysis, which can only be achieved using multi-collector instruments. MC-TIMS and MC-ICP-MS can currently be considered comparable in terms of analytical performance, as mass spectrometric sensitivity is of minor concern for measurements in urine and blood. Whereas double-spiking techniques are the only available choice for correction of mass bias in MC-TIMS measurements [101, 102], standard–sample bracketing is the most suitable option in MC-ICP-MS [103]. For either technique, separation of calcium from the sample matrix is essential, which mostly neutralizes possible advantages of MC-ICP-MS in terms of sample throughput, especially when matrix separation can be automated [104]. In TIMS, matrix separation is essential, both to reduce the sample load on the filament and for efficient ionization. In ICP-MS, it is necessary to reduce both spectral and non-spectral interferences. In argon-based ICP-MS, $^{40}Ca^+$ is excluded from analysis owing to an unresolvable interference from $^{40}Ar^+$, even with high-resolution instruments. Because of the high $^{40}Ar^+$ intensity, other masses in the calcium mass spectrum can potentially be affected by scattering ions. Argon ionization can be reduced by plasma desolvation or by using "cool plasma" conditions, that is, low radiofrequency power and central gas flow rate [105]. Although analyte–matrix separation will not reduce $^{40}Ar^+$ interferences, it (largely) avoids interference from $^{86}Sr^{2+}$ ($^{43}Ca^+$) and $^{88}Sr^{2+}$ ($^{44}Ca^+$) and interferences arising from polyatomic ions such as MgO^+, CaH^+, KH^+, AlO^+, SiO^+, and SiN^+ [99].

Calcium and iron are possibly the most promising elements for the possible use of isotope fractionation effects in studies of element metabolism and for identification of impairments of clinical relevance. Skulan *et al.* [55] demonstrated in a landmark study that the calcium isotope fractionation effect increases with trophic level in aquatic food chains and in terrestrial organisms [106]. Chu *et al.* [107] followed the hypothesis that calcium isotope fractionation effects in mammals are associated with diet and suggested that calcium isotope fractionation effects could potentially be used as a marker of dairy product intake. Skulan *et al.* [55] suggested that calcium isotope fractionation effects in urine can be used to study bone calcium balance and turnover, a hypothesis that has been supported by a study in bedrest patients subjected to bisphosphonate treatment and a physical exercise regimen [108], and also by a study in mice showing differences in calcium isotopic composition of bone and plasma [109]. Heuser and Eisenhauer [110] more recently suggested mass fractionation during calcium reabsorption in the kidneys as another mechanism.

Current findings indicate the challenges associated with the possible use of isotope fractionation effects as clinical markers. In contrast to biochemical markers, which are associated with specific reactions or physiological processes, any of the numerous chemical reactions in which a given element is involved can contribute to its isotopic signature. Although this makes interpretation of a specific isotopic signature more difficult, it may offer the opportunity to study element homeostasis at the systemic level, which is currently difficult using conventional biomarkers.

16.7.2
Iron

It is well known and confirmed that heme-bound iron from animal-based foods is absorbed differently from iron in plant-based foods [111]. This requires intrinsic labeling of the diet/test meal if iron absorption from meat needs to be assessed. A separate pathway of iron absorption has been suggested for ferritin-bound iron [112], but it appears to be of minor relevance to iron nutrition. Ferritin is the body's most important iron storage protein and can trap iron in a hollow protein sphere, but it is effectively released during digestion [113]. Therefore, it should enter the same common chemical pool from which all other non-heme iron is absorbed.

Iron consists of four stable isotopes, ^{54}Fe, ^{56}Fe, ^{57}Fe, and ^{58}Fe, with average natural abundances of 5.84, 91.75, 2.1, and 0.28%, respectively. Because most functional body iron is present in blood, which is easily accessible to sampling, studies on iron absorption and utilization are nowadays mostly based on the incorporation and subsequent isotopic analysis of iron in red blood cells [9]. As explained earlier, the amount of iron isotopic label in circulation that has been absorbed can be assessed either by estimating blood volume and circulating iron [44] or by injection of a second isotopic label after test meal intake that serves as a reference [45–47].

The major challenges associated with the erythrocyte incorporation technique arise from the relatively high dose requirements. Red blood cells, as the target tissue, are the pool that contains most body iron. In addition, substantial interference from the polyatomic ions ArO^+, $ArOH^+$, and ArN^+ makes iron one of the most difficult elements to characterize isotopically by ICP-MS, in terms of both precision and accuracy. This refers specifically to biological samples, where additional interferences from $^{40}CaO^+$, $^{40}CaOH^+$, and $^{42}CaO^+$ ions from the matrix can affect the measurement and make matrix separation advisable. Unless iron uptake from large amounts of iron, such as from iron supplements, is the subject of investigation, standard quadrupole-based ICP-MS systems are less suitable. High-resolution or collision/reaction cells are required for overcoming spectral interferences, possibly supported by cool plasma conditions, iron separation from matrix elements, and/or the use of a desolvating nebulizer [114, 115]. Under optimum operating conditions, the precisions in iron isotope ratio measurements that can be achieved using single-collector instruments are of the order of 0.1–0.3% RSD. This is not ideal, but can be sufficient for the study purpose if higher isotope doses are acceptable. In combination with an MC set-up for simultaneous ion beam detection, precisions of the order of 0.01% RSD can be achieved in whole-blood samples [116–119], which usually allows the use of isotope doses well below average daily iron intakes. This makes MC-ICP-MS and MC-TIMS, the most advanced techniques for isotopic analysis of metals, the tool of choice and often the single possible option for this methodology. However, it must be noted that such precisions can only be achieved following rigorous sample preparation and quality control protocols to ensure that data are both precise and accurate [73].

MC-ICP-MS instruments are still not widely available and owners are often reluctant to accept isotopically enriched samples for measurements in case of possible carryover effects. Single-collector instruments are more widely accessible and attempts have been made to reduce the dose requirements by studying incorporation of isotopic label into reticulocytes, that is, newly formed red blood cells, rather than the entire red blood cell population [120]. Because freshly absorbed label occurs first in the reticulocyte fraction, iron isotopic enrichments for a given dose are much higher and permit working with about one-third of the original dose. Fecal balance studies remain an option if measurement precision and, therefore, isotope dosages are the limiting factor, but they are less favored owing to technical inconveniences both for the volunteer and the researcher, and also possible inaccuracies associated with incomplete feces collection [29].

Because urine is not a major pathway of iron loss, urinary monitoring is not a relevant choice for iron, simply for physiological reasons. Plasma appearance using either one or two isotopic labels, however, is a viable alternative to the erythrocyte incorporation technique [45]. By injection of a known amount of iron isotopic label, a reference for plasma appearance is introduced, which is comparable to complete absorption of the oral label. Iron absorption can theoretically be determined from the amount ratio of oral to intravenous label in serum at a certain time point, but owing to differences in the plasma kinetics of both labels, it is advisable to compare areas under the curve resulting from the time-dependent changes in plasma appearance of the labels [121]. Apart from lower requirements in terms of measurement precision, plasma-based techniques can be very useful in situations where the volume of blood (to estimate circulating iron) and erythrocyte incorporation efficiency of isotopic labels are highly variable and difficult to predict. This is the case during pregnancy [45–47] or in situations where red blood cell uptake of iron is impaired due to other micronutrient deficiencies.

Compartmental modeling techniques were used in the 1960s and 1970s to assess sizes and exchange rates of major body iron compartments, including blood cells, liver, muscle, bone marrow, spleen, and plasma [122]. Ferrokinetic studies are still used as a diagnostic tool in clinical settings using iron radioisotopes [123]. It is unlikely that stable isotopes will replace radioisotopes in these applications for the simple reason that the latter can be analyzed much more easily and faster, which is decisive in patient care. The potential of iron stable isotopes for ferrokinetic studies, however, has yet not been fully made use of in research settings. Sarria *et al.* [121] introduced a single compartmental model to assess iron absorption by kinetic modeling. Studies by Fomon and co-workers [53, 54] and Hunt *et al.* [52] determined iron losses and turnover in long-term studies based on the fate of the iron isotopic label in the body. Modeling techniques would permit observational time spans to be shortened and are waiting to be explored for retrieving this information.

As a result of the relatively large relative mass differences of its naturally occurring isotopes, iron has always been a candidate element for natural isotope fractionation effects. Attempts to measure these effects have failed over decades mainly because of mass spectrometric limitations. It was only in the 1990s that evidence for natural variations in the isotopic composition of iron could be generated, mainly owing to analytical developments. Whereas first investigations employed MC-TIMS and double-spiking techniques to correct for instrumental fractionation [124], MC-ICP-MS is now widely accepted as the method of choice owing to a usually better precision at often higher mass spectrometric sensitivity [125, 126]. Early reports of iron isotope fractionation effects led to the hypothesis that they could be used as tracers of biological activity [58, 127], even in extraterrestrial material. The past decade, however, has shown that abiotic processes such as redox reactions, precipitation, diffusion, and complexation can likewise account for such effects. Because such processes also occur in the human body, the observation of Walczyk and von Blanckenburg [117] that iron in human blood differs from diet and geosphere in its isotopic composition was not unexpected. In a series of follow-up studies, the authors provided further evidence that the efficiency of iron absorption is primarily responsible for this effect [128]. In agreement with this hypothesis, it was found that hemochromatotic patients have a different iron isotopic composition of blood [129]. Preliminary data from an independent trial confirmed this finding and indicated a similar effect for zinc [130], which requires further investigation. In a recent experiment in a pig, iron isotopic analysis of the pig's mucosa, dietary iron sources, and tissues of all organs involved in iron metabolism provided direct evidence that iron uptake by the intestinal mucosa is mass-sensitive [131].

Because the iron turnover of the body is low (<0.03% of body iron is lost/replaced per day), it should take years until even a drastic change in iron isotopic composition of absorbed iron has a measurable impact on the iron isotopic composition of blood. Ohno *et al.* [132] verified this hypothesis by determining the iron isotopic composition of blood in a healthy individual over 12 months. On the one hand, its stability disqualifies iron isotope fractionation effects *a priori* as a biomarker of acute conditions affecting iron metabolism. The stability of iron isotope fractionation effects in blood, however, can potentially be used to discriminate between individuals with a genetic predisposition or persistent dietary habits that result in an inefficient absorption of iron in the intestine. Such differences remain elusive to date, apart from disorders with a strong impact on iron absorption such as hemochromatosis [133] or thalassemia [134]. At present, it is hardly possible to identify impairments in iron absorption efficiency using conventional tracer techniques, as intestinal iron absorption varies substantially within and between days in the individual.

16.7.3 Zinc

The adult human body contains about 2 g of zinc with about 60% being located in muscle tissue, 30% in bone and 10% in other organs. Zinc is an intracellular ion with only 0.1% of body zinc circulating in plasma. As opposed to iron or calcium, there is no specific storage organ for zinc in the body. Excess zinc is deposited inside the cell in metallothioneins (low molecular weight and cysteine-rich proteins). Zinc is involved in a multitude of chemical reactions in the body, primarily in the form of zinc-dependent enzymes. Zinc plays a functional role in DNA and RNA replication, heme synthesis, free radical defense, and the metabolization of fats, proteins, and carbohydrates. Disorders associated with zinc deficiency include impaired growth, development, and immune function. Poor zinc status affects blood clotting and wound healing, thyroid function, taste perception, sperm development, and pregnancy outcomes [74].

Zinc consists of five stable isotopes, ^{64}Zn, ^{66}Zn, ^{67}Zn, ^{68}Zn, and ^{70}Zn, with average natural abundances of 48.63, 27.90, 4.10, 18.75, and 0.62%, respectively. The most ideal stable isotopic tracers are ^{67}Zn and ^{70}Zn owing to their low natural abundance and ^{68}Zn as an optional tracer. Because of the ease of extrinsic labeling, many isotope studies have been conducted over the past decades to evaluate whether zinc absorption efficiency from extrinsically and intrinsically labeled foods is similar or, ideally, non-distinguishable. Although there are no indications for distinctly different pools from which zinc is absorbed, as is the case for heme and non-heme iron, the common pool theory could not be consistently validated [135–139]. This indicates that extrinsic labeling must be considered with caution in zinc studies. Differences in zinc absorption were also observed for labeled ZnO, as a water-insoluble compound, and water-soluble $ZnCl_2$ [140]. This illustrates the need for the synthesis of fortification compounds from isotopically labeled zinc for assessment of bioavailability, which should be considered a general recommendation for any element under investigation.

As for iron, administered doses need to be kept as low as possible in order not to disturb plasma homeostasis and/or to ensure that the added label does not alter the efficiency of element absorption [141]. This affects, to some extent, the choice of methodology and instrumental set-up for isotopic analysis. The most straightforward methodology, as for any other element, is the assessment of apparent zinc absorption as the difference between administered isotope dose and dose recovered in feces following oral administration of the isotope. Examples in which this technique has been used successfully include the effect of zinc intake on fractional zinc absorption from the diet [141, 142], studies on the effect of diets high in animal products [143] or high in cereals, vegetables, and fruits [144], and the effect of contraceptive use [145], copper [146], and age [147]. Incomplete collection of feces after dose administration is a well-known source of bias for this methodology. For zinc absorption studies, ytterbium [148] and dysprosium [26, 149] have been suggested and successfully validated as quantitative markers for completeness of fecal collection. Rare earth elements are practically not absorbed in the human intestine. Given together with the zinc isotopic label in aqueous solution, their incomplete recovery is indicative of incomplete collection of fecal matter by the volunteer. In the above studies, their appearance in stool was found to follow closely that of the extrinsic zinc isotopic label. Given in parallel with the isotope dose, they can potentially be used to shorten fecal collection periods by extrapolating recovered amounts of isotopic label and rare earth elements in stool to 100% of the administered dose for the rare earth element.

It is no coincidence that most studies that rely solely on apparent absorption using a single isotopic label date back to the early years of using isotopic labels in nutrition studies. First, the attainable precision in zinc isotopic analysis of available instrumentation poses minor restrictions on isotope doses. Most of the ingested label is excreted in feces and zinc possesses a stable isotope of very low abundance (^{70}Zn) for which dose requirements to obtain a well-resolvable isotopic enrichment in the stool are relatively low in most conditions. Second, it was only in the 1990s that it was widely recognized that endogenous losses of zinc into the intestine through digestive juices are not only significant, but also highly variable, being a central mechanism of human zinc homeostasis [150, 151]. Endogenous zinc losses can be assessed by administration of an intravenous dose of a zinc isotopic label and measuring the appearance of the isotopic label in stool [152–155]. For intravenous administrations, ^{70}Zn is the isotope of choice owing to its low abundance. As discussed earlier, minimization and slow administration of the intravenous dose are crucial in order not to trigger a homeostatic response by the body, which would ultimately result in bias.

When endogenous losses are known, it is possible to calculate true zinc absorption as the amount of isotopic label crossing the intestinal mucosa [141, 156, 157]. This is done by correcting the fecal recovery of the oral isotopic label for the fraction of oral label that has been absorbed and readily excreted into the intestine. The latter equals the percentage of intravenous isotopic label that has been recovered in the collected stool. Alternatively, plasma deconvolution analysis can be used for assessment of true zinc absorption based on the appearance and disappearance of the oral and intravenous label in plasma [158–161]. Here, appearance of the intravenous label is used as a reference, as it enters circulation directly, which is comparable to 100% absorption if the label were to have been taken up through the intestine.

While fecal collection is cumbersome, plasma deconvolution analysis is more invasive as it requires venipuncture not only for label administration, but also for collection of plasma samples. As for calcium, it has therefore been explored whether the double-isotope urinary monitoring technique can be employed for zinc. Here, true absorption is calculated from the amount ratio of oral to intravenous label in urine, which, ideally, should equal that in plasma and would make plasma sampling and fecal collection superfluous. This approach has been validated in several studies [159, 162–164] and is now commonly accepted as the method of choice, provided that intravenous tracer administration is possible and ethically acceptable. Examples include studies during pregnancy [158, 165–167], investigation of the effect of zinc intake on zinc absorption [168], and studies to assess zinc absorption from maize [169] and biofortified wheat [170]. As for calcium, the accuracy of measuring true zinc absorption by urinary monitoring depends mainly on two factors. First, the intravenous dose must not disturb plasma homeostasis, and second, the kinetics of appearance and disappearance of the tracer have to mimic closely those of the oral tracer. If the latter is given, the amount ratio of both labels in plasma, and consequently urine, would be invariable over time and could be retrieved from a single spot urine sample. However, both preconditions are rarely fulfilled, which demands longer collection periods and possibly an empirical identification of the most suitable time point and duration of urine collection. Approaches include the collection and analysis of several spot urine samples from the same subject, collected over several days [165, 167, 168], 24 h urine collections on days 3–6 post-isotope administration [157, 163] or spot urine samples on days 6–8 post-dose [170]. To avoid the need for an intravenous injection, Yeung *et al.* [171] suggested the use of an oral isotopic label only and the use of a constant to translate oral tracer recovery in urine to true zinc absorption. The basic thinking is the same as in the single isotope approach suggested for calcium [50, 86], but both come with the same restriction that they can only be considered a surrogate measure as they may yield accurate results only in the population group and experimental setting in which they have been validated.

A major obstacle in understanding homeostatic mechanisms for zinc is the difficulty in measuring zinc status and to assess zinc deficiency reliably in the individual. In this domain, stable isotope techniques have contributed substantially to our understanding of zinc metabolism, through the insights gained by kinetic studies and compartmental analysis as reviewed by Krebs and Hambidge [172]. Over the past three decades, more than a dozen studies have been conducted to develop compartmental models of human zinc metabolism, based on the administration of an oral and an intravenous label (see Wastney *et al.* [173], Miller *et al.* [174], Griffin *et al.* [175], King *et al.* [176], and Donangelo *et al.* [177]). Identified models differ in complexity and range from relatively simple three-compartment models to models with as many as 10 compartments. Apart from their important contributions to our understanding of zinc physiology and homeostasis, which go beyond the scope of this book, a major contribution of these investigations lies in the evolution of a possible biomarker of zinc status. By measuring the disappearance of an intravenous tracer over 10 days, the exchangeable zinc pool (EZP) can be estimated, that is, the zinc pool that exchanges with zinc in plasma within 3 days [178]. The size of the EZP can also be derived from isotopic data for urine [179, 180]. EZP was found to correlate well with endogenous losses, and also with absorbed zinc, which illustrates its value as a potential marker of zinc status [181].

The first studies employing zinc stable isotopes for metabolic studies used fast atom bombardment (FAB) as the ionization technique for mass spectrometric analysis [182]. FAB-MS was not selected at that time as the method of choice; it was selected because of availability, rather than performance characteristics [183]. With the wider availability of ICP-MS by the end of the 1980s, ICP-MS became the preferred analytical technique for zinc tracer studies. While single-collector ICP-MS instruments at that time compared well with single-collector TIMS in terms of analytical precision, its capability of higher sample throughput made the difference. At relatively poor external precisions of the order of 0.3–1.0% for zinc isotopic analysis, simple digestion without further matrix separation was found to give sufficiently accurate data for zinc kinetic studies [184]. This perception has now changed, however, mainly because of a better recognition of matrix-related interferences affecting zinc. Spectral interferences may affect all zinc isotopes [185] and are a result of the occurrence of isobaric, doubly charged, and polyatomic ions: ^{64}Zn ($^{64}Ni^+$, $^{40}Ar^{24}Mg^+$, $^{36}Ar^{12}C^{16}O^+$, $^{32}S^{16}O_2^+$, $^{27}Al^{37}Cl^+$, $^{48}Ti^{16}O^+$, $^{48}Ca^{16}O^+$), ^{66}Zn ($^{132}Ba^{2+}$, $^{40}Ar^{26}Mg^+$, $^{49}Ti^{16}O^1H^+$, $^{34}S^{16}O_2^+$), ^{67}Zn ($^{134}Ba^{2+}$, $^{35}ClO_2^+$), and ^{68}Zn ($^{70}Ge^+$, $^{140}Ce^{2+}$, $^{40}Ar^{14}N^{16}O^+$, $^{40}Ar^{14}N^{16}N^+$, $^{35}Cl_2^+$, $^{54}Fe^{16}O^+$). Although they cannot be fully eliminated by matrix separation, their effects on analytical accuracy and precision can be reduced in this way. Although chelation/extraction techniques have been explored for clinical samples [184, 186], ion-exchange chromatography is now considered preferable for high-precision isotopic analysis [187, 188].

Although external precisions for zinc isotopic analyses of the order of 0.5% RSD are sufficient for simple fecal balance studies, they are less satisfactory for applications that require the intravenous administration of a zinc isotopic label. Using MC-ICP-MS, external precisions better than 0.1% RSD can be achieved routinely for isotopically enriched samples. This has been demonstrated by Tanimizu *et al.* [189], who measured Zn isotope ratios at external precisions of 0.05% RSD for the $^{70}Zn/^{64}Zn$ ratio and down to 0.01% for the $^{66}Zn/^{64}Zn$ ratio using double-spiking techniques and double-focusing MC-ICP-MS. Similar precisions were achieved by Ingle *et al.* [190], who compared a quadrupole-based TIMS set-up with a magnetic sector MC-ICP-MS system equipped with a collision cell and a double-focusing MC-ICP-MS instrument for zinc isotopic analysis of fecal samples. This compares with reported external precisions of ~0.02% RSD for $^{67}Zn/^{66}Zn$ analysis by MC-TIMS using internal normalization techniques for correction of instrumental fractionation [191] and 0.1% combined relative uncertainty for the $^{68}Zn/^{67}Zn$ isotope ratio [192]. A major difference, however, lies in the zinc requirements for isotopic analysis. Zinc has a very high first ionization energy (9.39 eV), which results in very low thermal ionization efficiencies of the order of 0.01%. This makes it hardly possible to reduce sample loadings below 500 ng zinc in TIMS with Faraday cup detection. This is different for MC-ICP-MS, which has been successfully used to measure zinc isotope ratios at sample loadings as low as 10 ng of zinc in carbonate samples [193]. Mass discrimination in these measurements has been corrected for by doping with a copper standard of known isotopic composition for parallel measurement of mass bias. This technique, known as "empirical external normalization," is unique to MC-ICP-MS and is based on the assumption that regularities and degree of mass discrimination in the ICP-MS instrument are very similar for different elements (see also chapter 5). Originally developed by Maréchal *et al.* [57], it has been further refined and validated for copper and zinc analysis over recent years, including a critical consideration of possible shortcomings [185, 194, 195].

To date, only MC-ICP-MS has been employed to study natural zinc isotope fractionation phenomena, although such measurements would also have been theoretically possible using MC-TIMS. While external normalization techniques can only be used in ICP-MS to correct for instrumental mass bias, double-spiking techniques are applicable for both types of instruments. For the measurement of relative differences of an isotope ratio relative to an accepted standard (δ-values), external precisions of the order of 0.005% RSD and better can be achieved by MC-ICP-MS [196]. Relative (differential) isotope ratio measurements can be conducted at a smaller measurement uncertainty than absolute isotope ratio measurements, but a major difficulty arises from the need to agree on a common laboratory standard for reporting such measurements to make measurements comparable between different laboratories (see also Chapters 6 and 7) [196].

Variations in Zn isotopic abundances in nature are now well confirmed and have been the subject of several investigations in the past decade, as reviewed by Cloquet *et al.* [187]. Although the focus has been largely on geological and environmental samples, including plants [196–198], attempts have also been made to detect zinc isotopic variations in animals and humans. In a seminal paper on natural copper and zinc isotope fractionation effects, Maréchal *et al.* [57] also analyzed a human blood sample. Later, Stenberg *et al.* [199] reported a significant difference in zinc isotopic composition between blood and hair, which was soon confirmed by Ohno *et al.* [200], who also found that the zinc isotopic signature in blood is not subject to seasonal variations. The most comprehensive study so far was reported recently by Balter *et al.* [201], who analyzed various body tissues of sheep for zinc isotopic composition. The data showed an enrichment of bone, muscle, serum, and urine in the heavier isotopes, whereas feces, red blood cells, kidney, and liver were found to be enriched in the light isotopes. Systematic investigations are now needed to elucidate the factors that determine the observed variations in zinc isotopic composition of mammals and how these variations can be used and under which conditions they may serve as biomarkers of human zinc metabolism.

16.7.4 Magnesium

Magnesium is an essential element that activates more than 300 enzyme systems. In addition to being an enzyme cofactor, major physiological roles of magnesium include bone and teeth mineralization, muscle contraction, nerve impulse transmission, and blood clotting The human body contains about 35 g of magnesium, two-thirds of which is located in bone as the major storage compartment. Magnesium deficiency is discussed as a risk factor for osteoporosis, diabetes mellitus type II, and hypertension [74].

Magnesium consists of three stable isotopes, ^{24}Mg, ^{25}Mg, and ^{26}Mg, with average natural abundances of 78.99, 10.00, and 11.01%, respectively. ^{25}Mg and ^{26}Mg can be used as tracers. About 20–60% of dietary magnesium is absorbed [202]. This makes it technically less challenging to determine apparent absorption as the difference between ingested isotope dose and fraction of dose recovered in feces is significant. The higher the tracer recovery in feces, the lower are the dose requirements to achieve a well-measurable isotopic enrichment. Apparent absorption has been successfully determined in several studies by administration of a single isotopic label and fecal monitoring [203–205]. In two of these studies, recovery of the oral label in urine was used to determine magnesium retention

[203, 204]. For green leafy vegetables, extrinsic and intrinsic labeling yielded similar data, which minimizes labeling efforts for these highly relevant magnesium sources in the diet [206, 207].

Apparent magnesium absorption can be determined even with basic quadrupole-based ICP-MS instruments, which can deliver precisions of 0.2–0.5% RSD for measuring the $^{25}Mg/^{24}Mg$ and $^{26}Mg/^{24}Mg$ isotope ratios [208, 209], which is similar to what can be achieved using sector field ICP-MS [210] or collision cell ICP-MS [211]. Such precisions require isotope doses of the order of 20–50 mg to reach enrichments in the percent range in the collected fecal material. This compares with a dietary magnesium intake of about 300 mg per day.

Fecal collections are cumbersome and deliver only apparent absorption or retention as information, and not the true absorption. Several attempts have been made in the past to explore whether tracer recovery of an oral label and a second intravenously administered isotope can be used to assess true absorption by urinary monitoring [22, 208, 212]. For validation, oral tracer recovery in feces has been used to assess apparent absorption, which yields true absorption if corrected for endogenous losses via fecal recovery of the intravenous tracer. Studies differ, however, in the suggested collection period for urine delivering similar results to the fecal monitoring technique. This can be explained by difficulties in administering the intravenous tracer slowly enough or in keeping doses low enough. Magnesium concentrations in serum are maintained in a narrow range, which may result in rapid discharge of the intravenous tracer via the kidneys if the burden of the intravenous dose is too high [22]. As for calcium, this may result in transient changes in the amount ratio of oral and intravenous tracer in urine, which translate into time-dependent changes in the calculated absorption value.

The parallel use of an oral and an intravenous label also permits the study of magnesium kinetics and the use of compartmental modeling techniques to assess basic parameters of magnesium metabolism [213–215]. In these studies, the branched three-compartmental model suggested by Avioli and Berman [216] was used, which was derived from a radiotracer study in humans using ^{28}Mg. Feillet-Coudray *et al.* [215] used these experiments to evaluate whether magnesium status can be assessed using the exchangeable pool concept suggested by Miller *et al.* [178]. Magnesium status is difficult to determine, as serum magnesium concentrations are kept in a narrow range, unless stores are substantially depleted. Low serum magnesium concentrations are therefore only observable in cases of severe magnesium deficiency. With a similar motivation, Wälti *et al.* [217] evaluated whether urinary excretion of an intravenously administered stable isotope dose can be used to assess status. While the study of Feillet-Coudray *et al.* [215] failed to detect a response of exchangeable magnesium pools to magnesium supplementation, no correlation between the isotopic information and magnesium concentration in muscle biopsies could be observed in the study by Wälti *et al.* [217].

Restrictions on the use of ICP-MS in magnesium metabolic studies arise primarily from the limits on precision. Single-collector instruments, irrespective of whether they are equipped with a quadrupole filter, with or without an additional collision/reaction cell, or a sector field mass spectrometer, are not capable of reaching precisions in magnesium isotopic analysis of better than 0.1% RSD, irrespective of possible limitations on accuracy. Analysis of organic matrices without prior magnesium separation can suffer from spectral overlap due to the occurrence of ions such as $^{23}Na^{1}H^{+}$, $^{12}C^{14}N^{+}$, $^{12}C_2^{+}$, and $^{48}Ca^{2+}$ [210, 211]. In consequence, single-collector instruments are most suitable for methodologies involving relatively high tracer recoveries. This includes the assessment of apparent absorption by fecal monitoring or plasma appearance/plasma deconvolution, in which the short-term appearance of the tracer in serum is monitored and absorption determined as the area under the resulting curve [208]. The same holds for single-collector TIMS. For applications that require the administration of an intravenous dose, dose minimization and, therefore, ultimate precision in isotopic analysis are pivotal. Required precisions of the order of 0.01% and better can only be achieved using MC-ICP-MS [218, 219] or MC-TIMS [22]. Although MC-TIMS and MC-ICP-MS are comparable in terms of precision for tracer experiments, MC-ICP-MS is certainly more sensitive. This allowed the determination of the amount ratio of oral and intravenously administered magnesium tracers in erythrocytes, similarly to the principle described earlier for measuring iron absorption. True magnesium absorption determined via erythrocyte incorporation of isotopic labels was not statistically different from true magnesium absorption determined by the double-isotope fecal monitoring technique [22].

Because magnesium has only three stable isotopes, double-spiking techniques cannot be employed to study magnesium isotopic fractionation by MC-TIMS. However, this is possible using MC-ICP-MS, which yielded clear evidence for isotope fractionation effects for magnesium in Nature [218–220]. However, no such measurements have been conducted so far in humans, although significant isotope fractionation effects similar to those observed for calcium can clearly be expected.

16.7.5
Selenium

The human body contains 15–30 mg of selenium. Selenium consists of six stable isotopes, ^{74}Se, ^{76}Se, ^{77}Se, ^{78}Se, ^{80}Se, and ^{82}Se, with average natural abundances of 0.89, 9.37, 7.63, 23.77, 49.61, and 8.73%, respectively. Selenium turnover is relatively low (about 0.1% of total body selenium per day), which comes with a current recommended dietary allowance (RDA) of only 15–70 μg per day, depending on age. Selenium deficiency may cause cardiac myopathy, induce arterial sclerosis, and negatively affect immune function. Selenium plays a central role in the anti-oxidant defense of the body in the form of glutathione peroxidase, which contains four selenocysteine residues. Its essential role, however, is not limited to the removal of reactive oxygen species (ROS). It is also a cofactor for other selenocysteine-containing enzymes, such as thioredoxin reductase and iodothyronine deiodinase, needed for activation and deactivation of thyroid hormones [74]. Selenium has been discussed as playing a role in preventing lung, colorectal, and prostate cancer through mitigation of oxidative stress. According to a recent Cochrane meta-analysis, however, there is no conclusive evidence yet that selenium supplementation reduces cancer risks, especially for individuals who are well nourished [221]. The safety range for dietary selenium intake is relatively small with a tolerable upper intake level of 400 μg per day. Excessive selenium intake can result in selenosis, which is characterized by a garlic odor in the breath, brittle nails, hair loss, and neurological damage.

Studies of selenium metabolism in humans using tracer techniques are challenging, for both physiological and analytical reasons. Selenium species both in food and in the human body are manifold, with new species still being identified [222–224]. Inorganic selenium (selenite SeO_3^{2-}, selenate SeO_4^{2-}, and selenide Se^{2-}) and organic forms of selenium, including selenoamino acids, are absorbed and metabolized differently, with selenite and selenate being apparently well absorbed, but less well retained by the body [225]. This makes selenium a prime example of an element for which absorption can differ significantly from bioavailability, that is, the fraction of an element in the diet that is both absorbed *and* utilized for physiological function.

Studies comparing selenium absorption from the same food labeled either intrinsically or extrinsically with a labeled inorganic selenium species, usually selenite or selenate, indicate that intrinsic labeling is a requirement for accurate assessment of selenium absorption from food. Significant differences between selenium absorption or retention from intrinsically and extrinsically labeled foods were found for poultry [226], egg white and yolk [227], trout and yeast [228], and garlic, wheat, and cod [229]. The requirement to label selenium in foods intrinsically makes non-isotopic, “cold” balance studies a viable alternative to isotopic techniques [230–232], despite the fact that these are now considered obsolete for other elements owing to their undisputed methodological limitations. For supplemental selenium, absorption and retention can be assessed using inorganic selenium species obtained by synthesis from isotopically enriched selenium [233]. For high-selenium yeast, as a widely available supplemental form of organic selenium, the selenium was labeled intrinsically by growing the yeast in a medium to which the isotopic label had been added in inorganic form [234].

In addition to the measurement of selenium absorption and retention, stable isotopic tracers have also been used for tracing plasma and urinary selenium kinetics for studying human selenium metabolism. Janghorbani *et al.* [235] used an intravenous dose of ^{74}Se-selenite and measured the appearance of the tracer in urine and serum to assess differences in selenium metabolism between Chinese males of different selenium status/intake. Kremer *et al.* [236] monitored *in vivo* methylation by giving an oral dose of ^{77}Se-selenite to a single human subject and quantifying the release of dimethyl selenide in the breath over time by GC–ICP-MS. In a series of rat studies, Suzuki and co-workers collected information on the metabolic conversion of various inorganic and organic selenium species by administering isotopically labeled selenium species orally and intravenously and by measuring their time-resolved appearance in urine and feces together with organ deposition [237–239]. In a recent human study, Wastney *et al.* [240]

administered ^{76}Se-selenite and ^{74}Se-selenomethionine orally and studied the tracer appearance in urine, feces, and blood over time to design a kinetic model of selenium metabolism that takes selenium speciation into account. Tracer kinetics in plasma and urine have also been suggested by Janghorbani *et al.* [241] to measure the selenite-exchangeable metabolic pool as a measure of selenium status. In contrast to zinc, however, the exchangeable pool concept has not been explored further.

The complexity of human selenium metabolism is one, but not the only, reason why selenium is among the essential elements that have been studied least using stable isotope techniques. Selenium isotopic analysis is also technically challenging for various reasons. Until the development of an NTIMS technique [242], mass spectrometric sensitivity was a major obstacle. At 9.75eV, the first ionization energy of selenium is forbiddingly high for efficient formation of positive ions in a thermal ion source. As for zinc, ICP-MS made a difference soon after becoming commercially available, simply because of the relative ease of obtaining high and stable ion currents [243]. ICP-MS analysis, however, comes with other challenges than TIMS, namely in terms of the accuracy of the measurement. The two most abundant selenium isotopes, ^{78}Se and ^{80}Se, suffer from spectral overlap with argon dimer ($^{40}Ar^{38}Ar^{+}$ and $^{40}Ar^{40}Ar^{+}$) signals, which can hardly be resolved even at a mass resolution of 10 000 (the current limit of commercially available sector field ICP-MS instrumentation). Interferences from doubly charged ions of the rare earth elements (Sm^{2+}, Gd^{2+}, Er^{2+}, Dy^{2+}), although not very abundant in biological samples, also need to be considered, apart from other polyatomic ion species.

Matrix separation, as in most cases, is the strategy of choice to improve the analytical accuracy in isotopic analysis by ICP-MS. Because selenium forms volatile hydrides, hydride generation (HG) ICP-MS is now widely employed to remove interfering matrix elements [244–247]. Various techniques have been developed to remove other hydride-forming elements, such as arsenic and germanium, that may form polyatomic ions interfering with the isotopic analysis of selenium. These include thiol-cellulose powder [248] or thiol-cotton [249] as elegant techniques for selenium sorption. Of course, HG is not suitable for avoiding the occurrence of argon dimer ions. This gives collision/reaction cell-based ICP-MS instruments an advantage over double-focusing sector field ICP-MS units. Using a suitable collision/reaction gas, the signals from the argon dimers are not resolved from the analyte signals, but the interfering ions are effectively removed from the ion beam [244, 250–252]. With basic quadrupole-based instrumentation, it is hardly possible to determine selenium isotope ratios at a precision/accuracy significantly below 1%. By using single-collector instruments equipped with a collision/reaction cell [248, 251] or double-focusing sector field instrumentation [248], it is possible to determine selenium isotope ratios at a precision and even accuracy of 0.2–0.5% RSD for most selenium isotope ratios. This level of performance can be considered sufficient for metabolic studies in humans and compares well with what can be achieved with NTIMS, although with much lower sample requirements.

As for many “heavy” elements in Nature, variations in natural isotopic abundances were supposed to exist for selenium, but were were considered for a long time to be immeasurable. This has changed with the advent of MC systems for ion detection, which improved the precision attainable in isotopic analysis substantially. Using either MC-NTIMS and double-spiking techniques [253–255] or double-focusing MC-ICP-MS [244, 247, 249, 256, 257], it has been demonstrated that relative differences in selenium isotope ratios can be determined at a precision of the order of 0.02% and better. At comparable precisions, isobaric interferences are virtually absent in MC-NTIMS, but the mass spectrometric sensitivity of MC-ICP-MS is more than 10 times higher. This analytical progress made the identification of significant variations in selenium isotopic composition in geological and microbial systems possible (see review by Johnson [258]). However, no attempts have been made so far to study isotope fractionation phenomena for selenium in mammals, including humans.

16.7.6 Copper

As for iron, differences in redox potentials are relatively small between the major oxidation states of copper, that is, Cu(I) and Cu(II). This gives copper its main function as a cofactor in enzymatic reactions involving electron-transfer processes. In the human body, most of the copper (about 40%) is present in muscle tissue with significant amounts also present in the liver, brain, and skeleton. About 5% of the copper can be found in serum, of which 80–90% is present as ceruloplasmin. Ceruloplasmin in serum and hephaestin at the basolateral side of the mucosa ensure oxidation of circulating Fe^{2+} to Fe^{3+} for iron binding to transferrin. Unbound Fe^{2+} is a major source of oxidative stress through Fenton/Haber–Weiss chemistry. Copper together with zinc is also a cofactor for superoxide dismutase, a key molecule in the anti-oxidant defense system of the body [74].

The adult human body contains 50–150 mg of copper. Copper consists of two stable isotopes: ^{63}Cu and ^{65}Cu, with average natural abundances of 69.15 and 30.85%, respectively. Because of its lower abundance, ^{65}Cu is the isotope of choice for tracer applications. As ^{65}Cu comprises about one-third of all copper atoms, dose requirements to obtain significant isotopic enrichment are relatively high compared with elements such as iron and calcium, which have stable isotopes with a natural abundance of less than 1%.

Because copper has only two stable isotopes, only one tracer isotope is available, which limits the choice of techniques for metabolic studies substantially. In most human studies, ^{65}Cu has been used to determine apparent absorption, as the difference between the amount of isotope ingested and the amount recovered in feces. To check for completeness of fecal collection, holmium has been suggested and successfully used as a non-absorbable, quantitative elemental marker in copper absorption studies [259]. Studies comparing copper absorption of foods intrinsically or extrinsically labeled with ^{65}Cu are limited. Johnson and co-workers [260, 261] found no statistically significant difference in copper absorption from intrinsically and extrinsically labeled goose liver, goose breast, peanut butter, and wheat. Harvey *et al.* [262] observed significant differences in copper absorption between intrinsically and extrinsically labeled sunflower seeds and soy beans, although they indicated that the observed differences could also have been due to differences in test meal size and, therefore, differences in molar ratios of copper to possible enhancers or inhibitors of copper absorption in the test meals.

Extrinsic labeling has been used in several studies to assess apparent copper absorption in humans. This includes studies to evaluate apparent copper absorption from plant- and animal-based diets [263, 264], and also the effect of ascorbic acid [265], age [146, 266, 267], copper intake [267–271], zinc intake [146], and iron supplementation [272] on copper absorption. By measuring urinary excretion of the tracer, copper retention has also been assessed from vitamin B6-deficient diets [271], in pregnant women [263], in young men consuming a low copper diet [268], and for high copper intakes [270].

Despite the fact that only one stable isotopic label can be used in metabolic studies, attempts have been made in the past to develop compartmental models of human copper metabolism, based on kinetic studies [273]. Scott and Turnlund [274] tried to overcome this limitation by giving an oral and an intravenous dose of ^{65}Cu to the same individual separated by 6 days and measuring the appearance of the tracer in plasma, urine, and feces following each of the two administrations. A similar approach was used by Harvey *et al.* [275] to determine the size of the exchangeable copper pool in humans and to estimate endogenous losses. Plasma kinetics were also explored as an alternative approach to assess copper absorption with the aim of avoiding cumbersome fecal collection and analysis [276, 277]. Because of the methodological limitations mentioned, radioisotopes and not stable isotopes are considered as the gold standard for studying copper metabolism. Using ^{67}Cu and whole-body counting, copper retention can be determined by direct counting without major limitations when radiation burdens are deemed tolerable, for example, in pathological conditions [28].

Both ICP-MS and TIMS have been used successfully for copper metabolic studies. A major advantage of using ICP-MS for copper isotopic analysis is the possibility of correcting for instrumental mass discrimination during isotopic analysis either by standard–sample bracketing or by "empirical external correction" using zinc as a monitor for instrumental mass bias [185]. Fractionation correction in TIMS is only possible for elements that have at least three stable isotopes and thus impossible for copper: attainable measurement repeatability can hardly be better than 0.1–0.5% RSD [191, 268]. A measurement repeatability of 0.03% RSD has been reported for a low-temperature TIMS technique [278], but this approach is difficult to apply for metabolic studies, as these usually come with demands for high sample throughput and sensitivity. For these reasons, ICP-MS is the method of choice although spectral interferences due to the occurrence of polyatomic ions, such as $^{40}Ar^{23}Na^{+}$ $^{32}S^{16}O_2{}^{1}H^{+}$, $^{33}S^{16}O_2{}^{+}$, $^{48}Ca^{16}O^{1}H^{+}$, $^{35}Cl^{14}N^{16}O^{+}$, and $^{40}Ar^{12}C_2{}^{1}H^{+}$, must be considered as a possible source of bias [279, 280]. This gives double-focusing instruments or instruments with collision/reaction cell technologies a clear advantage over basic quadrupole-based instruments.

Diemer *et al.* [281] compared the performances of different types of ICP-MS instrumentation for copper isotopic analysis, namely a quadrupole-based single-collector instrument and two double-focusing instruments (single- and multi-collector). Reported combined uncertainties for isotopic analysis were of the order of 0.3, 0.2, and 0.07% ($k = 1$), respectively, for absolute isotope ratio measurements of synthetic isotope mixtures [281]. A similar performance of 0.1–0.2% RSD for the $^{63}Cu/^{65}Cu$ isotope ratio was reported for a single-collector double-focusing sector field ICP-MS instrument [279]. This compares with a measurement repeatability of less than 0.01% RSD for isotope ratio measurements relative to a standard to study natural isotope fractionation phenomena using MC-ICP-MS [57, 194, 195]. Although copper has the potential for significant isotopic variations in clinical samples, no systematic investigations have yet been reported.

16.7.7
Molybdenum

Molybdenum is considered an ultra-trace element with an approximate amount of 5 mg in the adult human body. It is a cofactor for at least three enzymes in humans (sulfite oxidase, xanthine oxidase, and aldehyde oxidase) and is involved in the catabolism of sulfur-containing amino acids, purine, and pyrimidine. A better understanding of human molybdenum metabolism is needed in order to give evidence-based recommendations regarding optimal nutrition, although molybdenum deficiency and associated pathological symptoms have not yet been observed in humans [74].

Molybdenum consists of seven stable isotopes, ^{92}Mo, ^{94}Mo, ^{95}Mo, ^{96}Mo, ^{97}Mo, ^{98}Mo, and ^{100}Mo, with average natural abundances of 14.77, 9.23, 15.90, 16.68, 9.56, 24.19, and 9.67%, respectively. Half-lives of molybdenum radioisotopes are in general not long enough (<67 h) for kinetic studies or intrinsic labeling experiments. So far, only a single study has been conducted to compare molybdenum absorption from extrinsically and intrinsically labeled foods [282]. Absorption of both labels was statistically not different for kale. For soy, however, the intrinsic label was less well absorbed and less well excreted than the extrinsic label given as ^{96}Mo-molybdate.

Molybdenum is known to be well absorbed and, accordingly, only relatively few stable isotope studies have been conducted focusing on intestinal absorption alone. This includes two consequent studies by Turnlund and co-workers, who studied absorption, excretion and retention of molybdenum in young men based on fecal and urinary recovery of orally administered ^{100}Mo and infused ^{97}Mo at different levels of habitual molybdenum intake [283, 284]. Later, Sievers *et al.* [285] assessed molybdenum absorption and retention in infants by measuring the recovery of the oral label in urine and feces and Giussani *et al.* [286] determined molybdenum absorption in seven healthy adults by plasma deconvolution analysis of ^{95}Mo and ^{96}Mo isotopic labels, given orally and intravenously, respectively. In three different studies, Turnlund and co-workers assessed the effect of habitual molybdenum intake on molybdenum tracer kinetics using an oral and intravenous isotopic label [287–289]. Key findings were that molybdenum plasma concentration can be used to assess dietary molybdenum intake [288] and that the body retains molybdenum more effectively when molybdenum intake is low and discharges molybdenum when molybdenum intake exceeds requirements [289]. Observations are complemented by studies describing the kinetics of urinary and plasma clearance of oral and intravenously administered molybdenum by Werner *et al.* [290] and Giussani *et al.* [291].

Similarly to selenium, the high first ionization energy of molybdenum (7.1 eV) limits the ionization efficiency in PTIMS. Despite this restriction, Turnlund *et al.* used PTIMS in their molybdenum work successfully [292]. Using internal normalization and an MC instrument, they obtained molybdenum isotope ratios in relevant biological matrices at precisions of 0.05–0.08% RSD using 1–2 μg of separated magnesium. Turnlund *et al.* [292] Lu and Masuda [293, 294] reported similar figures of merit for single-collector PTIMS using 20 μg of molybdenum, separated from geological samples. Measurement of molybdenum isotope ratios is facilitated by the generation of negatively charged MoO_3^- ions, which permits sample amounts to be reduced to the low nanograms range [295–297], at comparable analytical precision to PTIMS. NTIMS has not only improved sensitivity, but also accuracy as possible isobaric interferences of ^{92}Zr, ^{94}Zr, ^{96}Zr, and ^{100}Ru become irrelevant at the higher mass of the MoO_3^- ions.

As for virtually any element subjected to ICP-MS analysis, also for molybdenum spectral interferences have to be considered. Nickel, manganese, iron, copper, cobalt, and zinc from the sample matrix can form polyatomic ions with oxygen, nitrogen, chlorine, or argon that can interfere with the isotopic analysis of molybdenum. Again, this makes matrix separation a precondition for precise and accurate isotopic analysis. Without matrix separation, Keyes and Turnlund [298] reported good reproducibilities of the order of 0.1–0.3% RSD for pure molybdenum standards using quadrupole-based ICP-MS, but with precisions in the percent range for digested and diluted plasma samples. No comments were made on the accuracy of the measured ratios *per se*, as the technique was designed for isotope dilution analysis of biological samples.

For ultimate precision and accuracy, either MC-TIMS or MC-ICP-MS is required. Achievable precisions for MC-ICP-MS are of the order of 0.002–0.01% [299–302]. These techniques have been developed and applied exclusively to study natural isotope fractionation phenomena in geological samples. While double-spiking techniques are required for TIMS to differentiate between natural and instrumental fractionation [297], measurement bias has been corrected for in MC-ICP-MS measurements by normalization to the isotope ratios of a doped zirconium or ruthenium standard [301], a palladium standard [302] or by standard sample bracketing using a molybdenum standard [303].

Acknowledgments

The author would like to thank Gina Chew and Nadia Nur Singh (National University of Singapore) and Dr. Barbara Mack for their assistance in preparing this chapter.

References

1 Allen, L. (2006) New approaches for designing and evaluating food fortification programs. *J. Nutr.*, **136**, 1055–1058.
2 Wienk, K.J., Marx, J.J., and Beynen, A.C. (1999) The concept of iron bioavailability and its assessment. *Eur. J. Nutr.*, **38**, 51–75.
3 Walczyk, T. (2001) The potential of inorganic mass spectrometry in mineral and trace element nutrition research. *Fresenius' J. Anal. Chem.*, **370**, 444–453.
4 Hoppe, M., Hulthén, L., and Hallberg, L. (2004) The validation of using serum iron increase to measure iron absorption in human subjects. *Br. J. Nutr.*, **92**, 485–488.
5 Hevesy, G. (1923) The absorption and translocation of lead by plants. A contribution to the application of the method of radioactive indicators in the investigation of the change of substance in plants. *Biochem. J.*, **17**, 439–445.
6 Lowman, J.T. and Krivit, W. (1963) New *in vivo* tracer method with the use of nonradioactive isotopes and activation analysis. *J. Lab. Clin. Med.*, **61**, 1042–1048.
7 Heumann, K. (1992) Isotope dilution mass spectrometry. *Int. J. Mass Spectrom. Ion Processes*, **118**, 575–592.
8 De Bièvre, P. (1994) Stable isotope dilution: an essential tool in metrology. *Fresenius' J. Anal. Chem.*, **350**, 277–283.
9 Walczyk, T., Davidsson, L., Zavaleta, N., and Hurrell, R. (1997) Stable isotope labels as a tool to determine the iron absorption by Peruvian school children from a breakfast meal. *Fresenius' J. Anal. Chem.*, **359**, 445–449.
10 Audi, G. (2003) The AME2003 atomic mass evaluation (II). Tables, graphs and references. *Nucl. Phys. A*, **729**, 337–676.
11 Fox, T.E., Fairweather-Tait, S.J., Eagles, J., and Wharf, S.G. (1991) Intrinsic labeling of different foods with stable isotope of zinc (^{67}Zn) for use in bioavailability studies. *Br. J. Nutr.*, **66**, 57–63.
12 Grusak, M. (1997) Intrinsic stable isotope labeling of plants for nutritional investigations in humans. *J. Nutr. Biochem.*, **8**, 164–171.
13 Motzok, I., Pennell, M.D., Davies, M.I., and Ross, H.U. (1975) Effect of particle size on the biological availability of reduced iron. *J. Assoc. Off. Anal. Chem.*, **58**, 99–103.
14 Wegmueller, R., Zimmermann, M.B., Moretti, D., Arnold, M., Langhans, W., and Hurrell, R.F.(2004) Particle size reduction and encapsulation affect the bioavailability of ferric pyrophosphate in rats. *J. Nutr.*, **134**, 3301–3304.
15 Beard, J. and Han, O. (2009) Systemic iron status. *Biochim. Biophys. Acta*, **1790**, 584–588.
16 Björn-Rasmussen, E., Hallberg, L., Isaksson, B., and Arvidsson, B. (1974) Food iron absorption in man. Applications of the two-pool extrinsic tag method to measure heme and nonheme iron absorption from the whole diet. *J. Clin. Invest.*, **53**, 247–255.
17 Turnlund, J.R. and Johnson, P.E. (1984) *Stable Isotopes in Nutrition*, American Chemical Society, Washington, DC.
18 Mellon, F.A. and Sandstrom, B. (1996) *Stable Isotopes in Human Nutrition: Inorganic Nutrient Metabolism*, Academic Press, London.
19 Lowe, N. and Jackson, M. (2001) *Advances in Isotope Methods for the Analysis of Trace Elements in Man*, CRC Press, Boca Raton, FL.
20 Abrams, S.A. and Wong, W.W. (2003) *Stable Isotopes in Human Nutrition: Laboratory Methods and Research Applications*, CAB International, Wallingford.
21 Wolfe, R.R. and Chinkes, D.L. (2004) *Isotopic Tracers in Metabolic Research, Principles and Practices of Kinetic Analysis*, John Wiley & Sons, Inc., Hoboken, NJ.

22 Bohn, T., Walczyk, T., Davidsson, L., Pritzkow, W., Klingbeil, P., Vogl, J., and Hurrell, R.F. (2004) Comparison of urinary monitoring, faecal monitoring and erythrocyte analysis of stable isotope labels to determine magnesium absorption in human subjects. *Br. J. Nutr.*, **91**, 113–120.

23 Simpson, F., Hall, G., Kelleher, J., and Losowsky, M. (1979) Radio-opaque pellets as faecal markers for faecal fat estimation in malabsorption. *Gut*, **20**, 581–584.

24 Schuette, S.A., Janghorbani, M., Young, V.R. and Weaver, C.M. (1993) Dysprosium as a nonabsorbable marker for studies of mineral absorption with stable isotope tracers in human subjects. *J. Am. Coll. Nutr.* **12**, 307–315.

25 Schuette, S.A., Janghorbani, M., Cohen, M.B., Krug, S., Schindler, T., Wagner, D.A., and Morris, S.J. (2003) Dysprosium chloride as a nonabsorbable gastrointestinal marker for studies of stable isotope-labeled triglyceride excretion in man. *J. Am. Coll. Nutr.*, **22**, 379–387.

26 Miller, L.V., Sheng, X.Y., Hambidge, K.M., Westcott, J.E., Sian, L., and Krebs, N.F. (2010) The use of dysprosium to measure endogenous zinc excretion in feces eliminates the necessity of complete fecal collections. *J. Nutr.*, **140**, 1524–1528.

27 Strickland, G.T., Beckner, W.M., and Leu, M.L. (1972) Absorption of copper in homozygotes and heterozygotes for Wilson's disease and controls: isotope tracer studies with ^{67}Cu and ^{64}Cu. *Clin. Sci.*, **43**, 617–625.

28 Sargent, T. III and Stauffer, H. (1979) Whole-body counting of retention of ^{67}Cu, ^{32}P and ^{51}Cr in man. *Int. J. Nucl. Med. Biol.*, **6**, 17–21.

29 Sarria, B. and Dainty, J.R. (2010) Comparison of faecal monitoring and area under the curve techniques to determine iron absorption in humans using stable isotope labelling. *J. Trace Elem. Med. Biol.*, **24**, 157–160.

30 Dainty, J.R., Roe, M.A., Teucher, B., Eagles, J., and Fairweather-Tait, S.J. (2003) Quantification of unlabelled non-haem iron absorption in human subjects: a pilot study. *Br. J. Nutr.*, **90**, 503–506.

31 Aubert, J.-P., Bronner, T.F., and Richelle, I.J. (1963) Quantitation of calcium metabolism. Theory. *J. Clin. Invest.*, **42**, 885–897.

32 Smith, D.L., Atkin, C., and Westenfelder, C. (1985) Stable isotopes of calcium as tracers: methodology. *Clin. Chim. Acta*, **146**, 97–101.

33 Yergey, A.L., Vieira, N.E., and Covell, D.G. (1987) Direct measurement of dietary fractional absorption using calcium isotopic tracers. *Biomed. Environ. Mass Spectrom.*, **14**, 603–607.

34 Miller, L.V., Krebs, N.F., and Hambidge, K.M. (1998) Human zinc metabolism: advances in the modeling of stable isotope data. *Adv. Exp. Med. Biol.*, **445**, 253–269.

35 Wastney, M.E., Patterson, B.H., Linares, O.A., Greif, P.C., and Boston, R.C. (1998) *Investigating Biological Systems Using Modeling: Strategies and Software*, Academic Press, San Diego, CA.

36 Dainty, J.R. (2001) Use of stable isotopes and mathematical modelling to investigate human mineral metabolism. *Nutr. Res. Rev.*, **14**, 295–315.

37 Denk, E., Hillegonds, D., Vogel, J., Synal, A., Geppert, C., Wendt, K., Fattinger, K., Hennessy, C., Berglund, M., Hurrell, R.F., and Walczyk, T. (2006) Labeling the human skeleton with ^{41}Ca to assess changes in bone calcium metabolism. *Anal. Bioanal. Chem.*, **386**, 1587–1602.

38 Dubach, R., Moore, C.V., and Minnich, V. (1946) Studies in iron transportation and metabolism V. Utilization of intravenously injected radioactive iron for hemoglobin synthesis, and an evaluation of the radioactive iron method for studying iron absorption. *J. Lab. Clin. Med.*, **31**, 1201–1222.

39 Wennesland, R., Brown, E., Hopper, J. Jr, Hodges, J. Jr, Guttentag, O., Scott, K., and Bradley, B. (1959) Red cell, plasma and blood volume in healthy

men measured by radiochromium (Cr^{51}) cell tagging and hematocrit: influence of age, somatotype and habits of physical activity on the variance after regression of volumes to height and weight combined. *J. Clin. Invest.*, **38**, 1065–1077.

40 Brown, E., Hopper, J. Jr, Hodges, J. Jr, Bradley, B., Wennesland, R., and Yamauchi, H. (1962) Red cell, plasma, and blood volume in healthy women measured by radiochromium cell-labeling and hematocrit. *J. Clin. Invest.*, **41**, 2182–2190.

41 Linderkamp, O., Versmold, H., Riegel, K., and Betke, K. (1977) Estimation and prediction of blood volume in infants and children. *Eur. J. Pediatr.*, **125**, 227–234.

42 Hosain, F., Marsaglia, G., and Finch, C. (1967) Blood ferrokinetics in normal man. *J. Clin. Invest.*, **46**, 1–9.

43 Janghorbani, M., Ting, B.T.G., and Fomon, S.J. (1986), Erythrocyte incorporation of ingested stable isotope of iron (^{58}Fe). *Am. J. Hematol.*, **21**, 277–288.

44 Kastenmayer, P., Davidsson, L., Galan, P., Cherouvrier, F., Hercberg, S., and Hurrell, R.F. (1994) A double stable isotope technique for measuring iron absorption in infants. *Br. J. Nutr.*, **71**, 411–424.

45 Whittaker, P., Lind, T., and Williams, J. (1991) Iron absorption during normal human pregnancy: a study using stable isotopes. *Br. J. Nutr.*, **65**, 457–463.

46 Barrett, J.F., Whittake, P.G., Williams, J.G. and Lind, T. (1994) Absorption of non-haem iron from food during normal pregnancy. *BMJ.*, **309**, 79–82.

47 Whittaker, P.G., Barrett, J.F., and Lind, T. (2001) The erythrocyte incorporation of absorbed non-haem iron in pregnant women. *Br. J. Nutr.*, **86**, 323–329.

48 Zlotkin, S.H., Lay, D.M., Kjarsgaard, J., and Longley, T. (1995) Determination of iron absorption using erythrocyte iron incorporation of two stable isotopes of iron (^{57}Fe and ^{58}Fe) in very low birthweight premature infants. *J. Pediatr. Gastroenterol. Nutr.*, **21**, 190–199.

49 Cercamondi, C., Egli, I., Ahouandjinou, E., Dossa, R., Zeder, C., Salami, L., Tjalsma, H., Wiegerinck, E., Tanno, T., and Hurrell, R.F. (2010) Afebrile *Plasmodium falciparum* parasitemia decreases absorption of fortification iron but does not affect systemic iron utilization: a double stable-isotope study in young Beninese women. *Am. J. Clin. Nutr.*, **92**, 1385–1392.

50 Lee, W.H., Wastney, M.E., Jackson, G.S., Martin, B.R., and Weaver, C.M. (2010) Interpretation of ^{41}Ca data using compartmental modeling in post-menopausal women. *Anal. Bioanal. Chem.*, **399**, 1613–1622.

51 Green, R., Charlton, R., Seftel, H., Bothwell, T., Mayet, F., Adams, B., Finch, C., and Layrisse, M. (1968) Body iron excretion in man: a collaborative study. *Am. J. Med.*, **45**, 336–353.

52 Hunt, J., Zito, C., and Johnson, L. (2009) Body iron excretion by healthy men and women. *Am. J. Clin. Nutr.*, **89**, 1792–1798.

53 Fomon, S.J., Drulis, J.M., Nelson, S.E., Serfass, R.E., Woodhead, J.C., and Ziegler, E.E. (2003) Inevitable iron loss by human adolescents, with calculations of the requirement for absorbed iron. *J. Nutr.*, **133**, 167–172.

54 Fomon, S.J., Nelson, S.E., Serfass, R.E., and Ziegler, E.E. (2005) Absorption and loss of iron in toddlers are highly correlated. *J. Nutr.*, **135**, 771–777.

55 Skulan, J., DePaolo, D., and Owens, T. (1997) Biological control of calcium isotopic abundances in the global calcium cycle. *Geochim. Cosmochim. Acta*, **61**, 2505–2510.

56 Walczyk, T. (1997) Iron isotopic variations in Nature. 20. Jahrtestagung der Arbeitsgemeinschaft Stabile Isotope, Weihenstephan, Book of Abstracts., p. 79.

57 Maréchal, C., Telouk, P., and Albarède, F. (1999) Precise analysis of copper and zinc isotopic compositions by plasma-source mass spectrometry. *Chem. Geol.*, **156**, 251–273.

58 Beard, B., Johnson, C., Cox, L., Sun, H., Nealson, K., and Aguilar, C. (1999) Iron isotope biosignatures. *Science*, **285**, 1889–1892.

59 Albarède, F. and Beard, B. (2004) Analytical methods for non-traditional isotopes. *Rev. Mineral. Geochem.*, **55**, 113–152.

60 Moody, J. and Epstein, M. (1991) Definitive measurement methods. *Spectrochim. Acta, B*, **46**, 1571–1575.

61 Sargent, M., Harrington, C., and Harte, R. (eds.) (2002) *Guidelines for Achieving High Accuracy in Isotope Dilution Mass Spectrometry (IDMS)*, Royal Society of Chemistry, Cambridge.

62 World Health Organization and Food and Agricultural Organization of the United Nations (2004) Iron, in *Vitamin and Mineral Requirements in Human Nutrition*, 2nd edn., WHO, Geneva, pp. 246–278.

63 Stürup, S. (2004) The use of ICPMS for stable isotope tracer studies in humans: a review. *Anal. Bioanal. Chem.*, **378**, 273–282.

64 Becker, J.S. (2007). *Inorganic Mass Spectrometry: Principles and Applications*, John Wiley & Sons, Ltd., Chichester.

65 Smith, S., Wastney, M., Nyquist, L., Shih, C.Y., Wiesmann, H., Nillen, J., and Lane, H.W. (1996) Calcium kinetics with microgram stable isotope doses and saliva sampling. *J. Mass Spectrom.*, **31**, 1265–1270.

66 Habfast, K. (1998) Fractionation correction and multiple collectors in thermal ionization isotope ratio mass spectrometry. *Int. J. Mass Spectrom.*, **176**, 133–148.

67 Vanhaecke, F., Balcaen, L., and Malinovsky, D. (2009) Use of single-collector and multi-collector ICP-mass spectrometry for isotopic analysis. *J. Anal. At. Spectrom.*, **24**, 863–863.

68 Hart, S. and Zindler, A. (1989) Isotope fractionation laws: a test using calcium. *Int. J. Mass Spectrom. Ion Processes*, **89**, 287–301.

69 Moore, L.J., Machlan, L.A., Shields, W.R., and Garner, E.L. (1974) Internal normalization techniques for high accuracy isotope dilution analyses. Application to molybdenum and nickel in standard reference materials. *Anal. Chem.*, **46**, 1082–1089.

70 Chu, Z.-Y., Yang, Y.-H., Guo, J.H., and Qiao, G.-S. (2011) Calculation methods for direct internal mass fractionation correction of spiked isotopic ratios from multi-collector mass spectrometric measurements. *Int. J. Mass Spectrom.*, **299**, 87–93.

71 Rudge, J.F., Reynolds, B.C., and Bourdon, B. (2009) The double spike toolbox. *Chem. Geol.*, **265**, 420–431.

72 Walczyk, T. (2004) TIMS versus multicollector-ICP-MS: coexistence or struggle for survival?. *Anal. Bioanal. Chem.*, **378**, 229–231.

73 Schoenberg, R. and von Blanckenburg, F. (2005) An assessment of the accuracy of stable Fe isotope ratio measurements on samples with organic and inorganic matrices by high-resolution multicollector ICP-MS. *Int. J. Mass Spectrom.*, **242**, 257–272.

74 Gropper, S.S., Smith, J.L., and Groff, J.L. (2009) *Advanced Nutrition and Human Metabolism*, 5th edn., Cengage/Wadsworth, Florence, KY.

75 Weaver, C., Heaney, R., Martin, B., and Fitzsimmons, M. (1992) Extrinsic vs intrinsic labeling of the calcium in whole-wheat flour. *Am. J. Clin. Nutr.*, **55**, 452–454.

76 Nickel, K.P., Martin, B.R., Smith, D.L., Smith, J.B., Miller, G.D., and Weaver, C.M. (1996) Calcium bioavailability from bovine milk and dairy products in premenopausal women using intrinsic and extrinsic labeling techniques. *J. Nutr.*, **126**, 1406–1411.

77 Weaver, C. and Heaney, R. (1991) Isotopic exchange of ingested calcium between labeled sources. Evidence that ingested calcium does not form a common absorptive pool. *Calcif. Tissue Int.*, **49**, 244–247.

78 Weaver, C., Heaney, R., Nickel, K., and Packard, P. (1997) Calcium bioavailability from high oxalate vegetables: Chinese vegetables, sweet potatoes and rhubarb. *J. Food Sci.*, **62**, 524–525.

79 Heaney, R., Dowell, M., Rafferty, K., and Bierman, J. (2000) Bioavailability of the calcium in fortified soy imitation

milk, with some observations on method. *Am. J. Clin. Nutr.*, **71**, 1166–1169.

80 Heaney, R., Dowell, M., and Barger-Lux, M. (1999) Absorption of calcium as the carbonate and citrate salts, with some observations on method. *Osteoporos. Int.*, **9**, 19–23.

81 Yergey, A.L., Abrams, S.A., Vieira, N.E., Aldroubi, A., Marini, J., and Sidbury, J.B. (1994) Determination of fractional absorption of dietary calcium in humans. *J. Nutr.*, **124**, 674–682.

82 Barger-Lux, M., Heaney, R., and Recker, R. (1989) Time course of calcium absorption in humans: evidence for a colonic component. *Calcif. Tissue Int.*, **44**, 308–311.

83 Cashman, K. (2003) Prebiotics and calcium bioavailability. *Curr. Issues Intest. Microbiol.*, **4**, 149–174.

84 Ceglia, L., Abrams, S.A., Harris, S.S., Rasmussen, H.M., Dallal, G.E., and Dawson-Hughes, B. (2010) A simple single serum method to measure fractional calcium absorption using dual stable isotopes. *Exp. Clin. Endocrinol. Diabetes*, **118**, 653–656.

85 Heaney, R.P. and Recker, R.R. (1985) Estimation of true calcium absorption. *Ann. Intern. Med.*, **103**, 516–521.

86 Lee, W., McCabe, G.P., Martin, B.R., and Weaver, C.M. (2011) Validation of a simple isotope method for estimating true calcium fractional absorption in adolescents. *Osteoporos. Int.*, **22**, 159–166.

87 World Health Organization and Food and Agricultural Organization of the United Nations (2004) Calcium, in *Vitamin and Mineral Requirements in Human Nutrition*, 2nd edn., WHO, Geneva, pp. 59–93.

88 Meacham, S., Grayscott, D., Chen, J.J., and Bergman, C. (2008) Review of the dietary reference intake for calcium: where do we go from here? *Crit. Rev. Food Sci. Nutr.*, **48**, 378–384.

89 Abrams, S., Sidbury, J., Muenzer, J., Esteban, N., Vieira, N., and Yergey, A. (1991) Stable isotopic measurement of endogenous fecal calcium excretion in children. *J. Pediatr. Gastroenterol. Nutr.*, **12**, 469–473.

90 Weaver, C., Wastney, M., and Spence, L. (2002) Quantitative clinical nutrition approaches to the study of calcium and bone metabolism. *Clin. Rev. Bone Miner. Metab.*, **1**, 219–232.

91 Smith, S.M., Wastney, M.E., O'Brien, K.O., Morukov, B.V., Larina, I.M., Abrams, S.A., Davis-Street, J.E., Oganov, V., and Shackelford, L.C. (2005) Bone markers, calcium metabolism, and calcium kinetics during extended-duration space flight on the Mir space station. *J. Bone Miner. Res.*, **20**, 208–218.

92 Denk, E., Hillegonds, D., Hurrell, R.F., Vogel, J., Fattinger, K., Häuselmann, H.J., Kraenzlin, M., and Walczyk, T. (2007) Evaluation of ^{41}calcium as a new approach to assess changes in bone metabolism: effect of a bisphosphonate intervention in postmenopausal women with low bone mass. *J. Bone Miner. Res.*, **22**, 1518–1525.

93 Weaver, C.M., Peacock, M., Martin, B.R., McCabe, G.P., Zhao, J., Smith, D.L., and Wastney, M.E. (1997) Quantification of biochemical markers of bone turnover by kinetic measures of bone formation and resorption in young healthy females. *J. Bone Miner. Res.*, **12**, 1714–1720.

94 Hill, P.A. (1998) Bone remodelling. *Br. J. Orthodont.*, **25**, 101–107.

95 Johnson, R., Berkovits, D., Boaretto, E., Gelbart, Z., Ghelberg, S., Meirav, O, Paul, M., Prior, J., Sossi, V., and Venczel, E. (1994) Calcium resorption from bone in a human studied by ^{41}Ca tracing. *Nucl. Instrum. Methods B*, **92**, 483–488.

96 Lin, Y., Hillegonds, D., Gertz, E., Van Loan, M., and Vogel, J. (2004) Protocol for assessing bone health in humans by tracing long-lived ^{41}Ca isotope in urine serum and saliva samples. *Anal. Biochem.*, **332**, 193–195.

97 Hillegonds, D., Fitzgerald, R., Herold, D., Lin, Y.-M., and Vogel, J.S. (2004) High-throughput measurement of ^{41}Ca by accelerator mass spectrometry to quantitate small changes in individual human bone turnover rates. *J. Assoc. Lab. Autom.*, **9**, 99–102.

98 Chen, Z. (2003) Inductively coupled plasma mass spectrometric analysis of calcium isotopes in human serum: a low-sample-volume acid-equilibration method. *Clin. Chem.*, **49**, 2050–2055.

99 Boulyga, S.F. (2010) Calcium isotope analysis by mass spectrometry. *Mass Spectrom. Rev.*, **29**, 685–716.

100 Bullen, T.D. and Walczyk, T. (2009) Environmental and biomedical applications of natural metal stable isotope variations. *Elements*, **5**, 381–385.

101 Russell, W. and Papanastassiou, D. (1978) Calcium isotope fractionation in ion-exchange chromatography. *Anal. Chem.*, **50**, 1151–1154.

102 Heuser, A., Eisenhauer, A., Gussone, N., Bock, B., Hansen, B., and Nagler, T. (2002) Measurement of calcium isotopes (δ^{44}Ca) using a multicollector TIMS technique. *Int. J. Mass Spectrom.*, **220**, 385–397.

103 Halicz, L., Galy, A., Belshaw, N.S., and O'Nions, R.K. (1999) High-precision measurement of calcium isotopes in carbonates and related materials by multiple collector inductively coupled plasma mass spectrometry (MC-ICP-MS). *J. Anal. At. Spectrom.*, **14**, 1835–1838.

104 Schmitt, A.D., Gangloff, S., Cobert, F., Lemarchand, D., Stille, P., and Chabaux, F. (2009) High performance automated ion chromatography separation for Ca isotope measurements in geological and biological samples. *J. Anal. At. Spectrom.*, **24**, 1089–1097.

105 Fietzke, J., Eisenhauer, A., Gussone, N., Bock, B., Liebetrau, V., Nagler, T., Spero, H.J., Bijma. J., and Dullo, C. (2004) Direct measurement of ^{44}Ca/^{40}Ca ratios by MC-ICP-MS using the cool plasma technique. *Chem. Geol.*, **206**, 11–20.

106 DePaolo, D. (2004) Calcium isotopic variations produced by biological kinetic radiogenic and nucleosynthetic processes. *Rev. Mineral. Geochem.*, **55**, 255–288.

107 Chu, N.C., Henderson, G.M., Belshaw, N.S., and Hedges, R.E. (2006) Establishing the potential of Ca isotopes as proxy for consumption of dairy products. *Appl. Geochem.*, **21**, 1656–1667.

108 Skulan, J., Bullen, T., Anbar, A.D., Puzas, J.E., Shackelford, L., LeBlanc, A. and Smiths, S.M. (2007) Natural calcium isotopic composition of urine as a marker of bone mineral balance. *Clin. Chem.*, **53**, 1155–1158.

109 Hirata, T., Tanoshima, M., Suga, A., Tanaka, Y.K., Nagata, Y., Shinohara, A., and Chiba, M. (2008) Isotopic analysis of calcium in blood plasma and bone from mouse samples by multiple collector-ICP-mass spectrometry. *Anal. Sci.*, **24**, 1501–1507.

110 Heuser, A. and Eisenhauer, A. (2010) A pilot study on the use of natural calcium isotope (^{44}Ca/^{40}Ca) fractionation in urine as a proxy for the human body calcium balance. *Bone*, **46**, 889–896.

111 Consaul, J. and Lee, K. (1983) Extrinsic tagging in iron bioavailability research: a critical review. *J. Agric. Food Chem.*, **31**, 684–689.

112 Lönnerdal, B. (2010) Alternative pathways for absorption of iron from foods. *Pure Appl. Chem.*, **82**, 429–436.

113 Hoppler, M., Schönbächler, A., Meile, L., Hurrell, R.F., and Walczyk, T. (2008) Ferritin-iron is released during boiling and *in vitro* gastric digestion. *J. Nutr.*, **138**, 878–884.

114 Vanhaecke, F., Balcaen, L., de Wannemacker, G., and Moens, L. (2002) Capabilities of inductively coupled plasma mass spectrometry for the measurement of Fe isotope ratios. *J. Anal. At. Spectrom.*, **17**, 933–943.

115 Dauphas, N. and Rouxel, O. (2006) Mass spectrometry and natural variations of iron isotopes. *Mass Spectrom. Rev.*, **25**, 515–550.

116 Walczyk, T. (1997) Iron isotope ratio measurements by negative thermal ionisation mass spectrometry using

FeF_4-molecular ions. *Int. J. Mass Spectrom. Ion Processes*, **161**, 217–227.

117 Walczyk, T. and von Blanckenburg, F. (2002) Natural iron isotope variations in human blood. *Science*, **295**, 2065–2066.

118 Stenberg, A., Malinovsky, D., Rodushkin, I., Andrén, H., Pontér, C., Öhlander, B., and Baxter, D.C. (2003) Separation of Fe from whole blood matrix for precise isotopic ratio measurements by MC-ICP-MS: a comparison of different approaches. *J. Anal. At. Spectrom.*, **18**, 23–28.

119 Benkhedda, K., Chen, H., Dabeka, R., and Cockell, K. (2008) Isotope ratio measurements of iron in blood samples by multi-collector ICP-MS to support nutritional investigations in humans. *Biol. Trace Elem. Res.*, **122**, 179–192.

120 Van Den Heuvel, E., Muys, T., Pellegrom, H., Bruyntjes, J., Van Dokkum, W., Spanhaak, S., and Schaafsma, G. (1998) A new method to measure iron absorption from the enrichment of ^{57}Fe and ^{58}Fe in young erythroid cells. *Clin. Chem.*, **44**, 649–654.

121 Sarria, B., Dainty, J.R., Fox, T.E., and Fairweather-Tait, S.J. (2005) Estimation of iron absorption in humans using compartmental modelling. *Eur. J. Clin. Nutr.*, **59**, 142–144.

122 Ricketts, C. and Cavill, I. (1978) Ferrokinetics: methods and interpretation. *Clin. Nucl. Med.*, **3**, 159–164.

123 Shreeve, W. (2007) Use of isotopes in the diagnosis of hematopoietic disorders. *Exp. Hematol.*, **35**, 173–179.

124 Johnson, C. and Beard, B. (1999) Correction of instrumentally produced mass fractionation during isotopic analysis of Fe by thermal ionization mass spectrometry. *Int. J. Mass Spectrom.*, **193**, 87–99.

125 Belshaw, N., Zhu, X., Guo, Y., and O'Nions, R. (2000) High precision measurement of iron isotopes by plasma source mass spectrometry. *Int. J. Mass Spectrom.*, **197**, 191–195.

126 Weyer, S. and Schwieters, J.B. (2003) High precision Fe isotope measurements with high mass resolution MC-ICPMS. *Int. J. Mass Spectrom.*, **226**, 355–368.

127 Bullen, T.D. and McMahon, P.M. (1998) Using stable Fe isotopes to assess microbially-mediated Fe^{3+} reduction in a jet-fuel contaminated aquifer. *Mineral. Mag.*, **62A**, 255–256.

128 Walczyk, T. and von Blanckenburg, F. (2005) Deciphering the iron isotope message of the human body. *Int. J. Mass Spectrom.*, **242**, 117–134.

129 Krayenbuehl, P.A., Walczyk, T., Schoenberg, R., von Blanckenburg, F., and Schulthess, G. (2005) Hereditary hemochromatosis is reflected in the iron isotope composition of blood. *Blood.*, **105**, 3812–3816.

130 Stenberg, A., Malinovsky, D., Öhlander, B., Andrén, H., Forsling, W., Engström, L.M., Anders Wahlin, A., Engström, E., Rodushkin, I., and Baxter, D.C. (2005) Measurement of iron and zinc isotopes in human whole blood: preliminary application to the study of HFE genotypes. *J. Bone Miner. Res.*, **19**, 55–60.

131 Hotz, K., Augsburger, H., and Walczyk. T. (2011) Isotopic signatures of iron in body tissues as a potential biomarker for iron metabolism. *J. Anal. At. Spectrom.*, **26**, 1347–1353

132 Ohno, T., Shinohara, A., Kohge, I., Chiba, M., and Hirata, T. (2004) Isotopic analysis of Fe in human red blood cells by multiple collector-ICP-mass spectrometry. *Anal. Sci.*, **20**, 617–621.

133 Camaschella, C. (2005) Understanding iron homeostasis through genetic analysis of hemochromatosis and related disorders. *Blood*, **106**, 3710–3717.

134 Hershko, C. (2010) Pathogenesis and management of iron toxicity in thalassemia. *Ann. N. Y. Acad. Sci.*, **1202**, 1–9.

135 Janghorbani, M., Istfan, N.W., and Pagounes, J.O. (1982) Absorption of dietary zinc in man: comparison of intrinsic and extrinsic labels using a triple stable isotope method. *Am. J. Clin. Nutr.*, **36**, 537–545.

136 Ketelsen, S., Stuart, M., Weaver, C., Forbes, R., and Erdman, J. (1984) Bioavailability of zinc to rats from defatted soy flour, acid-precipitated soy concentrate and neutralized soy concentrate as determined by intrinsic and extrinsic labeling techniques. *J. Nutr.*, **114**, 536–542.
137 Fairweather-Tait, S.J., Fox, T.E., Wharf, S.G., and Eagles, J. (1991) Apparent zinc absorption by rats from foods labelled intrinsically and extrinsically with ^{67}Zn. *Br. J. Nutr.*, **66**, 65–71.
138 Fox, T.E., Fairweather-Tait, S.J., Eagles, J., and Wharf, S.G. (1994) Assessment of zinc bioavailability: studies in rats on zinc absorption from wheat using radio- and stable isotopes. *Br. J. Nutr.*, **71**, 95–101.
139 Donangelo, C.M., Woodhouse, L.R., King, S.M., Toffolo, G., Shames, D.M., Viteri, F.E., Fernando, E., Cheng, Z., Welch, R.M., and King, J.C. (2003) Iron and zinc absorption from two bean (*Phaseolus vulgaris* L.) genotypes in young women. *J. Agric. Food. Chem.*, **51**, 5137–5143.
140 de Romaña, D.L., Lönnerdal, B., and Brown, K.H. (2003) Absorption of zinc from wheat products fortified with iron and either zinc sulfate or zinc oxide. *Am. J. Clin. Nutr.*, **78**, 279–283.
141 Sian, L., Hambidge, K.M., Westcott, J.L., Miller, L.V., and Fennessey, P.V. (1993) Influence of a meal and incremental doses of zinc on changes in zinc absorption. *Am. J. Clin. Nutr.*, **58**, 533–536.
142 Wada, L., Turnlund, J.R., and King, J. C. (1985) Zinc utilization in young men fed adequate and low zinc intakes. *J. Nutr.*, **115**, 1345–1354.
143 Swanson, C.A., Turnlund, J.R., and King, J.C. (1983) Effect of dietary zinc sources and pregnancy on zinc utilization in adult women fed controlled diets. *J. Nutr.* **113**, 2557–2567.
144 Mason, P.M., Judd, P.A., Fairweather-Tait, S.J., Eagles, J., and Minski, M.J. (1990) The effect of moderately increased intakes of complex carbohydrates (cereals, vegetables and fruit) for 12 weeks on iron and zinc metabolism. *Br. J. Nutr.* **63**, 597–611.
145 King, J.C., Raynolds, W.L., and Margen, S. (1978) Absorption of stable isotopes of iron, copper, and zinc during oral contraceptives use. *Am. J. Clin. Nutr.*, **31**, 1198–1203.
146 August, D., Janghorbani, M., and Young, V.R. (1989) Determination of zinc and copper absorption at three dietary Zn–Cu ratios by using stable isotope methods in young adult and elderly subjects. *Am. J. Clin. Nutr.*, **50**, 1457–1463.
147 Turnlund, J.R., Durkin, N., Costa, F., and Margen, S. (1986) Stable isotope studies of zinc absorption and retention in young and elderly men. *J. Nutr.*, **116**, 1239–1247.
148 Ulusoy, U. and Whitley, J.E. (1999) Determination of intestinal uptake of iron and zinc using stable isotopic tracers and rare earth markers. *Nutr. Res.*, **19**, 675–688.
149 Sheng, X.Y., Hambidge, K.M., Krebs, N.F., Lei, S., Westcott, J.E., and Miller, L.V. (2005) Dysprosium as a nonabsorbable fecal marker in studies of zinc homeostasis. *Am. J. Clin. Nutr.*, **82**, 1017–1023.
150 Hambidge, M. and Krebs, N.F. (2001) Interrelationships of key variables of human zinc homeostasis: relevance to dietary zinc requirements. *Annu. Rev. Nutr.*, **21**, 429–452.
151 Hambidge, K.M., Miller, L.V., Tran, C. D. and Krebs, N.F. (2005) Measurements of zinc absorption: application and interpretation in research designed to improve human zinc nutriture. *Int. J. Vitam. Nutr. Res.*, **75**, 385–393.
152 Ziegler, E.E., Serfass, R.E., Nelson, S.E., Figueroa-Colon, R., Edwards, B.B., Houk, R.S., and Thompson, J.J. (1989) Effect of low zinc intake on absorption and excretion of zinc by infants studied with ^{70}Zn as extrinsic tag. *J. Nutr.*, **119**, 1647–1653.
153 Jalla, S., Krebs, N.F., Rodden, D., and Hambidge, K.M. (2004) Zinc homeostasis in premature infants does

not differ between those fed preterm formula or fortified human milk. *Pediatr. Res.* **56**, 615–620.

154 Krebs, N.F., Reidinger, C.J., Miller, L.V., and Hambidge, K.M. (1996) Zinc homeostasis in breast-fed infants. *Pediatr. Res.*, **39**, 661–665.

155 Krebs, N.F., Reidinger, C.J., Miller, L.V., and Borschel, M.W. (2000) Zinc homeostasis in healthy infants fed a casein hydrolysate formula. *J. Pediatr. Gastroenterol. Nutr.*, **30**, 29–33.

156 Fairweather-Tait, S.J., Fox, T.E., Wharf, S.G., Eagles, J., and Kennedy, H. (1992) Zinc absorption in adult men from a chicken sandwich made with white or wholemeal bread, measured by a double-label stable-isotope technique. *Br. J. Nutr.*, **67**, 411–419.

157 Harvey, L.J., Dainty, J.R., Hollands, W.J., Bull, V.J., Hoogewerff, J.A., Foxall, R.J., McAnena, L., Strain, J.J., and Fairweather-Tait, S.J. (2007) Effect of high-dose iron supplements on fractional zinc absorption and status in pregnant women. *Am. J. Clin. Nutr.*, **85**, 131–136.

158 O'Brien, K.O., Zavaleta, N., Caulfield, L.E., Wen, J., and Abrams, S.A. (2000) Prenatal iron supplements impair zinc absorption in pregnant Peruvian women. *J. Nutr.*, **130**, 2251–2255.

159 Shames, D.M., Woodhouse, L.R., Lowe, N.M., and King, J.C. (2001) Accuracy of simple techniques for estimating fractional zinc absorption in humans. *J. Nutr.*, **131**, 1854–1861.

160 Lowe, N.M., Woodhouse, L.R., Sutherland, B., Shames, D.M., Burri, B.J., Abrams, S.A., Turnlund, J.R., Jackson, M.J., and King, J.C. (2004) Kinetic parameters and plasma zinc concentration correlate well with net loss and gain of zinc from men. *J. Nutr.*, **134**, 2178–2181.

161 Meunier, N., Feillet-Coudray, C., Rambeau, M., Andriollo-Sanchez, M., Brandolini-Bunlon, M., Coulter, S.J., Cashman, K.D., Mazur, A., and Coudray, C. (2005) Impact of micronutrient dietary intake and status on intestinal zinc absorption in late middle-aged men: the ZENITH study. *Eur. J. Clin. Nutr.*, **59**, (Suppl. 2), S48–S52.

162 Friel, J.K., Naake, V.L., Miller, L.V., Fennessey, P.V., and Hambidge, K.M. (1992) The analysis of stable isotopes in urine to determine the fractional absorption of zinc. *Am. J. Clin. Nutr.*, **55**, 473–477.

163 Friel, J.K., Andrews, W.L., Simmons, B.S., Miller, L.V., and Longerich, H.P. (1996) Zinc absorption in premature infants: comparison of two isotopic methods. *Am. J. Clin. Nutr.*, **63**, 342–347.

164 Lowe, N.M., Woodhouse, L.R., Matel, J.S., and King, J.C. (2000) Comparison of estimates of zinc absorption in humans by using 4 stable isotopic tracer methods and compartmental analysis. *Am. J. Clin. Nutr.*, **71**, 523–529.

165 Fung, E.B., Ritchie, L.D., Woodhouse, L.R., Roehl, R., and King, J.C. (1997) Zinc absorption in women during pregnancy and lactation: a longitudinal study. *Am. J. Clin. Nutr.*, **66**, 80–88.

166 King, J.C. (2001) Effect of reproduction on the bioavailability of calcium, zinc and selenium. *J. Nutr.*, **131**, 1355S–1358S.

167 Hambidge, K.M., Abebe, Y., Gibson, R.S., Westcott, J.E., Miller, L.V., Lei, S., Stoecker, B.J., Arbide, I., Teshome, A., Bailey, K.B., and Krebs, N.F. (2006) Zinc absorption during late pregnancy in rural southern Ethiopia. *Am. J. Clin. Nutr.*, **84**, 1102–1106.

168 Tran, C.D., Miller, L.V., Krebs, N.F., Lei, S., and Hambidge, K.M. (2004) Zinc absorption as a function of the dose of zinc sulfate in aqueous solution. *Am. J. Clin. Nutr.*, **80**, 1570–1573.

169 Hambidge, K.M., Huffer, J.W., Raboy, V., Grunwald, G.K., Westcott, J.L., Sian, L., Miller, L.V., Dorsch, J.A., and Krebs, N.F. (2004) Zinc absorption from low-phytate hybrids of maize and their wild-type isohybrids. *Am. J. Clin. Nutr.*, **79**, 1053–1059.

170 Rosado, J.L., Hambidge, K.M., Miller, L.V., Garcia, O.P., Westcott, J.,

Gonzalez, K., Conde, J., Hotz, C., Pfeiffer, W., and Ortiz-Monasterio, I. (2009) The quantity of zinc absorbed from wheat in adult women is enhanced by biofortification. *J. Nutr.*, **139**, 1920–1925.

171 Yeung, G., Schauer, C., and Zlotkin, S. (2001) Fractional zinc absorption using a single isotope tracer. *Eur. J. Clin. Nutr.*, **55**, 1098–1103.

172 Krebs, N.F. and Hambidge, K.M. (2001) Zinc metabolism and homeostasis: the application of tracer techniques to human zinc physiology. *Biometals*, **14**, 397–412.

173 Wastney, M., House, W., Barnes, R., and Subramanian, K. (2000) Kinetics of zinc metabolism: variation with diet, genetics and disease. *J. Nutr.*, **130**, 1355S–1359S.

174 Miller, L.V., Krebs, N.F., and Hambidge, K.M. (2000) Development of a compartmental model of human zinc metabolism: identifiability and multiple studies analyses. *Am. J. Physiol. Regul. Integr. Comp. Physiol.*, **279**, R1671–R1684.

175 Griffin, I., King, J., and Abrams, S. (2000) Body weight-specific zinc compartmental masses in girls significantly exceed those reported in adults: a stable isotope study using a kinetic model. *J. Nutr.*, **130**, 2607–2612.

176 King, J., Shames, D., Lowe, N., Woodhouse, L., Sutherland, B., Abrams, S., Turnlund, J.R., and Jackson, M.J. (2001) Effect of acute zinc depletion on zinc homeostasis and plasma zinc kinetics in men. *Am. J. Clin. Nutr.*, **74**, 116–124.

177 Donangelo, C.M., Zapata, C.L.V., Woodhouse, L.R., Shames, D.M., Mukherjea, R., and King, J.C. (2005) Zinc absorption and kinetics during pregnancy and lactation in Brazilian women. *Am. J. Clin. Nutr.*, **82**, 118–124.

178 Miller, L.V., Hambidge, K.M., Naake, V.L., Hong, Z., Westcott, J.L., and Fennessey, P.V. (1994) Size of the zinc pools that exchange rapidly with plasma zinc in humans: alternative techniques for measuring and relation to dietary zinc intake. *J. Nutr.*, **124**, 268–276.

179 Lei, S., Mingyan, X., Miller, L.V., Tong, L., Krebs, N.F., and Hambidge, K.M. (1996) Zinc absorption and intestinal losses of endogenous zinc in young Chinese women with marginal zinc intakes. *Am. J. Clin. Nutr.*, **63**, 348–353.

180 Manary, M.J., Hotz, C., Krebs, N.F., Gibson, R.S., Westcott, J.E., Arnold, T., Broadhead, R.L., and Hambidge, K.M. (2000) Dietary phytate reduction improves zinc absorption in Malawian children recovering from tuberculosis but not in well children. *J. Nutr.*, **130**, 2959–2964.

181 Krebs, N., Hambidge, K., Westcott, J., Miller, L., Sian, L., Bell, M., and Grunwald, G. (2003) Exchangeable zinc pool size in infants is related to key variables of zinc homeostasis. *J. Nutr.*, **133**, 1498S–1501S.

182 Peirce, P., Hambidge, K., Goss, C., Miller, L., and Fennessey, P. (1987) Fast-atom-bombardment mass spectrometry for the determination of zinc stable isotopes in biological samples. *Anal. Chem.*, **59**, 2034–2037.

183 Eagles, J., Fairweather-Tait, S., Portwood, D., Self, R., Goetz, A., and Heumann, K. (1989) Comparison of fast atom bombardment mass spectrometry and thermal ionization quadrupole mass spectrometry for the measurement of zinc absorption in human nutrition studies. *Anal. Chem.*, **61**, 1023–1025.

184 Ramanujam, V., Yokoi, K., Egger, N., Dayal, H., Alcock, N., and Sandstead, H. (1999) Polyatomics in zinc isotope ratio analysis of plasma samples by inductively coupled plasma-mass spectrometry and applicability of nonextracted samples for zinc kinetics. *Biol. Trace Elem. Res.*, **68**, 143–158.

185 Mason, T.F.D., Weiss, D.J., Horstwood, M., Parrish, R.R., Russell, S.S., Mullane, E., and Coles, B.J. (2004) High-precision Cu and Zn isotope analysis by plasma source mass spectrometry. Part 1: spectral interferences and their correction. *J. Anal. At. Spectrom.*, **19**, 209–217.

186 Durrant, S.F., Krushevska, A., Amarasiriwardena, D., Argentine, M.D., Romon-Guesnier, S., and Barnes, R.M. (1994) Matrix separation by chelation to prepare biological materials for isotopic zinc analysis by inductively coupled plasma mass spectrometry. *J. Anal. At. Spectrom.*, **9**, 199–204.

187 Cloquet, C., Carignan, J., Lehmann, M.F., and Vanhaecke, F. (2008) Variation in the isotopic composition of zinc in the natural environment and the use of zinc isotopes in biogeosciences: a review. *Anal. Bioanal. Chem.*, **390**, 451–463.

188 Shiel, A., Barling, J., Orians, K., and Weis, D. (2009) Matrix effects on the multi-collector inductively coupled plasma mass spectrometric analysis of high-precision cadmium and zinc isotope ratios. *Anal. Chim. Acta*, **633**, 29–37.

189 Tanimizu, M., Asada, Y., and Hirata, T. (2002) Absolute isotopic composition and atomic weight of commercial zinc using inductively coupled plasma mass spectrometry. *Anal. Chem.*, **74**, 5814–5819.

190 Ingle, C.P., Langford, N., Harvey, L.J., Dainty, J.R., Turner, P.J., Sharp, B.L., and Lewis, D.J. (2004) Comparison of three different instrumental approaches to the determination of iron and zinc isotope ratios in clinical samples. *J. Anal. At. Spectrom.*, **19**, 404–406.

191 Yamakawa, A., Yamashita, K., Makishima, A., and Nakamura, E. (2009) Chemical separation and mass spectrometry of Cr, Fe, Ni, Zn, and Cu in terrestrial and extraterrestrial materials using thermal ionization mass spectrometry. *Anal. Chem.*, **81**, 9787–9794.

192 Ghidan, O.Y. and Loss, R.D. (2010) Accurate and precise elemental abundance of zinc in reference materials by an isotope dilution mass spectrometry TIMS technique. *Geostand. Geoanal. Res.*, **34**, 185–191.

193 Pichat, S., Douchet, C., and Albarède, F. (2003) Zinc isotope variations in deep-sea carbonates from the eastern equatorial Pacific over the last 175 ka. *Earth Planet. Sci. Lett.*, **210**, 167–178.

194 Archer, C. and Vance, D. (2004) Mass discrimination correction in multiple-collector plasma source mass spectrometry: an example using Cu and Zn isotopes. *J. Anal. At. Spectrom.*, **19**, 656–665.

195 Peel, K., Weiss, D., Chapman, J., Arnold, T., and Coles, B. (2008) A simple combined sample–standard bracketing and inter-element correction procedure for accurate mass bias correction and precise Zn and Cu isotope ratio measurements. *J. Anal. At. Spectrom.*, **23**, 103–110.

196 Arnold, T., Schönbächler, M., Rehkämper, M., Dong, S., Zhao, F.J., Kirk, G.J.D., Coles, B.J., and Weiss, D.J. (2010) Measurement of zinc stable isotope ratios in biogeochemical matrices by double-spike MC-ICPMS and determination of the isotope ratio pool available for plants from soil. *Anal. Bioanal. Chem.*, **398**, 3115–3125.

197 Arnold, T., Kirk, G.J.D., Wissuwa, M., Frei, M., Zhao, F.J., Mason, T.F.D., and Weiss, D.J. (2010) Evidence for the mechanisms of zinc uptake by rice using isotope fractionation. *Plant Cell Environ.*, **33**, 370–381.

198 Moynier, F., Pichat, S., Pons, M.L., Fike, D., Balter, V., and Albarède, F. (2009) Isotopic fractionation and transport mechanisms of Zn in plants. *Chem. Geol.*, **267**, 125–130.

199 Stenberg, A., Andrén, H., Malinovsky, D., Engström, E., Rodushkin, I., and Baxter, D.C. (2004) Isotopic variations of Zn in biological materials. *Anal. Chem.*, **76**, 3971–3978.

200 Ohno, T., Shinohara, A., Chiba, M., and Hirata, T. (2005) Precise Zn isotopic ratio measurements of human red blood cell and hair samples by multiple collector-ICP-mass spectrometry. *Anal. Sci.*, **21**, 425–428.

201 Balter, V., Zazzo, A., Moloney, A.P., Moynier, F., Schmidt, O., Monahan, F.J., and Albarède, F. (2010) Bodily variability of zinc natural isotope abundances in sheep. *Rapid Commun. Mass Spectrom.*, **24**, 605–612.

202 Bohn, T. (2008) Dietary factors influencing magnesium absorption in humans. *Curr. Nutr. Food Sci.*, **4**, 53–72

203 Sabatier, M., Arnaud, M.J., Kastenmayer, P., Rytz, A., and Barclay, D.V. (2002) Meal effect on magnesium bioavailability from mineral water in healthy women. *Am. J. Clin. Nutr.*, **75**, 65–71.

204 Wälti, M.K., Zimmermann, M.B., Walczyk, T., Spinas, G.A., and Hurrell, R.F. (2003) Measurement of magnesium absorption and retention in type 2 diabetic patients with the use of stable isotopes. *Am. J. Clin. Nutr.*, **78**, 448–453.

205 Seki, N., Hamano, H., Iiyama, Y., Asano, Y., Kokubo, S., Yamauchi, K., Tamura, Y., Uenishi, K., and Kudou, H. (2007) Effect of lactulose on calcium and magnesium absorption: a study using stable isotopes in adult men. *J. Nutr. Sci. Vitaminol.*, **53**, 5–12.

206 Schwartz, R., Grunes, D.L., Wentworth, R.A., and Wien, E.M. (1980) Magnesium absorption from leafy vegetables intrinsically labeled with the stable isotope ^{26}Mg. *J. Nutr.*, **110**, 1365–1371.

207 Schwartz, R., Spencer, H., and Welsh, J.J. (1984). Magnesium absorption in human subjects from leafy vegetables, intrinsically labeled with stable ^{26}Mg. *Am. J. Clin. Nutr.*, **39**, 571–576.

208 Sabatier, M., Keyes, W.R., Pont, F., Arnaud, M.J., and Turnlund, J.R. (2003) Comparison of stable-isotope-tracer methods for the determination of magnesium absorption in humans. *Am. J. Clin. Nutr.*, **77**, 1206–1212.

209 Dombovari, J., Becker, J.S., and Dietze, H.J. (2000) Isotope ratio measurements of magnesium and determination of magnesium concentration by reverse isotope dilution technique on small amounts of ^{26}Mg-spiked nutrient solutions with inductively coupled plasma mass spectrometry. *Int. J. Mass Spectrom.*, **202**, 231–240.

210 De Wannemacker, G., Ronderos, A., Moens, L., Vanhaecke, F., Bijvelds, M.J.C., and Kolar, Z.I. (2001) Use of double-focusing sector field ICP-mass spectrometry in tracer experiments, aiming at the quantification of Mg^{2+} transport across the intestine of tilapia fish. *J. Anal. At. Spectrom.*, **16**, 581–586.

211 Becker, J.S., Füllner, K., Seeling, U.D., Fornalczyk, G., and Kuhn, A.J. (2008) Measuring magnesium, calcium and potassium isotope ratios using ICP-QMS with an octopole collision cell in tracer studies of nutrient uptake and translocation in plants. *Anal. Bioanal. Chem.*, **390**, 571–578.

212 Schwartz, R., Spencer, H., and Wentworth, R.A. (1978) Measurement of magnesium absorption in man using stable ^{26}Mg as a tracer. *Clin. Chim. Acta*, **87**, 265–273.

213 Sojka, J., Wastney, M., Abrams, S., Froese, S., Martin, B., Weaver, C., and Peacock, M. (1997) Magnesium kinetics in adolesecent girls determined using stable isotopes: effects of high and low calcium intake. *Am. J. Physiol.*, **273**, R710–R715.

214 Abrams, S.A. (1998) The relationship between magnesium and calcium kinetics in 9- to 14-year-old children. *J. Bone Miner. Res.*, **13**, 149–153.

215 Feillet-Coudray, C., Coudray, C., and Tressol, J.C. (2002) Exchangeable magnesium pool masses in healthy women: effects of magnesium supplementation. *Am. J. Clin. Nutr.*, **75**, 72–78.

216 Avioli, L.V. and Berman M. (1966) Mg^{28} kinetics in man. *J. Appl. Physiol.*, **2**, 1688–1694.

217 Wälti, M.K., Walczyk, T., Zimmermann, M.B., Fortunato, G., Weber, M., Spinas, G.A., and Hurrell, R.F. (2006) Urinary excretion of an intravenous ^{26}Mg dose as an indicator of marginal magnesium deficiency in adults. *Eur. J. Clin. Nutr.*, **60**, 147–154.

218 Galy, A., Belshaw, N.S., Halicz, L., and O'Nions, R.K. (2001) High-precision measurement of magnesium isotopes by multiple-collector inductively coupled plasma mass spectrometry. *Int. J. Mass Spectrom.*, **208**, 89–98.

219 Wombacher, F., Eisenhauer, A., Heuser, A., and Weyer, S. (2009) Separation of Mg, Ca and Fe from geological reference materials for stable isotope ratio analyses by MC-ICP-MS and double-spike TIMS. *J. Anal. At. Spectrom.*, **24**, 627–636.

220 Chang, V.T.-C., WIlliams, R.J.P., Makishima, A., Belshawl, N.S., and O'Nions, R.K. (2004) Mg and Ca isotope fractionation during $CaCO_3$ biomineralisation. *Biochem. Biophys. Res. Commun.*, **323**, 79–85.

221 Dennert, G., Zwahlen, M., Brinkman, M., Vinceti, M., Zeegers, M.P.A., and Horneber, M., *et al.* (2011) Selenium for preventing cancer. *Cochrane Database Syst. Rev.*, **5**, CD005195.

222 Birringer, M., Pilawa, S., and Flohé, L. (2002) Trends in selenium biochemistry. *Nat. Prod. Rep.*, **19**, 693–718.

223 Dumont, E., Vanhaecke, F., and Cornelis, R. (2006) Selenium speciation from food source to metabolites: a critical review. *Anal. Bioanal. Chem.*, **385**, 1304–1323.

224 Gammelgaard, B., Jackson, M.I., and Gabel-Jensen, C. (2011) Surveying selenium speciation from soil to cell – forms and transformations. *Anal. Bioanal. Chem.*, **399**, 1743–1763.

225 Fairweather-Tait, S., Collings, R., and Hurst, R. (2010) Selenium bioavailability: current knowledge and future research requirements. *Am. J. Clin. Nutr.*, **91**, 1484S–1491S.

226 Christensen, M., Janghorbani, M., Steinke, F., Istfan, N., and Young, V. (1983) Simultaneous determination of absorption of selenium from poultry meat and selenite in young men: application of a triple stable-isotope method. *Br. J. Nutr.*, **50**, 43–50.

227 Sirichakwal, P., Young, V., and Janghorbani, M. (1985) Absorption and retention of selenium from intrinsically labeled egg and selenite as determined by stable isotope studies in humans. *Am. J. Clin. Nutr.*, **41**, 264–269.

228 Fox, T.E., van den Heuvel, E.G.H.M., Atherton, C.A., Dainty, J.R., Lewis, D.J., Langford, N.J., Crews, H.M., Luten J.B., Lorentzen, M., Sieling, F. W., van Aken-Schneyder, P., Hoek, M., Kotterman, M.J.J., van Dael, P., and Fairweather-Tait, S.J. (2004) Bioavailability of selenium from fish, yeast and selenate: a comparative study in humans using stable isotopes. *Eur. J. Clin. Nutr.*, **58**, 343–349.

229 Fox, T.E., Atherton, C., Dainty, J.R., Lewis, D.J., Langford, N.J., Baxter, M.J., Crews, H.M., and Fairweather-Tait, S.J. (2005) Absorption of selenium from wheat, garlic, and cod intrinsically labeled with Se-77 and Se-82 stable isotopes. *Int. J. Vitam. Nutr. Res.*, **075**, 0179–0186.

230 Bügel S.H., Sandstrom, B., and Larsen E.H. (2001) Absorption and retention of selenium from shrimps in man. *J. Trace Elem. Med. Biol.*, **14**, 198–204.

231 Hawkes, W.C., Alkan, F.Z., and Oehler, L. (2003) Absorption, distribution and excretion of selenium from beef and rice in healthy North American men. *J. Nutr.*, **133**, 3434–3442.

232 Bügel, S., Sandström, B., and Skibsted, L.H. (2004) Pork meat: a good source of selenium? *J. Trace Elem. Med. Biol.*, **17**, 307–311.

233 Ehrenkranz, R.A., Gettner, P.A., Nelli, C.M., Sherwonit, E.A., Williams, J.E., Ting, B.T., and Janghorbani, M. (1991) Selenium absorption and retention by very-low-birth-weight infants: studies with the extrinsic stable isotope tag ^{74}Se. *J. Pediatr. Gastroenterol. Nutr.*, **13**, 125–133.

234 Bügel, S., Larsen, E.H., Sloth, J.J., Flytlie, K., Overvad, K., Steenberg, L.C., and Moesgaard, S.(2008) Absorption, excretion, and retention of selenium from a high selenium yeast in men with a high intake of selenium. *J. Food Nutr. Res.*, **52**, doi: 10.3402/fnr.v52i0.1642.

235 Janghorbani, M., Xia, Y., Ha, P., Whanger, P., Butler, J., and Olesik, J. (1999) Metabolism of selenite in men with widely varying selenium status. *J. Am. Coll. Nutr.*, **18**, 462–469.

236 Kremer, D., Ilgen, G., and Feldmann, J. (2005) GC–ICP-MS determination of dimethylselenide in human breath after ingestion of ^{77}Se-enriched selenite: monitoring of *in-vivo* methylation of selenium. *Anal. Bioanal. Chem.*, **383**, 509–515.

237 Suzuki, K. and Itoh, M. (1997) Metabolism of selenite labelled with enriched stable isotope in the bloodstream. *J. Chromatogr. B*, **692**, 15–22.

238 Suzuki, K.T., Ohta, Y., and Suzuki, N. (2006) Availability and metabolism of ^{77}Se-methylseleninic acid compared simultaneously with those of three related selenocompounds. *Toxicol. Appl. Pharmacol.*, **217**, 51–62.

239 Suzuki, K.T., Somekawa, L., Kurasaki, K., and Suzuki, N. (2006) Simultaneous tracing of ^{76}Se-selenite and ^{77}Se-selenomethionine by absolute labeling and speciation. *Toxicol. Appl. Pharmacol.*, **217**, 43–50.

240 Wastney, M., Combs, G., Canfield, W., Taylor, P., Patterson, K., Hill, A., Moler, J.E., and Patterson, B.H. (2011) A human model of selenium that integrates metabolism from selenite and selenomethionine. *J. Nutr.*, **141**, 708–717.

241 Janghorbani, M., Martin, R., Kasper, L., Sun, X., and Young, V. (1990) The selenite-exchangeable metabolic pool in humans: a new concept for the assessment of selenium status. *Am. J. Clin. Nutr.*, **51**, 670–677.

242 Wachsmann, M. and Heumann, K.G. (1992) Negative thermal ionization mass spectrometry of main group elements, Part 2. 6th group: sulfur, selenium, and tellurium. *Int. J. Mass Spectrom. Ion Processess*, **114**, 209–220.

243 Ting, B.T.G., Mooers, C.S., and Janghorbani, M. (1989) Isotopic determination of selenium in biological materials with inductively coupled plasma mass spectrometry. *Analyst*, **114**, 667–667.

244 Rouxel, O., Ludden, J., Carignan, J., Marin, L., and Fouquet, Y. (2002) Natural variations of Se isotopic composition determined by hydride generation multiple collector inductively coupled plasma mass spectrometry. *Geochim. Cosmochim. Acta*, **66**, 3191–3199.

245 Vandael, P. (2004) Stable isotope-enriched selenite and selenate tracers for human metabolic studies: a fast and accurate method for their preparation from elemental selenium and their identification and quantification using hydride generation atomic absorption spectrometry. *J. Trace Elem. Med. Biol.*, **18**, 75–80.

246 Elwaer, N. and Hintelmann, H. (2007) Comparative performance study of different sample introduction techniques for rapid and precise selenium isotope ratio determination using multi-collector inductively coupled plasma mass spectrometry (MC-ICP/MS). *Anal. Bioanal. Chem.*, **389**, 1889–1899.

247 Far, J., Bérail, S., Preud'homme, H., and Lobinski, R. (2010) Determination of the selenium isotopic compositions in Se-rich yeast by hydride generation-inductively coupled plasma multicollector mass spectrometry. *J. Anal. At. Spectrom.*, **25**, 1695–1695.

248 Elwaer, N. and Hintelmann, H. (2008) Comparing the precision of selenium isotope ratio measurements using collision cell and sector field inductively coupled plasma mass spectrometry. *Talanta*, **75**, 205–214.

249 Zhu, J.M., Johnson, T.M., Clark, S.K., and Zhu, X.K. (2008) High precision measurement of selenium isotopic composition by hydride generation multiple collector inductively coupled plasma mass spectrometry with a ^{74}Se–^{77}Se double spike. *Chin. J. Anal. Chem.*, **36**, 1385–1390.

250 Reyes, L.H., Gayón, J.M.M., Alonso, J.I.G., and Sanz-Medel, A. (2003) Determination of selenium in biological materials by isotope dilution analysis with an octapole reaction system ICP-MS. *J. Anal. At. Spectrom.*, **18**, 11–16.

251 Schaumlöffel, D., Bierla, K., and Lobiński, R. (2007) Accurate determination of selenium in blood serum by isotope dilution analysis using inductively coupled plasma collision cell mass spectrometry with xenon as collision gas. *J. Anal. At. Spectrom.*, **22**, 318–321.

252 Al-Saad, K.A., Amr, M.A. and Helal, A. I. (2011) Collision/reaction cell ICP-MS with shielded torch and sector field ICP-MS for the simultaneous determination of selenium isotopes in biological matrices. *Biol. Trace Elem. Res.*, **140**, 103–113.

253 Johnson, T., Herbel, M., Bullen, T., and Zawislanski, P. (1999) Selenium isotope ratios as indicators of selenium sources and oxyanion reduction. *Geochim. Cosmochim. Acta*, **63**, 2775–2783.

254 Herbel, M., Johnson, T., Oremland, R., and Bullen, T. (2000) Fractionation of selenium isotopes during bacterial respiratory reduction of selenium oxyanions. *Geochim. Cosmochim. Acta*, **64**, 3701–3709.

255 Ellis, A., Johnson, T.M., Herbel, M.J., and Bullen, T.D. (2003) Stable isotope fractionation of selenium by natural microbial consortia. *Chem. Geol.*, **195**, 119–129.

256 Carignan, J. and Wen, H. (2007) Scaling NIST SRM 3149 for Se isotope analysis and isotopic variations of natural samples. *Chem. Geol.*, **242**, 347–350.

257 Wen, H. and Carignan, J. (2011) Selenium isotopes trace the source and redox processes in the black shale-hosted Se-rich deposits in China. *Geochim. Cosmochim. Acta*, **75**, 1411–1427.

258 Johnson, T. (2004) A review of mass-dependent fractionation of selenium isotopes and implications for other heavy stable isotopes. *Chem. Geol.*, **204**, 201–214.

259 Harvey, L.J., Majsak-Newman, G., Dainty, J.R., Wharf, S.G., Reid, M.D., Beattie, J.H., and Fairweather-Tait, S.J. (2002) Holmium as a faecal marker for copper absorption studies in adults. *Clin. Sci.*, **102**, 233–240.

260 Johnson, P.E. and Lykken, G.I. (1988) Copper-65 absorption by men fed intrinsically and extrinsically labeled whole wheat bread. *J. Agric. Food Chem.*, **36**, 537–540.

261 Johnson, P.E., Stuart, M.A., Hunt, J.R., Mullen, L., and Starks, T.L. (1988) ^{65}Copper absorption by women fed intrinsically and extrinsically labeled goose meat, goose liver, peanut butter and sunflower butter. *J. Nutr.*, **118**, 1522–1528.

262 Harvey, L.J., Dainty, J.R., Beattie, J.H., Majsak-Newman, G., Wharf, S.G., Reid, M.D., and Fairweather-Tait, S.J. (2005) Copper absorption from foods labelled intrinsically and extrinsically with Cu-65 stable isotope. *Eur. J. Clin. Nutr.*, **59**, 363–368.

263 Turnlund, J.R., Swanson, C.A., and King, J.C. (1983) Copper absorption and retention in pregnant women fed diets based on animal and plant proteins. *J. Nutr.*, **113**, 2346–2352.

264 Hunt, J.R. and Vanderpool, R.A. (2001) Apparent copper absorption from a vegetarian diet. *Am. J. Clin. Nutr.*, **74**, 803–807.

265 Jacob, R.A., Skala, J.H., Omaye, S.T., and Turnlund, J.R. (1987) Effect of varying ascorbic acid intakes on copper absorption and ceruloplasmin levels of young men. *J. Nutr.*, **117**, 2109–2115.

266 Turnlund, J.R., Reager, R.D., and Costa, F. (1988) Iron and copper absorption in young and elderly men. *Nutr. Res.*, **8**, 333–343.

267 Olivares, M., Lönnerdal, B., Abrams, S.A., Pizarro, F., and Uauy, R. (2002) Age and copper intake do not affect copper absorption measured with the use of ^{65}Cu as a tracer in young infants. *Am. J. Clin. Nutr.*, **76**, 641–645.

268 Turnlund, J.R., Keyes, W.R., Peiffer, G.L., and Scott, K.C. (1998) Copper absorption, excretion, and retention by young men consuming low dietary copper determined by using the stable isotope ^{65}Cu. *Am. J. Clin. Nutr.*, **67**, 1219–1225.

269 Harvey, L.J., Majsak-Newman, G., Dainty, J.R., Lewis, D.J., Langford, N.J., Crews, H.M., and Fairweather-Tait, S.J. (2003) Adaptive responses in men fed low- and high-copper diets. *Br. J. Nutr.*, **90**, 161–168.

270 Turnlund, J.R., Keyes, W.R., Kim, S.K., and Domek, J.M. (2005) Long-term high copper intake: effects on copper absorption, retention, and homeostasis in men. *Am. J. Clin. Nutr.*, **81**, 822–828.

271 Turnlund, J.R., Keyes, W.R., Hudson, C.A., Betschart, A.A., Kretsch, M.J., and Sauberlich, H.E. (1991) A stable-isotope study of zinc, copper, and iron absorption and retention by young women fed vitamin B-6-deficient diets. *Am. J. Clin. Nutr.*, **54**, 1059–1064.

272 Domellöf, M., Hernell, O., Abrams, S.A., Chen, Z., and Lönnerdal, B. (2009) Iron supplementation does not affect copper and zinc absorption in breastfed infants. *Am. J. Clin. Nutr.*, **89**, 185–190.

273 Turnlund, J.R. (1998) Human whole-body copper metabolism. *Am. J. Clin. Nutr.*, **67**, 960S–964S.

274 Scott, K.C. and Turnlund, J.R. (1994) Compartmental model of copper metabolism in adult men. *J. Nutr. Biochem.*, **5**, 342–350.

275 Harvey, L.J., Dainty, J.R., Hollands, W.J., Bull, V.J., Beattie, J.H., Venelinov, T.I., Hoogewerff, J.A., Davies, I.M., and Fairweather-Tait, S.J. (2005) Use of mathematical modeling to study copper metabolism in humans. *Am. J. Clin. Nutr.*, **81**, 807–813.

276 Venelinov, T., Beattie, J., Dainty, J., Hollands, W., Fairweather-Tait, S., and Harvey, L. (2007) Stable isotope pilot study of exchangeable copper kinetics in human blood plasma. *J. Trace Elem. Med. Biol.*, **21**, 138–140.

277 Beattie, J.H., Reid, M.D., Harvey, L.J., Dainty, J.R., Majsak-Newman, G., and Fairweather-Tait, S.J. (2001) Selective extraction of blood plasma exchangeable copper for isotope studies of dietary copper absorption. *Analyst*, **126**, 2225–2229.

278 Hosoe, M., Fujii, Y., and Okamoto, M. (1988) Mass-spectrometric measurements of copper isotopic-ratios using low-temperature thermal ionization. *Anal. Chem.*, **60**, 1812–1815.

279 Vanhaecke, F., Moens, L., and Dams, R. (1998) The accurate determination of copper in two groundwater candidate reference materials by means of high resolution inductively coupled plasma mass spectrometry using isotope dilution for calibration. *J. Anal. At. Spectrom.*, **13**, 1189–1192.

280 Mason, T.F.D., Weiss, D.J., Horstwood, M., Parrish, R.R., Russell, S.S., Mullane, E., and Coles, B.J. (2004) High-precision Cu and Zn isotope analysis by plasma source mass spectrometry. Part 2: correcting for mass discrimination effects. *J. Anal. At. Spectrom.*, **19**, 218–226.

281 Diemer, J., Quétel, C.R., and Taylor, P.D.P. (2002) Comparison of the performance of different ICP-MS instruments on the measurement of Cu in a water sample by ICP-IDMS. *J. Anal. At. Spectrom.*, **17**, 1137–1142.

282 Turnlund, J., Weaver, C., Kim, S., Keyes, W., Gizaw, Y., Thompson, K., and Peiffer, G.L. (1999) Molybdenum absorption and utilization in humans from soy and kale intrinsically labeled with stable isotopes of molybdenum. *Am. J. Clin. Nutr.*, **69**, 1217–1223.

283 Turnlund, J., Keyes, W., Peiffer, G., and Chiang, G. (1995) Molybdenum absorption, excretion, and retention studied with stable isotopes in young men during depletion and repletion. *Am. J. Clin. Nutr.*, **61**, 1102–1109.

284 Turnlund, J., Keyes, W., Peiffer, G., and Chiang, G. (1995) Molybdenum absorption, excretion, and retention studied with stable isotopes in young men at five intakes of dietary molybdenum. *Am. J. Clin. Nutr.*, **62**, 790–796.

285 Sievers, E., Dörner, K., Garbe-Schönberg, D., and Schaub, J. (2001) Molybdenum metabolism: stable isotope studies in infancy. *J. Trace Elem. Med. Biol.*, **15**, 185–191.

286 Giussani, A., Arogunjo, A., Claire Cantone, M., Tavola, F., and Veronese, I. (2006) Rates of intestinal absorption of molybdenum in humans. *Appl. Radiat. Isot.*, **64**, 639–644.

287 Thompson, K. and Turnlund, J. (1996) Kinetic model of molybdenum metabolism developed from dual stable isotope excretion in men consuming a low molybdenum diet. *J. Nutr.*, **126**, 963–972.

288 Turnlund, J., Keyes, W.R. (2004) Plasma molybdenum reflects dietary molybdenum intake. *J. Nutr. Biochem.*, **15**, 90–95.

289 Novotny, J. and Turnlund, J. (2006) Molybdenum intake influences molybdenum kinetics in men. *J. Nutr.*, **137**, 37–42.

290 Werner, E., Roth, P., Heinrichs, U., Giussani, A., Cantone, M.C., Zilker, T., Felgenhauer, N., and Greim, H. (2000) Internal biokinetic behaviour of molybdenum in humans studied with stable isotopes as tracers. *Isot. Environ. Health Stud.*, **36**, 123–132.

291 Giussani, A., Cantone, M.C., Hollriegl, V., Oeh, U., Tavola, F., and Veronese, I., (2007) Modelling urinary excretion of molybdenum after oral and intravenous administration of stable tracers. *Radiat. Protect. Dosimetry*, **127**, 136–139.

292 Turnlund, J., Keyes, W., and Peiffer, G. (1993) Isotope ratios of molybdenum determined by thermal ionization mass spectrometry for stable isotope studies of molybdenum metabolism in humans. *Anal. Chem.*, **65**, 1717–1722.

293 Lu, Q. and Masuda, A. (1992) High accuracy measurement of isotope ratios of molybdenum in some terrestrial molybdenites. *J. Am. Soc. Mass Spectrom.*, **3**, 10–17.

294 Lu, Q. and Masuda, A. (1994) Isotopic composition and atomic weight of molybdenum. *Int. J. Mass Spectrom. Ion Processes*, **130**, 65–72.

295 Giussani, A. and Hansen, C. (1995) Application of thermal ionization mass spectrometry to investigations of molybdenum absorption in humans. *Int. J. Mass Spectrom. Ion Processes*, **148**, 171–178.

296 Wieser, M. and de Laeter, J. (2000) Thermal ionization mass spectrometry of molybdenum isotopes. *Int. J. Mass Spectrom.*, **197**, 253–261.

297 Wieser, M. and de Laeter, J.R. (2003) A preliminary study of isotope fractionation in molybdenites. *Int. J. Mass Spectrom.*, **225**, 177–183.

298 Keyes, W.R. and Turnlund, J.R. (2002) Determination of molybdenum and enriched Mo stable isotope concentrations in human blood plasma by isotope dilution ICP-MS. *J. Anal. At. Spectrom.*, **17**, 1153–1156.

299 Lee, D. and Halliday, A. (1995) Precise determinations of the isotopic compositions and atomic weights of molybdenum, tellurium, tin and tungsten using ICP magnetic sector multiple collector mass spectrometry. *Int. J. Mass Spectrom. Ion Processes*, **146**, 35–46.

300 Dauphas, N., Reisberg, L., and Marty, B. (2001) Solvent extraction, ion chromatography, and mass spectrometry of molybdenum isotopes. *Anal. Chem.*, **73**, 2613–2616.

301 Anbar, A.D., Knab, K.A., and Barling, J. (2001) Precise determination of mass-dependent variations in the isotopic composition of molybdenum using MC-ICPMS. *Anal. Chem.*, **73**, 1425–1431.

302 Malinovsky, D., Rodushkin, I., Baxter, D., Ingri, J., and Ohlander, B. (2005) Molybdenum isotope ratio measurements on geological samples by MC-ICPMS. *Int. J. Mass Spectrom.*, **245**, 94–107.

303 Wen, H., Carignan, J., Cloquet, C., Zhu, X., and Zhang, Y. (2010) Isotopic delta values of molybdenum standard reference and prepared solutions measured by MC-ICP-MS: proposition for delta zero and secondary references. *J. Anal. At. Spectrom.*, **25**, 716–721.

17
Isotopic Analysis via Multi-Collector Inductively Coupled Plasma Mass Spectrometry in Elemental Speciation

Vladimir N. Epov, Sylvain Berail, Christophe Pécheyran, David Amouroux, and Olivier F.X. Donard

17.1
Introduction

Isotope fractionation occurs during physical, chemical, and biological processes (see also Chapter 1). During (bio)chemical reactions, isotopes of an element are distributed between different chemical species of that element. According to the IUPAC definition, a chemical species is "an ensemble of chemically identical molecular entities that can explore the same set of molecular energy levels on the time scale of the experiment" [1]. Another definition given by IUPAC describes a chemical species of an element as "a specific form of an element defined as to isotopic composition, electronic or oxidation state, and/or complex or molecular structure" [1]. In other words, chemical species are atoms, molecules, molecular fragments, or ions taking part in a chemical process or being subjected to a measurement. The uptake, accumulation, transport, and interaction of the different elements in nature are influenced by their species.

For example, Figure 17.1a graphically exemplifies the presence and distribution of different chemical species (Sp-1, Sp-2, Sp-3, and Sp-4) of a trace element in the environment. The main environmental processes involved are volatilization, deposition/input, reduction, oxidation, precipitation/adsorption, dissolution/desorption, complexation/alkylation, dissociation/dealkylation, bioaccumulation/uptake, and release. According to this figure, the element enters the atmosphere from natural and/or anthropogenic sources in the form of gaseous species Sp-2. In the atmosphere, the species Sp-2 can be oxidized to ionic species Sp-1, which can either be precipitated via wet precipitation or be re-reduced back to Sp-2. In aquatic ecosystems, the species Sp-1 can undergo three main processes: (i) reduction to species Sp-2, (ii) chelation with organic ligands or by bacteria to form species Sp-3, or (iii) precipitation in the form of solid species Sp-4. Organic species Sp-3 can bioaccumulate in biological organisms [2] or dissociate back to species Sp-1.

Isotopic Analysis: Fundamentals and Applications Using ICP-MS, First Edition. Edited by Frank Vanhaecke and Patrick Degryse.

Published 2012 by WILEY-VCH Verlag GmbH & Co. KGaA

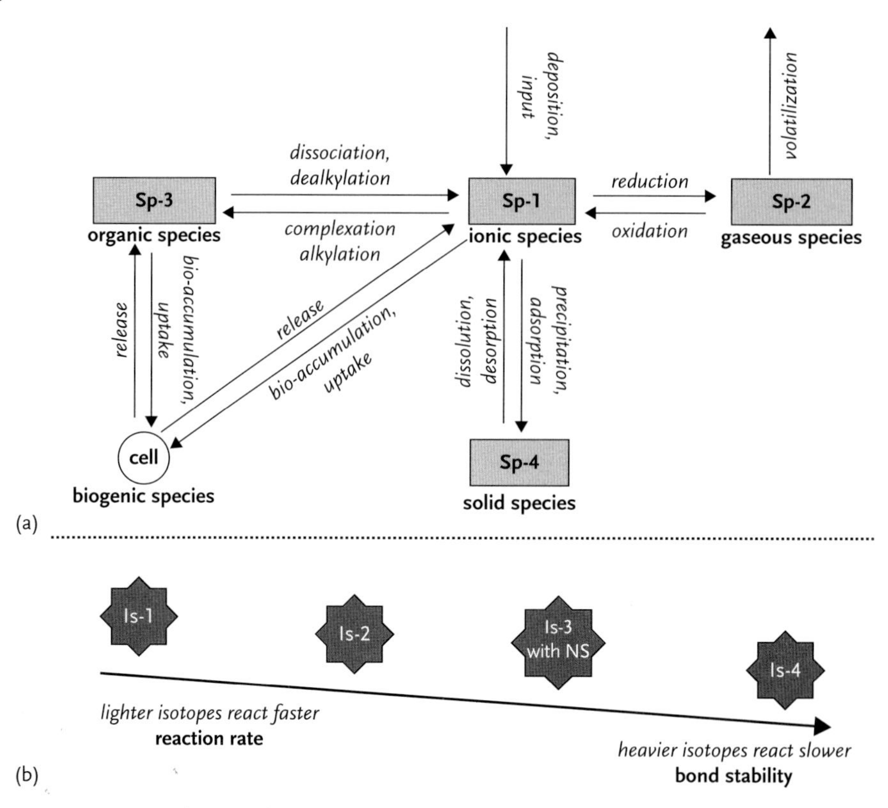

Figure 17.1 (a) Schematic diagram summarizing the fate of an element species (Sp-1, Sp-2, Sp-3, Sp-4) in the environment. (b) Example of an element with several stable isotopes (Is-1, Is-2, Is-3, Is-4). NS represents a (non-zero) nuclear spin.

During these environmental processes, isotopes of this element are fractionated between different species. As presented in Figure 17.1b, several isotopes of the element (Is-1, Is-2, Is-3, and Is-4) can fractionate because in chemical reactions lighter isotopes react faster than heavier isotopes. This kinetic phenomenon has a mass-dependent nature or is governed by the nuclear mass isotope effect, causing "traditional" mass-dependent isotope fractionation. However, under specific conditions, some isotopes of this element (Is-3) can also fractionate due to one of the mass-independent mechanisms: (i) the nuclear volume effect, causing mass-independent isotope fractionation due to the different nuclear volumes of the isotopes, or (ii) the nuclear spin (NS) effect, causing mass-independent isotope fractionation due to hyperfine Fermi interaction between the electrons and a nucleus having a nuclear spin [3] (see also Chapter 1). For example, both mass-dependent and mass-independent isotope fractionation have recently been demonstrated experimentally for Hg in environmental samples and during chemical and biological processes [4].

In the previous paragraph, a general example of isotope fractionation between different species (Sp) of an element was described. It is clear, however, that these types of processes and fractionation will depend on the specific physicochemical properties of the element. To determine the isotopic composition of the target element in the species, development of analytical techniques is required. These can rely on either on-line or off-line separation of the species using chromatography or another technique, followed by the measurement of the isotopic composition of the target element that it contains by mass spectrometry.

17.2
Advantage of On-Line versus Off-Line Separation of Elemental Species

To separate different chemical species of an element (oxidation states, organometallic compounds, coordination compounds or metal- and heteroatom-containing biomolecules) from one another, several analytical techniques are used [5, 6]. These separation techniques include (i) specific sample preparation methods, such as complexation, different extraction and derivatization methods and precipitation, and (ii) instrumental techniques, such as electrophoresis [7], liquid chromatography (LC) [8], gas chromatography (GC), ion chromatography (IC), flow injection [9], chemical vapor generation [10], and X-ray methods. The majority of methods in speciation analysis include a separation step before the detection of the different species. A literature survey revealed that high performance liquid chromatography (HPLC) [8] is the dominant separation technique in elemental speciation, followed by IC.

Chromatographic techniques with a liquid mobile phase can be used for the separation of different chemical species, in either off-line or on-line mode. However, off-line separation of the species can have several disadvantages in comparison with on-line separation: (i) it is more time-consuming, (ii) it is more labor-intensive, (iii) the amount of the collected species after off-line separation can be small and therefore additional sample preparation steps, such as preconcentration, may be required, and (iv) it can be difficult to optimize the off-line separation method, as the elution profile can vary and cannot be controlled on-line. In addition, GC separation of different species can, in most cases, only be used in on-line mode, since the gaseous species are more difficult to collect off-line. When combining isotopic analysis with chromatographic on-line separation of species, multi-collector (MC) inductively coupled plasma mass spectrometry (ICP-MS) is the most suitable type of instrumentation owing to several advantages, including sensitivity, selectivity, the high ionization efficiency and robustness of the ICP source, which allows for the hyphenation of chromatography or flow injection, and the simultaneous monitoring of the isotopes of interest, providing high isotope ratio precision. Also, on-line coupling of chromatographic separation with MC-ICP-MS could overcome some of the issues related to off-line separation, since it provides improved sample throughput and allows the use of alternative instrumental mass bias correction procedures. Details of the instrumentation for on-line determination of isotope ratios in species are described in Section 17.3, and applications relying on isotopic analysis in species via MC-ICP-MS are described for different elements in Section 17.4. Table 17.1 summarizes the published applications of isotopic analysis via MC-ICP-MS in species of different elements [11–23]. Unfortunately, to date, only GC has been coupled with MC-ICP-MS.

17.3 Coupling Chromatography with MC-ICP-MS

As discussed previously, the development of MC-ICP-MS instruments has produced a powerful technique for measuring isotope ratios precisely and accurately and therefore the technique has achieved growing success since its conception in the early 1990s [24]. One of the main advantages compared with other techniques capable of providing information on the isotopic composition of the elements, such as thermal ionization mass spectrometry (TIMS), is the ability to analyze a wide range of elements, even those with a high ionization energy (such as Hg and Hf) owing to the high ionization efficiency of the ICP source [25]. Furthermore, another important characteristic of the ICP ionization source is that it works at atmospheric pressure, which allows the introduction of samples as either a liquid or a gas. Hence this makes it possible to couple different introduction systems to MC-ICP-MS, including laser ablation (LA) (see also Chapter 4) and different types of chromatography. The latter provides accurate and precise isotope ratios in different elemental species. The main issues that need to be considered when deploying such a hyphenated technique are (i) technical issues related to the coupling of the chromatography to the ICP source for specific applications, (ii) proper treatment of the transient signal data, and (iii) the often low concentrations of the analytes.

17.3.1 Instrumentation: LC, GC, HPLC, and IC Coupled with MC-ICP-MS

When coupling a chromatographic technique to MC-ICP-MS, one should keep in mind that it has the same ion source as other types of ICP-MS instruments (quadrupole-based, single-collector sector field and time-of-flight ICP-MS). It is therefore possible to use the same strategy for the coupling of chromatographic columns to the plasma source of an MC-ICP-MS instrument. However, measuring isotope ratios with high precision requires dealing with additional issues, such as the instrumental mass bias, the transient nature of the signals, and suitable sample preparation. For MC-ICP-MS, the mass bias is significant and can reach the percent range [25]. Hence proper correction for mass bias when using a hyphenation technique is one of the challenges for the isotopic analysis in elemental species. In addition, on-line chromatography implies dealing with fairly short transient signals, corresponding to the elution of the chromatographic peak (from a few seconds up to more than 1 min). This complicates the data treatment

Table 17.1 Publications on the measurement of isotope ratios in elemental species using the combination of chromatography and MC-ICP-MS.

Element	Species measured	Isotopes measured	Samples	Hyphenation method	Ref.
Hg	$MeHg^+$, IHg^{2+}	^{198}Hg, ^{199}Hg, ^{200}Hg, ^{201}Hg, ^{202}Hg	Standards	GC–MC-ICP-MS	[11]
	$MeHg^+$, IHg^{2+}	^{198}Hg, ^{199}Hg, ^{200}Hg, ^{201}Hg, ^{202}Hg	Standards and candidate NRC CRMs	GC–MC-ICP-MS	[12]
	$MeHg^+$, IHg^{2+}	^{198}Hg, ^{199}Hg, ^{200}Hg, ^{201}Hg, ^{202}Hg	Standards and CRMs IAEA-085, BCR 464	GC–MC-ICP-MS	[13]
	$MeHg^+$, IHg^{2+}	^{198}Hg, ^{199}Hg, ^{200}Hg, ^{201}Hg, ^{202}Hg, ^{204}Hg	IHg^{2+} methylation by anaerobic bacteria	GC–MC-ICP-MS	[14]
	Hg^0, Me_2Hg, $MeHg^+$, IHg^{2+}	^{196}Hg, ^{198}Hg, ^{199}Hg, ^{200}Hg, ^{201}Hg, ^{202}Hg, ^{204}Hg	Standards, Me_2Hg and CRMs IAEA-085, IAEA-086, BCR 464	GC–MC-ICP-MS	[15]
Pb	$PbEt_4$	^{204}Pb, ^{206}Pb, ^{207}Pb, ^{208}Pb	Tetraethyllead synthesized from NIST NBS 981	GC–MC-ICP-MS	[11, 16, 17]
S	SF_6	^{32}S, ^{33}S, ^{34}S	Certified SF_6 gas	GC–MC-ICP-MS	[18]
	Organosulfur species	^{32}S, ^{33}S, ^{34}S	Eight organosulfur compounds	GC–MC-ICP-MS	[19]
Sb	Me_3Sb	^{121}Sb, ^{123}Sb	Me_3Sb generated by anaerobic bacteria	GC–MC-ICP-MS	[20]
Cl	C_2HCl_3, C_2Cl_4	^{35}Cl, ^{37}Cl	Trichloroethene, tetrachloroethene	GC–MC-ICP-MS	[21]
Br	Organobromine species	^{79}Br, ^{81}Br	BDE-47, BDE-99, BDE-100	GC–MC-ICP-MS	[22]
	Organobromine species	^{79}Br, ^{81}Br	C_6H_4ClBr, $C_6H_3OBr_3$, C_2H_4Br, C_7H_7Br, $C_6H_3Br_3$	GC–MC-ICP-MS	[23]

strategy. Also, possible coelution of matrix components needs to be taken into account as this might cause spectral interferences or affect the instrumental mass discrimination. Finally, the sample preparation steps should not alter either the original isotope ratio or the original speciation of the sample [26].

17.3.1.1 Liquid Chromatography

To date, LC coupled on-line with MC-ICP-MS (LC–MC-ICP-MS) has been mainly used for applications other than isotopic analysis in species. These applications include sample preconcentration, separation of the target element from matrix components or interfering elements [27–32], and species quantification relying on isotope dilution [33, 34]. Although the description of these applications is beyond the scope of this chapter, they are technically helpful for the development of methods for isotopic analysis in species using LC–MC-ICP-MS. Most of the time, LC requires minimum sample preparation and several separation modes (e.g., adsorption, partition, ion exchange, and size exclusion) are available. Another advantage of LC is the relative simplicity of the coupling to the ICP–MS system. In addition, the precision of isotope ratios obtained using LC–MC-ICP-MS can be similar to that attainable using continuous sample introduction systems, as discussed elsewhere [27]. Technically, almost all types of LC could be used for this coupling. However, one of the main limitations is the use of the eluent; only mobile phases with limited salt concentration and within a certain pH range may be used, and it is better to avoid the use of organic solvents [26]. So far, two different types of separation have been used in LC–MC-ICP-MS coupling: for most applications: ion-exchange columns were utilized for the purpose of target element–matrix separation [27–31], and in other cases reversed-phase columns have been used [32–34].

LC is typically operated at room temperature. The eluent flow rate in LC corresponds well with the sample uptake rate of the traditional introduction system (pneumatic nebulizer) used for MC-ICP-MS [26]. As a consequence, LC–MC-ICP-MS coupling can be simply accomplished by connecting the column outlet of the chromatographic system with the nebulizer of the MC-ICP-MS unit (Figure 17.2).

17.3.1.2 Gas Chromatography

In GC, volatile species of an element (e.g., S, Se, Sb, or Hg) are separated from one another depending on their affinity towards a stationary phase and their vapor pressure. Separation of the volatile forms can be achieved either at a constant temperature or (even more efficiently) by using a temperature ramp, while an inert carrier gas (such as He) is used to transport the species through the column. Gas chromatography often requires that the compounds of interest are derivatized to generate volatile species. For example, ethylation (Et) or propylation (Pr) of an element (El) species with $NaBEt_4$ or $NaBPr_4$, respectively, can be used for some elements, thus creating the ethylated or propylated forms of the element or an organic salt thereof (e.g., Pb, Hg, or Sn). These ethylated or propylated forms are

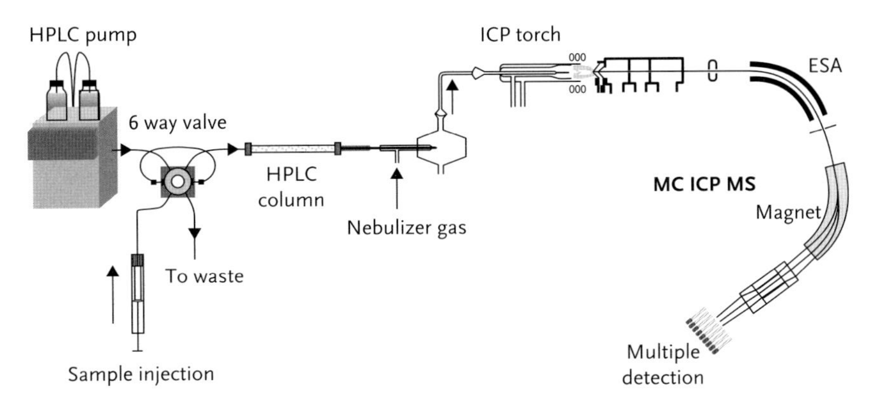

Figure 17.2 HPLC–MC-ICP-MS coupling.

volatile under GC conditions. During the reaction, the derivatized species are simultaneously trapped and concentrated in an organic solvent (e.g., hexane or isooctane) prior their injection on to the GC column.

$$
\begin{aligned}
n\mathrm{NaBEt_4} + \mathrm{El}^{n+} &\rightarrow n\mathrm{BEt_3} + \mathbf{ElEt}_n + n\mathrm{Na}^+ \\
(n-k)\mathrm{NaBEt_4} + \mathrm{R}_k\mathrm{El}^{(n-k)+} &\rightarrow (n-k)\mathrm{BEt_3} + \mathbf{R}_k\mathbf{ElEt}_{n-k} + (n-k)\mathrm{Na}^+ \\
n\mathrm{NaBPr_4} + \mathrm{El}^{n+} &\rightarrow n\mathrm{BEt_3} + \mathbf{ElPr}_n + n\mathrm{Na}^+ \\
(n-k)\mathrm{NaBPr_4} + \mathrm{R}_k\mathrm{El}^{(n-k)+} &\rightarrow (n-k)\mathrm{BEt_3} + \mathbf{R}_k\mathbf{ElPr}_{n-k} + (n-k)\mathrm{Na}^+
\end{aligned}
$$

where R is an alkyl group (methyl, ethyl, propyl, butyl, etc.), $k = 1$–4, and El = target element (Hg, Pb, Sn, etc.). In addition, coupling of GC to MC-ICP-MS (GC–MC-ICP-MS) is not as straightforward as for LC. However, in most previous studies in which both isotopic and speciation information were required, the hyphenation of GC with MC-ICP-MS was used. The most noticeable advantage of GC–MC-ICP-MS is better sensitivity compared with LC–MC-ICP-MS. This is mainly due to the fact that (nearly) 100% of the introduced analyte reaches the plasma with GC–MC-ICP-MS [26], whereas with LC–MC-ICP-MS, the analyte introduction efficiency is limited to values $<5\%$. In addition, GC is a more stable sample introduction system because the mobile phase is an inert gas. However, the precision of isotope ratio measurements accomplished using GC–MC-ICP-MS is generally somewhat worse (Table 17.2). RSDs are about 2–10 times higher than when using continuous sample introduction, owing to the short duration of a chromatographic peak in GC (seconds) and the smaller amount of analyte introduced.

Depending on the application, the derivatized sample can either be introduced directly into the GC unit [11, 13, 15, 16] or be preconcentrated on a Tenax or cryogenic trap, followed by thermal desorption for introduction into the column [12, 20]. Although capillary columns are now the most widely applied [11, 13, 15, 16, 20], the use of packed columns has also been reported [12].

Table 17.2 Precision for delta values measured using GC–MC-ICP-MS.

Study	Year	Species	Measured delta value	External precision (2SD) (‰)
Van Acker *et al.* [21]	2006	Trichloroethene	$\delta^{37/35}Cl$	0.5
Epov *et al.* [13]	2008	$MeHg^+$	$\delta^{202/198}Hg$	0.56
Amrani *et al.* [19]	2009	Dihexyl sulfide	$\delta^{34/32}S$	0.16
Epov *et al.* [15]	2010	IHg^{2+}	$\delta^{202/198}Hg$	0.28

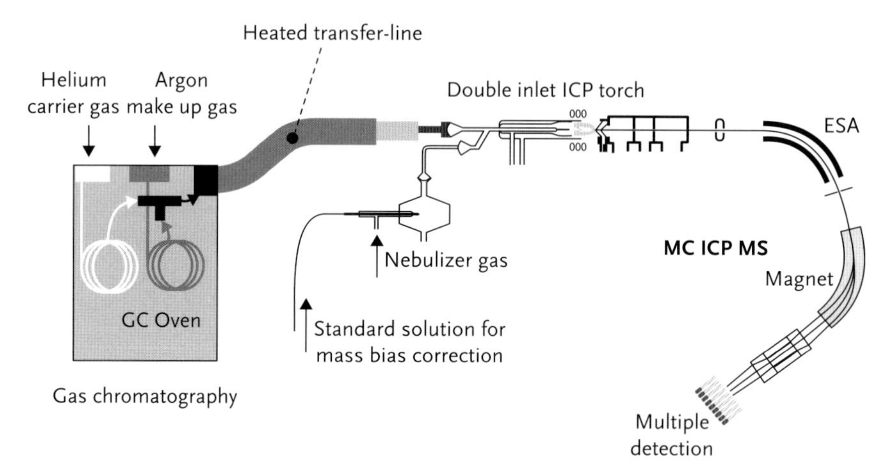

Figure 17.3 GC–MC-ICP-MS coupling in wet plasma configuration [17].

When using a GC unit coupled to an ICP–MS system, two different strategies have been reported:

1. Operation under wet plasma conditions is the most commonly used approach [11, 13, 15, 16, 20]. With this approach, a standard solution is introduced for tuning of MC-ICP-MS parameters, for monitoring the plasma conditions during the isotope ratio measurements, and for mass bias correction. The standard solution used for these purposes is introduced as a wet aerosol via a pneumatic nebulizer connected to the spray chamber (Figure 17.3). This wet plasma configuration provides a more stable signal [13] and the excess of oxygen atoms from the water molecules helps to burn the carbon from the organic solvent. This reduces the deposition of elemental carbon on the cones of the MC-ICP-MS instrument.
2. Operation under dry plasma conditions is the second option. The standard solution is introduced as a dry aerosol through a desolvating system (such as an Aridus, DSN-100 or Apex unit) [12, 23]. The use of this configuration significantly reduces oxide-based interferences. This is essential for some elements, such as sulfur [19].

For both these configurations, the standard, in wet or dry aerosol form, is mixed with the column effluent, either via a simple T-piece or by using a special double-inlet ICP torch (Figure 17.3). To obtain the best results, several MC-ICP-MS parameters have to be specifically optimized and the type of interface cones has to be properly chosen for each configuration. Since the flow rate of He ($\sim$0.025 l min^{-1}) at the exit of the GC column is not high enough to accomplish sample introduction into the plasma, an additional Ar gas flow (make-up gas) with a flow rate of $\sim$0.4 l min^{-1} is admixed (Figure 17.3). For connecting the GC column and the ICP torch, a heated transfer line ($\sim$300 °C) is used to avoid any condensation of the separated species between the column exit and the plasma.

It should be noted that GC parameters can play a crucial role in improving the precision of the measured isotope ratios. Some authors recommend the optimization of the GC temperature program and gas flow rate to increase the width of chromatographic peaks [13], because, as will be further discussed, isotope ratio precision and peak width are interdependent.

17.3.2 Acquisition, Mass Bias Correction, and Data Treatment Strategy

Coupling of chromatography to MC-ICP-MS requires dealing with relatively short transient signals, which makes the data evaluation more complex. The simultaneous detection of the signals of different isotopes is one of the main differences between MC and other types of ICP-MS instrumentation. Hence the acquisition and treatment of the signals are a crucial aspect in GC–MC-ICP-MS.

17.3.2.1 Signal Acquisition

The main condition for signal acquisition is that the total measurement window must be wide enough to record the chromatogram in its entirety. Two strategies can be used, depending on the type of MC-ICP-MS instrument deployed: (i) subdivision of the total acquisition time into blocks and cycles similar to that typical for conventional continuous analysis, in which case it is important that the number of cycles is sufficiently high to acquire the entire chromatogram (e.g., 300 cycles [12]), or (ii) time-resolved analysis (TRA) [15] (Figure 17.4), in which the acquisition is started and ended on demand.

An important aspect related to the acquisition of a GC–MC-ICP-MS trace is the integration time. This is a key parameter, as it has a large impact on the precision of the isotope ratios. The integration time defines the number of points describing a given chromatographic peak. Between 15 and 40 points with an integration time from 0.05 s [16] to 1 s [12, 13, 23] each is typical for GC–MC-ICP-MS [11, 12] (for comparison, the integration time when dealing with continuous signals is usually between 3 and 10 s). Since the Faraday detectors used in MC-ICP-MS produce a more stable response for longer integration times, it is necessary to find a compromise between the number of points per chromatographic peak and the integration time. Also, it should be noted that the available range of integration times depends on the type of mass spectrometer used and its software capabilities.

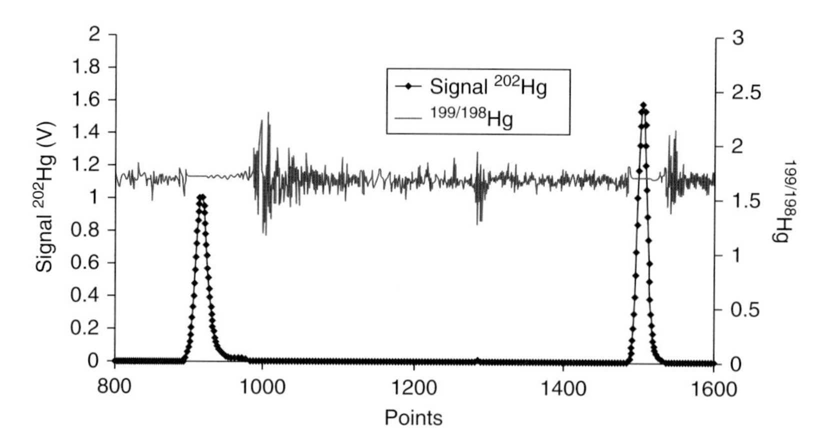

Figure 17.4 Example of a GC–MC-ICP-MS Hg species isotope analysis (first peak $MeHg^+$ and second peak IHg^{2+}) in a STREM/NIST standard ([Hg] = 500 μg l^{-1}). TRA acquisition mode, integration time = 0.5 s.

17.3.2.2 Mass Bias Correction

Depending on the isotopes monitored and the nature of the sample, several types of mass bias correction models can be used when coupling chromatography to MC-ICP-MS. Some studies report the use of internal mass bias correction with an exponential law when an invariant isotope ratio is available (as is the case, for example, for Sr or Nd) [27, 29–31]. It is the simplest model, as merely another isotope ratio of the element of interest is used for the correction.

It is also possible to use external mass bias correction. In this case, another element, which is absent in the sample and which is located in the same mass range as the analyte, is simultaneously and continuously introduced [33]. In this case the exponential law is preferentially used for correction of the measured ratios [20].

Finally, another approach to correct for the mass bias is the "standard–sample–standard" bracketing technique [13, 15, 28, 34], in which a certified isotopic standard is measured before and after each sample. Generally, for isotopic analysis in species, this type of correction should be applied "species by species," which means that the standard must contain the element of interest in the same species as the sample.

To improve the precision and accuracy of results, both external mass bias correction and standard–sample–standard bracketing techniques can be used simultaneously. This has been done, for example, for Hg isotopic studies in elemental species [13, 15].

17.3.2.3 Data Treatment Strategy

Several methods for the calculation of isotope ratios in species have been developed in order to improve the precision and accuracy of the results. So far, four calculation approaches have been reported (Figure 17.5):

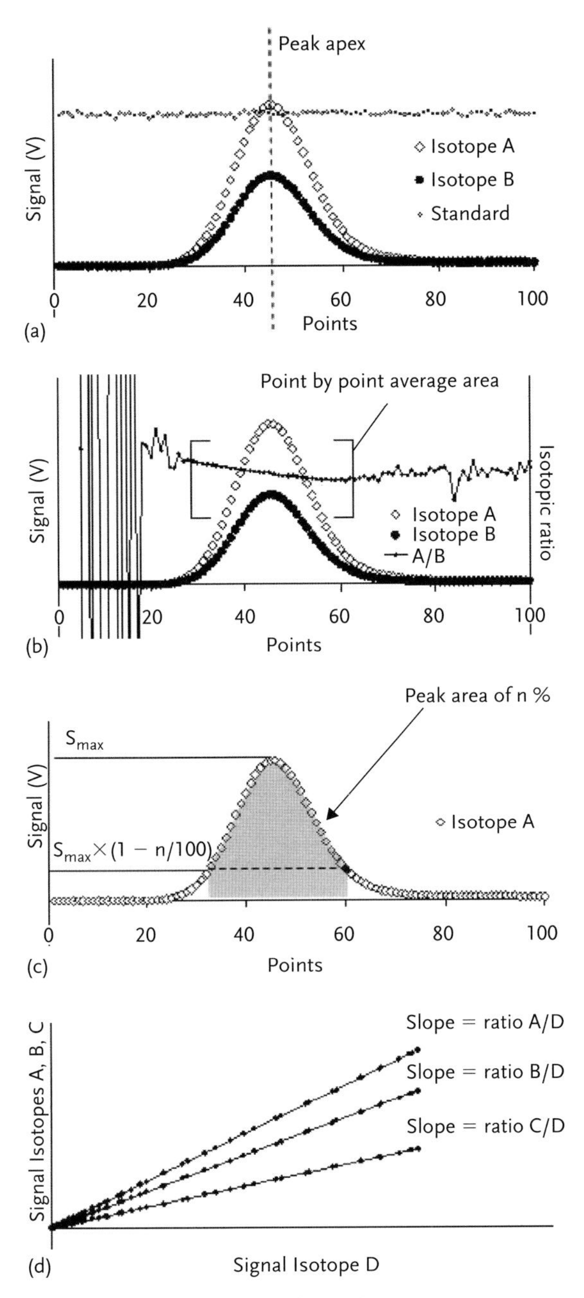

Figure 17.5 Illustration of different data treatment strategies: (a) peak apex; (b) point by point; (c) peak integration; (d) linear regression slope.

1. **Peak apex method.** A single data point which corresponds to the maximum of the transient signal peak is used for calculating the isotope ratios [12] (Figure 17.5a). This calculation is used both for the sample and for the bracketing standard.
2. **Point-by-point (or average peak ratio) method.** In this method, the isotope ratio is determined for several points of a selected peak area (usually 15–40 points, centered around the maximum of the peak) and the average value is calculated (Figure 17.5b). The area of calculation is selected based on the stability of the measured isotope ratio. Mass bias is corrected for by using a standard also measured via this point-by-point approach. [11–13, 15, 16].
3. **Peak integration (or peak area) method.** In this case, the isotope ratio is obtained by dividing the integrated peak areas for the two isotopes. As shown in Figure 17.5c, different peak areas (from 5 to 100% of the peak width, centered around the maximum of the signal) can be selected for integration of the peak. Results show that the best performance is obtained for a peak area >90% [13, 15]. The average value for the isotope ratio of the internal standard, calculated using the same peak area, is used for the mass bias correction.
4. **Linear regression slope.** Recently, a new approach has been developed for calculating isotope ratios in transient signals. This approach was first developed for sample introduction via LA [35] and was then transferred to GC–MC-ICP-MS, where it was used for measuring Hg isotope ratios in species [15]. When using this simple method, the isotope ratio is given by the slope of the linear regression line, that is, the best fitting line through the data points obtained by plotting the signal intensities of the two isotopes as a function of one another (Figure 17.5d). All the points of the GC peak are used, including those of the baseline, and no background correction is needed. The average isotope ratio of the internal standard covering the same peak area is used for the mass bias correction.

The performances of these different methods of data treatment have been evaluated and compared [12, 13, 15]. Dzurko *et al.* [12] compared the first three methods (Figure 17.5a–c) and demonstrated that the point-by-point method provides the best external precision. In another study [15], the newer linear regression slope method was compared with the other three methods of data treatment. This new method was found to give the best internal and external precision, because it provides a natural weighting of the points, favoring those originating from the most intense signal zone over those originating from the edges of the peaks. It also allows the measurement of samples and bracketing standards of different concentrations and comparison to isotope ratios in different species.

There is also no common approach for background correction or subtraction when using transient signal measurement. Some authors observed that the subtraction of the background did not change the isotope ratio noticeably and therefore they did not take the background into account [12]. However, for isotope ratio measurements in species using transient signals, different approaches to the background correction have also been tried:

1. **Subtraction of the electronic background.** The electronic noise of the detectors is measured and subtracted from the signal at each point in the chromatogram. Depending on the MC-ICP-MS instrument type used, measuring the electronic noise can be accomplished either by closing the gate valve or by deflecting the ion beam [27].
2. **Subtraction of the analytical background.** The average signal of the baseline of the chromatogram, measured before the elution of the peak, is subtracted from the signal of the peak of the species [13, 15].
3. **No background subtraction.** When using the linear regression slope method of data treatment, it is not necessary to subtract any background since the background intensities are taken into account when calculating the slope.

17.3.3 Consequences of the Transient Nature of the Signal

17.3.3.1 Shape and Width of the Peak

In contrast to the coupling of chromatography with single-collector ICP-MS, the measurement of isotope ratios in species using the transient signals registered by MC-ICP-MS requires a maximum number of measurements, hence wider and flatter chromatographic peaks are preferred. This requires a compromise with a suitable chromatographic resolution to prevent the overlapping of different species during elution. The precision of the measured isotope ratios is strongly influenced by the shape and width of the corresponding peaks. According to Poisson counting statistics, the wider and higher the chromatographic signal is, the better is the isotope ratio precision achieved.

Under traditional conditions, the peak width for LC can easily achieve 60 s, whereas it is generally only about 5 s for GC [27]. These short transient signals with GC jeopardize the precision required for most environmental applications. Hence GC–MC-ICP-MS has to be specifically optimized in order to increase the peak width [13].

17.3.3.2 Drift of the Isotope Ratios During Peak Elution

During elution of the chromatographic peak, drift of the measured isotope ratio has been observed (Figure 17.6). This is one of the main analytical issues for chromatography coupled with MC-ICP-MS, since this variation could influence the precision and accuracy of the measurement. This drift of the isotope ratio has been observed for different elements, different types of sample introduction systems, and different types of MC-ICP-MS instrumentation [11, 12, 20, 27, 36, 37] (see also Chapter 4). In addition, this drift has an unpredictable behavior, with either a positive or a negative slope (Figure 17.6).

Different hypotheses have been formulated to explain this drift, but none of them provides a clear explanation for this phenomenon. Below, the main suggestions are summarized.

1. In the first studies related to the coupling of chromatography and MC-ICP-MS, it was assumed that this drift was due to isotope fractionation occurring inside the

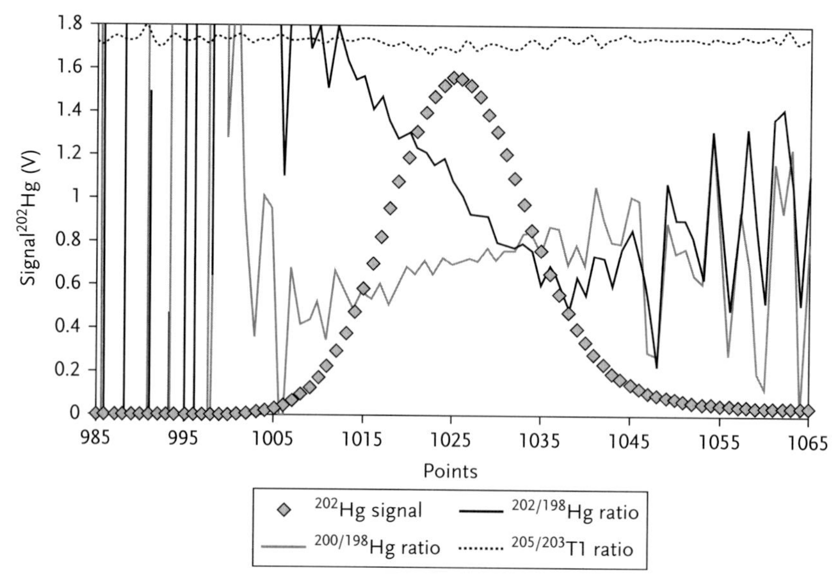

Figure 17.6 Example of the drift in Hg isotope ratios observed during the measurement of a transient signal using GC–MC-ICP-MS. A Tl solution is simultaneously and continuously introduced as a mass bias correction standard.

chromatographic column [20]. However, as demonstrated later, this is not the predominant nature of the drift, since the same drift was also observed for other sample introduction systems coupled with MC-ICP-MS: flow injection analysis (FIA) [27] and LA [37]. Also, it was observed that the slope of the drift is not dependent on the mass of isotopes [11] and can have either a positive or a negative direction for the same instrument set-up and the same pair of isotopes [12].

2. A general variation in the instrumental mass bias could be an explanation. However, this is not the case, since the ratio measured for the internal standard, which is introduced simultaneously and continuously into the ICP (Figure 17.3), does not demonstrate this drift during the elution of the peak [11].
3. Other suggestions, such as the influence of the analyte concentration or the drift of the background, have also been excluded [11].

The most credible explanation for this drift is a slow response of the detection system of the MC-ICP-MS to a short transient signal [11, 12, 37]. The geometry of the collector array of an MC-ICP-MS instrument and, in particular, the design of the Faraday cup preamplifier are only perfectly fit for integrating continuous signals. This hypothesis is backed up by the observation that the slope of the drift could change in direction when using different detector configurations for the same pair of isotopes [12]. Hence the response of the detection system is the most probable origin of this isotope ratio drift. A mathematical correction has been proposed to take this behavior into account [37]. However, to avoid the influence of this drift on the isotope ratio measurement, 100% of the chromatographic peak should be taken into account for calculation (see Section 17.3.2.3 for details).

17.4 Environmental and Other Applications

Recently, several new methods have been developed for measuring isotope ratios in elemental species of some real-world samples, standards, and reference materials using chromatographic separation coupled on-line to MC-ICP-MS. So far, there are about 15 published papers which describe the application of on-line chromatography coupled with MC-ICP-MS to isotopic analysis in elemental species of Hg, Pb, S, Sb, Cl, and Br. In this section, these applications, including sample preparation, analytical characteristics of the technique, instrumentation, species and isotopes monitored, and some useful results, are discussed.

17.4.1 Mercury

Mercury is the element the isotopic composition of which in species has been investigated most intensively (Table 17.1). Mercury has seven stable isotopes (^{196}Hg, ^{198}Hg, ^{199}Hg, ^{200}Hg, ^{201}Hg, ^{202}Hg, and ^{204}Hg) and is present in four main environmental chemical entities: elemental mercury (Hg^0), inorganic ionic mercury species (IHg^{2+}), organic methylmercury species ($MeHg^+$), and dimethylmerury (Me_2Hg). Owing to the toxicity of Hg, especially of the organic forms, isotopic analysis of Hg is an important means to understand the complex biogeochemical cycle of this element. During the last decade, several studies have shown that Hg isotopic signatures allow the identification of several Hg transformation pathways, and also the identification of sources of Hg in the environment. Furthermore, the importance of Hg speciation in the environment has led to the development of techniques allowing the study of species-specific Hg isotopic composition and signature using MC-ICP-MS [11–15].

GC has been used for the separation of Hg species and MC-ICP-MS to measure Hg isotope ratios. Krupp and Donard [11] coupled capillary GC with different MC-ICP-MS instruments via a home-made transfer line for the measurement of the Hg isotope ratios in two Hg species, $MeHg^+$ and IHg^{2+} (Table 17.1). A Tl standard was added continuously to the plasma by using a T-piece for the connection of the GC unit and the spray chamber with the plasma torch. The five most abundant Hg isotopes were measured simultaneously in the SPEX (IHg^{2+} species) and STREM ($MeHg^+$ species) mercury standards. These species were first derivatized to Et_2Hg and EtMeHg, respectively. Then the species were separated from one another using GC and introduced directly into the MC-ICP-MS instrument, with which isotope ratios were measured based on the short transient signals obtained.

Dzurko *et al.* [12] determined compound-specific Hg isotope ratios using GC interfaced with MC-ICP-MS (Table 17.1). The procedure consists of (i) species extraction based on ethylation, followed by preconcentration of the species thus obtained on a Tenax trap, (ii) thermal desorption of these Hg species and separation using a home-made GC unit, and (iii) registration of the isotope ratios in Hg species using a Neptune MC-ICP-MS instrument. Thallium was introduced simultaneously with Hg into the mass spectrometer via an Apex desolvating sample introduction system. The authors also discussed and compared calculation approaches for achieving an improvement in analytical precision. The isotopic composition of Hg was measured in two species of two NRC candidate reference materials.

Epov *et al.* [13] optimized the GC separation of Hg species by using wide transient signals, which improved the precision and accuracy of isotope ratio measurement in Hg species by MC-ICP-MS (Table 17.1). They used either ethylation or propylation for the derivatization of IHg^{2+} and $MeHg^{+}$. Similarly to previous studies, Tl was used for the correction for instrumental mass bias. Another novel aspect of the method was the bracketing of the sample by the IHg^{2+} species of NIST SRM 3133 (Hg solution) reference material and the $MeHg^{+}$ species of the STREM standard, which enabled the relative isotopic composition of these species to be obtained. It was demonstrated that integration of the peak gives accurate and precise results. Using the technique developed, the Hg isotopic composition was measured in IHg^{2+} species of the F65A, RL24H and UM-Almaden secondary standards. The isotopic composition of Hg in the $MeHg^{+}$ species was measured in BCR CRM 464 Tuna Fish and IAEA 085 Human Hair reference materials.

In a subsequent investigation, the same group [14] utilized the technique developed to study species-specific stable isotope fractionation of mercury during methylation of IHg^{2+} by anaerobic bacteria (*Desulfobulbus propionicus*) in the dark (Table 17.1). The isotopic composition of Hg in IHg^{2+} and $MeHg^{+}$ species was measured on-line from the same sample. The authors demonstrated mass-dependent fractionation of Hg isotopes during the experiment.

Finally, Epov *et al.* [15] presented a new approach for calculating isotope ratios in species using chromatography coupled with MC-ICP-MS (Table 17.1) and demonstrated the precise and accurate measurement of the isotopic composition of Hg in four species (Hg^{0}, IHg^{2+}, $MeHg^{+}$, and Me_2Hg). An example of the calculation of Hg isotope ratios ($^{204}Hg/^{198}Hg$, $^{202}Hg/^{198}Hg$, $^{201}Hg/^{198}Hg$, $^{200}Hg/^{198}Hg$, $^{199}Hg/^{198}Hg$, and $^{196}Hg/^{198}Hg$) for NIST SRM 3133 standard is shown in Figure 17.5d. The *x*-axis represents the signal intensity (in volts) for the ^{198}Hg isotope, which is the denominator of the isotope ratios. The *y*-axis represents the signal intensities for the other Hg isotopes (^{204}Hg, ^{202}Hg, ^{201}Hg, ^{200}Hg, ^{199}Hg, and ^{196}Hg), which are the nominators of the isotope ratios. The Hg isotope ratios are given by the slopes of the regression lines. The Hg isotopic composition (expressed as $\delta^{204}Hg$, $\delta^{202}Hg$, $\delta^{201}Hg$, $\delta^{200}Hg$, $\delta^{199}Hg$, and $\delta^{196}Hg$,) was measured using the new method for (i) the secondary standards RL24H and F65A, representing IHg^{2+} species and having an isotopic composition influenced by strongly positive and strongly negative mass-dependent fractionation, respectively, (ii) the STREM methylmercury standard, representing $MeHg^{+}$ species and having an isotopic composition affected by negative mass-dependent isotope fractionation, (iii) synthesized dimethylmercury (Me_2Hg), having an isotopic composition affected by negative mass-dependent fractionation, and (iv) fish (ERM CE464) and hair (IAEA 085, IAEA 086) reference materials, representing $MeHg^{+}$ and IHg^{2+} species, the isotopic composition of which seems to be affected by both mass-dependent and mass-independent isotope fractionation. Isotope fractionation during degradation of IHg^{2+} to Hg^0 was also studied and negative mass-dependent fractionation was observed. With the new methodology, it is possible to (i) measure the isotopic composition when a sample and a bracketing standard have significantly different concentrations, (ii) measure the isotopic composition of different species in samples versus that of single species in a bracketing standard, and (iii) measure the isotope ratios for low-abundant isotopes. This method could become the preferred approach for studying the isotopic composition of species, since it can also be applied to the coupling of other chromatographic techniques with MC-ICP-MS (GC–MC-ICP-MS, LC–MC-ICP-MS, HPLC–MC-ICP-MS or IC–MC-ICP-MS).

17.4.2
Lead

Organolead compounds are pollutants in air, water, soil, and sediments. The toxicity of organolead species depends on the organic groups bound to the Pb atom. Lead has four stable isotopes (^{204}Pb, ^{206}Pb, ^{207}Pb, and ^{208}Pb) and isotopic analysis of Pb can be used for geological dating and to track environmental processes and sources of Pb species.

To date, there are three publications demonstrating the on-line determination of the Pb isotopic composition in lead species by means of GC–MC-ICP-MS [11, 16, 17]. Tetraethyllead species were obtained by Krupp and co-workers [11, 16, 17] by derivatization of Pb^{2+} from an isotopically certified lead solution, by dissolving NIST SRM 981 Lead Wire (Table 17.1). Tetraethyllead was dissolved in isooctane and introduced into an MC-ICP-MS instrument through a gas chromatograph for isotopic analysis based on a short transient signal. The authors discussed the precision, accuracy, and other analytical parameters for optimization of the measurement.

However, no environmental applications have been published yet for on-line lead isotope ratio measurement in species.

17.4.3
Sulfur

Along with inorganic species of different oxidation states (−2, 0, +2, +4, and +6), sulfur has a number of volatile organic compounds with different toxicities, which can be produced both in nature and anthropogenically. Of the four stable isotopes of sulfur (^{32}S, ^{33}S, ^{34}S, and ^{36}S), the three most abundant can be monitored by MC-ICP-MS (the signals of $^{36}S^+$ and $^{36}Ar^+$ show isobaric overlap), hence the corresponding isotope ratios can be used to study isotope fractionation during different processes and the isotopic composition in different sulfur species can be determined. Sulfur isotope ratio measurement is complicated as a result of interference from O_2^+, which can be overcome either by using a higher mass resolution or by using dry plasma conditions.

The first study of the isotopic composition in sulfur species was reported by Krupp *et al.* [18]. They introduced SF_6 gas into a GC unit coupled to an MC-ICP-MS instrument and the corresponding data were recorded in transient signal mode (Table 17.1). The optimization of the measurements and the reproducibility of the isotope ratios involving the three most abundant S isotopes in SF_6 species were discussed.

Amrani *et al.* [19] developed a highly sensitive and robust method for the determination of $\delta^{34}S$ in individual organic species by GC coupled with MC-ICP-MS (Table 17.1). After the optimization of the GC–MC-ICP-MS hyphenation, the isotopic composition of S in eight organosulfur species (dihexyl sulfide, dibenzothiophene, 1-octadecanethiol, 3-octylthiophene, 1-dodecanethiol, 4,6-diethyldibenzothiophene, 3-hexylthiophene, and benzothiophene) was compared with that of the reference gas SF_6. The isotopic composition of sulfur ($\delta^{34}S$) was found to vary from 30.71‰ for 1-octadecanethiol to −7.00‰ for 1-dodecanethiol.

17.4.4 Antimony

Antimony is a metalloid and its presence in the environment is increasing owing to extensive anthropogenic utilization. The toxicity of antimony depends on its chemical form, and the main species of interest include Sb(III), Sb(V), monomethylstibonic acid, and dimethylstibonic acid. Sb has two stable naturally abundant isotopes, ^{123}Sb and ^{125}Sb, with natural relative abundances of 57.2 and 42.8%, respectively.

There has been only one study on the development of isotope ratio measurement in Sb species (Table 17.1). Wehmeier *et al.* [20] used GC coupled with MC-ICP-MS to measure the isotopic composition of Sb in trimethylstibine [Me_3Sb]. The instrumental mass bias was corrected for using the standard–sample–standard bracketing technique with an Me_3Sb standard, as a suitable internal standard was not found for this purpose. For the calculation, the authors used the peak integration method. The isotopic composition of trimethylstibine generated by anaerobic bacteria as measured by the authors showed positive $\delta^{123}Sb$ values relative to the gaseous Me_3Sb standard.

17.4.5 Halogens

Both inorganic and organic species of bromine and chlorine are present in different types of samples, including environmental, biological, and food samples. The organic species can be separated from one another by GC. Both halogens have two stable isotopes, ^{35}Cl (75.8%) and ^{37}Cl (24.2%) and ^{79}Br (50.7%) and ^{81}Br (49.3%), which can be used to study the fractionation processes. Three papers have described attempts to measure the isotopic composition of these halogen species using GC hyphenated with MC-ICP-MS [21–23].

Van Acker *et al.* [21] developed a method for the isotopic analysis of Cl in chlorinated aliphatic hydrocarbons (Table 17.1). Trichloroethene (TCE) and perchloroethene (PCE) were tested as models for chlorinated aliphatic hydrocarbons by the GC–MC-ICP-MS method. TCE and PCE were injected directly into a GC column coupled to an MC-ICP-MS instrument, operated in high-resolution mode to separate the signals of $^{37}Cl^+$ and $^{36}Ar^1H^+$. The data from the transient signals were integrated and instrumental mass bias was corrected using sample–standard comparison. The results obtained were reported as $\delta^{37}Cl$ in permil versus the SMOC (Standard Mean Ocean Chloride), and $\delta^{37}Cl$ values for TCE and PCE were not statistically different from zero.

The first on-line GC–MC-ICP-MS measurement of brominated organic compounds was reported by Sylva *et al.* [22]. They demonstrated that after optimization of this hyphenation, the background signals, the formation of hydrated Ar dimers ($^{38}Ar^{40}Ar^1H^+$ and $^{40}Ar^{40}Ar^1H^+$), and the solvent peak did not affect the results. A precision of 0.3‰ was achieved for three brominated benzenes: 2,2′,4,4′-tetrabromodiphenyl ether (BDE-47), 2,2′,4,4′,5-pentabromodiphenyl ether (BDE-99), and 2,2′,4,4′,6-pentabromodiphenyl ether (BDE-100). Samples were bracketed by monobromobenzene to calculate $\delta^{81}Br$ (Table 17.1).

Recently, Gelman and Halicz presented a new GC–MC-ICP-MS method to study the isotopic composition of bromine species [23]. Instrument parameters were optimized to measure ^{79}Br and ^{81}Br simultaneously (Table 17.1), A solution of NIST SRM 987 isotopic reference material was introduced continuously for mass bias correction (via $^{84}Sr/^{86}Sr$ monitoring), and the $^{83}Kr^+$ signal (Kr is a contaminant in Ar) was used to correct for the contributions of Kr at m/z ratios of 84 and 86. The Br isotopic composition was measured for the following bromine species: dibromoethane, 2-bromochlorobenzene, 3-bromotoluene, tribromophenol, and tribromobenzene (listed from the lowest to the highest $^{81}Br/^{79}Br$ isotope ratio).

17.5 Conclusion and Future Trends

As can be seen from this review of studies on isotopic analysis in elemental species by chromatography coupled to MC-ICP-MS published to date, this research area is still in an early stage of development. Most of the publications describe the development of techniques and analysis of standards, and only a few demonstrate the application of the methods to the analysis of environmental samples or to study different processes. A main aspect to be taken into account during this type of analysis is the required quantitative yield of the species (at least close to 100%).

Based on the discussion in this chapter, the typical procedure for isotopic analysis in species using GC–MC-ICP-MS can be summarized as presented in Figure 17.7: (i) an external standard (bracketing standard) containing a single species with known isotopic composition is measured, and the internal standard is measured simultaneously using continuous sample introduction (see Figures 17.2 and 17.3 for details); (ii) a sample containing several species with unknown isotopic composition is measured, and also here the internal standard is measured

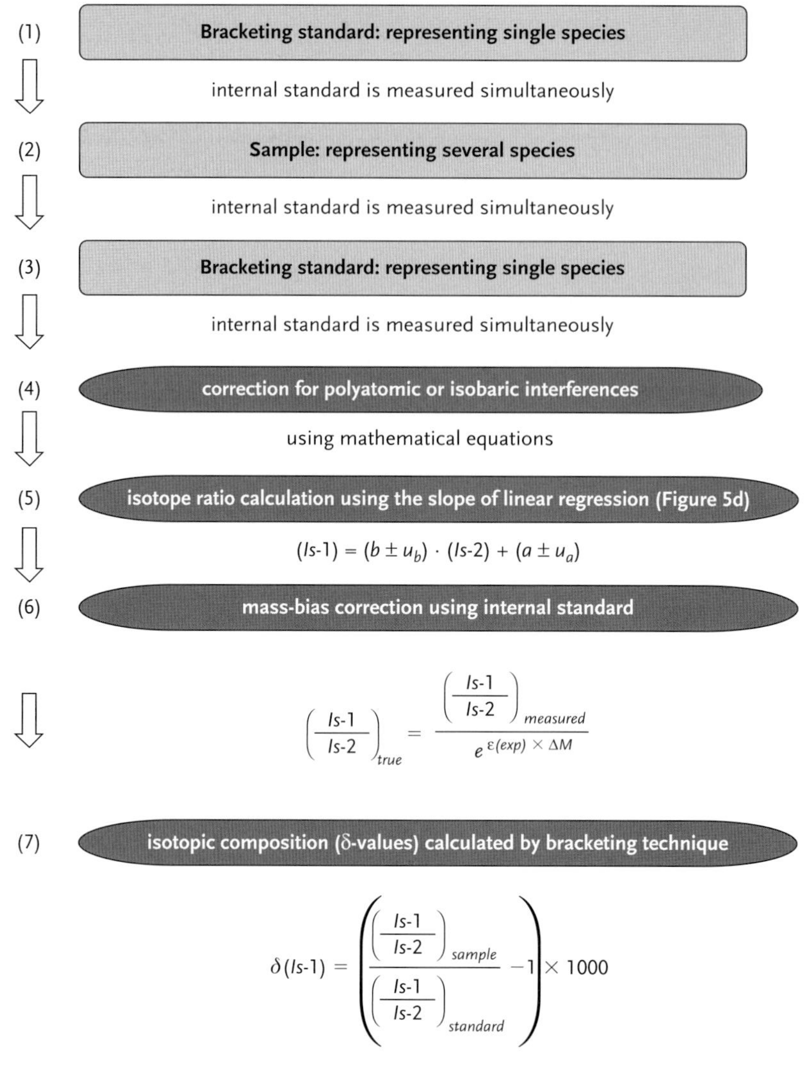

Figure 17.7 Schematic diagram of the method for determination of the isotopic composition of the target element in different species via chromatography coupled to MC-ICP-MS.

simultaneously; (iii) the measurement of the external standard is repeated; (iv) correction for interference from polyatomic ions or isobars is accomplished for each isotope and for each sample/standard using mathematical corrections; (v) the isotope ratio (*Is-1*)/(*Is-2*) is calculated for each sample/standard using the linear regression slope $b \pm u_b$ (see Figure 17.5d for details); (vi) the instrumental mass bias is corrected for every isotope ratio, using the exponential law, whereby the internal standard with known isotope ratio is used to obtain the fractionation factor [$\varepsilon(exp)$] used in the corresponding equations; and (vii) finally, the isotopic composition is calculated (often expressed as δ values versus the bracketing standard). This method of calculation should produce statistically accurate and precise results.

So far, isotope ratio determination in species using the hyphenation of chromatography and MC-ICP-MS has been tried for only a few elements: Hg, Pb, S, Sb, Cl, and Br. However, the approach to measure the isotopic composition in elemental species described above can be applied to other elements and be used for the combination of other chromatographic techniques with MC-ICP-MS also. On-line hyphenation is especially advantageous for GC, since GC is fairly difficult to use in off-line mode. The list of the potential candidate elements for the isotopic analysis in elemental species using on-line chromatography is not limited to those already discussed in the literature (Table 17.1). These studies can also be applied to the elements which have been investigated previously for their speciation using chromatography coupled with single-collector ICP-MS: Cd, Cr, Cu, Ge, In, Fe, Mo, Ni, Se, Si, Te, Sn, W, V, and Zn.

References

1 IUPAC (1997) *IUPAC Compendium of Chemical Terminology*, 2nd edn. (the "Gold Book"), compiled by A.D. McNaught and A. Wilkinson, Blackwell Scientific, Oxford. XML on-line corrected version: http://goldbook.iupac.org (2006–), created by M. Nic, J. Jirat, and B. Kosata, updates compiled by A. Jenkins. ISBN 0-9678550-9-8; doi: 10.1351/goldbook.

2 Watras, C.J., Back, R.C., Halvorsen, S., Hudson, R.J.M., Morrison, K.A., and Wente, S.P. (1998) Bioaccumulation of mercury in pelagic freshwater food webs. *Sci. Total Environ.*, **219**, 183–208.

3 Buchachenko, A.L. (2009) Mercury isotope effects in the environmental chemistry and biochemistry of mercury-containing compounds. *Russ. Chem. Rev.*, **78** (4), 319–328.

4 Bergquist, B.A. and Blum, J.D. (2009) The odds and evens of mercury isotopes: applications of mass-dependent and mass-independent fractionation. *Elements*, **5**, 353–357.

5 Harrington, C.F., Clough, R., Hansen, H.R., Hill, S.J., Pergantis, S.A., and Tyson, J.F. (2009) Atomic spectrometry update. Elemental speciation. *J. Anal. At. Spectrom.*, **24** (8), 999–1025.

6 Feldmann, J., Salaun, P., and Lombi, E. (2009) Critical review perspective: elemental speciation analysis methods in environmental chemistry – moving towards methodological integration. *Environ. Chem.*, **6** (4), 275–289.

7 Timerbaev, A.R. (2009) Capillary electrophoresis coupled to mass spectrometry for biospeciation analysis: critical evaluation. *Trends Anal. Chem.*, **28** (4), 416–425.

8 Montes-Bayon, M., DeNicola, K., and Caruso, J.A. (2003) Liquid chromatography–inductively coupled plasma mass spectrometry. *J. Chromatogr. A*, **1000** (1–2), 457–476.

9 Van Staden, J.F. and Stefan, R.I. (2004) Chemical speciation by sequential injection analysis: an overview. *Talanta*, **64** (5), 1109–1113.

10 Rouxel, O., Ludden, J., and Fouquet, Y. (2003) Antimony isotope variations in natural systems and implications for their use as geochemical tracers. *Chem. Geol.*, **200** (1–2), 25–40.

11 Krupp, E.A. and Donard, O.F.X. (2005) Isotope ratios on transient signals with GC–MC-ICP-MS. *Int. J. Mass Spectrom.*, **242** (2–3), 233–242.

12 Dzurko, M., Foucher, D., and Hintelmann, H. (2009) Determination

of compound-specific Hg isotope ratios from transient signals using gas chromatography coupled to multicollector inductively coupled plasma mass spectrometry (MC-ICP/MS). *Anal. Bioanal. Chem.*, **393** (1), 345–355.

13 Epov, V.N., Rodriguez-Gonzalez, P., Sonke, J.E., Tessier, E., Amouroux, D., Maurice Bourgoin, L., and Donard, O.F.X. (2008) Simultaneous determination of species-specific isotopic composition of Hg by gas chromatography coupled to multicollector ICPMS. *Anal. Chem.*, **80** (10), 3530–3538.

14 Rodriguez-Gonzalez, P., Epov, V.N., Bridou, R., Tessier, E., Guyoneaud, R., Monperrus, M., and Amouroux, D. (2009) Species-specific stable isotope fractionation of mercury during Hg(II) methylation by an anaerobic bacteria (*Desulfobulbus propionicus*) under dark conditions. *Environ. Sci. Technol.*, **43** (24), 9183–9188.

15 Epov, V.N., Berail, S., Jimenez-Moreno, M., Perrot, V., Pecheyran, C., Amouroux, D., and Donard, O.F.X. (2010) New approach to measure isotopic ratios in species using MC-ICP-MS coupled with chromatography. *Anal. Chem.*, **82** (13), 5652–5662.

16 Krupp, E.M., Pecheyran, C., Meffan-Main, S., and Donard, O.F.X. (2001) Precise isotope-ratio measurements of lead species by capillary gas chromatography hyphenated to hexapole multicollector ICP-MS. *Fresenius' J. Anal. Chem.*, **370** (5), 573–580.

17 Krupp, E.M., Pecheyran, C., Pinaly, H., Motelica-Heino, M., Koller, D., Young, S.M.M., Brenner, I.B., and Donard, O.F.X. (2001) Isotopic precision for a lead species ($PbEt_4$) using capillary gas chromatography coupled to inductively coupled plasma-multicollector mass spectrometry. *Spectrochim. Acta B*, **56** (7), 1233–1240.

18 Krupp, E.M., Pecheyran, C., Meffan-Main, S., and Donard, O.F.X. (2004) Precise isotope-ratio determination by CGC hyphenated to ICP-MCMS for speciation of trace amounts of gaseous sulfur, with SF_6 as example compound. *Anal. Bioanal. Chem.*, **378** (2), 250–255.

19 Amrani, A., Sessions, A.L., and Adkins, J.F. (2009) Compound-specific $\delta^{34}S$ analysis of volatile organics by coupled GC/multicollector-ICPMS. *Anal. Chem.*, **81** (21), 9027–9034.

20 Wehmeier, S., Ellam, R., and Feldmann, J. (2003) Isotope ratio determination of antimony from the transient signal of trimethylstibine by GC–MC-ICP-MS and GC–ICP-TOF-MS. *J. Anal. At. Spectrom.*, **18** (9), 1001–1007.

21 Van Acker, M.R.M.D., Shahar. A., Young, E.D., and Coleman, M.L. (2006) GC/multiple collector-ICPMS method for chlorine stable isotope analysis of chlorinated aliphatic hydrocarbons. *Anal. Chem.*, **78** (13), 4663–4667.

22 Sylva, S.P., Ball, L., Nelson, R.K., and Reddy, C.M. (2007) Compound-specific Br-81/Br-79 analysis by capillary gas chromatography/multicollector inductively coupled plasma mass spectrometry. *Rapid Commun. Mass Spectrom.*, **21** (20), 3301–3305.

23 Gelman, F. and Halicz, L. (2010) High precision determination of bromine isotope ratio by GC–MC-ICPMS. *Int. J. Mass Spectrom.*, **289** (2–3), 167–169.

24 Douthitt, C.B. (2008) The evolution and applications of multicollector ICPMS (MC-ICPMS). *Anal. Bioanal. Chem.*, **390** (2), 437–440.

25 Walczyk, T. (2004) TIMS versus multicollector-ICP-MS: coexistence or struggle for survival? *Anal. Bioanal. Chem.*, **378** (2), 229–231.

26 Moldovan, M., Krupp, E.M., Holliday, A.E., and Donard, O.F.X. (2004) High resolution sector field ICP-MS and multicollector ICP-MS as tools for trace metal speciation in environmental studies: a review. *J. Anal. At. Spectrom.*, **19** (7), 815–822.

27 Günther-Leopold, I., Wernli, B., Kopajtic, Z., and Gunther, D. (2004) Measurement of isotope ratios on transient signals by MC-ICP-MS. *Anal. Bioanal. Chem.*, **378** (2), 241–249.

28 Günther-Leopold, I., Waldis, J.K., Wernli, B., and Kopajtic, Z. (2005) Measurement of plutonium isotope ratios in nuclear fuel samples by HPLC–MC-ICP-MS. *Int. J. Mass Spectrom.*, **242** (2–3), 197–202.

29 Caruso, S., Günther-Leopold, I., Murphy, M.F., Jatuff, F., and Chawla, R. (2008) Comparison of optimised germanium gamma spectrometry and multicollector inductively coupled plasma mass spectrometry for the determination of ^{134}Cs, ^{137}Cs and ^{154}Eu single ratios in highly burnt UO_2. *Nucl. Instrum. Methods Phys. Res. A*, **589** (3), 425–435.

30 Garcia-Ruiz, S., Moldovan, M., and Alonso, J.I.G. (2008) Measurement of strontium isotope ratios by MC-ICP-MS after on-line Rb–Sr ion chromatography separation. *J. Anal. At. Spectrom.*, **23** (1), 84–93.

31 Garcia-Ruiz, S., Moldovan, M., and Garcia Alonso, J.I. (2007) Large volume injection in ion chromatography separation of rubidium and strontium for on-line inductively coupled plasma mass spectrometry determination of strontium isotope ratios. *J. Chromatogr. A*, **1149** (2), 274–281.

32 Santamaria-Fernandez, R., Hearn, R., and Wolff, J.C. (2008) Detection of counterfeit tablets of an antiviral drug using delta S-34 measurements by MC-ICP-MS and confirmation by LA–MC-ICP-MS and HPLC–MC-ICP-MS. *J. Anal. At. Spectrom.*, **23** (9), 1294–1299.

33 Fietzke, J., Liebetrau, V., Günther, D., Gürs, K., Hametner, K., Zumholz, K., Hansteen, H., and Eisenhauer, A. (2008) An alternative data acquisition and evaluation strategy for improved isotope ratio precision using LA–MC-ICP-MS applied to stable and radiogenic strontium isotopes in carbonates. *J. Anal. At. Spectrom.*, **23** (7), 955–961.

34 Evans, D., Hintelmann, H., and Dillon, P. (2001) Measurement of high precision isotope ratios for mercury from coals using transient signals. *J. Anal. At. Spectrom.*, **16** (9), 1064–1069.

35 Hirata, T., Hayano, Y., and Ohno, T. (2003) Improvements in precision of isotopic ratio measurements using laser ablation–multiple collector-ICP-mass spectrometry: reduction of changes in measured isotopics ratios. *J. Anal. At. Spectrom.*, **18** (10), 1283–1288.

36 Clough, R., Belt, S.T., Evans, E.H., and Catterick, T. (2003) Uncertainty contributions to species specific isotope dilution analysis. Part 2. Determination of methylmercury by HPLC coupled with quadrupole and multicollector ICP-MS. *J. Anal. At. Spectrom.*, **18** (9), 1039–1046.

37 Clough, R., Belt, S.T., Fairman, B., Catterick, T., and Evans, E.H. (2005) Uncertainty contributions to single and double isotope dilution mass spectrometry with HPLC–CV–MC-ICP-MS for the determination of methylmercury in fish tissue. *J. Anal. At. Spectrom.*, **20** (10), 1072–1075.

Index

Note: Page numbers in *italics* refer to figures and tables.

Isotopic Analysis: Fundamentals and Applications Using ICP-MS, First Edition. Edited by Frank Vanhaecke and Patrick Degryse.

Published 2012 by WILEY-VCH Verlag GmbH & Co. KGaA